Antioxidants in Animal Production, Reproduction, Health and Welfare

Antioxidants in Animal Production, Reproduction, Health and Welfare

Editors

Youssef A. Attia
Woo Kyun Kim
Nesrein Hashem
Maria de Olivera

MDPI

Basel • Beijing • Wuhan • Barcelona • Belgrade • Novi Sad • Cluj • Manchester

Editors

Youssef A. Attia
Department of Agriculture
King Abdulaziz University
Jeddah
Saudi Arabia

Woo Kyun Kim
Poultry Science Department
University of Georgia
Athens
United States

Nesrein Hashem
Department of Animal and Fish Production
Alexandria University
Alexandria
Egypt

Maria de Olivera
Faculty of Veterinary Medicine
Rio Verde University
Rio Verde
Brazil

Editorial Office
MDPI
St. Alban-Anlage 66
4052 Basel, Switzerland

This is a reprint of articles from the Special Issue published online in the open access journal *Animals* (ISSN 2076-2615) (available at: www.mdpi.com/journal/animals/special_issues/Animal_Antioxidants).

For citation purposes, cite each article independently as indicated on the article page online and as indicated below:

Lastname, A.A.; Lastname, B.B. Article Title. *Journal Name* **Year**, *Volume Number*, Page Range.

ISBN 978-3-0365-8969-5 (Hbk)
ISBN 978-3-0365-8968-8 (PDF)
doi.org/10.3390/books978-3-0365-8968-8

Contents

About the Editors

Youssef A. Attia

Professor Dr. Youssef A. Attia is a professor of Poultry Nutrition at King Abdulaziz University, Jeddah, Saudi Arabia. He gained a Doctorate of Philosophy in Poultry Production, University of Zagazig, Egypt, University of Georgia, USA. His job career has included Department Head, Vice Dean, and Dean in Charge. Professor Attia has published 330 articles, including 16 books + book chapters + translated book+ student books). Dr. Attia is a reviewer, and editorial board member and has gained many awards.

Woo Kyun Kim

Prof. Dr. Woo K. Kim is a professor/poultry nutritionist in the Department of Poultry Science, University of Georgia. His research has focused on the roles of feed additives, amino acids, vitamins, and minerals in nutrient utilization, gut health, and bone health in poultry. He has published over 200 peer-reviewed papers and has given 88 national/international invited talks and mentored 24 graduate students. He received the 2020 PSA American Feed Industry Association Nutrition Research Award and the 2021 UGA-CAES Outstanding Mentor Award.

Nesrein Hashem

Prof. Dr. Nesrein M. Hashem is a professor of Animal Physiology at the Animal and Fish Production Dept., Faculty of Agriculture, Alexandria University-AU, Egypt. She has professional experience in research, teaching, and society service; published about 120 scientific articles; three book chapters on the field of animal production; contributed to research projects funded by STDF and ASRT; received the shield of AU of excellent scientific publication in 2022, the State Encouragement Award for Women in 2019, and AU Award for Scientific Encouragement in 2017.

Maria de Olivera

Professor Dr. Maria Cristina de Oliveira is a staff member at the Faculty of Veterinary Medicine, University of Rio Verde, in Brazil. She started her career in 1996 as a Professor of Animal Physiology. In addition, she has professional experience in research in the field of animal production, as well as teaching. She has supervised six M.Sc. students and twenty-two students in undergraduate research projects, and all the dissertations and monographs were published in high-ranked journals.

Preface

One of the most important biological processes increasingly attracting biologists is redox homeostasis and its relationship with the efficiency and maintenance of different biological systems. The redox status imbalance in livestock production is associated with low animals' productive and reproductive performance and can be considered a drive for many health problems. This is mainly due to the negative impacts of imbalanced levels between antioxidant action and free radical formation in the body. Elevated levels of free radicals can adversely affect cell functions, stimulate apoptotic pathways, and damage different functioning and structural molecules.

Based on these facts, the editors of the present Special Issue have discussed the importance of this topic and subsequently approved unanimously to initiate this distinguished Special Issue entitled "Antioxidants in Animal Production, Reproduction, Health and Welfare". This is to present to the scientific community a piece of art and cutting-edge knowledge on maintaining redox homeostasis and minimizing oxidative stress in food-producing animals.

Moreover, this Special Issue presents different innovative strategies/solutions to maintain the redox status of animals, particularly when animals face harsh environmental conditions and/or improper physiological processes, which can make animals more vulnerable to negative oxidative stress impacts. In this Special Issue, 25 research articles and review articles were published. The articles reflect the international research trend on this topic, as articles from various research groups worldwide were published. Finally, we hope that by the release of this Special Issue, we can pave the way for more research work that enables us to understand more clearly the role of redox homeostasis in the farm animals' health and productivity to meet both animal welfare standards and the need for mass production of animal-based foods.

Youssef A. Attia, Woo Kyun Kim, Nesrein Hashem, and Maria de Olivera
Editors

animals

MDPI

Article

Effect of Folic Acid Supplements on Progesterone Profile and Blood Metabolites of Heat-Stressed Holstein Cows during the Early Stage of Pregnancy

Abdelrahman A. Kilany [1], Abdel-Halim A. El-Darawany [2], Akram A. El-Tarabany [1,*] and Khaled M. Al-Marakby [2]

[1] Radioisotopes Applications Division, Department of Biological Applications, NRC, Egyptian Atomic Energy Authority, Inshas, Cairo P.O. Box 13759, Egypt; elkelany_86@icloud.com

[2] Department of Animal Production, Faculty of Agriculture, Zagazig University, Zagazig P.O. Box 44511, Egypt; a.a.eldarawany@hotmail.com (A.-H.A.E.-D.); jnakhaled303@yahoo.com (K.M.A.-M.)

* Correspondence: eltarabanyakram@yahoo.com

Simple Summary: The fertility of a dairy cow can be described as the female's ability to conceive and maintain pregnancy when insemination is performed at the proper time relative to ovulation. In this context, poor estrous detection and embryonic or fetal losses are common causes for reduced reproductive performance in dairy cows. Furthermore, heat stress severely deteriorates conception and pregnancy rates in dairy farms. The aim of this study was to see how oral folic acid (FA) supplements affect progesterone levels, blood metabolites, endocrine patterns, and blood biochemical concentrations in heat-stressed pregnant cows. Oral FA supplementation (10 μg kg^{-1}) in the first month of gestation improved the progesterone profile, as well as blood folates, pregnancy associated glycoprotein (PAG), growth hormone (GH), and Insulin-like growth factor-1 (IGF-1) concentrations in heat-stressed Holstein cows.

Abstract: The aim was to elucidate the impact of oral folic acid (FA) supplements on progesterone profile, blood metabolites and biochemical indices of heat-stressed Holstein cows during the early stage of pregnancy. The study lasted from the day of artificial insemination through the end of the fourth week of pregnancy. The first group (CON, n = 17) received 0 μg of FA/kg BW as a control. The second and third groups received oral FA doses of 5 (FA_5, n = 19) and 10 (FA_{10}, n = 20) μg kg^{-1} BW, respectively. At the 2nd and 3rd weeks of pregnancy, the FA_{10} group had greater progesterone levels than the CON group ($p < 0.05$). The FA_{10} group had a greater progesterone level than the FA_5 and CON groups at the fourth week of pregnancy ($p < 0.01$). The FA_{10} group had higher folate levels than CON group during the first three weeks of pregnancy ($p < 0.01$). Both FA-supplemented groups had significantly greater serum folates than the CON group by the end of the fourth week of pregnancy ($p < 0.01$). At the 2nd and 4th weeks of pregnancy, the FA_{10} group had greater levels of serum glucose and globulin than the CON group (p = 0.028 and 0.049, respectively). Both FA-supplemented groups had greater serum growth hormone (GH) levels at the 4th week of pregnancy (p = 0.020). Additionally, the FA_{10} group showed significantly higher levels of IGF-1 at the 2nd and 4th week of gestation (p = 0.040 and 0.001, respectively). FA supplementation decreased the levels of non-esterified fatty acid (NEFA) at the 2nd and 4th week of gestation (p = 0.020 and 0.035, respectively). Additionally, the FA_{10} group showed significantly higher pregnancy-associated glycoprotein (PAG) levels at the 2nd and 4th week of gestation (p = 0.005 and 0.001, respectively). In conclusion, oral FA supplementation (10 mcg kg^{-1}) in the first month of gestation improved the progesterone profile, as well as blood folates, PAG, GH, and IGF-1 concentrations in heat-stressed Holstein cows. These findings could be useful in developing practical strategies to keep dairy cows' regular reproductive patterns under heat stress conditions.

Keywords: fertility; Holstein; folic acid; pregnancy

Citation: Kilany, A.A.; El-Darawany, A.-H.A.; El-Tarabany, A.A.; Al-Marakby, K.M. Effect of Folic Acid Supplements on Progesterone Profile and Blood Metabolites of Heat-Stressed Holstein Cows during the Early Stage of Pregnancy. *Animals* **2022**, *12*, 1872. https://doi.org/10.3390/ani12151872

Academic Editors: Clive J. C. Phillips and Rosalia Crupi

Received: 23 March 2022
Accepted: 30 June 2022
Published: 22 July 2022

1. Introduction

The fertility of dairy cows can be described as the female's ability to conceive and maintain pregnancy when insemination is performed at the proper time relative to ovulation [1]. In this context, poor estrous detection and embryonic or fetal losses are common causes for reduced reproductive performance in dairy cows [2]. Furthermore, heat stress severely deteriorates the conception and pregnancy rates in dairy farms [3]. This can be attributed to the negative effects of hyperthermia on the biological functions of female reproductive tract [4]. Jordan [5] claimed that the unfavorable effects of thermal stress could be observed from day 42 before insemination to day 40 after insemination. Furthermore, the influence of maternal heat stress on embryos varies depending on the stage of development; nonetheless, the initial days after fertilization are the most essential for the embryo [6].

Folic acid (FA) is a B-complex vitamin with the sole biochemical role of mediating the transfer of 1-C units in animals [7]. FA also has a variety of activities, including regulating neurotransmission and gene expression [8,9]. Folate metabolism appears to have a significant impact on the synthesis of milk proteins in the epithelial cells of mammary glands [10,11]. Although rumen microorganisms can generate folates, it has not been proven that such amounts are sufficient for dairy cows to perform at their best [12]. To some extent, the forage:concentrate ratio may influence the amount of folates produced in the rumen [13]. In this context, rumen breakdown of orally supplied FA appears to be very high, as FA supplement hardly raises folate concentrations in the digesta [14]. Total serum folates in dairy cows declined by 40% from two months postpartum until parturition [15]. Additionally, supplemental FA in gestation and lactation enhanced milk yield and milk protein yields in multiparous cows but not in primiparous cows [16].

It is widely acknowledged that a successful pregnancy is the most crucial aspect in ensuring a profitable dairy cow. In this regard, FA has attracted a lot of interest in animal research, especially during pregnancy. In multiparous dairy cows, Duplessis et al. [17] found that FA supplementation improved reproduction indices. The use of a periconceptional FA supplement slightly increased the number of twin pregnancies in humans [18]. Only a small number of studies have looked into the impact of FA supplements on the reproductive performance of heat-stressed cows [9,19], to the best of our knowledge. As a result, the aim of this study was to see how oral FA supplements affect progesterone levels, blood metabolites, and blood biochemical concentrations in heat-stressed pregnant cows.

2. Materials and Methods

The current research was performed at dairy station number 2 (30°36′16.0″ N, 31°54′21.5″ E), El Salhiya company for investment and development, New Salhiya, Sharkia governorate, Egypt.

2.1. Animals, Management and Experimental Design

In total, 120 multiparous cows (4–8 years of age) from El Salhiya Company for investment and development were used in the first stage of this experiment. The selected cows had homogenous body condition scores (3.4 ± 0.2). All cows were regularly monitored during the early post-partum period to exclude the reproductive disorders and then vaccinated against diseases that impede the reproductive traits by veterinarians. The cows were equally divided into three groups (each 40 cows) and housed in three yards (50 m × 60 m) with open stalls. The animals were fed a total mixed ration (TMR). The TMR was adjusted to achieve the optimal requirements of nutrients for milk production and body condition [20]. The chemical composition of TMR was described in Table 1. All cows were milked twice daily at 12 h intervals. The first group received 0 µg of folic acid/kg BW and served as control (CON). Orally, the second and third groups received 5 (FA_5) and 10 (FA_{10}) µg of folic acid/kg BW two times weekly, respectively. The folic acid tablets (Folate 1.000 µg folic acid) were purchased from Nutricost Co., Vineyard, UT, USA. The tablets were administered using a long cylindrical bolus gun, which is a tube made of stainless steel that has a spring-action handle. The experiment started at the day of artificial insemination (AI) to the end of fourth week of gestation period. When pregnancy was confirmed at day 28 following the AI, only cows that conceived were included in this study, and hence the actual sizes of the experimental groups were 17, 18 and 20 cows in CON, FA5 and FA10 groups, respectively.

Table 1. Chemical composition of the total mixed ration (TMR).

Item	Dry Matter Basis (%)
Crude Protein	17.96
NDF	26.32
Cellulose	14.93
Hemicellulose	8.78
Lignin	2.62
Crude Fat	5.1
Ash	10.72
TDN	71.74
DE (Mcal/kg)	3.23
ME (Mcal/kg	2.58

TDN: total digestible nutrients; NDF: neutral detergent fiber; DE: digestible energy; ME: metabolizable energy.

2.2. Reproductive Management and Experimental Procedures

During the first two weeks post partum, all cows having an abnormal puerperium period were eliminated from the study (dystocia, twinning, retained placenta, ovarian cysts, primary metritis or ketonuria). Cows with an average body condition score of 3.5 ± 0.3 were considered to be in a good condition. The Ovsynch synchronization protocol was initiated at 45 ± 4 day postpartum in all groups. All cows were given 10 g GnRH (day 0; Buserelin; Receptal; Intervet), 25 mg PGF2α (day 7; dinoprost; Lutylase; pfizer), and one dosage of PGF2α 56 h before the second dose of GnRH. Artificial insemination (AI) was conducted 18 h following the second dosage of GnRH. The insemination processes were practiced by three expert technicians. Diagnosis of pregnancy was conducted 28 days following the AI (transrectal ultrasonography; Pie Medical, Maastricht, The Netherlands), and only cows that conceived to AI were included in this study (CON, n = 17; FA_5, n = 19; FA_{10}, n = 20). Hence, the samples from those pregnant animals were included in the current study.

2.3. Estimation of Progesterone Profile and Serum Folates

On a weekly basis, blood (3 mL) was sampled from the jugular vein for four consecutive weeks to evaluate the concentrations of progesterone and folates. Centrifugation at 1008× g for 15 min separated the serum, which was then kept at −20 °C. The serum progesterone levels were determined using a radioimmunoassay method (commercial kits, Diagnostic Product Corporation, Los Angeles, CA, USA). In antibody-coated tubes, unknown samples or standards are treated with I125-labeled hormone. The liquid contents of the tube were aspirated after incubation, and the radioactivity was measured in a computerized gamma counter at the Egyptian Atomic Energy Authority, Nuclear Research Centre. The intra- and inter-assay coefficients of variation were 4.9 and 7.5 percent, respectively. The concentrations of serum folic acid were determined using chemical commercial kits and quantitative enzymatic colorimetric techniques (MDSS GmbH, Hannover, Germany).

2.4. Estimation of Blood Metabolites and Biochemical Indices

Blood was taken from the jugular vein at the end of the second and fourth weeks of pregnancy to assess the concentrations of various blood metabolites and biochemical markers. Centrifugation at 1008× g for 15 min was performed to separate the serum, which was then frozen and stored at −20 °C until analysis. Total protein, urea, triglycerides, cholesterol, glucose, and albumin concentrations were determined using chemical commercial kits and quantitative enzymatic colorimetric techniques (MDSS GmbH, Hannover, Germany). The concentration of pregnancy associated glycoprotein (PAG) was detected by enzyme-linked immune sorbent assay (ELIZA) technique, with Bovine PAG antibody (Bioassay technology laboratory Co, Ningguo, Shanghai, China, Cat No MBS1602347), and the standard curve range was 0.05 ng/mL–20 ng/mL, sensitivity 0.02 ng/mL. The levels of serum non-esterified fatty acid (NEFA) were measured by ELIZA technique with bovine NEFA antibody (Bioassay technology laboratory Co, Ningguo, Shanghai, China, Cat No E0021Bo), with a standard curve range of 2 μmol/L–600 μmol/L and a sensitivity of 1.17 μmol/L. The concentrations of Beta-Hydroxybutric Acid (BHB) and insulin-like growth factor-1 (IGF-1) were determined also by ELIZA technique (Boster Biomedical Technology, Valley Ave, Pleasanton, CA, USA). The concentration of serum cortisol was determined (Diagnostic Product Corporation, Los Angeles, CA, USA).

2.5. Metrological Data

The temperature–humidity index indicates heat stress by representing the interaction impact of both temperature and humidity. The records of daily temperature and humidity during this study were procured from the New Salhiya metrological station, approximately 9 km away from the farm area. Values of the THI were calculated as follows [21]:

$$\text{THI} = [(1.8 \times \text{AT}) + 32] - [0.55 - (0.0055 \times \text{RH})] \times [(1.8 \times \text{AT}) - 26]$$

where AT is ambient temperature (°C) and RH is relative humidity (%). The THI is considered to be high when the value exceeds 75 (Table 2).

Table 2. Average air temperature, humidity, and temperature humidity index (THI) during the early gestation period.

Experimental Weeks	Air Temperature, °C	Humidity, %	THI
1st week	37	75	81
2nd week	37	73	82
3rd week	36	76	81
4th week	36	75	80

2.6. Statistical Analysis

The General Linear Model procedures (GLM) of the IBM SPSS software application were used to examine the data (Version 16.0; IBM Corp., Armonk, NY, USA). For blood metabolites and biochemical indices (2nd and 4th week), the statistical model was constructed as follows:

$$Y_{ijk} = \mu + T_i + P_j + e_{ijk}$$

where:

Y_{ijk}: An observation of each trait.
μ: The overall mean.
T_i: The fixed effect of treatment ith (1, 2, 3).
P_j: The fixed effect of parity jth.
e_{ijk}: The random effect of error.

Data repeatedly measured (progesterone profile and serum folates) were subjected to ANOVA with repeated measures (IBM SPSS software). Duncan's Multiple Range Test was used to compare the means. Significance was stated at $p < 0.05$.

3. Results

3.1. Progesterone Profile and Serum Folates

The effects of FA supplements on progesterone profile during the early stage of pregnancy in heat-stressed Holstein cows are presented in Figure 1. The oral FA supplements significantly improved the serum progesterone level ($p < 0.001$). Throughout the first week of pregnancy, there were no significant differences between the experimental groups. Meanwhile, the FA_{10} group had significantly greater progesterone levels than the CON group at the 2nd and 3rd weeks of pregnancy ($p < 0.05$). By the end of the fourth week of pregnancy, the FA_{10} group had significantly higher progesterone levels than the FA_5 and CON groups ($p < 0.01$).

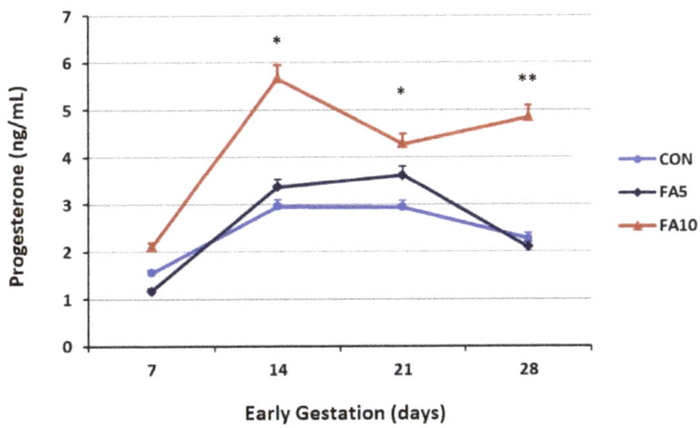

Figure 1. Effects of folic acid supplements (FA_5 and FA_{10}) on progesterone profile of heat-stressed Holstein cows during early gestation period. The concentrations of progesterone were affected by treatment ($p < 0.001$) and time ($p = 0.003$), but not treatment by time ($p = 0.315$). *, ** Significant differences at $p < 0.05$ and 0.01, respectively.

The effects of FA supplements on the levels of serum folates during the early stage of pregnancy in heat-stressed Holstein cows are shown in Figure 2. The concentrations of serum folates were significantly affected by FA supplements ($p = 0.009$). During the first three weeks of gestation, the FA_{10} group exhibited significantly higher folates level than the CON group ($p < 0.01$). Both FA-supplemented groups had significantly greater serum folates than the CON group by the end of the fourth week of pregnancy ($p < 0.01$).

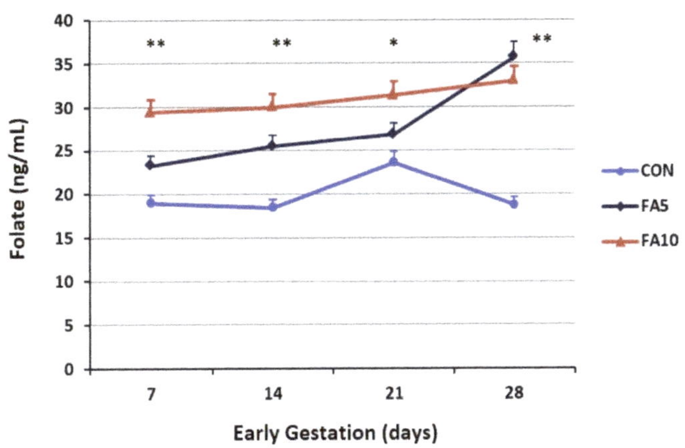

Figure 2. Effect of folic acid supplements (FA_5 and FA_{10}) on serum folate profile in heat-stressed Holstein cows during the early gestation period. The concentration of folate was affected by treatment ($p = 0.009$) and time ($p = 0.002$), but not treatment by time ($p = 0.147$). *, ** Significant differences at $p < 0.05$ and 0.01, respectively.

3.2. Blood Metabolites and Biochemical Indices

Table 3 shows the impact of FA supplementation on blood metabolite concentrations in heat-stressed Holstein cows. At the 2nd and 4th weeks of pregnancy, the FA_{10} group had greater levels of serum glucose and globulin than the CON group (p = 0.028 and 0.049, respectively). When compared to the CON group, FA supplements significantly reduced serum urea at the 4th week of pregnancy (p = 0.028). Meanwhile, there were no significant differences in total protein, cholesterol, triglycerides, HDL, and LDL concentrations across the experimental groups.

Table 3. Effect of folic acid supplements on the concentrations of blood metabolites in heat-stressed Holstein cows during the early stage of pregnancy.

Parameter	Gestation (Week)	Experimental Groups				
		[1] CON	[2] FA_5	[3] FA_{10}	[4] SEM	p-Value
Glucose, mmol/L	2nd week	2.46 [b]	2.35 [b]	2.90 [a]	0.18	0.041
	4th week	2.55	2.16	2.22	0.16	0.118
Urea, mmol/L	2nd week	4.52	4.85	4.46	0.28	0.856
	4th week	5.21 [a]	4.17 [b]	4.45 [ab]	0.33	0.028
Total protein, g/dL	2nd week	7.27	6.73	5.91	0.75	0.622
	4th week	4.44	6.56	6.97	0.67	0.087
Albumin, g/dL	2nd week	4.56 [a]	3.96 [a]	2.16 [b]	0.25	0.010
	4th week	3.65	3.29	4.57	0.32	0.134
Globulin, g/dL	2nd week	3.33	2.56	2.88	0.27	0.404
	4th week	3.28 [b]	2.99 [b]	4.20 [a]	0.23	0.049
Cholesterol, mmol/L	2nd week	8.63	8.99	9.12	0.36	0.496
	4th week	8.54	9.14	8.06	0.32	0.405
Triglycerides, mmol/L	2nd week	5.31	6.38	5.36	0.32	0.095
	4th week	6.97	6.57	6.27	0.34	0.677
[5] HDL, mmol/L	2nd week	4.06	2.61	4.24	0.22	0.053
	4th week	3.95	3.72	5.68	0.28	0.209
[6] LDL, mmol/L	2nd week	3.30	3.51	2.90	0.17	0.488
	4th week	3.09	3.11	2.22	0.15	0.404

[1] Control group; [2] folic acid supplements 5 $\mu g\ kg^{-1}$; [3] folic acid supplements 10 $\mu g\ kg^{-1}$; [4] standard error of mean; [5] high-density lipoprotein; [6] low-density lipoprotein. [a,b] Values within a row with different superscripts differ significantly.

Table 4 shows the effects of FA supplementation on cortisol, GH, and IGF-1 levels in heat-stressed Holstein cows during the early stage of pregnancy. Although cortisol levels did not change significantly, both FA-supplemented groups had significantly greater levels of GH than the CON group at the 4th week of pregnancy (p = 0.020). Additionally, the FA_{10} group had significantly higher IGF-1 levels at the 2nd and 4th week of gestation (p = 0.040 and 0.001, respectively).

Table 5 shows the impacts of FA supplementation on NEFA, BHB, and PAG levels in heat-stressed Holstein cows during the early stage of pregnancy. FA supplements significantly decreased the levels of NEFA at the 2nd and 4th week of gestation (p = 0.020 and 0.035, respectively). Meanwhile, there were no significant changes in BHB concentrations at the 4th week of pregnancy (p = 0.069). At the 2nd and 4th week of gestation, the FA_{10} group had significantly higher PAG levels compared to the FA_5 and CON groups (p = 0.005 and 0.001, respectively).

Table 4. Effect of folic acid supplements on the concentrations of cortisol, GH and IGF-1 in heat-stressed Holstein cows during the early stage of pregnancy.

Parameter	Gestation (Week)	Experimental Groups				
		[1] CON	[2] FA_5	[3] FA_{10}	[4] SEM	*p*-Value
Cortisol, µg/dL	2nd week	4.13	3.20	3.26	0.31	0.189
	4th week	3.92	3.16	4.29	0.24	0.323
[5] GH, pg/mL	2nd week	0.280	0.250	0.254	0.040	0.889
	4th week	0.220 [b]	0.310 [a]	0.303 [a]	0.033	0.020
[6] IGF-1, pg/mL	2nd week	396 [b]	301 [b]	786 [a]	30.4	0.040
	4th week	349 [c]	486 [b]	729 [a]	31.7	0.001

[1] Control group; [2] folic acid supplements 5 µg kg^{-1}; [3] folic acid supplements 10 µg kg^{-1}; [4] standard error of mean; [5] growth hormone; [6] insulin-like growth factor-1. [a,b,c] Values within a row with different superscripts differ significantly.

Table 5. Effect of folic acid supplements on the concentrations of NEFA, BHB and PAG in heat-stressed Holstein cows during the early stage of pregnancy.

Parameter	Gestation (Week)	Experimental Groups				
		[1] CON	[2] FA_5	[3] FA_{10}	[4] SEM	*p*-Value
[5] NEFA, µmol/L	2nd week	173.5 [a]	130.5 [b]	109.0 [b]	6.92	0.020
	4th week	100.9 [a]	96.3 [ab]	82.2 [b]	4.15	0.035
[6] BHB, mmol/L	2nd week	0.535 [a]	0.405 [ab]	0.377 [b]	0.04	0.013
	4th week	0.349	0.379	0.374	0.03	0.069
[7] PAG, ng/mL	2nd week	1.82 [b]	2.42 [b]	5.21 [a]	0.19	0.005
	4th week	3.23 [b]	2.44 [b]	5.93 [a]	0.18	0.001

[1] Control group; [2] folic acid supplements 5 µg kg^{-1}; [3] folic acid supplements 10 µg kg^{-1}; [4] standard error. [5] Non-esterified fatty acid, [6] Beta-Hydroxybutric Acid, [7] pregnancy-associated glycoprotein. [a,b] Values within a row with different superscripts differ significantly.

4. Discussion

The purpose of this study was to see how FA supplements affected the progesterone profile and blood chemistry of heat-stressed Holstein cows during the first month of pregnancy. Progesterone is thought to have an important role in the implantation and growth of embryos in the early stages of pregnancy [22]. Furthermore, cows exposed to summer heat stress often have a considerable reduction in progesterone levels [23]. This could be attributed to decreased progesterone production or perhaps to disruption of corpus luteum (CL) development during thermal stress [24]. Herein, FA supplements (10 mcg kg^{-1}) significantly improved the progesterone profile in heat-stressed cows during the first month of gestation. Hence, this may enhance the implantation process of embryos under chronic thermal stress. El-Tarabany et al. [22] suggested that FA supplements considerably increased serum progesterone concentrations throughout the early stages of pregnancy, which is in line with our findings. In this regard, Starbuck et al. [25] found a positive relationship between plasma progesterone levels and pregnancy maintenance at week five of pregnancy. In heat-stressed mice, FA supplementation scavenges the reactive oxygen species in embryos and improves the cell number in blastocyst [26]. Shin and Shiota [27] also suggested that FA supplements decreased the incidence of neural tube abnormalities in heat-stressed mouse embryos. Furthermore, at high levels of THI, using progesterone- releasing intra-vaginal inserts (CIDRsynch protocol) maintains an acceptable and consistent pregnancy rate [3]. On day 16, uterine luminal concentrations of interferon tau were associated with progesterone levels on days 4 and 5, preventing luteolisis and thereby embryonic losses in pregnant cows [28].

Previous research has shown that the serum folate levels in dairy cows reduced by 40% from the second month post partum (around AI) through parturition [15], which is usually independent of an exogenous FA supplements [29]. Accordingly, ruminal synthesis of folates appears to be insufficient to restore serum folates throughout early pregnancy and lactation. In this context, the drastic decline of serum folates in multiparous dairy cows could reflect an elevated requirement for FA by the maternal–fetal complex. In the current study, the FA_{10} group exhibited significantly greater progesterone levels than the control group during the first three weeks of gestation. Both FA-supplemented groups had significantly higher levels of serum folates than the CON group by the end of the fourth week of pregnancy. Li et al. [30] found that serum folate concentrations increased linearly with increasing rumen protected FA (RPFA) supplementation in multiparous Holstein cows, which is consistent with our findings. Similarly, supplemental FA at 2.6 g/day significantly improved plasma folate concentrations [31]. Girard and Matte [16] also claimed that FA supplementation enhanced the concentrations of serum folate, which peaked in supplemented cows during the early lactation period.

Glucose, as the principal metabolic fuel for maintenance, fetal growth, and milk production, is an essential component of dairy cows' protection against negative energy balance [32]. Due to a negative energy balance after calving, which is reflected by changes in blood metabolic and hormone profiles, high rates of body condition score reductions occur [33]. The large energy demand of dairy cows is partially supplied by gluconeogenesis, and glucose plays a vital role in metabolism and homeostasis [34]. In the current study, FA supplementation (FA_{10}) increased the serum glucose in heat-stressed cows during the early gestation period. Consistent with these findings, Li et al. [30] observed significant increases in plasma glucose levels when multiparous dairy cows were fed RPFA-supplemented diets. Similarly, supplementing nursing cows with FA and vitamin B12 improves metabolic efficiency by raising plasma glucose or glucose irreversible loss rate in the early stages of lactation [31]. They discovered that amino acid gluconeogenesis, such as glycine, serine, threonine, and total sulphur amino acids, can promote serum glucose level. In a more recent study, Li et al. [35] found that adding RPFA to maternal diets boosted serum glucose levels in growing lambs. Duplessis et al. [36] also demonstrated that weekly intramuscular injections of FA raised plasma glucose in postpartum dairy cows (320 mg). The discrepancy in the results could be explained by the fact that the current study used a lower dose of FA than Duplessis et al. [36]. When compared to the CON group, FA supplementation reduced serum urea levels at the 4th week of pregnancy. In Holstein cows, Rastani et al. [37] observed that the concentration of urea depends on the DM intake and the variation of urea may be influenced by the levels of NEFA. Meanwhile, others reported that the RPFA supplements did not affect the levels of serum urea in growing lambs [35]. They also support our findings that FA supplements did not affect the levels of serum triglycerides and cholesterol.

The catabolic and anabolic activities of lipid in the animal body are reflected in serum lipid metabolites. In this scenario, negative energy balance triggers fat mobilization, resulting in a rise in blood NEFA and BHB concentrations [38]. The reduced serum NEFA and BHB in FA-supplemented groups (mostly FA10) may imply that body fat catabolism was dramatically reduced in heat-stressed pregnant cows. As a result of these findings, FA supplements improved energy balance and lowered body fat mobilization, indicating positive changes in BW of pregnant cows. Li et al. [30] observed that during the early lactation stage of Holstein cows, blood concentrations of NEFA and BHB dropped linearly as RPFA supplementation increased. Meanwhile, others suggested that serum NEFA and BHB were not influenced by the addition of RPFA to the diets of growing Hu lambs [35].

IGF-1 and GH play important roles in cell metabolism, adipogenesis, glucose and energy metabolism [39]. Hence, IGF-1 and GH are useful indicators for assessing the metabolic condition of animals. In the present study, both FA-supplemented groups exhibited significantly higher levels of GH anf IGF-1 at the 4th week of gestation. Similarly, the addition of RPFA to the lamb diets increased the levels of serum GH and IGF-1 at days 120 and 180 of age [35]. In this context, the increased levels of serum GH and IGF-1 in the RPFA-supplemented groups may be attributed to the elevated ruminal VFA concentrations or dietary protein availability [40,41]. Wang et al. [42] found that dietary FA supplementation raised IGF-1 levels at the 90th day of pregnancy in multiparous pregnant ewes. Nonetheless, they revealed that dietary FA supplementation had no effect on GH concentrations. In dairy heifers, weekly FA injections (40 mg) had no effect on serum GH and IGF-1 secretion [43].

The placenta of ruminants such as cows, ewes, and goats expresses a vast family of inactive aspartic proteinases known as PAG [44]. Furthermore, PAG concentration is a valid marker for placental growth and function that could be utilized to predict embryonic loss [45]. The FA_{10} group had considerably greater PAG levels than the FA_5 and CON groups throughout the first month of pregnancy. This seems to be correlated with the higher progesterone levels in the FA_{10} group. Ayad et al. [46] found a favorable association between progesterone and PAG concentrations in dairy cattle during the first three months of pregnancy, which supports our findings. In this context, Ricci et al. [47] claimed that average PAG concentrations in dairy cows increase from day 15 to day 35 of pregnancy. During the time from day 29 to day 36 following AI, Nanas et al. [48] reported no variation in PAG levels between summer and winter pregnancies.

5. Conclusions

Finally, FA supplements (10 $\mu g\ kg^{-1}$) given to heat-stressed Holstein cows during the first month of pregnancy enhanced their progesterone profile, as well as serum folates, PAG, GH, and IGF-1 levels. In heat-stressed pregnant cows, FA supplementation decreased NEFA and BHB levels. These findings could be useful in developing practical strategies for dairy cows to maintain their regular reproductive patterns under heat stress conditions.

Author Contributions: Conceptualization, A.A.K., A.-H.A.E.-D., A.A.E.-T. and K.M.A.-M.; methodology A.A.K., A.-H.A.E.-D., A.A.E.-T. and K.M.A.-M.; formal analysis, A.A.K., A.-H.A.E.-D., A.A.E.-T. and K.M.A.-M.; investigation, A.A.K., A.-H.A.E.-D., A.A.E.-T. and K.M.A.-M.; resources, A.A.K., A.-H.A.E.-D., A.A.E.-T. and K.M.A.-M.; data curation, A.A.K., A.-H.A.E.-D., A.A.E.-T. and K.M.A.-M.; writing original draft preparation, A.A.K., A.-H.A.E.-D. and K.M.A.-M.; writing review and editing, A.A.K., A.A.E.-T. and K.M.A.-M.; visualization, A.A.K., A.-H.A.E.-D., A.A.E.-T. and K.M.A.-M.; supervision, A.A.K., A.-H.A.E.-D., A.A.E.-T. and K.M.A.-M. All authors have read and agreed to the published version of the manuscript.

Funding: This research received no external funding.

Institutional Review Board Statement: The study was conducted according to the guidelines of the Declaration of Helsinki and approved by the Animal Ethics Committee of Zagazig University, Egypt (approval no. ZU-IACUC/2/F/68/2022).

Data Availability Statement: Data available on reasonable request from the corresponding author.

Acknowledgments: The authors are grateful to Mohammed Abdel-Hameed, the dairy farm (station 2) manager, El Salhiya Company for investment and development, New Salhiya, Sharkia governorate, Egypt, and his team for their cooperation in this study.

Conflicts of Interest: The authors declare no conflict of interest.

References

1. Darwash, A.O.; Lamming, G.E.; Woolliams, J.A. Estimation of genetic variation in the interval from calving to postpartum ovulation of dairy cows. *J. Dairy Sci.* **1997**, *80*, 1227–1234. [CrossRef]
2. Sturman, H.; Oltenacu, E.A.B.; Foote, R.H. Importance of inseminating only cows in estrus. *Theriogenology* **2000**, *53*, 1657–1668. [CrossRef]
3. El-Tarabany, M.S.; El-Tarabany, A.A. Impact of thermal stress on the efficiency of ovulation synchronization protocols in Holstein cows. *Anim. Reprod. Sci.* **2015**, *160*, 138–145. [CrossRef] [PubMed]
4. Wolfenson, D.; Roth, Z.; Merdan, R. Impaired reproduction in heat-stressed dairy cattle: Basic and applied aspects. *Anim. Reprod. Sci.* **2000**, *60–61*, 535–547. [CrossRef]
5. Jordan, E.R. Effects of heat stress on reproduction. *J. Dairy Sci.* **2003**, *86*, E104–E114. [CrossRef]

6. Hansen, P.J.; Aréchiga, C.F. Strategies for managing reproduction in the heat-stressed dairy cow. *J. Anim. Sci.* **1999**, *77* (Suppl. 2), 36–50. [CrossRef]
7. Choi, S.-W.; Mason, J.B. Folate and carcinogenesis: An integrated scheme. *J. Nutr.* **2000**, *130*, 129–132. [CrossRef]
8. Pogribny, I.P.; Karpf, A.R.; James, S.R.; Melnyk, S.; Han, T.; Tryndyak, V.P. Epigenetic alterations in the brains of Fisher 344 rats induced by long-term administration of folate/methyl-deficient diet. *Brain Res.* **2008**, *1237*, 25–34. [CrossRef]
9. Khan, M.Z.; Khan, A.; Xiao, J.; Dou, J.; Liu, L.; Yu, Y. Overview of folic acid supplementation alone or in combination with vitamin B12 in dairy cattle during periparturient period. *Metabolites* **2020**, *10*, 263. [CrossRef]
10. Menzies, K.K.; Lefèvre, C.; Sharp, J.A.; Macmillan, K.L.; Sheehy, P.A.; Nicholas, K.R. A novel approach identified the FOLR1 gene, a putative regulator of milk protein synthesis. *Mamm. Genome* **2009**, *20*, 498–503. [CrossRef]
11. Schwab, E.C.; Schwab, C.G.; Shaver, R.D.; Girard, C.L.; Putnam, D.E.; Whitehouse, N.L. Dietary forage and nonfiber carbohydrate contents influence B-vitamin Intake, duodenal flow, and apparent ruminal synthesis in lactating dairy cows. *J. Dairy Sci.* **2006**, *89*, 174–187. [CrossRef]
12. Ghoshal, K.; Li, X.; Datta, J.; Bai, S.; Pogribny, I.; Pogribny, M.; Huang, Y.; Young, D.; Jacob, S.T. A folate and methyl-deficient diet alters the expression of DNA methyltransferases and methyl CpG binding proteins involved in epigenetic gene silencing in livers of F344 rats. *J. Nutr.* **2006**, *136*, 1522–1527. [CrossRef] [PubMed]
13. Santschi, D.E.; Chiquette, J.; Berthiaume, R.; Martineau, R.; Matte, J.J.; Mustafa, A.F.; Girard, C.L. Effects of the forage to concentrate ratio on B-vitamin concentrations in different ruminal fractions of dairy cows. *Can. J. Anim. Sci.* **2005**, *85*, 389–399. [CrossRef]
14. Collier, R.J.; Dahl, G.E.; VanBaale, M.J. Major advances associated with environmental effects on dairy cattle. *J. Dairy Sci.* **2006**, *89*, 1244–1253. [CrossRef]
15. Girard CLMatte, J.J.; Tremblay, G.F. Serum folates in gestating and lactating dairy cows. *J. Dairy Sci.* **1989**, *72*, 3240–3246. [CrossRef]
16. Girard, C.L.; Matte, J.J. Dietary supplements of folic acid during lactation: Effects on the performance of dairy cows. *J. Dairy Sci.* **1998**, *81*, 1412–1419. [CrossRef]
17. Duplessis, M.; Girard, C.L.; Santschi, D.E.; Laforest, J.P.; Durocher, J.; Pellerin, D. Effects of folic acid and vitamin B12 supplementation on culling rate, diseases, and reproduction in commercial dairy herds. *J. Dairy Sci.* **2014**, *97*, 2346–2354. [CrossRef]
18. Czeizel, A.E.; Vargha, P. Periconceptional folic acid/multivitamin supplementation and twin pregnancy. *Am. J. Obstet. Gynecol.* **2004**, *191*, 790–794. [CrossRef]
19. Gagnon, A.; Khan, D.R.; Sirard, M.A.; Girard, C.L.; Laforest, J.P.; Richard, F.J. Effects of intramuscular administration of folic acid and vitamin B12 on granulosa cells gene expression in postpartum dairy cows. *J. Dairy Sci.* **2015**, *98*, 7797–7809. [CrossRef]
20. National Research Council. *Nutrient Requirements of Dairy Cattle*, 7th ed.; National Academies Press: Washington, DC, USA, 2001.
21. Kendall, P.E.; Webster, J.R. Season and physiological status affects the circadian body temperature rhythm of dairy cows. *Livest. Sci.* **2009**, *125*, 155–160. [CrossRef]
22. El-Tarabany, M.S.; Atta, M.A.; Emara, S.S.; Mostafa, M.M. Folic acid and flaxseed oil supplements in Ossimi ewes: Effect on body weight changes, progesterone profile, blood chemistry, and litter traits. *Trop. Anim. Health Prod.* **2020**, *52*, 301–308. [CrossRef] [PubMed]
23. Wolfenson, D.; Sonego, H.; Bloch, A.; Shaham-Albalancy, A.; Kaim, M.; Folman, Y.; Meidan, R. Seasonal differences in progesterone production by luteinized bovine thecal and granulosa cells. *Domest. Anim. Endocrinol.* **2002**, *22*, 81–90. [CrossRef]
24. Wolfenson, D.; Roth, Z. Impact of heat stress on cow reproduction and fertility. *Anim. Front.* **2019**, *9*, 32–38. [CrossRef] [PubMed]
25. Starbuck, M.J.; Dailey, R.A.; Inskeep, E.K. Factors affecting retention of early pregnancy in dairy cattle. *Anim. Reprod. Sci.* **2004**, *84*, 27–39. [CrossRef] [PubMed]
26. Koyama, H.; Ikeda, S.; Sugimoto, M.; Kume, S. Effects of folic acid on the development and oxidative stress of mouse embryos exposed to heat stress. *Reprod. Domest. Anim.* **2012**, *47*, 921–927. [CrossRef]
27. Shin, J.-H.; Shiota, K. Folic acid supplementation of pregnant mice suppresses heat-induced neural tube defects in the offspring. *J. Nutr.* **1999**, *129*, 2070–2073. [CrossRef]
28. Mann, G.E.; Lamming, G.E. The influence of progesterone during early pregnancy in cattle. *Reprod. Domen. Anim.* **1999**, *34*, 269–274. [CrossRef]
29. Girard, C.L.; Matte, J.J.; Tremblay, G.F. Gestation and lactation of dairy cows: A role for folic acid? *J. Dairy Sci.* **1995**, *78*, 404–411. [CrossRef]
30. Li, H.Q.; Liu, Q.; Wang, C.; Yang, Z.M.; Guo, G.; Huo, W.J.; Pei, C.X.; Zhang, Y.L.; Zhang, S.L.; Wang, H.; et al. Effects of dietary supplements of rumen-protected folic acid on lactation performance, energy balance, blood parameters and reproductive performance in dairy cows. *Anim. Feed Sci. Technol.* **2016**, *213*, 55–63. [CrossRef]
31. Graulet, B.; Matte, J.J.; Desrochers, A.; Doepel, L.; Palin, M.F.; Girard, C.L. Effects of dietary supplements of folic acid and vitamin B12 on metabolism of dairy cows in early lactation. *J. Dairy Sci.* **2007**, *90*, 3442–3455. [CrossRef]
32. De Koster, J.D.; Opsomer, G. Insulin resistance in dairy cows. *Vet. Clin. Food Anim. Pract.* **2013**, *29*, 299–322. [CrossRef] [PubMed]
33. Wathes, D.C.; Cheng, Z.; Chowdhury, W.; Fenwick, M.A.; Fitzpatrick, R.; Morris, D.G.; Patton, J.; Murphy, J.J. Negative energy balance alters global gene expression and immune responses in the uterus of postpartum dairy cows. *Physiol. Genom.* **2009**, *39*, 1–13. [CrossRef] [PubMed]

34. Herdt, T.H. Variability characteristics and test selection in herd-level nutritional metabolic profle testing. *Vet. Clin. N. Am. Food Anim. Pract.* **2000**, *16*, 387–403. [CrossRef]
35. Li, H.Q.; Wang, B.; Li, Z.; Luo, H.L.; Wang, Y.J.; Zhang, C.; Jian, L.Y.; Gao, Y.F.; Lu, W.; Liu, M.; et al. Effects of rumen-protected folic acid addition in maternal and post-weaning diets on growth performance, total tract digestibility, ruminal fermentation and blood metabolites in lambs. *Anim. Feed Sci. Technol.* **2020**, *260*, 114364. [CrossRef]
36. Duplessis, M.; Lapierre, H.; Pellerin, D.; Laforest, J.P.; Girard, C.L. Effects of intramuscular injections of folic acid, vitamin B12, or both, on lactational performance and energy status of multiparous dairy cows. *J. Dairy Sci.* **2017**, *100*, 4051–4064. [CrossRef] [PubMed]
37. Rastani, R.R.; Lobos, N.E.; Aguerre, M.J.; Grummer, R.R.; Wattiaux, M.A. Relationships between Blood Urea Nitrogen and Energy Balance or Measures of Tissue Mobilization in Holstein Cows During the Periparturient Period. *Profess Anim. Sci.* **2006**, *22*, 382–385. [CrossRef]
38. Xu, C.; Wang, Z.; Liu, G.W.; Li, X.B.; Xie, G.H.; Xia, C.; Zhang, H.Y. Metabolic characteristic of the liver of dairy cows during ketosis based on comparative proteomics. *Asian-Aust. J. Anim. Sci.* **2008**, *21*, 1003–1008. [CrossRef]
39. Blüher, S.; Kratzsch, J.; Kiess, W. Insulin-like growth factor I, growth hormone and insulin in white adipose tissue. *Best Pract. Res. Clin. Endocrinol. Metab.* **2005**, *19*, 577–587. [CrossRef]
40. Zhao, G.Y.; Sun, Y.B. Effects of volatile fatty acids on IGF-I, IGFBP-3, GH, insulin and glucagon in plasma, and IGF-I and IGFBP-3 in different tissues of growing sheep nourished by total intragastric infusions. *Asian-Aust. J. Anim. Sci.* **2010**, *23*, 366–371. [CrossRef]
41. Smith, J.M.; Amburgh, M.E.V.; Diaz, M.C.; Lucy, M.C.; Bauman, D.E. Effect of nutrient intake on the development of the somatotropic axis and its responsiveness to GH in Holstein bull calves. *J. Anim. Sci.* **2002**, *80*, 1528–1537. [CrossRef] [PubMed]
42. Wang, B.; Li, Z.; Li, H.; Luo, H.; Blair, H.T.; Jian, L.; Diao, Z. Effect of dietary folic acid supplementation during pregnancy on blood characteristics and milk composition of ewes. *Animals* **2020**, *10*, 433. [CrossRef] [PubMed]
43. Petitclerc, D.; Dumoulin, P.; Ringuet, H.; Matte, J.; Girard, C. Plane of nutrition and folic acid supplementation between birth and four months of age on mammary development of dairy heifers. *Can. J. Anim. Sci.* **1999**, *79*, 227–234. [CrossRef]
44. Haugejorden, G.; Waage, S.; Dahl, E.; Karlbert, K.; Beckers, J.F.; Ropstad, E. Pregnancy associated glycoproteins (PAG) in postpartum cows, ewes, goats and their offspring. *Theriogenology* **2006**, *66*, 1976–1984. [CrossRef] [PubMed]
45. Pohler, K.G.; Pereira, M.H.C.; Lopes, F.R.; Lawrence, J.C.; Keisler, D.H.; Smith, M.F.; Vasconcelos, J.L.M.; Green, J.A. Circulating concentrations of bovine pregnancy-associated glycoproteins and late embryonic mortality in lactating dairy herds. *J. Dairy Sci.* **2016**, *99*, 1584–1594. [CrossRef] [PubMed]
46. Ayad, A.; Sousa, N.M.; Sulon, J.; Hornick, J.L.; Watts, J.; Lopez-Gatius, F.; Iguer-Ouada, M.; Beckers, J.F. Influence of progesterone concentrations on secretory functions of trophoblast and pituitary during the first trimester of pregnancy in dairy cattle. *Theriogenology* **2007**, *67*, 1503–1511. [CrossRef]
47. Ricci, A.; Carvalho, P.D.; Amundson, M.C.; Fourdraine, R.H.; Vincenti, L.; Fricke, P.M. Factors associated with pregnancy-associated glycoprotein (PAG) levels in plasma and milk of Holstein cows during early pregnancy and their effect on the accuracy of pregnancy diagnosis. *J. Dairy Sci.* **2015**, *98*, 2502–2514. [CrossRef]
48. Nanas, I.; Chouzouris, T.M.; Dovolou, E.; Dadouli, K.; Stamperna, K.; Kateri, I.; Barbagianni, M.; Amiridis, G.S. Early embryo losses, progesterone and pregnancy associated glycoproteins levels during summer heat stress in dairy cows. *J. Therm. Biol.* **2021**, *98*, 102951. [CrossRef]

animals

MDPI

Article

Modulation of Antioxidant Defense, Immune Response, and Growth Performance by Inclusion of Propolis and Bee Pollen into Broiler Diets

Saad N. AL-Kahtani [1], Abdulaziz A. Alaqil [2,*] and Ahmed O. Abbas [2,3,*]

1 Department of Arid Land Agriculture, College of Agricultural and Food Sciences, King Faisal University, P.O. Box 420, Al-Ahsa 31982, Saudi Arabia; salkahtani@kfu.edu.sa
2 Department of Animal and Fish Production, College of Agricultural and Food Sciences, King Faisal University, P.O. Box 420, Al-Ahsa 31982, Saudi Arabia
3 Department of Animal Production, Faculty of Agriculture, Cairo University, 7 Gamma St., Giza 12613, Egypt
* Correspondence: aalaqil@kfu.edu.sa (A.A.A.); aabbas@kfu.edu.sa (A.O.A.)

Simple Summary: Broiler meat provides a considerable nutritional source of animal protein in the human diet. However, the intensive and accelerating growth in broiler breeding programs requires a continuous or intermittent use of antibiotics to improve the health and performance of broilers. Antibiotic resistance and residues problems cause a general limitation on the use of antibiotics in poultry production worldwide, and hence, prompt researchers and feed companies to find natural alternatives. In the present study, we investigated the possible impact of propolis (PR) and bee pollen (BP) in nutritional strategies on the performance and immunity of broiler chickens. The obtained results display the ability of PR and BP contained within the broiler diets to enhance the antioxidant defense system and improve several immunological parameters. These beneficial effects coincided with an increase in the growth performance of broilers. Thus, supplementation of PR and BP separately or in combination could be recommended into broiler diets for their positive impacts as natural products on the performance and health of broilers.

Abstract: (1) Background: Propolis and bee pollen have natural bioactive compounds that may support the performance and immunological response of broilers. (2) Methods: The study included 300 1 d old Cobb-500 broiler chicks. Starting from 22–42 d of age, chicks were divided according to a 2 × 2 factorial design into one of the four treatment groups (5 replicates × 15 chicks per replicate); a basal diet without supplementation (CONT) or supplemented with 1 g/kg of propolis (PR) or bee pollen (BP) separately or in an even combination (PR + BP). (3) Results: A significant ($p < 0.05$) increase was obtained in the body-weight gain of broilers treated with PR, BP, and PR + BP compared to the CONT. The total antioxidant capacity and superoxide dismutase were highly ($p < 0.05$) activated in all treated groups compared to the CONT. Immunological parameters, especially the leukocyte cell viability, T- and B-lymphocyte proliferation, immunoglobulins (IgA and IgM), antibody titers, and wattle-swelling test were significantly ($p < 0.05$) enhanced in the treated broilers with PR and/or BP compared to the CONT. (4) Conclusions: The dietary supplementation of PR and/or BP could be beneficial for broiler growth through maximizing the antioxidant- and immune-system defenses.

Keywords: propolis; bee pollen; productive performance; antioxidants; immune response; broilers

Citation: AL-Kahtani, S.N.; Alaqil, A.A.; Abbas, A.O. Modulation of Antioxidant Defense, Immune Response, and Growth Performance by Inclusion of Propolis and Bee Pollen into Broiler Diets. *Animals* **2022**, *12*, 1658. https://doi.org/10.3390/ani12131658

Academic Editor: Raffaella Rossi

Received: 10 May 2022
Accepted: 24 June 2022
Published: 28 June 2022

1. Introduction

The intensive and rapid production in broiler selection programs require continuous or intermittent courses of antibiotic therapy to improve broiler well-being and performance. However, antibiotic resistance and residues problems lead to a general limitation on the use of antibiotics in poultry production worldwide, and prompt researchers and feed companies to seek natural alternatives [1]. Recently, the natural products have been widely considered in nutritional strategies for optimizing the health and performance of poultry during intensive production. Among the regarded candidates of natural products are those produced by honey-bee workers, such as propolis (PR) and bee pollen (BP) [2].

Propolis consists of resinous substances from various plants in addition to essential oils and waxes gathered by the bees [3], while BP consists of pollen grains mixed with nectar and the hypopharyngeal-glands secretion of the bees [4]. Both products are considered to be feed supplements in animal nutrition due to their abundant sources of nutrients [5,6], flavonoids [7,8], antioxidants [4,9], digestive enzymes [10,11], and antimicrobial compounds [12,13].

The beneficial effects of PR and BP on poultry production and health have been documented in research. It was reported that dietary PR supplementation reduced the oxidative stress induced by paraquat herbicides in turkey [14] or by heat stress in Japanese quail [15]. Bee pollen supplementation into broiler diets promotes some immunological traits, such as increasing leukocytes, decreasing heterophil/lymphocyte ratio, speeding the antibody production, and reinforcing the immune-organs formation [16,17]. In addition, broiler growth aspects were enhanced by dietary PR and/or BP supplementation through morphological and bacterial regulation in the gastrointestinal tract [13]. Further studies concluded that honeybee products, including PR and BP, improve the growth performance and immune functions in Japanese quail [18]. However, the information available in the literature about the addition of bee products into broiler diets remain scant, especially regarding the physiological mechanism of action such as immunological and antioxidant status. The objective of this study was to highlight the possible impact of PR and BP inclusion, alone or together, into broiler diets on their performance along with the antioxidant and immunological defense system.

2. Materials and Methods

2.1. PR and BP Preparation and Analysis

The PR and BP were obtained from a collection of beehives situated in the Agricultural and Veterinary Research Station, King Faisal University, Saudi Arabia. Propolis was obtained in the form of yellow-brown powder, while dry BP was obtained as small yellow pellets. Samples of PR and BP were subjected to a chemical analysis based on the methods of International AOAC [19]. The total phenolic contents of both PR and BP were estimated according to the Folin–Ciocalteu colorimetric methods [20], considering gallic acid as the standard and obtaining the optical density by a CE1010-Spectrophotometer (Cecil Instruments Limited, Cambridge, UK) at 765 nm. The total flavonoids in PR and BP were also determined according to the aluminum calorimetric methods [21], considering catechol as the standard and measuring the absorbance at 435 nm by the spectrophotometer. In addition, the scavenging activities of PR and BP samples against 2,2-diphenyl-1-picrylhydrazyl (DPPH)-free radicals were measured by the spectrophotometer at 520 nm [22]. Table 1 represents the chemical characteristics of the PR and BP used in the experiment.

Table 1. The chemical characteristics of propolis and bee pollen.

Item	Propolis	Bee Pollen
Dry matter (%)	90.8	90.5
Carbohydrate (g) [1]	1.9 g	67.6
Crude fiber (g) [1]	68.7	1.2
Total lipids (g) [1]	9.2 g	3.7
Crude protein (g) [1]	2.6 g	17.1
Total ash (g) [1]	0.9 g	2.9
Phenolic content (mg GAE/g) [2]	2.8	2.4
Flavonoid content (mg CAT/g) [2]	1.4	0.9
DPPH-free radical scavenging activity (%) [3]	89.3	84.5

[1] Results of chemical analyses calculated per 100 g dry matter. [2] Calculated as mg gallic acid equivalent (GAE) or catechol equivalent (CAT), respectively, per g dry weight of the sample. [3] Calculated as % of the scavenging activity against 2,2-diphenyl-1-picrylhydrazyl (DPPH) radicals.

2.2. Birds and Treatments

A total of 300 male broiler chicks (Cobb500™) were obtained from a local hatchery at one day of age and raised in an open-system floor house. During the experiment, all the chicks were maintained within the same optimum conditions of temperature, humidity, and lighting as recommended by the manufacturer's guideline of Cobb-500 broiler management (available at: https://www.cobb-vantress.com/en_US/products/cobb500/; accessed on 1 March 2022). According to the guidelines, basal diets of soybean–corn mixture were formulated to meet the standard requirements of Cobb-500 broilers (Table 2). Birds were given unlimited access to food and water during the experiment.

Table 2. The chemical characteristics of propolis and bee pollen.

Ingredients (g/kg)	Starter (0–8 d)	Grower (9–21 d)	Finisher (22–42 d)
Corn	607.0	654.0	693.0
Gluten meal	70.0	50.0	50.0
Soybean meal, 48% CP	289.0	243.0	203.0
Soybean oil	0.0	20.0	22.0
Di-calcium phosphate	4.0	4.0	4.0
Limestone	20.0	19.0	18.0
Salt	4.5	4.5	4.5
Vitamin–Mineral Premix [1]	5.5	5.5	5.5
Nutritional composition			
Dry matter (g/kg) [2]	906.0	901.0	908.9
Total ash (g/kg) [2]	55.0	53.0	39.1
Crude protein (g/kg) [2]	229.8	199.8	184.6
Crude fat (g/kg) [2]	58.3	77.5	83.4
Crude fiber (g/kg) [2]	32.0	35.0	35.8
Metabolizable energy (MJ/kg) [3]	12.6	13.1	13.3
L-lysine (g/kg) [3]	12.1	11.6	10.4
DL-Methionine (g/kg) [3]	4.8	4.7	4.3
Calcium (g/kg) [3]	9.1	8.6	8.1
Available phosphorus (g/kg) [3]	4.5	4.2	4.1

[1] Premix provides the following components per kg of the basal diet: vitamins A 10 KIU, D_3 5 KIU, E 65 IU, K 3 mg, B_1 3 mg, B_2 9 mg, B_6 4 mg, B_{12} 0.02 mg, biotin 0.20 mg, niacin 20 mg, pantothenic acid 15 mg, folic acid 2 mg, and choline chloride 500 mg; and minerals Mn 100 mg, Fe 40 mg, Zn 100 mg, Cu 15 mg, Se 0.35 mg, and Iodine 1 mg. [2] Determined values. [3] Calculated values.

From 22–42 d of age, birds were randomly assigned into 4 treatment groups according to a 2 × 2 factorial design with five replicates of 15 birds each (75 chicks per group). The birds in each replicate were raised on 5 cm-deep litter of wood-shavings in a floor yard area of 1.35 × 1.35 m^2. The first group served as a control and was fed on a basal diet without supplementation (CONT). The other experimental groups were fed on a basal diet supplemented with either 1 g/kg propolis (PR group), 1 g/kg bee pollen (BP group), or an even mixture of PR and BP at 1 g/kg each (PR + BP group), respectively. Both PR and BP were crushed using a grinder (Moulinex Type LM201, Mayenne, France) and mixed daily with the basal diet before they were introduced to the broilers. The productive performance of chicks was evaluated during the experimental period length of 22–42 d of age. Furthermore, blood samples were taken from birds when the treatments ended at 42 days of age for the purpose of studying some antioxidant indicators and immunological response, as posteriorly described in detail.

2.3. Productive Performance

Individual body weights in each group were recorded at days 22 and 42 of age to determine the initial body weight (IBW), final body weight (FBW), and body weight gain (BWG) for each experimental group. Feed intake (FI) was calculated by taking the leftover feed from the total amount of feed that was given for each replicate in the treatment group.

The feed conversion ratio (FCR) was then determined for each replicate in the treatment group based on FI per unit of BWG.

2.4. Antioxidant Indicators

As soon as the treatments were over, two blood samples from the brachial vein were taken for each replicate per experimental group (n = 10) and immediately transferred into heparinized tubes. Plasma was separated by centrifuging the blood samples for 10 min at 2000× *g* at 4 °C and stored to figure out the antioxidant indicators. The total antioxidant capacity (TAC), total superoxide dismutase (T-SOD), and catalase (CAT) assays were performed using an automated microplate scanner and the available colorimetric kits (MBS2540515, MBS2563691, and MBS2540413, respectively; MyBioSource, San Diego, CA, USA) according to the manufacturer's instructions. Table S1 summarizes the detection limits, sensitivity, intra-assay CV%, and inter-assay CV% for each assay.

2.5. Immunological Parameters

2.5.1. Leukocyte's Count, Differentiation, and Viability

Two blood samples per replication in each experimental group (n = 10) were taken at the conclusion of the treatments (42 d of age) and relocated into heparinized tubes. Ten µL of the fresh sample was diluted with 490 µL brilliant cresyl blue stain solution. A drop of the mixture was mounted on a hemocytometer slide, and the total white blood cells (TWBC) count was then detected under a microscope at 200× magnification [23]. Another 10 µL of the blood sample was smeared on a glass slide, then fixed and stained using Hema-3 solutions (Fisher Scientific, Pittsburg, PA, USA). Differentiation of approximately 200 leukocytes was performed under a microscope at 1000× magnification with oil immersion, and the heterophil-to-lymphocyte (H/L) ratio was then detected [24].

The remaining blood samples were allocated to determine leucocyte cell viability (LCV) according to the methods described by Abbas et al. [25]. First, peripheral blood mononuclear cells (PBMC) were separated by centrifuging the blood samples with Histopaque-1077 medium (Sigma Chemical Co., St. Louis, MO, USA) at 1030× *g* for 20 min at 4 °C. The PBMCs were washed twice using RPMI-1640 culture medium (Thermo Fisher Scientific, Waltham, MA, USA), then resuspended with phosphate-buffered saline (PBS) (pH 7.2). Thereafter, 100 µL of cell suspension were pipetted with 25 µL of MTT solution (5 mg of tetrazolium salt; MTT, Serva, Heidelberg, Germany; dissolved in 1 mL of AIM-V medium; Thermofisher) in a 96-well plate. The plates were incubated for 4 h at 37 °C, then centrifuged at 600× *g* for 10 min. The incubation medium was discarded, and each well was refilled with 100 µL of acidified isopropyl alcohol solution (0.04 N HCl). Finally, the absorbance of formazan was measured at 570 nm using an automated ELISA reader (Bio-Rad Laboratories Inc., Hercules, CA, USA).

2.5.2. Lymphocyte Proliferation

Proliferation of T- and B-lymphocyte cells was evaluated in blood samples obtained from 2 broilers per replicate (n = 10 per treatment group) according to the methodology described by Alaqil et al. [26]. The protocol, in brief, started with the isolation of the PBMCs, washing twice, and resuspending in RPMI-1640 medium as previously stated. The viable lymphocytes in each sample were plated in triplicates at a constant concentration of 1×10^6 cells/mL in a 96-well plate. The experimental wells were supplemented with 50 µL of 5% Concanavalin-A mitogen or 1% Lipopolysaccharide to stimulate T- or B-lymphocytes, respectively, whereas the control wells were filled with 50 µL RPMI medium. The samples were incubated for 48 h at 42 °C, 5% CO_2, and saturated humidity, then appended with 15 µL of MTT solution and incubated again for 4 h, and finally complemented with 100 µL of 10% sodium dodecyl sulfates in 0.04 M HCl. The optical density at 570 nm (OD570) was recorded for the experimental wells against the control wells using an automated ELISA. Stimulation index (SI) of T- and B-lymphocytes was computed as the OD570 ratio for stimulated to unstimulated cells in each sample.

2.5.3. Immunoglobulins Assay

Blood samples were collected into heparinized tubes from 2 chicks per replicate in each treatment group (n = 10). The plasma was separated by centrifuging at 2000× *g* for 10 min at 4 °C then kept at −20 °C to be used in immunoglobulin (Ig) assay. The IgA, IgM, and IgG were analyzed in accordance with the manufacturer instructions of commercial ELISA kits specific for chickens (MBS564152, MBS706158 and MBS260043, respectively; MyBioSource). In brief, 100 µL of the diluted samples (1:5000 in the sample-conjugate diluent) or standards were pipetted (in duplicate) into predesignated wells in a microtiter plate. After incubation, the plates were washed twice, and the contents were removed by sharp striking on an absorbent paper. The plate was then incubated in the dark after the addition of 100 µL of an appropriate dilution of enzyme–antibody conjugate. The contents were removed after being washed three times, then 100 µL of Chromogen-substrate solution were pipetted to each well and incubated in the dark. Finally, the reaction was stopped by the addition of 100 µL sulfuric acid (0.3 M) and the absorbance was read at 450 nm using ELISA microplate-reading scanner. The IgA, IgM, and IgG levels were computed using a 4-parameter logistic curve fit generated from the chicken reference serum absorbance. The specifications of the immunoglobulin ELISA assays are summarized in Table S2.

2.5.4. Humoral and Cellular Immunity Assay

The sheep red blood cells antibody (SRBC-AB) titer was evaluated to point out the humoral immunity of broilers in this study. One week before the end of the experiment (at 35 d of age), 2 birds per replicate per treatment group (n = 10) received an IV injection with 1 mL of 5% SRBC. After that, blood samples were taken from the birds, allowed for clotting at RT for 2 h, then centrifuged at 4 °C for 10 min at 400× *g* to separate the sera. Serial doubling dilutions of sera samples (25 µL each) were pipetted in a 96-well plate, and 25 µL of 2% SRBC solution was added to each dilution. The plates were gently vortexed and kept overnight for agglutination at RT. The antibody titer was calculated as $\log_2$ value of the inverse of the last dilution with positive agglutination in the well's bottom [27].

The cellular immunity of broilers was assessed using procedures outlined in a prior study [28]. At 42 d of age, 2 birds per replication in each treatment group (n = 10) were assigned for the test. Briefly, birds were intradermally injected with 0.1 mL sterile PBS supplemented with 0.5 mg mitogenic phytohemagglutinin (PHA) (Thermo Fisher Scientific) in a marked area of the wattle. Twenty-four hours later, the increase in the wattle thickness was measured as a positive reaction to the PHA-wattle immune test.

2.6. Statistical Analysis

Data for all variables were analyzed using two-way analysis of variance (ANOVA) and explored with a General Linear Model (GLM) procedure of SPSS software (version 22.0; IBM Corp., Armonk, NY, USA, 2013). The main factors were propolis supplementation (−PR versus + PR) and bee pollen supplementation (−BP versus + BP). The interaction between the two main factors (PR × BP) was tested and represented as CONT, PR, BP, and PR + BP, respectively. The experimental unit was considered to be the number of observations per treatment group for each test performed (n = 5 for the productive performance traits, and n = 10 for the other antioxidant indicators and immunological parameters). The mean differences were tested at 0.05 level of significance using the post hoc Duncan's test.

3. Results

3.1. Productive Performance

The effect of dietary supplementations of PR, BP, and their interaction on the broiler performance is shown in Table 3. The broiler productive performances from 1–21 d of age were nearly similar between groups and within the normal ranges of the Cobb-500 broiler's performance guideline. From 22–42 d of age, PR and BP treatments substantially ($p < 0.05$) increased the FBW, BWG, and FI of broilers. In contrast, no significant differences were obtained in the FCR among PR or BP groups ($p > 0.05$). No interaction effect was observed in the performance traits of broilers except for the BWG, indicating the highest BWG in PR + BP combination group followed by PR, BP, and CONT group, respectively.

Table 3. Effect of propolis (PR), bee pollen (BP), and their interaction (PR + BP) on the productive performance of broiler chickens.

Groups		Traits [1]				
		IBW (g)	FBW (g)	BWG (g)	FI (g)	FCR
PR effect	−PR	755	2489 [b]	1734 [b]	154 [b]	1.86
	+PR	745	2666 [a]	1920 [a]	165 [a]	1.81
	n	10	10	10	10	10
	SEM	7.8	16.4	8.9	2.4	0.021
	p-value	0.376	<0.001	<0.001	0.003	0.095
BP effect	−BP	754	2528 [b]	1774 [b]	155 [b]	1.84
	+BP	746	2628 [a]	1881 [a]	164 [a]	1.83
	n	10	10	10	10	10
	SEM	7.8	16.4	8.9	2.4	0.021
	p-value	0.476	0.001	<0.001	0.024	0.669
Interaction effect	CONT	760	2421	1661 [d]	148	1.87
	PR	749	2636	1887 [b]	163	1.82
	BP	751	2559	1808 [c]	160	1.86
	PR + BP	742	2697	1955 [a]	168	1.80
	n	5	5	5	5	5
	SEM	10.9	23.1	12.5	3.4	0.030
	p-value	0.900	0.118	0.006	0.295	0.921

Means with dis-similar superscripts (a, b, c, d) in the same column significantly differ at $p < 0.05$. SEM: standard error of means. n: number of observations in the treatment group. [1] Traits: IBW, initial body weight 22 d of age; FBW, final body weight at 42 d of age; BWG, body weight gain; FI, feed intake; and FCR, feed conversion ratio.

3.2. Antioxidant Indicators

The effect of PR, BP, and their interaction treatments on the plasma antioxidant indicators of broiler chickens is represented in Table 4. A significant increase in the TAC, T-SOD, and CAT activity was obtained in broilers treated with PR or BP. There was a significant interaction effect between PR and BP. In comparison with the CONT, the TAC, T-SOD, and CAT were significantly ($p < 0.05$) higher when PR or BP was added to the broiler diets, especially in the PR + BP group ($p < 0.05$).

Table 4. Effect of propolis (PR), bee pollen (BP), and their interaction (PR + BP) on the antioxidant indicators of broiler chickens.

Groups		Traits [1]		
		TAC (U/mL)	T-SOD (U/mL)	CAT (U/mL)
PR effect	−PR	4.81 [b]	4.57 [b]	0.81 [b]
	+PR	5.94 [a]	6.70 [a]	0.86 [a]
	n	20	20	20
	SEM	0.046	0.114	0.009
	p-value	<0.001	<0.001	<0.001
BP effect	−BP	5.01 [b]	4.33 [b]	0.79 [b]
	+BP	5.74 [a]	6.95 [a]	0.87 [a]
	n	20	20	20
	SEM	0.046	0.114	0.009
	p-value	<0.001	<0.001	<0.001

Table 4. *Cont.*

Groups		Traits [1]		
		TAC (U/mL)	T-SOD (U/mL)	CAT (U/mL)
Interaction effect	CONT	4.18 ^{d}	3.08 ^{d}	0.78 ^{c}
	PR	5.84 ^{b}	5.58 ^{c}	0.80 ^{bc}
	BP	5.43 ^{c}	6.06 ^{b}	0.83 ^{b}
	PR + BP	6.04 ^{a}	7.83 ^{a}	0.91 ^{a}
	n	10	10	10
	SEM	0.065	0.161	0.013
	p-value	<0.001	0.030	0.029

Means with dis-similar superscripts (a, b, c, d) in the same column significantly differ at $p < 0.05$. SEM: standard error of means. n: number of observations in the treatment group. [1] Traits: TAC, total antioxidant capacity; T-SOD, total superoxide dismutase; and CAT, catalase.

3.3. Immunological Parameters

3.3.1. Leukocytes Count, Differentiation, and Viability

The effect of PR, BP, and their interaction on the TWBC count, H/L ratio, and leukocyte cell viability of broiler chickens are presented in Table 5. TWBC was significantly ($p < 0.05$) increased by the PR treatment. The addition of PR or BP to broiler diets significantly ($p < 0.05$) decreased the H/L ratio and increased the LCV. There were no interaction effects for PR and BP on TWBC and H/L ratio. In contrast, the LCV was substantially ($p < 0.05$) enhanced by 13% and 10% in the PR and BP groups, respectively, while the highest level of LCV (18%) was obtained in the PR + BP combination group, compared to the CONT group.

Table 5. Effect of propolis (PR), bee pollen (BP), and their interaction (PR + BP) on leukocytes count, differentiation, and viability of broiler chickens.

Groups		Traits [1]		
		TWBC (10^3/mL)	H/L Ratio	LCV (%)
PR effect	−PR	44.09 ^{b}	0.43 ^{a}	105 ^{b}
	+PR	48.60 ^{a}	0.38 ^{b}	115 ^{a}
	n	20	20	20
	SEM	0.455	0.005	0.4
	p-value	<0.001	<0.001	<0.001
BP effect	−BP	45.76	0.42 ^{a}	106 ^{b}
	+BP	46.93	0.39 ^{b}	114 ^{a}
	n	20	20	20
	SEM	0.455	0.005	0.4
	p-value	0.076	<0.001	<0.001
Interaction effect	CONT	43.24	0.45	100 ^{d}
	PR	48.28	0.39	113 ^{b}
	BP	44.94	0.41	110 ^{c}
	PR + BP	48.93	0.37	118 ^{a}
	n	10	10	10
	SEM	0.644	0.008	0.5
	p-value	0.422	0.194	<0.001

Means with dis-similar superscripts (a, b, c, d) in the same column significantly differ at $p < 0.05$. SEM: standard error of means. n: number of observations in the treatment group. [1] Traits: TWBC, total white blood cells; H/L ratio, heterophils to lymphocytes (H/L) ratio; and LCV, leukocyte cell viability.

3.3.2. Lymphocyte Proliferation and Humoral and Cellular Immunity

The results of lymphocyte proliferation and humoral and cellular immunity of broilers as affected by PR, BP, and their interaction are shown in Table 6. The addition of PR or BP into broiler diets significantly($p < 0.05$) improved the T- and B-lymphocyte SI, SRBC-AB titer, and PHA-wattle immune reaction test. Except for B-lymphocyte proliferation, there was a significant interaction effect on the other traits ($p < 0.05$). The T-lymphocyte SI was elevated in the PR and BP groups compared to the CONT group by 40% and 53%, respectively, while it was elevated by 66% in broilers treated with PR + BP in combination. The broilers treated with PR, BP, or PR + BP exhibited a significant ($p < 0.05$) rise in the anti-SRBC-AB titer by approximately 16% than that in the CONT group. In the PHA-reaction test, the wattle thickness in broilers treated with PR or BP alone was significantly ($p < 0.05$) incremented by approximately 0.09 and 0.11 mm in comparison with the CONT broilers, while the highest wattle swelling (0.23 mm thicker than CONT) occurred in the broilers treated with PR + BP together.

Table 6. Effect of propolis (PR), bee pollen (BP), and their interaction (PR + BP) on lymphocyte proliferation and humoral and cellular immunity of broiler chickens.

Groups		Traits [1]			
		T-lymphocytes SI	B-lymphocytes SI	SRBC-AB Titer (log2)	PHA-Wattle Test (mm)
PR effect	−PR	4.72 ^{b}	2.27 ^{b}	7.02 ^{b}	0.51 ^{b}
	+PR	5.70 ^{a}	3.43 ^{a}	7.51 ^{a}	0.62 ^{a}
	n	20	20	20	20
	SEM	0.034	0.031	0.046	0.003
	p-value	<0.001	<0.001	<0.001	<0.001
BP effect	−BP	4.47 ^{b}	2.31 ^{b}	7.06 ^{b}	0.50 ^{b}
	+BP	5.96 ^{a}	3.39 ^{a}	7.48 ^{a}	0.63 ^{a}
	n	20	20	20	20
	SEM	0.034	0.031	0.046	0.003
	p-value	<0.001	<0.001	<0.001	<0.001
Interaction effect	CONT	3.73 ^{d}	1.71	6.51 ^{b}	0.46 ^{d}
	PR	5.21 ^{c}	2.90	7.60 ^{a}	0.55 ^{c}
	BP	5.72 ^{b}	2.83	7.53 ^{a}	0.57 ^{b}
	PR + BP	6.20 ^{a}	3.95	7.43 ^{a}	0.69 ^{a}
	n	10	10	10	10
	SEM	0.048	0.043	0.065	0.004
	p-value	<0.001	0.500	<0.001	0.001

Means with dis-similar superscripts (a, b, c, d) in the same column significantly differ at $p < 0.05$. SEM: standard error of means. n: number of observations in the treatment group. [1] Traits: SI, stimulation index of T- and B-lymphocytes; SRBC-AB titer, sheep red blood cells antibody titer; and PHA-wattle test, phytohemagglutinin wattle test.

3.3.3. Immunoglobulin Assay

The effect of PR, BP, and their interaction on the immunoglobulin concentration in broilers is summarized in Table 7. A significant increase in the IgA, IgM, and IgG concentrations were obtained in the broilers treated with PR. The BP treatment significantly increased the levels of IgA and IgM in broilers. There was an interaction effect of PR and BP on the IgA concentration, showing the highest IgA levels in broilers treated with PR + BP in combination compared to the other groups.

Table 7. Effect of propolis (PR), bee pollen (BP), and their interaction (PR + BP) on immunoglobulins of broiler chickens.

Groups		Traits [1]		
		IgA (μg/mL)	IgM (μg/mL)	IgG (mg/mL)
PR effect	−PR	162.11 ^{b}	440.48 ^{b}	1.78 ^{b}
	+PR	175.02 ^{a}	502.62 ^{a}	1.87 ^{a}
	n	20	20	20
	SEM	0.490	2.426	0.042
	p-value	<0.001	<0.001	0.042

Table 7. *Cont.*

Groups		Traits [1]		
		IgA (μg/mL)	IgM (μg/mL)	IgG (mg/mL)
BP effect	−BP	162.97 ^{b}	456.83 ^{b}	1.85
	+BP	174.17 ^{a}	486.27 ^{a}	1.81
	n	20	20	20
	SEM	0.490	2.426	0.042
	p-value	<0.001	<0.001	0.506
Interaction effect	CONT	159.15 ^{c}	427.57	1.84
	PR	166.79 ^{b}	486.09	1.86
	BP	165.08 ^{b}	453.39	1.73
	PR + BP	183.26 ^{a}	519.16	1.89
	n	10	10	10
	SEM	0.693	3.430	0.060
	p-value	<0.001	0.298	0.248

Means with dis-similar superscripts (a, b, c, d) in the same column significantly differ at $p < 0.05$. SEM: standard error of means. n: number of observations in the treatment group. [1] Traits: IgA, immunoglobulin A; IgM, immunoglobulin M; and IgG, immunoglobulin G.

4. Discussion

There is a global consent that feeding antibiotics as growth promoters in livestock production can negatively affect human and animal health in the long run. Hence, seeking natural alternatives has been widely welcomed in the feed industry and research, especially in the field of poultry production [1]. In the present study, PR and BP were supplemented into broiler diets during 22–42 d of age at 1 g/kg either alone or in combination, and the results were investigated. The PR and BP levels applied in the previous studies were in a range between 25–5000 mg of PR and 300–45,000 mg of BP per kg of broiler diets [2,13,18,29]. The levels applied in the current work were justified according to a preliminary study, which include a diet with PR at a range of 0.025–5.0 g and BP at a range of 0.3–4.5 g per kg. Based on the results of the preliminary study, 1 g/kg of PR or BP into broiler diets was chosen considering the physiological effects and the economical prospective.

Results show that dietary supplementation of PR or BP at a dose of 1 g per kg could improve some traits of the broiler performance such as FBW and BWG. These results, along with other reports [2], indicate that PR and BP could be successfully applied in broiler diets as natural growth promoters. The significant improvement in BWG of broilers treated with PR + BP, compared to the CONT and other groups, may be due to the nutritive value of PR and BP as additional sources of protein, lipids, and carbohydrates (Table 1). Similar results were also confirmed in broilers [17,30] and in other poultry species, such as turkey [14] and quail [15,18], and in rabbits [8]. In contrast, there is no obvious interaction effect between PR and BP on the FI and FCR of broilers in the present study. Moreover, Attia et al. [2] recorded a reduction in FI in the broiler groups supplemented from 0–35 d of age with 300 mg/kg diets of PR, PB, or their combination compared to the control broilers. Based on the relative weights of body organs (gizzard, liver, heart, intestine, pancreas, and abdominal fat) measured in their study, Attia et al. [2] concluded that reduced FI in PR- and BP-treated groups did not affect the development of the gastrointestinal tract.

As shown in Table 1, PR and BP contain many components with antioxidant activity, such as phenolic acids and flavonoids, and both have scavenging activity against DPPH-free radicals of approximately 85–90%. Compared with the CONT birds, the plasma TAC T-SOD and CAT levels were significantly increased by at least 30%, 81%, and3%, respectively, when PR and BP were added separately or in combination to the broiler diets (Table 4). The improvement of antioxidant defense system in broilers treated with PR and BP may be attributed to the direct capability of polyphenols and flavonoids that existed in the honeybee products to eliminate free radicals [2,31–33]. However, CAT activity seems to be less affected by PR and/or BP in the present study. This could be explained by the finding that the role of CAT enzyme as an antioxidant becomes more substantial during the high levels of oxidative stress [34], which did not occur in the present study.

Results of several immunological parameters in broilers fed with PR and BP are intrinsically discussed in the current study. It was found that TWBC count increased, while H/L ratio decreased in broilers treated with individual PR or BP treatment. Similarly, Attia et al. [35] indicated that PR and/or BP supplemented at 300 mg/kg to Arbor Acers broiler diets significantly ($p < 0.05$) increased TWBC and decreased H/L ratio at 36 d of age compared to the control broilers. In contrast, the PR and BP treatments increased the leukocyte cell viability by 13% and 10%, respectively, compared to the control–untreated birds; moreover, this increment reached 18% over the control when PR and BP were supplemented in combination into broiler diets (Table 5). Similar to the results of the current study, a 40% increment in the leukocyte viability was demonstrated earlier in laying hens treated with PR compared to the untreated birds [25]. It was also suggested that antioxidant properties of propolis can contribute to the increasing leukocyte cell viability through the control of the fork-head box (Foxo) genes pathway involved in cellular apoptosis and oxidative stress resistance [36,37].

Compared to the CONT group, the lymphocyte proliferation was improved in broiler groups fed with PR or BP separately, while the maximum stimulation in T-lymphocytes was obtained in the PR + BP broiler group (Table 6). It was reported that immunomodulatory properties of PR and BP are associated with its high contents of flavonoids and phenolic acids [14,35,38]. Our results are in line with the documented findings that the bioactive and antioxidant compounds in the honeybee products, including PR and BP, sustain the thymus and bursa tissues to consequently generate active T and B lymphocyte cells, respectively, and augment the immune modulation via influencing the lymphocyte proliferation in birds [15,39,40]. In addition, B-lymphocytes are responsible for processing and presenting natural immunoglobulin antibodies [41]. It is known that introducing bee products into birds' nutrition stimulates immuno-competence and triggers antibody production [18].

The present study also proved the beneficial effects of PR and BP on the humeral and cellular immunity in broilers. Results display remarkable increases in the antibody titer against SRBC and wattle swelling against PHA in those broilers supplemented with PR and/or BP (Table 6). In line with our results, the humeral and cellular immune responses in Japanese quail were enhanced in birds supplemented with ethanolic extract of PR or BP powder compared to the control birds [18]. The mechanisms underline these effects were slightly discussed in the literature. The stimulation of humoral immunity in the PR and BP groups may be attributed to the redistribution of peripheral blood leukocytes towards an augmentation in the lymphocyte populations compared to the other components [42]. Other reports suggest that antioxidant properties of flavones and phenols, which exist in PR and BP, may inhibit the synthesis of immunosuppressor, prostaglandin, and thus contribute to a higher humoral response [43]. It is also possible that these compounds prompt macrophages and lymphocytes to release interleukins, such as IL-1 and IL-2, which enhance the T- and B-cells' mitogenesis [44]. Furthermore, PR and BP could stimulate the B-lymphocytes indirectly by increasing the anti-inflammatory cytokines then turning into plasma cells which in turn produce effective antibodies [45].

In a specific study on immunoglobulin titers against Newcastle vaccination in broilers [16], a significant increase in the IgM titer at 21 d of age was observed, but not in the IgG titer, in the BP-treated group vs. the control. In the current study, including PR or BP individually into broiler diets led to higher levels of IgM and IgA, which are partially considered as a natural, first-line defense in birds [46]. In contrast, there were no effects of PR and/or BP on the plasma IgM and IgG concentration, while a significant increase in IgA was obtained in the PR + BP group compared to the other groups (Table 7). These results agree with the fact assuming that IgM and IgG is more active after being challenged by infection [47].

5. Conclusions

Dietary supplementation with PR and BP separately or in combination could improve the growth performance and yield a higher body weight gain of broiler chickens. In addition, the total antioxidant capacity and superoxide dismutase activity were obviously increased in the treated broilers. Moreover, several immune functions were also enhanced by the PR and/or BP, such as leukocyte viability, lymphocyte proliferation, immunoglobulin concentration, and humoral and cellular immunity. Therefore, inclusion of PR and BP into broiler diets could be beneficial for broiler performance through improving the antioxidant and immune systems.

Supplementary Materials: The following supporting information can be downloaded at: https://www.mdpi.com/article/10.3390/ani12131658/s1, Table S1: Antioxidants assay specification according to the colorimetric kits' manufacturer; Table S2: Immunoglobulins (Ig) assay specification according to the ELISA kits' manufacturer.

Author Contributions: S.N.A.-K., A.A.A. and A.O.A. have the same contribution to conceptualization, methodology, formal analysis, investigation, data curation, writing of original draft, review and editing, supervision, and project administration. All authors have read and agreed to the published version of the manuscript.

Funding: This research was funded by Deanship of Scientific Research, Vice Presidency for Graduate Studies and Scientific Research, King Faisal University, Saudi Arabia, grant number AN000552.

Institutional Review Board Statement: The animal study protocol was approved by the Research Ethics Committee at King Faisal University (protocol code: KFU-REC-2022-MAR-EA000545).

Informed Consent Statement: Not applicable.

Data Availability Statement: The data presented in this study are available in the manuscript.

Acknowledgments: The author would like to thank the Deanship of Scientific Research at King Faisal University for the administrative and technical support provided to this work. The author is very grateful to all the personnel from the Department of Arid Land Agriculture and the Department of Animal and Fish Production, College of Agricultural and Food Sciences, King Faisal University, for their assistance in broilers' management and sample preparation throughout the experimental period.

Conflicts of Interest: The authors declare no conflict of interest. The funders had no role in the design of the study; in the collection, analyses, or interpretation of data; in the writing of the manuscript, or in the decision to publish the results.

References

1. Attia, Y.A.; Zeweil, H.S.; Alsaffar, A.A.; El-Shafy, A.S. Effect of non-antibiotic feed additives as an alternative to flavomycin on productivity, meat quality and blood parameters in broilers. *Arch. Geflugelkd.* **2011**, *75*, 40–48.
2. Attia, Y.A.; Al-Hamid, A.E.A.; Ibrahim, M.S.; Al-Harthi, M.A.; Bovera, F.; Elnaggar, A.S. Productive performance, biochemical and hematological traits of broiler chickens supplemented with propolis, bee pollen, and mannan oligosaccharides continuously or intermittently. *Livest. Sci.* **2014**, *164*, 87–95. [CrossRef]
3. Bankova, V.S.; de Castro, S.L.; Marcucci, M.C. Propolis: Recent advances in chemistry and plant origin. *Apidologie* **2000**, *31*, 3–15. [CrossRef]
4. Carpes, S.T.; Prado, A.; Moreno, I.A.M.; Mourão, G.B.; De Alencar, S.M.; Masson, M.L. Screening of the antioxidant potential of bee pollen produced in the southern region of Brazil. *Quim. Nova* **2008**, *31*, 1660–1664. [CrossRef]
5. Taha, E.K.A. Chemical composition and amounts of mineral elements in honeybee-collected pollen in relation to botanical origin. *J. Apic. Sci.* **2015**, *59*, 75–81. [CrossRef]
6. Viuda-Martos, M.; Ruiz-Navajas, Y.; Fernández-López, J.; Pérez-Álvarez, J.A. Functional properties of honey, propolis, and royal jelly. *J. Food Sci.* **2008**, *73*, 117–124. [CrossRef]
7. Prytzyk, E.; Dantas, A.P.; Salomão, K.; Pereira, A.S.; Bankova, V.S.; De Castro, S.L.; Neto, F.R.A. Flavonoids and trypanocidal activity of Bulgarian propolis. *J. Ethnopharmacol.* **2003**, *88*, 189–193. [CrossRef]
8. Attia, Y.A.; Al-Hanoun, A.; Bovera, F. Effect of different levels of bee pollen on performance and blood profile of New Zealand White bucks and growth performance of their offspring during summer and winter months. *J. Anim. Physiol. Anim. Nutr.* **2011**, *95*, 17–26. [CrossRef]

9. Chen, Y.J.; Huang, A.C.; Chang, H.H.; Liao, H.F.; Jiang, C.M.; Lai, L.Y.; Chan, J.T.; Chen, Y.Y.; Chiang, J. Caffeic acid phenethyl ester, an antioxidant from propolis, protects peripheral blood mononuclear cells of competitive cyclists against hyperthermal stress. *J. Food Sci.* **2009**, *74*, 162–167. [CrossRef]
10. Wang, J.; Li, S.; Wang, Q.; Xin, B.; Wang, H. Trophic effect of bee pollen on small intestine in broiler chickens. *J. Med. Food* **2007**, *10*, 276–280. [CrossRef]
11. Prakatur, I.; Miskulin, M.; Pavic, M.; Marjanovic, K.; Blazicevic, V.; Miskulin, I.; Domacinovic, M. Intestinal Morphology in Broiler Chickens Supplemented with Propolis and Bee Pollen. *Animals* **2019**, *9*, 301. [CrossRef] [PubMed]
12. Castaldo, S.; Capasso, F. Propolis, an old remedy used in modern medicine. *Fitoterapia* **2002**, *73* (Suppl. 1), S1–S6. [CrossRef]
13. Kročko, M.; Čanigová, M.; Bezeková, J.; Lavová, M.; Haščík, P.; Ducková, V. Effect of Nutrition with Propolis and Bee Pollen Supplements on Bacteria Colonization Pattern in Gastrointestinal Tract of Broiler Chickens. *Anim. Sci. Biotechnol.* **2012**, *45*, 63–67.
14. Abbas, A.O.; Kamel, N.N.; Khalifa, W.H.; Gouda, G.F.F.; El-Manylawi, M.A.F.A.F.; Mehaisen, G.M.K.K.; Mashaly, M.M.M. Propolis supplementation attenuates the negative effects of oxidative stress induced by paraquat injection on productive performance and immune function in Turkey poults. *Poult. Sci.* **2017**, *96*, 4419–4429. [CrossRef] [PubMed]
15. Mehaisen, G.M.K.K.; Ibrahim, R.M.; Desoky, A.A.; Safaa, H.M.; El-Sayed, O.A.; Abass, A.O. The importance of propolis in alleviating the negative physiological effects of heat stress in quail chicks. *PLoS ONE* **2017**, *12*, e0186907. [CrossRef] [PubMed]
16. Oliveira, M.d.; Silva, D.d.; Loch, F.; Martins, P.; Dias, D.; Simon, G. Effect of Bee Pollen on The Immunity and Tibia Characteristics in Broilers. *Braz. J. Poult. Sci.* **2013**, *15*, 323–328. [CrossRef]
17. Farag, S.A.; El-Rayes, T.K. Effect of bee-pollen supplementation on performance, carcass traits and blood parameters of broiler chickens. *Asian J. Anim. Vet. Adv.* **2016**, *11*, 168–177. [CrossRef]
18. Babaei, S.; Rahimi, S.; Amir, M.; Torshizi, K.; Tahmasebi, G. Effects of propolis, royal jelly, honey and bee pollen on growth performance and immune system of Japanese quails. *Vet. Res. Forum* **2016**, *7*, 13–20.
19. AOAC. *Official Methods of Analysis of AOAC International*, 18th ed.; Association of Official Analysis Chemists International: Washington, DC, USA, 2005; ISBN 0935584544.
20. Singleton, V.L.; Rossi, J.A. Colorimetry of Total Phenolics with Phosphomolybdic-Phosphotungstic Acid Reagents. *Am. J. Enol. Vitic.* **1965**, *16*, 144–158.
21. Quettier-Deleu, C.; Gressier, B.; Vasseur, J.; Dine, T.; Brunet, C.; Luyckx, M.; Cazin, M.; Cazin, J.C.; Bailleul, F.; Trotin, F. Phenolic compounds and antioxidant activities of buckwheat (Fagopyrum esculentum Moench) hulls and flour. *J. Ethnopharmacol.* **2000**, *72*, 35–42. [CrossRef]
22. Oktay, M.; Gülçin, İ.; Küfrevioğlu, Ö.İ. Determination of in vitro antioxidant activity of fennel (Foeniculum vulgare) seed extracts. *LWT-Food Sci. Technol.* **2003**, *36*, 263–271. [CrossRef]
23. Gehad, A.E.; Mehaisen, G.M.K.; Abbas, A.O.; Mashaly, M.M. The Role of Light Program and Melatonin on Alleviation of Inflammation Induced by Lipopolysaccharide Injection in Broiler Chickens. *Int. J. Poult. Sci.* **2008**, *7*, 193–201. [CrossRef]
24. Zhang, L.; Yue, H.Y.; Zhang, H.J.; Xu, L.; Wu, S.G.; Yan, H.J.; Gong, Y.S.; Qi, G.H. Transport stress in broilers: I. Blood metabolism, glycolytic potential, and meat quality. *Poult. Sci.* **2009**, *88*, 2033–2041. [CrossRef] [PubMed]
25. Abbas, A.O.; Alaqil, A.A.; El-Beltagi, H.S.; Abd El-Atty, H.K.; Kamel, N.N. Modulating laying hens productivity and immune performance in response to oxidative stress induced by e. Coli challenge using dietary propolis supplementation. *Antioxidants* **2020**, *9*, 893. [CrossRef]
26. Alaqil, A.A.; Abbas, A.O.; El-Beltagi, H.S.; Abd El-Atty, H.K.; Mehaisen, G.M.K.; Moustafa, E.S. Dietary supplementation of probiotic lactobacillus acidophilus modulates cholesterol levels, immune response, and productive performance of laying hens. *Animals* **2020**, *10*, 1588. [CrossRef]
27. Loa, C.C.; Lin, T.L.; Wu, C.C.; Bryan, T.; Thacker, H.L.; Hooper, T.; Schrader, D. Humoral and cellular immune responses in turkey poults infected with turkey coronavirus. *Poult. Sci.* **2001**, *80*, 1416–1424. [CrossRef]
28. Al-Khalifa, H. Immunological techniques in avian studies. *Worlds. Poult. Sci. J.* **2016**, *72*, 573–584. [CrossRef]
29. Kleczek, K.; Majewska, K.; Makowski, W.; Michalik, D. The effect of diet supplementation with propolis and bee pollen on the physicochemical properties and strength of tibial bones in broiler chickens. *Arch. Anim. Breed.* **2012**, *55*, 97–103. [CrossRef]
30. Haščík, P.; Elimam, I.O.E.; Garlík, J. the Effect of Addition Bee Pollen To Feed Mixtures on Internal Fat of Broiler Ross 308. *J. Microbiol. Biotechnol. Food Sci.* **2012**, *2*, 246–252.
31. Ferreira, I.C.F.R.; Aires, E.; Barreira, J.C.M.; Estevinho, L.M. Antioxidant activity of Portuguese honey samples: Different contributions of the entire honey and phenolic extract. *Food Chem.* **2009**, *114*, 1438–1443. [CrossRef]
32. Fatrcová-Šramková, K.; Nôžková, J.; Kačániová, M.; Máriássyová, M.; Rovná, K.; Stričík, M. Antioxidant and antimicrobial properties of monofloral bee pollen. *J. Environ. Sci. Health. B.* **2013**, *48*, 133–138. [CrossRef] [PubMed]
33. Kaur, R.; Kalia, P.; Kumar, N.R.; Harjai, K. Preliminary studies on different extracts of some honey bee products. *J. Appl. Nat. Sci.* **2013**, *5*, 420–422. [CrossRef]
34. Yan, H.; Harding, J.J. Glycation-induced inactivation and loss of antigenicity of catalase and superoxide dismutase. *Biochem. J.* **1997**, *328*, 599–605. [CrossRef] [PubMed]
35. Attia, Y.A.; Al-Khalaifah, H.; Ibrahim, M.S.; Al-Hamid, A.E.A.; Al-Harthi, M.A.; El-Naggar, A. Blood Hematological and Biochemical Constituents, Antioxidant Enzymes, Immunity and Lymphoid Organs of Broiler Chicks Supplemented with Propolis, Bee Pollen and Mannan Oligosaccharides Continuously or Intermittently. *Poult. Sci.* **2017**, *96*, 4182–4192. [CrossRef] [PubMed]

36. Bi, Y.; Xu, L.; Qiu, L.; Wang, S.; Liu, X.; Zhang, Y.; Chen, Y.; Zhang, Y.; Xu, Q.; Chang, G.; et al. Reticuloendotheliosis virus inhibits the immune response acting on lymphocytes from peripheral blood of chicken. *Front. Physiol.* **2018**, *9*, 4. [CrossRef]
37. Cabrera-Ortega, A.A.; Feinberg, D.; Liang, Y.; Rossa, C.; Graves, D.T. The Role of Forkhead Box 1 (FOXO1) in the Immune System: Dendritic Cells, T Cells, B Cells, and Hematopoietic Stem Cells. *Crit. Rev. Immunol.* **2017**, *37*, 1–13. [CrossRef] [PubMed]
38. Seven, I.; Aksu, T.; Seven, P.T. The Effects of Propolis on Biochemical Parameters and Activity of Antioxidant Enzymes in Broilers Exposed to Lead-Induced Oxidative Stress. *Asian-Australas. J. Anim. Sci.* **2010**, *23*, 1482–1489. [CrossRef]
39. Sforcin, J.M.; Orsi, R.O.; Bankova, V. Effect of propolis, some isolated compounds and its source plant on antibody production. *J. Ethnopharmacol.* **2005**, *98*, 301–305. [CrossRef]
40. Yuan, J.; Liu, J.; Hu, Y.; Fan, Y.; Wang, D.; Guo, L.; Nguyen, T.L.; Zhao, X.; Liu, X.; Liu, C.; et al. The immunological activity of propolis flavonoids liposome on the immune response against ND vaccine. *Int. J. Biol. Macromol.* **2012**, *51*, 400–405. [CrossRef]
41. Thornton, B.P.; Větvicka, V.; Ross, G.D. Natural antibody and complement-mediated antigen processing and presentation by B lymphocytes. *J. Immunol.* **1994**, *152*, 1727–1737.
42. Saki, A.A.; Khoramabadi, V.; Nourian, A.R.; Zamani, P. Immune response, blood parameters and growth performance in broiler fed reduced protein diet supplemented with β-hydroxy-β-methyl butyrate and conjugated linoleic acid. *Acta Sci. Anim. Sci.* **2018**, *40*, 1–6. [CrossRef]
43. Taheri, H.R.; Rahmani, H.R.; Pourreza, J. Humoral immunity of broilers is affected by oil extracted propolis (OEP) in the diet. *Int. J. Poult. Sci.* **2005**, *4*, 414–417. [CrossRef]
44. Havsteen, B.H. The biochemistry and medical significance of the flavonoids. *Pharmacol. Ther.* **2002**, *96*, 67–202. [CrossRef]
45. Oršolić, N.; Bašić, I. Immunomodulation by water-soluble derivative of propolis: A factor of antitumor reactivity. *J. Ethnopharmacol.* **2003**, *84*, 265–273. [CrossRef]
46. Parmentier, H.K.; Lammers, A.; Hoekman, J.J.; De Vries Reilingh, G.; Zaanen, I.T.A.; Savelkoul, H.F.J. Different levels of natural antibodies in chickens divergently selected for specific antibody responses. *Dev. Comp. Immunol.* **2004**, *28*, 39–49. [CrossRef]
47. Härtle, S.; Magor, K.E.; Göbel, T.W.; Davison, F.; Kaspers, B. Structure and evolution of avian immunoglobulins. *Avian Immunol.* **2022**, *1*, 101–119. [CrossRef]

animals

MDPI

Article

Fructooligosaccharide Supplementation Boosts Growth Performance, Antioxidant Status, and Cecal Microbiota Differently in Two Rabbit Breeds

Ayman H. Abd El-Aziz [1], Mahmoud M. Abo Ghanima [1], Walaa F. Alsanie [2,3], Ahmed Gaber [3,4], Abd El-Wahab Alsenosy [5], Ahmed A. Easa [6], Sherif A. Moawed [7], Sayed Haidar Abbas Raza [8], Ahmed Elfadadny [9], Hany Abo Yossef [10], Wafaa M. Ghoneem [11], Mustafa Shukry [12,*], Amin Omar Hendawy [6] and Khalid Mahrose [13]

1 Animal Husbandry and Animal Wealth Development Department, Faculty of Veterinary Medicine, Damanhour University, Damanhour 22511, Egypt; ayman.sadaka@vetmed.dmu.edu.eg (A.H.A.E.-A.); aboghoneima.mmyvet2@vetmed.dmu.edu.eg (M.M.A.G.)
2 Department of Clinical Laboratories Sciences, The Faculty of Applied Medical Sciences, Taif University, P.O. Box 11099, Taif 21944, Saudi Arabia; w.alsanie@tu.edu.sa
3 Centre of Biomedical Sciences Research (CBSR), Deanship of Scientific Research, Taif University, P.O. Box 11099, Taif 21944, Saudi Arabia; a.gaber@tu.edu.sa
4 Department of Biology, College of Science, Taif University, P.O. Box 11099, Taif 21944, Saudi Arabia
5 Biochemistry Department, Faculty of Veterinary Medicine, Damanhour University, Damanhour 22511, Egypt; dr_alsenosy_2010@yahoo.com
6 Department of Animal and Poultry Production, Faculty of Agriculture, Damanhour University, Damanhour 22516, Egypt; ahmed.allam@agr.dmu.edu.eg (A.A.E.); amin.hendawy@gmail.com (A.O.H.)
7 Department of Animal Wealth Development, Biostatistics Division, Faculty of Veterinary Medicine, Suez Canal University, Ismailia 41522, Egypt; sherifmoawed@vet.suez.edu.eg
8 College of Animal Science and Technology, Northwest A&F University, Xianyang 712100, China; haiderraza110@nwafu.edu.cn
9 Department of Animal Internal Medicine, Faculty of Veterinary Medicine, Damanhour University, Damanhour 22511, Egypt; ahmed.elfadadny@vetmed.dmu.edu.eg
10 Food Hygiene Department, Faculty of Veterinary Medicine, Alexandria University, Edfina 22758, Egypt; Hanyyossef84@gmail.com
11 Animal Production Department, Faculty of Agriculture, Cairo University, Giza 12613, Egypt; wafaaghoneem@agr.cu.edu.eg
12 Department of Physiology, Faculty of Veterinary Medicine, Kafrelsheikh University, Kafrelsheikh 22511, Egypt
13 Animal and Poultry Production Department, Faculty of Technology and Development, Zagazig University, Zagazig 44511, Egypt; ostrichkhalid@zu.edu.eg
* Correspondence: mostafa.ataa@vet.kfs.edu.eg

Citation: Abd El-Aziz, A.H.; Abo Ghanima, M.M.; Alsanie, W.F.; Gaber, A.; Alsenosy, A.E.-W.; Easa, A.A.; Moawed, S.A.; Raza, S.H.A.; Elfadadny, A.; Yossef, H.A.; et al. Fructooligosaccharide Supplementation Boosts Growth Performance, Antioxidant Status, and Cecal Microbiota Differently in Two Rabbit Breeds. *Animals* **2022**, *12*, 1528. https://doi.org/10.3390/ani12121528

Academic Editors: Rosalia Crupi and Raffaella Rossi

Received: 8 April 2022
Accepted: 9 June 2022
Published: 13 June 2022

Simple Summary: Rapidly rising incomes are dependent on animal protein production and the worldwide demand for livestock. It is expected that moving towards more intensive production systems to sustain this increased demand will depend on growth promoters. Some growth promoters, such as prebiotics, might be considered alternative non-antibiotic feed supplementation as they enhance performance without any side effects on the consumer's health. The present study inspected the influence of supplementation of β-fructan® (a commercial fructooligosaccharide; FOS) in the drinking water of growing rabbits on growth performance, carcass traits, hematological and biochemical indices, antioxidant status, and cecal microbiota of the NZW- and APRI-line rabbits (Animal Production Research Institute Line). FOS supplementation in rabbits enhanced growth carcass characteristics, significantly improving hematological parameters and antioxidant status, and minimized pathogenic Escherichia coli bacteria (from 3.45 in control groups to 2.89 and 2.24 (Log10 CFU g^{-1}) in 0.5 mL and 1 mL FOS-treated rabbits, respectively.

Abstract: The present study examined the effects of fructooligosaccharide (FOS) supplementation in drinking water on the growth performance, carcass characteristics, hematological and biochemical parameters, antioxidant status, and cecal microbiota of New Zealand White (NZW) and APRI rabbits. A total of 180 male NZW and APRI rabbits (aged five weeks; average live body weight

700 ± 39 g) were divided into six groups (30 rabbits/group; 5 replicates/group) in a two × three factorial arrangement. Rabbits of each breed were randomly assigned to one of three treatments of FOS (control; 0.00, FOS-0.5, and FOS-1.0). Results showed that rabbits' final body weight, FBWG, and carcass traits were considerably enhanced compared to those in the control group. The interaction effect of the supplement with the rabbit breed increased the growth, carcass traits, and hematobiochemical and antioxidant parameters with increasing FOS levels. In the cecum of both rabbit breeds, the total bacterial count and *Escherichia coli* population were considerably low, with a substantial increase in the number of *Lactobacilli* supplemented by FOS. In conclusion, FOS supplementation enhanced growth and carcass traits by improving the hematobiochemical parameters and antioxidant status and reducing cecal pathogenic bacteria in both breeds.

Keywords: antioxidant status; carcass; fructooligosaccharide; growth; haemato-biochemical parameters; cecal microbiota

---

1. Introduction

Improving animal productivity and boosting immunity using natural substances is a primary goal in animal breeding [1–7]. Recent studies have shown that immunostimulants, such as probiotics and prebiotics, have the potential to be used as protective and environment-friendly substitutions to antibiotics in mammals and poultry species [8–11]. These compounds are a possible method to enhance animal health and performance without antibiotics [12]. Prebiotics, such as inulin-type fructans and galacto-oligosaccharides, reveal immune-stimulating properties to the host through selective promoting of growth and/or encouraging the growth of some beneficial bacteria (i.e., probiotics) [13,14]. Fructooligosaccharides (FOSs) are considered the popular forms of prebiotics that consist of short-chain and undigested carbohydrates [9,15] because the β-linkages between fructose monomers cannot be hydrolyzed by the endogenous enzymes [16]. FOS is derived from the cell wall of the yeast, *Saccharomyces cerevisiae*, and has been reported to possess the ability to improve growth performance, decrease pathogenic bacterial count, and enhance immunity in two different rabbit breeds (New Zealand White and V-line rabbits) [1].

FOS may accelerate the gut fermentation of beneficial microorganisms, such as *Lactobacillus* and *Bifidobacterium*, and reduce the accumulation of pathogenic bacteria, such as *Clostridium perfringens* and *Escherichia coli* [15,17–19], thus enhancing the general health of animals [1,6,17]. Therefore, FOS is considered to be a prebiotic [20]. Dietary FOSs was reported to improve the growth traits (body weight, weight gain, and feed conversion ratio) and immune responses of broilers [21–23].

To our knowledge, there are no reports on adding FOSs to the drinking water of growing rabbits. Hence, this study was conducted to detect the possible effects of adding FOS (β-fructan®, a commercial FOS) in drinking water on the productive performance, carcass characteristics, hematobiochemical parameters, oxidative stress, and cecal microbiota of New Zealand White (NZW) and APRI rabbits. We hypothesized that oral FOS supplementation in combination with rabbit breed would enhance growth, improve blood biochemistry and antioxidant status, and improve microflora population diversity to alleviate the weaning stress of the rabbits.

2. Materials and Methods

2.1. Ethical Declaration

This research was performed after the approval of the Ethics of the Institutional Committee of Animal Husbandry and Animal Wealth Development Department, Faculty of Veterinary Medicine, Damanhour University, Egypt (DMU/VetMed-2019-/0145).

2.2. Animal Rearing and Study Design

APRI rabbit was produced by crossing Baladi Red bucks with a V line to create F1 ($\frac{1}{2}$B$\frac{1}{2}$V) stock, and it was continued for two generations of intersex mating to attain performing constancy. A total of 180 weaned APRI and NZW rabbits (male, aged five weeks, weighing 700 ± 39 g) were collected and allocated to six groups (30 rabbits per group), and each group was divided into five replicates, each with six rabbits. The rabbits were assigned at random using a two × three completely factorial design (NZW and APRI-line with three treatments of a commercial FOS known as β-fructan®). The control group was not treated with FOS, and the first and second groups were supplemented with FOS-0.5 mL and FOS-0.1 mL, respectively. The experimental groups received 0.5 and 1.0 mL β-fructan (1,3 pharmaceutical grade 10%) per liter of drinking water for three sequential days per week (Glencore Company, Ann Arbor, MI, USA). Each rabbit in the 0.5 mL β-fructan-treated group was supplemented with 349.8 mg of β-fructan during the eight-weeks experimental period, while in the 1 mL β-fructan-treated group, each rabbit was supplemented with 699.75 mg of β-fructan. Rabbits were housed in galvanized wire batteries with standard dimensions (60 × 35 × 35 cm). All cages were supplied with galvanized steel feeding hoppers and automatic drinkers (nipples). Plastic ear tags identified rabbits. Freshwater was provided *ad libitum*, and a standard pelleted ration was provided *ad libitum* twice daily at 8 am and 2 pm. The pellets measured 1 cm in length and 0.4 cm in diameter. Rabbit cages were regularly cleaned and disinfected. Urine and feces dropped beneath the batteries were removed every morning.

2.3. Experimental Feed Diet Preparation

Diet was prepared following the NRC [24] and Lebas [25] recommendations (Table 1). The analysis of the ingredients was performed according to AOAC [26].

Table 1. Ingredients and chemical composition (%) of the basal diet.

Ingredients	%
Yellow corn	9.5
Soybean meal (44%)	15
Wheat bran	17
Barley	21.7
Barley hay	34.5
Dicalcium phosphate [1]	1.2
Ground limestone [2]	0.25
DL-Methionine	0.05
Common salt	0.5
Vitamin + mineral premix [3]	0.3
Total	100
Chemical composition	
Dry matter	87.8
Moisture	12.2
Crude protein	17.9
Crude fiber	13.75
Ether extract	3.6
Nitrogen-free extract [4]	42.75
Ash	9.8
DE (kcal /kg) [5]	2677.97

[1] Dicalcium phosphate: 20% phosphorus and 25% calcium; [2] limestone: 34% calcium. [3] Amounts per kg: Vitamin A—12,000 and 900 IU of vitamin A and D3, respectively. While 2 mg of each vitamin K3, B1, and B6. 50 mg of vitamin E, 6 mg vitamin B2, 0.01 mg vitamin B12, 0.2 mg biotin, 20 mg pantothenic, 50 mg niacin, 5 mg folic acid, 8.5 mg manganese, 70 mg zinc, 75 mg iron, 5 mg copper, 0.75 mg iodine, 0.1 mg selenium. [4] Nitrogen free extract (NFE) was calculated by difference = 100 − (moisture % + CP% + EE% + CF% + Ash %). [5] Digestible energy (DE) was calculated according to values given in the feed composition tables of the NRC [24].

2.4. Productive Performance and Carcass Characteristics

At the start of the fifth week, the animals were weighed individually until the end of the experiment (13 weeks of age). The rabbit's daily feed consumption was calculated every week to evaluate the feed conversion ratio (FCR). Final body weight (FBW), body weight gain (BWG), and total feed consumption (TFC) were determined. Fifteen rabbits from each group were randomly selected to evaluate carcass characteristics at the end of the experiment (13th week). Rabbits were fasted for 12 h before being slaughtered. After removing the skin and bones, the carcasses were measured individually to evaluate the weight and percentage of the dressed animals. The offal weight includes blood, viscera, lungs, skin, arms, and tail. The obtained results were presented as the % of live weight [27]. The dressing % was calculated as hot carcass weight × 100/fasting weight. The carcass was divided into three cuts, viz., (1) the two forelegs (including the thoracic muscle inserting system), (2) the loin (the abdominal wall and the riveting after the seventh thoracic rib), and (3) the hind legs (including the sacral bone and the lumbar vertebra after the sixth lumbar vertebra).

2.5. Hematology and Biochemical and Serum Oxidative Stress Evaluations

Two blood samples were collected from the lateral ear vein (30 rabbits) during the slaughter. One sample contained an anticoagulant and was used to determine the count of white blood cells (WBCs), red blood cells (RBCs), lymphocytes, monocytes, and mean corpuscular hemoglobin (MCH), red cell distribution width (RDW), platelet count, hematocrit %, and hemoglobin concentration [28]. The other blood sample was centrifuged (15 min, 3000× *g*) at 15–24 °C for plasma separation and stored at −20 °C until analysis. Total protein, albumin, cholesterol, alanine aminotransferase (ALT), aspartate aminotransferase (AST), and creatinine levels were measured in plasma using commercial kits. Moreover, the levels of glutathione peroxidase (GPX), superoxide dismutase (SOD), and total antioxidant capacity (T-AOC) were evaluated using the colorimetric method (kits obtained from Bio-diagnostic, Cairo, Egypt).

2.6. Bacterial Count

Total bacteria, *E. coli*, and *lactobacilli* were all counted using the ring-plate method in the rabbit cecum sample [29,30].

2.7. Data Analysis

The attained results were statistically analyzed with the general linear model procedure of SAS® (Cary, NC, USA) [31]. Homogeneity of variances among studied groups was verified [32]. The analysis was performed using this model: $Y_{ijk} = \mu + S_i + E_j + SE_{ij} + e_{ijK}$, where μ = observed mean for the concerned treatment, S_i = breed effect, E_j = treatment effect, SE_{ij} = interaction effect of breed and treatment, and e_{ijk} = the error related to individual observation using Duncan's multiple range test [32]. According to Ahrens et al. [33], the percentages were converted into arcsine values. Results were considered statistically significant at $p \leq 0.05$.

3. Results

The FBW, FBWG, TFC, and FCR of rabbits supplemented with oral FOS were considerably enhanced compared to those in the control group (Table 2). The NZW rabbits treated with 1% FOS in drinking water showed the largest FBW, followed by APRI rabbits in which drunk water increased with the same level of FOS. BWG was more significant in NZW rabbits that consumed 1% and 0.5% FOS and APRI rabbits that consumed water supplemented with 1% FOS than their counterparts. Increasing FOS concentrations in drinking water decreased ($p < 0.001$) the amount of feed consumed in both breeds. The NZW rabbits that consumed water supplemented with 1% and 0.5% FOS and the APRI rabbits that consumed 1% FOS-supplemented water showed the lowest ($p < 0.001$) FCR compared with the other groups (Table 2).

Table 2. Growth performance of rabbits as affected by breed and supplementation of the diets with fructooligosaccharide (FOS).

Items		Initial Body Weight (g)	Final Body Weight (g)	Body Weight Gain (g)	Total Feed Consumption (g)	Feed Conversion Ratio (g Feed/g Gain)
			Breed			
NZW		756.45	2261.45	1464.50	4740.24	3.278
APRI		706.25	2207.91	1461.54	4736.51	3.281
			FOS supplementation			
FOS 0.5 mL/L DW		721.25	2160.62	1470.00	4647.45	3.19
FOS 1 mL/L DW		737.18	2409.06	1561.56	4609.54	2.96
Control		735.62	2134.37	1357.50	4958.14	3.68
			Breed x Treatment interaction			
NZW	Control	768.12	2139.37 ^{c}	1389.37 ^{cd}	4914.37 ^{b}	3.65 ^{a}
	FOS 0.5 mL	738.12	2176.25 ^{c}	1519.62 ^{abc}	4699.77 ^{c}	3.11 ^{bc}
	FOS 1 mL	763.12	2500.00 ^{a}	1583.75 ^{a}	4606.58 ^{d}	2.92 ^{c}
APRI	Control	703.12	2129.37 ^{c}	1325.62 ^{d}	5001.91 ^{a}	3.71 ^{a}
	FOS 0.5 mL	704.37	2145.00 ^{c}	1420.37 ^{bcd}	4595.12 ^{d}	3.26 ^{b}
	FOS 1 mL	711.25	2318.12 ^{b}	1539.37 ^{ab}	4612.50 ^{d}	3.007 ^{bc}
SEM		18.383	19.472	18.249	8.151	0.044
			Two-way ANOVA (*p-value*)			
Breed		0.179	0.176	0.936	0.820	0.979
Treatment		0.927	<0.001	<0.001	<0.001	<0.001
Interaction		0.827	<0.001	0.001	<0.001	<0.001

Means within each column for each division with no common superscript letters are significantly different ($p < 0.05$). SEM = standard error of means.

Total giblets, gastrointestinal tract, liver, and dressing % in NZW and APRI rabbits that consumed FOS-supplemented water were significantly enhanced due to FOS and the interaction between FOS and breed ($p < 0.001$) compared with the control groups (Table 3). The difference between FOS levels was insignificant.

Table 3. Carcass traits of rabbits as affected by breed and supplementation of the diets with fructooligosaccharide (FOS) (%).

Items		Forequarter	Loin	Hindquarter	Giblets	Gastrointestinal Tract	Liver	Dressing
				Breed				
NZW		33.48	27.10	39.21	3.67 ^{a}	25.59	5.04	55.24
APRI		33.28	27.01	39.19	3.20 ^{b}	27.24	4.64	54.74
				FOS supplementation				
FOS 0.5 mL/L DW		33.22	27.26	39.26	3.64 ^{a}	26.38 ^{ab}	5.02 ^{a}	55.18 ^{a}
FOS 1 mL/L DW		33.73	27.60	39.68	3.71 ^{a}	24.13 ^{b}	5.17 ^{a}	56.61 ^{a}
Control		33.20	26.32	38.66	2.97 ^{b}	28.73 ^{a}	4.34 ^{b}	53.19 ^{b}
				Breed x Treatment interaction				
NZW	Control	33.16	26.93	38.43	3.11 ^{b}	28.32 ^{a}	4.15 ^{b}	53.13 ^{b}
	FOS 0.5 mL	33.28	27.38	39.74	3.28 ^{b}	25.77 ^{a}	4.92 ^{ab}	56.28 ^{a}
	FOS 1 mL	34.23	27.81	40.08	4.20 ^{a}	21.86 ^{b}	5.67 ^{a}	56.93 ^{a}
APRI	Control	32.23	25.72	37.87	2.66 ^{b}	29.14 ^{a}	4.15 ^{b}	53.06 ^{b}
	FOS 0.5 mL	33.23	27.01	39.43	3.21 ^{b}	26.99 ^{a}	4.53 ^{ab}	53.98 ^{b}
	FOS 1 mL	34.06	27.51	39.62	4.16 ^{a}	26.39 ^{a}	5.12 ^{ab}	56.37 ^{a}
SEM		0.214	0.250	0.243	0.081	0.490	0.128	0.269
				Two-way ANOVA (*p-value*)				
Breed		0.645	0.850	0.976	0.010	0.110	0.137	0.413
Treatment		0.350	0.127	0.255	0.003	0.005	0.038	0.001
Interaction		0.185	0.262	0.119	<0.001	0.008	0.047	0.003

Means within each column for each division with no common superscript letters are significantly different ($p < 0.05$).

Substantial improvements ($p < 0.001$) were found in the two breeds in hematobiochemical and antioxidant parameters (Tables 4 and 5), which were enhanced with increasing FOS levels (FOS-1.0), with no detrimental effects on the kidney and liver. However, rabbits consuming 1% FOS had higher blood biochemicals and antioxidant parameters values than those receiving 0.5% FOS-supplemented water.

Table 4. Hematological parameters of rabbits as affected by breed and supplementation of the diets with fructooligosaccharide (FOS).

Item		WBC $10^3/\mu L$	Lymphocytes $10^3/\mu L$	Monocytes $10^3/\mu L$	RBC $10^6/\mu L$	Hgb %	MCV ft	HCT %	MCH pg	RDW %	Platelets $10^3/\mu L$
Breed											
NZW		6.11	4.36 ^{a}	0.59	4.47	11.07	60.88	32.22	24.39	25.37 ^{a}	152.07
APRI		5.94	3.71 ^{b}	0.58	4.41	11.06	61.60	34.83	24.80	24.95 ^{b}	153.07
FOS supplementation											
FOS 0.5 mL/L DW		5.95 ^{b}	3.62 ^{b}	0.57 ^{b}	4.46	11.12 ^{a}	61.55 ^{b}	34.55 ^{ab}	24.90 ^{a}	25.27 ^{b}	153.60 ^{b}
FOS 1 mL/L DW		7.57 ^{a}	5.81 ^{a}	0.65 ^{a}	4.44	11.24 ^{a}	64.53 ^{a}	36.46 ^{a}	25.319 ^{a}	26.52 ^{a}	160.40 ^{a}
Control		4.56 ^{c}	2.69 ^{c}	0.53 ^{c}	4.43	10.83 ^{b}	57.65 ^{c}	29.57 ^{b}	23.70 ^{b}	23.69 ^{c}	143.70 ^{c}
Breed × Treatment interaction											
NZW	Control	4.72 ^{e}	3 ^{d}	0.532 ^{c}	4.42	10.84 ^{b}	58.18 ^{d}	33.12 ^{ab}	24 ^{ab}	23.80 ^{d}	144.80 ^{c}
	FOS 0.5 mL	6.12 ^{c}	3.71 ^{c}	0.586 ^{b}	4.44	11.14 ^{a}	62.24 ^{bc}	34.96 ^{a}	24.98 ^{ab}	25.40 ^{c}	153.80 ^{b}
	FOS 1 mL	7.83 ^{a}	6.38 ^{a}	0.664 ^{a}	4.49	11.24 ^{a}	64.68 ^{a}	36.50 ^{a}	25.40 ^{a}	26.90 ^{a}	160.60 ^{a}
APRI	Control	5.78 ^{d}	2.38 ^{e}	0.528 ^{c}	4.44	10.82 ^{b}	57.12 ^{d}	26.02 ^{b}	23.40 ^{b}	23.58 ^{d}	142.60 ^{c}
	FOS 0.5 mL	7.32 ^{b}	3.52 ^{c}	0.566 ^{b}	4.42	11.10 ^{a}	60.86 ^{c}	34.14 ^{a}	24.80 ^{ab}	25.14 ^{c}	153.40 ^{b}
	FOS 1 mL	7.32 ^{b}	5.24 ^{b}	0.646 ^{a}	4.49	11.24 ^{a}	64.38 ^{ab}	36.42 ^{a}	25 ^{ab}	26.14 ^{bc}	160.20 ^{a}
SEM		0.047	0.047	0.004	0.037	0.024	0.324	1.05	0.215	0.084	0.840
Breed		0.101	0.001	0.205	0.380	0.891	0.281	0.226	0.355	0.021	0.557
Treatment		0.001	0.001	0.001	0.958	0.001	0.001	0.036	0.022	0.001	0.001
Interaction		0.001	0.001	0.001	0.951	0.001	0.001	0.077	0.115	0.001	0.001

Means within each column for each division with no common superscript letters are significantly different ($p < 0.05$). SEM: standard error of the means; MCV: mean corpuscular volume; HCT; hematocrit; MCH: mean corpuscular hemoglobin; RDW: red cell distribution width.

In both rabbit breeds that consumed FOS-supplemented water, the cecum, total bacterial, and *E. coli* populations (Table 6) were considerably lower ($p < 0.001$), with a substantial increase in the Lactobacillus population compared with the control groups. Rabbits that consumed 1% FOS showed the most significant count of beneficial bacteria and a lower count of pathogenic ones compared with the other treatments.

Table 5. Blood biochemical parameters and some selected oxidative stress biomarkers of rabbits as affected by breed and supplementation with fructooligosaccharide (FOS).

Item		Serum Biochemical Parameters								Oxidative Stress Biomarkers		
		Total Protein (g/dL)	**Albumin (g/dL)**	**Globulin (g/dL)**	**A/G Ratio**	**Cholesterol (mg/dL)**	**ALT (U/L)**	**AST (U/L)**	**Creatinine (mg/dL)**	**GPX (U/L)**	**SOD (U/L)**	**T-AOC (mmol/L)**
Breed												
NZW		6.45	3.70	2.75	1.48	50.80	35	31.87	1.74	27.27^{a}	79.40	1.32
APRI		6.42	3.69	2.72	1.43	54.80	35.66	31.93	1.74	25.80^{b}	78.60	1.31
FOS supplementation												
FOS 0.5 mL/L DW		6.59^{b}	3.69	2.89^{b}	1.28^{b}	48^{b}	35.10	32.60	1.79	27.70^{b}	80.30^{b}	1.33^{b}
FOS 1 mL/L DW		7.14^{a}	3.74	3.39^{a}	1.11^{b}	33.30^{c}	36.10	31.40	1.69	31.40^{a}	88.40^{a}	1.36^{a}
Control		5.57^{c}	3.65	1.91^{c}	1.98^{a}	77.10^{a}	34.80	31.70	1.74	20.50^{c}	68.30^{c}	1.27^{c}
Breed × Treatment interaction												
NZW	Control	5.60^{c}	3.67	1.97^{c}	1.87^{a}	76.6^{a}	34.4	31.6	1.74	20.80^{d}	68.60^{c}	1.28^{c}
	FOS 0.5 mL	6.60^{b}	3.70	2.90^{b}	1.27^{b}	44^{c}	34.8	32.6	1.73	28.20^{bc}	81.40^{b}	1.33^{b}
	FOS 1 mL	7.20^{a}	3.78	3.50^{a}	1.06^{b}	31.80^{d}	35.8	31.4	1.67	32.80^{a}	88.60^{a}	1.36^{a}
APRI	Control	5.54^{c}	3.62	1.85^{c}	2.09^{a}	77.6^{a}	35.2	31.8	1.81	20.20^{d}	68^{c}	1.27^{c}
	FOS 0.5 mL	6.58^{b}	3.68	2.89^{b}	1.28^{b}	52^{b}	35.4	32.7	1.78	27.20^{c}	79.20^{b}	1.33^{b}
	FOS 1 mL	7.08^{a}	3.71	3.29^{ab}	1.15^{b}	34.80^{d}	36.4	31.4	1.71	30^{b}	88.20^{a}	1.34^{ab}
SEM		0.049	0.020	0.054	0.048	1.018	0.293	0.374	0.023	0.248	0.728	0.004
Two-way ANOVA (*p-value*)												
Breed		0.789	0.960	0.774	0.638	0.061	0.257	0.930	0.989	0.016	0.588	0.518
Treatment		0.001	0.221	0.001	0.001	0.001	0.188	0.409	0.183	0.001	0.001	0.001
Interaction		0.001	0.367	0.001	0.001	0.001	0.450	0.860	0.571	0.001	0.001	0.001

Means within each column for each division with no common superscript letters are significantly different ($p < 0.05$). SEM: standard error of the means; ALT; alanine aminotransferase; AST; aspartate aminotransferase; GPX, glutathione peroxidases; SOD, superoxide dismutase; T-AOC, total antioxidant capacity.

Table 6. Cecal bacterial counts of rabbits as affected by breed and supplementation with fructooligosaccharide (FOS) (Log10 CFU g^{-1}).

Items		Total Bacterial Count (TBC)	*Cecal Escherichia coli (E. coli)*	*Cecal Lactobacilli*
Breed				
NZW		7.09	2.82^{b}	7.57
APRI		7.14	2.96^{a}	7.60
FOS supplementation				
FOS 0.5 mL/L DW		7.03^{b}	2.89^{b}	7.86^{a}
FOS 1 mL/L DW		6.05^{c}	2.24^{c}	7.95^{a}
Control		8.26^{a}	3.54^{a}	6.94^{b}
Breed × Treatment interaction				
NZW	Control	8.24^{a}	3.50^{a}	6.90^{b}
	FOS 0.5 mL	7^{b}	2.84^{b}	7.80^{a}
	FOS 1 mL	5.98^{c}	2.12^{d}	7.92^{a}
APRI	Control	8.28^{a}	3.58^{a}	6.98^{b}
	FOS 0.5 mL	7.06^{b}	2.94^{b}	7.90^{a}
	FOS 1 mL	6.13^{c}	2.36^{c}	8^{a}
SEM		0.027	0.029	0.081
Two-way ANOVA (*p*-value)				
Breed		0.697	0.024	0.839
Treatment		<0.001	<0.001	0.001
Interaction		<0.001	<0.001	0.001

Means within each column for each division with no common superscript letters are significantly different ($p < 0.05$). SEM = standard error of means.

4. Discussion

This study investigated the possible effects of adding FOS in drinking water on the growth performance, carcass characteristics, hematobiochemical parameters, oxidative stress biomarkers, and cecal microbiota of NZW and APRI rabbits. Our results showed that supplementing water with FOS significantly enhanced the growth performance traits of the two rabbit breeds.

The beneficial effects of adding FOS to the drinking water of growing rabbits may be due to the augmentation of feed efficiency and absorption, which improves anabolic metabolism, enhances the intestinal response to pathogens, and increases serum protein levels, thereby encouraging rabbit growth [1,6,7]. Prebiotics provide suitable environments for the growth of helpful microflora and inhibit the growth of pathogenic bacteria, which may explain the improvement in growth performance [7].

Consistent with our findings, the rabbits' growth was enhanced with *Bacillus subtilis* and FOS with a more significant average daily BWG than the control [22]. In addition, Inmunair17.5® (*Propionibacterium acnes* and coli lipopolysaccharides) as a prebiotic in the drinking water of fattening NZW rabbits resulted in an enhancement of BW at marketing, BWG, and FCR [11]. Comparable findings reported that a diet supplemented with *S. cerevisiae* and probiotics accelerated the BWG and FCR of NZW rabbits [34,35]. By contrast, Rotolo et al. [36,37] found that the dietary supplementation of *S. cerevisiae* did not affect rabbits' BW, BWG, and FCR. Additionally, Zarei et al. [38] reported that dietary prebiotics did not modify FCR in laying hens. In broilers, Xu et al. [14] concluded that supplementation with 4 g of FOS/kg diet increased BWG and improved FCR.

Regarding carcass characteristics, our findings were consistent with those observed by Abd El-Aziz et al. [1], Mahrose et al. [6], and Abo Ghanima et al. [2]. Similarly, Mousa et al. [11] showed that dressing and giblet percentages were significantly higher in the carcasses of rabbits that drank water supplemented with 1 mL Inmunair17.5®/litter. However, Rotolo et al. [36] found nonsignificant changes in the carcass characteristics of growing rabbits treated with dietary prebiotics. Moreover, Juśkiewicz et al. [39] concluded that increasing turkeys fed with a diet supplemented with FOS showed no differences from the control group. There were no significant changes between the two rabbit breeds regarding the breed impact on carcass traits in the present study. Such an absence of significant differences in carcass traits between genetic breeds has also been confirmed in previous studies [1,7].

Hematological measurements are valuable indicators for evaluating the animals' health statuses [4]. In our study, most hematological parameters were altered by the water supplemented with FOS in the two rabbit breeds. Our findings are consistent with those of Akrami et al. [13], who found that WBC counts were increased in fish fed with 1% FOS compared with the control group. They also found a nonsignificant elevation of RBCs, MCV, HCT, Hgb, and lymphocytes in the fish fed with a diet supplemented with 1% FOS.

In a study on birds, FOS supplementation resulted in low heterophil counts, indicating that FOS may reduce stress reactions and alleviate the possible damaging consequences on growth performance [9]. Moreover, broilers supplemented with FOS had more significant monocyte counts than broilers fed with the control diet. Monocytes comprise 5%–10% of peripheral blood leukocytes and can migrate rapidly in response to diseases, release cytokines, and differentiate into macrophages and dendritic cells to assist the innate immune response [40]. FOS supplementation increased monocyte %, suggesting that dietary FOS supplementation in broilers augments cytokine release and alleviates pathogenic infections rapidly [9]. This effect is probably due to the alteration in the gut microbiota, such as variations in the *Lactobacillus* profile, which shows diverse patterns for dendritic cell activation [41,42]. The findings concerning hematological indices revealed that these measurements were increased in rabbits that consumed water supplemented with FOS-0.5 and FOS-0.1. Hoseinifar et al. [43] mentioned that WBC count, primarily lymphocytes, was significantly increased in belugas fed with 1 and 2 g kg^{-1} dietary oligofructose. The high leukocyte count may increase activity and improve defense mechanisms during feeding.

Leukocytes are imperative cells that stimulate the immune responses of fish. They produce antibodies and may exhibit macrophage activities [44]. Saha et al. [19] obtained similar results, where the MCH in broilers receiving a water-soluble organic additive at different doses fluctuated from that in the control.

The total protein and globulin levels were increased in the experimental groups treated with varying levels of FOS in their diets, indicating a more robust innate immune response. Globulin is believed to be the main protein that plays a significant role in immune response [5]. Moreover, FOS was found to have the potential to control enteric pathogens and alter immunity [1]. This result was also previously supported by Abd El-Gawad et al. [17] who concluded that ALT and AST activities were diminished with dietary FOS than in the control fish group. Our results failed to show significant differences in AST and ALT activities with FOS supplementation in drinking water.

Interestingly, our findings showed an increase in SOD, GPX, and T-AOC values in the supplemented groups of the two rabbit breeds. These results suggest that the FOS-supplemented drinking water could alleviate oxidative stress in the two breeds of growing rabbits and maintain their healthy. The first line of antioxidant enzymatic defense is believed to involve GPX, SOD, and T-AOC [5], which act as biomarkers of oxidative stress due to the inequality between the production and elimination of reactive oxygen species. The enhancement of antioxidant enzymatic activities in the present study with FOS supplementation in the drinking water of growing rabbits was also previously reported by Guerreiro et al. [45] and Zhang et al. [46] as FOS supplementation may relieve oxidative stress [17].

In the present study, FOS supplementation in drinking water caused a stimulatory impact on the growth of health-supporting bacterial species (*Lactobacillus*). Moreover, FOS supplementation decreased the total bacterial count and harmful or potential pathogens (*E. coli*) in the two rabbit breeds. Our results are consistent with those reported by Xu et al. [16] who examined the effects of FOS at doses of 0, 2, 4, and 8 g/kg diet on intestinal microbiota. The inclusion of FOS at a 4 g/kg diet resulted in a beneficial effect on *Bifidobacterium* and *Lactobacillus*, with an immediate reduction of *E. coli* growth in the broilers' gastrointestinal tract. Saminathan et al. [47] evaluated the impact of applying various oligosaccharides by isolating 11 *Lactobacillus* species from the gastrointestinal tract of fowls. The in vitro data revealed that Lactobacillus species utilized FOS more competently than mannan oligosaccharides. The increased availability of FOS may be related to particular enzymatic actions and the oligosaccharide conveyance technique of *Lactobacillus* species. Nevertheless, broilers' intestinal microbiota is further complicated than in vitro examinations. Prebiotics may be fermented not only by *Lactobacillus* species but also by other microbes in the gastrointestinal tracts of animals [23].

5. Conclusions

FOS supplementation in the drinking water of rabbits improved most growth performance parameters, carcass characteristics, hematobiochemical parameters, antioxidant status, and cecal microbiota in NZW and APRI rabbits. Moreover, the response of NZW rabbits to FOS supplementation was more significant than that of APRI rabbits.

Author Contributions: Conceptualization, A.H.A.E.-A., W.M.G., A.E.-W.A. and M.M.A.G.; methodology, W.F.A., A.G., A.E.-W.A. and K.M.; software, W.M.G. and A.A.E.; validation, S.A.M., K.M. and A.A.E.; formal analysis, S.H.A.R.; investigation., A.H.A.E.-A.; resources, K.M., A.E.-W.A. and M.S.; data curation, A.O.H., A.E., H.A.Y., W.M.G. and K.M.; writing—original draft preparation, M.S., A.O.H., M.M.A.G., and K.M.; writing—review and editing, W.F.A., A.G., S.H.A.R. and M.S.; visualization, H.A.Y., W.F.A., W.M.G. and A.G.; supervision, W.F.A., A.O.H. and A.G.; project administration, M.S. and K.M.; funding acquisition, W.F.A., A.G., M.S., A.E.-W.A. and A.H.A.E.-A. All authors have read and agreed to the published version of the manuscript..

Funding: The research was partially funded by Taif University Researchers Supporting Project number (TURSP2020/53), Taif University, Taif, Saudi Arabia.

Institutional Review Board Statement: All investigations in this study were carried out following the Native Experimental Animal Care Committee and approved by the Ethics of the Institutional Committee of Animal Husbandry and Animal Wealth Development Department, Faculty of Veterinary Medicine, Damanhour University, Egypt (DMU/VetMed-2019-/0145).

Informed Consent Statement: Not applicable.

Data Availability Statement: The data that support the findings of this study are available on request from the corresponding author. The data are not publicly available due to privacy or ethical restrictions.

Acknowledgments: The authors appreciate Taif University Researchers Supporting Project number (TURSP2020/53), Taif University, Taif, Saudi Arabia.

Conflicts of Interest: The authors declare no conflict of interest. The funder had no role in the study's design or interpretation of data.

References

1. Abd El-Aziz, A.H.; Mahrose, K.M.; El-Kasrawy, N.I.; Alsenosy, A.E.A. Yeast as growth promoter in two breeds of growing rabbits with special reference to its economic implications. *Acad. Bras. Cienc.* **2021**, *93*, e20190274. [CrossRef] [PubMed]
2. Abo Ghanima, M.M.; Abd El-Aziz, A.H.; Noreldin, A.E.; Atta, M.S.; Mousa, S.A.; El-Far, A.H. β-glucan administration improves growth performance and gut health in New Zealand White and APRI rabbits with different breed responses. *PLoS ONE* **2020**, *15*, e0234076. [CrossRef] [PubMed]
3. Alagawany, M.; Mahrose, K.M. Influence of different levels of certain essential amino acids on the performance, egg quality criteria and economics of Lohmann brown laying hens. *Asian J. Poult. Sci.* **2014**, *8*, 82–96. [CrossRef]
4. Awad, A.; Fahim, H.; El-Shhat, A.E.; Mahrose, K.; Shazly, S. Dietary Echinacea purpurea administration enhanced egg laying performance, serum lipid profile, antioxidant status and semen quality in duck breeders during summer season. *J. Anim. Physiol. Anim. Nutr.* **2021**, *105*, 757–765. [CrossRef] [PubMed]
5. Hassan, F.A.; Mahrose, K.M.; Basyony, M.M. Effects of grape seed extract as a natural antioxidant on growth performance, carcass characteristics and antioxidant status of rabbits during heat stress. *Arch. Anim. Nutr.* **2016**, *70*, 141–154. [CrossRef] [PubMed]
6. Mahrose, K.M.; Elhack, M.E.A.; Mahgoub, S.A.; Attia, F.A.M. Influences of stocking density and dietary probiotic supplementation on growing Japanese quail performance. *Acad. Bras. Cienc.* **2019**, *91*, e20180616. [CrossRef] [PubMed]
7. Abd El-Aziz, A.H.; El-Kasrawy, N.I.; Abd El-Hack, M.E.; Kamel, S.Z.; Mahrous, U.E.; El-Deeb, E.M.; Atta, M.S.; Amer, M.S.; Naiel, M.A.E.; Khafaga, A.F.; et al. Growth, immunity, relative gene expression, carcass traits and economic efficiency of two rabbit breeds fed prebiotic supplemented diets. *Anim. Biotechnol.* **2022**, *33*, 417–428. [CrossRef] [PubMed]
8. Jha, R.; Das, R.; Oak, S.; Mishra, P. Probiotics (Direct-Fed Microbials) in Poultry Nutrition and Their Effects on Nutrient Utilization, Growth and Laying Performance, and Gut Health: A Systematic Review. *Animals* **2020**, *10*, 1863. [CrossRef]
9. Shang, Y.; Kumar, S.; Thippareddi, H.; Kim, W.K. Effect of Dietary Fructooligosaccharide (FOS) Supplementation on Ileal Microbiota in Broiler Chickens. *Poult. Sci.* **2018**, *97*, 3622–3634. [CrossRef]
10. Shang, Y.; Kim, W.J.I.J.P.S. Roles of fructooligosaccharides and phytase in broiler chickens: Review. *Int. J. Poult. Sci.* **2017**, *16*, 16–22. [CrossRef]
11. Mousa, Y.I.; Abdel-Monem, U.; Bazid, A. Effect of Spirulina And Prebiotic (Inmunair 17.5®) on New-Zealand White Rabbits Performance. *Zagazig J. Agric. Res.* **2018**, *97*, 3622–3634. [CrossRef]
12. Kim, G.B.; Seo, Y.M.; Kim, C.H.; Paik, I.K. Effect of dietary prebiotic supplementation on the performance, intestinal microflora, and immune response of broilers. *Poult. Sci.* **2011**, *90*, 75–82. [CrossRef] [PubMed]
13. Akrami, R.; Iri, Y.; Rostami, H.K.; Razeghi Mansour, M. Effect of dietary supplementation of fructooligosaccharide (FOS) on growth performance, survival, lactobacillus bacterial population and hemato-immunological parameters of stellate sturgeon (Acipenser stellatus) juvenile. *Fish Shellfish Immunol.* **2013**, *35*, 1235–1239. [CrossRef] [PubMed]
14. Sohail, M.U.; Hume, M.E.; Byrd, J.A.; Nisbet, D.J.; Ijaz, A.; Sohail, A.; Shabbir, M.Z.; Rehman, H. Effect of supplementation of prebiotic mannan-oligosaccharides and probiotic mixture on growth performance of broilers subjected to chronic heat stress. *Poult. Sci.* **2012**, *91*, 2235–2240. [CrossRef] [PubMed]
15. Swiatkiewicz, S.; Arczewska-Wlosek, A. Prebiotic fructans and organic acids as feed additives improving mineral availability. *World's Poult. Sci. J.* **2012**, *68*, 269–279. [CrossRef]
16. Xu, Z.R.; Hu, C.H.; Xia, M.S.; Zhan, X.A.; Wang, M.Q. Effects of dietary fructooligosaccharide on digestive enzyme activities, intestinal microflora and morphology of male broilers. *Poult. Sci.* **2003**, *82*, 1030–1036. [CrossRef]
17. Abd El-Gawad, E.A.; Abd El-latif, A.M.; Amin, A.A.; Abd-El-Azem, M. Effect of dietary fructooligosaccharide on bacterial Infection, oxidative stress and histopathological alterations in Nile tilapia (*Oreochromis niloticus*). *Glob. Vet.* **2015**, *15*, 339–350.
18. Bogusławska-Tryk, M.; Piotrowska, A.; Burlikowska, K. Dietary fructans and their potential beneficial influence on health and performance parametrs in broiler chickens. *J. Cent. Eur. Agric.* **2012**, *13*, 272–291. [CrossRef]
19. Saha, M.; Chowdhury, S.; Hossain, M.; Islam, M.; Roy, B. Effect of a Water Soluble Organic Additive on Growth Performances, Hematological Parameters and Cost Effectiveness in Broiler Production. *J. Progress. Agric.* **2010**, *21*, 73–82. [CrossRef]

20. Meksawan, K.; Chaotrakul, C.; Leeaphorn, N.; Gonlchanvit, S.; Eiam-Ong, S.; Kanjanabuch, T. Effects of Fructo-Oligosaccharide Supplementation on Constipation in Elderly Continuous Ambulatory Peritoneal Dialysis Patients. *Perit. Dial. Int.* **2016**, *36*, 60–66. [CrossRef]
21. Al-Rawashdeh, O.; Gumaa, A.; Saeed, M.; Orban, J.; Patterson, J.; Nour, A. Effects of sucrose thermal oligosaccharide caramel and feed restriction on the performance, hematological values and cecal bacteriological counts of broiler chickens. *Acta Vet.* **2000**, *50*, 225–239.
22. Liu, L.; Xu, Y.; Xu, X. Effect of supplementation with two combinations of alternative to antimicrobials by stages on cecal fermentation in rabbits. *Czech J. Anim. Sci.* **2018**, *63*, 419–427. [CrossRef]
23. Teng, P.Y.; Kim, W.K. Review: Roles of Prebiotics in Intestinal Ecosystem of Broilers. *Front. Vet. Sci.* **2018**, *5*, 245. [CrossRef] [PubMed]
24. Council, N.R. *Nutrient Requirements of Rabbits: Second Revised Edition, 1977*; The National Academies Press: Washington, DC, USA, 1977; p. 30.
25. Lebas, F. Feeding strategy for small and medium scale rabbit units. In Proceedings of the 3rd Conference of Asian Rabbit Production Association, Bali, Indonesia, 27–29 August 2013; pp. 27–29.
26. Association of Official Analytical Chemists. *Official Methods of Analysis: Changes in Official Methods of Analysis Made at the Annual Meeting. Supplement*; Association of Official Analytical Chemists: Washington, DC, USA, 1990; Volume 15.
27. Blasco Mateu, A.; Ouhayoun, J.; Masoero, G. Harmonization of criteria and terminology in rabbit meat research. *World Rabbit Sci.* **1993**, *1*, 3–10. [CrossRef]
28. Drabkin, D.L.; Austin, J.H. Spectrophotometric studies: I. Spectrophotometric constants for common hemoglobin derivatives in human, dog, and rabbit blood. *J. Biol. Chem.* **1932**, *98*, 719–733. [CrossRef]
29. Dierick, N.; Decuypere, J.; Molly, K.; Van Beek, E.; Vanderbeke, E.J.L.P.S. The combined use of triacylglycerols (TAGs) containing medium chain fatty acids (MCFAs) and exogenous lipolytic enzymes as an alternative to nutritional antibiotics in piglet nutrition: II. In vivo release of MCFAs in gastric cannulated and slaughtered piglets by endogenous and exogenous lipases; effects on the luminal gut flora and growth performance. *Livest. Prod. Sci.* **2002**, *76*, 1–16.
30. Van Nevel, C.J.; Decuypere, J.A.; Dierick, N.; Molly, K. The influence of Lentinus edodes (Shiitake mushroom) preparations on bacteriological and morphological aspects of the small intestine in piglets. *Arch. Anim. Nutr.* **2003**, *57*, 399–412. [CrossRef]
31. Levene, H. Robust tests for equality of variances. In *Contributions to Probability and Statistics. Essays in Honor of Harold Hotelling*; Olkin, I., Ed.; Stanford University Press: Palo Alto, CA, USA, 1960; pp. 278–292.
32. Duncan, D.B. Multiple Range and Multiple F Tests. *Biometrics* **1955**, *11*, 1–42. [CrossRef]
33. Ahrens, W.H.; Cox, D.J.; Budhwar, G. Use of the arcsine and square root transformations for subjectively determined percentage data. *Weed Sci.* **1990**, *38*, 452–458. [CrossRef]
34. Shehata, S.; Mahrose, K.M.; Ismail, E. Effect of amino yeast addition on growth performance, digestion, carcass traits and economical efficiency of growing rabbit. *Egypt J. Nutr. Feed* **2012**, *15*, 75–80.
35. El-Badawi, A.Y. Growth performance of male NZW rabbits fed diets supplemented with beneficial bacteria or live yeast. *Agric. Eng. Int. CIGR J.* **2017**, *19*, 220–226.
36. Rotolo, L.; Gai, F.; Peiretti, P.; Ortoffi, M.; Zoccarato, I.; Gasco, L. Live yeast (*Saccharomyces cerevisiae* var. *boulardii*) supplementation in fattening rabbit diet: Effect on productive performance and meat quality. *Livest. Sci.* **2014**, *162*, 178–184. [CrossRef]
37. Foad, K.A.E.E. Evaluation of baker's yeast (*Saccharomyces cerevisiae*) supplementation on the feeding value of hydroponic barley sprouts for growing rabbits. *Egypt. Poult. Sci. J.* **2017**, *37*, 833–854. [CrossRef]
38. Zarei, M.; Ehsani, M.; Torki, M. Dietary inclusion of probiotics, prebiotics and synbiotic and evaluating performance of laying hens. *Am. J. Agric. Biol. Sci.* **2011**, *6*, 249–255. [CrossRef]
39. Juśkiewicz, J.; Zduńczyk, Z.; Jankowski, J.; Juśkiewicz, M. Occurrence of mild-diarrhoea in turkeys fed a FOS-rich diet and its impact on caecal parameters. *Med. Weter.* **2007**, *63*, 290–293.
40. Beal, R.K.P.C.; Davison, T.F.; Smith, A.L. Immunological development of the avian gut. In *Avian Gut Function in Health and Disease*; Perry, G.C., Ed.; CABI: Oxfordshire, UK, 2006; pp. 85–103.
41. Christensen, H.R.; Frøkiær, H.; Pestka, J. Lactobacilli differentially modulate expression of cytokines and maturation surface markers in murine dendritic cells. *J. Immunol.* **2002**, *168*, 171–178. [CrossRef]
42. Janardhana, V.; Broadway, M.M.; Bruce, M.P.; Lowenthal, J.W.; Geier, M.S.; Hughes, R.J.; Bean, A.G. Prebiotics modulate immune responses in the gut-associated lymphoid tissue of chickens. *J. Nutr.* **2009**, *139*, 1404–1409. [CrossRef]
43. Hoseinifar, S.H.; Mirvaghefi, A.; Merrifield, D.L.; Amiri, B.M.; Yelghi, S.; Bastami, K.D. The study of some haematological and serum biochemical parameters of juvenile beluga (*Huso huso*) fed oligofructose. *Fish Physiol. Biochem.* **2011**, *37*, 91–96. [CrossRef]
44. Jalali, M.A.; Ahmadifar, E.; Sudagar, M.; Takami, G.A. Growth efficiency, body composition, survival and haematological changes in great sturgeon (*Huso huso* Linnaeus, 1758) juveniles fed diets supplemented with different levels of Ergosan. *Aquac. Res.* **2009**, *40*, 804–809. [CrossRef]
45. Guerreiro, I.; Pérez-Jiménez, A.; Costas, B.; Oliva-Teles, A. Effect of temperature and short chain fructooligosaccharides supplementation on the hepatic oxidative status and immune response of turbot (*Scophthalmus maximus*). *Fish Shellfish Immunol.* **2014**, *40*, 570–576. [CrossRef]

46. Zhang, C.N.; Li, X.F.; Jiang, G.Z.; Zhang, D.D.; Tian, H.Y.; Li, J.Y.; Liu, W.B. Effects of dietary fructooligosaccharide levels and feeding modes on growth, immune responses, antioxidant capability and disease resistance of blunt snout bream (*Megalobrama amblycephala*). *Fish Shellfish Immunol.* **2014**, *41*, 560–569. [CrossRef] [PubMed]
47. Saminathan, M.; Sieo, C.C.; Kalavathy, R.; Abdullah, N.; Ho, Y.W. Effect of prebiotic oligosaccharides on growth of Lactobacillus strains used as a probiotic for chickens. *Afr. J. Microbiol. Res.* **2011**, *5*, 57–64.

animals

MDPI

Review

Potential Use of Tannin Extracts as Additives in Semen Destined for Cryopreservation: A Review

Mohammed S. Liman [1,2], Abubeker Hassen [3], Lyndy J. McGaw [4], Peter Sutovsky [5] and Dietmar E. Holm [1,*]

1 Department of Production Animal Studies, Faculty of Veterinary Science, University of Pretoria, Pretoria 0110, South Africa; limanms69@gmail.com
2 Niger State Livestock and Fisheries Institute, Ministry of Livestock and Fisheries Development, Minna 920001, Niger State, Nigeria
3 Department of Animal and Wildlife Sciences, Faculty of Natural and Agricultural Science, University of Pretoria, Pretoria 0028, South Africa; abubeker.hassen@up.ac.za
4 Phytomedicine Programme, Department of Paraclinical Sciences, Faculty of Veterinary Science, University of Pretoria, Pretoria 0110, South Africa; lyndy.mcgaw@up.ac.za
5 Division of Animal Sciences, Department of Obstetrics, Gynecology and Women's Health, University of Missouri, Columbia, MO 65211-5300, USA; sutovskyp@missouri.edu
* Correspondence: dietmar.holm@up.ac.za

Simple Summary: Freezing of semen used for artificial reproductive technologies (ART) affects the survival and vigour of sperm cells due to excessive production of reactive oxygen species (ROS) during the freezing and thawing processes. ROS plays a physiological role in sperm function but excessive ROS production from damaged sperm cells can hinder sperm's motility and their ability to fertilise an oocyte. Tannins, a class of water-soluble plant polyphenols, are known to have antioxidant and other health-promoting effects and may serve as binders/acceptors to reduce the deleterious effects of excessive ROS produced during the freezing and thawing process. This review is the first to analyse the available data supporting the use of tannins as additives to semen extenders to improve the survival of cryopreserved spermatozoa during storage and after thawing. It is concluded that tannins and their derivatives have naturally protective properties with the potential to ameliorate sperm cell survival after freezing.

Abstract: Cryopreservation and storage of semen for artificial insemination (AI) result in excessive accumulation of reactive oxygen species (ROS). This leads to a shortened life span and reduced motility of spermatozoa post-thawing, with consequent impairment of their function. However, certain levels of ROS are essential to facilitate the capacitation of spermatozoa required for successful fertilisation. Tannins, as well-known antioxidant compounds, may act as ROS binders/acceptors/scavengers to inhibit the damaging effects of ROS. This review comprises an analysis of the semen cryopreservation protocol and health functions of tannins, as well as the effects of ROS on fresh and cryopreserved semen's longevity and fertilisation. Additionally, we surveyed available evidence of the effects of tannin extract feed supplementation on male fertility. We furthermore interrogated existing theories on tannin use as a potential additive to semen extenders, its relationship with semen quality, and to what degree existing theories have been investigated to develop testable new hypotheses. Emphasis was placed on the effects of tannins on ROS, their involvement in regulating sperm structure and function during cryopreservation, and on post-thaw sperm motility, capacitation, and fertilising ability. The diverse effects of tannins on the reproductive system as a result of their potential metal ion chelation, protein precipitation, and biological antioxidant abilities have been identified. The current data are the first to support the further investigation of the incorporation of tannin-rich plant extracts into semen extenders to enhance the post-thaw survival, motility, and fertilising ability of cryopreserved spermatozoa.

Keywords: cryopreservation; spermatozoa; tannin; polyphenols; semen additives; antioxidant

Citation: Liman, M.S.; Hassen, A.; McGaw, L.J.; Sutovsky, P.; Holm, D.E. Potential Use of Tannin Extracts as Additives in Semen Destined for Cryopreservation: A Review. *Animals* **2022**, *12*, 1130. https://doi.org/10.3390/ani12091130

Academic Editor: Eva Bussalleu

Received: 28 February 2022
Accepted: 22 April 2022
Published: 28 April 2022

1. Introduction

Cryopreservation reduces the functional and structural integrity of spermatozoa due to the development of reactive oxygen species (ROS) [1,2]. ROS are produced during numerous chemical reactions in different parts of the mammalian body [1]. In the testes, ROS are produced during spermatogenesis within the seminiferous tubules and steroidogenesis in the interstitium [3]. Cryopreservation and storage of semen lead to changes in the sperm mitochondrial membrane and the resident electron transport chain [3], which result in the excessive release of ROS, hydrogen peroxide (H_2O_2), nitric oxide (NO), or superoxide anion (O_2^-), with consequences on sperm capacitation and the acrosome reaction [2]. Cryoprotectants are important for the cryo-survival of spermatozoa [4], and these may include egg yolk, glycerol [4], dimethyl sulphoxide (DMSO) [5], ethylene glycol [6], Triladyl® (a commercially available semen extender) [7], and butylated hydroxytoluene (BHT) [8–11]. Combinations of cryoprotectants such as glycerol and ethylene glycol [7,12] and acetamide together with lactamide [13] may also be employed. Antioxidant substances may reduce the impact of oxidative stress and thereby improve the quality of semen post-thawing [14]. Cryoprotectants are important for the cryo-survival of spermatozoa [15].

Low levels of ROS are, however, associated with increased sperm motility, viability, increased capacity for successful fertilisation during sperm–oocyte interactions, and fertility in mammalian species [16]. Antioxidant additives in semen diluents for cryopreservation should therefore not aim to eliminate ROS [17]. When ROS occur in small concentrations, they act as mediators of normal sperm function, whereas when present in excess, they are toxic to spermatozoa [14].

Sperm capacitation normally occurs in the oviduct and involves biochemical and structural changes that make the spermatozoa competent to attach to the zona pellucida of the oocyte, penetrate it, and fuse with the oolemma [18]. The cellular changes that occur include the activation of soluble adenyl cyclase that produces cAMP, the influx of Ca^{2+} ions, Zn ions [19,20], efflux of cholesterol from the plasma membrane, leading to its fluidity/fuseability, and the generation of more ROS, with a consequent increase in intracellular pH [7]. Additionally, activation of protein kinase A and downstream protein tyrosine kinases results in the protein phosphorylation of numerous proteins on tyrosine residues [21]. This process results in the hyperactivation of sperm tail motility, which is necessary for sperm detachment from the oviductal sperm reservoir and the penetration of the egg vestment at fertilisation. It was reported that controlled and low ROS generation plays a physiological role during the capacitation and acquisition of sperm's fertilising ability, with ROS-specific scavengers inhibiting the process [14,22,23]. These processes of ROS affecting the spermatozoa have been reviewed previously [24].

Thus, ROS homeostasis appears to be equally important for timely sperm capacitation within the female oviduct, and for the prevention of premature capacitation during semen processing and cryopreservation for artificial insemination (AI).

Plants contain combinations of complex polymeric phenols, which are amongst the most studied phytochemicals because of their diverse array of useful biological functions and health-promoting effects [14]. Consequently, their antioxidation effects on the production of ROS, sperm longevity, and fertilising potential were reviewed using the available peer-reviewed data on tannin extract supplementation for male fertility. The aim was to document the utilisation of the biological and reproductive health benefits of tannins, with a view to exploiting their potential for use as additives to improve the cryopreservation of semen. To our knowledge, this review is the first to recommend further structured evaluation of the value of tannin extracts or compounds as additives into semen destined for cryopreservation [14,25].

2. Methodology

This theoretical literature review (TLR) focused firstly on the existing evidence of the biological and health benefits of tannins, specifically with regard to their antioxidant properties and resultant inhibitory effects on lipid peroxidation, as well as their antiviral, antibacterial, and anti-inflammatory effects in terms of protecting spermatozoa against microbial infections during semen processing, cryopreservation, and distribution. This first section is divided into three subsections addressing the cryopreservation of semen using tannins, and their relevant biological and health functions, respectively. Secondly, we investigated the current evidence on the effect of ROS on sperm viability/semen longevity, and on the requirement for low levels of ROS in semen fertility.

Articles used in this review had a concise hypothesis, with keywords searched on databases including Google Scholar, Scopus/ScienceDirect, and Web of Science/Pubmed (which includes CABAbstracts, Medline and Zoological Records). Inclusion words: "additives"; "cryopreservation", "tannin-extracts" or "fractions", or "compound" and "health" or "biological" with emphasis on specific functions, namely "antiviral", "antibiotic", "antioxidant", and "protection against lipid peroxidation", excluding anticancer and antidiabetes. Database results for Google Scholar (645, 76%); Scopus/ScienceDirect (94, 11%); and Web of Science/Pubmed (105, 13%) are all published reports, respectively (Figure 1). The relevant reports used are represented as cited in this study.

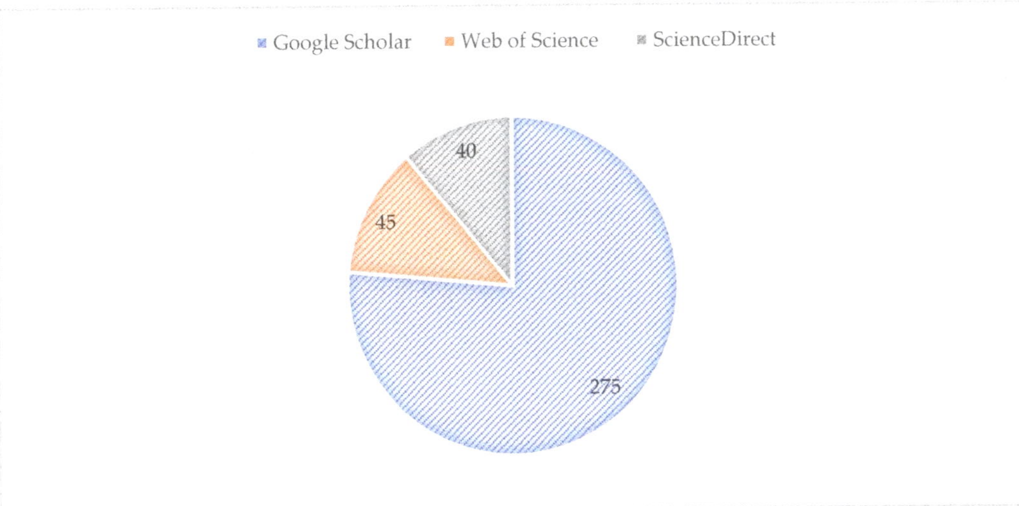

Figure 1. Database of the number of reports from Google Scholar, Scopus, and Web of Science on tannin additives for cryopreservation (n = 844).

3. Effect of ROS on Cryopreserved Spermatozoa

3.1. ROS Effect on Sperm Cryopreservation and Longevity

ROS are a group of molecules (free radicals, oxygen ions, peroxides, etc.) that are produced during aerobic metabolism in the mitochondria of cells, and are important components of physiological processes and cellular signalling events [1]. The liquid or frozen semen preservation and its effect on semen quality were reviewed previously [14]. Oxidative damage in semen impairs spermatozoal function, resulting in a loss of motility, loss of mitochondrial activity, increase in deoxyribonucleic acid (DNA) damage, and lack of activation of apoptotic pathways [26]. Consequently, unresolved issues affecting fertility are encountered in artificially collected semen samples, such as infections, inadequate constituents of semen extenders and protocols adopted during cryopreservation processes, and the overall need for highly skilled intra-uterine insemination. Mammalian spermatozoa naturally contain antioxidants and ROS scavenging enzymes, such as glutathione (GSH), superoxide dismutase, and catalase (CAT) [27]. These endogenous antioxidants often are not sufficient to prevent lipid peroxidation during cryopreservation [28]. Excess ROS that develop during the storage of spermatozoa are largely responsible for damage to spermatozoa. The damage of the sperm plasma membrane due to the effect of ROS consequently exposes semen to lipid peroxidation, resulting from the high content of polyunsaturated acids, and DNA damage [29]. Thus, the cryopreservation of semen is dependent on the reversible reduction of the survival and metabolic activity of spermatozoa [30]. This could be achieved by the provision of an effective environment for the steady cooling of semen, with a focus on the development of extenders that maintain membrane integrity, increase motility, maximise sperm's ability to capacitate, prevent oxidative stress, and minimise the generation of reactive oxygen species (ROS) during cryopreservation and storage [31–33]; see Figure 2.

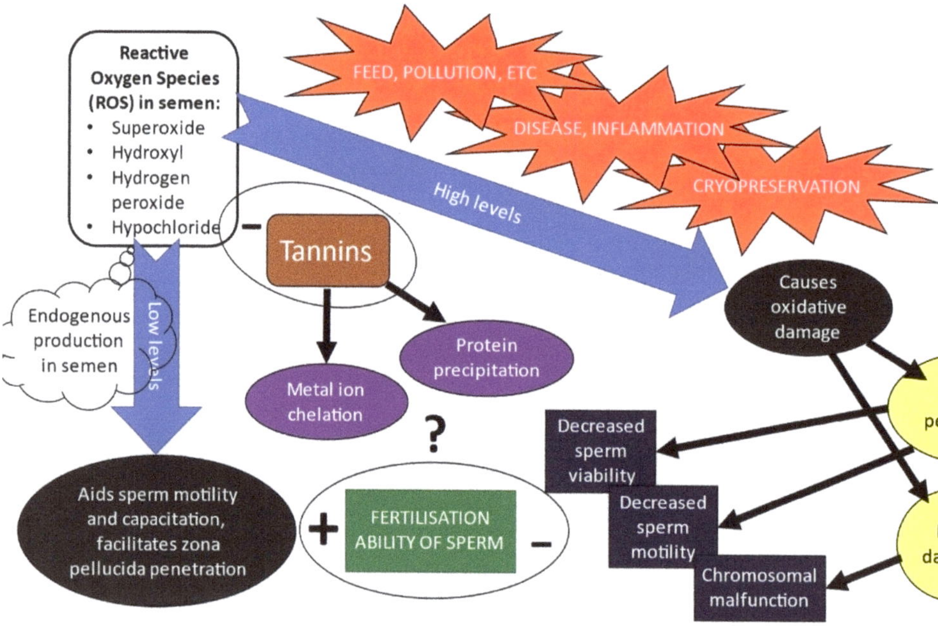

Figure 2. Demonstration of the homeostatic effects of ROS on sperm's fertilisation ability, and the potential benefits of tannins to ameliorate its effects, particularly following cryopreservation.

The cytotoxic action of ROS on spermatozoa is mediated by high concentrations of phospholipid-bound polyunsaturated fatty acids (PUFA) in the sperm plasma membrane, especially docosahexaenoic acid (DHA), with six double bonds per molecule, which makes them susceptible to free radical attack [34]. Additionally, spermatozoa lack an enzyme apurinic/apyrimidinic endonuclease (APEI), which plays a significant role in DNA repair and base excision repair pathways [24]. Furthermore, the sperm DNA is hyper-condensed; thus, it is not easily accessible to repair mechanisms.

3.2. *Effects of ROS on Sperm's Fertilising Potential*

Low levels of ROS are associated with increased sperm motility, viability, increased capacity for fertilisation during sperm–oocyte fusion, and general male fertility in mammalian species [16]. When ROS are in low concentrations, they act as mediators of normal sperm function, whereas in excess, they are toxic to spermatozoa. Sperm capacitation is a complex process by which spermatozoa acquire the ability to fertilise the mature oocyte. This occurs within the oviductal sperm reservoir and involves the biochemical and morphological changes that make the spermatozoon competent to attach to the zona pellucida of the oocyte, penetrate it, and fuse with the oolemma [18]. Conception rates in livestock AI depend on the quality of semen, which is generally low post-thawing, with the capacitation and fertilisation processes being dependent on the effect of the sub-lethal dysfunction of spermatozoa [35]. Premature sperm capacitation brought about by cryopreservation and thawing is referred to as cryocapacitation [36] and, similarly to physiological capacitation, is irreversible and terminal, leading to a shortened sperm lifespan and eventual death before spermatozoa can reach the oviductal fertilisation site following AI. The selection of animals with good-quality semen for cryopreservation and AI is a critical step in improving the fertility levels of frozen–thawed semen [37,38]. Despite having satisfactory fertility testing in terms of fresh-stored semen, the frozen–thawed semen of some animal species does not meet standards of acceptable fertilisation results suitable for commercial AI programmes [38,39]. Accumulated evidence indicates that inherent male progeny variability is one of the factors in semen cryopreservation responsible for the marked differences in sperm cryo-survival [37–40]. Individual differences in sperm quality and cryo-survival are addressed by ongoing efforts to identify gene variants and differentially expressed sperm proteins associated with either high or low sperm cryotolerance in livestock species [41,42].

4. Tannins

4.1. Properties of Tannins

Tannins are sourced from a multitude of trees and shrubs. Notable for industrial importance are black wattle or black mimosa (Mimosa tannin, *Acacia mearnsii*), quebracho wood (*Schinopsis lorentzhii*), oak bark (*Quercus robur*), chestnut wood (*Castanea sativa*), mangrove wood (*Algarobilla chilena*), gambir (*Uncaria gambir*), the bark of several species of pines and firs, such as *Pinus radiata* and *Pinus nigra*, as well as many other plants harbouring extractable tannins [43–46]. Tannins are a renewable resource used in several fields, ranging from the traditional application of tanning to producing heavy-duty leather and as wood adhesives up until the 1960s and 1970s, whereafter new applications were investigated [44], such as the proposed use of chestnut tannin as an antimicrobial and to reduce mycotoxins [47]. Tannins dissolve in water to form colloidal solutions, with their solubility dependent on the degree of polymerisation [48]. They are soluble in alcohol and acetone, and react with ferric chloride [49]. They have moderate stability in aqueous solutions, especially during extraction with boiling water (decoctions), in which they decompose in 30 min into gallic acid, ellagic acid, and corilagin [44]. At the centre of hydrolysable tannins is a polyol carbohydrate (D-glucose), which is partially or completely esterified with a phenolic group such as gallic acid (gallotannins) or ellagic acid (ellagitannins). Hydrolysable tannins are hydrolysed by weak acids or weak bases to produce carbohydrates and phenolic acids. Condensed tannins (proanthocyanidins) are polymers of 2–50 (or more) flavonoid units joined by carbon-to-carbon bonds, which are not easily cleaved by hydrolysis.

4.2. Extraction of Tannins

Tannins, both hydrolysable and condensed, are commonly extracted with a mixture of water and acetone. Optimal yield may be obtained from fresh, frozen, or lyophilised material. Some tannin-rich extracts are available from varied sources and are used as supplements to improve reproduction.

4.3. Medicinal Properties and Biological Functions of Tannins

The health benefits of tannins include antioxidant, anti-carcinogenic, cardioprotective, antimutagenic, antiviral, antibacterial, haemostatic, and anti-inflammatory properties, as well as inhibition of lipid perioxidation [45,46] Hydrolysable tannins are often cited for their antimicrobial activity [46] and chemopreventive properties against degenerative diseases [50]. These multi-functional properties of tannins are utilised in the treatment of human diseases [51]. Hydrolysable tannins are also inhibitors of α-glucosidase, which is an enzyme known to be involved in the modulation of the absorption of glucose in tissues [48].

Antioxidants have been used in semen extenders, including cysteamine, taurine, trehalose, and selenium, to improve the motility, viability, and membrane integrity of post-thawed semen [52,53], with significant results. Some other antioxidants, such as Vitamin C and E and catalase, have been used to supplement human, cattle, boar, rabbit, and stallion semen [54,55] In a study of the α-glucosidase inhibition and antioxidant activity of an oenological commercial tannin (Tan'Activ® toasted oak wood *Quercus robur*), the extraction and fractionation process yielded four fractions, with one of the fractions generating a sub-fraction with enhanced α-glucosidase inhibitory activity with an inhibitory concentration (IC_{50}) of 6.15 μg/mL [56]. The oak wood is used for barrel staves in the winemaking process and the polyphenols are not only used in the ageing of wine but in maintaining aroma/flavour, as well as contributing useful health properties [57,58].

Synthetic water-soluble polymers such as polyvinyl pyrrolidone (PVP) and polyethylene glycol (PEG) are used as tannin-binding agents for quantification and to neutralise the negative effect of tannins in animal diets [49]. The PVP is used to bind hydrolysable tannins, while PEG is used for condensed tannins. These groups of tannins both contain sufficient oxygen molecules in their chains to form strong hydrogen bonds, with the phenolic and hydroxyl groups in tannins serving to precipitate them from solutions [49].

4.4. Use of Tannins as Supplements to Improve Reproduction Outomces or as Semen-Protective Agents

Tannin extracts or compounds are extracted using ethanol or water into powdered substances and stored at −20 °C [56,59,60] for later use as supplements (feed) (Table 1) or added into semen extenders (Table 2), etc., after optimisation.

Table 1. Reported effects of tannin extracts used as food/feed supplements on reproduction outcomes in humans or animals.

Plant from Which Tannins Were Extracted (References)	Compounds Identified	Extraction Method	Subject (Animal Species and Gender)	Effect on Reproduction
Zingiber officinale (ginger) root extract [61]	High content of total flavonoids, tannins, alkaloids, and total phenolic components	Ethanol	*Rattus rattus* (Rat)—male	Restored testis histopathological alterations, reduced arsenic, and improved sperm parameters.
Spondias mombin leaf extract [62]	Leaves contain saponins, alkaloids, flavonoids, tannins, steroids, phenolics, phlobatannins, cardiac glycosides, cardenolides, and dienolides with saponins	Ethanol	*Cavia porcellus* (Guinea pig)—male	Induced infertility in males via endocrine dysregulation, anti-spermatogenic activity, testicular dysfunction, and antioxidative stress.
Allium triquetrum (wild garlic) bulb and leaf extract [63]	Tannins (leaves have higher concentration)	Water	*Rattus norvegicus* (Wistar rat)—male	Used in the treatment of reproductive toxicity of lead acetate by reducing lead testicular injury by boosting sperm characteristics and ameliorating oxidative sperm markers.
Azadirachta indica leaf and fruit extracts [64]	Not reported	Methanol	*Rattus norvegicus* (Long Evans rat)—male	At 200 μg/mL, increased percentage of morphological defects. (Cellular detachment in the seminiferous epithelium with sperm death without decrease in number of sperm).

Table 1. *Cont.*

Plant from Which Tannins Were Extracted (References)	Compounds Identified	Extraction Method	Subject (Animal Species and Gender)	Effect on Reproduction
Acacia mearnsii (Black Wattle) bark extract [65]	Condensed tannins average MW 1250 (500 to 3000), non-tannin polyphenols, salts, sugars, and organic acids. Total tannins (65.5%), tannic acid, and condensed tannin (30.5%) as leucocyanidin	Water	*Ovis aries* (Sheep: mutton merino)—male	Increase in testicular length, semen volume, semen concentration, and reduction in sperm with morphological defects.
Phoenix dactylifera (Date palm) fruit extract [66]	Review study		*Homo sapiens* (Man)—male and female	It has a potent effect on male hormones, seminal vesicle parameters, and sperm motility and viability.
Turraea fischeri bark extract [67]	20 compounds including several isomers of flavonolignan cinchonain-I and dominant bis-dihydrophenoxyl propanoid-substituted catechins hexsoides	Methanol	*Rattus norvegicus* (Wistar rat)—male	Enhanced reduction in the elevated levels of aspartate aminotransferase (AST), malondialdehyde (MDA), and increased glutathione (GSH) content in the liver.
Mucuna pruriens (Thai (T-MP)) seed extract [68]	Not reported	Water	*Rattus rattus* (Rat)—male and female	Exhibit antioxidation capacity, phytoestrogenic effect on females, and increased testicular and sperm markers of male fertility.
Vitis vinifera (Grape) seed tannin extract (GPE) [69]	GPE has a 95% purity coefficient (56.5% condensed tannins)	Not reported	*Ovis aries* (Hu lambs)	Improved the seminiferous tubules' development, diameter, and increase in Sertoli cells. Also increase in superoxide dismutase (SOD).
Caesalpinia pulcherrima bark extracts [70]	Alkaloids, flavonoids, steroids, and triterpenes	Water and ethanol	*Rattus rattus* (Rat)—female	Reduced ovarian size and increased uterine weight.

MW = Molecular weight.

Certain tannin concentrations have exerted efficiency in fertilisation, but with no effect on sperm kinematic parameters, acrosome integrity, mitochondrial membrane integrity, lipid perioxidation, or capacitation status or its viability [73]. The ethanol extract of a commercial oenological tannin (*Quercus robur*, toasted oak wood Tan'Activ®) had a biological effect at a concentration of 10 μg/mL, stimulating an increase ($p < 0.001$) in in vitro swine sperm capacitation at the tail principal piece (B pattern) and increased ($p < 0.001$) oocyte fertilisation rate [60]. However, at 100 μg/mL, the opposite effect was recorded on both sperm capacitation (B pattern) and fertilising ability, associated with higher sperm viability [60]. Where 5% crude tannin was added to the semen of the Bali breed of cattle for 14 days, it increased ($p > 0.001$) motility and viability, with a decrease in abnormal semen [76]. Guava (*Psidium guajava*) leaf extract, comprising 3% crude tannin, was added to liquid semen (stored for 15 days at 4–5 °C) of Ettawa crossbred Boer goats and improved ($p < 0.001$) the motility and viability and maintained intact plasma membranes of the spermatozoa, while a concentration of 24% of the crude tannin reduced viable sperm content [71]. Altogether, it appears that tannins may benefit extended semen through ROS scavenging and microbial growth limitation. It is yet to be determined if tannins may also convey cryotolerance during semen preservation.

Table 2. Tannin-rich extracts used as semen-protective agents in humans and various domestic animals.

Plants from Which Tannins Were Extracted (References)	Compounds Identified	Extraction Methods	Subject (Animal Species and Gender)	Effect on Sperm
Psidium guajava (Crude guava) leaf tannin extract [71]	2.41% of tannin, 20.80% of phenols per 17.825 g of extract	Methanol, ethyl acetate, and acetone	*Capra aegagrus hircus* (Etawa crossbred goat)	At 3%, increase in sperm motility, viability, and maintained intact plasma membrane integrity.
Aspalathus linearis (Rooibos) extracts [72]	Major flavonoids, flavols, and low tannins	Water	*Sus scrofa domesticus* (Pig)	Enhanced the sperm velocity, protected acrosome integrity, and preserved membrane integrity during 96 h of storage.
Mixture of chestnut and Quebracho wood (60/40) tannin-rich vegetal extract [73]	94.2% tannin content	Filter Freiberg-hide powder method	*Sus scrofa domesticus* (Pig)	Increased penetration rate with oocytes inseminated with thawed sperm pretreated with vegetal extract, and at 5 μg/mL, it exerts total efficiency on fertilisation.
Entada abyssinica (Splinter bean) bark extract [74]	28 compounds including tannins and gallic acid derivatives	Methanol	*Ovis aries* (Sheep)	Increased post-thaw progressive sperm motility, plasma membrane integrity, % of intact sperm increased with decrease in apoptotic/necrotic sperm.
Quercus robur (Toasted oak wood) (Tan activ®) [56,60]	Monogalloyl glucose (332.2), Glucose esterified by hexahydroxydiphenic acid (482.2), Gallic acid (170.1), Ellagitannins, castalin (632.4), Vescalsgin (934.6), Grandinin or its isomer roburin E (1066.7)	Ethanol	*Sus scrofa domesticus* (Pig)	Stimulated the sperm capacitation and oocyte fertilisation rate in a swine in vitro fertilisation trial.
Capparis spinosa leaf extract [59]	Flavones and flavanols, total flavonoids, total phenolic content, tannins, and the total carbohydrates	Water and ethanol	*Homo sapiens* (Man)	Increased progressive, total in vitro motility, viability, and maintained sperm DNA integrity.
Avena sativa (Oats) seed extract [75]	Phenols—93.2 mg/g, Flavonoids—67 mg/g, Saponins—5.9%, Glycosides—17.6%, Terpenoids—4.6%, Rutin—179 ppm, Kaemperol—513 ppm, Quercetin 409 ppm, Gallic acid—348 ppm	Water	*Bos taurus* (Bovine: Holstein)	Improved sperm individual motility, viability, plasma membrane integrity, and acrosome integrity.

5. Conclusions

To our knowledge, this is the first review recommending the addition of tannin-rich extract or compounds into semen destined for cryopreservation, exploiting their diverse effects on biological systems due to their potential for metal ion chelation and biological antioxidation. The varied biological roles, however, together with the enormous structural variations of these compounds, make it difficult to develop a model that allows accurate prediction of the role of tannins in any biological system. Therefore, it becomes imperative for studies to be conducted on tannin biological activities by determining their chemical structure, biological activity, and structure–activity relationships so that potential applications can be explored. While the inquiry into the biological activities of tannins is still in its infancy, it holds a promise of utility in livestock-assisted reproductive technology and human reproductive therapy. The addition of plant tannin extracts, extract fractions, or purified/synthetic compounds derived therefrom to semen may elevate the quality and viability of semen intended for cryopreservation. Beyond sperm cryopreservation, protocols for semen collection, processing, and liquid semen distribution in relevant livestock species could benefit from judicious, experiment-validated tannin supplementation, taking advantage of the antioxidant properties of tannins.

Author Contributions: Conceptualisation, A.H., D.E.H. and M.S.L.; methodology, A.H., L.J.M., D.E.H. and M.S.L.; validation, M.S.L.; A.H., D.E.H., L.J.M. and P.S.; formal analysis, M.S.L.; investigation, M.S.L.; resources, L.J.M., D.E.H. and A.H.; data curation, D.E.H. and M.S.L.; writing—original draft preparation, M.S.L.; writing—review and editing, M.S.L., D.E.H., A.H., L.J.M. and P.S.; visualisation, L.J.M.; supervision, D.E.H., L.J.M. and A.H.; project administration, D.E.H.; funding acquisition, M.S.L. and D.E.H. All authors have read and agreed to the published version of the manuscript.

Funding: This research was funded by the Red Meat Research and Development Trust of South Africa project committee (RMRD SA). P.S. was supported by grant number 2021-67015-33404 from the USDA National Institute of Food and Agriculture, grant number 1R01HD084353 from NIH/NICHD, and a travel grant from the University of Missouri South African Education Program. The project was also supported by the "Translational Medicine Research Theme" of the Faculty of Veterinary Sciences, University of Pretoria.

Institutional Review Board Statement: The study was approved by the University of Pretoria research committee and Animal Ethics Committee of the University of Pretoria (REC 193-19 of 8 October 2020).

Informed Consent Statement: Not applicable.

Data Availability Statement: The data presented in this study are available upon request from the corresponding author.

Acknowledgments: We would like to thank Susan Marsh (susanmarsh@up.ac.za), Faculty Library Manager: Jotello F. Soga Veterinary Science, for guidance in the use of search tools.

Conflicts of Interest: The authors declare no conflict of interest. The funders had no role in the design of the study; in the collection, analyses, or interpretation of data; in the writing of the manuscript, or in the decision to publish the results.

References

1. Gao, S.; Li, C.; Chen, L.; Zhou, X. Actions and mechanisms of reactive oxygen species and antioxidative system in semen. *Mol. Cell. Toxicol.* **2017**, *13*, 143–154. [CrossRef]
2. Aitken, R.J. Free radicals, lipid peroxidation and sperm function. *Reprod. Fertil. Dev.* **1995**, *7*, 659–668. [CrossRef] [PubMed]
3. Leahy, T.; Gadella, B.M. Sperm surface changes and physiological consequences induced by sperm handling and storage. *Reproduction* **2011**, *142*, 759. [CrossRef] [PubMed]
4. Van Wagtendonk-De Leeuw, A.; Den Daas, J.; Kruip, T.A.; Rall, W.F. Comparison of the efficacy of conventional slow freezing and rapid cryopreservation methods for bovine embryos. *Cryobiology* **1995**, *32*, 157–167. [CrossRef]
5. El-Harairy, M.; Eid, L.N.; Zeidan, A.; Abd El-Salaam, A.; El-Kishk, M. Quality and fertility of the frozen-thawed bull semen as affected by the different cryoprotectants and glutathione levels. *J. Am. Sci.* **2011**, *7*, 791–801.
6. Rodriguez-Martinez, H.; Barth, A.D. In vitro evaluation of sperm quality related to in vivo function and fertility. *Soc. Reprod. Fertil. Suppl.* **2007**, *64*, 39–54. [CrossRef]
7. Taşdemir, U.; Büyükleblebici, S.; Tuncer, P.B.; Coşkun, E.; Özgürtaş, T.; Aydın, F.N.; Büyükleblebici, O.; Gürcan, I.S. Effects of various cryoprotectants on bull sperm quality, DNA integrity and oxidative stress parameters. *Cryobiology* **2013**, *66*, 38–42. [CrossRef] [PubMed]
8. Watson, P. Recent developments and concepts in the cryopreservation of spermatozoa and the assessment of their post-thawing function. *Reprod. Fertil. Dev.* **1995**, *7*, 871–891. [CrossRef]
9. Sahashi, Y.; Otsuki, T.; Higaki, S.; Nagano, M.; Yamashita, Y.; Hishinuma, M. Effect of butylated hydroxytoluene on dog sperm longevity in chilling storage and cryopreservation. *J. Vet. Med. Sci.* **2011**, *73*, 1103090461. [CrossRef]

10. Naijian, H.R.; Kohram, H.; Shahneh, A.Z.; Sharafi, M.; Bucak, M.N. Effects of different concentrations of BHT on microscopic and oxidative parameters of Mahabadi goat semen following the freeze–thaw process. *Cryobiology* **2013**, *66*, 151–155. [CrossRef]
11. Ghorbani, M.; Amiri, I.; Khodadadi, I.; Fattahi, A.; Atabakhsh, M.; Tavilani, H. Influence of BHT inclusion on post-thaw attributes of human semen. *Syst. Biol. Reprod. Med.* **2015**, *61*, 57–61. [CrossRef] [PubMed]
12. Tuncer, P.B.; Buyukleblebici, S.; Eken, A.; Tasdemir, U.; Durmaz, E.; Büyükleblebici, O.; Coskun, E. Comparison of Cryoprotective Effects of Lycopene and Cysteamine in Different Cryoprotectants on Bull Semen and Fertility Results. *Reprod. Domest. Anim.* **2014**, *49*, 746–752. [CrossRef] [PubMed]
13. Nagase, H.; Tomizuka, T.; Hanada, A.; Hosoda, T.; Morimoto, H. Cryoprotective action of some amide solutions on spermatozoa of domestic animals. I. Effects of formamide, acetamide and lactamide on the motility of bovine spermatozoa after pellet freezing. *Jpn. J. Anim. Reprod.* **1972**, *18*, 15–21. [CrossRef]
14. Ros-Santaella, J.L.; Pintus, E. Plant Extracts as Alternative Additives for Sperm Preservation. *Antioxidants* **2021**, *10*, 772. [CrossRef]
15. Clément, C.; Witschi, U.; Kreuzer, M. The potential influence of plant-based feed supplements on sperm quantity and quality in livestock: A review. *Anim. Reprod. Sci.* **2012**, *132*, 1–10. [CrossRef]
16. Baumber, J.; Ball, B.A.; Gravance, C.G.; Medina, V.; Davies-Morel, M.C. The effect of reactive oxygen species on equine sperm motility, viability, acrosomal integrity, mitochondrial membrane potential, and membrane lipid peroxidation. *J. Androl.* **2000**, *21*, 895–902.
17. Sutovsky, P. New approaches to boar semen evaluation, processing and improvement. *Reprod. Domest. Anim.* **2015**, *50*, 11–19. [CrossRef]
18. Senger, P.L. Endocrinology of the male and spermatogenesis. In *Pathways to Pregnancy and Parturation*, 3rd ed.; Current Conceptions: Pullman, WA, USA, 2012; pp. 202–227.
19. Kerns, K.; Zigo, M.; Drobnis, E.Z.; Sutovsky, M.; Sutovsky, P. Zinc ion flux during mammalian sperm capacitation. *Nat. Commun.* **2018**, *9*, 2061. [CrossRef]
20. Kerns, K.; Sharif, M.; Zigo, M.; Xu, W.; Hamilton, L.E.; Sutovsky, M.; Ellersieck, M.; Drobnis, E.Z.; Bovin, N.; Oko, R.; et al. Sperm cohort-specific zinc signature acquisition and capacitation-induced zinc flux regulate sperm-oviduct and sperm-zona pellucida interactions. *Int. J. Mol. Sci.* **2020**, *21*, 2121. [CrossRef]
21. Ecroyd, H.W.; Jones, R.C.; Aitken, R.J. Endogenous redox activity in mouse spermatozoa and its role in regulating the tyrosine phosphorylation events associated with sperm capacitation. *Biol. Reprod.* **2003**, *69*, 347–354. [CrossRef]
22. O'Flaherty, C.; de Lamirande, E.; Gagnon, C. Positive role of reactive oxygen species in mammalian sperm capacitation: Triggering and modulation of phosphorylation events. *Free Radic. Biol. Med.* **2006**, *41*, 528–540. [CrossRef] [PubMed]
23. Aitken, R.J.; Nixon, B. Sperm capacitation: A distant landscape glimpsed but unexplored. *Mol. Hum. Reprod.* **2013**, *19*, 785–793. [CrossRef] [PubMed]
24. Aitken, R.J.; Smith, T.B.; Jobling, M.S.; Baker, M.A.; De Iuliis, G.N. Oxidative stress and male reproductive health. *Asian J. Androl.* **2014**, *16*, 31. [CrossRef]
25. Khojasteh, S.M.B.; Khameneh, R.J.; Houresfsnd, M.; Yaldagard, E. A review on medicinal plants used for improvement of spermatogenesis. *Biol. Med.* **2016**, *8*, 1. [CrossRef]
26. Aitken, R.J.; Gordon, E.; Harkiss, D.; Twigg, J.P.; Milne, P.; Jennings, Z.; Irvine, D.S. Relative impact of oxidative stress on the functional competence and genomic integrity of human spermatozoa. *Biol. Reprod.* **1998**, *59*, 1037–1046. [CrossRef]
27. Kantola, M.; Saaranen, M.; Vanha-Perttula, T. Selenium and glutathione peroxidase in seminal plasma of men and bulls. *Reproduction* **1988**, *83*, 785–794. [CrossRef] [PubMed]
28. Aurich, J.; Schönherr, U.; Hoppe, H.; Aurich, C. Effects of antioxidants on motility and membrane integrity of chilled-stored stallion semen. *Theriogenology* **1997**, *48*, 185–192. [CrossRef]
29. De Lamirande, E.; Leclerc, P.; Gagnon, C. Capacitation as a regulatory event that primes spermatozoa for the acrosome reaction and fertilization. *Mol. Hum. Reprod.* **1997**, *3*, 175–194. [CrossRef]
30. Çoyan, K.; Başpınar, N.; Bucak, M.N.; Akalın, P.P.; Ataman, M.B.; Ömür, A.D.; Güngör, Ş.; Küçükgünay, S.; Özkalp, B.; Sarıözkan, S. Influence of methionine and dithioerythritol on sperm motility, lipid peroxidation and antioxidant capacities during liquid storage of ram semen. *Res. Vet. Sci.* **2010**, *89*, 426–431. [CrossRef]
31. Paulenz, H.; Söderquist, L.; Ådnøy, T.; Fossen, O.H.; Berg, K.A. Effect of milk-and TRIS-based extenders on the fertility of sheep inseminated vaginally once or twice with liquid semen. *Theriogenology* **2003**, *60*, 759–766. [CrossRef]
32. Hollinshead, F.; Evans, G.; Evans, K.; Catt, S.L.; Maxwell, W.; O'Brien, J. Birth of lambs of a pre-determined sex after in vitro production of embryos using frozen–thawed sex-sorted and re-frozen–thawed ram spermatozoa. *Reproduction* **2004**, *127*, 557–568. [CrossRef] [PubMed]
33. Salvador, I.; Yániz, J.; Viudes-de-Castro, M.; Gómez, E.; Silvestre, M. Effect of solid storage on caprine semen conservation at 5 C. *Theriogenology* **2006**, *66*, 974–981. [CrossRef] [PubMed]
34. Da Ros, V.G.; Maldera, J.A.; Willis, W.D.; Cohen, D.J.; Goulding, E.H.; Gelman, D.M.; Rubinstein, M.; Eddy, E.M.; Cuasnicu, P.S. Impaired sperm fertilizing ability in mice lacking Cysteine-RIch Secretory Protein 1 (CRISP1). *Dev. Biol.* **2008**, *320*, 12–18. [CrossRef] [PubMed]
35. Watson, P.F. The causes of reduced fertility with cryopreserved semen. *Anim. Reprod. Sci.* **2000**, *60–61*, 481–492. [CrossRef]
36. Pini, T.; Leahy, T.; de Graaf, S.P. Sublethal sperm freezing damage: Manifestations and solutions. *Theriogenology* **2018**, *118*, 172–181. [CrossRef]

37. Yeste, M. Sperm cryopreservation update: Cryodamage, markers, and factors affecting the sperm freezability in pigs. *Theriogenology* **2016**, *85*, 47–64. [CrossRef]
38. Roca, J.; Broekhuijse, M.; Parrilla, I.; Rodriguez-Martinez, H.; Martinez, E.; Bolarin, A. Boar differences in artificial insemination outcomes: Can they be minimized? *Reprod. Domest. Anim.* **2015**, *50*, 48–55. [CrossRef]
39. Yeste, M. Recent advances in boar sperm cryopreservation: State of the art and current perspectives. *Reprod. Domest. Anim.* **2015**, *50*, 71–79. [CrossRef]
40. Abdalla, H.; Ali, M.A.E.; El-Tarabany, M.S. Fertility of commercial sexed semen and the economic analyses of its application in Holstein heifers. *Adv. Anim. Vet. Sci.* **2014**, *2*, 535–542. [CrossRef]
41. Yánez-Ortiz, I.; Catalán, J.; Rodríguez-Gil, J.E.; Miró, J.; Yeste, M. Advances in sperm cryopreservation in farm animals: Cattle, horse, pig and sheep. *Anim. Reprod. Sci.* **2021**, 106904. [CrossRef]
42. Brym, P.; Wasilewska-Sakowska, K.; Mogielnicka-Brzozowska, M.; Mańkowska, A.; Paukszto, Ł.; Pareek, C.S.; Kordan, W.; Kondracki, S.; Fraser, L. Gene promoter polymorphisms in boar spermatozoa differing in freezability. *Theriogenology* **2021**, *166*, 112–123. [CrossRef]
43. Koleckar, V.; Kubikova, K.; Rehakova, Z.; Kuca, K.; Jun, D.; Jahodar, L.; Opletal, L. Condensed and hydrolysable tannins as antioxidants influencing the health. *Mini Rev. Med. Chem.* **2008**, *8*, 436–447. [CrossRef] [PubMed]
44. Pizzi, A. *Tannins: Major Sources, Properties and Applications. Monomers, Polymers and Composites from Renewable Resources*; Elsevier: Amsterdam, The Netherlands, 2008; pp. 179–199.
45. Okuda, T.; Ito, H. Tannins of constant structure in medicinal and food plants—Hydrolyzable tannins and polyphenols related to tannins. *Molecules* **2011**, *16*, 2191–2217. [CrossRef]
46. Martinez, J.; Sasse, F.; Brönstrup, M.; Diez, J.; Meyerhans, A. Antiviral drug discovery: Broad-spectrum drugs from nature. *Nat. Prod. Rep.* **2015**, *32*, 29–48. [CrossRef] [PubMed]
47. Pizzi, A. Tannins medical/pharmacological and related applications: A critical review. *Sustain. Chem. Pharm.* **2021**, *22*, 100481. [CrossRef]
48. Yang, Y.; Lian, G.; Yu, B. Naturally occurring polyphenolic glucosidase inhibitors. *Isr. J. Chem.* **2015**, *55*, 268–284. [CrossRef]
49. Silanikove, N.; Perevolotsky, A.; Provenza, F.D. Use of tannin-binding chemicals to assay for tannins and their negative postingestive effects in ruminants. *Anim. Feed Sci. Technol.* **2001**, *91*, 69–81. [CrossRef]
50. Cerdá, B.; Tomás-Barberán, F.A.; Espín, J.C. Metabolism of antioxidant and chemopreventive ellagitannins from strawberries, raspberries, walnuts, and oak-aged wine in humans: Identification of biomarkers and individual variability. *J. Agric. Food Chem.* **2005**, *53*, 227–235. [CrossRef]
51. Kumari, M.; Jain, S. Tannins: An antinutrient with positive effect to manage diabetes. *Res. J. Recent Sci. ISSN* **2012**, *2277*, 2502.
52. Aisen, E.; Medina, V.; Venturino, A. Cryopreservation and post-thawed fertility of ram semen frozen in different trehalose concentrations. *Theriogenology* **2002**, *57*, 1801–1808. [CrossRef]
53. Akalin, P.P.; Bucak, M.N.; Güngör, Ş.; Başpinar, N.; Coyan, K.; Dursun, Ş.; Ili, P.; Aksoy, A.; Karaşör, F.; Bilgili, A.; et al. Influence of lycopene and cysteamine on sperm and oxidative stress parameters during liquid storage of ram semen at 5 C. *Small Rumin. Res.* **2016**, *137*, 117–123. [CrossRef]
54. Aitken, J.; Fisher, H. Reactive oxygen species generation and human spermatozoa: The balance of benefit and risk. *Bioessays* **1994**, *16*, 259–267. [CrossRef] [PubMed]
55. Aitken, R.J.; Bennetts, L.E.; Sawyer, D.; Wiklendt, A.M.; King, B.V. Impact of radio frequency electromagnetic radiation on DNA integrity in the male germline. *Int. J. Androl.* **2005**, *28*, 171–179. [CrossRef] [PubMed]
56. Muccilli, V.; Cardullo, N.; Spatafora, C.; Cunsolo, V.; Tringali, C. α-Glucosidase inhibition and antioxidant activity of an oenological commercial tannin. Extraction, fractionation and analysis by HPLC/ESI-MS/MS and 1H NMR. *Food Chem.* **2017**, *215*, 50–60. [CrossRef]
57. Zhang, X.-K.; He, F.; Zhang, B.; Reeves, M.J.; Liu, Y.; Zhao, X.; Duan, C.-Q. The effect of prefermentative addition of gallic acid and ellagic acid on the red wine color, copigmentation and phenolic profiles during wine aging. *Food Res. Int.* **2018**, *106*, 568–579. [CrossRef]
58. Zhang, B.; Cai, J.; Duan, C.-Q.; Reeves, M.J.; He, F. A review of polyphenolics in oak woods. *Int. J. Mol. Sci.* **2015**, *16*, 6978–7014. [CrossRef] [PubMed]
59. Rad, M.K.; Ghani, A.; Ghani, E. In vitro effects of Capparis spinosa L. extract on human sperm function, DNA fragmentation, and oxidative stress. *J. Ethnopharmacol.* **2021**, *269*, 113702.
60. Spinaci, M.; Muccilli, V.; Bucci, D.; Cardullo, N.; Gadani, B.; Tringali, C.; Tamanini, C.; Galeati, G. Biological effects of polyphenol-rich extract and fractions from an oenological oak-derived tannin on in vitro swine sperm capacitation and fertilizing ability. *Theriogenology* **2018**, *108*, 284–290. [CrossRef]
61. Seif, M.; Abd El-Aziz, T.; Sayed, M.; Wang, Z. Zingiber officinale ethanolic extract attenuates oxidative stress, steroidogenic gene expression alterations, and testicular histopathology induced by sodium arsenite in male rats. *Environ. Sci. Pollut. Res.* **2021**, *28*, 19783–19798. [CrossRef]
62. Ogunro, O.B.; Yakubu, M.T. Antifertility effects of 60-day oral gavage of ethanol extract of Spondias mombin leaves in guinea pigs: A biochemical, reproductive and histological study. *Asian Pac. J. Reprod.* **2021**, *10*, 56.
63. Kahalerras, L.; Otmani, I.; Abdennour, C. Wild Garlic Allium triquetrum L. Alleviates Lead Acetate-Induced Testicular Injuries in Rats. *Biol. Trace Elem. Res.* **2022**, *200*, 2205–2222. [CrossRef] [PubMed]

64. Irais, C.-M.; Claudia, B.-R.; David, P.-E.; Ashutosh, S.; Rubén, G.-G.; Agustina, R.-M.; del Carmen, V.-M.M.; Mario-Alberto, R.-G.; Luis-Benjamín, S.-G. Leaf and Fruit Methanolic Extracts of Azadirachta indica Exhibit Antifertility Activity on Rats' Sperm Quality and Testicular Histology. *Curr. Pharm. Biotechnol.* **2021**, *22*, 400–407. [CrossRef] [PubMed]
65. Ahmed, O.; Lehloenya, K.; Mphaphathi, M.; Hassen, A. Effect of Acacia mearnsii Tannin Extract Supplementation on Reproductive Performance and Oxidative Status of South African Mutton Merino Rams. *Animals* **2021**, *11*, 3266. [CrossRef] [PubMed]
66. Shehzad, M.; Rasheed, H.; Naqvi, S.A.; Al-Khayri, J.M.; Lorenzo, J.M.; Alaghbari, M.A.; Manzoor, M.F.; Aadil, R.M. Therapeutic Potential of Date Palm against Human Infertility: A Review. *Metabolites* **2021**, *11*, 408. [CrossRef] [PubMed]
67. Sobeh, M.; Mahmoud, M.F.; Sabry, O.M.; Adel, R.; Dmirieh, M.; El-Shazly, A.M.; Wink, M. HPLC-PDA-MS/MS characterization of bioactive secondary metabolites from Turraea fischeri bark extract and its antioxidant and hepatoprotective activities in vivo. *Molecules* **2017**, *22*, 2089. [CrossRef]
68. Iamsaard, S.; Arun, S.; Burawat, J.; Yannasithinon, S.; Tongpan, S.; Bunsueb, S.; Lapyuneyong, N.; Choowong-In, P.; Tangsrisakda, N.; Chaimontri, C.; et al. Evaluation of antioxidant capacity and reproductive toxicity of aqueous extract of Thai Mucuna pruriens seeds. *J. Integr. Med.* **2020**, *18*, 265–273. [CrossRef]
69. Li, W.; Yao, R.; Xie, L.; Liu, J.; Weng, X.; Yue, X.; Li, F. Dietary supplementation of grape seed tannin extract stimulated testis development, changed fatty acid profiles and increased testis antioxidant capacity in pre-puberty hu lambs. *Theriogenology* **2021**, *172*, 160–168. [CrossRef]
70. Shaik, R.; Mohamad, S.; Rao, N.V. Evaluation of Anti Fertility Activities of Bark Extracts of *Caesalpinia pulcherrima* Linn (Caesalpiniaceae) in Rats. *Indian J. Pharm. Sci.* **2021**, *83*, 393–397.
71. Wurlina, W.; Mas'ud Hariadi, E.S.; Susilowati, S.; Meles, D.K. The effect of crude guava leaf tannins on motility, viability, and intact plasma membrane of stored spermatozoa of Etawa crossbred goats. *Vet. World* **2020**, *13*, 530. [CrossRef]
72. Ros-Santaella, J.L.; Pintus, E. Rooibos (*Aspalathus linearis*) extract enhances boar sperm velocity up to 96 h of semen storage. *PLoS ONE* **2017**, *12*, e0183682. [CrossRef]
73. Galeati, G.; Bucci, D.; Nerozzi, C.; Gadani, B.; Tamanini, C.; Mislei, B.; Spinaci, M. Improvement of in vitro fertilization by a tannin rich vegetal extract addition to frozen thawed boar sperm. *Anim. Reprod.* **2020**, *17*. [CrossRef]
74. Sobeh, M.; Hassan, S.A.; Hassan, M.A.; Khalil, W.A.; Abdelfattah, M.A.; Wink, M.; Yasri, A. A polyphenol-rich extract from Entada abyssinica reduces oxidative damage in cryopreserved ram semen. *Front. Vet. Sci.* **2020**, *956*, 604477. [CrossRef] [PubMed]
75. Ali, M.M.; Banana, H.J. Effect of adding n-acetylcystiene and avena sativa extract to tris extender on post-cryopreservative semen characteristics of holstein bulls. *Plant Arch.* **2020**, *20*, 1209–1216.
76. Fitriyah, A.; Said, D.O.; Harianto, H. Improvement of Sperm Quality of Bali Cattle by Supplementation of Crude Tannin in the Semen. In *Proceeding of the 1st International Conference on Tropical Agriculture*; Springer: Cham, Switzerland, 2017.

animals

MDPI

Article

Effects of Different Selenium Sources on the Laying Performance, Egg Quality, Antioxidant, and Immune Responses of Laying Hens under Normal and Cyclic High Temperatures

Weihan Wang [†], Ruifen Kang [†], Meiling Liu, Zhong Wang, Lihong Zhao, Jianyun Zhang, Shimeng Huang and Qiugang Ma [*]

State Key Laboratory of Animal Nutrition, College of Animal Science and Technology, China Agricultural University, Beijing 100193, China; w18833192466@163.com (W.W.); ruifenkang@cau.edu.cn (R.K.); 17861509728@163.com (M.L.); wangzh@cau.edu.cn (Z.W.); lihongzhao100@126.com (L.Z.); jyzhang@cau.edu.cn (J.Z.); shimengh@cau.edu.cn (S.H.)
* Correspondence: maqiugang@cau.edu.cn
† These authors contributed equally to this work.

Simple Summary: Heat stress is a major environmental stressor in livestock production, which seriously threatens the global livestock industry and causes huge economic losses. Strategies to address heat stress are mainly divided into three categories: genetic regulation, environmental regulation, and nutritional diet regulation. Among them, nutritional regulation is one of the current research hotspots. Selenium is an essential trace element for animals and humans. As a component of some anti-oxidant enzymes and selenoproteins, it plays an important role in improving anti-oxidation, anti-stress, immune function, and reproduction performance. This study investigated the effects of different selenium sources on the laying performance, egg quality, antioxidant, and immune responses of laying hens under different temperatures. The results showed that supplementation with different selenium sources can increase the egg yolk color, antioxidant capacity, and immune status, and can alleviate heat stress damage. Additionally, the effect of organic selenium is better than that of inorganic selenium.

Abstract: The aim of this study was to evaluate the effects of different selenium (Se) sources on the laying performance, egg quality, antioxidant, and immune responses of laying hens under different temperatures. In an 8-week experiment, a total of 480 44-week-old laying hens were randomly divided into 8 groups, with 6 replicates for each group and 10 hens per replicate, and fed with a basal diet (BK), basal diet with 0.3 mg/kg of Se from sodium selenite (SS), from Se yeast (SY), or from selenium-enriched yeast culture (SYC) under normal temperature (NT, 26 ± 2 °C) and cyclic high temperature (CHT, 26 ± 2 °C~33 ± 2 °C). CHT decreased the laying performance and serum levels of Se, immunoglobulin G (IgG), and interleukin-10 (IL-10), and significantly increased the serum free triiodothyronine (FT3), deiodinase-I (DI-I), and heat stress protein (HSPs) ($p < 0.05$). In addition, SYC increased the egg yolk color, and SS increased serum IgG level. SS, SY, and SYC reduced the level of interleukin-6 (IL-6) ($p < 0.05$). In conclusion, Se can increase egg yolk color, antioxidant capacity, and immune capacity under heat stress, and the effect of organic Se is better than that of inorganic Se.

Keywords: trace element; laying hen performance; high temperature; inflammation; oxidative stress

Citation: Wang, W.; Kang, R.; Liu, M.; Wang, Z.; Zhao, L.; Zhang, J.; Huang, S.; Ma, Q. Effects of Different Selenium Sources on the Laying Performance, Egg Quality, Antioxidant, and Immune Responses of Laying Hens under Normal and Cyclic High Temperatures. *Animals* **2022**, *12*, 1006. https://doi.org/10.3390/ani12081006

Academic Editors: Youssef A. Attia, Woo Kyun Kim, Nesrein Hashem and Maria de Olivera

Received: 14 February 2022
Accepted: 11 April 2022
Published: 13 April 2022

1. Introduction

With the global climate crisis, the frequency of high temperature weather has increased, and heat stress has become one of the important factors restricting the production of laying hens [1,2]. The high temperature environment will cause changes in the physiological functions of laying hens such as White Leghorn hens and Hy-Line Brown, and reduce their egg production, egg weight, eggshell thickness, and reproductive performance, while also altering the immunity of laying hens [3–5]. Thus, it is more urgent to develop new strategies to diminish negative effects of heat stress caused by high temperature and lighten economic loss caused by heat stress.

Selenium (Se) is an essential trace element for animals and humans. As a component of some antioxidant enzymes and selenoproteins, it plays an important role in improving anti-oxidation, anti-stress, immune function, and reproduction performance [6]. In addition, Se deficiency could cause a variety of diseases in humans and animals, for example Kaschin–Beck in humans; exudative quality, muscular trophic disease, and pancreatic trophic atrophy in poultry [7,8]; and white muscle disease in sheep [9]. Therefore, it is important to supplement Se to foods and feeds to prevent Se-deficiency diseases.

Se yeast (SY) is an organic Se source with high Se content that is transformed by microorganisms [10]. As a feed supplement Se product, SY has significant advantages over inorganic Se sources [11]. It has been reported that the bioavailability and retention of organic Se is higher than that of inorganic Se in broiler chickens [12,13]. The addition of Se in feed will affect the Se content in eggs, and SY has a higher deposition efficiency for Se than inorganic Se [11]. Compared with inorganic Se, SY can significantly increase the daily feed intake and eggshell strength of laying hens [14]. SY can improve the oxidative stress caused by high temperature and enteric bacterial infection, and that may be due to the fact that SY improves the redox state of the chickens [15]. More importantly, SY has the characteristics of strong activity, low toxicity, high absorption rate, low environmental pollution, and wide safety range compared to inorganic Se [16,17]. Thus, the reasonable use of SY is essential to improve chicken production performance.

At present, most studies are about the effect of different levels of the same Se source in the diet on the production performance of laying hens [5,14,18]; there are few reports on whether adding Se to feed can resist the adverse consequences of laying hens caused by heat stress. The purpose of this study was to investigate the effects of different sources of Se in the diet on the laying performance, egg quality, anti-oxidation, immune function, and heat stress associated indicators of laying hens under normal temperature (NT) and cyclic high temperatures (CHT). We evaluated the feeding effect of different sources of Se under heat stress, and provide a theoretical basis for the utilization of Se in the production practice of laying hens.

2. Materials and Methods

2.1. Animal Experimental Ethics

The experiment was approved by the China Agricultural University Animal Care and Use Committee (A0041011202-1-1, Beijing, China).

2.2. Selenium Source

Common yeast culture and selenium-enriched yeast culture (SYC) were obtained from Hebei Feimote Biotechnology Co., Ltd., and both use the same yeast strain (preservation number: ACCC20060). The SYC was produced by solid fermentation of the yeast strains in a selenium-containing medium, and the Se content was 30 mg/kg; Se yeast (SY) named Alkosel (inactivated whole cell yeast containing 1000 mg/kg of organic Se, from Lallemand Inc., Montreal, QC, Canada). Sodium selenite premix (SS) containing 1% of inorganic Se was purchased from Hebei Yuanda Zhongzheng Biotechnology Co., Ltd. (Shijiazhuang, China). The Se content in the feed and serum were analyzed by using a fluorescence spectrophotometer (Hitachi 850, Hitachi Ltd., Tokyo, Japan) according to the protocol of Liao et al. [19].

2.3. Animals and Treatments

A total of 480 44-week-old Peking Pink laying hens were purchased from Beijing Huadu Yukou Poultry Industry Co., Ltd. The chickens were housed in wire cages and randomly divided into 8 groups using a 4 × 2 double factorial design, with 6 replicates in each group of 10 birds each, with 2 hens per cage, which dimensions were

45 cm × 45 cm × 45 cm. Each cage was equipped with one nest and two nipples. The layer of normal temperature (NT) and cyclic high temperature (CHT) were reared in two houses with four groups of chickens in each house, and the temperatures were controlled by using an electric heating tube and a thermostat. The experimental hens were fed a basal diet without Se addition, or a basal diet with 0.3 mg/kg of Se from SS, from SY, or from SYC. The procedure of heat stress was as follows (Figure 1). During the 4-week normal feeding period (weeks 45–48), the house temperature was maintained at 26 ± 2 °C. Then, during the 2-week heat stress period (weeks 49–50), the house temperature was maintained at 33 ± 2 °C from 09:00 to 18:00, then cooled to 26 ± 2 °C, and maintained until 07:00 the next day, with ventilation for 2 min every 100 min. During the two-week convalescence (weeks 51–52), the house temperature was maintained at 26 ± 2 °C. NT groups keep the temperature at 26 ± 2 °C [4] during these three stages. A corn-soybean-meal diet was formulated as a basal diet according to Chicken Feeding Standards (except selenium) (NY/T33-2004). The diet composition and nutritional level are shown in Table 1. Feed and water were provided ad libitum during the 8-week experiment period. There was no mortality during the experiment.

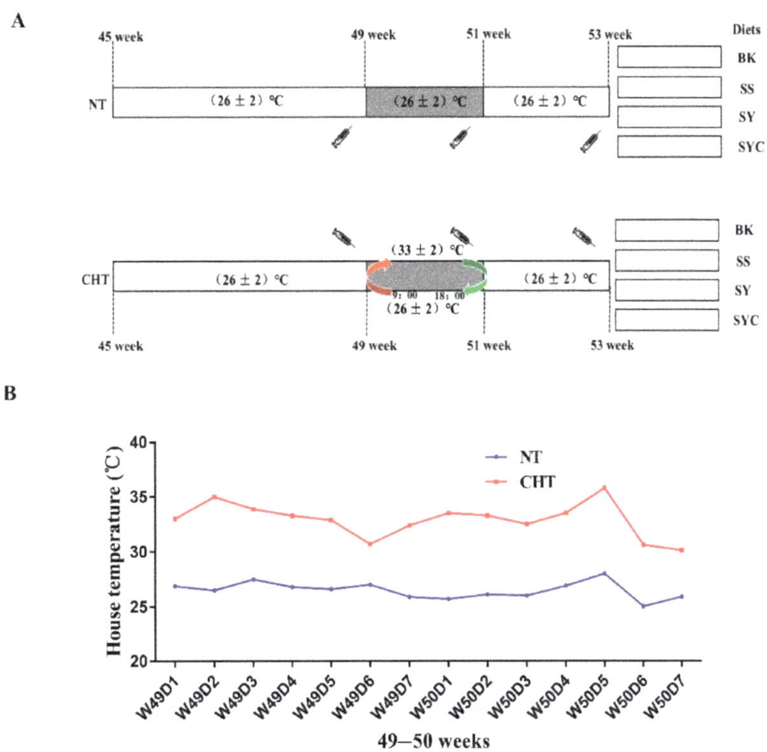

Figure 1. Experimental treatments and timeline of temperature control. (**A**) : blood sampling; BK: basal diet; SS, SY, and SYC mean basal diet with 0.3 mg/kg of Se from sodium selenite, from Se yeast, and from selenium-enriched yeast culture, respectively; NT: normal temperature, the temperature maintained at (26 ± 2) °C from weeks 45 to 52, CHT: cyclic high temperature, the house temperature maintained at (26 ± 2) °C during the weeks of 45–48 and 51–52, however, from weeks 49 to 50, the house temperature was maintained at (33 ± 2) °C from 09:00 to 18:00, then cooled to (26 ± 2) °C. (**B**) Ambient temperature measured in the bird house from 9:00 to 18:00 during the period from 49 to 50 weeks. The letter W means week and D means day; for example, W49D1 means week 49 day 1.

Table 1. Diet composition and nutritional levels.

	BK	SS	SY	SYC
Ingredients		**Composition (%)**		
Corn	63.93	63.93	63.93	63.93
Soybean meal	23.60	23.60	23.60	23.60
Ordinary yeast culture	1.00	1.00	1.00	0.00
Selenium-enriched yeast culture	0.00	0.00	0.00	1.00
Limestone	9.00	9.00	9.00	9.00
Dicalcium Phosphate	1.60	1.60	1.60	1.60
Salt	0.30	0.30	0.30	0.30
DL-methionine	0.11	0.11	0.11	0.11
L-lysine hydrochloride	0.08	0.08	0.08	0.08
Threonine	0.02	0.02	0.02	0.02
Tryptophan	0.02	0.02	0.02	0.02
Vitamins [1]	0.04	0.04	0.04	0.04
Trace minerals [2]	0.30	0.30	0.30	0.30
Nutrient		Levels		
ME [3], MJ/kg	14.91	14.91	14.91	14.91
Crude protein, %	15.40	15.40	15.40	15.40
Calcium, %	3.64	3.64	3.64	3.64
Phosphorus, %	0.35	0.35	0.35	0.35
Lysine, %	0.81	0.81	0.81	0.81
Total sulfur amino acids %	0.58	0.58	0.58	0.58
Threonine, %	0.61	0.61	0.61	0.61
Selenium [4], mg/kg	0.049	0.352	0.373	0.368

BK: basal diet, SS: sodium selenite, SY: selenium yeast, and SYC: selenium-enriched yeast culture. [1] The vitamins provided per kg of diet: vitamin A, 8000 IU; vitamin D3, 3600 IU; vitamin E, 21 IU; vitamin K3, 4.2 mg; vitamin B1, 3 mg; vitamin B2, 10.2 mg; folic acid, 0.9 mg; calcium pantothenate, 15 mg; smoke acid, 45 mg; vitamin B6, 5.4 mg; vitamin B12, 0.024 mg; and biotin, 0.15 mg. [2] The trace minerals provided per kg of diet: iron, 60 mg; manganese, 60 mg; copper, 8 mg; zinc, 80 mg; and iodine, 0.35 mg. [3] ME: metabolizable energy. [4] Except for the selenium content, all others are calculated values.

2.4. *Laying Performance*

During the whole experimental period, the number of eggs and egg weight in each replicate were recorded daily. The feed consumption in each replicate was recorded weekly. The rate of egg production, mean weight of eggs, daily egg production, the ratio of feed:egg, and the average feed intake were then calculated.

2.5. *Rectal Temperature*

On the last day of normal feeding, heat stress, and convalescence (12:00 pm), two chickens were randomly selected from each repetition of each group to determine the rectal temperature by using a portable digital thermometer (Tianjin Jinming Instrument Co., Ltd., JM222L, Tianjin, China).

2.6. *Egg Quality*

On the last day of normal feeding, heat stress, and convalescence, three eggs were randomly selected from each repetition of each group to determine the egg quality. Egg quality mainly includes Haugh units (HU), egg yolk color, eggshell thickness, egg shell percent, and egg yolk percent. Egg HU and egg yolk color were determined by egg analyzer TM (Orka Teachnology Ltd., Ramat Hasharon, Israel). The eggshell thickness was calculated by a vernier caliper.

2.7. *Blood Collection and Serum Analysis*

One day before the end of normal feeding, heat stress, and convalescence, fasting for 12 h, two chickens were selected randomly for each replication, and wing vein blood sampling was performed. The blood samples were clotted in the heparin treated tubes

(Jiangsu Rongsheng Jiamei Bioreagent Co., Ltd., Jiangsu, China) for 3 h and then centrifuged at 999× g, 20 min at 4 °C to obtain sera. The serum samples were stored at −20 °C until further analysis. Serum malondialdehyde (MDA, A003-1-2), glutathione peroxidase (GSH-Px, A005-1-2), vitamin E (VE, A008-1-1), nuclear factor-E2- related factor-2 (Nrf-2, KT81122-B), Kelch-like ECH-associated protein-1 (Keap-1, KT81121-B), immunoglobulin A (IgA, H108-1-2), immunoglobulin G (IgG, H106), immunoglobulin M (IgM, H109), interleukin-1β (IL-1β, H002), interleukin-6 (IL-6, H007-1-2), interleukin-10 (IL-10, H009-1), free triiodothyro- nine (FT3, H224), free thyroxine (FT4, H225), deiodinase-I (DI-I, KT7342-A), corticosterone (Cort, H205), and heat stress protein (HSPs, KT7345-A) were determined according to the manufacturer's instructions of corresponding assay kits (Nanjing Jiancheng Bioengineering Institute, Nanjing, China). The samples were examined in duplicate and the results were measured using a microplatereader (MultiskanTM GO, ThermoFisher, Waltham, MA, USA).

2.8. Data Statistics and Analysis

The experimental data were analyzed using SPSS 20.0 software. The data before CHT were analyzed by a one-way ANOVA followed by Duncan's multiple comparison, which was for normally distributed data to determine the differences in means among treatments, whereas the Kruskal-Wallis test was used for non-normally distributed data. After CHT, the data were analyzed according to a two-way ANOVA for a 2 × 4 factorial design using the general linear model (GLM) procedure. The difference of different Se sources and levels, temperatures, and their interactive effects were determined by Duncan's multiple comparison. $p < 0.05$ was considered significant.

In the supplementary material, the relative fold change of indexes associated with rectal temperature, laying performance, egg quality, antioxidant, immune response, and heat stress were analyzed by using GraphPad Prism version 7.01 (GraphPad Software, Inc., San Diego, CA, USA). The relative fold change means the relative change fold of the period of heat stress (weeks 49–50) and convalescence (weeks 51–52) to normal feeding period (weeks 45–48). The data were analyzed by a one-way ANOVA followed by Duncan's multiple comparison. Significance compared with weeks 45–48, * $p < 0.05$, ** $p < 0.01$, and *** $p < 0.001$; # indicated that weeks 51–52 were compared with weeks 49–50, # $p < 0.05$., ## $p < 0.01$, and ### $p < 0.001$.

3. Results

3.1. Rectal Temperature

As shown in Table 2, at week 48, compared with the BK, different Se sources had no significant effect on the rectal temperature (RT) of laying hens (p = 0.119), but Se supplementation significantly decreased the RT of laying hens ($p < 0.05$). At week 50, the RT increased significantly as the temperature increased ($p < 0.05$). In addition, the RT decreased in the SYC group compared to BK and SS ($p < 0.05$), but had no difference to SY ($p > 0.05$). At week 52, CHT significantly increased the RT of laying hens, compared with the normal temperature (NT) ($p < 0.05$); Se supplementation also had a significant effect on the RT of laying hens ($p < 0.05$). There were interactive effects between temperature and dietary Se sources and concentrations on RT at weeks 50 and 52 ($p < 0.05$). In addition, in NT, the RT of hens among BK, SS, SY, and SYC groups were decreased at week 50 compared to week 48, while the RT increased in week 52 compared to week 50 ($p < 0.05$). CHT significantly increased the RT among BK + HS, SS + HS, SY + HS, and SYC + HS groups during the heat stress period compared to the normal feeding period ($p < 0.05$), while the RT significantly decreased in the convalescence period compared to the heat stress period ($p < 0.001$) (Figure S1).

Table 2. Effects of different Se sources in diet on rectal temperature of laying hens.

Diets	HS [1]	Rectal Temperature (°C)		
		Week 48 [2]	Week 50 [2]	Week 52 [2]
BK	−	40.13	39.92	40.02
BK	+	40.15	40.80	40.18
SS	−	40.08	39.90	40.05
SS	+	40.08	40.77	40.12
SY	−	40.15	39.85	40.00
SY	+	40.07	40.73	40.12
SYC	−	40.13	39.78	40.02
SYC	+	40.07	40.68	40.10
SEM		0.026	0.035	0.036
BK		40.14	40.36 ^{a}	40.10
SS		40.11	40.33 ^{a}	40.08
SY		40.11	40.30 ^{ab}	40.06
SYC		40.08	40.23 ^{b}	40. 06
SEM		0.018	0.024	0.026
	−	40.11	39.86 ^{b}	40.02 ^{b}
	+	40.10	40.75 ^{a}	40.13 ^{a}
SEM		0.013	0.017	0.018
p-values	HS	0.651	<0.001	<0.001
	Se L	0.048	0.640	0.015
	Se S	0.364	0.872	0.771
	Se L × HS	0.207	<0.001	<0.001
	Se S × HS	0.453	<0.001	0.004

Different letters indicate statistically significant differences ($p < 0.05$). BK: basal diet, SS: sodium selenite, SY: selenium yeast, SYC: selenium-enriched yeast culture, Se L: selenium levels, and Se S: selenium sources. [1] HS− = Normal temperature (26 ± 2 °C), HS+ = Circulating high temperature (26 ± 2 °C~33 ± 2 °C). [2] Weeks 48, 50, and 52: each group has six replicates ($n = 6$).

3.2. Laying Performance

As shown in Table 3, during the period from weeks 45 to 48, different Se sources and concentrations had no significant effect on the laying performance of laying hens ($p > 0.05$). From weeks 49 to 50, CHT significantly decreased the rate of egg production, daily egg production, mean weight of eggs, and the average feed intake, compared with the NT ($p < 0.05$). Different Se sources had no significant effect on the rate of egg production, daily egg production, mean weight of eggs, the average feed intake and the ratio of feed: egg of laying hens ($p > 0.05$). From weeks 51 to 52, CHT significantly decreased the rate of egg production, daily egg production, and mean weight of eggs and increased the ratio of feed:egg of laying hens compared with the NT ($p < 0.05$), but had no significant effect on the average feed intake ($p > 0.05$). Different Se sources had no significant effects, but had an upward trend in the rate of egg production, daily egg production, and average feed intake ($p > 0.05$). In addition, in NT, no obvious changes of laying performance were observed among BK, SS, SY, and SYC groups ($p > 0.05$). CHT significantly decreased the egg production and average feed intake among BK + HS, SS + HS, SY + HS, and SYC + HS groups during heat stress period compared to normal feeding ($p < 0.001$), while the egg production and average feed intake significantly increased in convalescence period compared with heat stress period ($p < 0.01$ and $p < 0.001$, respectively) (Figure S2).

Table 3. Effects of different Se sources in diets on performance of laying hens.

Diets	HS [1]	Egg Production Rate (%)			Egg Production (g/hen/d)			Mean Weight of Eggs (g)			Average Feed Intake (g)			Feed: Egg Ratio (g:g)		
		Weeks 45–48 [2]	Weeks 49–50 [2]	Weeks 51–52 [2]	Weeks 45–48	Weeks 49–50	Weeks 51–52	Weeks 45–48	Weeks 49–50	Weeks 51–52	Weeks 45–48	Weeks 49–50	Weeks 51–52	Weeks 45–48	Weeks 49–50	Weeks 51–52
BK	−	87.38	82.73	87.44	50.54	48.76	51.32	57.83	58.90	58.69	106.00	106.59	108.02	2.09	2.19	2.11
BK	+	91.91	71.66	82.03	53.52	40.21	46.59	58.24	56.12	56.81	106.38	83.11	103.94	1.99	2.09	2.24
SS	−	88.09	83.92	84.70	51.27	49.31	49.66	58.18	58.77	58.66	103.85	106.45	104.03	2.03	2.16	2.10
SS	+	90.00	72.61	84.12	52.56	40.94	48.85	58.39	56.43	58.06	111.04	87.33	106.56	2.12	2.14	2.19
SY	−	88.81	87.99	90.73	51.09	52.83	53.03	57.53	59.97	58.42	107.38	109.87	109.67	2.10	2.10	2.07
SY	+	88.81	73.46	85.64	51.43	41.87	48.55	57.92	56.99	56.67	111.09	81.67	107.81	2.16	1.97	2.24
SYC	−	89.76	87.26	89.63	51.33	50.77	51.83	57.20	58.20	57.83	106.14	107.07	105.62	2.06	2.11	2.04
SYC	+	90.00	75.59	85.55	51.59	42.00	48.68	57.32	55.53	56.88	107.81	83.41	108.54	2.09	2.04	2.25
SEM		1.420	2.222	1.918	0.898	1.481	1.131	0.359	0.790	0.444	2.044	2.010	2.109	0.047	0.066	0.048
BK		89.64	77.20	84.74	52.03	44.49	48.96	58.04 ^{a,b}	57.52	57.75	106.19	94.86	105.98	2.04	2.14	2.18
SS		89.05	78.27	84.41	51.92	45.13	49.26	58.29 ^{a}	57.61	58.36	107.45	96.89	105.29	2.07	2.15	2.15
SY		88.81	80.73	88.19	51.26	47.36	50.79	57.73 ^{a,b}	58.48	57.55	109.24	95.78	108.74	2.13	2.03	2.16
SYC		89.88	81.43	87.59	51.46	46.39	50.26	57.26 ^{b}	56.87	57.36	106.98	95.24	107.08	2.08	2.08	2.15
SEM		0.500	1.571	1.356	0.320	1.047	0.800	0.132	0.559	0.314	0.750	1.421	1.491	0.017	0.047	0.034
	−	88.51	85.48 ^{a}	88.13 ^{a}	51.06	50.42 ^{a}	51.46 ^{a}	57.69	58.96 ^{a}	58.41 ^{a}	105.85	107.49 ^{a}	106.83	2.08	2.15 ^{a}	2.08 ^{b}
	+	90.18	73.34 ^{b}	84.34 ^{b}	52.28	41.26 ^{b}	48.17 ^{b}	57.97	56.27 ^{b}	57.11 ^{b}	109.08	83.88 ^{b}	106.71	2.09	2.04 ^{b}	2.23 ^{a}
SEM		0.078	1.111	0.959	0.449	0.741	0.566	0.179	0.395	0.222	1.022	1.005	1.054	0.024	0.023	0.024
p-values	HS	0.104	<0.001	0.008	0.062	<0.001	<0.001	0.271	<0.001	<0.001	0.051	<0.001	0.936	0.691	0.049	<0.001
	Se L	0.733	0.113	0.209	0.516	0.144	0.224	0.352	0.833	0.982	0.315	0.502	0.542	0.187	0.348	0.539
	Se S	0.731	0.339	0.120	0.754	0.331	0.395	0.024	0.137	0.069	0.512	0.705	0.273	0.408	0.220	0.952
	Se L × HS	0.107	0.696	0.494	0.118	0.737	0.305	0.775	0.926	0.283	0.261	0.955	0.134	0.039	0.787	0.759
	Se S × HS	0.766	0.732	0.473	0.817	0.646	0.271	0.932	0.922	0.425	0.402	0.091	0.460	0.817	0.720	0.458

Different letters indicate statistically significant differences ($p < 0.05$). BK: basal diet, SS: sodium selenite, SY: selenium yeast, SYC: selenium-enriched yeast culture, Se L: selenium levels, and Se S: selenium sources. [1] HS− = Normal temperature (26 ± 2 °C), HS+ = Circulating high temperature (26 ± 2 °C~33 ± 2 °C). [2] Weeks 45–48, 49–50, and 51–52: each group has six replicates ($n = 6$).

3.3. Egg Quality

As shown in Table 4, during the period from 45 to 48 weeks, there was no obvious difference among different groups on the egg quality ($p > 0.05$). From weeks 49 to 50, CHT had no significant effects on the HU, the egg yolk color, eggshell thickness, eggshell percent, or egg yolk percent compared with the NT ($p > 0.05$). Se supplementation significantly increased the egg yolk color compared with the BK ($p < 0.05$). Different Se sources had a significant effect on eggshell percent and eggshell thickness ($p < 0.05$). From weeks 51 to 52, Se supplementation had a significant effect on egg yolk color; different Se sources had significant effects on egg yolk color and egg yolk percent ($p < 0.05$); the color of egg yolk in the SYC group was significantly darker than that of the SS and SY groups ($p < 0.05$). The egg yolk percent in SY was significantly higher than that in the SS group ($p < 0.05$). There were interactive effects between temperature and dietary Se sources on eggshell thickness from weeks 51 to 52 ($p < 0.05$). There were interactive effects between temperature and dietary Se concentrations on HU, egg yolk color, and eggshell thickness from weeks 51 to 52, and eggshell percent and egg yolk percent from weeks 45–48 ($p < 0.05$). In addition, in NT, the egg yolk color in the BK group decreased in both the weeks 49–50 and 51–52 compared to the weeks 45–48 ($p < 0.001$), while in SS and SYC groups, the color only decreased in week 51–52 compared to weeks 45–48 ($p < 0.01$ and $p < 0.001$, respectively), no obvious difference was observed in SYC group ($p > 0.05$). The eggshell thickness of week 51–52 significantly increased in BK and SS group compared to weeks 45–48 and 49–50 ($p < 0.01$), CHT significantly decreased egg yolk color among BK + HS, SS + HS, SY + HS, and SYC + HS groups during both the heat stress and convalescence period compared to normal feeding period ($p < 0.05$, $p < 0.01$, and $p < 0.001$, respectively) (Figure S3).

3.4. Serum Se and Vitamin E Content

As shown in Table 5, at weeks 48, 50, and 52, Se supplementation significantly increased the serum Se content compared with the BK ($p < 0.05$); no difference was observed among SS, SY, and SYC groups. In addition, CHT significantly decreased the serum Se content ($p < 0.05$) at weeks 50 and 52. The VE content in CHT was lower than that in NT at week 50 ($p < 0.05$), and SYC increased the VE content of layers at week 52 compared with the BK ($p < 0.05$). There were interactions between dietary Se levels and temperatures in Se content at week 52 and VE content at both weeks 48 and 50 ($p < 0.05$). Additionally, there were interactions between dietary Se sources and temperatures in VE content at weeks 48, 50, and 52 ($p < 0.05$). In addition, in NT, serum Se in BK group decreased in week 52 compared to week 48 and week 50 ($p < 0.05$ and $p < 0.001$, respectively). No obvious changes of Se were observed among SS, SY, and SYC groups ($p > 0.05$). Serum VE in BK and SYC group decreased in week 50 compared to week 48 ($p < 0.01$), and increased in BK in week 52 compared to week 50 ($p < 0.001$). Additionally, the VE level in SS group increased in week 52 compared to week 48 and week 50 ($p < 0.001$). No obvious changes of VE were observed in SY groups ($p > 0.05$). CHT significantly decreased the serum Se among BK + HS, SS + HS, SY + HS, and SYC + HS groups during heat stress period compared to normal feeding ($p < 0.001$), Furthermore, the serum Se among BK + HS, SS + HS, and SY + HS groups in week 52 were also lower than those in week 48 ($p < 0.05$ and $p < 0.001$, respectively), while no obvious changes of Se between week 48 and week 52 were observed the SYC + HS group ($p > 0.05$). CHT significantly decreased VE in the BK + HS group and increased VE in SYC + HS group in week 52 compared to week 48 ($p < 0.05$ and $p < 0.001$, respectively) (Figure S4).

Table 4. Effects of different Se sources in diets on egg quality of laying hens.

Diets	HS [1]	Egg Haugh Unites			Egg Yolk Color			Eggshell Percent (%)			Egg yolk Percent (%)			Eggshell Thickness (mm)		
		Weeks 45–48 [2]	Weeks 49–50 [2]	Weeks 51–52 [2]	Weeks 45–48	Weeks 49–50	Weeks 51–52	Weeks 45–48	Weeks 49–50	Weeks 51–52	Weeks 45–48	Weeks 49–50	Weeks 51–52	Weeks 45–48	Weeks 49–50	Weeks 51–52
BK	–	67.71	73.08	64.50	4.55	3.33	3.38	10.43	10.25	10.75	27.54	27.31	27.05	0.39	0.39	0.44 [a]
BK	+	70.51	73.86	72.49	4.66	3.33	2.89	10.05	10.53	10.90	28.70	28.28	28.34	0.39	0.38	0.41 [a,b]
SS	–	68.17	68.41	74.56	4.50	3.88	2.77	10.28	10.15	10.89	27.63	27.90	26.75	0.38	0.38	0.41 [a,b]
SS	+	72.29	74.89	67.38	4.67	3.89	3.66	10.16	10.03	10.77	26.49	26.62	26.79	0.39	0.37	0.43 [a]
SY	–	66.24	73.96	68.23	4.56	4.44	3.66	10.24	10.46	10.60	28.18	28.39	28.03	0.38	0.39	0.40 [a,b]
SY	+	72.28	74.75	68.97	4.83	3.72	3.88	10.60	10.65	10.96	27.59	27.41	28.78	0.40	0.39	0.43 [a]
SYC	–	67.08	73.73	69.73	4.44	4.00	3.89	10.25	10.47	10.78	28.24	27.52	28.26	0.38	0.38	0.39 [b]
SYC	+	71.14	77.35	66.87	4.56	3.94	3.83	10.54	10.42	11.15	27.48	28.34	27.03	0.40	0.41	0.43 [a]
SEM		2.299	2.310	2.987	0.130	0.223	0.245	0.166	0.169	0.191	0.405	0.459	0.526	0.006	0.007	0.008
BK		69.11	73.47	68.50	4.61	3.33 [b]	3.14 [b]	10.24	10.39	10.82	28.12 [a]	27.80	27.69 [a,b]	0.39	0.38	0.42
SS		70.23	71.66	70.98	4.58	3.88 [a]	3.22 [a,b]	10.22	10.09	10.83	27.07 [b]	27.26	26.77 [b]	0.39	0.37	0.42
SY		69.26	74.36	68.60	4.70	4.08 [a]	3.78 [a,b]	10.42	10.56	10.78	27.89 [a,b]	27.90	28.41 [a]	0.39	0.39	0.42
SYC		69.11	75.55	68.30	4.50	3.97 [a]	3.86 [a]	10.39	10.44	10.96	27.87 [a,b]	27.93	27.64 [a,b]	0.39	0.39	0.41
SEM		0.819	1.633	2.112	0.045	0.158	0.173	0.059	0.119	0.135	0.174	0.325	0.372	0.004	0.005	0.005
	–	67.29 [b]	72.30	69.26	4.51	3.92	3.43	10.30	10.33	10.75	27.90	27.78	27.52	0.38	0.38	0.41
	+	71.55 [a]	75.22	68.93	4.68	3.72	3.57	10.33	10.41	10.94	27.57	27.67	27.74	0.39	0.39	0.42
SEM		1.150	1.155	1.494	0.065	0.112	0.122	0.083	0.084	0.095	0.228	0.230	0.263	0.003	0.004	0.004
p-values	HS	0.012	0.082	0.879	0.077	0.226	0.428	0.759	0.529	0.166	0.304	0.771	0.566	0.006	0.342	0.077
	Se L	0.823	0.840	0.746	0.862	<0.001	0.021	0.453	0.870	0.837	0.173	0.792	0.836	0.815	0.858	0.207
	Se S	0.869	0.237	0.622	0.333	0.683	0.025	0.422	0.023	0.615	0.135	0.265	0.013	0.627	0.023	0.152
	Se L × HS	0.609	0.454	0.028	0.722	0.485	0.040	0.044	0.325	0.848	0.011	0.059	0.103	0.147	0.506	<0.001
	Se S × HS	0.887	0.475	0.422	0.816	0.210	0.153	0.309	0.623	0.346	0.825	0.056	0.177	0.907	0.024	0.481

Different letters indicate statistically significant differences ($p < 0.05$). BK: basal diet, SS: sodium selenite, SY: selenium yeast, SYC: selenium-enriched yeast culture, Se L: selenium levels, and Se S: selenium sources. [1] HS− = Normal temperature (26 ± 2 °C), HS+ = Circulating high temperature (26 ± 2 °C~33 ± 2 °C). [2] Weeks 45–48, 49–50, and 51–52: each group has six replicates ($n = 6$).

Table 5. Effects of different Se sources in diet on Se and VE content in serum of laying hens.

Diets	HS [1]	Se (μg/mL)			VE (μg/mL)		
		Week 48 [2]	Week 50 [2]	Week 52 [2]	Week 48	Week 50	Week 52
BK	−	0.08	0.07	0.06 [d]	75.94	81.51 [c]	81.71 [b,c]
BK	+	0.07	0.04	0.06 [d]	113.38	34.91 [a]	70.60 [c]
SS	−	0.18	0.17	0.18 [a,b]	61.61	84.97 [b,c]	106.15 [a,b]
SS	+	0.19	0.13	0.15 [c]	75.92	42.99 [a]	94.54 [a,b,c]
SY	−	0.19	0.18	0.20 [a]	90.81	83.39 [a,b]	79.20 [b,c]
SY	+	0.18	0.15	0.16 [b,c]	75.46	66.38 [a]	102.32 [a,b]
SYC	−	0.19	0.18	0.19 [a,b]	123.97	77.77 [a]	87.85 [a,b,c]
SYC	+	0.18	0.15	0.16 [b,c]	61.37	75.40 [a]	116.36 [a]
SEM		0.004	0.007	0.005	10.800	5.016	8.114
BK		0.08 [b]	0.06 [b]	0.06 [b]	94.66	58.21	76.16 [b]
SS		0.18 [a]	0.15 [a]	0.17 [a]	68.77	63.98	100.34 [a,b]
SY		0.19 [a]	0.17 [a]	0.18 [a]	83.14	74.89	90.76 [a,b]
SYC		0.18 [a]	0.16 [a]	0.18 [a]	92.67	76.59	102.10 [a]
SEM		0.008	0.005	0.003	7.636	3.547	5.737
	−	0.16	0.15 [a]	0.16 [a]	88.08	81.91 [a]	88.73
	+	0.16	0.12 [b]	0.13 [b]	81.53	54.92 [b]	95.95
SEM		0.002	0.003	0.002	5.400	2.508	4.057
p-values	HS	0.137	0.008	<0.001	0.397	<0.001	0.350
	Se L	<0.001	<0.001	<0.001	0.146	0.073	0.002
	Se S	0.421	0.125	0.018	0.099	0.241	0.335
	Se L × HS	0.595	0.991	<0.001	0.002	0.003	0.074
	Se S × HS	0.264	0.642	0.549	0.004	0.002	0.039

Different letters indicate statistically significant differences ($p < 0.05$). BK: basal diet, SS: sodium selenite, SY: selenium yeast, SYC: selenium-enriched yeast culture, VE: Vitamin E, Se L: selenium levels, and Se S: selenium sources. [1] HS− = Normal temperature (26 ± 2 °C), HS+ = Circulating high temperature (26 ± 2 °C~33 ± 2 °C). [2] Weeks 48, 50, and 52: each group has six replicates (n = 6).

3.5. Serum Antioxidant Status

As shown in Table 6, at week 48, compared with the BK, Se supplementation significantly decreased the serum MDA content and increased the serum GSH-PX level ($p < 0.05$), but had no significant effect on Nrf-2 and Keap-1 contents ($p > 0.05$). Compared to the BK, SS and SY significantly decreased MDA level ($p < 0.05$), and SS, SY, and SYC significantly increased GSH-PX level ($p < 0.05$). At week 50, compared with the BK, Se supplementation decreased serum MDA content ($p < 0.05$), and significantly increased GSH-PX level ($p < 0.05$), but had no significant effect on Nrf-2 ($p > 0.05$). At week 52, CHT increased the content of GSH-PX ($p < 0.05$), but had no significant effect on MDA, Nrf-2, or Keap-1, compared to the NT ($p > 0.05$). SS and SY significantly decreased serum MDA, and significantly increased GSH-PX level, compared with the BK ($p < 0.05$). In addition, in NT, the GSH-PX and Keap-1 in BK group decreased in both week 50 and 52 compared to week 48 ($p < 0.05$ and $p < 0.01$, respectively). Additionally, the GSH-PX in SYC group in week 52 was lower than that in week 48, and Keap-1 in SY and SYC decreased in both week 50 and 52 compared to week 48 ($p < 0.05$ and $p < 0.001$, respectively). No obvious difference of Keap-1 was observed in the SS group ($p > 0.05$). CHT significantly decreased GSH-PX level in the BK + HS group in both the heat stress period and convalescence period ($p < 0.05$ and $p < 0.001$, respectively); no obvious changes of GSH-PX were observed in the SS + HS, SY + HS, or SYC + HS groups ($p > 0.05$). Additionally, CHT significantly decreased Keap-1 in the BK + HS and SY + HS groups in the convalescence period compared to the normal feeding period ($p < 0.01$ and $p < 0.001$, respectively), and increased the MDA level among BK + HS, SS + HS, and SY + HS groups compared to normal feeding. No obvious changes of GSH-PX in the SS + HS and SY + HS groups of MDA in SYC + HS group were observed ($p > 0.05$) (Figure S5).

Table 6. Effects of dietary different Se sources on serum antioxidant status of laying hens.

Diets	HS [1]	MDA (nmol/mL)			GSH-Px (U/mL)			Nrf-2(ng/mL)			Keap-1(pg/mL)		
		Week 48 [2]	Week 50 [2]	Week 52 [2]	Week 48	Week 50	Week 52	Week 48	Week 50	Week 52	Week 48	Week 50	Week 52
BK	–	3.99	5.42	4.30	368.37	283.82	292.40	3.32	2.46	2.67	141.44	105.42	93.27
BK	+	3.89	5.32	4.14	406.02	294.61	319.71	3.30	2.36	2.60	169.22	117.00	113.48
SS	–	2.62	3.91	2.78	445.43	387.30	449.84	3.27	2.25	2.33	113.67	94.82	97.95
SS	+	2.00	4.06	3.26	452.70	417.47	492.77	3.24	2.19	1.90	142.77	107.21	89.39
SY	–	2.72	4.08	3.28	442.80	406.90	418.35	3.77	2.29	2.08	168.31	89.75	99.52
SY	+	2.40	4.20	3.60	443.23	389.73	431.34	3.47	1.69	2.41	128.98	76.82	102.09
SYC	–	3.03	4.31	4.09	438.39	375.12	219.30	3.44	1.92	2.40	122.21	91.24	96.33
SYC	+	3.50	4.39	4.36	437.87	365.91	335.56	2.79	2.63	2.38	146.08	114.43	100.39
SEM		0.317	0.270	0.296	11.027	23.078	26.699	0.193	0.273	0.313	13.703	9.594	6.326
BK		3.94^{a}	5.37^{a}	4.22^{a}	387.19^{b}	289.21^{b}	306.06^{b}	3.31	2.41	2.63	155.33	111.21^{a}	103.37
SS		2.31^{c}	3.98^{b}	3.02^{b}	449.06^{a}	402.39^{a}	471.30^{a}	3.25	2.22	2.12	128.22	101.02^{ab}	93.67
SY		2.56^{c}	4.14^{b}	3.44^{b}	443.01^{a}	398.31^{a}	424.84^{a}	3.62	1.99	2.25	148.65	83.28^{b}	100.81
SYC		3.26^{ab}	4.35^{b}	4.22^{a}	438.13^{a}	370.51^{a}	277.43^{b}	3.11	2.28	2.39	134.15	102.84^{ab}	98.36
SEM		0.148	0.191	0.209	5.514	16.319	18.879	0.074	0.193	0.221	5.317	6.784	4.473
	–	3.09	4.43	3.61	423.74	363.28	344.97^{b}	3.45	2.23	2.37	136.41	95.31	96.77
	+	2.94	4.49	3.84	434.95	366.93	394.84^{a}	3.20	2.22	2.32	146.76	103.87	101.34
SEM		0.158	0.135	0.148	5.513	11.539	13.349	0.096	0.137	0.156	6.852	4.797	3.163
p-values	HS	0.520	0.752	0.159	0.160	0.845	0.020	0.077	0.965	0.887	0.293	0.281	0.385
	Se L	<0.001	<0.001	0.010	<0.001	<0.001	<0.001	0.895	0.278	0.146	0.111	0.057	0.273
	Se S	0.015	0.406	0.001	0.615	0.336	<0.001	0.054	0.549	0.676	0.322	0.095	0.525
	Se L × HS	0.921	0.636	0.297	0.059	0.802	0.495	0.339	0.797	0.949	0.307	0.799	0.052
	Se S × HS	0.222	0.990	0.929	0.929	0.553	0.155	0.279	0.068	0.483	0.695	0.171	0.558

Different letters indicate statistically significant differences ($p < 0.05$). BK: basal diet, SS: sodium selenite, SY: selenium yeast, SYC: selenium-enriched yeast culture, MDA: malondialdehyde:, GSH-Px: glutathione peroxidase, Nrf-2: nuclear factor-E2-related factor-2, Keap-1: Kelch-like ECH-associated protein-1, Se L: selenium levels, and Se S: selenium sources. [1] HS− = Normal temperature (26 ± 2 °C), HS+ = Circulating high temperature (26 ± 2 °C~33 ± 2 °C). [2] Weeks 48, 50, and 52: each group has six replicates (n = 6).

3.6. Serum Immune Indexes

As shown in Table 7, at week 48, SYC significantly decreased serum IL-6 level compared with the BK and SY ($p < 0.05$), but had no significant difference with SS ($p > 0.05$), while the level of IgM in SYC was significantly higher than that in SS ($p < 0.05$). At week 50, CHT significantly decreased serum IgG compared with the BK ($p < 0.05$), but had no significant effect on IgA, IgM, IL-1β, IL-6, and IL-10 ($p > 0.05$). SY significantly increased the level of IgM compared to SYC ($p < 0.05$). At week 52, CHT significantly decreased IL-1β and IL-10 in the sera compared with the NT ($p < 0.05$), but had no significant effect on IgA, IgG, IgM, or IL-6 ($p > 0.05$). Se supplementation significantly decreased IL-6 content ($p < 0.05$), and SS significantly increased serum IgG compared with the BK ($p < 0.05$). In addition, in NT, the IL-10 in the BK group decreased in week 50 compared to week 48 ($p < 0.001$), and increased in week 52 compared to week 50 ($p < 0.001$). No obvious changes of IgA, IgG, IgM, IL-1β, or IL-6 were observed in the BK group ($p > 0.05$). The levels of IgM in SYC group in week 52 increased compared to week 50 ($p < 0.001$), and the level of IL-6 decreased in SY group in both week 50 and 52 compared to week 48 ($p < 0.05$). CHT reduced the levels of IgG in BK + HS and SYC + HS in the heat stress period compared to normal feeding period ($p < 0.05$) and also decreased the level of the IL-10 in BK + HS, SS + HS, and SYC + HS groups ($p < 0.01$ and $p < 0.001$, respectively). No obvious changes of IgG in the SS + HS and SY + HS groups, or of IL-10 in the SY + HS groups, were observed ($p > 0.05$) (Figure S6).

Table 7. Effects of different Se sources in diet on serum immune indexes of laying hens.

Diets	HS [1]	IgA (g/L)			IgG (g/L)			IgM (g/L)			IL-1β (pg/mL)			IL-6 (pg/mL)			IL-10 (pg/mL)		
		Week 48 [2]	Week 50 [2]	Week 52 [2]	Week 48	Week 50	Week 52	Week 48	Week 50	Week 52	Week 48	Week 50	Week 52	Week 48	Week 50	Week 52	Week 48	Week 50	Week 52
BK	−	1.06	0.92	1.11	6.71	6.12	6.42	0.73	0.55	0.74	26.85	31.99	32.48	86.00	86.69	85.90	32.82	19.95	29.97
BK	+	1.04	0.86	1.09	7.58	5.23	5.19	0.68	0.56	0.78	28.88	33.39	27.43	89.56	86.35	89.65	32.85	22.16	28.58
SS	−	0.96	0.91	1.08	6.28	6.51	8.91	0.58	0.60	0.73	27.92	31.08	28.19	85.19	86.97	75.14	31.28	20.41	26.80
SS	+	1.05	0.84	1.31	6.92	6.12	7.04	0.71	0.56	0.94	25.15	26.20	24.80	82.30	84.89	78.74	28.27	22.31	26.45
SY	−	0.99	1.03	1.03	7.69	7.33	6.53	0.66	0.68	0.79	22.18	28.87	30.66	91.11	75.18	77.74	32.90	21.04	28.16
SY	+	1.25	0.99	1.08	7.08	5.94	5.97	0.69	0.59	0.66	28.61	28.33	29.40	86.03	83.85	74.44	27.66	23.83	26.09
SYC	−	1.03	1.06	1.03	6.71	6.05	6.77	0.76	0.55	0.88	25.21	27.49	28.14	82.74	87.08	79.76	29.30	25.36	27.93
SYC	+	1.11	0.83	1.04	7.23	4.76	6.87	0.80	0.45	0.67	26.30	30.36	25.61	79.53	88.28	75.67	27.22	21.33	23.35
SEM		0.094	0.067	0.098	0.568	0.544	0.674	0.049	0.050	0.065	1.711	1.979	1.599	2.855	3.889	3.835	1.726	1.159	1.369
BK		1.05	0.89	1.10	7.15	5.67	5.81 [b]	0.71 [a,b]	0.55 [a,b]	0.76	27.87	32.69	29.95	87.78 [a]	86.52	87.78 [a]	32.83	21.05	29.27
SS		1.00	0.87	1.19	6.60	6.31	7.97 [a]	0.65 [b]	0.58 [ab]	0.83	26.53	28.64	26.50	83.74 [a,b]	85.93	76.94 [b]	29.77	21.36	26.63
SY		1.12	1.01	1.06	7.38	6.63	6.25 [a,b]	0.67 [a,b]	0.63 [a]	0.72	25.40	28.60	30.03	88.57 [a]	79.52	76.09 [b]	30.28	22.43	27.12
SYC		1.07	0.94	1.03	6.97	5.41	6.82 [a,b]	0.78 [a]	0.50 [b]	0.77	25.75	28.93	26.87	81.14 [b]	87.68	77.72 [b]	28.26	23.35	25.64
SEM		0.033	0.047	0.069	0.195	0.385	0.477	0.018	0.035	0.046	0.641	1.399	1.131	1.078	2.750	2.712	0.664	0.819	0.968
	−	1.01	0.98	1.06	6.85	6.51 [a]	7.16	0.68	0.59	0.78	25.54	29.86	29.87 [a]	86.26	83.98	79.64	31.58	21.69	28.22 [a]
	+	1.11	0.88	1.13	7.20	5.51 [b]	6.27	0.72	0.54	0.76	27.24	29.57	26.81 [b]	84.36	85.84	79.63	29.00	22.41	26.11 [b]
SEM		0.047	0.033	0.049	0.284	0.272	0.337	0.025	0.025	0.033	0.856	0.990	0.800	1.427	1.944	1.918	0.863	0.579	0.684
p-values	HS	0.140	0.226	0.239	0.385	0.012	0.196	0.305	0.312	0.525	0.171	0.860	0.013	0.353	0.284	0.997	0.043	0.387	0.037
	Se L	0.826	0.338	0.911	0.727	0.325	0.036	0.865	0.675	0.769	0.168	0.020	0.109	0.167	0.503	0.001	0.022	0.171	0.017
	Se S	0.490	0.146	0.222	0.393	0.080	0.046	0.035	0.041	0.280	0.795	0.984	0.067	0.042	0.103	0.914	0.482	0.245	0.551
	Se L × HS	0.301	0.579	0.486	0.469	0.880	0.683	0.176	0.289	0.468	0.875	0.491	0.317	0.128	0.647	0.429	0.226	0.299	0.676
	Se S × HS	0.576	0.303	0.499	0.482	0.604	0.342	0.584	0.780	0.513	0.257	0.162	0.800	0.918	0.378	0.551	0.647	0.842	0.312

Different letters indicate statistically significant differences ($p < 0.05$). BK: basal diet, SS: sodium selenite, SY: selenium yeast, SYC: selenium-enriched yeast culture, IgA: immunoglobulin A, IgG: immunoglobulin G, IgM: immunoglobulin M, IL-1β: interleukin-1β, IL-6: interleukin-6, IL-10: interleukin-10, Se L: selenium levels, and Se S: selenium sources. [1] HS− = Normal temperature (26 ± 2 °C), HS+ = Circulating high temperature (26 ± 2 °C~33 ± 2 °C). [2] Weeks 48, 50, and 52: each group has six replicates (n = 6).

3.7. Serum Heat Stress Associated Indexes

As shown in Table 8, at week 48, different Se sources had no significant effects on the serum heat stress associated indexes ($p > 0.05$). At week 50, compared with the BK, different Se sources had no significant effects on HSPs, FT3, FT4, DI-I, and Cort ($p > 0.05$). CHT significantly increased the serum FT3 content of laying hens compared with the NT ($p < 0.05$), but had no significant effect on HSPs, FT4, DI-I, and Cort ($p > 0.05$). At week 52, CHT significantly increased the levels of serum DI-I and HSPs ($p < 0.05$), but had no significant effect on FT3, FT4, and CORT compared to NT ($p > 0.05$). The levels of DI-I in SY and SYC groups were significantly higher than that in SS ($p < 0.05$). In addition, SY significantly increased the content of FT3 compared with the BK and SYC ($p < 0.05$). There were interactions between dietary Se sources and temperatures in DI-I and FT4 at week 52 ($p < 0.05$); additionally, there were interactions between dietary Se sources and temperatures in CORT at week 50 ($p < 0.05$). In addition, in NT, the CORT in BK group decreased in both week 50 and 52 compared to week 48 ($p < 0.05$ and $p < 0.01$, respectively). The HSPs in BK in week 50 increased compared to week 48 ($p < 0.01$) and decreased in week 52 compared to week 50 ($p < 0.001$). No obvious changes of FT3, FT4, and DI-I were observed in BK group ($p > 0.05$). In the heat stress period, CHT increased the levels of FT3 in SY + HS and SYC + HS groups ($p < 0.05$ and $p < 0.001$, respectively), FT4 in SS + HS and SY + HS groups ($p < 0.05$ and $p < 0.\ 01$, respectively), and HSPs among BK + HS, SS + HS, SY + HS, and SYC + HS groups compared to normal feeding period ($p < 0.001$). Compared to the heat stress period, the levels of FT3 and FT4 were significantly decreased in SYC + HS group ($p < 0.05$) and SS + HS group ($p < 0.01$), respectively (Figure S7).

Table 8. Effects of different Se sources in diets on serum heat stress indexes of laying hens.

Diets	HS [1]	FT3 (pmol/L)			FT4 (pmol/L)			FT4 (pmol/L)			CORT (nmol/L)			HSPs (ng/L)		
		Week 48 [2]	Week 50 [2]	Week 52 [2]	Week 48	Week 50	Week 52	Week 48	Week 50	Week 52	Week 48	Week 50	Week 52	Week 48	Week 50	Week 52
BK	−	1.88	1.98	2.01	3.77	4.08	4.78	9.23	8.89	8.80 [b]	130.29	112.52	103.97	329.02	418.15	274.16
BK	+	2.09	2.72	1.78	4.05	4.61	3.7	10.56	9.65	14.03 [a]	118.44	115.89	121.52	286.24	404.79	302.08
SS	−	2.06	2.26	2.31	3.54	5.06	4.65	11.91	10.34	8.41 [b]	124.32	121.89	118.74	324.65	359.57	253.50
SS	+	1.72	2.76	1.88	3.66	5.99	3.35	9.69	8.14	12.30 [a]	117.12	108.13	108.58	258.34	370.66	267.52
SY	−	1.81	2.62	2.64	3.68	6.15	4.74	10.08	8.47	13.79 [a]	112.01	105.75	100.91	274.20	365.90	222.89
SY	+	2.16	3.31	2.95	3.69	6.33	4.63	9.2	8.11	13.53 [a]	125.63	123.89	112.38	273.45	407.63	273.09
SYC	−	1.75	2.30	2.14	4.06	5.08	4.08	9.12	8.90	12.86 [a]	112.76	116.08	103.84	251.98	352.42	245.70
SYC	+	1.63	3.15	1.96	4.18	5.60	6.27	9.10	8.90	13.06 [a]	113.24	118.52	107.81	233.90	392.82	314.50
SEM		0.194	0.311	0.276	0.515	0.803	0.645	0.735	0.590	0.553	5.326	4.970	4.737	19.836	28.604	17.023
BK		1.99	2.35	1.89 [b]	3.91	4.35	4.24	9.90	9.27	11.41 [a,b]	124.37	114.20	112.74	307.63	411.47	288.12
SS		1.89	2.51	2.10 [a,b]	3.60	5.53	4.00	10.80	9.24	10.35 [b]	120.72	115.01	113.66	291.49	365.11	260.51
SY		1.98	2.97	2.79 [a]	3.69	6.24	4.69	9.64	8.29	13.66 [a]	118.82	114.82	106.64	273.82	386.77	247.99
SYC		1.69	2.72	2.05 [b]	4.12	5.34	5.18	9.11	8.90	12.96 [a]	113.00	117.30	105.83	242.94	372.62	280.10
SEM		0.137	0.220	0.195	0.364	0.568	0.456	0.520	0.417	0.391	3.766	3.514	3.349	14.026	20.226	12.037
	−	1.88	2.29 [b]	2.28	3.76	5.09	4.56	10.08	9.15	10.96 [b]	119.84	114.06	106.86	294.96 [a]	374.01	249.07 [b]
	+	1.90	2.99 [a]	2.14	3.90	5.64	4.49	9.64	8.70	13.23 [a]	118.61	116.60	112.57	262.98 [b]	393.98	289.30 [a]
SEM		0.097	0.156	0.138	0.257	0.401	0.323	0.368	0.295	0.277	2.663	2.485	2.368	9.918	14.302	8.511
p-values	HS	0.848	0.005	0.488	0.707	0.377	0.867	0.396	0.357	0.004	0.745	0.474	0.128	0.029	0.400	0.015
	Se L	0.858	0.143	0.071	0.819	0.046	0.476	0.161	0.347	0.052	0.101	0.713	0.305	0.292	0.127	0.079
	Se S	0.905	0.361	0.018	0.219	0.502	0.203	0.268	0.279	<0.001	0.808	0.858	0.208	0.285	0.746	0.180
	Se L × HS	0.928	0.905	0.787	0.649	0.991	0.211	0.090	0.104	<0.001	0.091	0.894	0.050	0.528	0.349	0.559
	Se S × HS	0.529	0.854	0.402	0.915	0.895	0.033	0.273	0.151	0.001	0.418	0.012	0.083	0.798	0.833	0.276

Different letters indicate statistically significant differences ($p < 0.05$). BK: basal diet, SS: sodium selenite, SY: selenium yeast, SYC: selenium-enriched yeast culture, FT3: free triiodothyronine, FT4: free thyroxine, DI-I: deiodinase-I, CORT: corticosterone, HSPs: heat stress protein, Se L: selenium levels, and Se S: selenium sources. [1] HS− = Normal temperature (26 ± 2 °C), HS+ = Circulating high temperature (26 ± 2 °C~33 ± 2 °C). [2] Weeks 48, 50, and 52: each group has six replicates ($n = 6$).

4. Discussion

4.1. Effects of Different Se Sources on Laying Performance of Heat-Stressed Laying Hens

High temperature is the main environmental factor affecting poultry production. Adult layer hens are prone to stress response due to plump surface feathers and underdeveloped sweat glands. Four weeks of continuous heat stress (35°C, 7 h/d) in 24-week-old Hy-Line Brown hens significantly decreased shell weight, feed intake, egg production, and body weight [20]. A previous study reported that heat stress reduced the feed conversion rate of laying hens by 31.6%, egg production by 36.4%, and egg weight by 3.41% [21]. The addition of organic selenium to the diet had no effect on the total weight, the white weight, shell weight, and shell thickness of eggs, but significantly increased egg yolk's weight [22]. In the present study, the rate of egg production, daily egg production, and mean weight of eggs of laying hens were significantly decreased after challenging with CHT, which was similar to previous studies [23,24], but the ratio of feed: egg was significantly increased during week 51–52. This may be due to the increase in endogenous loss in layer hens repairing damage caused by CHT in weeks 51–52. A previous study has found that high temperature can increase the consumption of endogenous nutrients in pigs to compensate for the intestinal damage caused by high temperature [25]. Therefore, we speculated that the increase in feed-to-egg ratio during the recovery period after heat stress may also be caused by the increased consumption of endogenous nutrients during the heat stress period for alleviating heat stress injury. In addition, adding inorganic or organic selenium had no significant effect on egg weight, egg shell thickness, or egg yolk percentage, which was partially inconsistent with the study of Januzi. This may be due to the different types and ages of laying hens.

4.2. Effects of Different Se Sources on Egg Quality of Heat-Stressed Laying Hens

Eggs were considered as a complete food and an excellent source of fatty acids, vitamins, minerals, and essential amino acids. High temperature (33.3 °C) can significantly reduce the protein height and shell weight of eggs, as well as the weight of egg yolks [26]. When laying hens were exposed to heat stress, egg weight, egg shell weight, egg shell thickness, and specific gravity will be significantly decreased [1]. Reduced feed intake was considered the main factor leading to lower productivity of birds under high temperatures [27], which was in line with the present study. The addition of SY and SS to layer diets had no significant effect on the HU of eggs. SY significantly increased egg yolk color [28]. In the present study, compared with the selenium-free diet group, SYC significantly increased the yolk color, which was consistent with the study of Paton et al.

4.3. Effects of Different Se Sources on Serum Antioxidant Status of Heat-Stressed Laying Hens

Animals in a high temperature environment will produce excessive free radicals, including reactive oxygen radicals (ROS), which disrupt the body's antioxidant system and cause oxidative stress in animals [29,30]. MDA can be used to evaluate the degree of lipid oxidative damage. GSH is an important antioxidant in the antioxidant system. It can not only remove peroxides in the enzymatic antioxidant system, but also help maintain the antioxidant capacity of vitamin C and vitamin E in the non-enzymatic antioxidant system [31]. The addition of inorganic Se and bacterial organic Se to the diet significantly increased serum Se levels and GSH-Px activity, and decreased MDA content of broilers [32]. Adding Se to the diet increased serum GSH-Px activity and liver SOD activity [33]. In the present study, the levels of GSH-PX and Se decreased with age. CHT also decreased the levels of GSH-PX and Se, and increased the level of MDA; however, Se supplementation could alleviate these adverse effects, which may be due to the Se being involved in the composition of glutathione peroxidase [34]. The CHT activated the body's antioxidant system, further promoting the synthesis of glutathione peroxidase and decreasing the level of serum Se. In addition, compared with selenium-free diets, SS, SY, and SYC significantly increased serum GSH-Px, VE, and Se level, and significantly decreased MDA content, which was consistent with the previous study. It showed that the body activated its own antioxidant defense system to resist the harm caused by CHT to the body, and Se improved the body's antioxidant capacity and can alleviate the body's oxidative damage.

4.4. Effects of Different Se Sources on Immune Function of Heat-Stressed Laying Hens

It has been reported that increased ambient temperature will reduce the performance and immune status of poultry [35]. A study found that the ratio of circulating antibodies IgG and IgM in chickens decreased when exposed to heat stress [36]. The change of serum immunoglobulin content is an important indicator of animal immune function. It has been reported that heat stress can reduce the level of serum immunoglobulin, thymus, and spleen weight of broiler chickens [37]. The supplementation of 0.3 mg of Se to the diet significantly increased the total IgY titer and gammaglobulins in the sera of broilers, as well as improving humoral immunity [38]. A study found that Se-enriched earthworm powder with 1.0 mg/kg of Se increased the serum albumin and IgG of laying hens [5]. In the present study, compared to the NT group, CHT significantly decreased the serum IgG content; adding SS, SY, and SYC increased the immunity of layers, which was consistent with previous studies. It showed that Se improved the body's immune function and can alleviate the body's damage caused by heat stress.

4.5. Effects of Different Se Sources on Serum Heat Stress Associated Indexes of Heat-Stressed Laying Hens

The central nervous system regulates the secretion of hormones and responds to external stimuli. Cort is a key indicator of the strength of stress. Previous studies have reported that, during heat stress, it stimulates the hypothalamus to synthesize corticotropin-releasing hormone (CRH), the release of corticotropin (ACTH) from the anterior pituitary, and the adrenal cortex cells to secrete Cort to achieve the purpose of alleviating stress [39–41]. When chickens were under heat stress, the hypothalamic-pituitary-adrenal (HPA) axis was activated, and the content of Cort in the sera increased [42]. In the present study, compared to the NT group, CHT increased the Cort content, though not significantly; the trend was consistent with previous studies. In addition, the temperature of an animal's body is also regulated by thyroid hormone, triiodothyroxine (T3), and thyroxine (T4). We found that the FT3 concentration significantly increased at week 50, which was in partial agreement with the report, which reported that after 14 days of continuous heat stress (35 °C) in 4-week-old broiler, the levels of FT3 and FT4 significantly increased [43], while the FT4 level had no significance in the present study. This difference may be related to the species, ages, and environment of the animals. When exposed to high temperatures, the body protects itself by synthesizing HSPs. The high concentrations of HSP70 can be found when poultry is exposed to heat stress. In the present study, compared to NT, CHT significantly increased the HSPs at week 52, which was consistent with the previous study. In addition, adding SS, SY, and SYC decreased the concentration of HSPs, which means that Se plays a certain role in helping the body recover from heat stress.

5. Conclusions

In conclusion, this study demonstrated that circulating high temperature (26 ± 2 °C~33 ± 2 °C) had no significant effect on the egg quality of laying hens, but significantly decreased the laying performance, serum IgG, and antioxidant capacity. Adding different Se had no obvious effect on laying performance, but can remarkably increase the color of egg yolk. In addition, Se supplementation significantly increased antioxidant capacity and serum IgG, further alleviating damage caused by heat stress.

Supplementary Materials: The following supporting information can be downloaded at: https://www.mdpi.com/article/10.3390/ani12081006/s1. Figure S1: Changes in rectal temperature of laying hens of different periods. BK: basal diet, SS: sodium selenite, SY: selenium yeast, SYC: selenium-enriched yeast culture. HS+ = under circulating high temperature (26 ± 2 °C~33 ± 2 °C). Significance compared with 48 week, * $p < 0.05$, ** $p < 0.01$ and *** $p < 0.001$, # indicated that 52 week was compared with 50 week, # $p < 0.05$, ## $p < 0.01$, ### $p < 0.001$. Figure S2: Changes in laying performance of laying hens of different periods. BK: basal diet, SS: sodium selenite, SY: selenium yeast, SYC: selenium-enriched yeast culture. HS+ = under circulating high temperature (26 ± 2 °C~33 ± 2 °C). Significance compared with 45–48 weeks, * $p < 0.05$, ** $p < 0.01$ and *** $p < 0.001$, # indicated that 51–52 weeks was compared with 49–50 weeks, # $p < 0.05$, ## $p < 0.01$, ### $p < 0.001$. Figure S3. Changes in egg quality of laying hens of different periods. BK: basal diet, SS: sodium selenite, SY: selenium yeast, SYC: selenium-enriched yeast culture. HS+ = under circulating high temperature (26 ± 2 °C~33 ± 2 °C). Significance compared with 45–48 weeks, * $p < 0.05$, ** $p < 0.01$ and *** $p < 0.001$, # indicated that 51–52 weeks was compared with 49–50 weeks, # $p < 0.05$, ## $p < 0.01$, ### $p < 0.001$. Figure S4. Changes in Se and VE content in serum of laying hens of different periods. BK: basal diet, SS: sodium selenite, SY: selenium yeast, SYC: selenium-enriched yeast culture. HS+ = under circulating high temperature (26 ± 2 °C~33 ± 2 °C). Significance compared with 48 week, * $p < 0.05$, ** $p < 0.01$ and *** $p < 0.001$, # indicated that 52 week was compared with 50 week, # $p < 0.05$, ## $p < 0.01$, ### $p < 0.001$. Figure S5. Changes in antioxidant status of laying hens of different periods. BK: basal diet, SS: sodium selenite, SY: selenium yeast, SYC: selenium-enriched yeast culture. HS+ = under circulating high temperature (26 ± 2 °C~33 ± 2 °C). Significance compared with 48 week, * $p < 0.05$, ** $p < 0.01$ and *** $p < 0.001$, # indicated that 52 week was compared with 50 week, # $p < 0.05$, ## $p < 0.01$, ### $p < 0.001$. Figure S6. Changes in serum immune indexes of laying hens of different periods. BK: basal diet, SS: sodium selenite, SY: selenium yeast, SYC: selenium-enriched yeast culture. HS+ = under circulating high temperature (26 ± 2 °C~33 ± 2 °C). Significance compared with 48 week, * $p < 0.05$, ** $p < 0.01$ and *** $p < 0.001$, # indicated that 52 week was compared with 50 week, # $p < 0.05$, ## $p < 0.01$, ### $p < 0.001$. Figure S7. Changes in serum heat stress indexes of laying hens of different periods. BK: basal diet, SS: sodium selenite, SY: selenium yeast, SYC: selenium-enriched yeast culture. HS+ = under circulating high temperature (26 ± 2 °C~33 ± 2 °C). Significance compared with 48 week, * $p < 0.05$, ** $p < 0.01$ and *** $p < 0.001$, # indicated that 52 week was compared with 50 week, # $p < 0.05$, ## $p < 0.01$, ### $p < 0.001$.

Author Contributions: Conceptualization, W.W.; Data curation, Z.W.; Formal analysis, R.K.; Funding acquisition, Q.M.; Investigation, W.W. and M.L.; Methodology, L.Z., J.Z. and S.H.; Project administration and funding acquisition, Q.M.; Supervision, Q.M.; Writing—original draft, R.K.; Writing—review and editing, R.K., S.H. and Q.M. All authors have read and agreed to the published version of the manuscript.

Funding: This research was funded by the National Key Research and Development Programs of China (2021YFD1300405) and a Special Fund for China Agricultural Research System program (Grant No. CARS-40-K08).

Institutional Review Board Statement: The experiment was approved by the China Agricultural University Animal Care and Use Committee (A0041011202-1-1, Beijing, China).

Data Availability Statement: The data presented in this study are available on request from the corresponding author.

Conflicts of Interest: The authors declare no conflict of interest.

References

1. Mashaly, M.M.; Hendricks, G.L.; Kalama, M.A.; Gehad, A.E.; Patterson, P.H. Effect of Heat Stress on Production Parameters and Immune Responses of Commercial Laying Hens. *Poult. Sci.* **2004**, *83*, 889–894. [CrossRef] [PubMed]
2. Battisti, D.S.; Naylor, R.L. Historical warnings of future food insecurity with unprecedented seasonal heat. *Science* **2009**, *323*, 240–244. [CrossRef] [PubMed]
3. Mack, L.A.; Felver-Gant, J.N.; Dennis, R.L.; Cheng, H.W. Genetic variations alter production and behavioral responses following heat stress in 2 strains of laying hens. *Poult. Sci.* **2013**, *92*, 285–294. [CrossRef] [PubMed]
4. Li, W.X.; Chen, Y.Q.; Zhao, L.H.; Ma, Q.G.; Zhang, J.Y.; Ji, C. No copper supplementation in a corn-soybean basal diet has no adverse effects on late-phase laying hens under normal and cyclic high temperatures. *Poult. Sci.* **2018**, *97*, 1352–1360. [CrossRef]
5. Sun, X.; Yue, S.; Qiao, Y. Dietary supplementation with selenium-enriched earthworm powder improves antioxidative ability and immunity of laying hens. *Poult. Sci.* **2020**, *99*, 5344–5349. [CrossRef]
6. Rayman, M.P. The importance of selenium to human health. *Lancet* **2000**, *356*, 233–241. [CrossRef]
7. Noguchi, T.; Cantor, A.H.; Scott, M.L. Mode of action of selenium and vitamin E in prevention of exudative diathesis in chicks. *J. Nutr.* **1973**, *103*, 1502–1511. [CrossRef]
8. Whitacre, M.E.; Combs, G.F. Selenium and mitochondrial integrity in the pancreas of the chick. *J. Nutr.* **1983**, *113*, 1972–1983. [CrossRef]

9. Whanger, P.D.; Weswig, P.H.; Schmitz, J.A.; Oldfield, J.E. Effects of selenium and vitamin E on blood selenium levels, tissue glutathione peroxidase activities and white muscle disease in sheep fed purified or hay diets. *J. Nutr.* **1977**, *107*, 1298–1307. [CrossRef]
10. Suhajda, A.; Hegoczki, J.; Janzso, B.; Pais, I.; Vereczkey, G. Preparation of selenium yeasts I. Preparation of selenium-enriched Saccharomyces cerevisiae. *J. Trace Elem. Med. Biol.* **2000**, *14*, 43–47. [CrossRef]
11. Payne, R.L.; Lavergne, T.K.; Southern, L.L. Effect of inorganic versus organic selenium on hen production and egg selenium concentration. *Poult. Sci.* **2005**, *84*, 232–237. [CrossRef] [PubMed]
12. Yoon, I.; Werner, T.M.; Butler, J.M. Effect of source and concentration of selenium on growth performance and selenium retention in broiler chickens. *Poult. Sci.* **2007**, *86*, 727–730. [CrossRef] [PubMed]
13. Lu, J.; Qu, L.; Ma, M.; Li, Y.F.; Wang, X.G.; Yang, Z.; Wang, K.H. Efficacy evaluation of selenium-enriched yeast in laying hens: Effects on performance, egg quality, organ development, and selenium deposition. *Poult. Sci.* **2020**, *99*, 6267–6277. [CrossRef] [PubMed]
14. Lu, J.; Qu, L.; Shen, M.M.; Wang, X.G.; Guo, J.; Hu, Y.P.; Dou, T.C.; Wang, K.H. Effects of high-dose selenium-enriched yeast on laying performance, egg quality, clinical blood parameters, organ development, and selenium deposition in laying hens. *Poult. Sci.* **2019**, *98*, 2522–2530. [CrossRef] [PubMed]
15. Mahmoud, K.Z.; Edens, F.W. Influence of organic selenium on hsp70 response of heat-stressed and enteropathogenic Escherichia coli-challenged broiler chickens (*Gallus gallus*). *Comp. Biochem. Physiol.* **2005**, *141*, 69–75. [CrossRef] [PubMed]
16. Kim, Y.Y.; Mahan, D.C. Comparative effects of high dietary levels of organic and inorganic selenium on selenium toxicity of growing-finishing pigs. *J. Anim. Sci.* **2001**, *79*, 942–948. [CrossRef]
17. Hoffman, D.J. Role of selenium toxicity and oxidative stress in aquatic birds. *Aquat. Toxicol.* **2002**, *57*, 11–26. [CrossRef]
18. Pan, C.; Zhao, Y.; Liao, S.F. Effect of selenium-enriched probiotics on laying performance, egg quality, egg selenium content, and egg glutathione peroxidase activity. *J. Agric. Food Chem.* **2011**, *59*, 11424–11431. [CrossRef]
19. Liao, X.D.; Lu, L.; Li, S.F.; Liu, S.B.; Zhang, L.Y.; Wang, G.Y.; Lu, A.; Luo, X.G. Effects of selenium source and level on growth performance, tissue selenium concentrations, antioxidation, and immune functions of heat-stressed broilers. *Biol. Trace Element Res.* **2012**, *150*, 158–165. [CrossRef]
20. Barrett, N.W.; Rowland, K.; Schmidt, C.J.; Lamont, S.J.; Rothschild, M.F.; Ashwell, C.M. Effects of acute and chronic heat stress on the performance, egg quality, body temperature, and blood gas parameters of laying hens. *Poult. Sci.* **2019**, *98*, 6684–6692. [CrossRef]
21. Star, L.; Juul-Madsen, H.R.; Decuypere, E.; Nieuwland, M.; de Vries Reilingh, G.; Van Den Brand, H.; Kemp, B.; Parmentier, H.K. Effect of early life thermal conditioning and immune challenge on thermotolerance and humoral immune competence in adult laying hens. *Poult. Sci.* **2009**, *88*, 2253–2261. [CrossRef] [PubMed]
22. Januzi, V.; Sena, L.; Elezi, X. The influence of layers feed supplementation with organic selenium on the eggs quality and selenium content in the egg. *Albanian J. Agric. Sci.* **2017**, 329–333. Available online: https://www.proquest.com/openview/9afb79ebb3015caf70eff1b3f14aa714/1?pq-origsite=gscholar&cbl=1316365 (accessed on 14 January 2022).
23. Ebeid, T.A.; Suzuki, T.; Sugiyama, T. High ambient temperature influences eggshell quality and calbindin-D28k localization of eggshell gland and all intestinal segments of laying hens. *Poult. Sci.* **2012**, *91*, 2282–2287. [CrossRef]
24. Song, Z.; Liu, L.; Sheikhahmadi, A. Effect of heat exposure on gene expression of feed intake regulatory peptides in laying hens. *J. Biomed. Biotechnol.* **2012**, *2012*, 484869. [CrossRef] [PubMed]
25. Morales, A.; Hernández, L.; Buenabad, L. Effect of heat stress on the endogenous intestinal loss of amino acids in growing pigs. *J. Anim. Sci.* **2016**, *94*, 165–172. [CrossRef] [PubMed]
26. Tadtiyanant, C.; Lyons, J.J.; Vandepopuliere, J.M. Influence of wet and dry feed on laying hens under heat stress. *Poult. Sci.* **1991**, *70*, 44–52. [CrossRef]
27. Tanor, M.A.; Leeson, S.; Summers, J.D. Effect of heat stress and diet composition on performance of White Leghorn hens. *Poult. Sci.* **1984**, *63*, 304–310. [CrossRef]
28. Paton, N.D. Organic Selenium in the Nutrition of Laying Hens: Effects on Egg Selenium Content, Egg Quality and Transfer to Developing Chick Embryo. Ph.D. Thesis, University of Kentucky, Lexington, KY, USA, 2001.
29. Lin, H.; Decuypere, E.; Buyse, J. Acute heat stress induces oxidative stress in broiler chickens. *Comp. Biochem. Physiol. A.* **2006**, *144*, 11–17. [CrossRef]
30. Belhadj Slimen, I.; Najar, T.; Ghram, A.; Dabbebi, H.; Ben Mrad, M.; Abdrabbah, M. Reactive oxygen species, heat stress and oxidative-induced mitochondrial damage. A review. *Int. J. Hyperth.* **2014**, *30*, 513–523. [CrossRef]
31. Jones, D.P. Redox potential of GSH/GSSG couple: Assay and biological significance. *Method Enzymol.* **2002**, *348*, 93–112.
32. Dalia, A.M.; Loh, T.C.; Sazili, A.Q.; Jahromi, M.F.; Samsudin, A.A. The effect of dietary bacterial organic selenium on growth performance, antioxidant capacity, and Selenoproteins gene expression in broiler chickens. *BMC Veter Res.* **2017**, *13*, 254. [CrossRef] [PubMed]
33. Han, X.J.; Qin, P.; Li, W.X.; Ma, Q.G.; Ji, C.; Zhang, J.Y.; Zhao, L.H. Effect of sodium selenite and selenium yeast on performance, egg quality, antioxidant capacity, and selenium deposition of laying hens. *Poult. Sci.* **2017**, *96*, 3973–3980. [CrossRef] [PubMed]
34. Rotruck, J.T.; Pope, A.L.; Ganther, H.E. Selenium: Biochemical role as a component of glutathione peroxidase. *Science* **1973**, *179*, 588–590. [CrossRef] [PubMed]

35. Tirawattanawanich, C.; Chantakru, S.; Nimitsantiwong, W.; Tongyai, S. The effects of tropical environmental conditions on the stress and immune responses of commercial broilers, Thai indigenous chickens, and crossbred chickens. *J. Appl. Poult. Res.* **2011**, *20*, 409–420. [CrossRef]
36. Lara, L.J.; Rostagno, M.H. Impact of heat stress on poultry production. *Animals* **2013**, *3*, 356–369. [CrossRef] [PubMed]
37. Park, S.O.; Hwangbo, J.H.; Ryu, C.M.; Park, B.S.; Choi, Y.H. Effects of Extreme Heat Stress on Growth Performance, Lymphoid Organ, IgG and Cecum Microflora of Broiler Chickens. *Int. J. Agric. Biol.* **2013**, *15*, 1204–1208.
38. Vakili, R.; Bahram, M. Effects of different dietary levels of selenium on metabolic parameters and humoral immunity in broiler chickens. *J. Vet. Res.* **2010**, *65*, 329–336.
39. Antoni, F.A. Hypothalamic control of adrenocorticotropin secretion: Advances since the discovery of 41-residue corticotropin-releasing factor. *Endocr. Rev.* **1986**, *7*, 351–378. [CrossRef]
40. Dallman, M.F.; Akana, S.F.; Jacobson, L.; Levin, N.; Cascio, C.S.; Shinsako, J. Characterization of Corticosterone Feedback Regulation of ACTH Secretion. *Ann. N. Y. Acad. Sci.* **1987**, *512*, 402–414. [CrossRef]
41. Vale, W.; Spiess, J.; Rivier, C.; Rivier, J. Characterization of a 41-residue ovine hypothalamic peptide that stimulates secretion of corticotropin and β-endorphin. *Science* **1981**, *213*, 1394–1397. [CrossRef]
42. Quinteiro-Filho, W.M.; Ribeiro, A.; Ferraz-de-Paula, V.; Pinheiro, M.L.; Sakai, M.; Sá, L.; Ferreira, A.; Palermo-Neto, J. Heat stress impairs performance parameters, induces intestinal injury, and decreases macrophage activity in broiler chickens. *Poult. Sci.* **2010**, *89*, 1905–1914. [CrossRef] [PubMed]
43. He, X.; Lu, Z.; Ma, B.; Zhang, L.; Li, J. Chronic heat stress alters hypothalamus integrity, the serum indexes and attenuates expressions of hypothalamic appetite genes in broilers. *J. Therm. Biol.* **2019**, *81*, 110–117. [CrossRef] [PubMed]

animals

MDPI

Article

Antioxidant Status, Blood Constituents and Immune Response of Broiler Chickens Fed Two Types of Diets with or without Different Concentrations of Active Yeast

Youssef A. Attia [1,*], Hanan Al-Khalaifah [2,*], Hatem S. Abd El-Hamid [3], Mohammed A. Al-Harthi [1], Salem R. Alyileili [4] and Ali A. El-Shafey [5]

1 Agriculture Department, Faculty of Environmental Science, King Abdulaziz University, P.O. Box 80208, Jeddah 21589, Saudi Arabia; malharthi@kau.edu.sa
2 Environment and Life Sciences Research Center, Kuwait Institute for Scientific Research (KISR), P.O. Box 24885, Safat-Kuwait 13109, Kuwait
3 Poultry and Fish Diseases Department, Faculty of Veterinary Medicine, Damanhour University, Damanhour 22511, Egypt; drhatem_deltavet@yahoo.co.uk
4 Department of Integrative Agriculture, College of Food and Agriculture, United Arab Emirates University, Al-Ain 15551, United Arab Emirates; salem7178767@gmail.com
5 Animal and Poultry Production Department, Faculty of Agriculture, Damanhour University, Damanhour 22511, Egypt; alielshafey2015@gmail.com
* Correspondence: yaattia@kau.edu.sa (Y.A.A.); hkhalifa@kisr.edu.kw (H.A.-K.); Tel.:+966-568575961 (Y.A.A.); +965-94949325 (H.A.-K.)

Simple Summary: Rations for broilers can be safely supplemented with probiotics such as active *Saccharomyces cerevisiae* (SC) yeast to stimulate oxidative reactions and immune response against stress and infectious agents. The current study suggested that SC yeast enhanced antioxidant capacity, growth rate, immune organ weights, immune response and the survival rate of broilers after Avian Influenza virus challenge at 38 days of age.

Abstract: Probiotics, such as active yeasts, are widely used to enhance poultry production and reduce feeding costs. This study aimed to investigate the antioxidant and immune responses of broilers to different concentrations of active *Saccharomyces cerevisiae* (SC) when supplemented to two types of diets. A total of 216 1-day-old Arbor Acres unsexed chicks were used in a factorial design, involving two feeds (regular- versus low-density diet) and three concentrations of SC (0%, 0.02% and 0.04%). The results revealed that the low-density diet reduced the body weight and production index of broilers. The addition of SC improved the production index more than the control diet. Total antioxidant capacity (TAC), alanine aminotransferase (ALT), aspartate aminotransferase (AST) and eosinophils were significantly higher in response to the regular-density diet than the low-density diet; however, phagocytic activity (PA), lymphocyte and lysozyme activity (LYS) were lower. *Saccharomyces cerevisiae* reduced ALT, AST, malondialdehyde (MAD) and TAC more than the standard set, but improved packed cell volume (PCV), hemoglobin (Hgb), red blood cells (RBCs), lymphocytes, monocytes, heterophils, phagocytic index (PI) and the immune response to Newcastle disease virus (NDV) and avian influenza (AI). In conclusion, supplementation of a regular- or low-density diet with SC at a concentration of 0.02% or 0.04% improved the antioxidant parameters, immune status and production index of broilers against stress and infectious agents.

Keywords: active yeast; antioxidant status; broilers; nutrient density; immune response

Citation: Attia, Y.A.; Al-Khalaifah, H.; Abd El-Hamid, H.S.; Al-Harthi, M.A.; Alyileili, S.R.; El-Shafey, A.A. Antioxidant Status, Blood Constituents and Immune Response of Broiler Chickens Fed Two Types of Diets with or without Different Concentrations of Active Yeast. *Animals* **2022**, *12*, 453. https://doi.org/10.3390/ani12040453

Academic Editor: Rosalia Crupi

Received: 9 November 2021
Accepted: 9 February 2022
Published: 12 February 2022

1. Introduction

There is a growing interest in the innovative biofortification of poultry feed rations through the use of functional ingredients to improve feed utilization and enhance production performance and the immune status of the flocks [1–8].

Saccharomyces cerevisiae (SC) has been used in poultry feed rations to enhance resistance to aflatoxicosis in poultry [9]. The antioxidant status and capacity of poultry have been shown to be significantly enhanced by supplementing poultry feed rations with SC, either alone or in combination with probiotics [10,11]. The weights of the primary and secondary immune organs in broilers have been shown to be increased after dietary supplementation with active yeast, potentially indicating immunocompetence in broilers [12]. In addition, Kiarie et al. [13] revealed that when added with feed enzymes, yeast derivatives can modulate cellular- and humoral-mediated immunity in broilers against intestinal coccidial infections. Zhou et al. [14] investigated the ability of yeast fractions to prevent pullorum disease and fowl typhoid in breeders. The results of the study revealed that dietary fortification of yeast fractions significantly reduced disease infection in the challenged breeders. In the same study, the positive culling rate of the pullets and their body weight were significantly reduced because of the addition of yeast fractions to the feed of birds challenged with *Salmonella* infection [14]. The intestinal microflora balance has also been shown to be improved in birds with dietary supplementation of yeast, due to the presence of mannan-oligosaccharides and fructo-oligosaccharides in the cell wall of yeast [15–19]. However, debate remains regarding the effects of active yeast on the production performance and immune status of chickens, particularly when they are fed diets containing different nutrient profiles or are placed under environmental stress [20–23]. Hayat et al. [24] suggested that this could be due to genetic differences, or differences in species, age, or environmental conditions. Thus, this study was carried out to fill this knowledge gap.

The current study aimed to fill the gap in the literature about the response of broilers fed two types of diets, with or without different concentrations of SC (active yeast) with respect to antioxidant status, blood constituents and immune status. Although previous studies investigated the effect of SC on the productive performance parameters in broiler chickens, there are relatively limited data in the literature on the direct effect of SC probiotic on the antioxidative status, blood constituents and immune status in broiler chickens fed low-density diet. The current literature does not yet adequately address the approaches of nutrient manipulation in broiler feed rations to shed light on the relationship between the effect of SC in low-density diets and the immune response of broiler chickens. Accordingly, this study was executed to elaborate on this vital relationship.

2. Materials and Methods

2.1. Birds, Dietary Treatments, Experimental Plan and Housing

This research work was approved by the Deanship of Scientific Research, King Abdulaziz University, Saudi Arabia, under protocol no: (FP-217-42 H). The protocol recommends general humane treatment of animals that did not cause animal (s) pain, suffering, distress, or lasting harm, according to the Royal Decree number M59 in 14/9/1431H.

A total of 216, 1-day-old Arbor Acres broilers (mixed sexes) were acquired. The chicks were marked randomly by way of wing-banding and were housed in 36 pens with 6 birds per pen (replicate). Each treatment involved 6 replicates. The body weight (BW) of all chicks was similar at the start of the experiment.

The chickens were provided mash feeds ad libitum, along with freely accessed waters. During the first 7 days, 23 h of light were used, followed by 20 h of light until the end of the experiment.

A factorial design (2 × 3) was applied to the experiment using two diets (a regular versus a low-density diet containing 10% fewer nutrients than the regular diet) and three levels of SC (an unfortified standard, 0.02%, or 0.04% SC). The broiler feed rations were formulated based on the Arbor Acres broilers guide [25]. There were six replications in each treatment and each replicate involved six birds. The SC yeast was purchased from China Way Co-operation, Taiwan, and had 12,000,000,000 active yeast per gram. The optimal dosage of SC was 200 to 400 g per ton of feed. Table 1 shows the composition of the dietary treatments fed to the chickens.

Table 1. Basal chemical composition of the experimental diets.

Ingredients (%)	Regular-Density Diet			Low-Density Diet		
	Starter	Grower	Finisher	Starter	Grower	Finisher
Maize	51.27	52.02	55.85	46.01	46.61	50.40
Rye	0	5	7	0	4.48	6.2
Wheat bran	0	0	0	10	10	10
Soybean meal (44% CP)	32.8	24.4	28.0	29.3	21.9	25.0
Vegetable oil	2.25	2.0	5.3	2.01	1.79	4.7
Full-fat soybean meal	10	13	1.6	9	11.64	1.42
Dicalcium phosphate	1.8	1.6	1.0	1.8	1.6	1.0
Limestone	1.0	1.0	0.15	1.0	1.0	0.15
L-Lysine	0.10	0.15	0.2	0.10	0.15	0.2
DL-Methionine	0.15	0.20	0.30	0.15	0.20	0.30
Vit + min premix [1]	0.30	0.30	0.30	0.30	0.30	0.30
NaCl	0.30	0.30	0.30	0.30	0.30	0.30
Washed building sand	0.03	0.03	0.00	0.03	0.03	0.03
Total	100	100	100	100	100	100
Calculated and deterimed analyses						
Metabolizable Energy (kcal/kg) [2]	3038	3100	3196	2735	2790	2876
Crude Protein (%) [3]	22.8	21.2	18.5	20.5	19.1	16.7
Lysine (%) [2]	1.33	1.23	1.04	1.20	1.11	0.94
Methionine (%) [2]	0.50	0.52	0.48	0.45	0.47	0.43
Meth + cysteine (%) [2]	0.87	0.87	0.78	0.78	0.78	0.70
Calcium (%) [2]	0.91	0.85	0.83	0.82	0.77	0.75
Av. P (%) [2]	0.46	0.41	0.41	0.414	0.369	0.369
Crude fat (%) [3]	6.09	6.45	6.8	5.48	5.81	6.12
Crude fibre (%) [3]	3.55	4.48	4.75	3.20	4.03	4.28
Ash (%) [3]	5.22	5.48	5.71	4.70	4.93	5.14

[1] Vit + Min mix. contains the following nutrients. Values are per kilogram of the diet: Vit. A, 12,000 IU; Vit. E (dl-α-tocopheryl acetate), 20 mg; menadione, 2.3 mg; Vit. D_3, 2200 ICU; riboflavin, 5.5 mg; calcium pantothenate, 12 mg; nicotinic acid, 50 mg; Choline, 250 mg; Vit. B_{12}, 10 µg; Vit. B_6, 3 mg; thiamine 3 mg; folic acid, 1 mg; d-biotin, 0.05 mg; Trace minerals (mg/kg of diet): Mn, 80; Zn, 60; Fe, 35; Cu, 8; Selenium, 0.1 mg, [2] Calculated analyses, [3] Determined analyses.

2.2. Data Gathering

Average pen body weight (g) was recorded at 1, 21 and 38 days of age and used to calculate the body weight gain (BWG, g/bird). The average pen feed intake (g/bird) was recorded and used to calculate the feed conversion rate (FCR, g feed/g gain) and survival rate (100–mortality rate) during the following periods: 1–21 days, 22–38 days and 1–38 days of age. The production index was calculated as follows: BW (kg) × survival rate Production index = ×100 production period in days × FCR.

2.3. Blood Sampling

Blood was collected from each treatment group (n = 6) before vaccination and again at 8 days post-vaccination. The serum was harvested by centrifuging the blood at 1500× g for 15 min.

2.4. Antioxidant Status and Biochemical Traits

Serum total antioxidant capacity (TAC) and malondialdehyde (MAD) were assayed as described in Erel [26] and Wyatt et al. [27], respectively. They were determined using commercial kits produced by Diamond Diagnostics (23 EL-Montazah St. Heliopolis, Cairo, Egypt, http://www.diamonddiagnostics.com (accessed on 1 February 2022). Total plasma protein and albumin concentrations were measured using the methods outlined in Armstrong and Carr [28] and Doumas and Peters [29], respectively. Subtracting albumin concentration from serum total protein gives an estimate of the globulin concentration, as described in Giangiacomo et al. [30]. Various kinds of globulin (α-, β- and γ-globulin) were determined based on methods described in Elias [31]. The activities of alanine aminotransferase (ALT, U/L) and aspartate aminotransferase (AST, U/L) were determined using techniques described in Reitman and Frankel [32]. Alkaline phosphatase (ALKP) enzyme action was measured in plasma, as described by Kim and Wyckoff [33].

2.5. Hematological Parameters

Wintrobe hematocrit tubes were used to measure the packed cell volume (PCV, %) by centrifugation for 20 min. at 2000× g. Hemoglobin (Hgb) level was estimated using the technique described in Eilers [34]. The mean corpuscular volume (MCV, μm 3), mean corpuscular hemoglobin (MCH, Pg) and mean corpuscular hemoglobin concentration (MCHC, g/dL) were measured using the equations described in Jain [35].

2.6. Immune Indices

The phagocytic activity (PA, % of phagocytic cells engulfing yeast cells) and phagocytic index (PI, number of yeast cell phagocytized/number of phagocytic cells) were determined as described in Kawahara et al. [36].

Broiler chickens were vaccinated according to the following schedule: inactivated avian influenza (AI) H5N2 at 10 days of age. Chickens were vaccinated with clone 30 eye drop on day 8 for Newcastle disease (NDV) and bivalent NDV vaccine was administered underneath the neck membrane, simultaneously with clone 30, at 8 days of age. The Gumboro intermediate vaccine and clone 30 were administered at 12 and 21 days of age, respectively (Nobilis, Intervet, Boxmeer, The Netherlands).

Blood samples (n = 6 per group) were taken just before vaccination (0 days post-vaccination) and again on 8th day post-vaccination. The samples were centrifuged at 1500× g for 15 min for serum separation, to determine antibody titers against NDV via the hemagglutination inhibition test (HI) test. This test was done using hemagglutination inhibition (HI) test according to OIE [37]. The geometric mean titer was calculated as recommended by the World Organization for Animal Health (OIE) [38].

Antibody responses were determined by the HI test, according to Seal et al. [39]. The assay measures antibodies attached to influenza antigen-coated plates [40]. Hemagglutination inhibition for NDV and AI were measured as described in Takatsy and Hamar [41].

A lymphocyte transformation test was performed, as described in Balhaa et al. [42]. Lympholyte-H (Cedarlane Laboratories Ltd., Burlington, ON, Canada) was used to layer the collected heparinized blood. After centrifugation, the lymphocytes in the interface layer were collected, washed and suspended in culture medium.

Serum bactericidal activity to the *Aeromonas hydrophila* strain was conducted following the protocols described in Rainger and Rowley [43]. The turbidimetric method was used to measure serum lysozyme activity [44]. The results were reflected as one unit of lysozyme activity as a reduction in absorbency of 0.001/min Lysozyme activity = (A0 − A)/A.

2.7. Challenge Test

The challenge test was conducted to study the impact of the diet on the survival rate of chicks [45,46] between 38 and 48 days of age. Six broiler chickens per treatment were randomly selected at 38 days of to represent all treatment replications. The chickens were vaccinated with inactivated avian influenza (AI) H5N2 at 10 days of age and then challenged with H5N1 at 38 days via the oculo-nasal route with 0.2 mL/bird (10^6/dose). The H5N1 was from research laboratory of Poultry Disease, Fac. Vet. Med., Damanhour University, where the challenge test was carried out following the regulations for animal welfare approved by the authorized ethics committee of the Egyptian Ministry of Agriculture according to Decree No. 27, 1967. The mortality was recorded daily during 38–48 days of age.

2.8. Histopathological Study

On day 38, 6 chickens from each treatment replicate were randomly selected and euthanized under anesthesia via an intravenous injection of sodium pentobarbital (50 mg/kg;

CAMEO chemicals, Tampa, FL, USA). Necropsies were performed for sample collection. Lymphoid organs (bursa of Fabricius, thymus and spleen) were weighed, and the body weight ratios of organs were calculated.

In addition, intestine, bursa of Fabricious, thymus and spleen specimens were collected from randomly collected broilers ($n = 6$ per treatment) at 38 days of age. Tissue specimens were prepared as previously described by Culling [47].

2.9. Morphometrical Study

An Optika binocular microscope, with an Optika imaging analyzer, was used to examine the morphological appearance of intestinal villi, determine the absorption surface and measure the longitudinal axis of the large follicle of the bursa. Five segments from each bird were used for this examination. In addition, quantitative measurements of the thymus cortical: medullary ratio were performed and the hyperplasia of the lymphoblastic cells was assayed by examining the spleen. The scale used was as follows: (−) for weak hyperplasia; (+) for moderate hyperplasia; (++) for severe hyperplasia.

2.10. Statistical Analyses

The data were analyzed using general linear models in SAS (SAS Institute, Cary, NC, USA [48]). A two-way factorial design (two kinds of diets × three concentrations of SC) was used to analyze the effects of the treatments on most of the parameters. An exception was survival rate in the challenge study, where age was included as a main effect only. The replicate was the experimental unit. Data were arcsine transformed prior to analysis to improve normality. Student–Newman–Keuls (SNK) post hoc tests were applied to evaluate differences between factor levels in the model. Differences were considered significant if $p \leq 0.05$.

3. Results

3.1. Growth Performance

Table 2 shows the impact of yeast concentrations on body weight and the European Production Efficiency Index (EPEI) of broilers fed regular- and low-density diets. During the experiment, the low-density diet was found to decrease the final body weight and EPEI, reaching 5% over the duration of the study. Diets supplemented with 0.02% and 0.04% SC resulted in a production index that was significantly enhanced relative to the control diet. However, the addition of SC in the diet at a concentration of 0.04% had a more substantial effect on the growth of 38-day-old chickens than the 0.02% level (Table 2). There was no significant relationship between the amount of SC and diet on the growth of broilers or EPEI between 1 and 38 days of age.

Table 2. Impact of different concentrations of *Saccharomyces cerevisiae* on body weight, survival rate, European Production Efficiency Index and blood hematological parameters of broilers fed regular- or low-density diets from days 1 to 38 of age.

Treatment	Body Weight at 38 Days of Age, g	Survival Rate, %	European Production Efficiency Index	PCV, %	Hemoglobin, g/dL	RBC, 10^6/uL	MCV, fL	MCH, pg	MCHC, g/dL
				Effect of diet					
Regular density	2099 ^{a}	100	314 ^{a}	30.4	9.91	1.65	186	60.6	32.7
Low density	2004 ^{b}	100	297 ^{b}	30.0	10.0	1.72	177	58.7	33.3
				Effect of yeast concentration					
Control	1865 ^{c}	100	281 ^{b}	28.3 ^{b}	9.43 ^{b}	1.60 ^{b}	180	59.7	33.2
0.02%	2106 ^{b}	100	318 ^{a}	31.6 ^{a}	10.5 ^{a}	1.76 ^{a}	182	60.8	33.5
0.04%	2183 ^{a}	100	318 ^{a}	30.7 ^{a}	9.87 ^{b}	1.70 ^{ab}	182	58.5	32.2

Table 2. *Cont.*

Treatment		Body Weight at 38 Days of Age, g	Survival Rate, %	European Production Efficiency Index	PCV, %	Hemoglobin, g/dL	RBC, 10^6/uL	MCV, fL	MCH, pg	MCHC, g/dL
Interaction between diet type and yeast concentration										
Regular density	Control	1946	100	288	18.7 [b]	8.75 [b]	1.58	172 [b,c]	55.7 [b,c]	32.4
	0.02%	2171	100	321	32.0 [a]	10.6 [a]	1.75	185 [a,b,c]	61.6 [a,b]	33.4
	0.04%	2179	100	334	32.3 [a]	10.3 [a]	1.62	202 [a]	64.5 [a]	32.2
Low density	Control	1784	100	274	29.7 [a]	10.1 [a]	1.61	188 [a,b]	63.8 [a,b]	34.0
	0.02%	2041	100	315	31.3 [a]	10.5 [a]	1.77	179 [a,b,c]	60.0 [a,b]	33.5
	0.04%	2187	100	303	29.1 [a]	9.37 [b]	1.78	163 [c]	52.4 [c]	32.3
SEM		41.30	ND	9.67	0.955	0.316	0.058	9.99	3.09	1.02
p value										
Diet type		0.0008	ND	0.0408	0.6336	0.7485	0.1473	0.256	0.466	0.458
Yeast		0.0001	ND	0.0006	0.0040	0.0038	0.0273	0.966	0.753	0.442
Interaction		0.1182	ND	0.4298	0.0121	0.0021	0.4088	0.029	0.009	0.715

[a,b,c] Means within a column with different superscripts are significantly different based on Student–Newman–Keuls (SNK) post hoc tests. MCV = Mean corpuscular volume; MCH = Mean corpuscular hemoglobin; MCHC = Mean corpuscular hemoglobin concentration; Number of observations was 6 replicates per interaction cell. ND = Not done.

3.2. Antioxidant Status and Biochemical Traits

Tables 3 and 4 show the impact of different yeast concentrations on the liver enzyme index, peroxidation index and blood serum biochemical constituents of broilers fed regular and low-density diets from 1 to 38 of age.

Table 3. Impact of different concentrations of *Saccharomyces cerevisiae* on liver enzymes and malondialdehyde (MDA) of broilers fed regular- and low-density diets from days 1 to 38 of age.

Treatment		AST (U/L)	ALT (U/L)	AST/ALT Ratio	Alkaline Phosphatase (U/L)	MDA.m Mol/dL
Effect of diet						
Regular density		64.7 [a]	56.4 [a]	1.14	11.4	1.44
Low density		63.7 [b]	55.1 [b]	1.15	11.6	1.45
Effect of yeast concentration						
Control		66.0 [a]	57.8 [a]	1.14	11.5	1.57 [a]
0.02%		63.43 [b]	54.9 [b]	1.15	11.2	1.32 [c]
0.04%		63.2 [b]	54.6 [b]	1.15	11.8	1.45 [b]
Interaction between diet and yeast concentration						
Regular density	Control	65.5 [a,b]	57.1 [a,b]	1.14	11.6	1.41 [b]
	0.02%	64.2 [b]	56.1 [b]	1.14	11.2	1.43 [b]
	0.04%	64.5 [b]	56.1 [b]	1.14	11.3	1.47 [b]
Low density	Control	66.6 [a]	58.6 [a]	1.13	11.3	1.73 [a]
	0.02%	62.6 [c]	53.7 [c]	1.16	11.2	1.21 [c]
	0.04%	62.0 [c]	53.1 [c]	1.16	12.2	1.42 [b]
SEM		0.537	0.701	0.015	0.587	0.058
p value						
Diet		0.0283	0.0299	0.4385	0.6668	0.7284
Yeast concentration		0.0001	0.0001	0.5854	0.6351	0.0005
Interaction		0.0046	0.0052	0.5946	0.6077	0.0001

[a,b,c] Means within a column with different superscripts are significantly different based on Student–Newman–Keuls (SNK) post hoc test. ALT = Alanine aminotransferase; AST = Aspartate aminotransferase; AST/ALT = Aspartate aminotransferase to alanine aminotransferase ratio; MAD = Malondialdehyde. Number of observations was 6 per interaction cell.

Table 4. Impact of different concentrations of *Saccharomyces cerevisiae* on the biochemical constituents of blood serum of broilers fed regular- and low-density diets from days 1 to 38 of age.

Treatment		Serum Biochemical Constituents (mg/dL)						
		Total Protein, g/dL	Albumin, g/dL	Globulin, g/dL	α-Gloulin, g/dL	β-Globulin, g/dL	γ-Globulin, g/dL	GAR
Effect of diet								
Regular density		5.00	2.70	2.24	0.912	0.730	0.595 [a]	0.846
Low density		4.80	2.73	2.11	0.958	0.720	0.437 [b]	0.784
Effect of yeast concentration								
Control		4.73 [b]	2.64	2.09	0.837	0.656 [b]	0.600	0.828
0.02%		5.01 [a]	2.70	2.17	0.956	0.768 [a]	0.450	0.792
0.04%		5.00 [a]	2.81	2.26	1.01	0.756 [a]	0.500	0.824
Interaction between diet and yeast concentration								
Regular density	Control	4.78	2.63	2.15	0.837	0.637	0.675	0.870
	0.02%	4.86	2.75	2.26	0.962	0.800	0.500	0.820
	0.04%	5.10	2.72	2.31	0.837	0.762	0.612	0.846
Low density	Control	4.68	2.65	2.03	0.837	0.675	0.525	0.786
	0.02%	4.86	2.65	2.08	0.950	0.737	0.400	0.763
	0.04%	5.02	2.91	2.22	1.087	0.750	0.387	0.802
SEM		0.137	0.153	0.113	0.072	0.044	0.067	0.075
p value								
Diet		0.191	0.792	0.1845	0.4459	0.7356	0.0066	0.3243
Yeast		0.0312	0.5149	0.3136	0.0612	0.0323	0.0908	0.8728
Interaction		0.8182	0.6448	0.9241	0.4702	0.5448	0.6516	0.9641

[a,b] Means within a column with different superscripts are significantly different based on Student–Newman–Keuls (SNK) post hoc tests. GAR = Globulin to albumin ratio. Number of observations was 6 chicks per interaction cell.

Data for the biochemical components of blood serum show no significant impact of diet type on serum biochemical constituents (total protein, albumin, α-, β-globulin, globulin and globulin/albumin ratio). However, γ-globulin, AST and ALT were significantly lower in birds fed a low-density diet than a regular-density diet (Tables 3 and 4). Supplementation of the diet with SC at 0.02 g/kg and 0.04% significantly lowered serum AST, ALT and MAD relative to the standard. Additionally, total serum protein and β- globulin were significantly greater in the groups that received SC supplementation compared with control groups without SC. There was no interaction effect between SC level and dietary treatment on blood biochemical constituents (ALT/AST and alkaline phosphatase, Table 3; and total protein, albumin, α, β and δ- globulin, G/A ratio, Table 4). However, there was a significant impact of the interaction on serum AST, ALT and MAD. Supplementation with SC at both concentrations significantly decreased serum AST and ALT of broilers on the low-density diet compared to the regular-density diet. Additionally, 0.02% of SC decreased MAD in broilers fed a low-density diet compared to a regular density diet.

3.3. Hematology of Blood

Table 5 shows the effects of different concentrations of yeast on white blood cells and its subpopulations of broilers fed regular- and low-density diets from 1 to 38 days of age. The results showed that lymphocytes of broilers fed a low-density diet were significantly higher than that of broilers fed a regular-density diet, but eosinophils were lower. The addition of SC to the diet greatly enhanced PCV, lymphocytes and monocytes; and 0.02% SC significantly increased Hgb and RBCs, but decreased heterophils and the H/L ratio, relative to control diet (Table 5). However, the addition of SC to the diet at 0.04% significantly increased heterophils.

Table 5. Impact of different concentrations of *Saccharomyces cerevisiae* on white blood cells (WBC) and its subpopulations in broilers fed regular- and low-density diets from 1 to 38 days of age.

Treatment		WBCs, 10^6/mm^3	Lymphocytes, %	Monocytes, %	Basophils, %	Eosinophils, %	Heterophils, %	H/L Ratio
Effect of diet								
Regular density		22.8	41.1 ^{b}	10.5	0.583	10.0 ^{a}	23.4	0.572
Low density		22.7	42.4 ^{a}	10.6	0.583	9.29 ^{b}	23.7	0.564
Effect of yeast concentration								
Control		22.8	40.0 ^{b}	9.68 ^{b}	0.562	9.87	23.1 ^{b}	0.583 ^{a}
0.02%		23.1	42.9 ^{a}	11.3 ^{a}	0.625	9.56	23.4 ^{b}	0.550 ^{b}
0.04%		22.4	42.4 ^{a}	10.8 ^{a}	0.562	9.56	24.1 ^{a}	0.571 ^{a,b}
Interaction between diet and yeast concentration								
Regular density	Control	21.7 ^{b}	39.7	9.25	0.625	10.6	23.0	0.585 ^{a}
	0.02%	23.2 ^{a}	41.7	11.5	0.750	9.87	23.7	0.571 ^{a}
	0.04%	23.5 ^{a}	42.0	11.0	0.375	9.62	23.5	0.560 ^{a}
Low density	Control	23.8 ^{a}	40.3	10.1	0.500	9.12	23.2	0.581 ^{a}
	0.02%	23.0 ^{a}	44.1	11.1	0.500	9.25	23.1	0.528 ^{b}
	0.04%	21.3 ^{b}	42.8	10.6	0.750	9.50	24.8	0.583 ^{a}
SEM		0.450	0.543	0.379	0.189	0.300	0.362	0.130
p value								
Diet		0.8219	0.0059	0.8939	1.0000	0.0041	0.2689	0.4844
Yeast		0.3213	0.0001	0.0004	0.9304	0.4936	0.0174	0.0453
Interaction		0.0001	0.2324	0.1783	0.2321	0.0815	0.3219	0.0485

^{a,b} Means within a column with different superscripts are significantly different based on Student–Newman–Keuls (SNK) post hoc tests. H/L = Heterophil to lymphocyte ratio. Number of observations was 6 per interaction cell.

There were significant effects of interaction between dietary treatments and SC concentrations on PCV, Hgb, MCV, MCH, white blood cells (WBCs) and the H/L ratio. Results showed that SC supplementation increased PCV and Hgb for birds fed a regular-density diet, but 0.4 g/kg SC significantly decreased Hgb in birds fed a low-density diet. In addition, MCV and MCH were increased dramatically by the addition of 0.04% SC to the regular-density diet, but they were reduced in birds fed the low-density diet. The results showed that using 0.02% and 0.04% of SC increased WBCs in birds fed a regular-density diet, but 0.04% of SC significantly decreased WBCs in those fed a low-density diet. On the other hand, 0.02% of SC significantly decreased H/L in birds fed the low-density diet only (Table 5).

3.4. Lymph Organs and Immune Response

Table 6 shows the effects of the experimental treatments on the lymphoid organs of broilers. The results showed no significant effect of diet density on lymph structures like the spleen, absolute weight of thymus and bursa of Fabricius. However, the percentage of thymus was significantly greater in birds fed a low-density diet than that in birds fed a regular-density diet. Immune responses to NDV and AI, as measured by HI, were not influenced by diet type. These organs, as well as the immune response to NDV and AI, were significantly greater in broilers fed a diet supplemented with 0.02% or 0.04% of SC than those fed a diet without SC supplementation. Moreover, the effects on the thymus, bursa of Fabricius and NDV and AI showed stepwise increases. There were significant effects of interactions between diet type and SC on the percentage of spleen weight, bursa weight and immune response to NDV. The results indicated that the absolute weights of the spleen and bursa of Fabricius significantly decreased in the group fed the regular-density diet supplemented with 0.04% of SC. Still, both levels of SC significantly increased immune response to NDV. On the other hand, both concentrations of SC significantly increased the absolute weights of the spleen and bursa of Fabricius in birds fed the low-density diet, but the response to NDV was stepwise (Table 6).

Table 6. Impact of different concentrations of *Saccharomyces cerevisiae* on immune organs and HI titer (log2) in response to avian influenza and Newcastle disease virus in 38-day-old broilers fed either a regular- or low-density diet from days 1 to 38 of age.

Treatment		Spleen, %	Thymus, %	Bursa of Fabricius, %	HI², Log2	
					NDV	AI
			Effect of diet			
Regular density		0.135	0.466 ^{b}	0.201	3.36	2.83
Low density		0.148	0.516 ^{a}	0.223	3.43	2.04
			Effect of yeast concentration			
Control		0.110 ^{b}	0.425 ^{c}	0.167 ^{c}	1.12 ^{c}	0.125 ^{c}
0.02%		0.153 ^{a}	0.490 ^{b}	0.213 ^{b}	3.54 ^{b}	2.69 ^{b}
0.04%		0.161 ^{a}	0.558 ^{a}	0.256 ^{a}	5.58 ^{a}	4.50 ^{a}
			Interaction between diet and yeast concentration			
Regular density	Control	0.112 ^{c}	0.366 ^{c}	0.162 ^{b}	1.17 ^{d}	0.250
	0.02%	0.136 ^{b}	0.486 ^{b}	0.189 ^{b}	3.92 ^{c}	3.37
	0.04%	0.155 ^{a,b}	0.545 ^{a,b}	0.252 ^{a}	5.00 ^{b}	4.87
Low density	Control	0.108 ^{c}	0.484 ^{b}	0.171 ^{b}	1.08 ^{d}	0.00
	0.02%	0.169 ^{a}	0.495 ^{b}	0.238 ^{a}	3.17 ^{c}	2.00
	0.04%	0.167 ^{a}	0.570 ^{a}	0.260 ^{a}	6.17 ^{a}	4.12
SEM		0.326	0.195	0.014	0.372	0.605
			p value			
Diet		0.0669	0.0214	0.0606	0.7161	0.1174
Yeast		0.0001	0.0001	0.0001	0.0001	0.0001
Interaction		0.1060	0.0828	0.2739	0.0393	0.6515

a,b,c,d Means within a column with different superscripts are significantly different based on Student–Newman–Keuls (SNK) post hoc tests. HI = Hemagglutination inhibition test; NDV = Newcastle disease virus; I = Influenza antigen. Number of observations was 6 chicks per interaction cell.

3.5. Immune Indices

Tables 7 and 8 show the impacts of the different experimental treatments on the immune parameters and survival rate of broilers, respectively. Diet type significantly influenced LYS, TAC and PA, revealing an increasing effect of a low-density diet on immune parameters, but lower TAC (Table 7).

Supplementation of the diet with SC significantly decreased TAC, but enhanced PI relative to the standard diet and had no impact on other traits related to the immunity, including serum LTT, BACT, LYS and PA (Table 7). In addition, the survival rate of broiler chickens fed a diet fortified with 0.04% of SC was significantly greater than that of chickens fed a dietary supplement of 0.02% SC, or a diet without SC (Table 8).

The interaction between SC concentration and diet type did not influence the LTT, BACT, LYS, TAC, PI, PA, or survival rate during the challenge experiment (Tables 7 and 8).

Table 7. Impact of different concentrations of *Saccharomyces cerevisiae* on immune indices of broilers fed a regular- or low-density diet from 1 to 38 days of age.

Treatment	LTT,%	BACT,%	LYS,%	TAC, mMol/dL	PI	PA,%
		Effect of diet				
Regular density	24.2	42.1	0.082 ^{b}	429 ^{a}	1.58	17.4 ^{b}
Low density	25.8	43.2	0.098 ^{a}	419 ^{b}	1.53	18.8 ^{a}
SEM	0.599	0.527	0.004	2.19	0.025	0.375
		Effect of yeast concentration				
Control	24.8	43.9	0.081	436 ^{a}	1.46 ^{b}	17.5
0.02%	25.1	41.8	0.095	420 ^{b}	1.61 ^{a}	18.3
0.04%	25.0	42.3	0.095	416 ^{b}	1.60 ^{a}	18.6
SEM	0.733	0.645	0.005	2.68	0.031	0.460

Table 7. *Cont.*

Treatment		LTT,%	BACT,%	LYS,%	TAC, mMol/dL	PI	PA,%
Interaction between diet and yeast concentration							
Regular density	Control	23.6	43.2	0.071	443	1.45	16.0
	0.02%	24.3	41.2	0.082	425	1.61	17.8
	0.04%	24.7	42.0	0.092	419	1.68	18.5
Low density	Control	26.1	44.6	0.091	430	1.47	19.0
	0.02%	26.0	42.5	0.107	414	1.61	18.8
	0.04%	25.3	42.7	0.097	413	1.52	18.7
SEM		1.03	0.913	0.007	3.39	0.044	0.650
p-value							
Diet		0.0693	0.1394	0.0128	0.0027	0.2240	0.0111
Yeast		0.9552	0.0746	0.1410	0.0001	0.0026	0.2054
Interaction		0.6676	0.9366	0.4206	0.6813	0.0942	0.1052

[a,b] Means within a column with different superscripts are significantly different based on Student–Newman–Keuls (SNK) post hoc tests. LTT = Lymphocyte transformation test; BACT = Bactericidal activity; LYS = Lysozyme activity; TAC = Total antioxidant capacity; PI = Phagocytic index; PA = Phagocytic activity. Number of observations was 6 per interaction cell.

Table 8. Impact of different concentrations of *Saccharomyces cerevisiae* on the survival rate of broiler chickens between 38 and 48 days of age that were fed a regular- or low-density diet and infected with HPAIV H5N1 at 38 days of age.

Treatment		Survival Rate after Infection of Influenza at 38 of Age,%
Effect of diet		
Regular density		64.8
Low density		59.2
Effect of yeast concentration		
Control		56.0 [b]
0.02%		56.0 [b]
0.04%		71.0 [a]
Effect of age		
39 d (1) challenge		100 [a]
40 d (2) challenge		100 [a]
41 d (3) challenge		70.0 [b]
42 d (4) challenge		56.6 [b,c]
43 d (5) challenge		50.0 [b,c]
44 d (6) challenge		50.0 [b,c]
45 d (7) challenge		50.0 [b,c]
46 d (8) challenge		45.0 [b,c]
47 d (9) challenge		40.0 [c]
48 d (10) challenge		30.0 [c]
Interaction between diet and yeast concentration		
Regular density	Control	56.0
	0.02%	64.0
	0.04%	70.0
Low density	Control	56.0
	0.02%	48.0
	0.04%	72.0
SEM		6.57
p value		
Diet		0.2947
Yeast		0.0001
Interaction		0.1380

[a,b,c] Means within a column with different superscripts are significantly different based on Student–Newman–Keuls (SNK) test. Number of observations was 6 broilers of 38-day-old per interaction cell.

3.6. Histology Study

Table 9 shows the impact of diet type and SC concentration on the morphology of the intestines and bursa of Fabricius and spleen, as shown in Figures 1–3. Diet type did not influence the length of intestinal villi or diameter of the large follicle of the bursa of Fabricius (Table 9). No changes were seen in the spleen and thymus due to diet type, SC concentration, or the interaction between these terms (Table 9; Figures 2 and 3).

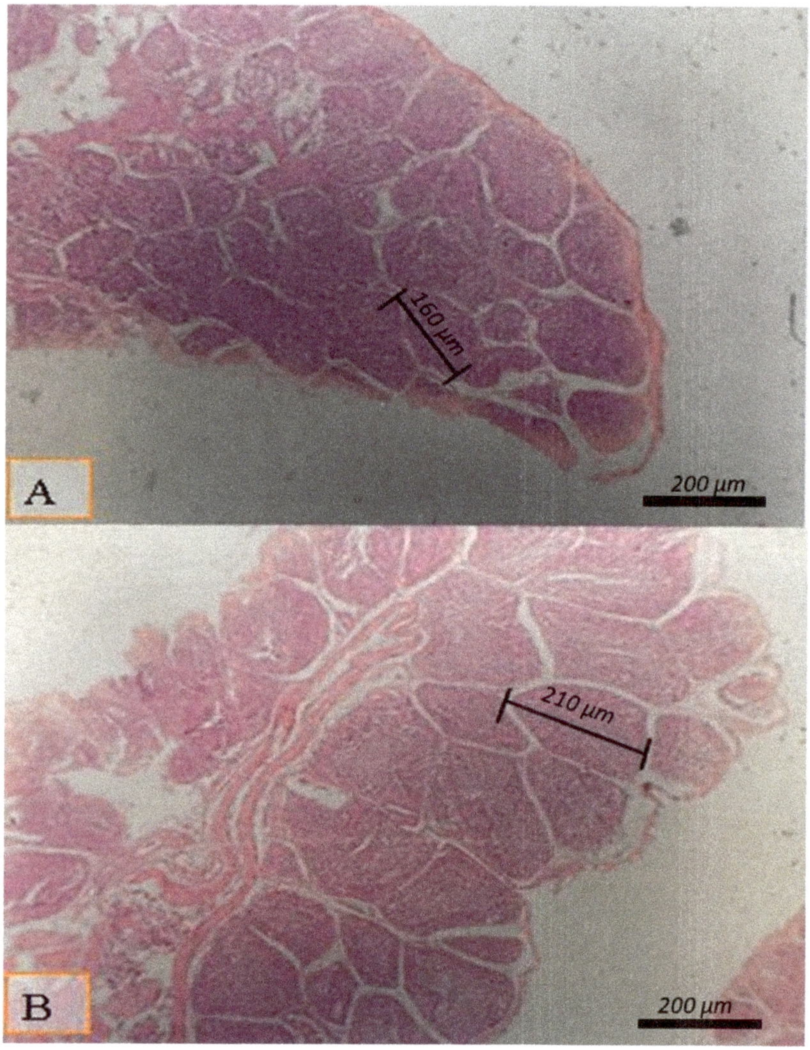

Figure 1. Micrograph of bursa of Fabricious of broiler at day 38 of age stained with HandE (X40) to investigate the follicle diameter in different groups; the distance between two follicular polar as presented all groups by lines: (**A**) Broilers supplemented with 0.02% *Saccharomyces cerevisiae;*(**B**) broilers supplemented with 0.04% *Saccharomyces cerevisiae*. Moderate enhancement in the follicular diameter was detected in broilers supplemented with 0.04% *Saccharomyces cerevisiae* (**B**).

Table 9. Impact of different concentrations of *Saccharomyces cerevisiae* on the morphology of the intestine, bursa of Fabricius and follicular cortical:medullary ratio of the thymus in 38-day-old broilers fed regular- or low-density diet.

Treatment		Length of Intestinal Villi, μm	L. Axis of Large Follicle of Bursa of Fabricius, μm	Follicular Cortical: Medullary Ratio of the Thymus
		Effect of diet		
Regular density		239	166	1:3
Low density		248	177	1:3
		Effect of yeast concentration		
Control		215 [b]	157 [b]	1:3
0.02%		227 [b]	164 [b]	1.3
0.04%		289 [a]	193 [a]	1:3
		Interaction between diet and yeast concentration		
Regular density	Control	211	157	1:3
	0.02%	226	159	1:3
	0.04%	279	181	1:3
Low density	Control	219	157	1:3
	0.02%	227	170	1:3
	0.04%	298	206	1:3
SEM		9.61	8.07	ND
		p value		
Diet		0.2389	0.0851	ND
Yeast		0.0001	0.0002	ND
Interaction		0.6707	0.2832	ND

[a,b] Means within a column with different superscripts are significantly different based on Student–Newman–Keuls (SNK) post hoc tests. Number of observations was 6 per interaction cell per age. ND = Not done.

Supplementation of the diet with 0.04% SC significantly enhanced the length of intestinal villi (Table 9) and the diameter of the large bursal follicle (Table 9 and Figure 1). The intestinal villi and bursa of Fabricious were enhanced by 29.8% and 22.9%, respectively. Both traits were significantly enhanced with the supplementation of SC at a concentration of 0.02%. The large bursal follicle's intestinal villi length and diameter were not affected by the interaction between diet type and SC concentration.

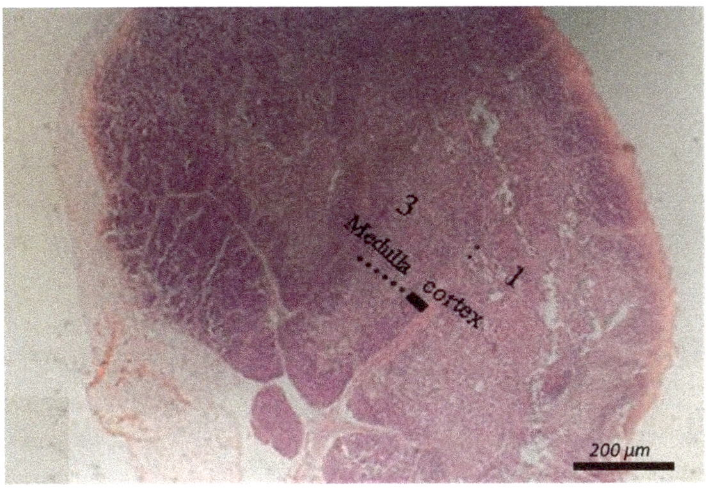

Figure 2. Micrograph of the thymus stained with HandE (X40) to explore the thymic cortical: medullary ratio.

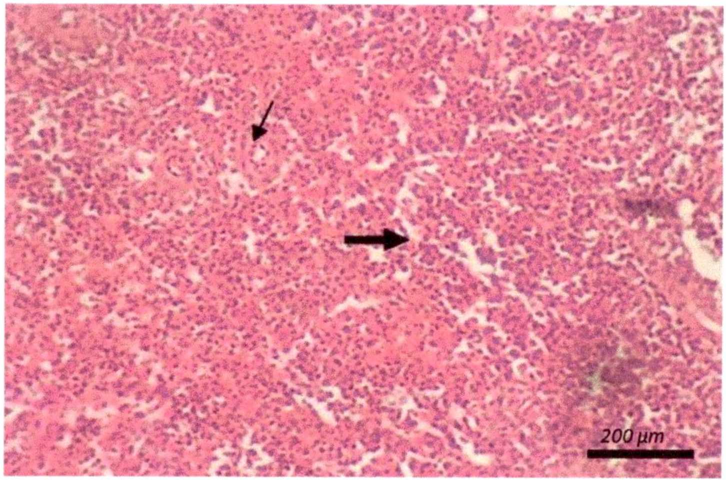

Figure 3. Micrograph of the spleen stained with HandE (X40) of regular density diet presented normal splenic histology featuring splenic arteriole (thin arrow) with white and red pulp (thick arrow). All groups presented the normal splenic histology as the control.

4. Discussion

The present work was conducted to fill the gap in knowledge regarding the impact of active SC yeast in relation to dietary composition on the antioxidant status, blood constituents and immune response of broiler chickens. Adding an SC product to the feed at concentrations of 0.02% or 0.04% improved the performance of broilers from 1 to 38 days of age.

To our knowledge, this is the first study investigating the interactive relationship between using yeast in low-density diets and its effect on immune response in broiler chickens. The percentage of thymus was significantly greater in birds fed a low-density diet than that of birds fed a regular-density diet. The immune response to NDV and AI were significantly greater in broilers fed a diet supplemented with 0.02% or 0.04% of SC than those fed a diet without SC supplementation. In addition, there were significant interactions between diet type and SC on the percentage of spleen weight, bursa weight and immune response to NDV. The results indicated that the absolute weights of the spleen and bursa of Fabricious were significantly high in the group fed the low-density diet supplemented with 0.04% of SC. Still, both levels of SC significantly increased immune response to NDV. Interestingly, both concentrations of SC significantly increased the absolute weights of the spleen and bursa of Fabricious in birds fed the low-density diet.

The increase in feed cost accompanied by the reduction in the availability of corn as a main feed ingredient will affect the production efficiency of poultry on the global level, especially during global pandemics such as the current coronavirus crisis. Nutritional manipulation by using the low-density diet supplemented with yeast could provide a great opportunity to improve the economic outcome by reducing the feed cost, which constitutes approximately 60–70% of the total poultry operation cost. Using low density-diet in broiler rations will provide a positive alternative to reduce feed cost. The other side of the coin is that using yeast in these diets compensated for the low-density contents of the diets by improving the antioxidant status and immune response of the broiler chickens. Any improvement in nutrition management and feed cost will have a direct impact on profitability and efficiency of poultry industry [49,50].

In addition, the group treated with 0.04% of SC had a significantly higher immune response than the other groups. These results contribute to the poultry industry important information that will improve production efficiency.

In addition, 0.02% of SC decreased MAD of broilers fed a low-density diet compared with a regular-density diet. These results indicate that SC enhances the oxidative status of broilers. Interestingly, Czech et al. [11] revealed that using 3% of *Yarrowia lipolytica* or SC yeast, in combination with *Bacillus* sp. probiotic, in the feed of turkeys from 7 to 112 days of age improved the antioxidant status of birds by preventing lipid peroxidation. This effect enhances the ability of poultry to handle stress and infectious agents. In another study [10], the authors investigated the mechanism by which SC enhances the oxidative status of broiler chickens. The authors included SC in either the feed or drinking water of stressed broilers and measured CYP1A2 and melanocortin-2 receptor (MC2R) gene expression in the adrenal glands and IL10 and AvBD1 in the spleen. The authors concluded that using SC in broilers' feed or drinking water for 40 days decreased stress and MC2R gene expression. They also showed that supplementation of SC fermentate in the feed was marginally more effective than adding it to drinking water in stimulating oxidative status and reducing stress in broiler chickens [10] and in detoxifying nitrate (21, 22) and aflatoxin (22, 23).

There has been an interest in using a low-density diet to feed broilers, to lower the growing pressure on the skeletal system of the bird and decrease skeletal diseases, the cost of feed and environmental pollution [51,52].

The body weight and EPEI were significantly decreased by 7.8% and 5.4%, respectively, over the study period for birds on the low-density diet compared to the regular-density diet. Indeed, the main effect of regular diet under the three SC levels was 2099 g while the mean body weight of the control diet without SC supplementations (1946 + 1784) was 1965 g. These findings indicate that the negative impact of diet structure persisted during the experimental period from 1 to 38 days of age [53].

The outcomes showed that the low-density diet improved liver function and increased the percentage of thymus and lymphocytes and PA, but decreased γ-globulin, eosinophils, TAC, ALT and AST. The current study revealed that supplementation of the feed with SC at 0.02 g/kg and 0.04% significantly lowered serum AST, ALT and MAD relative to the standard diet. Gheisari and Kholeghipour [9] showed that using live yeast had no significant impact on hematological indices such as RBCs, WBCs and PCV. On the other hand, in another study, there was a positive association between supplementation of feed with SC and hematological traits of chickens, such as RBCs, WBCs and PCV [51]. In the same study, probiotics had no effect ($p > 0.05$) on hemoglobin and WBCs at the finisher phase. Yet, a significant effect ($p < 0.05$) was observed for RBCs and packed cell volume.

Gut morphology was modulated because of the addition of 10% wheat bran in the low-density diet. The results indicated that the diets had no impact on the length of intestinal villi. Previous studies have shown that dietary cereal with a high nonstarch polysaccharides (NSP) level could enhance the dimension of the gastrointestinal tract [54]. Steenfeldt [55] observed that the arabinoxylan level in wheat is significantly and positively correlated to the relative masses of the duodenum, jejunum and ileum. It has been stated that dietary supplementation may modulate the morphology of the intestinal mucosa. Accordingly, NSP in the diet can also impact the morphology of the gastrointestinal tract [56]. Iji [57] showed that the crypt deepness of the jejunum and ileum was significantly enhanced by dietary addition of guar gum and xanthin gum. This finding demonstrated that NSP can improve cell turnover in the gastrointestinal tract. Enhanced crypt deepness indicates enhanced villus cell proliferation and in turn improved utilization of the nutrients by the gastrointestinal tract. This suggests that these cereals impact the mass of the gastrointestinaltract and morphology of the intestine [58].

These results suggest that the dilution of nutrients in feed via adding 10% wheat bran improves the immune response and production index of broilers 1 to 38 days of age. These outcomes agree with earlier studies by Abudabos [53] and Attia [54]. Wheat bran polysaccharides have been shown to act as antioxidants and immunostimulators and have anti-inflammatory, antitussive, anticancerous and antimutagenic properties [59–62]. Furthermore, wheat bran arabinoxylans have been shown to to enhance macrophage phagocytosis in animals [63]. They are immunostimulants of the antibody response in chickens by enhancing the total IgG and IgM anti-SRBC antibody titers on 7th and 14th day post primary antibody response (PPI) and post-secondary inoculation (PSI) of sheep red blood cells (SRBCs) compared to the control. Additionally, Korte et al. [64] stated that supplementing feed with arabinoxylans significantly induced anti-SRBC antibody titers, representing enhanced humoral immunity in chickens.

The results indicated that supplementing feed with SC at either 0.02% or 0.04% significantly affected growth and EPEI relative to the control dietary treatment. The effect persisted throughout all tested periods. Furthermore, supplementing the low-density diet with 0.02% or 0.04% of SC resulted in an enhanced production index compared with the regular-density diet lacking the addition.

These results agree with other studies that have investigated similar effects of SC on growth performance [65–68]. In addition, the positive effect of a higher dose of SC are in line with results reported by Valdivie [69], who found that the growth performance of broilers significantly improved with an increased supplementation dose of SC. The improved gastrointestinal health and growth performance of broilers supplemented with SC may be due to the presence of effective ingredients in SC such as Vitamin B, cellulostic enzymes, phytase, monooligosaccharides (MOS) and glucomannan [70].

A diet supplemented with 0.04% SC increased the length of the villi and SC supplementation enhanced the production index and the body weight of broiler from 1 to 38 days of age in a dose-dependent manner.

There were significant improvements in blood serum biochemistry and liver function due to SC supplementation. Consistent with these results, Paryad and Mahmoudi [71] and Hosseini [72] showed that SC at 1.5% significantly enhanced total plasma protein, albumin and globulin and WBCs and decreased the H/L ratio. Furthermore, Zhang et al. [64] revealed that SC supplementation to broiler chickens significantly lowered the 2-thiobarbituric acid-reactive substances (TBARS) in the breast and drumstick meats and increased villus height, compared to the control group.

Supplementation of SC significantly enhanced the spleen, thymus, bursa of Fabricious and HI in response to NDV and AI, with a positive concentration-dependent impact of SC on the thymus, bursa of Fabricious and immune response to NDV and AI. The diameter of the bursal follicle significantly enhanced at 0.04% SC, indicating an improvement in the number of B-lymphoblasts, leading to an increase in the B-lymphocytes responsible of humoral immunity stimulation through antibody production. Further evidence of this effect was reflected by the increased survival rate of broilers challenged with AI at 38 d of age. In addition, SC supplementation was associated with improvements in β-globulin and hematological traits such as PCV, Hgb, RBCs, lymphocytes, monocytes and PI. These data provide more evidence for an improved health status of broilers fed a diet supplemented with SC. The effect of SC on the relative weights of the thymus and bursa and immune response to NDV and AI, was dose-dependent. Similarly, Newman [73], Spring et al. [74] and Zhang et al. [70] showed that SC supplementation of the diet enhanced production performance by improving the immune status, intestinal lumen health and digestion and nutrient utilization of birds. In addition, Gheisari and Kholeghipour [9] found that broiler chickens fed SC at a concentration of 0.02% had higher antibody titers against NDV than the control at 38 d of age, but it did not affect AI titers. The positive effect of SC on immune response could be attributed to its cell wall constituents, including chitin, mannan and glucan, which have immunostimulant effects [2,71,75–78].

5. Conclusions

The antioxidant status and total antioxidant capacity of broiler chickens were improved by supplementation of the diet with SC. Supplementation of either a regular-density diet or a low-density diet with SC at either 0.02% or 0.04% enhanced the BWG and EPEI of broilers during 1 to 38 days of age. Additionally, broilers fed a low-density diet supplemented with either 0.02% or 0.04% SC had a greater body weight and EPEI than birds fed the control diet with no supplementary SC. However, fortification of the diet with 0.04% resulted in significantly enhanced immune organs and a higher immune response.

Author Contributions: Y.A.A., H.S.A.E.-H. and A.A.E.-S. conducted the experimental design and first draft of the manuscript; H.S.A.E.-H. and A.A.E.-S. carried out the experiment, data collection and contribution to the experimental set-up and laboratory work; H.A.-K. and S.R.A. carried out the statistical analyses and revision; Y.A.A., M.A.A.-H. and H.S.A.E.-H. proofread the manuscript. All authors have read and agreed to the published version of the manuscript.

Funding: The authors would like to thank the Kuwait Institute for Scientific Research, Kuwait, for providing the publications fees (APC).

Institutional Review Board Statement: The Animal Care and Use Committee (ACUC) at King Fahd Medical Research Center has approved the referenced protocol used in the current study, numbered ACUC-22-1-2, dated 1-2-2022.

Informed Consent Statement: Not applicable.

Data Availability Statement: Data were presented in the manuscript.

Acknowledgments: The Deanship of Scientific Research (DSR) at King Abdulaziz University, Jeddah, Saudi Arabia, has funded this project, under grant no. (FP-217-43H). The authors acknowledged the administrative, technical and financial support by DSR, King Abdulaziz University, Jeddah, Saudi Arabia.

Conflicts of Interest: The authors declare no conflict of interest.

References

1. Wang, W.; Li, Z.; Han, Q.; Guo, Y.; Zhang, B.; D'Inca, R. Dietary live yeast and mannan-oligosaccharide supplementation attenuate intestinal inflammation and barrier dysfunction induced by Escherichia coli in broilers. *Br. J. Nutr.* **2016**, *116*, 1878–1888. [CrossRef] [PubMed]
2. Attia, Y.A.; Al-Khalaifah, H.; Ibrahim, M.S.; Al-Hamid, A.E.A.; Al-Harthi, M.A.; El-Naggar, A. Blood Hematological and Biochemical Constituents, Antioxidant Enzymes, Immunity and Lymphoid Organs of Broiler Chicks Supplemented with Propolis, Bee Pollen and Mannan Oligosaccharides Continuously or Intermittently. *Poult. Sci.* **2017**, *96*, 4182–4192. [CrossRef] [PubMed]
3. Al-Khalifa, H.; Givens, D.I.; Rymer, C.; Yaqoob, P.; Juniper, D. The effect of increasing levels of Fish Oil on Immune Responses of Broiler Chickens. Advances in Animal Bio-Sciences Journal. *Proc. Br. Soc. Anim. Sci. Annu. Conf.* **2009**, *2009*, 235. [CrossRef]
4. Al-Khalifa, H.; Givens, D.I.; Rymer, C.; Yaqoob, P. Effect of n-3 fatty acids on immune function in broiler chickens. *Poult. Sci.* **2012**, *91*, 74–88. [CrossRef]
5. Al-Khalifa, H.; Ragheb, G.; Al-Nasser, A.; Al-Bahouh, M.; Khalil, F. Immunomodulation of Black Seed in Two Strains of Laying Hens. *Int. J. Poult. Sci.* **2013**, *12*, 451–455. [CrossRef]
6. Attia, Y.; Al-khalaifah, H.S.; Abd El-Hamid, H.; Al-Harthi, M.; El-Shafey, A. Multi-enzymes and immune response Effect of different levels of multi-enzymes on immune response, blood hematology and biochemistry, antioxidants status and organs histology of broiler chicks fed standard and low-density diets. *Front. Vet. Sci.* **2019**, *6*, 510. [CrossRef]
7. Flees, J.; Greene, E.; Ganguly, B.; Dridi, S.J.N. Phytogenic feed- and water-additives improve feed efficiency in broilers via modulation of (an)orexigenic hypothalamic neuropeptide expression. *Neuropeptides* **2020**, *81*, 102005. [CrossRef]
8. Suresh, G.; Santos, D.U.; Rouissi, T.; Hegde, K.; Brar, S.K.; Mehdi, Y.; Godbout, S.; Chorfi, Y.; Ramirez, A.A.J.A.F.S. Technology, In-field poultry tests to evaluate efficacy of bioformulation consisting of enzymes and yeast biomass. *Anim. Feed. Sci. Technol.* **2020**, *262*, 114398. [CrossRef]
9. Gheisari, A.A.; Kholeghipour, B. *Effect of Dietary Inclusion of Live Yeast (Saccharomyces cerevisiae) on Growth Performance, Immune Responses and Blood Parameters of Broiler Chickens*; World's Poultry Science Association (WPSA): Beekbergen, The Netherlands, 2006; p. 10.
10. Nelson, J.R.; Sobotik, E.B.; Athrey, G.; Archer, G.S.J.P.S. Effects of supplementing yeast fermentate in the feed or drinking water on stress susceptibility, plasma chemistry, cytokine levels, antioxidant status, and stress- and immune-related gene expression of broiler chickens. *Poult. Sci.* **2020**, *99*, 3312–3318. [CrossRef]
11. Czech, A.; Merska-Kazanowska, M.; Całyniuk, Z. Redox Status, Biochemical Parameters and Mineral Elements Content in Blood of Turkey Hens Fed a Diet Supplemented with Yarrowia lipolytica Yeast and Two Bacillus Species. *Animals* **2020**, *10*, 459. [CrossRef]
12. Leung, H.; Patterson, R.; Barta, J.; Karrow, N.; Kiarie, E. Nucleotide-rich yeast extract fed to broiler chickens challenged with Eimeria: Impact on growth performance, jejunal histomorphology, immune system, and apparent retention of dietary components and caloric efficiency. *Poult. Sci.* **2019**, *98*, 4375–4383. [CrossRef] [PubMed]

13. Kiarie, E.G.; Leung, H.; Kakhki, R.A.M.; Patterson, R.; Barta, J.R. Utility of Feed Enzymes and Yeast Derivatives in Ameliorating Deleterious Effects of Coccidiosis on Intestinal Health and Function in Broiler Chickens. *Front. Veter. Sci.* **2019**, *6*, 473. [CrossRef] [PubMed]
14. Zhou, C.; Liang, J.; Jiang, W.; He, X.; Liu, S.; Wei, P. The effect of a selected yeast fraction on the prevention of pullorum disease and fowl typhoid in commercial breeder chickens. *Poult. Sci.* **2020**, *99*, 101–110. [CrossRef] [PubMed]
15. Mehdi, A.; Hasan, G. Immune response of broiler chicks fed yeast derived mannan oligosaccharides and humate against Newcastle disease. *WASJ* **2012**, *18*, 779–785.
16. Al-Khalifa, H. Production of added-value poultry meat: Enrichment with n-3 polyunsaturated fatty acids. *World's Poult. Sci. J.* **2015**, *71*, 319–326. [CrossRef]
17. Al-Khalifa, H.; Al-Nasser, A.; Al-Bahouh, M.; Ragheb, G.; Al-Qalaf, S.; Al-Omani, N.; Ahmad, A. The effect of polyunsaturated fatty acids on avian immune cell subpopulations in peripheral blood, spleen, and thymus. *World's Poult. Sci. J.* **2016**, *72*, 531–534. [CrossRef]
18. Al-Khalaifah, H.; Al-Nasser, A.; Givens, D.; Rymer, C.; Yaqoob, P. Comparison of different dietary sources of n-3 polyunsaturated fatty acids on immune response in broiler chickens. *Heliyon* **2020**, *6*, e03326. [CrossRef]
19. Al-Nasser, A.; Al-Khalaifah, H.; Al-Mansour, H.; Ahmad, A.; Ragheb, G. Evaluating farm size and technology use in poultry production in Kuwait. *World's Poult. Sci. J.* **2020**, *76*, 365–380. [CrossRef]
20. Attia, Y.A.; Abd El Hamid, A.E.; Ismaiel, A.M.; El-Naggar, A.S. The detoxication of nitrate by two antioxidants or a probiotic and the effects on blood and seminal plasma profiles and reproductive function of NZW rabbit bucks. *Animal* **2013**, *7*, 591–601. [CrossRef]
21. Attia, Y.A.; Abd El Hamid, A.E.; Ismaiel, A.M.; de Oliveira Maria, C.; Al-Harthi, M.A.; El-Naggar, A.S.; Simon, G.A. Nitrate detoxification using antioxidants and probiotics in the water of rabbits. *Rev. Colomb. Cienc. Pecu. (RCCP)* **2018**, *31*, 130–138. [CrossRef]
22. Attia, Y.; Al-Hamid, A.A.; Allakany, H.; Al-Harthi, M.; Mohamed, N. Necessity of continuing of supplementation of non-nutritive feed additive during days 21–42 of age following 3 weeks of feeding aflatoxin to broiler chickens. *J. Appl. Anim. Res.* **2015**, *44*, 87–98. [CrossRef]
23. Attia, Y.A.; Allakany, H.F.; Abd Al-Hamid, A.E.; Al-Saffar, A.A.; Hassan, R.A.; Mohamed, N.A. Capability of different non-nutritive feed additives on improving productive and physiological traits of broiler chicks fed diets with or without aflatoxin during the first 3 weeks of life. *J. Anim. Physiol. Anim. Nutr.* **2013**, *97*, 754–772. [CrossRef] [PubMed]
24. Hayat, J.; Savage, T.F.; Mirosh, L.W. The reproductive performance of two genetically distinct lines of medium white turkey hens when fed breeder diets with and without a yeast culture containing Saccharomyces cerevisiae. *Anim. Feed. Sci. Technol.* **1993**, *43*, 291–301. [CrossRef]
25. Arbor Acres Broiler Management Handbook. Available online: https://eu.aviagen.com/assets/Tech_Center/AA_Broiler/AA-BroilerHandbook2018-EN.pdf (accessed on 1 February 2022).
26. Erel, O. A novel automated direct measurement method for total antioxidant capacity using a new generation, more stable ABTS radical cation. *Clin. Biochem.* **2004**, *37*, 277–285. [CrossRef]
27. Wyatt, T.A.; Kharbanda, K.K.; Tuma, D.J.; Sisson, J.H. Malondialdehyde–acetaldehyde-adducted bovine serum albumin activates protein kinase C and stimulates interleukin-8 release in bovine bronchial epithelial cells. *Alcohol* **2001**, *25*, 159–166. [CrossRef]
28. Armstrong, W.D.; Carr, C.W. Estimation of serum total protein. In *Physiological Chemistry Laboratory Directions*, 3rd ed.; Burges Publishing Co.: Minneapolis, MN, USA, 1964.
29. Doumas, B.T.; Peters, T. Serum and urine albumin: A progress report on their measurement and clinical significance. *Clin. Chim. Acta* **1997**, *258*, 3–20. [CrossRef]
30. Giangiacomo, J.; Cleary, T.G.; Cole, B.R.; Hoffsten, P.; Robson, A.M. Serum Immunoglobulins in the Nephrotic Syndrome. *N. Engl. J. Med.* **1975**, *293*, 8–12. [CrossRef]
31. Elias, M. Serum cortisol, testosterone, and testosterone-binding globulin responses to competitive fighting in human males. *Aggress. Behav.* **1981**, *7*, 215–224. [CrossRef]
32. Reitman, S.; Frankel, S. A Colorimetric Method for the Determination of Serum Glutamic Oxalacetic and Glutamic Pyruvic Transaminases. *Am. J. Clin. Pathol.* **1957**, *28*, 56–63. [CrossRef]
33. Kim, E.E.; Wyckoff, H.W. Reaction mechanism of alkaline phosphatase based on crystal structures: Two-metal ion catalysis. *J. Mol. Biol.* **1991**, *218*, 449–464. [CrossRef]
34. Eilers, R.J. Notification of Final Adoption of an International Method and Standard Solution for Hemoglobinometry Specifications for Preparation of Standard Solution. *Am. J. Clin. Pathol.* **1967**, *47*, 212–214. [CrossRef] [PubMed]
35. Jain, S.K. Membrane lipid peroxidation in erythrocytes of the newborn. *Clin. Chim. Acta* **1986**, *161*, 301–306. [CrossRef]
36. Kawahara, E.; Ueda, T.; Nomura, S. In Vitro Phagocytic Activity of White-Spotted Char Blood Cells after Injection with Aeromonas salmonicida Extracellular Products. *Fish Pathol.* **1991**, *26*, 213–214. [CrossRef]
37. OIE—World Organisation for Animal Health (OIE). Manual of Diagnostic Tests and Vaccines for Terrestrial Animals. 2021. Available online: https://www.oie.int/en/what-we-do/standards/codes-and-manuals/terrestrial-manual-online-access/ (accessed on 1 February 2022).
38. OIE—World Organisation for Animal Health (OIE). *OIE Terrestrial Manual 2015*; Chapter 2.3.4 Avian Influenza; OIE: Paris, France, 2015.

39. Seal, B.S.; King, D.J.; Sellers, H.S. The avian response to Newcastle disease virus. *Dev. Comp. Immunol.* **2000**, *24*, 257–268. [CrossRef]
40. Cosgrove, A.S. An Apparently New Disease of Chickens: Avian Nephrosis. *Avian Dis.* **1962**, *6*, 385. [CrossRef]
41. Takatsy, G.; Hamar, M. The antibody content of immune sera absorbed with influenza virus, as determined by different serological tests. *Acta Microbiol. Acad. Sci. Hung.* **1956**, *3*, 299–306.
42. Balhaa, R.L.; Hinz, H.H.; Luders, H.; Siegmann, O. linical experiences with the drugs for lymphocyte transformation in chickens and turkey flocks. *Tierarztl. Umsch.* **1985**, *43*, 507–508.
43. Rainger, G.; Rowley, A. Antibacterial activity in the serum and mucus of rainbow trout, Oncorhynchus mykiss, following immunisation with Aeromonas salmonicida. *Fish Shellfish. Immunol.* **1993**, *3*, 475–482. [CrossRef]
44. Engstad, R.E.; Robertsen, B.; Frivold, E. Yeast glucan induces increase in lysozyme and complement-mediated haemolytic activity in Atlantic salmon blood. *Fish Shellfish. Immunol.* **1992**, *2*, 287–297. [CrossRef]
45. Allan, J.; Mitchell, T.; Harborne, N.; Bohm, L.; Crane-Robinson, C. Roles of H1 domains in determining higher order chromatin structure and H1 location. *J. Mol. Biol.* **1986**, *187*, 591–601. [CrossRef]
46. Biswas, H.R.; Haoque, M.M.; Oxley, M.; Prodhan, M.A.M. A comparative study on the protection of indigenous chickens against Newcastle disease induced by Australian NDV4HR and locally produced conventional vaccines in Bangladesh. *Prev. Vet. Med.* **1996**, *26*, 157–164. [CrossRef]
47. Culling, C.F. *Handbook of Histopathological and Histochemical Staining Techniques*, 3rd ed.; Butterworth: London, UK, 1983.
48. SAS Intstitute Inc. *SAS/STAT Software, Version 9*; SAS Institute, Inc.: Cary, NC, USA, 2002.
49. Hewson, K.; Liu, S.; Morgan, N.; Selle, P.; Wu, S.; Townsend, G. Agrifutures Australia Chicken Meat–Nutrition Program of Research. In Proceedings of the 31st Annual Australian Poultry Science Symposium, Sydney, Australia, 16–19 February 2020; p. 58.
50. Al-Khalaifah, H.; Al-Nasser, A.; Abdulmalek, N.; Al-Mansour, H.; Ahmed, A.; Ragheb, G. Impact of SARS-Con-V2 on the poultry industry in Kuwait: A case study. *Front. Vet. Sci.* **2002**, *7*, 656. [CrossRef]
51. Onifade, A.A. Growth performance, carcass characteristics, organs measurement and haematology of broiler chickens fed a high fibre diet supplemented with antibiotics or dried yeast. *Food/Nahrung* **1997**, *41*, 370–374. [CrossRef]
52. Attia, Y.; Hassan, R.A.; El-Din, A.E.T.; Abou-Shehema, B.M. Effect of ascorbic acid or increasing metabolizable energy level with or without supplementation of some essential amino acids on productive and physiological traits of slow-growing chicks exposed to chronic heat stress. *J. Anim. Physiol. Anim. Nutr.* **2010**, *95*, 744–755. [CrossRef] [PubMed]
53. Abudabos, A.M. Effect of Enzyme Supplementation to Normal and Low Density Broiler Diets Based on Corn-soybean Meal. *Asian J. Anim. Veter. Adv.* **2012**, *7*, 139–148. [CrossRef]
54. Attia, Y.A. Value of rice bran, its maximal utilization, and upgrading by phytase and other enzymes and diet-formulation based on available amino acid for broiler chicks. *Arch. Geflugelkd.* **2003**, *67*, 157–166.
55. Steenfeldt, S. The dietary effect of different wheat cultivars for broiler chickens. *Br. Poult. Sci.* **2001**, *42*, 595–609. [CrossRef]
56. Yamauchi, K.-E. Review on Chicken Intestinal Villus Histological Alterations Related with Intestinal Function. *J. Poult. Sci.* **2002**, *39*, 229–242. [CrossRef]
57. Iji, P. The impact of cereal non-starch polysaccharides on intestinal development and function in broiler chickens. *World's Poult. Sci. J.* **1999**, *55*, 375–387. [CrossRef]
58. Thomas, D.V.; Ravindran, V. Effect of Cereal Type on the Performance, Gastrointestinal Tract Development and Intestinal Morphology of the Newly Hatched Broiler Chick. *J. Poult. Sci.* **2008**, *45*, 46–50. [CrossRef]
59. Akhtar, M.; Tariq, A.F.; Awais, M.M.; Iqbal, Z.; Muhammad, F.; Shahid, M.; Hiszczynska-Sawicka, E. Studies on wheat bran Arabinoxylan for its immunostimulatory and protective effects against avian coccidiosis. *Carbohydr. Polym.* **2012**, *90*, 333–339. [CrossRef] [PubMed]
60. Lin, W.C.; Lee, M.T.; Lo, C.T.; Chang, S.C.; Lee, T.-T. Effects of dietary supplementation of Trichoderma pseudokoningii fermented enzyme powder on growth performance, intestinal morphology, microflora and serum antioxidantive status in broiler chickens. *Ital. J. Anim. Sci.* **2018**, *17*, 153–164. [CrossRef]
61. Li, B.; Leblois, J.; Taminiau, B.; Schroyen, M.; Beckers, Y.; Bindelle, J.; Everaert, N. The effect of inulin and wheat bran on intestinal health and microbiota in the early life of broiler chickens. *Poult. Sci.* **2018**, *97*, 3156–3165. [CrossRef]
62. Al-Khalaifa, H.; Al-Nasser, A.; Al-Surayee, T.; Al-Kandari, S.; Al-Enzi, N.; Al-Sharrah, T.; Ragheb, G.; Al-Qalaf, S.; Mohammed, A. Effect of dietary probiotics and prebiotics on the performance of broiler chickens. *Poult. Sci.* **2019**, *98*, 4465–4479. [CrossRef] [PubMed]
63. Cao, Y.; Xu, Y.; Auchoybur, M.L.; Chen, W.; He, S.; Qin, W.; Su, C.; Huang, F.; Qiu, Z.; Li, L.; et al. Regulatory role of IKKα in myocardial ischemia/reperfusion injury by the determination of M1 versus M2 polarization of macrophages. *J. Mol. Cell. Cardiol.* **2018**, *123*, 1–12. [CrossRef]
64. Korte, J.; Fröhlich, T.; Kohn, M.; Kaspers, B.; Arnold, G.J.; Härtle, S. 2D DIGE analysis of the bursa of Fabricius reveals characteristic proteome profiles for different stages of chicken B-cell development. *Proteomics* **2013**, *13*, 119–133. [CrossRef]
65. Nilson, A.P.; Miazzo, R.D. *Use of Brewers Yeast S. cerevisiae to Replace Part of the Vitamin Mineral Premix in Finisher Broiler Diets*; XXII Worlds Poult. Cong.: Istanbul, Turkey, 2004.
66. Shareef, A.M.; Al-Dabbagh, A.S.A. Effect of probiotic (Saccharomyces cerevisiae) on performance of broiler chicks. *Iraqi J. Vet. Sci.* **2009**, *23* (Suppl. S1), 23–29.

67. Koc, F.; Samli, H.; Okur, A.; Ozduven, M.; Akyurek, H.; Senkoylu, N. Effects of Saccharomyces cerevisiae and/or mannano-ligosaccharide on performance, blood parameters and intestinal microbiota of broiler chicks. *Bulg. J. Agric. Sci.* **2010**, *16*, 643–650.
68. Abou El-Naga, M.K. Effect of dietary yeast supplementation on broiler performance. *Egypt. Poult. Sci. J.* **2012**, *32*, 95–106.
69. Valdivie, M. Saccharomyces yeast as a by-product from alcohol production on final molasses in diets for broilers. *Cuban J. Agric. Sci.* **1975**, *9*, 327–331.
70. Zhang, A.; Lee, B.; Lee, S.; Lee, K.; An, G.; Song, K.; Lee, C. Effects of yeast (*Saccharomyces cerevisiae*) cell components on growth performance, meat quality, and ileal mucosa development of broiler chicks. *Poult. Sci.* **2005**, *84*, 1015–1021. [CrossRef] [PubMed]
71. Paryad, A.; Mahmoudi, M. Effect of different levels of supplemental yeast Saccharomyces cerevisiae on performance, blood constituents and carcass characteristics of broiler chicks. *Afr. J. Agric. Res.* **2008**, *3*, 835–842.
72. Hosseini, S. The Effect of Saccharomysec cerevisiae on blood parameters of broiler chicken's. *Glob. Vet.* **2011**, *74*, 411–414.
73. Newman, K. Mannan-oligosaccharides: Natural polymers with significant impact on the gastrointestinal microflora and the immune system. *Biotechnol. Feed. Ind.* **1994**, *10*, 167–180.
74. Spring, P.; Wenk, C.; Dawson, K.A.; Newman, K.E. The effects of dietary mannanoligosaccharides on cecal parameters and the concentrations of enteric bacteria in the ceca of salmonella-challenged broiler chicks. *Poult. Sci.* **2000**, *79*, 205–211. [CrossRef] [PubMed]
75. Saleh El-Din, S.H.S.; Abd El-Hamid, A.E. Effect of different levels of vitamin E and mannanoligosaccharide on productive performance and immune response of broiler chicks. *Poult. Sci.* **2012**, *91*, 30.
76. Attia, Y.; Al-Khalaifah, H.; El-Hamid, H.A.; Al-Harthi, M.; El-Shafey, A. Growth performance, digestibility, intestinal morphology, Carcass traits and meat quality of broilers fed marginal nutrients deficiency-diet supplemented with different levels of active Yeast. *Livest. Sci.* **2020**, *233*, 103945. [CrossRef]
77. Bilal, R.M.; Hassan, F.U.; Saeed, M.; Rafeeq, M.; Zahra, N.; Fraz, A.; Saeed, S.; Khan, M.A.; Mahgoub, H.A.M.; Farag, M.R.; et al. Role of Yeast and Yeast-Derived Products as Feed Additives in Broiler Nutrition. *Anim. Biotechnol.* **2021**, 1–10. [CrossRef]
78. He, T.; Mahfuz, S.; Piao, X.; Wu, D.; Wang, W.; Yan, H.; Ouyang, T.; Liu, Y. Effects of live yeast (Saccharomyces cerevisiae) as a substitute to antibiotic on growth performance, immune function, serum biochemical parameters and intestinal morphology of broilers. *J. Appl. Anim. Res.* **2021**, *49*, 15–22. [CrossRef]

animals

MDPI

Article

Effects of *Eimeria tenella* Infection on Key Parameters for Feed Efficiency in Broiler Chickens

Janghan Choi [1], Hanseo Ko [1], Yuguo Hou Tompkins [1], Po-Yun Teng [1], Jeferson M. Lourenco [2], Todd R. Callaway [2] and Woo Kyun Kim [1,*]

[1] Department of Poultry Science, University of Georgia, Athens, GA 30602, USA; choij@uga.edu (J.C.); hsko@uga.edu (H.K.); yuguot@uga.edu (Y.H.T.); pyteng@uga.edu (P.-Y.T.)

[2] Department of Animal and Dairy Science, University of Georgia, Athens, GA 30602, USA; jefao@uga.edu (J.M.L.); todd.callaway@uga.edu (T.R.C.)

* Correspondence: wkkim@uga.edu; Tel.: +1-706-542-1346

Simple Summary: Coccidiosis, which can be induced by *Eimeria* spp., causes tremendous economic losses in the poultry production. *Eimeria tenella* (*E. tenella*) is one of the poultry *Eimeria* spp. that damage cecal tissue. Broilers infected with *E. tenella* can have reduced body weight, feed efficiency, and gut health because ceca are the main site for producing volatile fatty acids (VFA; important energy sources) and ceca accommodate diverse pathogens. To find appropriate strategies to cope with *E. tenella* infection, modes of actions of *E. tenella* infection on broiler growth and health should be investigated, and experimental infection model should be established. In the study, different levels of sporulated *E. tenella* oocysts were inoculated to the broilers, and the inoculation dosages induced mild infection in the ceca of broilers. The current study showed that *E. tenella* infection damaged feed efficiency and small intestinal health in broilers, mainly by reducing cecal volatile fatty acids (VFA) production. Different inoculation levels modulated the tendency of fecal moisture content and fecal oocyst shedding at different time points. Based on the results, energy supplementation and/or modulation of cecal microbiota potentially ameliorates negative effects of *E. tenella* infection in broilers.

Abstract: The purpose of the study was to investigate effects of different inoculation dosages of *E. tenella* on growth performance, gastrointestinal permeability, oocyst shedding, intestinal morphology, fecal consistency, ileal apparent digestibility, antioxidant capacity, and cecal VFA profile in broiler chickens. Five different dosages (T0: 0, T1: 6250, T2: 12,500, T3: 25,000, and T4: 50,000) of *E. tenella* oocysts were inoculated via oral gavage to fourteen-day-old broilers. Inoculation of *E. tenella* linearly increased FCR ($p < 0.05$), and feed intake was quadratically increased on 6 days post-infection (dpi; $p = 0.08$) and 7 dpi ($p = 0.09$). Cecal lesion score of each treatment was T0: 0; T1: 0.39 ± 0.14; T2: 0.93 ± 0.21; T3: 1.25 ± 0.16; and T4: 1.58 ± 0.2. Cecal total VFA production was linearly reduced due to *E. tenella* infection on 6 dpi ($p < 0.01$). *E. tenella* infection deepened cecal crypts depth on 6 dpi (CD; $p < 0.05$). Gastrointestinal permeability tended to be linearly increased ($p = 0.07$). *E. tenella* infection tended to linearly reduce duodenal VH ($p = 0.1$) and jejunal VH on 9 dpi ($p = 0.09$). Different dosages of *E. tenella* modulated the tendency of fecal moisture content and oocyst shedding. Therefore, *E. tenella* infection impaired feed efficiency and small intestinal health mainly by reducing cecal VFA production and deepening cecal CD in broilers.

Keywords: *Eimeria tenella*; broiler chickens; oocyst shedding; volatile fatty acids; feed efficiency; cecal health

Citation: Choi, J.; Ko, H.; Tompkins, Y.H.; Teng, P.-Y.; Lourenco, J.M.; Callaway, T.R.; Kim, W.K. Effects of *Eimeria tenella* Infection on Key Parameters for Feed Efficiency in Broiler Chickens. *Animals* **2021**, *11*, 3428. https://doi.org/10.3390/ani11123428

Academic Editor: Raffaella Rossi

Received: 8 November 2021
Accepted: 29 November 2021
Published: 1 December 2021

1. Introduction

Coccidiosis causes tremendous economic losses in broiler production by impairing gut health and depressing growth performance and feed efficiency of broiler chickens, and expensive anti-coccidial treatments also increase the overall production cost [1,2].

Avian coccidiosis are induced by *Eimeria* spp., which are protozoan parasites, and there are 7 known *Eimeria* spp. that can infect chickens: *Eimeria acervulina*, *E. maxima*, *E. tenella*, *E. brunetti*, *E. necatrix*, *E. mitis*, and *E. praecox* [3]. Each species resides at the different section of the intestinal tract of broiler chickens, and thereby it has different modes of actions to affect growth performance and gut health of broilers [4]. Teng et al. [5] reported that *E. maxima* decreased digestibility of crude proteins and amino acids in broilers. *Eimeria* spp. can be transmitted via the fecal–oral route. The infection is initiated by ingestion of sporulated (infectious) oocysts, and after the asexual and sexual replications, un-sporulated oocysts are excreted with feces [6]. In an appropriate environment, the oocysts can be sporulated and become infectious, and this life cycle can be repeated with poultry growth cycle.

Eimeria tenella (*E. tenella*) resides in the mucus membrane of ceca, and during its replications, epithelial cells in ceca are damaged, resulting in hemorrhagic diarrhea and impaired growth performance and intestinal health in broilers [7]. The ceca, the main intestinal compartment for bacterial fermentation, can be reservoirs for pathogenic bacteria and their toxins that can cause oxidative stress after entering the blood stream of broilers [8]. However, ceca also play crucial roles in producing beneficial bacterial metabolites including vitamins, volatile fatty acids (VFA), lactic acid, and antimicrobial compounds via bacterial fermentation [9]. The VFA are not only inhibit the growth of pathogenic bacteria, but also are energy substrates for the host and induce gut development of chickens by accelerating gut epithelial cell proliferation [10]. Moreover, VFA interact with fat metabolism via mitogen-activated protein kinase (MAPK) pathway [11]. These suggest that VFA are closely associated with feed efficiency by providing extra energy to the host or influencing metabolism of chickens.

To cope with *Eimeria* spp. infection in broiler production, anti-coccidial drugs, and vaccination has been used in the broiler industry. However, the use of anti-coccidial drugs has been restricted by inhibiting the use of old anti-coccidial drugs and requiring Veterinary Feed Directives (VFD) registrations because of the spread of resistant *Eimeria* strain and consumer pressure [12,13]. Furthermore, vaccination is expensive and can prevent spread of *Eimeria* spp. [14]. Recently, a lot of attention has been paid to find nutritional interventions to control *Eimeria* spp. infection in broilers. Diverse bioactive compounds, including essential oils [15], probiotics [16], sodium butyrate [17], and plant extracts [18], were studied to control or to ameliorate negative effects of *E. tenella* infection and in broilers. The modes of actions of those bioactive compounds may include damaging cell wall of *E. tenella*, modulating cecal microbiota, and/or enhancing the immunity of broilers. To find suitable nutritional interventions, it is important to understand mode of actions of *E. tenella* on the growth of chickens and to set up appropriate experimental infection models to test novel nutritional interventions. Therefore, the hypothesis of this study was that impaired cecal health due to *E. tenella* infection may result in reduced growth performance and impaired intestinal health because of reduced VFA production and increased oxidative stress. The purpose of the study was to investigate the effects of different inoculation dosages of *E. tenella* on growth performance, gastrointestinal permeability, oocyst shedding, fecal consistency, intestinal morphology, ileal apparent digestibility, antioxidant capacity, and cecal VFA in broiler chickens.

2. Materials and Methods

2.1. Experimental Design, Diets, and Growth Performance

This study was approved by the Institutional Animal Care and Use committee of the University of Georgia, and this experiment was conducted at the Poultry Research Center, University of Georgia, Athens, GA. A total of 360 fourteen-day-old male Cobb500 broiler chickens were distributed to 5 treatments with 6 replicates (12 birds per battery cage) in a completely randomized design. The experimental treatments were (1) treatments 0 (T0): administration with 1 mL of phosphate-buffered solution (PBS) as a sham-challenged group; (2) treatments 1 (T1): administration with 6250 sporulated oocysts of *E. tenella*; treatment 2 (T2): administration with 12,500 sporulated oocysts of *E. tenella*; treatments (T3): administration with 25,000 sporulated oocysts of *E. tenella*; and treatment (T4), administration with 50,000 sporulated oocysts of *E. tenella*. The sham-challenged groups were placed on the top of the cages to minimize the cross-infection. The *E. tenella* used in the study was a wild-type strain. To each bird, 1 mL of inoculum was administrated by oral gavage. As shown in Table 1, the experimental diet (D 14 to 23) was formulated to meet or exceed Cobb 500 nutrient requirements (2018) and included 3 g/kg of titanium (IV) oxide (Acros Organics, Morris Plains, NJ, USA) as an indigestible marker to determine the apparent ileal digestibility (AID). In-feed anticoccidials were not included in the experimental diet (D 14 to 23) and in the pre-experimental diet (D 0 to 14).

Table 1. Diet composition and calculated analysis of the broiler diet (g/kg, as fed basis).

Ingredients	D 14 to 23
Corn	700.8
Soybean meal (480 g crude protein/kg)	241.73
Soybean oil	15.84
Deflour phosphate	13.99
Sand	7.00
Limestone	6.11
Titanium dioxide	3.00
DL-Methionine 99%	2.86
L-Lysine HCl 78%	2.80
Vitamin Premix [1]	2.50
Common Salt	1.79
L-threonine	0.77
Mineral Premix [2]	0.80
Total	1,000
Calculated energy and nutrient value, %	
Metabolizable energy, Mcal/kg	3,100
Crude protein	18.375
SID [3] Methionine	0.552
SID Total sulfur amino acids	0.8
SID Lysine	1.02
SID Threonine	0.66
Total calcium	0.76
Available phosphate	0.38

[1] Vitamin mix provided the following in mg/100 g diet: thiamine-HCl, 1.5; riboflavin 1.5; nicotinic acid amide 15; folic acid 7.5; pyridoxine-HCl, 1.2; d-biotin 3; vitamin B-12 (source concentration, 0.1%) 2; d-calcium pantothenate 4; menadione sodium bisulfite, 1.98; α-tocopherol acetate (source 500,000 IU/g), 22.8; cholecalciferol (source 5,000,000 IU/g) 0.09; retinyl palmitate (source 500,000 IU/g), 2.8; ethoxyquin, 13.34; I-inositol, 2.5; dextrose, 762.2; [2] Mineral mix provided the following in g/100 g diet: $Ca(H_2PO_4)_2 \cdot H_2O$, 3.62; $CaCO_3$, 1.48; KH_2PO_4, 1.00; Na_2SeO_4, 0.0002; $MnSO_4 \cdot H_2O$, 0.035; $FeSO_4 \cdot 7H_2O$, 0.05; $MgSO_4 \cdot 7H_2O$, 0.62; KIO_3, 0.001; NaCl, 0.60; $CuSO_4 \cdot 5H_2O$, 0.008; $ZnCO_3$, 0.015; $CoCl_2 \cdot 6H_2O$, 0.00032; $NaMoO_4 \cdot 2H_2O$, 0.0011; KCl, 0.10; dextrose, 0.40; [3] SID: standard ileal digestible amino acid.

During the entire experiment period, birds had free access to water and feed, and temperature was controlled according to the recommendation of Cobb Broiler Management Guide. Body weight (BW) of the birds per cage were recorded on 6 days post-infection (dpi) and 9 dpi, and feed disappearance were recorded daily to calculate average daily gain (ADG), average daily feed intake (ADFI), and feed conversion ratio (FCR). The acute phase was 0 to 6 dpi, and recovery phase was 6 to 9 dpi.

2.2. Gastrointestinal Permeability, Oocyst Shedding, and Fecal Consistency

Gastrointestinal permeability was measured according to Teng et al. [19] with minor modifications. On 5, 6, and 7 dpi, one bird per cage was administrated with 1 mL of 2.2 mg/mL of fluorescein isothiocyanatedextran 4 kDa (FITC-D4; Sigma-Aldrich Co., St. Louis, MO, USA) dissolved in PBS. After 2 h, birds were euthanized and blood from the heart was collected into heparin-free vacutainer tubes. The tubes stood in a dark container at room temperature for 1 h for clotting and centrifuged at 1000× *g* for 15 min to recover serum. The collected serum samples were transferred to a 96 black well plate (Greiner Bio-one, Monroe, NC, USA) in duplicate, and the fluorescence was measured at an excitation wavelength of 485 nm and an emission wavelength of 528 nm by using a ICTOR3 Multilabel Plate Reader (Perkin Elmer, Waltham, MA, USA). The FITC-D4 concentration in the serum was quantified by prepared standard solution using pooled serum from birds that were not inoculated with FITC-D4 (not part of the study) and expressed relative to the T0 group.

Fresh fecal samples were collected on 5 to 6 dpi, 6 to 7 dpi, and 7 to 8 dpi to measure fecal moisture content and fecal oocyst shedding. Fecal samples were put in a 60 °C oven until constant weight, and the weights before and after drying were recorded to calculate moisture content. Oocysts in fecal samples were counted using a McMaster chamber to calculate oocyst shedding per gram of feces. Briefly, 3 to 5 g of fecal samples were put in 50 mL tubes and mixed thoroughly with 25 mL of distilled water to ensure a uniform suspension. Afterwards, 1 mL of the suspension were mixed with 9 mL of the saturated salt solutions. The mixed solution was loaded into a McMaster chamber, and number of oocysts were counted using a microscope. Oocyst shedding were expressed as $\log_{10}$ (oocysts/g feces).

2.3. Sample Collection and Lesion Score

On 6 and 9 dpi, 4 birds per cage were euthanized by the cervical dislocation method for sample collection and lesion scoring (6 dpi). Cecal lesion scoring from 4 birds per cage was conducted in a blind fashion according to the 4-score scale method [20]. Around 3 to 5 cm of intestinal sections of mid-duodenum, mid-jejunum, mid-ileum, and mid-ceca were collected, and then rinsed with PBS to remove digesta, and stored in 10% neutral-buffered formalin for further steps. From 4 birds per cage, ileal digesta (from the Meckel's diverticulum to 15 cm upper from the ileo-cecal-colic junction) were collected and dried in a 60 °C oven until constant weight to determine ileal moisture content and for digestibility analysis. Liver and ceca samples were collected and snap-frozen and stored at −80 °C.

2.4. Intestinal Morphology

The fixed tissues in 10% neutral-buffered formalin were embedded in paraffin and cut into 4 μm, and hematoxylin and eosin (H&E) staining was conducted. The H&E-stained slides were read using a microscope (BZ-X810; Keyence, Osaka, Japan). The villus height (VH) and crypts depth (CD) of five well-oriented villi per section and their corresponding crypts for the five villi were measured for duodenum, jejunum, and ileum samples, and CD was measured for ceca samples by using ImageJ (National Institutes of Health, Bethesda, MD, USA). The ratios of VH to CD were calculated for each villus and crypt.

2.5. Apparent Ileal Digestibility

The concentrations of titanium dioxide in oven-dried samples (0.3 g for ileal digesta samples and 0.5 g for the feed sample) were analyzed according to Short et al. [21]. Dry matter (DM), organic matter (OM), and ash apparent ileal digestibility were determined according to Lin and Olukosi [22].

2.6. Liver Total Antioxidant Capacity (TAC)

Total antioxidant capacity (TAC) in the liver was measured using a commercial kit (QuantiCromAntioxidant Assay Kit; DTAC-100) (BioAssay Systems, Hayward, CA, USA). Approximately 100 mg of frozen liver samples were homogenized in 1 mL of PBS for 45 s using a beads beater and centrifuged at 10,000× g for 10 min. Aliquots of supernatants were taken for the analyses of protein content using the Pierce™ BCA Protein Assay Kit (Thermo Fisher Scientific, Cleveland, OH, USA) after 20 times dilution. Afterwards, TAC was measured according to the manufacture's protocol without further dilutions. The absorbance was measured using SpectraMax® ABS Plus microplate reader (Molecular devices, San Jose, CA, USA). The TAC values were expressed as nM trolox quivalents/mg protein.

2.7. VFA Concentrations in Cecal Digesta

Concentrations of VFA in cecal digesta were analyzed according to Lourenco et al. [23]. Cecal samples were collected from birds on 6 and 9 dpi, and the samples were snap-frozen in liquid nitrogen and stored at −80 °C for further analyzes. Once thawed, samples were diluted and homogenized by placing 0.5 g into 3 mL of distilled water. The samples were vigorously homogenized for 1 min and subsequently frozen at −20 °C. After the samples were thawed, these were centrifuged at 10,000× g for 10 min, and 850 μL of supernatant were collected and mixed with 170 μL of the fresh 25% (wt/vol) meta-phosphoric acid solution, and immediately frozen at −20 °C overnight. The samples were centrifuged at 10,000× g for 10 min, and 800 μL of supernatant was collected and mixed with 1600 μL ethyl acetate. Samples were vigorously homogenized for 10 s and allowed to settle for 5 min. The top layer was transferred to a screw-thread vial and analyzed in a gas chromatograph (Shimadzu GC-2010 plus; Shimadzu Corporation, Kyoto, Japan) equipped with an autoinjector (AOC-20i; Shimadzu Corporation, Kyoto, Japan). A capillary column (Zebron ZB-FFAP; 30 m × 0.32 mm × 0.25 μm; Phenomenex Inc., Torrance, CA, USA) was used for the separation of the VFA. The sample injection volume was set at 1 mL, and helium was used as the carrier gas. The column temperature was initially set at 110 °C, and gradually increased to 200 °C over the course of 6 min. The flame ionization detector was set at 350 °C.

2.8. Statistical Analyses

Statistical analyses were performed using SAS (version 9.4; SAS Inst. Inc., Cary, NC, USA). Data normality was checked using proc univariate except for lesion score data. All groups were compared using proc mixed in a completely randomized design followed by Tukey's comparison test. Kruskal–Wallis test followed by the Dwass, Steel, Critchlow-Fligner post hoc test was used to analyze lesion score data. Orthogonal polynomial contrasts were utilized to evaluate the significance of linear or quadratic effects of different *E. tenella* inoculation dosages, and the inoculation dosages of *E. tenella* were normalized by using the base 2 logarithm of the number of sporulated *E. tenella* number for orthogonal polynomial contrasts [19]. Statistical significance was set at $p < 0.05$, and trends ($0.05 \leq p \leq 0.1$) were also presented.

3. Results

3.1. Growth Performance and Lesion Score

As shown in Table 2, no significant differences were observed in BW, ADG, and ADFI in the acute phase ($p > 0.1$) among the treatments. However, daily feed intake tended to quadratically increase on 6 dpi ($p = 0.08$) and 7 dpi ($p = 0.09$) due to *E. tenella* infection (Figure 1). In the acute phase, FCR was linearly increased due to the inoculation of *E. tenella* ($p < 0.05$). There were no significant differences in growth performance among the treatments in the recovery phase.

Table 2. Growth performance parameters including body weight (BW; g), average daily gain (ADG; g/d), average daily feed intake (ADFI; g/d), and feed conversion ratio (FCR; g/g) of broiler chickens infected with different dosages of *Eimeria tenella* during the acute phase [0 to 6 days post-infection (dpi)] and recovery phase (6 to 9 dpi) [1].

Items	*Eimeria tenella*-Challenged							Polynomial Contrast	
	T0	T1	T2	T3	T4	SEM	*p* Value	Lin.	Quad.
Initial BW	359.4	360.3	356	359.5	358	3.72	0.38		
0 to 6 dpi									
BW	703.4	705.5	697	699.5	680.77	27.28	0.55	0.16	0.43
ADG	57.33	57.53	57.01	56.54	53.80	4.32	0.56	0.16	0.37
ADFI	84.92	87.24	88.63	86.05	84.31	4.61	0.55	0.69	0.11
FCR	1.48	1.52	1.55	1.52	1.57	0.06	0.19	0.04	0.64

Table 2. *Cont.*

Items	*Eimeria tenella*-Challenged					SEM	*p* Value	Polynomial Contrast	
	T0	T1	T2	T3	T4			Lin.	Quad.
6 to 9 dpi									
BW	881.8	859	876.9	885.73	889.19	79.63	0.97	0.69	0.72
ADG	65.81	57.40	61.68	63.83	69.70	13.74	0.62	0.43	0.22
ADFI	115.4	120.9	123.77	123.1	123.1	14.86	0.87	0.4	0.55
FCR	1.9	2.17	2.03	1.97	1.77	0.39	0.50	0.38	0.16
0 to 9 dpi									
ADG	60.16	57.49	58.83	58.97	59.10	6.47	0.97	0.94	0.66
ADFI	95.09	98.47	100.6	98.16	97.24	6.81	0.72	0.65	0.22
FCR	1.62	1.74	1.71	1.67	1.64	0.14	0.54	0.85	0.15

[1] T0: treatment 0 (Sham-challenged with phosphate-buffered saline); T1, treatment 1 (challenged with 6250 sporulated oocysts of *Eimeria tenella*); T2, treatment 2 (challenged with 12,500 sporulated oocysts of *Eimeria tenella*); T3, treatment 3 (challenged with 25,000 sporulated oocysts of *Eimeria tenella*); T4, treatment 4 (challenged with 50,000 sporulated oocysts of *Eimeria tenella*). At each time point, orthogonal polynomial contrasts analysis was conducted to see linear pattern (L) and quadratic pattern (Q) among the treatments.

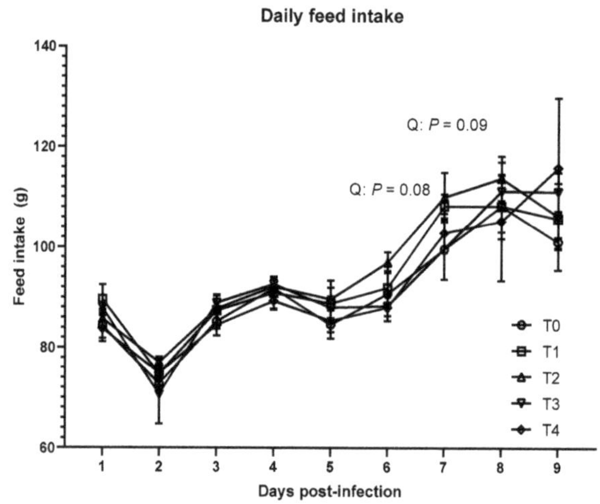

Figure 1. Daily feed intake of broiler chickens infected with different dosages of *Eimeria tenella*. Daily feed intake was measured in the T0 (treatment 0; Sham-challenged with phosphate-buffered saline); T1 (treatment 1; challenged with 6250 sporulated oocysts of *E. tenella*); T2 (treatment 2; challenged with 12,500 sporulated oocysts of *E. tenella*); T3 (treatment 3; challenged with 25,000 sporulated oocysts of *E. tenella*); T4 (treatment 4; challenged with 50,000 sporulated oocysts of *E. tenella*) groups during 1 to 9 days post-infection. At each time point, orthogonal polynomial contrasts analysis was conducted to see linear pattern (L) and quadratic pattern (Q) among the treatments.

Cecal lesion due to *E. tenella* infection was not detected in the T0 group on 6 dpi (Figure 2). The T4 group had a higher lesion score compared to the T1 group ($p < 0.05$); lesion scores > 0 and <2 were also obtained in T1, T2, and T3 groups.

3.2. Oocyst Shedding and Fecal/ileal Moisture Content

As shown in Figure 3, *E. tenella* was not detected in the feces of the sham-challenged (T0) group at all time points. On 5 to 6 dpi, the T4 group had significantly higher oocyst shedding compared to T1 group, and *E. tenella* infection linearly increased oocyst shedding. However, there were no significant differences in oocyst shedding among the treatments on 6 to 7 dpi and 7 to 8 dpi.

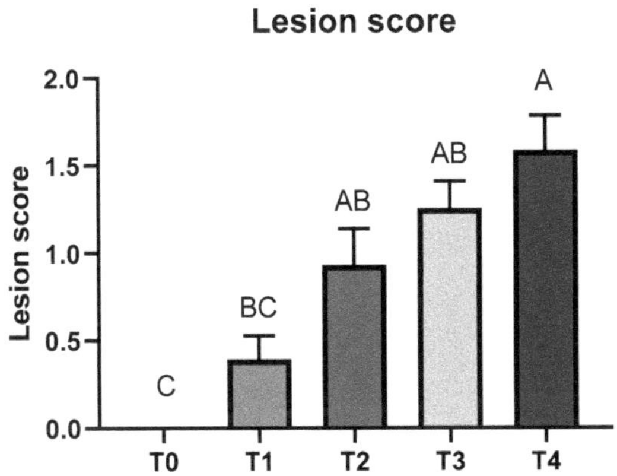

Figure 2. Cecal lesion score of broiler chickens infected with different dosages of *Eimeria tenella* on 6 days post-infection. Cecal lesion score was measured in the T0 (treatment 0; Sham-challenged with phosphate-buffered saline); T1 (treatment 1; challenged with 6250 sporulated oocysts of *E. tenella*); T2 (treatment 2; challenged with 12,500 sporulated oocysts of *E. tenella*); T3 (treatment 3; challenged with 25,000 sporulated oocysts of *E. tenella*); T4 (treatment 4; challenged with 50,000 sporulated oocysts of *E. tenella*) groups on 6 days post-infection. Different letters at the same time point represent significantly different ($p < 0.05$) by utilizing the Kruskal–Wallis test followed by the Dwass, Steel, Critchlow-Fligner post hoc test.

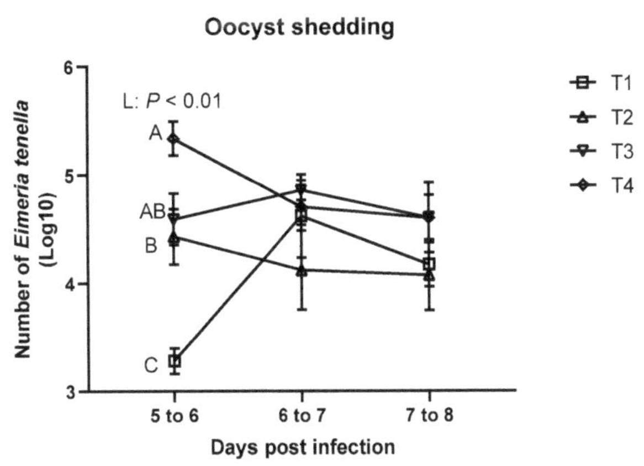

Figure 3. Oocyst shedding of broiler chickens infected with different dosages of *Eimeria tenella*. Fecal *oocyst shedding* was measured using McMaster chamber in the T0 (treatment 0; Sham-challenged with phosphate-buffered saline); T1 (treatment 1; challenged with 6250 sporulated oocysts of *E. tenella*); T2 (treatment 2; challenged with 12,500 sporulated oocysts of *E. tenella*); T3 (treatment 3; challenged with 25,000 sporulated oocysts of *E. tenella*); T4 (treatment 4; challenged with 50,000 sporulated oocysts of *E. tenella*) groups on 5 to 6, 6 to 7, and 7 to 8 days post-infection. Number of *E. tenella* oocysts is shown as $\log_{10}$ (oocysts/g feces). At each time point, orthogonal polynomial contrasts analysis was conducted to see linear pattern (L) and quadratic pattern (Q) among the treatments. Different letters at the same time point represent significantly different ($p < 0.05$) by proc mixed followed by the Tukey's multiple comparison test among the treatment groups.

As shown in Table 3, Fecal moisture content was modulated due to *E. tenella* infection on 7 to 8 dpi ($p < 0.05$). Fecal moisture content was quadratically decreased on 5 to 6 dpi ($p < 0.05$), and quadratically increased on 6 to 7 dpi ($p < 0.05$) due to *E. tenella* infection. *E. tenella* infection linearly increased fecal moisture content on 7 to 8 dpi ($p < 0.05$). On

6 dpi, *E. tenella* infection altered ileal moisture content, and ileal moisture content was linearly decreased on 6 dpi due to *E. tenella* infection ($p < 0.05$).

Table 3. Fecal moisture content (fecal consistency) and ileal moisture content of broiler chickens infected with different dosages of *Eimeria tenella on* 5 to 6 days post-infection (dpi), 6 to 7 dpi, and 7 to 8 dpi (fecal moisture content), and at 6 and 9 dpi (ileal moisture content) [1].

Items	*Eimeria tenella*-Challenged							Polynomial Contrast	
	T0	T1	T2	T3	T4	SEM	*p* Value	Lin.	Quad.
Feces									
5 to 6 dpi	78.23	76.76	75.91	76.56	76.99	1.66	0.22	0.22	0.04
6 to 7 dpi	68.14	69.68	73.38	74.64	68.66	4.87	0.12	0.35	0.03
7 to 8 dpi	70.81	71.87	73.09	75.07	74.86	2.52	0.03	<0.01	0.65
Ileum									
6 dpi	81.75	81.83	80.6	80.94	80.81	0.76	0.03	0.01	0.33
9 dpi	80.34	80.1	81.24	80.62	79.93	1.22	0.42	0.9	0.18

[1] T0: treatment 0 (Sham-challenged with phosphate-buffered saline); T1, treatment 1 (challenged with 6250 sporulated oocysts of *E. tenella*); T2, treatment 2 (challenged with 12,500 sporulated oocysts of *E. tenella*); T3, treatment 3 (challenged with 25,000 sporulated oocysts of *E. tenella*); T4, treatment 4 (challenged with 50,000 sporulated oocysts of *E. tenella*). At each time point, orthogonal polynomial contrasts analysis was conducted to see linear pattern (L) and quadratic pattern (Q) among the treatments.

3.3. Gastrointestinal Permeability

No significant differences in gastrointestinal permeability were observed on 5 dpi and 6 dpi (Figure 4). However, gastrointestinal permeability tended to be linearly increased ($p = 0.07$). However, the numerically highest group, T4, had only 1.16 folds FITC fluorescence compared to the T0 group.

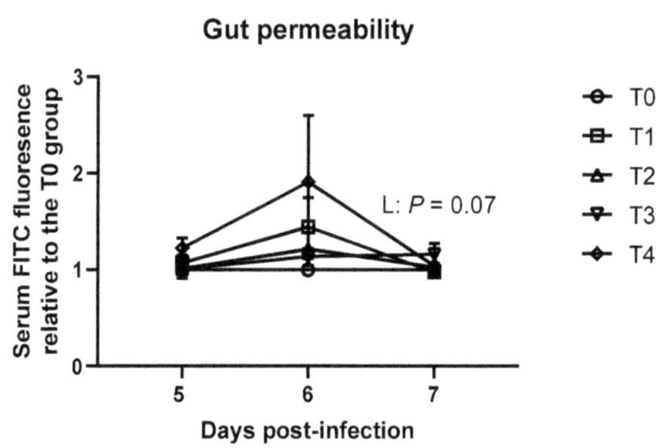

Figure 4. In vivo gastrointestinal permeability of broiler chickens infected with different dosages of *Eimeria tenella*. In vivo gastrointestinal permeability was measured using fluorescein isothiocyanate dextran 4 kDa in the T0 (treatment 0; Sham-challenged with phosphate-buffered saline); T1 (treatment 1; challenged with 6250 sporulated oocysts of E. tenella); T2 (treatment 2; challenged with 12,500 sporulated oocysts of *E. tenella*); T3 (treatment 3; challenged with 25,000 sporulated oocysts of *E. tenella*); T4 (treatment 4; challenged with 50,000 sporulated oocysts of *E. tenella*) groups on 5, 6, and 7 days post-infection. At each time point, orthogonal polynomial contrasts analysis was conducted to see linear pattern (L) and quadratic pattern (Q) among the treatments.

3.4. Intestinal Morphology and Apparent Ileal Digestibility

As shown in Table 4, *E. tenella* infection linearly ($p < 0.05$) and quadratically ($p < 0.05$) increased cecal CD, whereas the sham-challenged group had the lowest cecal CD compared to the *E. tenella* challenged groups on 6 dpi ($p < 0.05$). On 6 dpi, ileal CD showed a tendency to be linearly reduced ($p = 0.06$), and ileal VH:CD tended to be increased ($p = 0.08$) due to *E. tenella* infection. On 9 dpi, the inoculation of *E. tenella* tended to linearly decrease duodenal VH (p = 0.1) and jejunal VH (p = 0.09). The T0 group had lower cecal CD compared to T2, T3, and T4 groups ($p < 0.05$), and *E. tenella* infection linearly ($p < 0.05$), and quadratically ($p < 0.05$) increased cecal CD. As shown in Figure 5, cecal CD were deepened, and gametocytes and developing oocysts were observed in the ceca of broiler chickens infected *E. tenella*.

Table 4. Duodenal, jejunal, ileal, and cecal morphology [villus height (VH; µm), crypts depth (CD; µm), and VH:CD] of broiler chickens infected with different dosages of *Eimeria tenella* on 6 days post-infection (dpi) and 9 dpi [1].

Items	*Eimeria tenella*-Challenged							Polynomial Contrast	
	T0	T1	T2	T3	T4	SEM	p Value	Lin.	Quad.
6 dpi									
Duodenum									
VH	2459.97	2585.52	2442.42	2154.12	2485.42	300.16	0.22	0.34	0.56
CD	282.46	245.17	253.82	227.51	255.55	35.36	0.17	0.13	0.08
VH:CD	9.05	10.8	9.86	10.39	9.85	1.78	0.53	0.61	0.26
Jejunum									
VH	1411.46	1375.62	1332.65	1321.93	1303.21	147	0.71	0.17	0.76
CD	234.92	222.03	223.93	221.29	225.43	34.52	34.52	0.66	0.57
VH:CD	6.38	6.56	6.08	6.28	5.94	1.22	0.91	0.47	0.85
Ileum									
VH	899.01	936.51	934.37	894.06	879.21	113.9	0.87	0.58	0.41
CD	194.11	181.49	183.02	184.97	161.46	24.03	0.24	0.06	0.59
VH:CD	4.92	5.28	5.18	5.09	5.72	0.59	0.22	0.08	0.54
Cecal CD	217.21 ^{b}	410.22 ^{a}	451.86 ^{a}	593.18 ^{a}	477.05 ^{a}	111.85	<0.01	<0.01	<0.01
9 dpi									
Duodenum									
VH	2475	2574.81	2452.6	2150.15	2375.54	278.63	0.17	0.1	0.87
CD	250.69	226.78	233.48	211.26	234.56	35.58	0.49	0.31	0.23
VH:CD	10.26	11.55	11.04	10.74	10.32	1.95	0.77	0.78	0.28
Jejunum									
VH	1410.12	1348.81	1320.96	1288.44	1257.51	159.32	0.54	0.09	0.82
CD	208.29	217.18	197.49	211.8	210.85	34.22	0.87	0.99	0.78
VH:CD	7.21	6.55	7.06	6.5	6.21	1.86	0.69	0.26	0.87
Ileum									
VH	888.68	917.4	879.2	885.43	876.32	108.22	0.97	0.69	0.85
CD	173.99	167.73	158.63	161.9	155.46	28.8	0.8	0.26	0.78
VH:CD	5.35	5.57	5.74	5.7	5.95	0.72	0.69	0.17	0.89
Cecal CD	87.57 ^{b}	163.55 ^{a,b}	182.47 ^{a}	224.19 ^{a}	196.3 ^{a}	47.71	<0.01	<0.01	0.02

[1] T0: treatment 0 (Sham-challenged with phosphate-buffered saline); T1, treatment 1 (challenged with 6250 sporulated oocysts of *E. tenella*); T2, treatment 2 (challenged with 12,500 sporulated oocysts of *E. tenella*); T3, treatment 3 (challenged with 25,000 sporulated oocysts of *E. tenella*); T4, treatment 4 (challenged with 50,000 sporulated oocysts of *E. tenella*). At each time point, orthogonal polynomial contrasts analysis was conducted to see linear pattern (L) and quadratic pattern (Q) among the treatments. Different letters at the same time point represent significantly different ($p < 0.05$) by proc mixed followed by the Tukey's multiple comparison test among the treatment groups.

In broiler chickens infected *E. tenella*, crypts were deepened and gametocytes and developing oocysts were observed.

There were no significant differences in the ileal apparent digestibility of DM, OM, and ash among treatments on 6 dpi and 9 dpi ($p > 0.1$; Table 5).

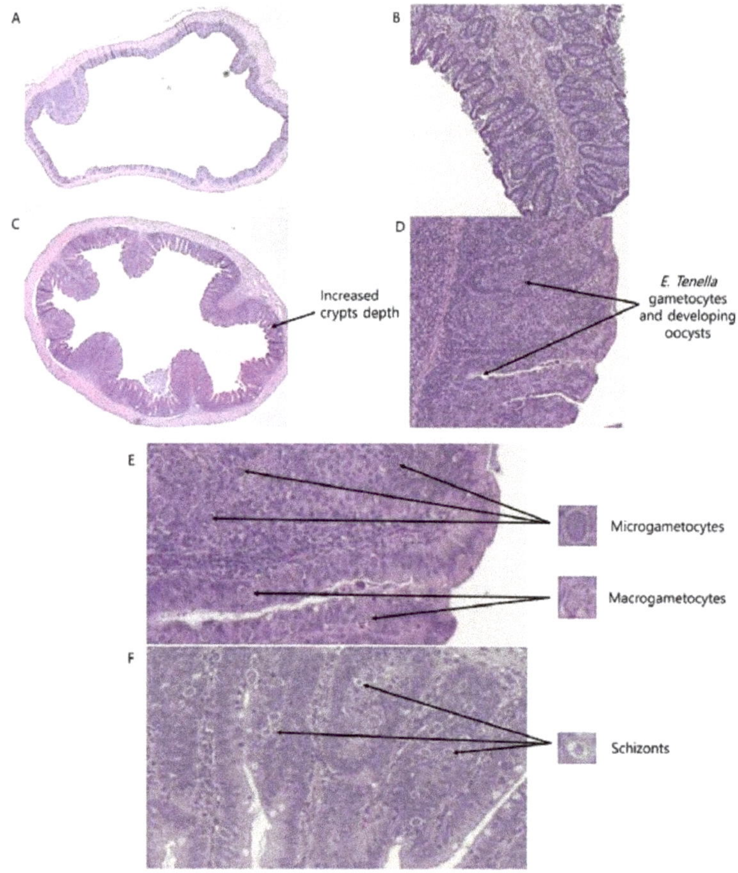

Figure 5. Ceca morphology of a non-challenged bird ((**A**) 2×; (**B**) 10×) and of a *Eimeria tenella* infected bird ((**C**) 2×; (**D**) 10×) in the T1 group (challenged with 6250 sporulated oocysts of *E. tenella*) on 6 days post-infection when tissues were stained with hematoxylin and eosin. Microgametocytes, macrogametocytes, and schizonts were observed in the ceca crypts (**E**,**F**).

Table 5. Apparent ileal digestibility (%) of dry matter (DM), organic matter (OM), and ash of broiler chickens infected with different dosages of *Eimeria tenella* on 6 days post-infection (dpi) and 9 dpi.

Items	*Eimeria tenella*-Challenged [1]							Polynomial Contrast	
	T0	T1	T2	T3	T4	SEM	*p* Value	Lin.	Quad.
6 dpi									
DM	69.16	70.31	68.35	71.53	70.19	1.96	0.09	0.21	0.96
OM	70.85	72.08	70.13	73.32	71.79	1.94	0.09	0.23	0.89
Ash	40.16	40	37.66	40.82	42.66	5.47	0.6	0.42	0.26
9 dpi									
DM	72.71	70.31	71.86	72.44	73.5	2.8	0.42	0.36	0.2
OM	75.02	72.29	74.32	74.68	75.8	2.69	0.29	0.31	0.18
Ash	33.03	36.24	29.47	33.86	33.82	9.49	0.81	0.95	0.76

[1] T0: treatment 0 (Sham-challenged with phosphate-buffered saline); T1, treatment 1 (challenged with 6250 sporulated oocysts of *E. tenella*); T2, treatment 2 (challenged with 12,500 sporulated oocysts of *E. tenella*); T3, treatment 3 (challenged with 25,000 sporulated oocysts of *E. tenella*); T4, treatment 4 (challenged with 50,000 sporulated oocysts of *E. tenella*). At each time point, orthogonal polynomial contrasts analysis was conducted to see linear pattern (L) and quadratic pattern (Q) among the treatments.

3.5. Liver Total Antioxidant Capacity (TAC)

As shown in Figure 6, liver TAC was not modulated due to *E. tenella* infection in broiler chickens ($p > 0.1$) on 6 dpi. However, different inoculation dosages of *E. tenella* quadratically increased liver TAC on 9 dpi ($p < 0.05$).

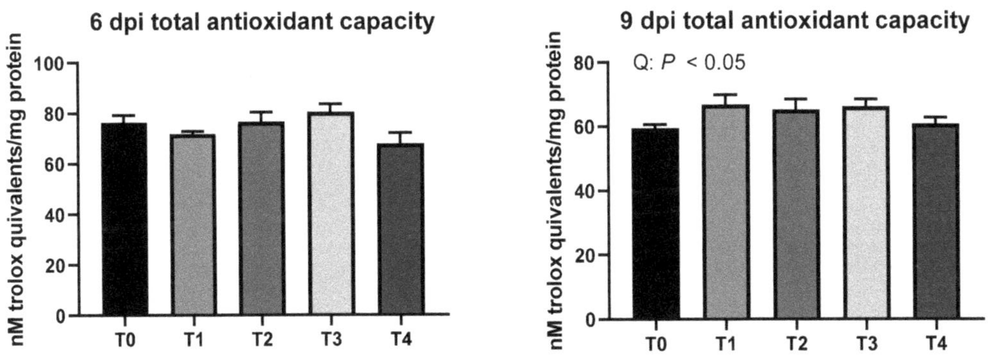

Figure 6. Total antioxidant capacity in the liver of broiler chickens infected with different dosages of *Eimeria tenella*. On 6 and 9 days post-infection (dpi), total antioxidant capacity in the liver was measured in the T0 (treatment 0; Sham-challenged with phosphate-buffered saline); T1 (treatment 1; challenged with 6250 sporulated oocysts of *E. tenella*); T2 (treatment 2; challenged with 12,500 sporulated oocysts of *E. tenella*); T3 (treatment 3; challenged with 25,000 sporulated oocysts of *E. tenella*); T4 (treatment 4; challenged with 50,000 sporulated oocysts of *E. tenella*) groups. At each time point, orthogonal polynomial contrasts analysis was conducted to see linear pattern (L) and quadratic pattern (Q) among the treatments.

3.6. VFA Concentrations in Cecal Digesta

As shown in Table 6, infection of *E. tenella* linearly decreased acetate concentration in ceca contents ($p < 0.05$), and the T4 group had significantly lower acetate concentration in ceca contents compared to the T0 group. Isobutyrate concentration in ceca contents was linearly decreased ($p < 0.05$) on 6 dpi as *E. tenella* dosage increased. Total VFA were linearly decreased due to *E. tenella* infection ($p < 0.05$), and the T4 group had significantly lower total VFA compared to the T0 group on 6 dpi. On 9 dpi, *E. tenella* tended to linearly decrease butyrate ($p = 0.09$), and significantly decreased valerate concentrations (Linear; $p < 0.05$) in ceca contents. The T0 group had significantly lower propionate compared to the T1 group, and T4 birds had significantly lower varelate compared to the T1 group.

Table 6. Concentrations of volatile fatty acids (mM) in ceca contents of broiler chickens infected with different dosages of *Eimeria tenella* at 6 days post-infection (dpi) and 9 dpi [1].

Items	*Eimeria tenella*-Challenged [1]							Polynomial Contrast	
	T0	T1	T2	T3	T4	SEM	*p* Value	Lin.	Quad.
6 dpi									
Acetate	48.21 [a]	35.2 [a,b]	35.64 [a,b]	36.94 [a,b]	26.91 [b]	10	0.02	<0.01	0.65
Propionate	8.37	9.03	0.84	7.27	6.1	2.95	0.5	0.11	0.45
Isobutyrate	0.26	0.17	0.19	0.15	0.12	0.08	0.07	<0.01	00.73
Butyrate	8.69	5.8	7.4	5.34	4.96	3.8	0.42	0.12	0.82
Isovalerate	0.42	0.42	0.45	0.32	0.37	0.17	0.72	0.35	0.82
Valerate	1.09	0.95	0.78	0.69	0.8	0.51	0.72	0.22	0.47
Total VFA	67.04 [a]	51.58 [a,b]	52.52 [a,b]	50.73 [a,b]	39.26 [b]	13.95	0.04	<0.01	0.8

Table 6. *Cont.*

Items	*Eimeria tenella*-Challenged [1]							Polynomial Contrast	
	T0	T1	T2	T3	T4	SEM	*p* Value	Lin.	Quad.
9 dpi									
Acetate	50.77	76.52	62.01	46.47	48.11	27.04	0.31	0.32	0.25
Propionate	6.14 [b]	11.87 [a]	8.17 [a,b]	7.82 [a,b]	7.07 [a,b]	3.34	0.06	0.62	0.07
Isobutyrate	0.16	0.32	0.15	0.16	0.16	0.14	0.22	0.39	0.55
Butyrate	8.64	13.04	8.58	6.24	6.3	4.94	0.16	0.09	0.4
Isovalerate	0.26	0.53	0.39	0.37	0.6	0.39	0.59	0.31	0.96
Valerate	0.96 [a,b]	1.65 [a]	0.86 [a,b]	0.8 [a,b]	0.57 [b]	0.55	0.03	0.03	0.21
Total VFA	66.94	103.94	68.72	61.86	62.83	35.97	0.26	0.29	0.43

[1] T0: treatment 0 (Sham-challenged with phosphate-buffered saline); T1, treatment 1 (challenged with 6250 sporulated oocysts of *E. tenella*); T2, treatment 2 (challenged with 12,500 sporulated oocysts of *E. tenella*); T3, treatment 3 (challenged with 25,000 sporulated oocysts of *E. tenella*); T4, treatment 4 (challenged with 50,000 sporulated oocysts of *E. tenella*). At each time point, orthogonal polynomial contrasts analysis was conducted to see linear pattern (L) and quadratic pattern (Q) among the treatments. Different letters at the same time point represent significantly different ($p < 0.05$) by proc mixed followed by the Tukey's multiple comparison test among the treatment groups.

4. Discussion

The purpose of the study was to investigate the effects of different inoculation dosages of *E. tenella* on growth performance, gastrointestinal permeability, oocyst shedding, intestinal morphology, fecal consistency, ileal apparent digestibility, antioxidant capacity, and cecal VFA profile in broiler chickens. Inoculation dosages were derived from our previous study [19], and the same strain of *E. tenella* was used. However, in the current study, milder infection (lesion score of T4: 1.6) was achieved compared to our previous study (lesion score of the T4 equivalent group: 2.7). The potential reason for milder infection would be that broilers were challenged with three different *Eimeria* spp. (*E. acervulina*, *E. maxima*, and *E. tenella*) in the previous study, which probably caused more severe infection by compensating the immune system in birds compared to the single species challenging in the current study. Mild-infection models are important to test a new bioactive compound because if the compounds do not show any beneficial effects at the mild infection model, they may not show any beneficial effects against coccidiosis at severe infection model either. In the present study, *E. tenella* infection decreased feed efficiency during the acute phase and increased feed intake on 6 and 7 dpi. These data are partially consistent with previous studies [24,25] which reported that FCR was increased along with reduced BW of broilers. In our current study, however, *E. tenella* infection increased FCR by increasing the feed intake of birds, whereas BW was also numerically ($p = 0.16$) decreased with a linear trend during the acute phase. Impaired feed efficiency in broilers infected with *E. tenella* during the acute phase in the present study would be associated with reduced acetate and total VFA production in the ceca. Acetate is the most abundant VFA, which are produced via bacterial fermentation in the ceca of broilers [26]. It is already well-established that *E. tenella* infection can negatively affect cecal microbiome, which results in depressed VFA production in broilers according to many previous studies [27,28]. Cecal VFA production affects the host's energy balance because VFA are important energy substrates for the host [29]. Gasaway [30] mentioned that VFA supply approximately 11% to 18% of the total energy production in chickens. In addition, *Eimeria* spp. compete for energy and nutrients for their asexual and sexual replications with the host [31]. Reduced available energy in the body can result in reduced BW or increased FCR by raising feed intake of the birds to decrease maintenance energy requirements or to meet the energy requirements, respectively [32]. In addition, reduced production of acetate, which can induce the secretion of satiety-stimulating hormones from the gut, could increase the feed intake of the birds [33]. Therefore, in this study, reduced VFA production in ceca due to *E. tenella* infection potentially reduced available energy levels in the body, and this increased feed intake and FCR to supply more energy to meet the energy requirements in broiler chickens infected with *E. tenella*.

This study showed that *E. tenella* infection deepened crypts depth of the ceca. The potential reasons for increased cecal crypts depth due to *E. tenella* infection are still unclear. It is proposed that *E. tenella* increased cecal CD (mucosa layer) to make their habitats in the ceca, or ceca crypts were deepened to increase VFA absorption because VFA production was restricted due to *E. tenella* infection. Nevertheless, deepened CD could inhibit the production and absorption of VFA in the ceca. Increased CD possibly turned in increased total goblet cells in the ceca and mucus secretion into the cecal contents. This possibly reduced concentration of VFA and modulated VFA production in ceca content by affecting microbiome of the broiler chickens. Ceca only have villi at the entrance of the ceca to filter large particles away and to act as an immunological detector of cecal contents [34,35], and middle and distal parts of the ceca only have smooth mucous membrane without villi [36]. Thereby, another possible reason for increased CD in broilers infected with *E. tenella* would be that birds increased CD to let crypts function like villi as a defensive mechanism in the proximal ceca because *E. tenella* infected ceca are vulnerable for further infections (e.g., bacterial infection).

We also hypothesized that impaired ceca health due to *E. tenella* infection may indirectly affect small intestinal health (the main area for nutrient digestion and absorption) by causing energy deficiency and inducing oxidative stress, and this may account for decreased feed efficiency in the current study. Nevertheless, no differences were observed in DM, OM, and ash apparent ileal digestibility among the treatments on 6 and 9 dpi in the current study. According to our previous study, *E. maxima* infection significantly decreased nutrient digestibility in broiler chickens, which indicates that different *Eimeria* spp. affect nutrient digestibility of broilers differently [5]. Liver total antioxidant capacity was same among the treatments on 6 dpi and even increased in broilers infected with *E. tenella* on 9 dpi, potentially because mild infection of *E. tenella* may allow birds to stimulate their antioxidant defensive system. Still, over-production of antioxidants (enzymatic and non-enzymatic) can result in over-use of energy and nutrients, which also may decrease available energy and nutrients for growth in chickens. However, a previous study by Georgieva et al. [25] reported that severe *E. tenella* infection model decreased antioxidant capacity of broilers. Ileal crypt depth was decreased, and ileal VH:CD was increased without affecting VH in *E. tenella* infected broilers in the current study. Probably, ileal morphology was enhanced with limited energy and nutrients sources as a compensation mechanism because cecal functionality was restricted due to *E. tenella* in the current study. There were negative effects of *E. tenella* on duodenal and jejunal morphology on 9 dpi and gastrointestinal permeability on 7 dpi. Energy and absorbed VFA play an important role in gut development in broiler chickens by stimulating gut epithelial cell proliferation [9]. In this study, reduced cecal VFA production would be the main factor that negatively affected gut health of *E. tenella* infected broilers rather than increased pathogens and toxin production because liver health was maintained in broilers infected with *E. tenella*. Whereas it cannot be concluded that impaired intestinal health caused impaired feed efficiency during the acute phase, energy deficiency due to poor VFA yield subsequently damaged intestinal health of broilers infected with *E. tenella*. Moreover, while differences were observed in gastrointestinal permeability, it was only less than two-fold difference. Our previous study [37] reported that more than 200 folds differences of gastrointestinal permeability were observed due to *E. maxima* infection. Hence, reduced VFA production in the ceca due to *E. tenella* infection caused energy deficiency in the body, which resulted in compensated gut health in broiler chickens.

In this study, different *E. tenella* inoculation dosages linearly increased oocyst shedding on 5 to 6 dpi, which implies that the higher inoculation dosages can results in higher oocyst shedding. Nonetheless, these data could be obtained in our study because our *E. tenella* strain induced mild-infection (lesion score below 2) in the ceca. Williams [38] demonstrated that higher inoculation dosage levels can linearly increase oocyst yields until reaching to the crowded dosages (e.g., maximally producing dosage), and higher dosages than the crowded dosages can decrease oocyst yields in broilers. However, no differences were observed on 6 to 7 dpi and 7 to 8 dpi in the current study. The current result also demonstrated that different challenge dosages of *E. tenella* have different peak point of shedding. The potential reason would be that higher number of *E. tenella* possibly decreased number of generation within the asexual and sexual stages of the life [35], which resulted in an earlier peak date for the highest dosage group in the study. Our study was the first to find that different *E. tenella* inoculation dosages resulted in different peak points for oocyst yields in the mildly infected broilers. Oocyst yields in severely infected (*E. tenella* lesion score 3 to 4) broilers at different time points should be investigated further.

Ileal and fecal moisture contents were modulated due to Eimeria infection in the current study. Ileal (6 dpi) and fecal moisture contents (5 to 6 dpi) were quadratically and linearly decreased, respectively. Afterwards, fecal moisture content was increased on 6 to 8 dpi. *E. tenella* are known to induce diarrhea containing mucus and blood. Although obvious bloody diarrhea was not achieved in the current study, fecal moisture content was increased after 6 dpi. Potentially, *E. tenella* damaged enterocytes in the ceca and modulated microbiota, and this caused electrolyte imbalance in the ceca which may explain the modulated ileal and fecal moisture contents in the current study [39]. Moreover, ceca play a crucial role in water absorption [40], and water absorption ability of the ceca of birds infected with *E. tenella* would be limited potentially due to thickened mucosa layer in the current study. Hypothetically, higher water loss due to *E. tenella* infection increased water intake [40] as birds increased their feed intake to compensate energy deficiency in the current study. In addition, increased fecal moisture content can result in increased litter moisture, and this can increase the incidence of food pad dermatitis in broilers [41].

This current study showed that mild infection of *E. tenella* impaired feed efficiency and gut health mainly through reducing VFA production in broilers. Thereby, supplementation of VFA or bioactive compounds that has high energy values and antimicrobial effects (e.g., medium chain fatty acids) can control mild-infection of *E. tenella* in broilers [11,42].

5. Conclusions

Orthogonal polynomial contrasts showed that *E. tenella* mild-infection reduced VFA production in the ceca, and this caused energy deficiency, which increased feed intake and impaired feed efficiency of broiler chickens. This suggests that the cecal VFA concentrations could be a key parameter to represent feed efficiency and *E. tenella* infection severity in broiler chickens. Furthermore, mild-infection of *E. tenella* modulated intestinal morphology, antioxidant capacity, and gastrointestinal permeability in the recovery phase. Different inoculation dosages of *E. tenella* changed oocyst shedding patterns and ileal/fecal moisture content. These current data showed that the mechanisms of *E. tenella* impair feed efficiency and gut health of broilers, which will be beneficial to study strategies to cope with *E. tenella* infection in broiler chickens. Based on the current study, 25,000 to 50,000 sporulated oocyst dosage range would be recommended as subclinical models for nutritional strategies.

Author Contributions: Conceptualization, J.C. and W.K.K.; methodology, J.C., P.-Y.T., J.M.L., T.R.C. and W.K.K.; validation, P.-Y.T., J.M.L., T.R.C. and W.K.K.; investigation, J.C., H.K. and Y.H.T.; data curation, J.C.; writing—original draft preparation, J.C.; writing—review and editing, H.K., Y.H.T., P.-Y.T., J.M.L. and W.K.K.; visualization, J.C. supervision, T.R.C. and W.K.K.; project administration, W.K.K. All authors have read and agreed to the published version of the manuscript.

Funding: This research received no external funding.

Institutional Review Board Statement: This study was approved by the Institutional Animal Care and Use committee of the University of Georgia (A2018 09-006, approved on 16 January 2019).

Informed Consent Statement: Not applicable.

Data Availability Statement: The data presented in this study are available on request from the corresponding author. The data are not publicly available due to further statistical analysis.

Acknowledgments: We appreciated Brett Marshall for his help with the gastrointestinal permeability measurement and Taylor Rae Krause for her help with the VFA analysis. The authors appreciate Alberta L. Fuller for providing the *E. tenella* strain for this study.

Conflicts of Interest: The authors declare no conflict of interest.

References

1. Blake, D.P.; Knox, J.; Dehaeck, B.; Huntington, B.; Rathinam, T.; Ravipati, V.; Ayoade, S.; Gilbert, W.; Adebambo, A.O.; Jatau, I.D. Re-calculating the cost of coccidiosis in chickens. *Vet. Res.* **2020**, *51*, 1–14. [CrossRef] [PubMed]
2. Choi, J.; Kim, W.K. Dietary application of tannins as a potential mitigation strategy for current challenges in poultry production: A review. *Animals* **2020**, *10*, 2389. [CrossRef]
3. Vermeulen, A.; Schaap, D.; Schetters, T.P. Control of coccidiosis in chickens by vaccination. *Vet. Parasitol.* **2001**, *100*, 13–20. [CrossRef]
4. López-Osorio, S.; Chaparro-Gutiérrez, J.J.; Gómez-Osorio, L.M. Overview of poultry Eimeria life cycle and host-parasite interactions. *Front. Vet. Sci.* **2020**, *7*, 384. [CrossRef] [PubMed]
5. Teng, P.-Y.; Yadav, S.; Shi, H.; Kim, W.K. Evaluating endogenous loss and standard ileal digestibility of amino acids in response to the graded severity levels of E. maxima infection. *Poult. Sci.* **2021**, *100*, 101426. [CrossRef]
6. Gaboriaud, P.; Sadrin, G.; Guitton, E.; Fort, G.; Niepceron, A.; Lallier, N.; Rossignol, C.; Larcher, T.; Sausset, A.; Guabiraba, R. The absence of gut microbiota alters the development of the apicomplexan parasite Eimeria tenella. *Front. Cell. Infect. Microbiol.* **2020**, *10*, 926. [CrossRef]
7. Zaman, M.A.; Iqbal, Z.; Abbas, R.Z.; Khan, M.N. Anticoccidial activity of herbal complex in broiler chickens challenged with Eimeria tenella. *Parasitology* **2012**, *139*, 237–243. [CrossRef] [PubMed]
8. Cason, J.; Hinton, A., Jr.; Northcutt, J.; Buhr, R.; Ingram, K.; Smith, D.; Cox, N. Partitioning of external and internal bacteria carried by broiler chickens before processing. *J. Food Prot.* **2007**, *70*, 2056–2062. [CrossRef] [PubMed]
9. Shang, Y.; Kumar, S.; Oakley, B.; Kim, W.K. Chicken gut microbiota: Importance and detection technology. *Front. Vet. Sci.* **2018**, *5*, 254. [CrossRef]
10. Józefiak, D.; Rutkowski, A.; Martin, S. Carbohydrate fermentation in the avian ceca: A review. *Anim. Feed Sci. Technol.* **2004**, *113*, 1–15. [CrossRef]
11. Zhang, J.-M.; Sun, Y.-S.; Zhao, L.-Q.; Chen, T.-T.; Fan, M.-N.; Jiao, H.-C.; Zhao, J.-P.; Wang, X.-J.; Li, F.-C.; Li, H.-F. SCFAs-induced GLP-1 secretion links the regulation of gut microbiome on hepatic lipogenesis in chickens. *Front. Microbiol.* **2019**, *10*, 2176. [CrossRef]
12. Broom, L.J. Evidence-based consideration of dietary 'alternatives' to anticoccidial drugs to help control poultry coccidial infections. *Worlds Poult. Sci. J.* **2021**, *77*, 43–54. [CrossRef]
13. Noack, S.; Chapman, H.D.; Selzer, P.M. Anticoccidial drugs of the livestock industry. *Parasitol. Res.* **2019**, *118*, 2009–2026. [CrossRef] [PubMed]
14. Blake, D.P.; Tomley, F.M. Securing poultry production from the ever-present Eimeria challenge. *Trends Parasitol.* **2014**, *30*, 12–19. [CrossRef]
15. Giannenas, I.; Florou-Paneri, P.; Papazahariadou, M.; Christaki, E.; Botsoglou, N.; Spais, A. Effect of dietary supplementation with oregano essential oil on performance of broilers after experimental infection with Eimeria tenella. *Arch. Anim. Nutr.* **2003**, *57*, 99–106. [CrossRef]
16. Memon, F.; Yang, Y.; Lv, F.; Soliman, A.; Chen, Y.; Sun, J.; Wang, Y.; Zhang, G.; Li, Z.; Xu, B. Effects of probiotic and Bidens pilosa on the performance and gut health of chicken during induced Eimeria tenella infection. *J. Appl. Microbiol.* **2021**, *131*, 425–434. [CrossRef]
17. Zhou, Z.; Nie, K.; Huang, Q.; Li, K.; Sun, Y.; Zhou, R.; Wang, Z.; Hu, S. Changes of cecal microflora in chickens following Eimeria tenella challenge and regulating effect of coated sodium butyrate. *Exp. Parasitol.* **2017**, *177*, 73–81. [CrossRef]
18. Song, X.; Li, Y.; Chen, S.; Jia, R.; Huang, Y.; Zou, Y.; Li, L.; Zhao, X.; Yin, Z. Anticoccidial effect of herbal powder "Shi Ying Zi" in chickens infected with Eimeria tenella. *Animals* **2020**, *10*, 1484. [CrossRef]
19. Teng, P.-Y.; Yadav, S.; de Souza Castro, F.L.; Tompkins, Y.H.; Fuller, A.L.; Kim, W.K. Graded Eimeria challenge linearly regulated growth performance, dynamic change of gastrointestinal permeability, apparent ileal digestibility, intestinal morphology, and tight junctions of broiler chickens. *Poult. Sci.* **2020**, *99*, 4203–4216. [CrossRef] [PubMed]
20. Johnson, J.; Reid, W.M. Anticoccidial drugs: Lesion scoring techniques in battery and floor-pen experiments with chickens. *Exp. Parasitol.* **1970**, *28*, 30–36. [CrossRef]
21. Short, F.; Gorton, P.; Wiseman, J.; Boorman, K. Determination of titanium dioxide added as an inert marker in chicken digestibility studies. *Anim. Feed Sci. Technol.* **1996**, *59*, 215–221. [CrossRef]
22. Lin, Y.; Olukosi, O.A. Qualitative and quantitative profiles of jejunal oligosaccharides and cecal short-chain fatty acids in broiler chickens receiving different dietary levels of fiber, protein and exogenous enzymes. *J. Sci. Food Agric.* **2021**, *101*, 5190–5201. [CrossRef] [PubMed]
23. Lourenco, J.M.; Kieran, T.J.; Seidel, D.S.; Glenn, T.C.; Silveira, M.F.D.; Callaway, T.R.; Stewart, R.L., Jr. Comparison of the ruminal and fecal microbiotas in beef calves supplemented or not with concentrate. *PLoS ONE* **2020**, *15*, e0231533. [CrossRef]

24. Christaki, E.; Florou-Paneri, P.; Giannenas, I.; Papazahariadou, M.; Botsoglou, N.A.; Spais, A.B. Effect of a mixture of herbal extracts on broiler chickens infected with Eimeria tenella. *Anim. Res.* **2004**, *53*, 137–144. [CrossRef]
25. Georgieva, N.; Koinarski, V.; Gadjeva, V. Antioxidant status during the course of Eimeria tenella infection in broiler chickens. *Vet. J.* **2006**, *172*, 488–492. [CrossRef] [PubMed]
26. Aljumaah, M.R.; Alkhulaifi, M.M.; Abudabos, A.M.; Alabdullatifb, A.; El-Mubarak, A.H.; Al Suliman, A.R.; Stanley, D. Organic acid blend supplementation increases butyrate and acetate production in Salmonella enterica serovar Typhimurium challenged broilers. *PLoS ONE* **2020**, *15*, e0232831. [CrossRef]
27. Macdonald, S.E.; Nolan, M.J.; Harman, K.; Boulton, K.; Hume, D.A.; Tomley, F.M.; Stabler, R.A.; Blake, D.P. Effects of Eimeria tenella infection on chicken caecal microbiome diversity, exploring variation associated with severity of pathology. *PLoS ONE* **2017**, *12*, e0184890. [CrossRef]
28. Kley, M.-V.; Oviedo-Rondón, E.; Dowd, S.; Hume, M.; Nalian, A. Effect of Eimeria infection on cecal microbiome of broilers fed essential oils. *ARS USDA Submiss.* **2012**, *11*, 747. [CrossRef]
29. LeBlanc, J.G.; Chain, F.; Martín, R.; Bermúdez-Humarán, L.G.; Courau, S.; Langella, P. Beneficial effects on host energy metabolism of short-chain fatty acids and vitamins produced by commensal and probiotic bacteria. *Microb. Cell. Fact.* **2017**, *16*, 1–10. [CrossRef]
30. Gasaway, W.C. Volatile fatty acids and metabolizable energy derived from cecal fermentation in the willow ptarmigan. *Comp. Biochem. Physiol. Part A Physiol.* **1976**, *53*, 115–121. [CrossRef]
31. Santos, T.S.D.; Teng, P.-Y.; Yadav, S.; Castro, F.L.D.S.; Gould, R.L.; Craig, S.W.; Chen, C.; Fuller, A.L.; Pazdro, R.; Sartori, J.R. Effects of Inorganic Zn and Cu Supplementation on Gut Health in Broiler Chickens Challenged with *Eimeria* spp. *Front. Vet. Sci.* **2020**, *7*, 230. [CrossRef]
32. Renaudeau, D.; Gilbert, H.; Noblet, J. Effect of climatic environment on feed efficiency in swine. In *Feed Efficiency in Swine*; Springer: Berlin/Heidelberg, Germany, 2012; pp. 183–210.
33. Canfora, E.E.; Blaak, E.E. Acetate: A diet-derived key metabolite in energy metabolism: Good or bad in context of obesity and glucose homeostasis? *Curr. Opin. Clin. Nutr. Metab. Care* **2017**, *20*, 477–483. [CrossRef] [PubMed]
34. Adji, A.V.; Plumeriastuti, H.; Ma'ruf, A.; Legowo, D. Histopathological Alterations of Ceca in Broiler Chickens (*Gallus gallus*) Exposed to Chronic Heat Stress. *World* **2019**, *9*, 211–217. [CrossRef]
35. Walker, R.A.; Sharman, P.A.; Miller, C.M.; Lippuner, C.; Okoniewski, M.; Eichenberger, R.M.; Ramakrishnan, C.; Brossier, F.; Deplazes, P.; Hehl, A.B. RNA Seq analysis of the Eimeria tenella gametocyte transcriptome reveals clues about the molecular basis for sexual reproduction and oocyst biogenesis. *BMC Genom.* **2015**, *16*, 94. [CrossRef]
36. Clench, M.H.; Mathias, J.R. The avian cecum: A review. *Wilson Bull.* **1995**, *107*, 93–121.
37. Teng, P.-Y.; Choi, J.; Tompkins, Y.; Lillehoj, H.; Kim, W. Impacts of increasing challenge with Eimeria maxima on the growth performance and gene expression of biomarkers associated with intestinal integrity and nutrient transporters. *Vet. Res.* **2021**, *52*, 1–12. [CrossRef] [PubMed]
38. Williams, R. Quantification of the crowding effect during infections with the seven Eimeria species of the domesticated fowl: Its importance for experimental designs and the production of oocyst stocks. *Int. J. Parasitol.* **2001**, *31*, 1056–1069. [CrossRef]
39. Chadwick, E.; Rahimi, S.; Grimes, J.; Pitts, J.; Beckstead, R. Sodium bisulfate feed additive aids broilers in growth and intestinal health during a coccidiosis challenge. *Poult. Sci.* **2020**, *99*, 5324–5330. [CrossRef]
40. Nakahiro, Y.; Isshiki, Y. Effect of cecal ligation on digestibility of crude fiber, cellulose and pentosan in chickens. *Jpn. Poult. Sci.* **1975**, *12*, 138–140. [CrossRef]
41. Van der Hoeven-Hangoor, E.; Van de Linde, I.; Paton, N.; Verstegen, M.; Hendriks, W. Effect of different magnesium sources on digesta and excreta moisture content and production performance in broiler chickens. *Poult. Sci.* **2013**, *92*, 382–391. [CrossRef] [PubMed]
42. Faradilla, Z.S.; Yunus, M.; Hermadi, H.A. The Effect of Dietary Administration of Virgin Coconut oil on Differential Leukocytes in Infected Chicken with Eimeria tenella. *J. World's Poult. Res.* **2020**, *10*, 615–622. [CrossRef]

animals

Article

Growth Performance, Antioxidant Activity, Immune Status, Meat Quality, Liver Fat Content, and Liver Histomorphology of Broiler Chickens Fed Rice Bran Oil

Shaimaa Selim [1,*], Eman Hussein [2], Nazema S. Abdel-Megeid [3], Sahar J. Melebary [4], Mohammad S. AL-Harbi [5] and Ahmed A. Saleh [6]

1 Department of Nutrition and Clinical Nutrition, Faculty of Veterinary Medicine, University of Menoufia, Shibin El-Kom 32514, Egypt
2 Department of Poultry and Fish Production, Faculty of Agriculture, University of Menoufia, Shibin El-Kom 32514, Egypt; eman.hussien@agr.menofia.edu.eg
3 Department of Cytology and Histology, Faculty of Veterinary Medicine, University of Sadat City, Sadat City 32897, Egypt; nazemah.abdelmageed@vet.usc.edu.eg
4 Department of Biology, College of Science, University of Jeddah, Jeddah 21493, Saudi Arabia; Sjmelebary@uj.edu.sa
5 Department of Biology, College of Science, Taif University, P.O. Box 11099, Taif 21944, Saudi Arabia; mharbi@tu.edu.sa
6 Department of Poultry Production, Faculty of Agriculture, Kafrelsheikh University, Kafrelsheikh 33516, Egypt; ahmed.saleh1@agr.kfs.edu.eg
* Correspondence: shimaaselim@vet.menofia.edu.eg

Citation: Selim, S.; Hussein, E.; Abdel-Megeid, N.S.; Melebary, S.J.; AL-Harbi, M.S.; Saleh, A.A. Growth Performance, Antioxidant Activity, Immune Status, Meat Quality, Liver Fat Content, and Liver Histomorphology of Broiler Chickens Fed Rice Bran Oil. *Animals* **2021**, *11*, 3410. https://doi.org/10.3390/ani11123410

Academic Editors: Woo Kyun Kim, Nesrein Hashem, Maria de Olivera and Youssef A. Attia

Received: 31 October 2021
Accepted: 27 November 2021
Published: 29 November 2021

Simple Summary: There are numerous approaches for enrichment of broiler's meat with valuable nutrients, for instance the enrichment with polyunsaturated fatty acids (PUFA). The addition of vegetable oils in the diets of broilers is an appropriate strategy to enrich the chicken meat with beneficial FA, however, this enrichment is accompanied by a lipid peroxidation with a resultant decrease in the nutritional value, quality, and shelf-life of the meat, and for that reason, the dietary supplementation with antioxidants becomes necessary. What places rice bran oil (RBO) on top of other vegetable oils is its antioxidant components and unique fatty acid profile and it is reported to induce substantial lipid-reducing effects and antioxidant properties. Therefore, this study was performed to determine the influence of RBO inclusion in the diets of broiler chickens on performance, carcass characteristics, blood parameters, meat quality, antioxidant activity, liver lipid content, and liver histological structure. RBO inclusion had a positive effect on the growing performance, dressing percentage, and immune status. Furthermore, RBO supplementation decreased the abdominal fat yield and EE content in the meat, while it increased the content of PUFA in the meat, which may be beneficial for consumers. RBO improved the antioxidant capacity of the meat and the liver, whereas it reduced the concentration of cholesterol and triglycerides in the blood, meat and liver. RBO could be used as an efficient ingredient in broiler chickens' diets to improve performance, immune status, antioxidant activity, blood lipid profile, and the nutritive value of meat.

Abstract: This trial was performed to determine the effect of rice bran oil (RBO) inclusion in diets of broiler chickens on performance, carcass characteristics, blood parameters, meat quality, antioxidant activity, liver lipid content, and liver histological structure. The 35-day feeding trial was conducted on 240 one-day-old Ross 308 broiler chickens, allocated to four treatment groups with six replicates each. RBO was examined at different inclusion levels, 0% (control), 1% ($RBO_{1\%}$), 1.5% ($RBO_{1.5\%}$), and 2% ($RBO_{2\%}$) in a completely randomized design. The results showed that at the end of the trial (35 days) the RBO supplementation had positive effects ($p < 0.001$) on the productivity parameters, but the feed intake was linearly decreased due to RBO inclusion. In addition, RBO supplementation linearly improved ($p < 0.05$) the dressing percentage, breast yield, immune organs relative weights, and meat glutathione concentration, while it decreased ($p < 0.01$) the abdominal fat yield and meat crude fat, triglycerides, cholesterol, and Malondialdehyde (MDA) contents in broiler's meat. Moreover, serum total protein, globulin, and high-density lipoprotein contents improved noticeably ($p < 0.01$)

due to offering an RBO-supplemented diet, but serum total lipids, total cholesterol, triglyceride, low-density lipoprotein, and aspartate aminotransferase concentrations linearly reduced ($p < 0.01$). The RBO supplementation augmented ($p < 0.05$) the phagocytic index, phagocytic activity, and antibody titer compared to control. On the other hand, RBO inclusion had no effect on the breast, thigh, or abdominal fat color parameters. Moreover, RBO supplementation reduced ($p < 0.01$) the content of total saturated FA (SFA), but increased ($p < 0.01$) the content of total monounsaturated FA (MUFA), and polyunsaturated FA in both breast and thigh meat. Chemical analysis of the liver tissue samples revealed that the inclusion of RBO linearly reduced ($p < 0.05$) hepatic cholesterol, triglyceride, and MDA contents. Histologically, the lipid percentage and number of lipid droplets ($p < 0.01$) were markedly lessened in the RBO-supplemented groups. The histological structure of the liver asses by light and electron microscope were normal in all groups without any pathological lesions. It is concluded that RBO could be used as a valuable ingredient in broiler chickens' diets to stimulate the growing performance and immune status, enhance the antioxidant activity and blood lipid profile, augment liver function, and improve the nutritive value of the meat.

Keywords: performance; carcass traits; blood biochemical constituents; abdominal fat color; meat fatty acid profile; liver histology; broilers

1. Introduction

A remarkable increase in worldwide meat consumption has been recorded during the last few decades, with a noteworthy increase in developing countries. The global production of chicken meat has been raised from 100.46 million tonnes (Mt) in 2014 to 118.02 Mt in 2019 [1]. Compared to other meat types, chicken meat is steadily and continuously raising worldwide owing to its low-price, health benefits, and sustainable production [2]. Meat quality is considered a valuable assessment criteria associated with the prerequisites that must be met to accomplish the consumers' demands and expectations. Consumers of the 21st Century are highly demanding ones with greater concern about meat quality, safety, and health benefits [3], which in turn necessitates the development of new approaches to enhance the meat nutritive value [4]. Currently, nutritional approaches are mainly depending on the application of natural constituents in poultry diets to enhance the nutritive value of the poultry products for better human health [2–4].

There are numerous approaches for enrichment of broiler's meat with valuable nutrients, for instance the enrichment with polyunsaturated fatty acids (PUFA). PUFA ($\omega 3$ and $\omega 6$ fatty acids, FA) are beneficial for the proper functions of the body, but these fatty acids cannot synthesize in the body and must be supplied by the diets. The addition of vegetable oils in the diets of broilers is an appropriate strategy to enrich the chicken meat with beneficial FA [5], which is in concordance with the consumers' interest and immunity perspectives [5–8]. However, enrichment of broiler's meat with PUFA is accompanied by a great susceptibility to lipid peroxidation with a resultant decrease in the nutritional value, quality, and shelf-life of the meat [5,8], and for that reason, the dietary supplementation with antioxidant becomes necessary [5]. Synthetic antioxidants are included in the diets of poultry to avoid or limit lipid peroxidation in the meat products [9]. Due to the consumers' awareness about consuming natural feed ingredients, nutritionists are seeking for natural antioxidants to replace synthetic ones [5,10]. These natural antioxidants have numerous advantages for instance lowering the incidence of metabolic diseases and improving the shelf life of meat, thus optimizing food safety and security [5].

Rice bran oil (RBO) is receiving great interest among other conventionally vegetable oils owing to its good quality, extended shelf-life, and well-proportioned fatty acid composition as well as the presence of numerous antioxidant components [11,12]. RBO is rich in tocopherols, tocotrienols, and other bioactive phytochemicals, including phytosterols, γ-oryzanol, squalene, and triterpene alcohols [13,14]. These natural bioactive components have been reported to reveal antioxidant, anti-inflammatory, and hypocholesterolaemic properties as well as boost the immune response [12]. RBO is one of the healthful edible oils because of its balanced fatty acids profile, with a ratio of 0.6:1.1:1 for saturated fatty acid (SFA), monounsaturated fatty acid (MUFA), and PUFA [11]. Oleic acid (C18:1; 42%), linoleic acid (C18:2; 32%), and palmitic acid (C16:0; 20%) are the three main fatty acids in RBO and represent about 90% of the total fatty acids in RBO [11,12]. Even though RBO has small proportion of α-linolenic acid, it is sufficient for de novo synthesis of other ω3-PUFA such as eicosapentaenoic acid and docosahexaenoic acid in tissue phospholipids [15].

What places RBO on top of other vegetable oils is its antioxidant components, which have been documented to have an outstanding nutritive significance and are recognized to induce substantial lipid-reducing effects and antioxidant properties based on research performed on mice and humans (see review, 12). Previous studies on the dietary supplementation of RBO in broilers showed enhanced growth performance, improved immune response, and reduced cholesterol concentration [16,17]. However, data concerning the impacts of dietary inclusion of RBO in the diets of broiler chickens on the antioxidant capacity, meat quality, meat fatty acid composition, hepatic lipid profile, and liver histomorphology are limited. Therefore, the present study aimed to investigate the effects of graded dietary inclusion levels of RBO on the growth performance, carcass traits, blood biochemical parameters, meat quality, abdominal fat color, lipid peroxidation, liver lipid content, and liver histological structure of broiler chickens. We hypothesized that the dietary inclusion of RBO might improve the meat quality, enhance the antioxidant capacity, and decrease meat and liver lipid content of broiler chickens owing to its unique FA profile and antioxidant constituents.

2. Materials and Methods

2.1. Ethical Approval

The care and procedure used for broiler chickens of the current trial were permitted by Institutional Animal Care and Use Committee (IACUC), Faculty of Veterinary Medicine, University of Sadat City (Ethical approval number: VUSC-018-1-20). The trial was complied with the European Union Council Directive 98/58/EC and Directive 2010/63/EU on the protection of animals used for experimental and other scientific purposes.

2.2. Experimental Design and Diets

This trial was performed on 240 one-day-old Ross 308 mixed sex broiler chickens (equal distribution between males and females in each group), obtained from a local commercial hatchery (Alwadi co., Sadat City, Menoufia, Egypt). Different levels of rice bran oil (RBO) were studied at 0% (control), 1% ($RBO_{1\%}$), 1.5% ($RBO_{1.5\%}$), and 2% ($RBO_{2\%}$) by replacing the vegetable oil (soybean oil) from the control basal diet. The broiler chickens were individually weighed and randomly allotted to four treatment groups of 60 broiler chicks each (six replicates/group) using a completely randomized design. The experiment lasted for 35 days and was allotted to three periods: starter (from 1–10 days), grower (from 11 to 24 days), and finisher periods (from 25 to 35 days). The experimental diets were formulated to be isocacloric and isonitrogenous and in accordance with Ross 308 broiler nutrition specifications [18]. Ingredients and chemical composition of the diets are presented in Table 1. Rice bran oil tested in the current experiment was obtained from Agricultural Research Center, Egypt. The fatty acid composition and γ-oryzanol of RBO and soybean oil are presented in Table 2. Feed and water were offered *ad libitum*. All broiler chicks were kept under the same management, hygienic, and environmental conditions. Broiler chickens in each replicate were housed in floor pens on wood shaves litter, at an optimum stocking density of 30 kg/m^2. The temperature and lighting programs were consistent with the recommendations of Ross breeding guide. The light schedule in all trial pens was kept at 23 L:1 D h for the first 7 days, followed by 20 L:4 D h till the end of the trial. The initial ambient temperature was approximately 32 °C during the first 7 days of life and then gradually decreased 2 °C per week till it reached 22 °C. The relative humidity was kept between 65 and 75%. No mortality was noted during the overall experimental period. The vaccination program was performed under the supervision of a veterinarian.

Table 1. Ingredients and chemical composition of the experimental diets [1].

Items, g/kg	Starter (d 1 to 10)				Grower (d 11 to 24)				Finisher (d 25 to 35)			
	CON	$RBO_{1\%}$	$RBO_{1.5\%}$	$RBO_{2\%}$	CON	$RBO_{1\%}$	$RBO_{1.5\%}$	$RBO_{2\%}$	CON	$RBO_{1\%}$	$RBO_{1.5\%}$	$RBO_{2\%}$
Yellow corn	560.0	560.0	560.0	560.0	602.6	602.6	602.6	602.6	647.9	647.9	647.9	647.9
SBM, 46% CP	319.0	319.0	319.0	319.0	284.0	284.0	284.0	284.0	230.0	230.0	230.0	230.0
Corn gluten, 62% CP	60.0	60.0	60.0	60.0	55.0	55.0	55.0	55.0	65.0	65.0	65.0	65.0
Soybean oil	20.0	15.0	10.0	0	20.0	15.0	10.0	0.0	20.0	15.0	10.0	0.0
RBO	0.0	5.0	10.0	20.0	0	5.0	10.0	20.0	0.0	5.0	10.0	20.0
Limestone	12.0	12.0	12.0	12.0	16.3	16.3	16.3	16.3	15.2	15.2	15.2	15.2
Dicalcium phosphate	17.3	17.3	17.3	17.3	11.1	11.1	11.1	11.1	10.5	10.5	10.5	10.5
L-Lysine [a]	2.34	2.34	2.34	2.34	1.86	1.86	1.86	1.86	2.41	2.41	2.41	2.41
DL-Methionine [b]	1.49	1.49	1.49	1.49	1.24	1.24	1.24	1.24	1.11	1.11	1.11	1.11
Common salt	2.5	2.5	2.5	2.5	2.5	2.5	2.5	2.5	2.5	2.5	2.5	2.5
Premix [c]	3.0	3.0	3.0	3.0	3.0	3.0	3.0	3.0	3.0	3.0	3.0	3.0
Na-bicarbonate	2.4	2.4	2.4	2.4	2.4	2.4	2.4	2.4	2.4	2.4	2.4	2.4
Calculated chemical composition												
ME, MJ/kg	12.55	12.55	12.55	12.55	12.95	12.95	12.95	12.95	13.30	13.30	13.30	13.30
CP	231.5	231.5	231.5	231.5	215.9	215.9	215.9	215.9	201.1	201.1	201.1	201.1
Calcium	9.6	9.6	9.6	9.6	8.7	8.7	8.7	8.7	8.1	8.1	8.1	8.1
Available P	4.8	4.8	4.8	4.8	4.4	4.4	4.4	4.4	4.1	4.1	4.1	4.1
Lysine	14.4	14.4	14.4	14.4	12.9	12.9	12.9	12.9	11.9	11.9	11.9	11.9
Methionine	5.6	5.6	5.6	5.6	5.1	5.1	5.1	5.1	4.8	4.8	4.8	4.8
Analyzed chemical composition												
DM	911.7	911.6	911.8	911.5	909.7	908.9	900.6	909.5	912.5	913.4	912.7	912.4
CP	230.6	230.4	230.5	230.8	214.9	215.3	215.0	215.5	200.0	199.9	200.8	200.6
EE	36.2	36.1	36.0	36.2	35.4	35.3	35.6	35.9	36.1	36.0	36.4	36.2
CF	45.4	45.7	45.8	45.3	44.9	45.1	44.9	45.6	45.2	45.0	45.3	45.7
Ash	59.8	59.8	60.2	60.0	59.8	59.2	59.5	59.8	61.0	59.9	61.2	61.1

[1] Dietary treatments: CON, basal diet without RBO; $RBO_{1\%}$, basal diet contained 1% RBO; $RBO_{1.5\%}$, basal diet contained 1.5% RBO; $RBO_{2\%}$, basal diet contained 2% RBO. [a] Supplied per kg of diet: vit. A, 12,000 IU; vit. D3, 3000 IU; vit. E, 40 mg; vit. K3, 3 mg; vit. B1, 2 mg; vit. B2, 6 mg; vit. B5, 12 mg; vi. B6, 5 mg; vitamin B12, 0.02 mg; vit. B3, 45 mg; vit. B7, 0.075 mg; vit. B9, 2 mg; Mn, 100 mg; Zn, 60 mg; Fe, 30 mg; Cu, 10 mg; I, 1 mg; Se, 0.2 mg; Co, 0.1 mg. [b] L-Lysine, lysine monohydrochloride (Feed Grade, 99%). [c] DL-Methionine, Met AMINO (DL-2-amino-4-(methyl-thio)-butane acid, DL-Met, α-amino-Y-methyl-oily acid) (Feed Grade, 99%).

Table 2. Fatty acid composition (weight %) and γ-oryzanol of rice bran oil (RBO) and soybean oil (SBO).

Items [1]		RBO	SBO
C14:0	Myristic	0.47	ND
C16:0	Palmitic	17.64	10.59
C17:0	Heptadecanoic	0.30	0.12
C18:0	Stearic	0.96	4.01
C20:0	Arachidic	0.67	0.40
C16:1	Palmitoleic	0.41	0.21
C17:1	Heptadecenoic	0.18	0.09
C18:1	Oleic	40.81	28.66
C20:1	Eicosenoic	0.71	0.3
C18:2	Linoleic	36.15	51.27
C18:3	Linolenic	1.85	4.98
SFA		20.04	15.12
MUFA		41.96	28.63
PUFA		38.00	56.25
γ-oryzanol (g/100 g oil)		3.58	-

[1] SFA, saturated fatty acids; MUFA, monounsaturated fatty acids; PUFA, polyunsaturated fatty acids. Data were collected from Agricultural Research Center, Egypt.

2.3. Broiler Performance and Carcass Traits

Throughout the trial period (1–35 days), the subsequent parameters were recorded: live weight (LW, g), weight gain (WG, g/chick), feed intake (FI, g feed/chick) and feed conversion ratio (FCR, g feed/g weight gain). Live weights of broiler chickens were recorded weekly by weighing the broiler chickens individually and FI was recorded on a daily basis. Average WG, FI, and FCR were calculated for the starter, grower, and finisher periods. At the end of the feeding trial period (35 days), 12 broiler chickens (six males and six females) of an average LW of each group were selected for slaughtering by cervical dislocation after a period of 12h fasting. The dressed carcass, breast and thigh yields, liver, gizzard, heart, spleen, thymus gland, bursa of Fabricius, and abdominal fat as a percentage of LW were recorded.

2.4. Sample Collection

At the end of the experiment (35 days), broilers were kept for blood sampling from the jugular vein, and then euthanized by cervical dislocation. After that, the breast and thigh meat samples were collected, divided into several parts, kept in plastic bags, labelled, and allocated for the subsequent analyses, involving pH determination, color measurements, proximate chemical analysis, lipid profile, lipid peroxidation, and fatty acid composition. Liver samples were also distributed into several parts for further determinations, including cholesterol and triglycerides concentrations, MDA, and histomorphology.

2.5. Blood Biochemical Parameters

At day 35, the collected blood samples were allowed to coagulate by centrifugation at 3000 rpm for 15 min. The serum samples were collected and kept at −20 °C for the clinical and biochemical analyses, involving protein and lipid metabolites, liver and kidney functions, and antibody titers. The serum biochemical constituents were analyzed spectrophotometrically (ultraviolet spectrophotometer UV4802, Unico Co., Dayton, OH, USA) using commercial analytical kits (Spectrum Diagnostics, Al Obour, Cairo, Egypt) following the manufacturer's manuals. Hemagglutination inhibition (HI) test was used to determine the antibody responses as described by Takatasy [19]. Another set of blood samples (12 blood samples/group) was collected into sterile vials having anticoagulant to determine phagocytic activity and phagocytic index, in accordance with the method of Kawahara et al. [20].

2.6. Meat Quality Measurements

The pH of the meat samples (breast and thigh) was measured in triplicate using a pH-meter (Beckman model 350, Beckman Coulter, Inc., East Lyme, CT, USA) at 24 h post-mortem as described by Egan et al. [21]. Measurements of meat samples and abdominal fat color parameters were performed at 24 h post-mortem by recording the following parameters: L* (lightness), a* (redness), and b* (yellowness) following the Commission International de l'Eclairage [22] using a model CR 410 Chroma meter (Konica Minolta, Tokyo, Japan) and documented the average of three measurements per each meat sample. Breast and thigh muscles were used in triplicate for measuring the proximate chemical analysis, including moisture (Method 950.46), crude protein (CP; Method 981.10), ether extract (EE; Method 960.39), and ash % (Method 920.153) following the procedures of AOAC [23].

Samples of breast and thigh meat were preserved in the refrigerator at a temperature of 4 °C for 7 days before measuring the lipid peroxidation. Malondialdehyde (MDA) was measured spectrophotometrically (UV4802, Unico Co., Dayton, OH, USA) following the method described by Ohkawa et al. [24] by using the analytical kits (Spectrum Diagnostics, Al Obour, Cairo, Egypt), and expressed as nmol/g of the meat. Glutathione peroxidase (GPx) activity was measured in accordance with Paglia and Valentine [25] using commercial analytical kits (Spectrum Diagnostics, Al Obour, Cairo, Egypt), and expressed as U/g of the meat. The meat cholesterol and triglyceride concentrations were performed spectrophotometrically (UV4802, Unico Co., Dayton, OH, USA) in accordance with the method of Bohac and Rhee [26]. The meat lipids were extracted according to Folch et al. [27] for FA analyses. The assessments of FA in the breast and thigh meat were done via the transformation of EE to FA methyl esters [28] using a gas chromatograph (Model GC-14A, Shimadzu Corporation, Kyoto, Japan) with a flame-ionization detector and a polar capillary column (BPX70, 0.25; SGE Incorporated, CA, USA).

2.7. Liver Lipid Content, MDA, and Histomorphology

The hepatic total cholesterol and triglycerides levels were measured using the analytical kits (Sigma-Aldrich St. Louis, MO, USA) and performed as per the manufacturer's manuals. Liver MDA levels were spectrophotometrically (UV4802, Unico Co., Dayton, OH, USA) determined according to Ohkawa et al. [24] using the analytical kits (Spectrum Diagnostics, Al Obour, Cairo, Egypt), and expressed as mg/kg of the liver.

For light microscopy, a liver sample (1 cm^2) was excised from each broiler chicken, immediately fixed in 10% neutral buffered formalin for 48 h. Afterwards, the samples were dehydrated in ascending grades series of alcohol, cleared in methyl benzoate, and embedded in paraffin wax. Sections of 5–7 µm thickness were performed using rotatory microtome and stained with Harri's Hematoxylin and eosin (H & E) for the routine histological examination [29]. The photomicrographs were taken using Leica digital camera connected with bilocular microscope for better demonstration of the results. For quantitative histological analyses, the hepatic lipid content was evaluated by assessing the percentage of the area occupied by lipid droplets inside the liver parenchyma with circularity filter to exclude artifacts such as sinusoids.

For electron microscopic examination, the liver samples were fixed in 2.5% buffered glutaraldehyde in 0.1 M phosphate buffer solution (pH 7.4) at 4 °C for 2 h, washing three times with phosphate buffer solution (PBS) (10 min, each), post fixed in 1% Osmic acid for 30 min, washing three times with PBS (10 min each), then dehydrated with ascending series of ethyl alcohol (30, 50, 70, 90% and absolute alcohol) infiltrated with acetone, for 30 min. After dehydration, samples were embedded in Araldite 502 resin. The plastic molds were cut in the Leica ultra-microtome, stained with 1% toluidine blue and photographed. After examination of semi-thin sections, ultra-thin sections were cut, stained with uranyl acetate. Then, counter stained with lead citrate [30]. The ultrathin (80 nm) sections were examined with a transmission electron microscope (JEOL-JEM-100 SX, Tokyo, Japan). For each broiler chicken, the diameters of 40 lipid droplets, the number and area of the lipid droplets in 10 randomly selected fields were determined manually. The number of lipid vacuoles in 1 mm^2 of the hepatic section as well as the area of the lipid droplets as a percent were measured.

2.8. Statistical Analysis

The trial was performed using a completely randomized design with four experimental groups of six replicates each. The normality of the data was checked by Kolmogorov Smirnov test before the statistical analysis. Data were analyzed by One-way ANOVA using IBM SPSS statistical package (version 22, SPSS Inc., Chicago, IL, USA) to evaluate the effect of treatment, along with a Tukey's test ($p < 0.05$). The replicate pen was used as an experimental unit for the performance data and the bird for the other variables. Orthogonal polynomial contrast test for linear, quadratic, and cubic to estimate the effects of RBO supplementation. Significance was set at $p < 0.05$ and values are presented as means ± standard error of the mean (SEM).

3. Results

3.1. Broilers Growth Performance

Growth performance data of broiler chickens fed the control and graded levels of RBO diets are shown in Table 3. Broiler chickens consuming the 1.5 and 2% RBO diets had a significantly higher ($p < 0.001$) LW during the starter and grower periods compared with those consuming the control diet, this was supported by linear and quadratic effects ($p < 0.001$) due to the inclusion of RBO in the diets of broiler chickens. Positive total weight gains ($p < 0.001$; linear, $p < 0.001$) were achieved for broiler chickens consuming the RBO diets than those fed on the control diet. However, broiler chickens fed the RBO diets had a significantly ($p < 0.001$; linear, $p < 0.001$; quadratic, $p < 0.001$) lower FI compared with those consuming the control diet. The mean FCR of the RBO treatment groups was significantly improved ($p < 0.001$) during the starter, grower, and whole the experimental periods compared to the control group, this was indicated by linear ($p < 0.001$) and quadratic ($p < 0.001$) responses as a result of the inclusion of RBO in their diets.

Table 3. Growth performance of broiler chickens fed diets contained graded level of rice bran oil (RBO) during the experiment (1–35 days of age).

Items [2]	Treatments [1]				SEM	p-Value	Contrasts		
	CON	$RBO_{1\%}$	$RBO_{1.5\%}$	$RBO_{2\%}$			Linear	Quadratic	Cubic
Initial BW (0 d), g	43.00	42.95	43.05	42.95	0.264	0.978	0.95	0.89	0.676
				1–21 day of age					
LW (21 d), g	735.79 ^{c}	775.06 ^{c}	853.67 ^{b}	948.04 ^{a}	17.932	<0.001	<0.001	0.03	0.679
Total WG, g	692.79 ^{c}	732.11 ^{c}	810.62 ^{b}	905.09 ^{a}	17.89	<0.001	<0.001	0.03	0.63
Daily WG, g	32.99 ^{c}	34.86 ^{c}	38.60 ^{b}	43.10 ^{a}	1.85	<0.001	<0.001	0.03	0.68
Daily FI, g	53.56 ^{a}	50.29 ^{b}	49.60 ^{b}	50.98 ^{b}	1.83	<0.001	<0.001	<0.001	<0.001
FCR	1.63 ^{a}	1.45 ^{b}	1.29 ^{c}	1.19 ^{c}	0.041	<0.001	<0.001	0.19	0.71
				22–35 day of age					
LW (35 d), g	2065.2 ^{c}	2137.5 ^{b}	2236.2 ^{a}	2264.5 ^{a}	16.47	<0.001	<0.001	0.06	0.06
Total WG, g	1329.4	1362.4	1382.5	1347.7	27.16	0.06	0.83	0.01	0.39
Daily WG, g	94.96	97.32	98.75	94.03	1.94	0.06	0.83	0.01	0.39
Daily FI, g	163.80 ^{a}	159.32 ^{a}	159.53 ^{a}	144.91 ^{b}	3.50	<0.001	<0.001	<0.001	<0.001
FCR	1.73 ^{a}	1.65 ^{ab}	1.62 ^{bc}	1.54 ^{c}	0.034	<0.001	<0.001	0.84	0.31
				1–35 day of age					
Total WG, g	2022.2 ^{c}	2094.5 ^{b}	2193.1 ^{a}	2221.6 ^{a}	16.42	<0.001	<0.001	0.06	0.06
Daily WG, g	57.77 ^{b}	59.84 ^{b}	62.66 ^{ab}	63.47 ^{a}	1.50	<0.001	<0.001	0.06	0.07
Daily FI, g	106.23 ^{a}	99.62 ^{b}	93.86 ^{c}	88.55 ^{c}	2.53	<0.001	<0.001	0.07	0.80
FCR	1.69 ^{a}	1.57 ^{b}	1.49 ^{c}	1.39 ^{d}	0.016	<0.001	<0.001	0.31	0.18

$^{a–d}$ Means within the same row having different letters are varied at $p < 0.05$; Tukey's tests were applied to compare means; SEM = Standard error of the mean. [1] Dietary treatments: CON, basal diet without RBO; $RBO_{1\%}$, basal diet contained 1% RBO; $RBO_{1.5\%}$, basal diet contained 1.5% RBO; $RBO_{2\%}$, basal diet contained 2% RBO. [2] LW, live weight; WG, weight gain; FI, feed intake; FCR, feed conversion ratio.

3.2. Carcass Traits

Carcass traits of broiler chickens are shown in Table 4. The inclusion of RBO in the diets of broiler chickens influenced the dressing per cent, with significantly greater values ($p < 0.001$; linear, $p < 0.001$; quadratic, $p < 0.05$; cubic, $p < 0.01$) being observed in the RBO group compared to the control group. RBO inclusion to the broiler diets at levels of 1.5 and 2% led to an increase ($p < 0.001$; linear, $p < 0.05$; quadratic, $p < 0.01$; cubic, $p < 0.001$) in the breast meat yield compared to those fed the control and $RBO_{1\%}$ diets. Non-significant difference ($p > 0.05$) was noted for the thigh, liver, gizzard, or heart yields between the treatment groups. However, the relative weight of abdominal fat decreased ($p < 0.05$; linear, $p < 0.01$, quadratic, $p < 0.01$) with the dietary inclusion of RBO compared to control. The immune organs' relative weights were significantly increased ($p < 0.01$) in the RBO groups compared to the control group, this was supported by linear response ($p < 0.001$).

3.3. Blood Parameters

The serum biochemical parameters of the broiler chickens fed the control and graded levels of RBO are presented in Table 5. Serum total protein and globulin showed linear ($p < 0.00$) and quadratic ($p < 0.05$) responses to the increasing dietary level of RBO, with a maximum corresponding to the $RBO_{1.5\%}$ and $RBO_{2\%}$ groups ($p < 0.01$), while there was non-significant difference in serum albumin concentration among the treatment groups. The serum concentrations of total lipids and triglycerides were linearly and quadratically ($p < 0.001$) decreased in the RBO groups compared to the control group. The RBO-supplemented groups exhibited significantly lower ($p < 0.001$) serum total cholesterol compared to control; this was characterized by linear ($p < 0.001$) and quadratic ($p < 0.05$) responses due to the inclusion of RBO in the diets of broiler chickens. RBO supplementation to broiler diets led to an increase ($p < 0.001$) in the serum HDL concentration and a decrease ($p < 0.05$) in the serum LDL level compared to the control group.

Table 4. Carcass traits (% of live body weight) and organs yields of broiler chickens fed diets contained graded level of rice bran oil (RBO) at 35 days of age.

Items	Treatments [1]				SEM	*p*-Value	Contrasts		
	CON	$RBO_{1\%}$	$RBO_{1.5\%}$	$RBO_{2\%}$			Linear	Quadratic	Cubic
				Slaughter parameter					
Dressing, %	73.34 [c]	75.65 [b]	76.07 [b]	77.98 [a]	0.460	<0.001	<0.001	0.039	0.006
				Muscle yields					
Breast yield, %	29.88 [c]	30.27 [bc]	31.49 [ab]	31.85 [a]	0.427	<0.001	0.01	0.007	<0.001
Thigh yield, %	21.70	21.85	21.37	21.91	0.509	0.743	0.96	0.33	0.65
				Organ yields					
Liver, %	2.51	2.56	2.55	2.53	0.140	0.50	0.23	0.12	0.38
Heart, %	0.47	0.50	0.47	0.49	0.031	0.11	0.73	0.42	0.30
Gizzard, %	2.55	2.56	2.58	2.59	0.210	0.62	0.52	0.18	0.29
Abdominal fat, %	2.38 [a]	1.75 [b]	1.72 [b]	1.70 [b]	0.190	0.014	0.003	0.002	0.72
				Immune organs					
Spleen, %	0.070 [b]	0.071 [b]	0.091 [a]	0.089 [a]	0.002	<0.001	<0.001	0.23	0.54
Thymus gland, %	0.097 [d]	0.102 [c]	0.112 [b]	0.126 [a]	0.003	<0.001	<0.001	0.06	0.63
Bursa of Fabricius, %	0.094 [d]	0.111 [c]	0.122 [b]	0.137 [a]	0.002	<0.001	<0.001	0.44	0.17

[a–d] Means within the same row having different letters are varied at $p < 0.05$; Tukey's tests were applied to compare means; SEM = Standard error of the mean. [1] Dietary treatments: CON, basal diet without RBO; $RBO_{1\%}$, basal diet contained 1% RBO; $RBO_{1.5\%}$, basal diet contained 1.5% RBO; $RBO_{2\%}$, basal diet contained 2% RBO.

Table 5. Blood biochemical constituents of broiler chickens fed the experimental diets at 35 days of age.

Items [2]	Treatments [1]				SEM	*p*-Value	Contrasts		
	CON	$RBO_{1\%}$	$RBO_{1.5\%}$	$RBO_{2\%}$			Linear	Quadratic	Cubic
				Protein metabolites					
Total protein, g/dL	4.18 [b]	4.50 [b]	5.06 [a]	4.96 [a]	0.137	<0.001	<0.001	0.04	0.05
Albumin, g/dL	3.06	3.16	3.40	3.46	0.148	0.06	0.11	0.85	0.50
Globulin, g/dL	1.12 [b]	1.34 [ab]	1.66 [a]	1.50 [a]	0.122	0.003	0.002	0.04	0.15
A/G	2.79	2.41	2.08	2.35	0.272	0.12	0.07	0.12	0.55
				Lipid metabolites					
Total lipids, mg/dL	664.95 [a]	518.30 [b]	424.67 [c]	405.68 [d]	4.794	<0.001	<0.001	<0.001	0.11
Total cholesterol, mg/dL	150.39 [a]	136.52 [b]	128.37 [c]	119.94 [d]	2.444	<0.001	<0.001	0.03	0.23
Triglycerides,	126.94 [a]	118.55 [b]	97.77 [d]	103.33 [c]	1.784	<0.001	<0.001	<0.001	<0.001
LDL, mg/dL	52.47 [a]	39.70 [b]	21.74 [d]	33.56 [c]	0.552	0.02	0.01	0.05	0.18
HDL, mg/dL	82.80 [c]	86.47 [b]	88.10 [b]	93.47 [a]	1.182	<0.001	<0.001	<0.001	<0.001
				Liver functions					
AST, U/dL	77.00 [a]	75.00 [ab]	71.80 [b]	73.60 [b]	1.183	0.003	0.002	0.04	0.12
ALT, U/dL	69.00	67.80	65.80	66.60	1.149	0.06	0.02	0.24	0.34
ALP, U/L	12.34	11.22	11.14	11.52	0.439	0.06	0.09	0.03	0.68
				Kidney functions					
Creatinine, mg/dL	0.52	0.53	0.47	0.50	0.190	0.20	0.51	0.46	0.22
Uric acid, mg/dL	5.56	5.52	5.60	5.70	0.332	0.12	0.34	0.34	0.31
				Antibody titer					
HIND, log^2	3.80 [c]	4.80 [b]	5.40 [b]	6.00 [a]	0.201	<0.001	<0.001	0.59	0.34
HIAI, log^2	4.00	3.80	4.60	4.40	0.351	0.15	0.09	0.24	0.09
HIBD, log^2	3.40	3.20	3.80	3.60	0.316	0.29	0.24	0.30	0.12
				Phagocytic activity					
Phagocytic index	3.57 [b]	3.68 [b]	4.18 [ab]	4.84 [a]	0.390	0.02	0.03	0.18	0.13
Phagocytic activity	61.80 [b]	65.80 [ab]	66.80 [ab]	71.40 [a]	2.577	0.01	0.02	0.13	0.17

[a–d] Means within the same row having different letters are varied at $p < 0.05$; Tukey's tests were applied to compare means; SEM = Standard error of the mean. [1] Dietary treatments: CON, basal diet without RBO; $RBO_{1\%}$, basal diet contained 1% RBO; $RBO_{1.5\%}$, basal diet contained 1.5% RBO; $RBO_{2\%}$, basal diet contained 2% RBO. [2] A/G, albumin to globulin ratio; LDL, low density lipoprotein; HDL, high density lipoprotein; AST, aspartate aminotransferase; ALT, alanine aminotransferase; ALP, alkaline phosphatase; HIND, Hemagglutination-inhibition test for Newcastle disease; HIAI, Hemagglutination-inhibition test for Avian Influenza; HIBD, Hemagglutination-inhibition test for bursa disease.

Serum AST showed linear and quadratic responses ($p < 0.05$) to the increasing dietary RBO levels, with lower values being noted for the $RBO_{1.5\%}$ and $RBO_{2\%}$ groups, while only a linear response ($p < 0.05$) for lower serum ALT and ALP concentrations were recorded. Serum creatinine and uric acid concentrations did not significantly vary ($p > 0.05$) among dietary treatments. The dietary inclusion of RBO did not affect the HI test for infectious bronchitis virus (HIBB) or bursa disease (HIBD). However, the RBO-supplemented groups exhibited significantly greater ($p < 0.001$) antibody titers to Newcastle Disease (HIND) compared to the CON group, this was indicated by a linear response ($p < 0.001$). Furthermore, a significant increase ($p < 0.05$; linear, $p < 0.05$) was also observed for the phagocytic activity and phagocytic index in the groups that included RBO in their diets compared to control.

3.4. Meat Quality Traits

Tables 6 and 7 showed the color parameters, pH, proximate chemical composition, and antioxidant capacity data of breast and thigh meat, respectively. Overall, there were non-significant differences in pH, L*, a*, or b* of breast and thigh meat of broiler chickens fed the RBO diets compared with those fed the control diet. The inclusion of RBO to the diets of broiler chickens linearly, quadratically, and cubically decreased the EE values ($p < 0.001$) but did not influence the moisture, CP, or ash contents of both breast and thigh meat when compared with the control ones. Dietary RBO supplementation linearly and quadratically, and cubically decreased the triglycerides and cholesterol contents of both breast and thigh meat ($p < 0.01$) compared to the control group. The concentration of meat MDA was significantly lower ($p < 0.001$) for broiler chickens consumed the RBO-diets than the control ones; this was strongly indicated by linear response ($p < 0.001$). On the other hand, the meat (breast and thigh) of broiler chickens consumed the RBO diets had a significantly greater GPx content than those consumed the control diet ($p < 0.001$; linear, $p < 0.001$; quadratic, $p < 0.05$, cubic, $p < 0.05$).

Table 6. Breast meat physical, chemical, lipid and antioxidant characteristics of broiler chickens fed the experimental diets at 35 days of age.

Items [2]	Treatments [1]				SEM	*p*-Value	Contrasts		
	CON	$RBO_{1\%}$	$RBO_{1.5\%}$	$RBO_{2\%}$			Linear	Quadratic	Cubic
Physical characteristics									
pH	6.10	6.07	6.07	6.11	0.019	0.11	0.56	0.20	0.97
Color									
L*	55.96	52.86	55.73	55.48	1.502	0.10	0.80	0.20	0.03
a*	3.02	3.27	3.16	3.40	0.497	0.22	0.28	0.12	0.56
b*	11.28	12.05	11.89	11.76	0.708	0.60	0.56	0.30	0.56
Proximate chemical composition									
Moisture, %	71.06	71.04	71.54	71.67	0.752	0.006	0.11	0.98	0.24
CP, %	24.40	24.50	24.30	24.21	0.591	0.12	0.17	0.13	0.27
EE, %	2.72 ^{a}	2.74 ^{a}	2.20 ^{b}	2.29 ^{b}	0.064	<0.001	<0.001	<0.001	<0.001
Ash, %	1.82	1.70	1.94	1.83	0.159	0.43	0.64	0.70	0.35
Lipid content									
Triglycerides, mg/100 g	57.84 ^{a}	44.15 ^{b}	40.44 ^{c}	40.39 ^{c}	0.581	<0.001	<0.001	<0.001	0.009
Cholesterol, mg/100 g	88.87 ^{a}	70.20 ^{b}	44.32 ^{c}	47.10 ^{c}	0.563	<0.001	<0.001	<0.001	<0.001
Antioxidant capacity									
GPx, U/g	0.77 ^{d}	1.13 ^{c}	1.40 ^{b}	2.02 ^{a}	0.057	<0.001	<0.001	0.01	0.04
MDA, mg/kg	0.86 ^{a}	0.71 ^{b}	0.62 ^{c}	0.38 ^{d}	0.013	<0.001	<0.001	0.001	0.003

$^{a–d}$ Means within the same row having different letters are varied at $p < 0.05$; Tukey's tests were applied to compare means; SEM = Standard error of the mean. [1] Dietary treatments: CON, basal diet without RBO; $RBO_{1\%}$, basal diet contained 1% RBO; $RBO_{1.5\%}$, basal diet contained 1.5% RBO; $RBO_{2\%}$, basal diet contained 2% RBO. [2] L*, lightness; a*, redness; b*, yellowness; CP, crude protein; EE, ether extract; GPx, glutathione peroxidase; MDA, malondialdehyde.

Table 7. Thigh meat physical, chemical, lipid and antioxidant characteristics of broiler chickens fed the experimental diets at 35 days of age.

Items [2]	Treatments [1]				SEM	*p*-Value	Contrasts		
	CON	$RBO_{1\%}$	$RBO_{1.5\%}$	$RBO_{2\%}$			Linear	Quadratic	Cubic
				Physical characteristics					
pH	6.25	6.22	6.15	6.22	0.023	0.25	0.42	0.34	0.13
				Color					
L*	55.59	54.34	54.53	55.77	1.428	0.73	0.89	0.27	0.93
a*	2.76	3.23	3.43	3.25	0.305	0.15	0.18	0.26	0.57
b*	11.73	12.31	11.81	12.34	0.690	0.36	0.95	0.72	0.07
				Proximate chemical composition					
Moisture, %	72.04	72.08	72.53	72.50	0.712	0.69	0.20	0.70	0.61
CP, %	21.62	21.86	22.22	22.58	0.802	0.19	0.25	0.36	0.65
EE, %	5.25 [a]	5.00 [a]	4.15 [b]	3.83 [b]	0.15	<0.001	<0.001	<0.001	<0.001
Ash, %	1.08	1.06	1.09	1.09	0.07	0.32	0.41	0.59	0.19
				Lipid content					
Triglycerides, mg/100 g	69.22 [a]	59.55 [b]	60.92 [b]	51.95 [c]	0.681	<0.001	<0.001	<0.001	<0.001
Cholesterol, mg/100 g	177.27 [a]	162.71 [b]	139.68 [c]	115.03 [d]	2.205	<0.001	<0.001	0.01	0.35
				Antioxidant capacity					
GPx, U/g	0.78 [d]	0.94 [c]	1.30 [b]	1.61 [a]	0.021	<0.001	<0.001	0.002	0.007
MDA, mg/kg	0.89 [a]	0.74 [b]	0.51 [c]	0.44 [c]	0.043	<0.001	<0.001	0.25	0.14

[a–d] Means within the same row having different letters are varied at $p < 0.05$; Tukey's tests were applied to compare means; SEM = Standard error of the mean. [1] Dietary treatments: CON, basal diet without RBO; $RBO_{1\%}$, basal diet contained 1% RBO; $RBO_{1.5\%}$, basal diet contained 1.5% RBO; $RBO_{2\%}$, basal diet contained 2% RBO. [2] L*, lightness; a*, redness; b*, yellowness; CP, crude protein; EE, ether extract; GPx, glutathione peroxidase; MDA, malondialdehyde.

Fatty acid composition of the breast and thigh muscles is shown in Table 8. The inclusion of RBO to broiler diets resulted in a decrease ($p < 0.01$, linear, $p < 0.01$) in SFA content and an increase in the MUFA, PUFA, n-6 FA and n-3 FA, in breast meat of the $RBO_{1.5\%}$ and $RBO_{2\%}$ groups, compared to the control group. Furthermore, a significant elevation ($p < 0.001$) was detected for the n-6:n-3 FA (linear and quadratic, $p < 0.001$), MUFA:SFA (linear and quadratic, $p < 0.001$), and PUFA:SFA (linear, $p < 0.001$) in the groups that contained RBO compared to control. Concerning the thigh meat, there was also a significant decrease ($p < 0.001$) in SFA content, while there was an enhancement ($p < 0.001$) for the MUFA, PUFA, and n-6 FA in the groups supplemented with RBO, this was supported by linear and quadratic responses ($p < 0.01$) compared to the control group. The ratios of n-6 to n-3 FA, MUFA to SFA, and PUFA to SFA were significantly greater ($p < 0.001$, linear, $p < 0.001$) in the thigh meat of the RBO groups than control.

3.5. Abdominal Fat Color Parameters

Abdominal fat coloration is presented in Table 9. Color parameters of abdominal fat were not influenced by the feeding of RBO for 35 days.

3.6. Liver Lipid Content and MDA

Liver cholesterol, triglycerides, and MDA contents of broiler chickens fed the experimental diets are presented in Table 10. Broiler chickens consumed the RBO diets had significantly lower ($p < 0.001$) liver cholesterol, triglycerides, and MDA contents when compared with those consumed the control diet, this was supported by linear ($p < 0.001$) and quadratic ($p < 0.05$) effects.

Table 8. Meat fatty acid profile (% total fatty acids) of broiler chickens fed the experimental diets at 35 days of age.

Items [2]	Treatments [1]				SEM	*p*-Value	Contrasts		
	CON	$RBO_{1\%}$	$RBO_{1.5\%}$	$RBO_{2\%}$			Linear	Quadratic	Cubic
				Breast meat					
SFA	35.00 ^{a}	34.29 ^{a}	32.12 ^{b}	31.22 ^{b}	0.286	<0.001	<0.001	0.65	0.02
MUFA	38.17 ^{c}	38.82 ^{bc}	39.83 ^{a}	39.37 ^{ab}	0.269	0.001	0.001	0.02	0.06
PUFA	26.83 ^{b}	26.89 ^{b}	28.05 ^{ab}	29.41 ^{a}	0.433	0.001	<0.001	0.07	0.53
n-6 FA	23.50 ^{c}	23.52 ^{c}	24.60 ^{b}	25.96 ^{a}	0.279	<0.001	<0.001	0.002	0.29
n-3 FA	3.33 ^{b}	3.37 ^{ab}	3.45 ^{a}	3.45 ^{a}	0.041	0.04	0.009	0.51	0.38
n6:n3	7.06 ^{c}	6.98 ^{d}	7.13 ^{b}	7.52 ^{a}	0.01	<0.001	<0.001	<0.001	0.003
MUFA:SFA	1.09 ^{d}	1.13 ^{c}	1.24 ^{b}	1.26 ^{a}	0.002	<0.001	<0.001	<0.001	<0.001
PUFA:SFA	0.77 ^{c}	0.78 ^{c}	0.87 ^{b}	0.94 ^{a}	0.02	<0.001	<0.001	0.11	0.19
				Thigh meat					
SFA	36.02 ^{a}	33.85 ^{b}	32.77 ^{c}	33.62 ^{bc}	0.286	<0.001	<0.001	0.001	0.38
MUFA	38.96 ^{c}	38.99 ^{c}	40.93 ^{a}	39.62 ^{b}	0.436	<0.001	<0.001	<0.001	<0.001
PUFA	25.02 ^{b}	27.16 ^{a}	26.30 ^{a}	26.76 ^{a}	0.327	0.001	0.003	0.007	0.003
n-6 FA	21.65 ^{b}	23.74 ^{a}	22.92 ^{a}	23.50 ^{a}	0.269	<0.001	0.001	0.004	0.001
n-3 FA	3.37	3.42	3.38	3.26	0.057	0.10	0.07	0.06	0.95
n6:n3	6.42 ^{d}	6.94 ^{b}	6.78 ^{c}	7.21 ^{a}	0.036	<0.001	<0.001	0.12	<0.001
MUFA:SFA	1.08 ^{c}	1.15 ^{b}	1.25 ^{a}	1.18 ^{b}	0.009	<0.001	<0.001	<0.001	<0.001
PUFA:SFA	0.69 ^{b}	0.80 ^{a}	0.80 ^{a}	0.79 ^{a}	0.015	<0.001	<0.001	0.001	0.06

$^{a–d}$ Means within the same row having different letters are varied at $p < 0.05$; Tukey's tests were applied to compare means; SEM = Standard error of the mean. [1] Dietary treatments: CON, basal diet without RBO; $RBO_{1\%}$, basal diet contained 1% RBO; $RBO_{1.5\%}$, basal diet contained 1.5% RBO; $RBO_{2\%}$, basal diet contained 2% RBO. [2] SFA, saturated fatty acids; MUFA, monounsaturated fatty acids; PUFA, polyunsaturated fatty acids; FA, fatty acids.

Table 9. Abdominal fat color parameters of broiler chickens fed the experimental diet at 35 days of age.

Items [2]	Treatments [1]				SEM	*p*-Value	Contrasts		
	CON	$RBO_{1\%}$	$RBO_{1.5\%}$	$RBO_{2\%}$			Linear	Quadratic	Cubic
L*	70.30	69.84	69.80	69.92	0.741	0.08	0.10	0.25	0.35
a*	6.40	6.81	6.71	6.59	0.365	0.20	0.31	0.16	0.89
b*	26.82	27.21	27.09	27.40	0.625	0.13	0.08	0.67	0.17

Tukey's tests were applied to compare means; SEM = Standard error of the mean. [1] Dietary treatments: CON, basal diet without RBO; $RBO_{1\%}$, basal diet contained 1% RBO; $RBO_{1.5\%}$, basal diet contained 1.5% RBO; $RBO_{2\%}$, basal diet contained 2% RBO. [2] L*, lightness; a*, redness; b*, yellowness.

Table 10. Liver triglycerides, cholesterol and Malondialdehyde (MDA) concentration of broiler chickens fed the experimental diets.

Items	Treatments [1]				SEM	*p*-Value	Contrasts		
	CON	$RBO_{1\%}$	$RBO_{1.5\%}$	$RBO_{2\%}$			Linear	Quadratic	Cubic
Cholesterol, mg/g	4.87 ^{a}	4.08 ^{b}	3.22 ^{c}	2.15 ^{d}	0.077	<0.001	<0.001	0.04	0.62
Triglycerides, mg/g	7.49 ^{a}	6.69 ^{b}	6.05 ^{c}	5.05 ^{d}	0.049	<0.001	<0.001	<0.001	0.35
MDA, mg/kg	1.78 ^{a}	0.64 ^{b}	0.51 ^{b}	0.38 ^{b}	0.021	<0.001	<0.001	<0.001	0.58

$^{a–d}$ Means within the same row having different letters are varied at $p < 0.05$; Tukey's tests were applied to compare means; SEM = Standard error of the mean. [1] Dietary treatments: CON, basal diet without RBO; $RBO_{1\%}$, basal diet contained 1% RBO; $RBO_{1.5\%}$, basal diet contained 1.5% RBO; $RBO_{2\%}$, basal diet contained 2% RBO.

3.7. *Liver Histomorphology*

3.7.1. Light Microscope

Figure 1 presents the photomicrographs of H & E stained liver sections and the histological measurements of broiler chickens fed the control and RBO-supplemented groups. Overall, the histological structures of the liver in all the treatment groups were normal. No pathological lesions or inflammatory changes were detected in the liver of the studied broiler chickens. The liver surface is covered by a layer of thin connective tissue. The hepatocytes were arranged in clusters. The cytoplasm was acidophilic without cytoplasmic vacuoles. Lymphocytes' aggregations were typically gathered around the portal area, the central veins, and among the hepatic plates. The parenchyma mainly composed of rows of conically shaped hepatocytes bordering sinusoids. The hepatocytes were appeared in a hexagonal design creating hepatic plates. The plates were looked irregularly from the edge of each hepatic lobe to the central vein. The hepatic sinusoids were lined by flat endothelial cells and large von kupffer cells. The percentage of hepatic lipid storage were linearly and quadratically decreased ($p < 0.01$) in the RBO-supplemented groups compared to the control ones.

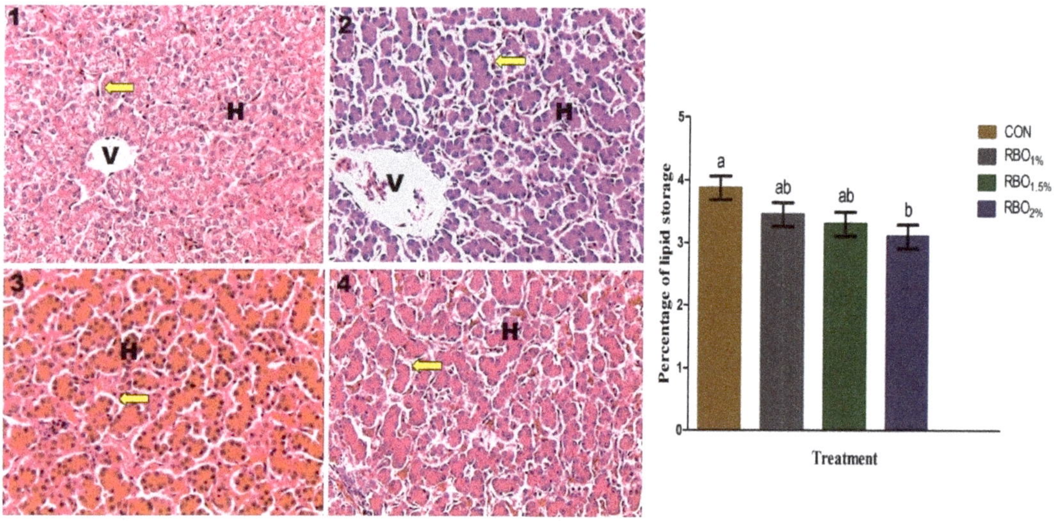

Figure 1. Histological structure of the liver at day 35 of age of broiler chickens fed the basal control diet without RBO (CON, 1), basal diet contained 1% RBO ($RBO_{1\%}$, 2), basal diet contained 1.5% RBO ($RBO_{1.5\%}$, 3), or basal diet contained 2% RBO ($RBO_{2\%}$, 4). Histomicrograph of the liver showing normal liver structure among the treatment groups. H, hepatocytes; V, central vein. Yellow arrows represent hepatic sinusoids between hepatocytes (H & E X100). Percentage of lipid storage estimated via histological method, ^{a,b} means with different letters are varied at $p < 0.05$. Data are shown as mean ± SEM.

3.7.2. Electron Microscope

Figure 2 shows the ultra-structure and the electron microscope measurements of the liver at day 35 of age of the broiler chickens. The ultra- structure of the liver of the ross chickens, at 35 days of age, showed that the hepatic parenchyma was arranged into anastomosing plates of two cell layers or in acinus like arrangement of four or more cells in cross sections. The free surface of the hepatocytes faced the sinusoids, while the central region of the plates contained the bile canaliculi formed by the involvement of two or more hepatocytes. The hepatocyte is appeared as a polygonal shape, with a round nucleus containing one or more nucleoli. The cytoplasm contains some cytoplasmic organelles as mitochondria are numerous, oval to elongate in shape, surrounded by flattened cisterna of rough endoplasmic reticulum. The smooth endoplasmic reticulum appeared as isolated circular area lying scattered throughout the cellular cytoplasm. Few glycogen granules appeared and some lysosome and lipid droplets were observed in the cellular cytoplasm. The diameters of the hepatic lipid droplets were similar in all the treatment groups. However, the percentage of hepatic lipid and the number of lipid droplets were linearly and quadratically reduced ($p < 0.001$) in the liver of broiler chickens fed the RBO diets compared to those fed the control diet.

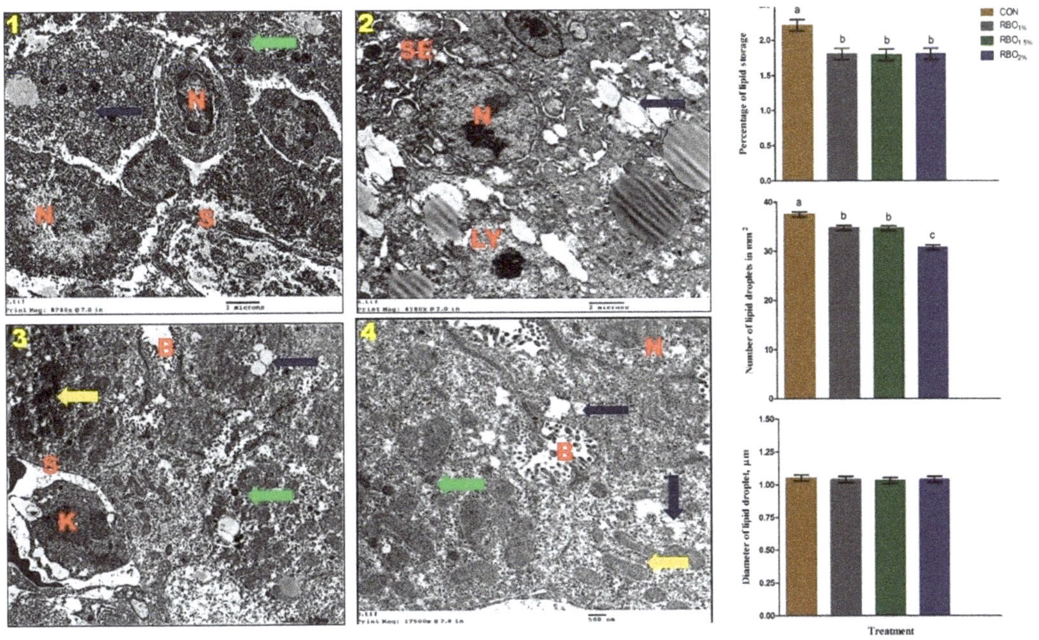

Figure 2. Transmission electron microscope of the liver at day 35 of broiler chickens fed the basal control diet without RBO (CON, 1), basal diet contained 1% RBO ($RBO_{1\%}$, 2), basal diet contained 1.5% RBO ($RBO_{1.5\%}$, 3), or basal diet contained 2% RBO ($RBO_{2\%}$, 4). B, bile canaliculi; S, sinusoids; K, Kupffer cell; N, nucleus of hepatocytes; SE, smooth endoplasmic reticulum; and Ly, lysosomes. Green, yellow, and blue arrows represent glycogen granules, shows mitochondria, and lipid droplet, respectively. Percentage of lipid storage, number and diameter of lipid droplets estimated via histological method, $^{a\text{–}c}$ means with different letters are varied at $p < 0.05$. Data are shown as mean ± SEM.

4. Discussion

The demands for healthy foods, such as nutrient-enriched animal products with PUFA and natural antioxidants, are rising. In addition, there is an increasing awareness among consumers about accessing the quality of animal products. RBO is considered one of the healthiest vegetable oils owing to its ideal FA composition and the presence of bioactive components such as γ-oryzanols, tocopherols, tocotrienols, polyphenols, sterols, and phenolic acids [11,12]. These bioactive components were reported to decrease the oxidative stress, blood cholesterol, and LDL levels in hyperlipidemic human subjects [31] or animal models [32,33], and they suggested that RBO may decrease the risk factors of cardiovascular diseases. Therefore, RBO can be an appropriate candidate as a natural source of PUFA and antioxidant constituents in the broilers' diets to enhance the production performance, health status, antioxidant capacity, and product quality.

In our study, there was an improvement of the final LW and total WG of the RBO groups compared to the control group. Our finding revealed that between the three RBO groups, a greater inclusion level ($RBO_{2\%}$) resulted in higher total WG and greater average daily WG as well as better FCR. However, daily FI of broilers during the entire experimental period was decreased by the inclusion of RBO in diets. The beneficial impacts of RBO on final LW, WG, and FCR were reported previously by Anitha et al. [34], which was enhanced when it included up to 3% in the diets of broilers. Similar findings concerning the FCR and LW enhancement of broilers were achieved by using 2% RBO [17]. However, these earlier studies reported no significant difference in FI among the experimental groups [17,34]. The beneficial effects of RBO on the broiler's performance may be due to its bioactive components, including γ-oryzanols, tocopherols, tocotrienols, polyphenols, sterols, and ferulic acid [12,17]. In previous studies, vegetable oil reduced the passage rate of feed in the gut, which permits more time for superior nutrient absorption and utilization [35,36], leading to a more effective utilization of nutrients from diet. The reduction in FI noticed herein of broiler chickens fed RBO may be attributed to PUFA and its high energy-yielding capacity [8,35,37]. Attia et al. [8] observed that the inclusion of vegetable oils rich in SFA, i.e., palmitic and stearic acids in the diets of broilers increased their FI when compared with the PUFA-enrich oils, i.e., linolenic and linoleic acids. Similarly, it was reported that SFA has low digestibility compared to UFA, especially during the initial stage of life [8,35,37].

The carcass parameters, determined at day 35 of age, showed that broiler chickens who were fed the RBO-enriched diets had a greater dressing percentage and breast yield, but a lower abdominal fat per cent compared to the control ones. These findings suggested that RBO augments the availability of energy for muscle development, whereas the reduce in abdominal fat per cent in the RBO groups indicates a shifting in energy expenditure for meat growth instead of an accumulation in the body, in particular the abdominal cavity [8]. It has been demonstrated that the sources and levels of PUFA considerably affected the carcass traits of broilers [8,35]. The dietary inclusion of oils rich in PUFA was reported to improve the oxidation and decrease FA synthesis, with a resultant reduction of abdominal fat accumulation in broiler chickens [38,39]. At a molecular level, Ahmed et al. [40] reported that RBO treatment overwhelmed the elevated hepatic de novo lipogenesis of the insulin resistance rats via downregulation of lipogenic genes, i.e., superoxide dismutase and catalase. Accordingly, the suppressing effect of RBO may be owing to its un-saponifiable components, which was also found to downregulate sterol regulatory element binding protein-1 [41]. All together, these results suggest that RBO may have a role in the lipid metabolism of broiler chickens; however, further investigations are required to confirm this mechanism.

Diet strategy is considered one of the documented approaches to augment the immune response in broiler chickens. The thymus gland, spleen, and bursa of Fabricius are the main lymphoid organs in poultry. The measurements of the lymphoid organs' weights and immune response have been usually applied to evaluate the general immune status of birds [42,43]. In the present study, RBO linearly increased the relative weights of the immune organs, improved serum total protein, and globulin, enhanced HIND antibody titer, and augmented phagocytic activity of the RBO-broilers, within the normal reported values, compared to control. It has been documented that the relative weights of lymphoid organs reveal the growth and function of the immune system, the humoral and cellular immunity [42,43]. Accordingly, an increase in the weights of these organs can imply an immunocompetence [42–44]. PUFA induces an immunomodulatory impact via influencing the intercellular signals which alter the leukocytes response due to antigenic stimulation [44]. The FA composition rather than the particular dietary lipids is necessary to augment the cellular and humoral immunity in broilers [8,45]. Linoleic acid, n-6 FA, was reported to be associated with an increase in the lymphocyte proliferation, immune organs size, and antibody titer in mice [46] and in broiler chickens [8,17]. RBO contain a considerable amount of linoleic acid (approximately 36%) and could be responsible for the observed immunomodulatory effect in the RBO groups. On the other hand, some plant components can augment the multiplication of advantageous microbiota, while diminishing the proliferation of harmful ones, consequently indirectly boosting host's immunity [47]. The flavonoids and phenolic compounds are known for their antioxidant, anti-inflammatory, and antimicrobial activities [48]. Since RBO is rich in flavonoids and phenolic compounds, in the present study, possibly dietary supplementation with RBO notably enhanced the immune status of broiler chickens.

Analysis of serum enzyme activities is an imperious procedure in the assessment of poultry health problems. Enzymes, such as ALT and AST, are generally existing with high concentrations in the liver [49] and are considered as particular indicators of hepatic cellular damage and inflammation [50]. ALP is an enzyme produced in all tissue types and is mainly responsible for dephosphorylation of a substrate and activated in alkaline pH. Its raised blood values are commonly noted in hepatic damages [49,50]. In our study, RBO supplementation linearly reduced the serum AST, ALT, and ALP when compared with the control group. The current findings indicate that RBO supplementation did not induce any detrimental effects on the liver. Lower serum AST, ALT, and ALP levels, within the normal values, may probably indicate that RBO decreased tissue damage and enhanced liver functions in the RBO broilers, since higher activities of these markers were reported to be associated with liver damage and inflammation [51]. Along with serum biochemical constituents, histological structure can be used to determine liver abnormalities and help in assessing disease diagnosis in poultry. In support of our findings, the general histological structure and ultrastructural of the liver of the experimental groups are normal and comparable to previous observations of the birds' liver [43,52]. No pathological lesions were observed in the broiler chickens' livers of the current experiment. The lipid content of the experimental groups whether assessed using images from light or electron microscope revealed that RBO feeding linearly decreased the percentage of hepatic lipid storage and number of lipid droplets compared to control. Regarding the hepatic chemical assessments, the broiler chickens consumed the RBO diets had significantly lower liver cholesterol, triglycerides, and MDA contents when compared with those consumed the control diet. The liver plays a central role in lipogenesis in birds, and contrary to mammals, the fat synthesis occurs mainly in the liver and is restricted in the adipose tissue [53]. Based on the obtained findings, we suggested that RBO can improve liver lipid profile and induce liver protective effects through its phenolic and antioxidant compounds [32,33]. The presence of phenolic compounds and antioxidants allows RBO to protect hepatic tissue from lipid peroxidation and thus decreased serum levels of ALT, AST, and ALP. Previous studies by Wang et al. [32] and Zhang et al. [33] showed that rice bran phenolic extract exerts its hypolipidemic effect through activation of AMP-activated protein kinase, wherein phenolic compounds found in rice bran such as p-coumaric acid, ferulic acid, and rutin play a fundamental role.

It is worth noting that RBO supplementation resulted in a reduction in serum levels of total lipids, total cholesterol, LDL, and triglycerides, while it increased serum HDL level. The favorable actions of RBO on total lipids, total cholesterol, and LDL are possibly due to its richness in non-saponifiable components, including sterols (b-sitosterol, campesterol and stigmasterol), c-oryzanol, and tocotrienols. Lichtenstien et al. [54] showed a higher content of b-sitosterol, campesterol and stigmasterol in RBO than other vegetable oils. Plant sterols are observed to be accountable for a 30–40% reduction in the cholesterol absorption [55]. Furthermore, oryzanol and tocotrienols found in RBO lessen the rate of endogenous cholesterol synthesis through diminishing the HMG-CoA reductase enzyme [32,56], and increasing the expression of cholesterol 7-alpha-hydroxylase, which is the rate-limiting enzyme in the synthesis of bile acids from cholesterol. Another suggested mechanism accountable for the reduction of total cholesterol due to feeding RBO is its omega-3 FA content which can inhibit the apolipoprotein-B100 and LDL synthesis and accordingly result in a decrease in total cholesterol [54,57]; however, the exact mechanism in broiler chickens is still unclear and necessitates further investigations.

Meat color is one of the critical indicators of freshness and wholesomeness of any meat product, relying on it, customers accomplish an initial impression of the product [58]. Numerous factors were reported to influence the color of broiler's meat including genetics, sex, age, diet composition, the heme pigments, and pre-slaughter condition [58]. In the current study, dietary RBO had no influence on L* (lightness), a* (redness), and b* (yellowness) values of breast or thigh meat at 24 h post-mortem among the experimental groups. The present findings are in line with the reports of Khatun et al. [59] and Jankowski et al. [60], who found that meat color was not influenced by the dietary oils supplementation. However, these results contradict the observations of Turcu et al. [5] and Qi et al. [61] and who found that dietary supplementation with various oils influenced the meat color of broiler chickens. The inconsistency between these studies may be due to various types of the studied plant oils. Measurement of the meat pH is vital because there is a relationship between pH and physicochemical properties of the meat, such as color and hardness [5]. In this study, the pH values of the breast and thigh muscles are within the ranges of normal pH limits [5]. Our results are in agreement with those of Khatun et al. [59] and Jankowski et al. [60] who observed no differences in the meat pH at 24 h post-mortem in broiler chickens fed various dietary oils. In goose, the inclusion of full-fat rice bran at 6–18% did not influence the meat quality traits, involving color parameters, and pH [62]. The lack of differences observed in meat color in the current trial might be owing to the similarity in the pH values for all groups. Furthermore, in this trial, the lack of variances between the experimental groups for the pH of the meat could be owing to the fact that RBO had no effects on the glycogen content of the meat. Glycogen plays a key role in the pH value of meat [63]; nevertheless, the meat glycogen content was not measured in the present trial, which necessitates further research. On the other hand, there was non-significant difference in broiler abdominal fat coloration in the present study, which is probably due to the low contents of beta-carotene, lutein, and zeaxanthin found in RBO [64]. To our knowledge, there have been no previous reports on the effects of dietary RBO on broiler meat or abdominal fat colorations.

It is well acknowledged that lipid peroxidation is one of the major processes that affect the quality of meat and its products [65]. The inclusion of RBO at different levels in the broiler diets has been performed to decrease the lipid content while increase the antioxidant contents of the meat in order to avoid its quality deterioration. In the present trial, the antioxidant effect of RBO on meat lipid peroxidation was noticed both in thigh and breast meat, compared to control samples. Dietary inclusion of RBO linearly and quadratically improved the meat GPx content, while reduced its EE %, triglycerides, cholesterol, and MDA contents. On the other hand, with the increasing RBO level there was no variation in CP content in the broiler meat, which may indicate that RBO had no influence on protein synthesis in the meat. In broiler chickens, in ovo injection of RBO increased the GPx content but decreased the MDA content in the breast and heart muscles [66]. Full fat rice bran inclusion linearly and quadratically reduced crude fat content in goose meat but did not affect the meat CP content [62]. The γ-oryzanol, tocopherols, and tocotrienols components of RBO are thought as the major ingredients responsible for the hypolipidemic and antioxidant activities of RBO [67]. Although the mechanism underlying this effect is not clear at the moment, oryzanol, and tocopherols of RBO are supposed to be the main cause for this beneficial outcome. Furthermore, phenolic compounds (e.g., polyphenols) induced the antioxidant potential which may diminish the formation of free radicals and subsequently counteract the peroxidation of PUFA [68].

Chicken meat is considered as good sources of protein and essential FA for humans. Diet manipulation has a considerable influence on the broiler meat composition. Meat FA composition is considered as a precious indicator of meat quality from the human health standpoint. Vegetable oils are valuable dietary constituents as a source of FA that can possibly be redirected in the poultry products [65]. Interestingly, our findings revealed that the inclusion of RBO in the broiler diets increased the meat MUFA, PUFA, n-6, and n-3 FA contents, but decreased the SFA level. The improvement of PUFA in the meat of RBO groups chiefly caused by the diminution of SFA and the elevation of C18:2n-6 and C18:3n-3 FA in the meat. The most effective of RBO supplementation was the enrichment of the breast meat with PUFA, n-6 FA, n-3 FA, and PUFA:SFA, and lowering the SFA content. PUFA show an important role in lessening the occurrence of cardiovascular and inflammatory diseases in humans [69]. To our knowledge, there is inadequate data in the literature regarding the dietary inclusion of RBO and the modification of the meat FA composition in poultry. According to Turcu, et al. [5] and Attia et al. [8], the dietary supplementation with PUFA-rich oils in broiler diets resulted in a modification in the meat FA profile, namely enrichment the meat with PUFA and a decrease in SFA level. Sun et al. [62] showed that full-fat rice bran inclusion decreased MUFA and increased PUFA level in the goose meat. In pigs, Jayaraman et al. [70] reported that the inclusion full-fat rice bran as a replacement of corn in their diets linearly decreased SFA (palmitic and stearic acids), but increased n-6 FA and n-3 FA in meat. The hypolipidemic action of the phenolic acids presented in RBO may be contributed to the enhancement of the FA profile of broiler meat through lipid homeostasis. Previous researchers have proposed that the FA content of poultry meat could be modified through dietary manipulation by adding plant oils and antioxidants [5,8,17,62,68], which in turn could extend the shelf-life and improve the quality of meat products.

5. Conclusions

Under the condition of the current experiment, RBO inclusion as a source of PUFA and natural antioxidants had a positive effect on the growing performance, dressing percentage, and immune status. Furthermore, RBO supplementation decreased the abdominal fat yield and EE content in the meat and increased the meat PUFA without influencing other meat quality traits such as color and pH, which may be beneficial for consumers. RBO improved the antioxidant capacity and lipid profile of the meat and the liver. In conclusion, RBO could be used as a valuable efficient ingredient in broiler chickens' diets to stimulate the growth performance and immune status, enhance the antioxidant activity and blood lipid profile, augment liver function, and improve the nutritive value of the meat.

Author Contributions: Conceptualization, S.S., E.H., N.S.A.-M. and A.A.S.; methodology, S.S., E.H., A.A.S. and N.S.A.-M.; software, S.S., S.J.M. and M.S.A.-H.; validation, S.S., E.H., N.S.A.-M. and A.A.S.; formal analysis, S.S., E.H. and N.S.A.-M.; investigation, S.S., E.H., N.S.A.-M., S.J.M. and M.S.A.-H.; resources, S.S., E.H. and M.S.A.-H.; data curation, S.S., E.H., N.S.A.-M., A.A.S., S.J.M. and M.S.A.-H.; writing—original draft preparation, S.S. and E.H.; writing—review and editing, S.S., E.H., N.S.A.-M., A.A.S., S.J.M. and M.S.A.-H. All authors have read and agreed to the published version of the manuscript.

Funding: The current work was funded by Taif University Researchers Supporting Project number (TURSP-2020/64), Taif University, Saudi Arabia.

Institutional Review Board Statement: The care and procedure used for broiler chickens of the current trial were permitted by Institutional Animal Care and Use Committee (IACUC), Faculty of Veterinary Medicine, University of Sadat City (Ethical approval number: VUSC-018-1-20).

Data Availability Statement: The data presented in this study are available on request from the corresponding author.

Acknowledgments: The authors are also grateful to Taif University Researchers Supporting Project number (TURSP-2020/64), Taif University, Saudi Arabia for the financial support and research facilities.

Conflicts of Interest: The authors declare no conflict of interest.

References

1. FAOSTAT. Food and Agriculture Organization of the United Nations. 2020. Available online: http://www.fao.org/faostat/en/#data (accessed on 28 September 2021).
2. Pinto da Rosa, P.; Pio Ávila, B.; Damé Veber Angelo, I.; Chesini, R.G.; Fernandes, T.A.; da Silva Camacho, J.; Bugoni, M.; Buttow Roll, V.F.; Gularte, M.A. Impact of different chicken meat production systems on consumers' purchase perception. *Br. Poult. Sci.* **2021**, *62*, 387–395. [CrossRef]
3. Sajdakowska, M.; Gębski, J.; Gutkowska, K.; Zakowska-Biemans, S. Importance of health aspects in Polish consumer choices of dairy products. *Nutrients* **2018**, *10*, 1007. [CrossRef] [PubMed]
4. Šrédl, K.; Prášilová, M.; Severová, L.; Svoboda, R.; Štěbeták, M. Social and economic aspects of sustainable development of livestock production and meat consumption in the Czech Republic. *Agriculture* **2021**, *11*, 102. [CrossRef]
5. Turcu, R.P.; Panaite, T.D.; Untea, A.E.; Vlaicu, P.A.; Badea, I.A.; Mironeasa, S. Effects of grape seed oil supplementation to broilers diets on growth performance, meat fatty acids, health lipid indices and lipid oxidation parameters. *Agriculture* **2021**, *11*, 404. [CrossRef]
6. Galli, C.; Calder, P.C. Effects of fat and fatty acid intake on inflammatory and immune responses: A critical review. *Ann. Nutr. Metab.* **2009**, *55*, 123–139. [CrossRef] [PubMed]
7. Swiatkiewicz, S.; Arczewska-Włosek, A.; Józefiak, D. The relationship between dietary fat sources and immune response in poultry and pigs: An updated review. *Livest. Sci.* **2015**, *180*, 237–246. [CrossRef]

8. Attia, Y.A.; Al-Harthi, M.A.; Abo El-Maaty, H.M. The effects of different oil sources on performance, digestive enzymes, carcass traits, biochemical, immunological, antioxidant, and morphometric responses of broiler chicks. *Front. Vet. Sci.* **2020**, *7*, 181. [CrossRef]
9. Salami, S.A.; Majoka, M.A.; Saha, S.; Garber, A.; Gabarrou, J.F. Efficacy of dietary antioxidants on broiler oxidative stress, performance and meat quality: Science and market. *J. Avian Biol. Res.* **2015**, *8*, 65–78. [CrossRef]
10. Selim, S.; Seleiman, M.F.; Hassan, M.M.; Saleh, A.A.; Mousa, M.A. Impact of dietary supplementation with *Moringa oleifera* leaves on performance, meat characteristics, oxidative stability, and fatty acid profile in growing rabbits. *Animals* **2021**, *11*, 248. [CrossRef]
11. Lai, O.M.; Jacoby, J.J.; Leong, W.F.; Lai, W.T. *Nutritional Studies of Rice-Bran Oil*; AOCS Press: Urbana, IL, USA, 2019; pp. 19–54.
12. Punia, S.; Kumar, M.; Sandhu, K.S. Rice-bran oil: An emerging source of functional oil. *J. Food Process. Preserv.* **2021**, *5*, e15318. [CrossRef]
13. McCaskill, D.R.; Zhang, F. Use of rice bran oil in foods. *Food Technol.* **1999**, *53*, 50–53.
14. Raghuram, T.C.; Rukmini, C. Nutritional significance of rice bran oil. *Indian J. Med. Res.* **1995**, *102*, 241–244.
15. Ali, A.; Devarajan, S. Nutritional and Health Benefits of Rice Bran Oil. In *Brown Rice*; Manickavasagan, A., Santhakumar, C., Venkatachalapathy, N., Eds.; Springer: Cham, Switzerland, 2017. [CrossRef]
16. Purushothaman, M.R.; Vasan, P.; Mohan, B.; Ravi, R. Utilization of tallow and rice bran oil in feeding broilers. *Ind. J. Poult. Sci.* **2005**, *40*, 175–178.
17. Kang, H.K.; Kim, C.H. Effects of dietary supplementation with rice bran oil on the growth performance, blood parameters, and immune response of broiler chickens. *J. Anim. Sci. Technol.* **2016**, *58*, 12. [CrossRef] [PubMed]
18. Aviagen, R. Ross Nutrition Specification. 2019. Available online: http://en.aviagen.com/assets/Tech_Center/Ross_Broiler/RossBroilerNutritionSpecs2019-EN.pdf (accessed on 10 December 2020).
19. Takatasy, G.Y. The use of 100 M in serological and virological micro methods. *Acta Microb. Acad. Sci.* **1955**, *3*, 191–202.
20. Kawahara, E.; Ueda, T.; Nomura, S. In vitro phagocytic activity of white spotted shark cells after injection with *Aeromoassalmonicida* extracellular products. *Fish Pathol.* **1991**, *26*, 213–214. [CrossRef]
21. Egan, H.; Kirk, R.S.; Sawyer, R.; Pearson, D. *Pearson's Chemical Analysis of Foods*, 8th ed.; Churchill Livingstone: Edinburgh, UK, 1981.
22. Commission Internationale de l'Eclairage, CIELab Colour System. *Commission Internationale de L'Eclairage*; CIE Publication: Paris, France, 1976.
23. AOAC. *Official Methods of Analysis of AOAC International*; AOAC International: Washington, DC, USA, 2005.
24. Ohkawa, K.; Ohishi, N.; Yagi, K. Assay for lipid peroxides in animal tissues by thiobarbituric acid reaction. *Anal. Biochem.* **1979**, *95*, 351–358. [CrossRef]
25. Paglia, D.E.; Valentine, W.N. Studies on the quantitative and qualitative characterization of erythrocyte glutathione peroxidase. *J. Lab. Clin. Med.* **1967**, *70*, 158–169. [PubMed]
26. Bohac, C.E.; Rhee, K.S. Influence of animal diet and muscle location on cholesterol content of beef and pork muscles. *Meat. Sci.* **1988**, *23*, 71–75. [CrossRef]
27. Folch, J.; Lees, M.; Sloane Stanley, G.H. A simple method for the isolation and purification of total lipides from animal tissues. *J. Biol. Chem.* **1957**, *226*, 497–509. [CrossRef]
28. Yang, C.J.; Yang, I.Y.; Oh, D.H.; Bae, I.H.; Cho, S.G.; Kong, I.G.; Uuganbayar, D.; Nou, I.S.; Choi, K.S. Effect of green tea by-product on performance and body composition in broiler chicks. *Asian-Australas. J. Anim. Sci.* **2003**, *16*, 867–872. [CrossRef]
29. Bancroft, J.D.; Gamble, M. *Theory and Practice of Histological Techniques*, 5th ed.; Churchill Livingston: London, UK, 2002.
30. Hayat, M.A. *Principles and Techniques of Electron Microscopy: Biological Applications*, 4th ed.; Cambridge University Press: Cambridge, UK, 2000.
31. Bumrungpert, A.; Chongsuwat, R.; Phosat, C.; Butacnum, A. Rice-bran oil containing gamma-oryzanol improves lipid profiles and antioxidant status in hyperlipidemic subjects: A randomized double-blind controlled trial. *J. Altern. Complement Med.* **2019**, *25*, 353–358. [CrossRef]
32. Wang, O.; Liu, J.; Cheng, Q.; Guo, X.; Wang, Y.; Zhao, L.; Zhou, F.; Baopang, J. Effects of ferulic acid and γ-oryzanol on high-fat and high-fructose diet-induced metabolic syndrome in rats. *PLoS ONE* **2015**, *10*, e0118135. [CrossRef] [PubMed]
33. Zhang, R.; Ma, Q.; Tong, X.; Liu, L.; Dong, L.; Huang, F.; Deng, Y.; Jia, X.; Chi, J.; Zhang, M. Rice bran phenolic extract supplementation ameliorates impaired lipid metabolism in high-fat-diet fed mice through AMPK activation in liver. *J. Funct. Foods* **2020**, *73*, 104131. [CrossRef]
34. Anitha, B.; Moorthy, M.; Viswanathan, K. Production performance of broilers fed with crude rice bran oil. *Int. J. Poult. Sci.* **2006**, *5*, 1046–1052.
35. Baiao, N.C.; Lara, L.J.C. Oil and fat in broiler nutrition. *Braz. J. Poult. Sci.* **2005**, *1*, 129–141. [CrossRef]
36. Poorghasemi, M.; Seidavi, A.; Qotbi, A.A.A.; Chambers, J.R.; Laudadio, V.; Tufarelli, V. Effect of dietary fat source on humoral immunity response of broiler chickens. *Eur. Poult. Sci.* **2015**, *79*. [CrossRef]
37. Ayed, H.B.; Attia, H.; Ennouri, M. Effect of oil supplemented diet on growth performance and meat quality of broiler chickens. *Adv. Tech. Biol. Med.* **2015**, *4*, 156. [CrossRef]
38. Abdulla, N.R. Phosphorus levels on growth performance, gut morphology and apparent digestibility of broiler chickens. *S. Afr. J. Anim. Sci.* **2016**, *46*, 42–53. [CrossRef]

39. Khatun, J.; Loh, T.C.; Akit, H.; Foo, H.L.; Mohamad, R. Fatty acid composition, fat deposition, lipogenic gene expression and performance of broiler fed diet supplemented with different sources of oil. *Anim. Sci. J.* **2017**, *88*, 1406–1413. [CrossRef]
40. Ahmed, M.A.; Mohamed, M.A.; Rashed, L.A.; Abd Elbast, S.A.; Ahmed, E.A. Rice bran oil improves insulin resistance by affecting the expression of antioxidants and lipid-regulatory genes. *Lipids* **2018**, *53*, 505–515. [CrossRef] [PubMed]
41. Ham, H.; Lee, Y.Y.; Park, J.Y.; Lee, C.; Kwak, J.; Kim, I.H.; Lee, J. Protective mechanisms of unsaponifiable matter from rice bran against tert-butyl hydroperoxide-induced oxidative damage in HepG2 cells. *J. Food Biochem.* **2016**, *40*, 526–534. [CrossRef]
42. Villagrán-de la Mora, Z.; Vázquez-Paulino, O.; Avalos, H.; Ascencio, F.; Nuño, K.; Villarruel-López, A. Effect of a symbiotic mix on lymphoid organs of broilers infected with Salmonella typhimurium and Clostridium perfringens. *Animals* **2020**, *10*, 886. [CrossRef]
43. Selim, S.; Abdel-Megeid, N.S.; Abou-Elnaga, M.K.; Mahmoud, S.F. Early nutrition with different diets composition versus fasting on immunity-related gene expression and histomorphology of digestive and lymphoid organs of layer-type chicks. *Animals* **2021**, *11*, 1568. [CrossRef] [PubMed]
44. Al-Khalaifah, H.; Al-Nasser, A. Dietary supplementation with various fat oils affect phytohemagglutinin skin test in broiler chickens. *Front. Immunol.* **2020**, *11*, 1735. [CrossRef]
45. Omidi, S.; Mohit, A.; Zadeh, N.G.H. Effect of dietary fat level and source on performance and immune system response of turkeys. *Acta Sci. Anim. Sci.* **2020**, *4*, e46775. [CrossRef]
46. Sierra, S.; Lara–Villoslada, F.; Olivares, M.; Jiménez, J.; Boza, J.; Xaus, J. Increased immune response in mice consuming rice bran oil. *Eur. J. Nutr.* **2005**, *44*, 509–516. [CrossRef]
47. Komiyama, Y.; Andoh, A.; Fujiwara, D.; Ohmae, H.; Araki, Y.; Fujiyama, Y.; Mitsuyama, K.; Kanauchi, O. New prebiotics from rice bran ameliorate inflammation in murine colitis models through the modulation of intestinal homeostasis and the mucosal immune system. *Scand. J. Gastroenterol.* **2011**, *46*, 40–52. [CrossRef]
48. Górniak, I.; Bartoszewski, R.; Króliczewski, J. Comprehensive review of antimicrobial activities of plant flavonoids. *Phytochem. Rev.* **2019**, *18*, 241–272. [CrossRef]
49. Jacobs, D.S. *Laboratory Test Handbook*, 4th ed.; Lexi-Comp: Cleveland, OH, USA, 1996.
50. Borsa, A.; Kohayagawa, A.; Boretti, L.P.; Saito, M.E.; Kuibida, K. Níveis séricos de enzimas de função hepática em frangos de corte de criação industrial clinicamente saudáveis. *Arq. Bras. Med. Vet. Zoo.* **2006**, *58*, 675–677. [CrossRef]
51. Franciscato, C. *Avaliação dos Minerais Séricos e da Função Hepática de Frangos de Corte Experimentalmente Intoxicados Com Aflatoxina e Submetidos a Diferentes Concentrações de Montmorilonita Sódica na Dieta. Degree Diss*; Universidade Federal de Santa Maria: Santa Maria, Brazil, 2006.
52. Alshamy, Z.; Richardson, K.C.; Harash, G.; Hünigen, H.; Röhe, I.; Hafez, H.M. Structure and age-dependent growth of the chicken liver together with liver fat quantification: A comparison between a dual-purpose and a broiler chicken line. *PLoS ONE* **2019**, *14*, e0226903. [CrossRef] [PubMed]
53. Hermier, D. Lipoprotein metabolism and fattening in poultry. *J. Nutr.* **1997**, *127*, 805S–808S. [CrossRef]
54. Lichtenstein, A.H.; Ausman, L.M.; Carrasco, W.; Gualtieri, L.J.; Jenner, J.L.; Ordovas, J.M.; Nicolosi, R.J.; Goldin, B.R.; Schaefer, E.J. Rice bran oil consumption and plasma lipid levels in moderately hypercholesterolemic humans. *Arterioscler. Thromb.* **1994**, *14*, 549–556. [CrossRef] [PubMed]
55. Ostlund, R.E., Jr. Phytosterols, cholesterol absorption and healthy diets. *Lipids* **2007**, *42*, 41–45. [CrossRef] [PubMed]
56. Minhajuddin, M.; Beg, Z.H.; Iqbal, J. Hypolipidemic and antioxidant properties of tocotrienol rich fraction isolated from rice bran oil in experimentally induced hyperlipidemic rats. *Food Chem. Toxicol.* **2005**, *43*, 747–753. [CrossRef]
57. Pourrajab, B.; Sohouli, M.H.; Amirinejad, A.; Fatahi, M.; Găman, M.; Shidfar, F. The impact of rice bran oil consumption on the serum lipid profile in adults: A systematic review and meta-analysis of randomized controlled trials. *Crit. Rev. Food Sci. Nutr.* **2021**, *10*, 1–11. [CrossRef]
58. Jaturasitha, S.; Chaiwang, N.; Kreuzer, M. Thai native chicken meat: An option to meet the demands for specific meat quality by certain groups of consumers; a review. *Anim. Prod. Sci.* **2017**, *57*, 1582–1587. [CrossRef]
59. Khatun, J.; Loh, T.C.; Akit, H.; Foo, H.L.; Mohamad, R. Influence of different sources of oil on performance, meat quality, gut morphology, ileal digestibility and serum lipid profile in broilers. *J. Appl. Anim. Res.* **2018**, *46*, 479–485. [CrossRef]
60. Jankowski, J.; Zdunczyk, P.; Mikulski, D.; Juskiewicz, J.; Naczmanski, J.; Pomianowski, J.F.; Zdunczyk, P. Fatty acid profile, oxidative stability, and sensory properties of breast meat from turkeys fed diets with a different n-6/n-3 PUFA ratio. *Eur. J. Lipid Sci. Technol.* **2012**, *114*, 1025–1035. [CrossRef]
61. Qi, K.; Chen, J.; Zhao, G.; Zheng, M.; Wen, J. Effect of dietary $\omega 6/\omega 3$ on growth performance, carcass traits, meat quality and fatty acid profiles of Beijing-you chicken. *J. Anim. Physiol. Anim. Nutr.* **2010**, *94*, 474–485. [CrossRef]
62. Sun, W.; Kang, P.; Xie, M.; Hou, S.S.; Wu, T.; Mei, H.M.; Liu, Y.L.; Hou, Y.Q.; Wu, L.Y. Effects of full-fat rice bran inclusion in diets on growth performance and meat quality of Sichuan goose. *Br. Poult. Sci.* **2016**, *57*, 655–662. [CrossRef]
63. Priolo, A.; Micol, D.; Agabriel, J. Effects of grass feeding systems on ruminant meat colour and flavour. A review. *Anim. Res.* **2001**, *50*, 185–200. [CrossRef]
64. Lamberts, L.; Delcour, J.A. Carotenoids in raw and parboiled brown and milled rice. *J. Agric. Food Chem.* **2008**, *56*, 11914–11919. [CrossRef]
65. Smet, K.; Raes, K.; Huyghebaert, G.; Haak, L.; Arnouts, S.; Smet, S. Lipid and protein oxidation of broiler meat as influenced by dietary natural antioxidant supplementation. *Poult. Sci.* **2008**, *87*, 1682–1688. [CrossRef]

66. Seifi, S.; Araghi, A.; Sayrafi, R.; Salehi, A. Effects of rice bran oil on qualitative properties of heart and breast muscle tissues in chicken embryo model. *Int. Food Res. J.* **2015**, *22*, 1894–1897.
67. Sengupta, A.; Ghosh, M.; Bhattacharyya, D.K. Antioxidative effect of rice bran oil and medium chain fatty acid rich rice bran oil in arsenite induced oxidative stress in rats. *J. Oleo Sci.* **2014**, *63*, 1117–1124. [CrossRef] [PubMed]
68. Jang, A.; Lin, X.D.; Shin, M.H.; Lee, B.D.; Lee, S.K.; Lee, J.H.; Jo, C. Antioxidative potential of raw breast meat from broiler chicks fed a dietary medicinal herb extract mix. *Poult. Sci.* **2008**, *87*, 2382–2389. [CrossRef] [PubMed]
69. Simopoulos, A.P.; Leaf, A.; Salem, N.J. Workshop statement on the essentiality of and recommended dietary intakes for omega-6 and omega-3 fatty acids. *Prostaglandins Leukot. Essent. Fat. Acids* **2000**, *63*, 119–121. [CrossRef]
70. Jayaraman, B.; Kinh, L.V.; Sy, P.V.; González-Vega, C.; Krishnan, P.; Channarayapatna, G. PSVI-12 full-fat rice bran as an alternative feed ingredient for corn in grow-finish pigs. *J. Anim. Sci.* **2021**, *99*, 205. [CrossRef]

animals

MDPI

Article

Use of Chemical Nano-Selenium as an Antibacterial and Antifungal Agent in Quail Diets and Its Effect on Growth, Carcasses, Antioxidant, Immunity and Caecal Microbes

Mahmoud Alagawany [1,*], Shaza Y. A. Qattan [2], Youssef A. Attia [3,*], Mohamed T. El-Saadony [4], Shaaban S. Elnesr [5], Mohamed A. Mahmoud [6], Mahmoud Madkour [7], Mohamed E. Abd El-Hack [1] and Fayiz M. Reda [1]

1 Poultry Department, Faculty of Agriculture, Zagazig University, Zagazig 44511, Egypt; dr.mohamed.e.abdalhaq@gmail.com (M.E.A.E.-H.); fayizreda@yahoo.com (F.M.R.)
2 Biological Sciences Department, Microbiology, Faculty of Science, King Abdulaziz University, Jeddah 80203, Saudi Arabia; syqattan@kau.edu.sa
3 Agriculture Department, Faculty of Environmental Sciences, King Abdulaziz University, Jeddah 21589, Saudi Arabia
4 Department of Agricultural Microbiology, Faculty of Agriculture, Zagazig University, Zagazig 44511, Egypt; m.elsadony@zu.edu.eg
5 Poultry Production Department, Faculty of Agriculture, Fayoum University, Fayoum 63514, Egypt; ssn00@fayoum.edu.eg
6 Department of Physiology, Faculty of Veterinary Medicine, New Valley University, New Valley 72511, Egypt; barhoma56@gmail.com
7 Animal Production Department, National Research Centre, Dokki, Giza 12622, Egypt; mahmoud.madkour9@gmail.com
* Correspondence: dr.mahmoud.alagwany@gmail.com (M.A.); yaattia@kau.edu.sa (Y.A.A.)

Citation: Alagawany, M.; Qattan, S.Y.A.; Attia, Y.A.; El-Saadony, M.T.; Elnesr, S.S.; Mahmoud, M.A.; Madkour, M.; Abd El-Hack, M.E.; Reda, F.M. Use of Chemical Nano-Selenium as an Antibacterial and Antifungal Agent in Quail Diets and Its Effect on Growth, Carcasses, Antioxidant, Immunity and Caecal Microbes. *Animals* **2021**, *11*, 3027. https://doi.org/10.3390/ani11113027

Academic Editor: Valentina Virginia Ebani

Received: 24 August 2021
Accepted: 19 October 2021
Published: 21 October 2021

Simple Summary: The chemical Nano-Selenium (Che-SeNPs) is a good example of applied nanotechnology used in the area of nutritional supplements due to its advantages and properties. From our results, dietary supplementation with Che-SeNPs could improve the performance of growing quails; the best level was 0.4 g Che-SeNPs/kg feed. Thus, this study supports the application of Che-SeNPs in quail diets in an effort to improve the productive and physiological performance. The results revealed that Che-SeNPs boosts the growth, blood biochemistry, antioxidant indices, immunity, and bacterial environment of the intestine of quail.

Abstract: Nano-minerals are used to enhance mineral bioavailability, which helps improve animal growth and health. The use of chemical nano-selenium (Che-SeNPs) has lately attracted great scientific interest, mainly due to its potential benefits for poultry. The current study was conducted to investigate the impact of the dietary supplementation of Che-SeNPs on the growth performance, carcass traits, blood constituents, antioxidant status, immunity, and gut microbiota of Japanese quails. A total of one week-old 180 Japanese quails were randomly distributed into four equal groups, and each group consisted of 45 unsexed birds with five replications (nine birds each). The first group was fed a basal diet without supplementation (0 g/kg Che-SeNPs), and the second, third, and fourth groups were fed diets containing 0.2, 0.4, and 0.6 g/kg Che-SeNPs, respectively. The results showed that the dietary supplementation of Che-SeNPs significantly ($p < 0.0001$) increased body weight, body weight gain, and feed conversion ratio, but decreased feed intake ($p < 0.0001$) compared to the control group. The highest values of growth performance were recorded in the group fed 0.4 g Che-SeNPs g/kg feed. Che-SeNPs levels did not affect the carcass traits, relative organs (except liver), or blood hematology (except platelet count and hemoglobin level) of quails. Plasma total protein, albumin, aspartate amino transferase (AST), and urea values were not affected by dietary Che-SeNPs, but alanine aminotransferase and lactate dehydrogenase values declined. Globulin and creatinine values were linearly increased with the inclusion of Che-SeNPs (0.4 and 0.6 g/kg) in quail diets compared to the control. The supplementation of Che-SeNPs in quail diets significantly improved ($p < 0.05$) the plasma lipid profile and activities of antioxidant enzymes compared to the control group. Immunoglobulin G values of Che-SeNPs (0.4 and 0.6 g/kg) were higher ($p < 0.05$)

than those in the control group. The groups fed diets supplemented with Che-SeNPs showed lower ($p < 0.0001$) total bacterial count, total yeast and molds count, *Coliform*, *Escherichia coli*, *Enterococcus* spp., and *Salmonella* spp. colonization, and higher ($p = 0.0003$ and 0.0048) lactic acid bacteria counts than those in the control group. In conclusion, Che-SeNPs supplemented up to 0.4 g/kg can improve the performance, lipid profile, antioxidant indices, and immunity, as well as decrease intestinal pathogens in quails during the fattening period (1–5 weeks of age).

Keywords: nano particles; selenium; performance; blood; pathogens; quails

1. Introduction

Selenium (Se) is one of the elements that can be used in diets as the chemical nano-selenium (Che-SeNPs). Se is required for the maintenance of physiological functions, growth, and health of birds. It also plays a crucial role in nutritional value and feed metabolism, leading to considerable growth [1]. Che-SeNPs has attracted more attention because of its strong adsorbing ability, high catalytic efficiency, high surface activity, and low toxicity compared to that of other chemical Se forms [2]. The high absorption of Che-SeNPs from the intestinal lumen into the body was observed. Shirsat et al. [3] highlighted that Che-SeNPs has antioxidant, anticancer, antibacterial, and antiprotozoal properties. El-Deep et al. [4] stated that dietary Che-SeNPs supplementation enhanced growth performance by improving immune or antioxidative properties in broiler chicks. Additionally, Ahmadi et al. [5] revealed that the dietary supplementation of Che-SeNPs improved growth performance and immune function without the deleterious effects on the internal organs of broiler chickens.

Previous investigations exhibited that Che-SeNPs augmented body weight gain and improved antioxidant functions of Arbor Acres broilers [3,6]. Se nanoparticles have also been utilized in food preservation methods such as packing food items and antiseptic coating over food materials. Studies have been conducted to highlight the disinfectant properties of Se nanoparticles against *Pseudomonas aeruginosa*, and *Proteus mirabilis* [7]. On the other hand, natural agents and trace elements including nanoparticles as feed additives may affect the diversity of gut microbiota and health [8]. Se is one of the important elements that can help microbiota to complete its action within the gut [9]. In this concern, the caecal counts of *Salmonella* and *E. coli* of quails were decreased in birds fed diets containing nano-curcumin when compared to the control diets [8].

Selenium can be considered an essential trace element and micronutrient for living creatures at low concentrations, but it becomes toxic and harmful at higher dose [2]. The extensive use of nano Se in nanotechnologies and medicine has increased the risk of their contamination in the environment, which could harm living species; however, it is useful to understand the assessment of Se-NPs toxicity to the biological ecosystem [10]. Nano-Se has lower toxicity than selenomethionine and is now the least toxic of all supplments of Se. Nano-Se has a threefold lower toxicity than organic Se and a sevenfold lower toxicity than inorganic Se [10].

The positive impacts of nanotechnology involving Se are well-known in many pathological conditions [11]. However, the inclusion of Che-SeNPs in quail diet during the growth period remains limited. It is hypothesized that the dietary addition of Che-SeNPs is expected to exert beneficial effects on growing quails. Therefore, the purpose of this study was to evaluate the antibacterial and antifungal activities of Che-SeNPs, and its beneficial effects on the growth, feed utilization, carcass traits, hematology, blood constituents, and cecal microbiota of growing quails.

2. Materials and Methods

2.1. Source of Selenium Nanoparticles

The study was carried out at Zagazig University, Zagazig, Egypt in conjunction with King Abdulaziz University, Jeddah, Saudi Arabia under protocol no: (FP-73-43). In this study, Che-SeNPs were prepared using wet chemicals. Sodium selenite (Na_2SeO_3) was used for producing Se nanoparticles with ascorbic acid ($C_6H_8O_6$) as a reducing agent. A stock of aqueous solution of 100 mM Na_2SeO_3 and 50 mM $C_6H_8O_6$ was prepared in a 1: 4 ratio. The solution was kept under a magnetic stirring condition at different rpm and ambient temperature for 30 min. The mixtures were allowed to react with each other in the concentrated form until the mixture changed from colorless to red. Next, the solution was centrifuged at 3000 rpm, pellets were collected, and Che-SeNPs was obtained [12,13]. Chemically synthesized nano-selenium was determined via UV–Vis spectroscopy using an automated spectrometer (Spectro UV–Vis double beam UVD 3500). The morphology and element percentage of selenium nanoparticles were measured using transmission electron microscopy and an energy dispersive X-ray analytical instrument. Fourier transform infrared spectroscopy (JASCO) was used to determine the properties of produced selenium nanoparticles including size, shape, charge, and stability. Characterization of Che-SeNPs; maximum UV absorbance at 300 nm, spherical shape by TEM, size (75.68 nm) and charge (−23.26 mV) by zeta seizer, and zeta potential, respectively.

2.2. Antibacterial Activity of Che-SeNPs

Listeria monocytogenes ATCC 15313, *Staphylococcus aureus* MTCC 1809, *Bacillus cereus* ATCC 11778, *E. coli* ATCC 25922, *P. aeruginosa* ATCC 27853, and *Salmonella enterica* MTCC 1253 were used in this study. The antibacterial activity of Che-SeNPs against animal and human pathogenic Gram-negative bacteria, *P. aeruginosa* ATCC 27853, *E. coli* ATCC 25922, and *S. enterica* MTCC 1253, and Gram-positive bacteria, *L. monocytogenes* ATCC 15313, *B. cereus* ATCC 11778, and *S. aureus* MTTC 1809, were estimated using the disc diffusion assay method. Mueller–Hinton agar medium consisting of peptone, beef extract, yeast extract, NaCl, and agar with 5, 3, 5, 5, and 20 g, respectively in 1 L of distilled water was prepared in slant to preserve all bacterial strains. Mueller–Hinton broth was used to activate bacterial cells; one hundred microliters of each bacterium (1×10^9 CFU/mL) were spread with sterile swabs in Mueller–Hinton agar plates. Freshly prepared selenium nanoparticles with different concentrations (50, 100, 200, 400, and 800 μg/mL) were loaded on paper discs (disc diameter was about 6 mm) and then were placed on the Muller–Hinton agar plates. Sodium selenite (50 μg/mL) and sterilized deionized water were loaded on paper discs and used as a positive and negative control, respectively. Mueller-Hinton agar plates were incubated for 24 h at 37 °C. After incubation, the obtained zones of inhibition surrounded the Che-SeNPs discs were measured and recorded as the mean ± standard deviation if they were greater than 6 mm. The minimum inhibitory concentration (MIC) of the Che-SeNPs was calculated based on a broth micro dilution method. Briefly, six pathogenic bacteria were cultured overnight at 37 °C in Mueller–Hinton broth and were adjusted to a final density of 10^9 CFU/mL by 0.5 McFarland standards. The Che-SeNPs (1 mg/mL) were homogenized with sterilized deionized water and dilutions of 50, 100, 200, 400, and 800 μg/mL were made. Next, 10 μL of different concentrations of Che-SeNPs was mixed in sterile test tubes contain 10 μL of bacterial inoculum and 90 μL of Mueller–Hinton broth. The test tubes were incubated for a day at 37 °C. The lower concentration of Che-SeNPs which inhibited bacterial strains growth or turbidity was considered the MIC. The lower concentration of Che-SeNPs which totally killed bacterial strains was defined as the minimum bactericidal concentration (MBC). The experiments were carried out in triplicate [8].

2.3. Antifungal Activity of Che-SeNPs

The antifungal activity of the Che-SeNPs was tested against animals and human pathogenic *Candida* strains. *Candida albicans* ATCC 4862, *C. glabrata* ATCC64677, *C. parapsilosis*

ATCC 22019, and C. *guilliermondii* ATCC 6260 were used in this study. The antifungal activity of Che-SeNPs against these four strains was evaluated via the disc diffusion method [14] using sterile cotton swab lawn cultures of selected fungi that were prepared on Sabouraud Dextrose agar (SDA) plates. Che-SeNPs was loaded on paper discs (disc diameter was about 6 mm) and then was placed in SDA surface. Selenium selenite and sterilized deionized water were used as the positive and negative controls, respectively. The plates were then incubated for 36h at 30 °C. The Che-SeNPs were tested for MIC using the broth dilution method [14]. Sabouraud broth was used as diluents for fungal species. About 10^6 CFU/mL cells could be inoculated. The Che-SeNPs levels (50 to 800 µg/mL) were prepared in sterilized deionized water and homogenized. Next, 10 µL of different concentrations of Che-SeNPs were mixed in sterile test tubes containing 10 µL of *Candida* inoculum and 90 µL of Sabouraud broth. The test tubes were incubated at 30 °C for 36 h. The obtained turbidity was estimated at 600 nm to determine the MIC values. The minimum concentration of Che-SeNPs that reduced fungi growth by 90% was considered the minimum inhibitory concentration (MIC). The concentration of Che-SeNPs at which complete fungal growth was not observed was defined as the minimum fungicidal concentration (MFC). The experiments were replicated in triplicate.

2.4. Animals, Design, and Diets

A total of 180 one-week-old Japanese quails with an average weight of 27.17 ± 0.075 g were used. Quail chicks were randomly allocated into four groups, and each group consisted of 45 unsexed birds with five replications (nine birds each). Quails were kept in conventional cages (90 × 40× 40 cm), and feed and water were open during the study (4 weeks). The treatments were as follows: the first group was fed the basal ration which containing 150 mg of Se as Se selenite, whereas the second, third, and fourth groups were fed diets supplemented with 0.2, 0.4, and 0.6 g/kg of Che-SeNP, respectively. The Che-SeNP was added at the top of the basal diet at the highest level and then diluted with the unsupplemented basal diet to achieve the desire concentration. The basal diet (Table 1) was formulated to meet the birds' requirements according to NRC [15]. The Ethics statement for Animal care and maintenance were in accordance with the guidelines of the Egyptian Research Ethics Committee and the guidelines for the Care and Use of Laboratory Animals by Zagazig University (ZU-IACUC/2/F/56/2021).

Table 1. Ingredients and nutrient contents of basal diet for growing Japanese quail.

Items	(g/kg)
Ingredient	
Maize 8.5%	518.0
Soybean meal 44%	367.0
Maize gluten meal 62%	52.1
Soybean oil	29.0
Limestone	7.0
Di-calcium phosphate	16.5
Salt	3.0
Premix [1]	3.0
L-Lysine	1.3
Dl-Methionine	1.1
Choline chloride	2.0
Total	1000
Calculated composition	
Metabolizable energy (MJ/kg)	12.53
Crude protein (g/kg)	240.0
Calcium (g/kg)	8.0
Nonphytate phosphorus (g/kg)	4.5
Lysine (g/kg)	13.0

Table 1. *Cont.*

Items	(g/kg)
Total sulphur amino acids (g/kg)	9.2

[1] Provides per kg of diet: Vitamin A, 12,000 I.U.; Vitamin D3, 5000 I.U.; Vitamin E, 130.0 mg; Vitamin K3, 3.605 mg; Vitamin B1 (thiamin), 3.0 mg; Vitamin B2 (riboflavin), 8.0 mg; Vitamin B6, 4.950 mg; Vitamin B12, 17.0 mg; Niacin, 60.0 mg; D-Biotin, 200.0 mg; Calcium D-pantothenate, 18.333 mg; Folic acid, 2.083 mg; manganese, 100 mg; iron, 80 mg; zinc, 80 mg; copper, 8 mg; iodine, 2 mg; cobalt, 500 mg; and selenium, 150 mg.

2.5. Growth Performance and Carcass Measurements

All growth parameters and feed utilization were measured at 1, 3, and 5 weeks of age. For carcass examinations, at 5weeks old, 20 birds (5 per treatment) were randomly selected, weighed, and euthanized. All edible parts were weighed and expressed as a percent of the live body weight before slaughter.

2.6. Microbiological Analysis

Ten grams of quail cecum samples (five samples per each treatment) were homogenized and transferred to a 250 mL conical flask containing 90 mL of sterile physiological saline solution consisting of 0.1% peptone and 0.85%NaCl; the mixture was well-mixed to obtain a 10^{-1} dilution. Serial dilutions from the previous dilution (10^{-1}) were prepared to obtain up to 10^{-6} dilution. The total bacterial count was counted using plate count agar medium at 30 °C for 24 h, and the total count of *Enterococcus* spp., was counted using Chromocult enterococci agar medium [16,17]. Total coliforms were enumerated by using MacConkey agar medium. Biochemical methods such as indole test, citrate reactions, methyl red, and Voges–Proskauer were used to identify *Escherichia coli*. DeManRogosa Sharpe agar was used to enumerate the lactic acid bacteria. *Salmonella Shigella* agar (SSA) media (Oxide CM 99) was used to count the *Salmonella* spp. The appearance of black colonies on SSA indicated the presence of *Salmonella* spp. SSA plates were incubated at 37 °C for 1 day. Sabouraud Dextrose agar (SDA) was used to count the molds and yeasts. SDA plates were incubated at 25 °C for 3–7 days. All the obtained microbiological results were then converted to logarithmic colony-forming units per gram (CFU/g) [8,18,19].

2.7. Blood Chemistry

After euthanization, blood samples were randomly collected from five quails per treatment into heparinized tubes. Hematological parameters were measured. Regarding biochemical parameters, we used a centrifuge (Janetzki, T32c, 5000 rpm, Germany) at 2146.56× *g* for 15 min to separate the plasma. The biochemical blood parameters were determined using commercial kits from Biodiagnostic Company (Giza, Egypt).

2.8. Statistics

The statistical analyses were carried out using SAS. The data of growth rate, feed efficiency, carcass parameters, hematology, blood chemistry, and microbiology were analyzed with a one-way analysis of variance using a normal distribution and the replicate as the experimental unit. Orthogonal polynomial contrasts were used to test the significance (linear and quadratic) of the gradual levels of dietary Che-SeNPs using the post-hoc Tukey's test ($p < 0.05$).

3. Results

3.1. Antibacterial Activity of Che-SeNPs

Three animal and human pathogenic Gram-negative bacteria (*E. coli, P. aeruginosa*, and *S. enterica*) and three Gram-positive bacteria (*L. monocytogenes*, *B. cereus*, and *S. aureus*) were selected to test Che-SeNPs antibacterial activity (Table 2). The maximum zones of inhibitions were observed in the three Gram-positive bacterial strains *L. monocytogenes* ATCC 15313, *B. cereus* ATCC 11778, and *S. aureus* MTTC 1809. The antibacterial activity of Che-SeNPs increased with increasing concentrations of Che-SeNPs. The effect of Che-

SeNPs was superior to that of sodium selenite as an antimicrobial agent against tested pathogenic microorganisms; moreover, the deionized water did not show any antimicrobial activity. The highest MIC of Che-SeNPs against *E. coli* ATCC 25922, *P. aeruginosa* ATCC 27853, and *S. enterica* MTCC 1253 was 45, 40, and 50μg/mL, respectively, whereas, the lowest MIC was 30, 35, and 25 μg/mL against *L. monocytogenes* ATCC 15313, *S. aureus* MTTC 1809, and *B. cereus* ATCC 11778, respectively (Table 2).

Table 2. Zone of inhibition produced by Sodium Selenite and selenium nanoparticles.

Item	Sod. Selenite (50 μg/mL)	Selenium Nanoparticles (μg/mL)					DI Water
		50	100	200	400	800	
Bacteria		Inhibition zones (mm)					
Listeria monocytogenes ATCC 15313	14 ± 0.2^{f}	15 ± 0.3^{e}	19 ± 0.1^{d}	23 ± 0.2^{c}	26 ± 0.2^{b}	32 ± 0.1^{a}	-
Staphylococcus aureus MTTC 1809	11 ± 0.4^{f}	13 ± 0.2^{e}	18 ± 0.2^{d}	20 ± 0.1^{c}	23 ± 0.3^{b}	28 ± 0.35^{a}	-
Bacillus cereus ATCC 11778	13 ± 0.2^{f}	16 ± 0.15^{e}	20 ± 0.1^{d}	23 ± 0.2^{c}	27 ± 0.15^{b}	33 ± 0.14^{a}	-
Escherichia coli ATCC 25922	9 ± 0.5^{f}	11 ± 0.45^{e}	15 ± 0.3^{d}	17 ± 0.4^{c}	21 ± 0.2^{b}	25 ± 0.2^{a}	-
Pseudomonas aeruginosa ATCC 27853	10 ± 0.5^{f}	11 ± 0.45^{e}	16 ± 0.4^{d}	20 ± 0.1^{c}	22 ± 0.3^{b}	27 ± 0.19^{a}	-
Salmonella enterica MTCC 1253	8 ± 0.5^{f}	11 ± 0.45^{f}	14 ± 0.5^{d}	17 ± 0.4^{c}	20 ± 0.5^{b}	24 ± 0.3^{a}	-
Fungi							
Candida albicans ATCC 4862	11 ± 0.4^{e}	12 ± 0.3^{d}	14 ± 0.2^{c}	15 ± 0.2^{b}	16 ± 0.15^{ab}	17 ± 0.1^{a}	-
Candida glabrata ATCC 64677	8 ± 0.5^{e}	9 ± 0.5^{d}	9 ± 0.5^{d}	10 ± 0.4^{c}	11 ± 0.4^{b}	13 ± 0.2^{a}	-
Candida parapsilosis ATCC 22019	10 ± 0.35^{d}	11 ± 0.4^{c}	12 ± 0.3^{b}	13 ± 0.3^{b}	14 ± 0.2^{a}	14 ± 0.3^{a}	-
Candida guilliermondii ATCC 6260	8 ± 0.5^{d}	8 ± 0.5^{d}	9 ± 0.5^{c}	9 ± 0.5^{c}	10 ± 0.3^{b}	11 ± 0.5^{a}	-

Mean ± SE, Means in the same row with a similar superscript letter following them are not significantly different ($p < 0.05$).

3.2. *Antifungal Activity of Che-SeNPs*

Che-SeNPs showed acceptable antifungal activity, ranging from 50 μg/mL to 800 μg/mL, against all the tested fungal strains. *C. albicans* ATCC 4862 was the most sensitive strain to Che-SeNP when compared to other tested strains. The MICs for *C. albicans* ATCC 4862, *C. glabrata* ATCC64677, *C. parapsilosis* ATCC 22019, and *C. guilliermondii* ATCC 6260 were 70, 90, 80, and 100 μg/mL, respectively (Table 3).

Table 3. The MIC (Minimum Inhibitory Concentration), MBC (Minimum Bactericidal Concentration), and MFC (Minimum fungicidal concentration) of the selenium nanoparticles.

Microorganisms	Selenium Nanoparticles	
	MIC μg/mL	MBC μg/mL
Bacteria		
Listeria monocytogenes ATCC 15313	30	60
Staphylococcus aureus MTTC 1809	35	70
Bacillus cereus ATCC 11778	25	50
Escherichia coli ATCC 25922	45	90
Pseudomonas aeruginosa ATCC 27853	40	80
Salmonella enterica MTCC 1253	50	100
Fungi		
Candida albicans ATCC 4862	70	140
Candida glabrata ATCC 64677	90	180

Table 3. *Cont.*

Microorganisms	Selenium Nanoparticles	
	MIC μg/mL	MBC μg/mL
Candida parapsilosis ATCC 22019	80	160
Candida guilliermondii ATCC 6260	100	200

3.3. Growth Performance

The effects of dietary Che-SeNPs supplementation on the growth performance of Japanese quails are presented in Table 4. From the results, quails fed diets containing Che-SeNPs had significantly higher body weight (BW) (linear, $p < 0.0001$ and quadratic, $p = 0.0004$) and body weight gain (BWG) (linear, $p < 0.0001$ and quadratic, $p = 0.0005$) during the whole experimental period. The group fed diets containing Che-SeNPs (0.4 g/kg diet) had the highest BW and BWG. Feed intake was decreased (linear and quadratic, $p < 0.001$) in the Che-SeNPs groups compared with that in the control group. The feed conversion ratio was linearly and quadratically improved with the addition of Che-SeNPs in quail diets during all periods. Generally, the best growth performance parameters were recorded in the group fed 0.4 g Che-SeNPs/kg feed.

Table 4. Growth performance of growing Japanese quail as affected by dietary treatments.

Items	Chemical Nano Selenium Levels (g/kg Diet)				SEM	*p* Value	
	0	0.2	0.4	0.6		Linear	Quadratic
Body weight (g)							
1 wk	27.1	27.2	27.2	27.2	0.049	0.812	0.598
3 wk	92.6 ^{b}	98.4 ^{a}	97.5 ^{a}	97.8 ^{a}	0.717	0.002	0.006
5 wk	178.5^{c}	190.8 ^{b}	198.1 ^{a}	193.7 ^{ab}	1.354	<0.0001	0.0004
Body weight gain (g/day)							
1–3 wk	4.67 ^{b}	5.09 ^{a}	5.02 ^{a}	5.04 ^{a}	0.051	0.003	0.006
3–5 wk	6.14 ^{c}	6.60 ^{b}	7.18 ^{a}	6.85 ^{ab}	0.101	0.001	0.008
1–5 wk	5.41 ^{c}	5.84 ^{b}	6.10 ^{a}	5.95 ^{ab}	0.051	<0.0001	0.0005
Feed intake (g/day)							
1–3 wk	14.1 ^{a}	12.9 ^{c}	13.2 ^{b}	13.1 ^{bc}	0.079	0.0002	0.0002
3–5 wk	23.7 ^{a}	19.9 ^{c}	21.0 ^{b}	20.2 ^{c}	0.202	<0.0001	0.0001
1–5 wk	18.9 ^{a}	16.4 ^{c}	17.1 ^{b}	16.7 ^{c}	0.083	<0.0001	<0.0001
Feed conversion ratio (g/g)							
1–3 wk	3.01 ^{a}	2.53 ^{b}	2.63 ^{b}	2.60 ^{b}	0.038	0.0002	0.0005
3–5 wk	3.86 ^{a}	3.02 ^{b}	2.93 ^{c}	2.95 ^{bc}	0.018	<0.0001	<0.0001
1–5 wk	3.49 ^{a}	2.81 ^{b}	2.81 ^{b}	2.80 ^{b}	0.015	<0.0001	<0.0001

Means in the same row with no superscript letters after them or a similar superscript letter following them are not significantly different ($p < 0.05$).

3.4. Carcass Traits

As indicated in Table 5, dietary Che-SeNPs levels did not affect carcass traits and relative organs (except liver) of Japanese quails. Supplementation of Che-SeNPs significantly increased the relative weight of the liver (linear and quadratic, $p < 0.05$) compared to the control group.

3.5. Blood Hematology

The effects of the addition of Che-SeNPs on the blood hematology of growing quails are presented in Table 6. Dietary Che-SeNPs levels did not affect ($p > 0.05$) white blood cells (WBCs), lymphocytes, mid-range, granulocytes, red blood cells (RBCs), hematocrit, and mean corpuscular volume of the growing quails. Platelet count and hemoglobin (Hb) levels were increased (linear, $p < 0.05$) by the addition of Che-SeNPs at 0.4 and 0.6 g/kg. The mean corpuscular volume value was augmented (linear, $p < 0.05$) by the addition of Che-SeNPs level compared to the control group (without Che-SeNPs). The supplementation of dietary

Che-SeNPs at levels of 0.6 g/kg decreased values of red blood cell distribution width linearly ($p = 0.0091$). In contrast, the dietary levels of 0.4 and 0.6 g/kg declined the values of red blood cell distribution volume linearly ($p = 0.0019$) compared to the control group.

Table 5. Carcass traits and relative organs of growing Japanese quail as affected by dietary treatments.

Items	Chemical Nano Selenium Levels (g/kg Diet)				SEM	p Value	
	0	0.2	0.4	0.6		Linear	Quadratic
Carcass %	73.7	76.1	71.7	72.8	1.087	0.199	0.566
Liver %	2.22 ^{b}	2.84 ^{a}	2.85 ^{a}	2.69 ^{a}	0.099	0.036	0.014
Gizzard %	2.54	2.30	2.10	2.35	0.208	0.531	0.372
Heart %	1.01	1.14	1.03	0.94	0.076	0.473	0.267
Giblets %	5.76	6.28	5.98	5.98	0.344	0.838	0.506
Dressing %	79.4	82.4	77.7	78.8	1.279	0.300	0.509

Means in the same row with no superscript letters after them or a similar superscript letter following them are not significantly different ($p < 0.05$).

Table 6. Hematological parameters of growing Japanese quail as affected by dietary treatments.

Items [1]	Chemical Nano Selenium Levels (g/kg Diet)				SEM	p Value	
	0	0.2	0.4	0.6		Linear	Quadratic
WBCs ($10^3/\mu L$)	22.4	23.3	23.7	23.7	1.972	0.575	0.741
LYM (%)	93.7	93.2	93.1	94.6	0.629	0.349	0.163
MID (%)	5.84	6.41	6.50	3.14	0.771	0.059	0.042
GRA (%)	0.32	0.44	0.36	0.17	0.084	0.269	0.175
RBCs ($10^6/\mu L$)	2.34	2.60	2.63	2.75	0.196	0.194	0.749
HGB (g/dL)	9.27 ^{b}	11.8 ^{ab}	12.4 ^{a}	13.4 ^{a}	0.822	0.001	0.399
HCT (%)	32.3	35.8	29.4	21.2	3.438	0.032	0.127
MCV (μm^3)	137.8	137.9	127.1	124.9	2.985	0.014	0.758
MCH (pg)	40.9 ^{c}	47.9 ^{b}	53.9 ^{a}	56.9 ^{a}	1.167	<0.0001	0.163
RDWSD	51.9 ^{a}	52.4 ^{a}	44.0 ^{ab}	41.5 ^{b}	2.420	0.009	0.574
RDWCV	13.1 ^{a}	13.3 ^{a}	11.9 ^{b}	11.2 ^{b}	0.349	0.002	0.222
PLT ($10^3/\mu L$)	5.67 ^{b}	8.67 ^{b}	16.7 ^{a}	15.0 ^{a}	2.771	0.032	0.476

Means in the same row with no superscript letters after them or with a common superscript letter following them are not significantly different ($p < 0.05$). [1] WBCs: white blood cells; LYM: lymphocytes; MID: mid-range; GRA: granulocytes; RBCs: red blood cells; HGB: hemoglobin; HCT: hematocrit; MCV: Mean corpuscular volume; MCH: Mean corpuscular hemoglobin; RDWSD: Red blood cell distribution width; RDWCV: Red blood cell distribution volume; PLT: Platelet count.

3.6. Blood Constituents

Liver and kidney function data are presented in Table 7. The total protein and albumin were not affected ($p > 0.05$) by dietary Che-SeNPs. The globulin value was lowest (linear, $p < 0.05$) in the Che-SeNPs levels of 0.4 and 0.6 g/kg compared with the Che-SeNPs levels of 0.2 g/kg and the control group. The quails fed diets containing Che-SeNPs had higher alanine aminotransferase (ALT) and lactate dehydrogenase ($p < 0.05$) than those in the control group. Dietary Che-SeNPs had no significant effect on AST and urea values. The inclusion of Che-SeNPs (0.4 and 0.6 g/kg) in quail diets increased the creatinine value (linear, $p < 0.05$) compared with that in the control and 0.2 g/kg Che-SeNPs groups.

Table 7. Liver and kidney function of growing Japanese quail as affected by dietary treatments.

Items [1]	Chemical Nano Selenium Levels (g/kg Diet)				SEM	p Value	
	0	0.2	0.4	0.6		Linear	Quadratic
TP (g/dL)	3.27	2.88	3.22	3.35	0.109	0.332	0.067
ALB (g/dL)	1.20	1.21	1.27	1.16	0.026	0.707	0.086
GLOB (g/dL)	1.63 ^{c}	1.72 ^{c}	1.95 ^{b}	2.19 ^{a}	0.066	0.0003	0.329
A/G (%)	0.74 ^{a}	0.71 ^{a}	0.65 ^{a}	0.53 ^{b}	0.022	0.0005	0.122
AST (IU/L)	221.2	229.0	237.8	238.3	4.207	0.031	0.500
ALT (IU/L)	10.8 ^{c}	12.9 ^{b}	13.7 ^{b}	16.67 ^{a}	0.493	<0.0001	0.415

Table 7. *Cont.*

Items [1]	Chemical Nano Selenium Levels (g/kg Diet)				SEM	p Value	
	0	0.2	0.4	0.6		Linear	Quadratic
LDH (IU/L)	119.5 c	143.5 a	133.2 b	144.0 a	1.328	<0.0001	0.0001
Creatinine (mg/dL)	0.33 b	0.33 b	0.42 a	0.44 a	0.017	0.0008	0.660
Urea (mg/dL)	6.86	7.03	7.12	7.24	0.098	0.050	0.839

Means in the same row with no superscript letters after them or a similar superscript letter following them are not significantly different ($p < 0.05$). [1] TP: total protein; ALB: albumin; GLOB: globulin; A/G: albumin/ globulin ratio; LDH: Lactate dehydrogenase, AST: aspartate aminotransferase and ALT: alanine aminotransferase.

The effects of Che-SeNPs inclusion in diets on the lipid profile of quails are presented in Table 8. Total cholesterol, triglyceride, and very-low-density lipoprotein were significantly decreased ($p < 0.05$) in Che-SeNPs-treated groups compared to those in control. The dietary supplementation of Che-SeNPs (0.2 and 0.4 g/kg) quadratically increased high-density lipoprotein (HDL) ($p = 0.0019$).

Table 8. Lipid profile of growing Japanese quail as affected by dietary treatments.

Items [1]	Chemical Nano Selenium Levels (g/kg Diet)				SEM	p Value	
	0	0.2	0.4	0.6		Linear	Quadratic
TC (mg/dL)	153.6 a	143.4 b	144.5 ab	125.3 c	1.912	0.0002	0.159
TG (mg/dL)	298.8 a	225.0 b	210.0 bc	192.0 c	5.882	<0.0001	0.002
HDL (mg/dL)	35.3 c	46.1 b	56.8 a	38.92 bc	3.092	0.164	0.002
LDL (mg/dL)	58.5 a	52.3 b	45.7 c	47.94 c	2.911	0.024	0.204
VLDL (mg/dL)	59.8 a	45.0 b	42.0 bc	38.40 c	1.176	<0.0001	0.002
SOD (U/mL)	0.12 c	0.22 b	0.29 a	0.22 b	0.004	<0.0001	<0.0001
MDA (nmol/mL)	0.33 a	0.24 b	0.22 b	0.13 c	0.007	<0.0001	0.792
GSH (ng/mL)	0.11 c	0.22 b	0.28 a	0.26 a	0.008	<0.0001	0.0001
GPX (mg/dL)	0.13 d	0.23 c	0.30 b	0.34 a	0.007	<0.0001	0.006
IgG (mg/dL)	0.89 b	1.33 a	1.19 a	1.17 a	0.042	0.013	0.002
IgM (mg/dL)	0.49 c	0.56 bc	0.64 b	0.90 a	0.026	<0.0001	0.024
IgA (mg/dL)	0.53 b	0.64 b	0.80	0.85 a	0.034	0.0007	0.547
Selenium	0.07 c	0.19 b	0.24 ab	0.27 a	0.016	0.0001	0.060

Means in the same row with no superscript letters after them or a similar superscript letter following them are not significantly different ($p < 0.05$). [1] TC: total cholesterol; TG: triglycerides; HDL: high-density lipoprotein; LDL: low-density lipoprotein. SOD: superoxide dismutase; MDA: malondialdehyde; GSH: reduced glutathione; GPX: glutathione peroxidase; IgG and M: immunoglobulin G.

3.7. Antioxidant Indices

The results of the antioxidant indices in the serum are given in Table 8. The activities of superoxide dismutase (SOD) and glutathione peroxidase, and the levels of reduced glutathione (GSH) were significantly increased (linear and quadratic, $p < 0.05$) by the dietary supplementation of Che-SeNPs compared with those in control. Dietary Che-SeNPs levels decreased malondialdehyde (MDA) levels linearly ($p < 0.0001$) compared to the control group. The values of immunoglobulin G (IgG) of Che-SeNPs-treated groups were higher (linear and quadratic, $p < 0.05$) than those in the control group. IgM and IgA values of Che-SeNPs (0.4 and 0.6 g/kg) were higher ($p < 0.05$) than those in the control group. The quails fed a diet supplemented with Che-SeNPs showed higher plasma selenium concentrations when compared to those fed the control diet (linear, $p = 0.0001$).

3.8. Microbiological Analysis

The different Che-SeNPs levels significantly affected the cecal microbiota of growing Japanese quails (Table 9). The groups fed a diet supplemented with Che-SeNPs showed lower total bacterial count, total yeast and molds count, *Coliform*, *E. coli*, *Enterococcus* ssp., and *Salmonella* spp. colonization than those in the control group (linear and quadratic,

$p < 0.0001$). However, the dietary supplementation of Che-SeNPs levels increased the lactic acid bacteria count (linear and quadratic, $p < 0.05$) compared to the control group.

Table 9. Caecal microbiota of growing Japanese quail as affected by dietary treatments.

Items	Chemical Nano Selenium Levels (g/kg Diet)				SEM	p Value	
	0	0.2	0.4	0.6		Linear	Quadratic
Microbiological count (Log CFU/g)							
TBC	6.05 [a]	5.32 [b]	5.08 [c]	5.31 [b]	0.006	<0.0001	<0.0001
TYMC	5.82 [a]	5.20 [b]	4.90 [d]	5.12 [c]	0.014	<0.0001	<0.0001
Coliform	5.96 [a]	5.26 [b]	5.08 [c]	4.92 [d]	0.035	<0.0001	0.0004
Escherichia coli	5.93 [a]	5.25 [b]	5.05 [c]	5.09 [c]	0.022	<0.0001	<0.0001
Lactic acid bacteria	5.31 [c]	5.58 [a]	5.40 [b]	5.53 [a]	0.017	0.0003	0.0048
Enterococcus spp.	5.86 [a]	5.23 [b]	4.97 [c]	5.08 [c]	0.026	<0.0001	<0.0001
Salmonella spp.	6.38 [a]	3.22 [b]	2.23 [c]	2.07 [c]	0.057	<0.0001	<0.0001

Means in the same row with no superscript letters after them or a similar superscript letter following them are not significantly different ($p < 0.05$). TBC: Total bacterial count. TYMC: total yeasts and molds count.

4. Discussion

Antimicrobial agents are critical in the pharmaceutical and textile industries, water purification, and food packaging. One notable disadvantage of organically synthesized compounds is toxicity in the body; therefore, the trend is to use inorganic nanoparticles such as Che-SeNPs with antimicrobial activity [20]. These Che-SeNPs have an inhibitory effect on many microorganisms. Currently, antimicrobial drugs are becoming less effective for many diseases globally because of the drug resistance capability of microbes. Microorganisms use their biofilm to resist antimicrobial drugs, and the membranes are the primary source of food contamination. Che-SeNPs have been used to control the growth and formation of biofilms of food spoilage bacteria, including *B. cereus*, *Enterococcus faecalis*, *S. aureus*, *E. coli* O157:H7, *S. typhimurium*, and *S. enterica* [21]. The development more effective antibacterial agents is vital for a wide range of applications in various diseases for better public health. However, the emergence of multiple antibiotic-resistant bacteria presents a public health threat. Many developed antimicrobial drugs have limited effective applications due to chemical imbalances, low biocompatibility, and poor long-term antibacterial efficiency. Che-SeNPs conjugated with quercetin and acetylcholine have shown a tremendous antimicrobial effect on the pathogen [22]. Probiotics are microorganisms that can improve intestinal microbial balance and benefit poultry health after consumption in adequate amounts. *Lactobacillus plantarum* and *L. johnsonii* cells are resistant against selenium dioxide, and their cell-free extracts were tested against *C. albicans* ATCC 14053 [7]. Selenium particles extracted from cultures of *S. carnosus* stabilized by their natural protein coating, for instance, show considerable activity against the nematode *Steinernema feltiae*, *Saccharomyces cerevisiae*, and *E. coli*. Natural SeNPs were found to be more active than mechanically generated selenium particles and can be applied as antimicrobial materials in medicine and agriculture [23]. Antimicrobial tests show SeNPs activity against *S. epidermidis*, but not against *E. coli* in a low Se concentration of 2 ppm. *S. aureus* is an important bacterium commonly found in numerous infections. *S. aureus* infections were difficult to treat due to their biofilm formation and defined antibiotic resistance. SeNPs were used effectively in the prevention and treatment of disease caused by *S. aureus* [8].

The antifungal activity of SeNPs was evaluated against *C. albicans* ATCC 4862, *C. glabrata* ATCC64677, *C. parapsilosis* ATCC 22019, and *C. guilliermondii* ATCC 6260 using the disk diffusion method [14] (Table 2). The common antifungal agents are enormously irritant and lethal, and it is necessary to formulate newer types of safe and cost-effective fungicidals. Accordingly, the present study illustrates that selenium nanoparticles have good antifungal activity against all pathogenic animals and human *Candida* species. Selenium nanoparticles showed better activity against *C. albicans* ATCC 4862 compared to other *Candida* species used in this study. In addition, it was proved that SeNPs ranging in size from 100 to 550 nm, with an average size of 245 nm, have low toxicity and high biological activities [24]. A similar observation was reported by Shakibaie et al. [7], who studied the antifungal activity of selenium nanoparticles against *Aspergillus fumigatus* and *C. albicans*, and found that the MICs for *A. fumigatus* and *C. albicans* were 100 and 70 µg/mL, respectively. However, the high surface-to-volume ratios and their nanoscale sizes provide better activity against biological materials. In addition, Che-SeNPs have significantly lower toxicity than other inorganic and organic forms of supplemental selenium [7].

The current data demonstrated that dietary supplementation with Che-SeNPs substantially affected BW, BWG, feed intake, and feed conversion ratio (FCR). A similar observation was stated by Zhou and Wang [25], who clarified a significant improvement in the FCR and growth performance by supplementation with Che-SeNP up to a 0.5-mg/kg basal diet. Khazraie and Ghazanfarpoor [26] illustrated that weight gain was significantly increased in quail chicks fed the Che-SeNPs-supplemented diet compared to the control. Selim et al. [27], using the Che-SeNPs form (0.15 to 0.30 ppm), showed a marked improvement in BW, BWG, and FCR of broiler chicks. Ibrahim et al. [28] indicated that dietary Che-SeNPs supplementation significantly improved BW, BWG, and FCR of broiler chicks compared to the control group. The improved performance may be due to (1) higher utilization of Che-SeNPs associated with the unique properties of the nano form, such as excellent bioavailability, higher solubility, high cellular uptake, and greater surface activity [2]; (2) the involvement of Se in regulating several enzymatic systems, which interfere in energy metabolism and metabolism of the essential fatty acid apurinic and apyrimidinic base; and (3) Che-SeNPs having high biological activity, immune regulation, and oxidation resistance [22]. In addition, the improved FCR can be elucidated by the Che-SeNPs role in enhancing the activity of intestinal microbiota to digest and absorb the nutrients via the intestinal barriers [9].

The results of the present study in carcass traits and relative organ weight of growing Japanese quails were in line with the study of Khazraie and Ghazanfarpoor [26], who stated that the supplementation of Che-SeNPs to the diet did not affect carcass traits of chicks. Additionally, Cai et al. [6] reported no significant effect of Che-SeNPs on the weights of carcass parts in broilers. Selim et al. [27] indicated that giblets were not affected due to the inclusion of Che-SeNPs in the diet. Recently, Bakhshalinejad et al. [29] reported that neither carcass yield nor carcass yield parts such as thigh and breast muscles and liver, gizzard, and heart of broilers were affected by different Che-SeNPs levels at 42 d of age. In the present study, the relative liver weight was significantly increased with Che-SeNPs; this increase (21–28% relative to control) may be due to the increase in live body weight in Che-SeNPs groups. However, the widespread use of Nano Se in medication and nanoelectronics has increased the risk of their environmental contaminations, which might affect animal species and humans, although it is useful to understand the assessment of the toxicity of Se-NPs to the biological ecosystem. It should be mentioned that the increase in WBCs was insignificant in the Che-SeNPs supplemented-groups; these change along with the change in liver percentage, even if not significant, warrant further investigation to confirm the safety of Che-SeNPs in animal and human nutrition.

Boostani et al. [30] exhibited that packed cell volume, RBCs and WBCs were not different between the birds supplemented with Che-SeNPs and the control birds, which is in line with the current results. Likewise, Chen et al. [31] revealed no significant difference in WBCs, RBCs, and packed cell volume of broilers fed different Se sources. Additionally, Mohamed et al. [32] illustrated that using Che-SeNPs in the diet of Sinai chicks did not significantly affect WBCs, eosinophils, and monocytes. However, our study indicated that Hb level was increased by adding Che-SeNPs, in agreement with Khazraie and Ghazanfarpoor [26], who reported a significant increase in Hb concentration in quails fed a diet containing Che-SeNPs. These findings may be caused by Se enhancing the activity of hemopoietic organs [33]. Se protects the neutrophils, RBCs, WBCs, and other blood components against peroxidative damage [34]. Deficiency of Se can increase ROS in body tissues, the significant adverse impacts on the consistency of immunity cells' performance and biological membranes [35].

The results of the current study on the blood biochemistry of quails were in agreement with previous studies. Serum total protein and albumin were not significantly affected due to Che-SeNPs supplementation to the broiler diet [27]. However, serum globulin levels were increased with the addition of Che-SeNPs in the diet [36]. Additionally, no significant difference in serum AST activity was observed of chicks fed a diet supplemented with Che-SeNPs [27]. However, our results are similar to the study of Elsaid [37], who reported increased serum ALT activity in birds fed a diet supplemented with Che-SeNPs. Selim et al. [27] found that increasing the Che-SeNPs level in broiler diets increased plasma creatinine levels compared to the control group. However, some studies showed that blood creatinine levels declined in birds fed a diet containing Che-SeNPs Elsaid [37]. The potential reason for these differences is possibly related to the dose and time of animal exposure. We conclude from the current study that the higher Che-SeNPs level is the cause of increased ALT and creatinine as indicators of liver and kidney oxidative damage, whereas lower levels showed less damage.

Selenium has a hypocholesterolemic activity. A significant reduction in plasma TC and an increase in HDL were detected in the Che-SeNPs-treated birds. The dietary addition of nano forms of selenium for hens caused substantial declines in serum levels of cholesterol as compared to that of the control [38]. Rizk [39] stated that Che-SeNPs addition in the chicken diet decreased cholesterol, triglycerides, and low-density lipoproteins and increased HDL compared with the control group. These results might be attributed to the lipolysis that increased with Se intake. Additionally, the reduction of cholesterol may be due to the role of Se in the activation of peroxisome proliferator-activated receptor-γ that can decrease sterol regulatory element-binding protein-2 level, resulting in decreased cholesterol synthesis [40].

The nutritional status of an animal greatly influences the antioxidant system. Se nanoparticles have vital roles in protecting the body cells from reactive oxygen species abundance by decreasing the production of free radicals and lipid peroxidation [41]. Se is well-known for its ability to boost the antioxidant capacity as it forms selenocysteine, a portion of the active center of GSH-peroxidase (Px) [42]. Therefore, a dietary supplementation of Se is essential to improve Se-dependent antioxidant enzymes. These enzymes can help in decreasing the concentration of lipid peroxides and hydrogen peroxide. Dietary Che-SeNPs enhanced oxidative stability and antioxidant ability in broilers [6]. Mohamed et al. [32] reported a positive effect on birds' plasma total antioxidant capacity when fed a diet containing Che-SeNPs. Aparna and Karunakaran [43] detected an increase in glutathione peroxidase and SOD cellular activity in birds fed Che-SeNPs compared to the control group. El-Deep et al. [4] displayed that Che-SeNPs enhanced the activities of SOD and GSH-Px and reduced MDA content in the liver of broilers. The improvement of antioxidant status in quails fed Che-SeNPs in the current study may be attributed to the fact that (1) Che-SeNPs had high antioxidant activity, because it has an augmented ability to trap free radicals with better antioxidant influence, (2) Che-SeNPs can act as a chemopreventive agent when administered at a smaller particle size, (3) Se plays a vital role as an antioxidant that could protect intestinal mucosa against pathogens and oxidative damage, and (4) Se has immunomodulation properties [44].

Nanominerals such as Che-SeNPs can increase immune parameters and disease resistance [4]. In the current study, we presented a potential approach to the application of Che-SeNPs to improve the immunity of quails. These findings can be due to the higher absorption of selenium nanoparticles. The present data are in harmony with the study of Cai et al. [6], who stated that dietary Che-SeNPs supplementation improved humoral immunity by increasing the levels of IgG and IgM of broiler chicks. Dietary Che-SeNPs supplementation showed immunostimulatory impacts in broiler chicks [45]. The improvement in serum immunoglobulins levels may be attributed to the essential biological role of Che-SeNPs in increasing T helper cells and enhancing the secretion of cytokines [46].

Additionally, Se plays a crucial role in the production of GSH-Px. Selenium inhibits arachidonic acid peroxidation and protects cells and tissues of the immune system from damage caused by free radicals. Therefore, it can be stated that Che-SeNPs boosts birds' immunity and antioxidant metabolites [45]. Studies have shown that the use of nanominerals in poultry production and its effect on performance and immunity, and reproduction is promising [47,48]. It has been suggested that the application of Se can help to strengthen immunity and decrease inflammation [49,50]. Se, according to Rooke et al. [51], may be involved in a variety of immune functions at the cellular and molecular levels, including lowering immunosuppressive markers such as glucocorticoids; reducing the duration and rate of intramammary infections; and regulating the function of lymphocytes, neutrophils, and natural killer cells. Our results suggest that feeding a diet enriched with Che-SeNPs might have immunostimulatory impacts on quails.

The regulation of microbiota in the gut can be achieved through dietary supplements that can encourage the growth of beneficial bacteria or selectively suppress pathogenic bacteria. Trace elements and natural agents as feed additives may affect the diversity of gut microbiota [8]. The present study found that supplementation of Che-SeNPs in quail diets declined harmful bacteria and increased beneficial bacteria. Se is one of the critical elements that can help microbiota complete its action within the gut [9]. Furthermore, Se supplementation augmented the population of caecum such as *Bifidobacterium* spp. and *Lactobacillus* spp. compared to the basal diet [9]. Therefore, using Che-SeNPs is one of the recommendations for reducing the population of harmful gut bacteria due to its inhibitory effect against many pathogenic bacteria.

Nanotechnology has been found to have advantageous uses in the food chain of humans, mainly through enhancing the bioavailability and delivering enough levels of vital nutrients, vitamins, and minerals in animal products used by humans [10,52–56]. Moreover, the consumers' demand for foods and their knowledge has been enhanced as consumers want safe and high-quality foods with high sensory quality, favorable health qualities, and prolonged shelf life [57]. Several studies proved the possibility of supplementing nanomaterials to improve mineral contents in animal products; nevertheless, most of these studies were carried out on chicken, meat, and eggs [58,59]. Therefore, more research is needed to analyze the ability of nanomaterials to affect the quality and nutritional content of meat and egg. In addition, the influence of nanomaterials on the environment and health needs further examination [60,61]. Thus, the application of nanoparticles in the poultry industry must be further investigated before they can be applied.

5. Conclusions

The current study's findings demonstrated that dietary supplementation with Che-SeNPs could improve the performance of growing quails. The highest values of growth performance were recorded in the group fed 0.4 g Che-SeNPs g/kg feed during the fattening periods (1–5 wk of age). Moreover, the dietary addition of Che-SeNPs improved the lipid profile, antioxidant indices, and immunity and decreased the intestinal pathogens of growing quails. The groups fed diets supplemented with Che-SeNPs showed lower total yeast and mold count, *Coliform*, *Escherichia coli*, *Enterococcus* spp., and *Salmonella* spp. colonization, and higher lactic acid bacteria counts than those in the control group. However, further studies are warranted to understand the effect of nanominerals and their mechanisms of action, sites of absorption, and transcript expression analysis of distribution.

Author Contributions: Conceptualization, M.A., M.T.E.-S., S.S.E. and F.M.R.; methodology, M.A., M.T.E.-S., S.S.E. and F.M.R.; formal analysis, M.A., M.T.E.-S. and F.M.R.; investigation, M.A., M.T.E.-S. and F.M.R.; data curation, M.A. and F.M.R.; writing—original draft preparation, M.A., S.Y.A.Q., Y.A.A., M.T.E.-S., S.S.E., M.A.M., M.M., M.E.A.E.-H. and F.M.R.; writing—review and editing, M.A., S.Y.A.Q., Y.A.A., M.T.E.-S., S.S.E., M.A.M., M.M., M.E.A.E.-H. and F.M.R. All authors have read and agreed to the published version of the manuscript.

Funding: The Deanship of Scientific Research (DSR) at King Abdulaziz University, Jeddah, Saudi Arabia has funded this project, under grant No. (FP-073-43).

Institutional Review Board Statement: The Ethics statement for Animal care and maintenance were in accordance with the guidelines of the Egyptian Research Ethics Committee and the guidelines for the Care and Use of Laboratory Animals by Zagazig University (ZU-IACUC/2/F/56/2021).

Informed Consent Statement: Not applicable.

Data Availability Statement: All data are published in the cited literature and reported in the text of this manuscript.

Acknowledgments: The authors acknowledged the administrative, technical, and financial support by DSR, King Abdulaziz University, Jeddah, Saudi Arabia.

Conflicts of Interest: The authors declare no conflict of interest.

References

1. Joshua, P.P.; Valli, C.; Balakrishnan, V. Effect of in ovo supplementation of nano forms of zinc, copper, and selenium on post-hatch performance of broiler chicken. *Vet. World* **2016**, *9*, 287–294. [CrossRef] [PubMed]
2. Zhang, J.; Wang, X.; Xu, T. Elemental Selenium at Nano Size (Nano-Se) as a Potential Chemopreventive Agent with Reduced Risk of Selenium Toxicity: Comparison with Se-Methylselenocysteine in Mice. *Toxicol. Sci.* **2007**, *101*, 22–31. [CrossRef] [PubMed]
3. Shirsat, S.; Kadam, A.; Naushad, M.; Mane, R.S. Selenium nanostructures: Microbial synthesis and applications. *RSC Adv.* **2015**, *5*, 92799–92811. [CrossRef]
4. El-Deep, M.H.; Shabaan, M.; Assar, M.H.; Attia, K.M.; Sayed, M.A.M. Comparative Effects of Different Dietary Selenium Sources on Productive Performance, Antioxidative Properties and Immunity in Local Laying Hens Exposed to High Ambient Temperature. *J. Anim. Poult. Prod.* **2017**, *8*, 335–343. [CrossRef]
5. Ahmadi, M.; Ahmadian, A.; Seidavi, A. Effect of different levels of nano-selenium on performance, blood parameters, immunity and carcass characteristics of broiler chickens. *Poult. Sci. J.* **2018**, *6*, 99–108.
6. Cai, S.J.; Wu, C.X.; Gong, L.M.; Song, T.; Wu, H.; Zhang, L.Y. Effects of nano-selenium on performance, meat quality, immune function, oxidation resistance, and tissue selenium content in broilers. *Poult. Sci.* **2012**, *91*, 2532–2539. [CrossRef] [PubMed]
7. Shakibaie, M.; Forootanfar, H.; Golkari, Y.; Mohammadi-Khorsand, T.; Shakibaie, M.R. Anti-biofilm activity of biogenic selenium nanoparticles and selenium dioxide against clinical isolates of Staphylococcus aureus, Pseudomonas aeruginosa, and Proteus mirabilis. *J. Trace Elem. Med. Biol.* **2015**, *29*, 235–241. [CrossRef]
8. Reda, F.M.; El-Saadony, M.T.; ElNesr, S.S.; Alagawany, M.; Tufarelli, V. Effect of Dietary Supplementation of Biological Curcumin Nanoparticles on Growth and Carcass Traits, Antioxidant Status, Immunity and Caecal Microbiota of Japanese Quails. *Animals* **2020**, *10*, 754. [CrossRef]
9. Yoon, I.; Werner, T.M.; Butler, J.M. Effect of Source and Concentration of Selenium on Growth Performance and Selenium Retention in Broiler Chickens. *Poult. Sci.* **2007**, *86*, 727–730. [CrossRef]
10. Abdelnour, S.; Alagawany, M.; Hashem, N.; Farag, M.; Alghamdi, E.; Hassan, F.; Bilal, R.; Elnesr, S.; Dawood, M.; Nagadi, S.; et al. Nanominerals: Fabrication Methods, Benefits and Hazards, and Their Applications in Ruminants with Special Reference to Selenium and Zinc Nanoparticles. *Animals* **2021**, *11*, 1916. [CrossRef]
11. Bhunchu, S.; Muangnoi, C.; Rojsitthisak, P. Curcumin diethyl disuccinate encapsulated in chitosan/alginate nanoparticles for improvement of its in vitro cytotoxicity against MDA-MB-231 human breast cancer celle. *Die Pharm.-Int. J. Pharm. Sci.* **2016**, *71*, 691–700.
12. El-Saadony, M.T.; Saad, A.M.; Najjar, A.A.; Alzahrani, S.O.; Alkhatib, F.M.; Shafi, M.E.; Selem, E.; Desoky, E.-S.M.; Fouda, S.E.-S.E.-S.; El-Tahan, A.M. The use of biological selenium nanoparticles in controlling Triticum aestivum L. crown root and rot diseases induced by Fusarium species and improve yield under drought and heat stress. *Saudi J. Biol. Sci.* **2021**, *28*, 4461–4471. [CrossRef]
13. El-Saadony, M.T.; Saad, A.M.; Taha, T.F.; Najjar, A.A.; Zabermawi, N.M.; Nader, M.M.; AbuQamar, S.F.; El-Tarabily, K.A.; Salama, A. Selenium nanoparticles from Lactobacillus paracasei HM1 capable of antagonizing animal pathogenic fungi as a new source from human breast milk. *Saudi J. Biol. Sci.* **2021**. [CrossRef]
14. Hariharan, H.; Al-Harbi, N.; Karuppiah, P.; Rajaram, S. Microbial synthesis of selenium nanocomposite using Saccharomyces cerevisiae and its antimicrobial activity against pathogens causing nosocomial infection. *Chalcogenide Lett.* **2012**, *9*, 509–515.
15. NRC. *Nutrient Requirements of Poultry*, 9th ed.; National Academy Press: Washington, DC, USA, 1994.
16. Alagawany, M.; El-Saadony, M.T.; Elnesr, S.; Farahat, M.; Attia, G.; Madkour, M.; Reda, F. Use of lemongrass essential oil as a feed additive in quail's nutrition: Its effect on growth, carcass, blood biochemistry, antioxidant and immunological indices, digestive enzymes and intestinal microbiota. *Poult. Sci.* **2021**, *100*, 101172. [CrossRef]
17. Alagawany, M.; Madkour, M.; El-Saadony, M.T.; Reda, F.M. Paenibacillus polymyxa (LM31) as a new feed additive: Antioxidant and antimicrobial activity and its effects on growth, blood biochemistry, and intestinal bacterial populations of growing Japanese quail. *Anim. Feed. Sci. Technol.* **2021**, *276*, 114920. [CrossRef]

18. Reda, F.M.; El-Saadony, M.T.; El-Rayes, T.K.; Attia, A.I.; El-Sayed, S.A.; Ahmed, S.Y.; Madkour, M.; Alagawany, M. Use of biological nano zinc as a feed additive in quail nutrition: Biosynthesis, antimicrobial activity and its effect on growth, feed utilisation, blood metabolites and intestinal microbiota. *Ital. J. Anim. Sci.* **2021**, *20*, 324–335. [CrossRef]
19. El-Saadony, M.T.; Alkhatib, F.M.; Alzahrani, S.O.; Shafi, M.E.; Abdel-Hamid, S.E.; Taha, T.F.; Aboelenin, S.M.; Soliman, M.M.; Ahmed, N.H. Impact of mycogenic zinc nanoparticles on performance, behavior, immune response, and microbial load in Oreochromis niloticus. *Saudi J. Biol. Sci.* **2021**, *28*, 4592–4604. [CrossRef]
20. Phull, A.R.; Abbas, Q.; Ali, A.; Raza, H.; Kim, S.J.; Zia, M.; Haq, I.-U. Antioxidant, cytotoxic and antimicrobial activities of green synthesized silver nanoparticles from crude extract of Bergenia ciliata. *Futur. J. Pharm. Sci.* **2016**, *2*, 31–36. [CrossRef]
21. Khiralla, G.M.; El-Deeb, B.A. Antimicrobial and antibiofilm effects of selenium nanoparticles on some foodborne pathogens. *LWT-Food Sci. Technol.* **2015**, *63*, 1001–1007. [CrossRef]
22. Huang, X.; Chen, X.; Chen, Q.; Yu, Q.; Sun, D.; Liu, J. Investigation of functional selenium nanoparticles as potent antimicrobial agents against superbugs. *Acta Biomater.* **2016**, *30*, 397–407. [CrossRef] [PubMed]
23. Estevam, E.C.; Griffin, S.; Nasim, M.J.; Denezhkin, P.; Schneider, R.; Lilischkis, R.; Dominguez-Alvarez, E.; Witek, K.; Latacz, G.; Keck, C.; et al. Natural selenium particles from Staphylococcus carnosus: Hazards or particles with particular promise? *J. Hazard. Mater.* **2017**, *324*, 22–30. [CrossRef] [PubMed]
24. Fesharaki, P.J.; Nazari, P.; Shakibaie, M.; Rezaie, S.; Banoee, M.; Abdollahi, M.; Shahverdi, A.R. Biosynthesis of selenium nanoparticles using Klebsiella pneumoniae and their recovery by a simple sterilization process. *Braz. J. Microbiol.* **2010**, *41*, 461–466. [CrossRef] [PubMed]
25. Zhou, X.; Wang, Y. Influence of dietary nano elemental selenium on growth performance, tissue selenium distribution, meat quality, and glutathione peroxidase activity in Guangxi Yellow chicken. *Poult. Sci.* **2011**, *90*, 680–686. [CrossRef] [PubMed]
26. Khazraie, S.; Ghazanfarpoor, R. Effect of nano-selenium particles and sodium selenite on performance, glutathione peroxidase and superoxide dismutase of quail under heat stress. *Int. J. Rev. Life. Sci.* **2015**, *5*, 875–882.
27. Selim, N.; Radwan, N.; Youssef, S.F.; Eldin, T.S.; Elwafa, S.A. Effect of Inclusion Inorganic, Organic or Nano Selenium Forms in Broiler Diets On: 2-Physiological, Immunological and Toxicity Statuses of Broiler Chicks. *Int. J. Poult. Sci.* **2015**, *14*, 144–155. [CrossRef]
28. Ibrahim, N.; Sabic, E.; Wakwak, M.; El-Wardany, I.; El-Homosany, Y.; Mohammad, N.E.-D. In-ovo and dietary supplementation of selenium nano-particles influence physiological responses, immunological status and performance of broiler chicks. *J. Anim. Feed. Sci.* **2020**, *29*, 46–58. [CrossRef]
29. Bakhshalinejad, R.; Hassanabadi, A.; Swick, R.A. Dietary sources and levels of selenium supplements affect growth performance, carcass yield, meat quality and tissue selenium deposition in broilers. *Anim. Nutr.* **2019**, *5*, 256–263. [CrossRef] [PubMed]
30. Boostani, A.; Sadeghi, A.A.; Mousavi, S.N.; Chamani, M.; Kashan, N. The effects of organic, inorganic, and nano-selenium on blood attributes in broiler chickens exposed to oxidative stress. *Acta Sci. Vet.* **2015**, *43*, 1–6.
31. Chen, G.; Wu, J.; Li, C. The effect of Different Selenium Levels on Production Performance and Biochemical Parameters of Broilers. *Ital. J. Anim. Sci.* **2013**, *12*, 79. [CrossRef]
32. Mohamed, H.S.; Rizk, Y.S.; Elslamony, A.E.; Soliman, A.A.; Ebrahim, A.F. Study the relationship between selenium and heat shock proteins under heat stress for local sinai chickens strain. *Egypt. Poult. Sci. J.* **2016**, *36*, 337–354. [CrossRef]
33. Tayeb, İ.; Qader, G. Effect of feed supplementation of selenium and vitamin E on production performance and some hematological parameters of broiler. *KSÜ Doğa Bilimleri Derg.* **2012**, *15*, 46–56.
34. Bickhardt, K.; Ganter, M.; Sallmann, P.; Fuhrmann, H. Investigations on manifestations of vitamin E and selenium deficiency in sheep and goats. *DTW Dtsch. Tierarztl. Wochenschr.* **1999**, *106*, 242–247. [PubMed]
35. Lessard, M.; Yang, W.C.; Elliott, G.S.; Rebar, A.H.; Van Vleet, J.F.; DesLauriers, N.; Brisson, G.J.; Schultz, R.D. Cellular immune responses in pigs fed a vitamin E- and selenium-deficient diet. *J. Anim. Sci.* **1991**, *69*, 1575–1582. [CrossRef]
36. Mahmoud, H.E.-D.; Ijiri, D.; Ebeid, T.A.; Ohtsuka, A. Effects of dietary nano-selenium supplementation on growth performance, antioxidative status, and immunity in broiler chickens under thermoneutral and high ambient temperature conditions. *J. Poult. Sci.* **2016**, *53*, 274–283.
37. Elsaid, A.E. Influence of In-Ovo Injection with Nanoparticles Selenium (SeNPs) on Growth Performance, Carcass Yield and Immune Status of Broiler Chicks. Ph.D. Thesis, Faculty of Agriculture, Damietta University, Damietta, Egypt, 2015.
38. Ismail, F.S.A.; Mostafa, M.Y.; Azzam, M.M.M.; Gorgy, M.A.L. Effect of Some Sources of Antioxiadnts on The Productive and Reproductive Performance of Turkey Hens. *J. Anim. Poult. Prod.* **2016**, *7*, 393–401. [CrossRef]
39. Rizk, Y.S. Effect of dietary source of selenium on productive and reproductive performance of Sinai laying hens under heat stress conditions. *Egypt. Poult. Sci. J.* **2018**, *37*, 461–489.
40. Klopotek, A.; Hirche, F.; Eder, K. PPARγ Ligand Troglitazone Lowers Cholesterol Synthesis in HepG2 and Caco-2 Cells via a Reduced Concentration of Nuclear SREBP-2. *Exp. Biol. Med.* **2006**, *231*, 1365–1372. [CrossRef]
41. Pilarczyk, B.; Jankowiak, D.; Tomza-Marciniak, A.; Pilarczyk, R.; Sablik, P.; Drozd, R.; Tylkowska, A.; Skólmowska, M. Selenium Concentration and Glutathione Peroxidase (GSH-Px) Activity in Serum of Cows at Different Stages of Lactation. *Biol. Trace Elem. Res.* **2012**, *147*, 91–96. [CrossRef]
42. Terova, G.; Rimoldi, S.; Izquierdo, M.; Pirrone, C.; Ghrab, W.; Bernardini, G. Nano-delivery of trace minerals for marine fish larvae: Influence on skeletal ossification, and the expression of genes involved in intestinal transport of minerals, osteoblast differentiation, and oxidative stress response. *Fish Physiol. Biochem.* **2018**, *44*, 1375–1391. [CrossRef]

43. Aparna, N.; Karunakaran, R. Effect of Selenium Nanoparticles Supplementation on Oxidation Resistance of Broiler Chicken. *Indian J. Sci. Technol.* **2016**, *9*, 1–5. [CrossRef]
44. Torres, S.K.; Campos, V.L.; León, C.G.; Rodríguez-Llamazares, S.M.; Rojas, S.M.; González, M.; Smith, C.; Mondaca, M.A. Biosynthesis of selenium nanoparticles by Pantoea agglomerans and their antioxidant activity. *J. Nanopart. Res.* **2012**, *14*, 1–9. [CrossRef]
45. Kadhim, Z.M. Study the Effect of Adding Two Levels of Organic Selenium and Nano-Selenium Particles to the Ration on Some Blood Traits and Lymphatic Organs of Broiler Chicks (Ross 308). Ph.D. Thesis, Animal Production Department, College of Agriculture, AlQasim Green University, Al Qasim, Iraq, 2018.
46. Shabani, R.; Fakhraei, J.; Yarahmadi, H.M.; Seidavi, A. Effect of different sources of selenium on performance and characteristics of immune system of broiler chickens. *Rev. Bras. Zootec.* **2019**, *48*, 48. [CrossRef]
47. Huang, S.; Wang, L.; Liu, L.; Hou, Y.; Li, L. Nanotechnology in agriculture, livestock, and aquaculture in China. A review. *Agron. Sustain. Dev.* **2015**, *35*, 369–400. [CrossRef]
48. Osama, E.; El-Sheikh, S.M.; Khairy, M.H.; Galal, A.A. Nanoparticles and their potential applications in veterinary medicine. *J. Adv. Vet. Res.* **2020**, *10*, 268–273.
49. Madkour, M.; Ali, H.; Yassein, S.; Abdel-Fatt, S.; El-Allawy, H.M. Effect of Dietary Organic Selenium Supplement on Growth and Reproductive Performance of Japanase Quail Breeders and Their Progeny and its Relation to Antioxidation and Thyroid Activity. *Int. J. Poult. Sci.* **2015**, *14*, 317–324. [CrossRef]
50. Rayman, M.P. The use of high-selenium yeast to raise selenium status: How does it measure up? *Br. J. Nutr.* **2004**, *92*, 557–573. [CrossRef]
51. Rooke, J.A.; Robinson, J.J.; Arthur, J.R. Effects of vitamin E and selenium on the performance and immune status of ewes and lambs. *J. Agric. Sci.* **2004**, *142*, 253–262. [CrossRef]
52. Korish, M.A.; Attia, Y.A. Evaluation of Heavy Metal Content in Feed, Litter, Meat, Meat Products, Liver, and Table Eggs of Chickens. *Animals* **2020**, *10*, 727. [CrossRef]
53. Attia, Y.A.; Abd Al-Hamid, A.E.; Zeweil, H.S.; Qota, E.M.; Bovera, F.; Monastra, G.; Sahledom, M.D. Effect of dietary amounts of inorganic and organic zinc on productive and physiological traits of White Pekin ducks. *Animals* **2013**, *7*, 895–900. [CrossRef]
54. El-Saadony, M.T.; Sitohy, M.Z.; Ramadan, M.F.; Saad, A.M. Green nanotechnology for preserving and enriching yogurt with biologically available iron (II). *Innov. Food Sci. Emerg. Technol.* **2021**, *69*, 102645. [CrossRef]
55. Attia, Y.; Abdalah, A.; Zeweil, H.; Bovera, F.; El-Din, A.T.; Araft, M. Effect of inorganic or organic selenium supplementation on productive performance, egg quality and some physiological traits of dual-purpose breeding hens. *Czech J. Anim. Sci.* **2010**, *55*, 505–519. [CrossRef]
56. Hassan, F.; Mobarez, S.; Mohamed, M.; Attia, Y.; Mekawy, A.; Mahrose, K. Zinc and/or Selenium Enriched Spirulina as Antioxidants in Growing Rabbit Diets to Alleviate the Deleterious Impacts of Heat Stress during Summer Season. *Animals* **2021**, *11*, 756. [CrossRef]
57. Konkol, D.; Wojnarowski, K. The Use of Nanominerals in Animal Nutrition as a Way to Improve the Composition and Quality of Animal Products. *J. Chem.* **2018**, *2018*, 5927058. [CrossRef]
58. Menon, S.; Shrudhi Devi, K.S.; Agarwal, H.; Shanmugam, V.K. Efficacy of Biogenic Selenium Nanoparticles from an Extract of Ginger towards Evaluation on Anti-Microbial and Anti-Oxidant Activities. *Colloid Interface Sci. Commun.* **2019**, *29*, 1–8. [CrossRef]
59. Alagawany, M.; Elnesr, S.S.; Farag, M.R.; Tiwari, R.; Yatoo, M.I.; Karthik, K.; Michalak, I.; Dhama, K. Nutritional significance of amino acids, vitamins and minerals as nutraceuticals in poultry production and health—A comprehensive review. *Vet. Q.* **2021**, *41*, 1–29. [CrossRef] [PubMed]
60. Prabhu, B.M.; Ali, S.F.; Murdock, R.C.; Hussain, S.M.; Srivatsan, M. Copper nanoparticles exert size and concentration dependent toxicity on somatosensory neurons of rat. *Nanotoxicology* **2010**, *4*, 150–160. [CrossRef] [PubMed]
61. Najafzadeh, H.; Ghoreishi, S.; Mohammadian, B.; Rahimi, E.; Afzalzadeh, M.; Kazemivarnamkhasti, M.; Ganjealidarani, H. Serum biochemical and histopathological changes in liver and kidney in lambs after zinc oxide nanoparticles administration. *Vet. World* **2013**, *6*, 6. [CrossRef]

animals MDPI

Article

Antioxidant and Antiapoptotic Effects of a *Turraea fischeri* Leaf Extract on Cryopreserved Goat Sperm

Soha A. Hassan [1,†], Wael A. Khalil [2,*,†], Mahmoud A. E. Hassan [3], Ahmed I. Yousif [3], Omar M. Sabry [4], Michael Wink [5] and Mansour Sobeh [6,*]

1 Basic Science Department, Faculty of Dentistry, October 6 University, Cairo 12566, Egypt; sohaahmed1999@yahoo.com
2 Department of Animal Production, Faculty of Agriculture, Mansoura University, Mansoura 35516, Egypt
3 Agriculture Research Center, Animal Production Research Institute, Ministry of Agriculture, Dokki, Giza 12619, Egypt; m.hassan55213@gmail.com (M.A.E.H.); ahmedyousif2788@yahoo.com (A.I.Y.)
4 Department of Pharmacognosy, College of Pharmacy, Cairo University, Cairo 11562, Egypt; omar.sabry@pharma.cu.edu.eg
5 Institute of Pharmacy and Molecular Biotechnology, Heidelberg University, 69120 Heidelberg, Germany; wink@uni-heidelberg.de
6 AgroBioSciences Research, Mohammed VI Polytechnic University, Lot 660–Hay MoulayRachid, Ben Guerir 43150, Morocco
* Correspondence: w-khalil@mans.edu.eg (W.A.K.); mansour.sobeh@um6p.ma (M.S.)
† These authors contributed equally to this work.

Citation: Hassan, S.A.; Khalil, W.A.; Hassan, M.A.E.; Yousif, A.I.; Sabry, O.M.; Wink, M.; Sobeh, M. Antioxidant and Antiapoptotic Effects of a *Turraea fischeri* Leaf Extract on Cryopreserved Goat Sperm. *Animals* **2021**, *11*, 2840. https://doi.org/10.3390/ani11102840

Academic Editors: Woo Kyun Kim, Youssef A. Attia, Maria de Olivera and Nesrein Hashem

Received: 26 July 2021
Accepted: 23 August 2021
Published: 29 September 2021

Simple Summary: The excessive production of reactive oxygen species (ROS) in cryopreservation and post-thawing affects sperm quality and subsequent fertilizing ability. Antioxidants of natural origin, such as plant extracts, rich in flavonoid and phenolic compounds, are of special interest in scavenging ROS. The supplementation of goat semen extender with 375 µg/mL *T. fischeri* leaf extract improved the functional and ultrastructural characteristics of cryopreserved sperm by maintaining antioxidant capacity, thus preventing membrane injury and reducing apoptosis.

Abstract: This study evaluated the efficacy of *Turraea fischeri* leaf extract for maintaining the viability of cryopreserved goat sperm. Ejaculated semen was collected from 5 mature Baladi bucks (50–60 kg, 2–4 years of age) and those samples with mass motility $\geq$ 70% and sperm concentration $\geq 2.5 \times 10^9$/mL were selected, pooled, and divided into 4 aliquots. Each aliquot was diluted in Tris-citric-soybean lecithin extender containing a different concentration of *T. fischeri* leaf extract (0, 125, 250, or 375 µg/mL). Treated semen samples were cooled to 5 °C, transferred to 0.25-mL French straws, and stored in liquid nitrogen (LN_2) at −196 °C. After thawing, membrane integrity was examined by transmission electron microscopy, apoptotic activity by Annexin/propidium iodide staining and flow cytometry, and both enzyme activities and antioxidant capacity by spectroscopic assays. The leaf extract at 375 µg/mL significantly improved semen quality as indicated by enhanced total antioxidant capacity, reduced H_2O_2 concentration, a greater proportion of structurally intact motile sperm, and concomitant reductions in apoptosis and necrosis. The extract also significantly increased the proportion of sperm with a contiguous plasma membrane and intact acrosome ($p < 0.05$). Furthermore, LC-MS revealed numerous secondary metabolites in the extract that may contribute to sperm cryopreservation.

Keywords: *Turraea fischeri*; polyphenolics; semen cryopreservation; sperm ultrastructure; antioxidant biomarker; apoptosis

1. Introduction

Artificial insemination (AI) is used widely in agriculture to optimize and spread commercially valuable genetic traits, including in goats. Effective cryopreservation of semen samples is critical for efficient AI. However, cryopreservation can reduce goat sperm quality, motility, and viability, resulting in lower fertility rates [1]. The reduced semen quality results in part from cellular oxidative stress, which causes the peroxidation of unsaturated fatty acids in the biomembrane. In healthy mammalian sperm cells, total antioxidant capacity (TAC) and reactive oxygen species (ROS) production remain in balance [2], but an increase in ROS production, a decrease in TAC, or both during cryopreservation can result in reduced motility and cell death [3]. Excessive ROS production during cryopreservation and after thawing not only degrade cellular membranes but may also damage DNA [4], further enhancing dysfunction [5]. To reduce oxidative stress, exogenous antioxidants are frequently added to cryopreservative solutions to sustain semen quality prior to AI [6]. Several studies have confirmed ROS scavenging capacity of various medicinal plant extracts to improve sperm motility and increase fertility rates [7–10].

The genus *Turraea* L., a genus of the Meliaceae (mahogany) family of tropical flowering trees and shrubs, includes 70 species distributed throughout Africa and Asia [11,12]. *Turraea fischeri* is widely used in east Africa to treat stomachache and infertility [13]. Previous studies have also documented antioxidant and hepatoprotective properties [14]. Furthermore, biochemical analyses of *T. fischeri* extracts have identified numerous potentially bioactive secondary metabolites such as limonoids as well as various sterols and flavonoids with antioxidant and anti-inflammatory activities [15].

To date, there are no studies available on the effect of the leaf extract from *T. fischeri* in semen extenders on semen cryopreservation. In the current study, the chemical profile of a leaf methanol extract from *T. fischeri* was characterized using HPLC-MS/MS, and its antioxidant activity was measured by in vitro spectroscopic assays. The effects of *T. fischeri* leaf extract on goat sperm viability, motility, and morphology, and antioxidant capacity were then examined after thawing by vital staining under light microscopy, flow cytometry following Annexin/propidium iodide staining, various spectroscopic assays, and transmission electron microscopy.

2. Materials and Methods

2.1. Plant Material, Extraction, LC-MS and Antioxidant Acctivities

Leaves of *Turraea fischeri* were collected from the Lupaga Site in Shinyanga, Tanzania, and stored at the Institute of Pharmacy and Molecular Biotechnology, Heidelberg University, under accession number P7336 [16].

The dried and ground leaves (100 g) were extracted three times in 100% methanol (3×500 mL) at ambient temperature and the three extracts combined. Pooled extracts were then filtered and dried under a vacuum at 40 °C. The obtained residue was lyophilized, yielding a fine dried powder (15 g). LC-MS analyses and in vitro antioxidant activities were performed according to Sobeh et al. [14]. Detailed methods are included in the Supplementary File.

2.2. Animals

Semen samples were collected from 5 mature, fertile Baladi bucks (50–60 kg LBW; 2–4 years of age) at the Animal Production Research Station, El-Karada, Kafrelsheikh, Animal Production Research Institute (APRI), Agricultural Research Center, Ministry of Agriculture, Egypt, using an artificial vagina. Sample collection was conducted in cooperation with the Physiology and Biotechnology Laboratory, Animal Production Department, Faculty of Agriculture, Mansoura University, Egypt, according to animal welfare guidelines of Mansoura University.

2.3. Animal Management

All 5 bucks were raised under the same environmental conditions. Feeding requirements were calculated according to the recommendations of Animal Production Research Institute, Ministry of Agriculture, Egypt. Each buck was fed 1.0 kg/day concentrate feed mixture containing 14% crude protein and 70% total digestible nutrients, plus 1.25 kg/day berseem hay from August to November or 5 kg/day Egyptian fresh berseem clover (*Trifolium alexandrinum*) from December to February. Animals had free access to trace mineralized salt and drinking water at all times.

2.4. Collection of Semen

Semen was collected from each buck by artificial vagina once weekly before feeding at 7–8 a.m. for five consecutive weeks. In total, 25 samples were obtained. Samples were transferred immediately into a water bath at 37 °C and only those with mass motility $\geq$ 70% and sperm concentration of at least 2.5×10^9/mL were retained for experiments. Samples were then pooled and divided into 4 aliquots for each treatment group.

2.5. Preparation of Extender

The Tris-citric-soybean lecithin extender contained 3.025 g/dL Tris (Sigma Chemical Co., St. Louis, MO, USA), 1.66 g/dL citric acid monohydrate (Sigma, Darmstadt, Germany), 1.25 g/dL glucose (Sigma Aldrich, St. Louis, MO, USA), 5% glycerol (Honeywell, Regen, Germany), 1% soybean lecithin (L-a-phosphatidylcholine, LAB: product number MC041), 100 IU/mL penicillin, and 100 µg/mL streptomycin. The components were mixed in a water bath at 37 °C and adjusted to 300 mOsm/kg in H_2O (pH 6.8) before addition of extract.

2.6. Cryopreservation

Pooled semen was diluted 1:10 in extender (*v*/*v*) containing the indicated extract concentration (0, 125, 250, and 375 µg/mL) and adjusted to a final sperm concentration of 2.5×10^8/mL. The mixture was then gradually cooled from 37 °C to 5 °C over 4 h (equilibration period) and transferred to 0.25-mL French straws (IMV Technologies, L'Aigle, France) for cryopreservation. Straws were first exposed to liquid nitrogen vapor for 10 min and then immersed in liquid nitrogen at −196 °C.

2.7. Thawing

After one month, the straws were thawed at 37 °C for 30 s in a water bath, and the various assessments conducted immediately.

2.8. Semen Evaluation

2.8.1. Progressive Motility

The proportion of sperm cells showing progressive motility was examined under a phase-contrast microscope (DM 500, Leica, Switzerland) with heated stage set to 37 °C. Briefly, aliquots of diluted sperm (10 µL) were placed on pre-warmed glass slides and sealed with coverslips. A total of 200 spermatozoa/slide from 3 randomly chosen fields were counted by the same investigator and the mean proportion (%) recorded.

2.8.2. Viability

Semen samples were double-stained with a mixture of 5% eosin (vital stain) and 10% nigrosin (background stain) to estimate the live: dead ratio [17]. The live: dead ratio was calculated by counting the unstained head area among 300 sperm at high magnification (400×) using a light microscope.

2.8.3. Gross Structural Abnormalities

Abnormalities in gross structure were assessed in 300 sperm cells during viability measurements using a light microscope. The following criteria were considered: (i) tail defects (abnormal tails), (ii) abnormal heads, and (iii) cytoplasmic droplets [18].

2.8.4. Plasma Membrane Integrity

The hypo-osmotic swelling test (HOS-t) was used to assess plasma membrane (PM) integrity according to a previously described protocol [19]. Briefly, semen (50 µL) was incubated at 37 °C for 30 min in a hypo-osmotic solution (500 μL at 75 mOsm/kg) containing fructose (6.75 g/L) and sodium citrate (3.67 g/L) in H_2O. A sample of the mixture was placed on a slide and covered with a coverslip. The number of spermatozoa with coiled or swollen tails (indicative of intact membranes under hypoosmotic conditions) among 300 sperm per slide was counted in each sample at 400× under phase-contrast microscopy.

2.8.5. Antioxidant Capacity and Enzyme Activities

The following biochemical parameters were measured in post-thawed extender: total antioxidant capacity (TAC, linearity up to 2 mM/L) [20], hydrogen peroxide concentration (H_2O_2, linearity up to 1.5 mM/L) [21], lactic dehydrogenase (LDH, linearity up to 1700 units/L) activity [22], aspartate transaminase (AST, linearity up to 150 units/mL) activity, and alanine transaminase (ALT, linearity up to 120 units/mL) activity [23]. All measurements were performed using a spectrophotometer (Spectro UV-VIS Auto, UV-2602, Labomed, Los Angeles, CA, USA) and commercial kits (Biodiagnostic, Giza, Egypt) according to the manufacturer's instructions.

2.8.6. Apoptosis and Necrosis

Semen samples were stained with Annexin-V (AV, calcium-dependent probe) for tracking phosphatidylserine (PS) externalization in the membrane and propidium iodide (PI) as an indicator of genomic DNA exposure using a commercial PS Detection Kit (IQP, Groningen, The Netherlands) according to the manufacturer's instructions. Briefly, semen samples were thawed and washed twice by centrifugation (300× *g* for 10 min at 4 °C) with phosphate-buffered saline. After the second centrifugation, the supernatant was removed, and the sperm pellet resuspended in binding buffer at 1×10^6 sperm cells/mL. Then, 100 μL of semen sample was transferred to culture tubes (5 mL) containing 5 μL AV (fluorescein isothiocyanate, FITC label, BD Biosciences, San Jose, CA, USA) and 5 μL PI (BD Biosciences). The mixed suspension was then incubated in the dark at room temperature (25 °C) for 15 min, followed by the addition of 400 μL binding buffer to each tube. Staining patterns were then evaluated by flow cytometry using an Accuri C6 Cytometer and Accuri C6 software (BD Biosciences) [24]. Cells negative for both AV and PI staining (A−/PI−) were classified as viable, those positive for AV and negative for PI (A+/PI−) as early apoptotic, those positive for both AV and PI (A+/PI+) as apoptotic, and those negative for AV and positive for PI (A−/PI+) as necrotic.

2.8.7. Ultrastructure

Sperm ultrastructure was examined by transmission electron microscopy (TEM) as described [25] with some modifications. Briefly, 500 μL of each semen sample was centrifuged and resuspended in cold (4 °C) fixative solution (2.5% glutaraldehyde in phosphate buffer) for 2 h. Samples were then washed and post-fixed in osmium tetroxide (1%) for 90 min at room temperature, dehydrated, cleared in gradient ethanol and propylene oxide, and embedded in Epon 812 (Fluka Chemie, AG, Buchs, Switzerland). Ultrathin sections (60–70 nm) were prepared using glass knives and observed using a JEOL-JEM 2100 TEM at 80 kV. Changes in PM and acrosome ultrastructure were examined from 300 sperm per sample.

2.9. Statistical Analysis

Arcsine transformation was performed before statistical analyses because that helps in dealing with percentage values for semen characteristics including progressive motility, viability, membrane integrity, acrosome integrity, structural abnormality, sperm viability by Annexin-V, plasma membrane integrity, acrosomal ultrastructure. Treatment group means were compared by one-way analysis of variance (ANOVA) [26] and Duncan's multiple range tests [27]. A $p < 0.05$ was considered statistically significant for all tests.

3. Results

3.1. Chemical Composition and In Vitro Antioxidant Activities

Analysis of a leaf methanol extract from *T. fischeri* by HPLC-MS/MS tentatively identified 17 compounds, including particularly high concentrations of phenolic acids (e.g., compounds 2, 3, and 4), flavonoids (compounds 10–17), and corresponding glycoside derivatives (Table 1). On the other hand, the bark extract from the same plant was rich in 20 secondary metabolites belonging to cinchonains and phenylpropanoid-substituted catechin [14]. The extract exhibited substantial antioxidant activity in two commonly used assays; FRAP and DPPH (Table 2). The observed results might be attributed to the existence of several phenolic acids (*p*-coumaroylquinic acid, feruloylquinic acid and caffeoylquinic acid) and flavonoids (Quercetin rutinoside, quercetin glucoside, kaempferol rutinoside, kaempferol glucoside and isorhamnetin glucoside). Comparable activities were reported from the bark extract [14].

Table 1. Characterization of secondary metabolites of *T. fischeri* by HPLC-MS/MS analyses. Precursor ions and corresponding fragment ions.

No.	Rt	M-H	MS/MS	Proposed Compound
1	1.49	133	115	Malic acid
2	2.34	353	191, 179	Caffeoylquinic acid
3	3.11	337	173	*p*-Coumaroylquinic acid
4	4.39	367	193, 173	Feruloylquinic acid
5	4.84	385	265, 137	Hydroxybenzoic acid rhamnoside derivative
6	9.03	411	265, 163, 119	*p*-Coumaric acid rhamnoside derivative
7	10.69	441	265, 193	Ferulic acid glucuronide derivative
8	12.44	505	463, 301, 179	Methyl quercetin glucoside
9	12.96	549	505, 301, 179	Dimethyl ether quercetin glucoside
10	21.72	741	609, 301, 271	Quercetin pentosyl-rutinoside
11	23.2	609	301, 179	Quercetin rutinoside
12	24.25	609	301, 179	Quercetin rutinoside
13	25.94	463	301, 179, 151	Quercetin glucoside
14	28.69	593	285	Kaempferol rutinoside
15	29.77	447	285	Kaempferol glucoside
16	31.2	477	315, 299	Isorhamnetin glucoside
17	32.34	477	315, 299	Isorhamnetin glucoside

Rt: retention time, M-H: pseudo molecular ion in the negative ion mode, MS/MS: mass fragmentation pattern.

Table 2. Antioxidant activities of a leaf methanol extract from *T. fischeri* as measured by FRAP, DPPH, and TPC assays. GAE = gallic acid equivalents.

Sample	FRAP (mM $FeSO_4$ Equivalent/mg Extract)	DPPH (EC_{50} µg/mL)	TPC mg GAE/g Extract
Leaf extract	9.84 ± 0.19	50.3 ± 3.5	236
Ascorbic acid	-	2.92 ± 0.29	-
Quercetin	24.04 ± 1.23	-	-

Ascorbic acid and quercetin are positive controls. DPPH, 2,2-diphenyl-1-picryl-hydrazyl-hydrate; FRAP, ferric-reducing antioxidant power assay; TPC, total phenolic content.

3.2. Effects of T. fischeri Leaf Extract on the Viability, Morphology, Function, and Membrane Integrity of Thawed Cryopreserved Goat Sperm

3.2.1. Effects on Sperm Motility, Viability, and Membrane Integrity

Post-thaw motility, viability, and membrane integrity were all significantly ($p < 0.05$) improved by addition of *T. fischeri* leaf extract (TFLE) to semen extender during cryopreservation, and the degree of improvement increased with TFLE concentration compared to semen extender alone, Table 3. Alternatively, acrosome integrity and the proportion of abnormal sperm were not affected significantly by TFLE ($p > 0.05$, Table 3).

Table 3. Effects of *T. fischeri* leaf extract on cryopreserved goat sperm characteristics (means ± SE, $n = 5$).

Concentration	Sperm Characteristic (% of Cells Examined)				
	Progressive Motility	Viability	Membrane Integrity	Acrosome Integrity	Structural Abnormality
Control	47.0 ± 2.55 [b]	45.0 ± 2.05 [b]	39.4 ± 0.87 [b]	86.8 ± 1.71	13.4 ± 1.33
Extract 125 μg/mL	54.0 ± 1.87 [a]	48.6 ± 1.94 [a,b]	48.6 ± 1.47 [a]	88.8 ± 1.50	14.4 ± 1.03
Extract 250 μg/mL	59.0 ± 1.87 [a]	53.8 ± 1.16 [a]	51.6 ± 2.54 [a]	86.4 ± 1.63	12.8 ± 0.37
Extract 375 μg/mL	60.0 ± 1.58 [a]	52.0 ± 1.58 [a]	48.8 ± 1.74 [a]	87.6 ± 1.21	14.0 ± 1.70

[a] and [b]: Values in the same column with different superscripts are significantly different at $p < 0.05$.

3.2.2. Effects on Apoptosis and Necrosis Rates

The proportion of viable sperm was significantly higher in TFLE-treated cryopreserved semen than control cryopreserved semen ($p < 0.05$), and the viable fraction increased progressively with TFLE concentration (Table 4). Consistent with this dose-dependent improvement in viability, the proportions of early apoptotic, total apoptotic, and necrotic spermatozoa decreased progressively with TFLE dose compared to controls. Further, at 375 μg/mL TFLE, the proportion of necrotic sperm was significantly lower than at all other concentrations.

Table 4. Effects of *T. fischeri* leaf extract on goat sperm viability (means ± SE, $n = 3$).

Concentration	Viable (%)	Early Apoptotic (%)	Apoptotic (%)	Necrotic (%)
	(A−/PI−)	(A+/PI−)	(A+/PI+)	(A−/PI+)
Control	39.7 ± 0.49 [d]	25.1 ± 0.69 [a]	31.8 ± 0.03 [a]	3.4 ± 0.23 [c]
Extract 125 μg/mL	51.9 ± 0.06 [c]	13.0 ± 0.15 [c]	25.2 ± 0.17 [b]	9.9 ± 0.03 [a]
Extract 250 μg/mL	58.9 ± 0.55 [b]	12.0 ± 0.30 [c]	20.6 ± 0.40 [c]	8.5 ± 0.17 [b]
Extract 375 μg/mL	62.4 ± 0.55 [a]	19.4 ± 0.55 [b]	17.3 ± 0.03 [d]	0.9 ± 0.03 [d]

[a], [b], [c] and [d]: Values in the same column with different superscripts are significantly different at $p < 0.05$.

3.2.3. Effects on Membrane Ultrastructure

Examination of sperm cell ultrastructure by TEM revealed that the plasma membrane (PM) was damaged by cryopreservation in semen extender alone, while supplementation with TFLE dose-dependently improved sperm ultrastructure after thawing. Table 5 summarizes the proportions of sperm cells in each treatment group demonstrating an intact, slightly swollen, swollen, or disrupted PM. Consistent with viability analyses, there was a TFLE dose-dependent increase in the portion of cells with an intact PM and decreases in the proportions with slightly swollen and swollen PM.

Table 5. Effect of *T. fischeri* leaf extract on sperm plasma membrane (PM) integrity (means ± SE, $n = 3$).

Concentration (mg/mL)	Intact PM (%)	Slightly Swollen PM (%)	Swollen PM (%)	Lost PM (%)
Control	39.3 ± 0.88 [c]	17.3 ± 0.88 [a]	33.7 ± 0.88 [a]	9.7 ± 0.88
Extract 125 μg/mL	43.0 ± 1.15 [c]	19.0 ± 1.00 [a]	30.0 ± 1.53 [b]	8.0 ± 1.15
Extract 250 μg/mL	54.3 ± 2.33 [b]	11.0 ± 1.53 [b]	24.7 ± 0.88 [c]	10.0 ± 1.53
Extract 375 μg/mL	64.4 ± 2.33 [a]	6.0 ± 2.31 [b]	21.3 ± 0.88 [c]	8.3 ± 1.20

[a], [b] and [c]: Values in the same column with different superscripts are significantly different at $p < 0.05$.

Figure 1A–G illustrates the different ultrastructural abnormalities resulting from cryopreservation. Four injury patterns were defined according to the degree of PM damage [28]. (i) Sperm with intact PM exhibited a normal head region with an intact acrosome (IA) and contiguous PM tightly surrounding the acrosomal ground substance (Figure 1A–C). In addition, the mid-sectional region of healthy sperm contained a contiguous mitochondrial sheath (MS) completely enclosing morphologically typical mitochondria. The axoneme also exhibited the normal 9 + 2 arrangement of microtubules (Figure 1B,C). (ii) Sperm with slightly damaged PMs exhibited an acrosomal physiological reaction (AR) characterized by the initial formation of small vesicles under a dilated and slightly separated plasma membrane (SP) (Figure 1A,B). (iii) In the third category, sperm exhibited a swollen PM (S) with a wavy appearance, and mitochondrial sheathes were also dilated (Figure 1D,F). (iv) Finally, another fraction of sperm showed discontinuous or disintegrated (DS) PMs, damaged membranes (DMs) around mitochondria, and various mitochondrial ultrastructural abnormalities. The axoneme structure at this stage also showed an abnormal microtubule arrangement. The distribution of these categories differed among treatment groups, particularly at higher TFLE concentrations. In the 375 µg/mL group, there was a significantly greater proportion of cells with an IA and a smaller proportion with atypical acrosomes ($p < 0.05$). There was also a trend for lower frequencies of typical AR and lost acrosome in TFLE-treated groups compared to controls cryopreserved in extender only (Table 6).

Table 6. Effect of *T. fischeri* leaf extract on acrosomal ultrastructure (means ± SE, $n = 3$).

Concentration (mg/mL)	Intact Acrosome (%)	Atypical AR (%)	Typical AR (%)	Lost Acrosome (%)
Control	69.3 ± 2.40 ^{b}	19.3 ± 0.88 ^{a}	8.4 ± 1.45	3.0 ± 1.15
Extract 125 µg/mL	70.3 ± 1.20 ^{b}	19.0 ± 0.58 ^{a}	8.0 ± 0.58	2.7 ± 0.88
Extract 250 µg/mL	76.3 ± 1.45 ^{a}	14.7 ± 1.45 ^{b}	6.7 ± 0.33	2.3 ± 1.20
Extract 375 µg/mL	79.7 ± 0.88 ^{a}	11.6 ± 1.20 ^{b}	7.0 ± 0.58	1.7 ± 0.33

^{a} and ^{b}: Values in the same column with different superscripts are significantly different at $p < 0.05$.

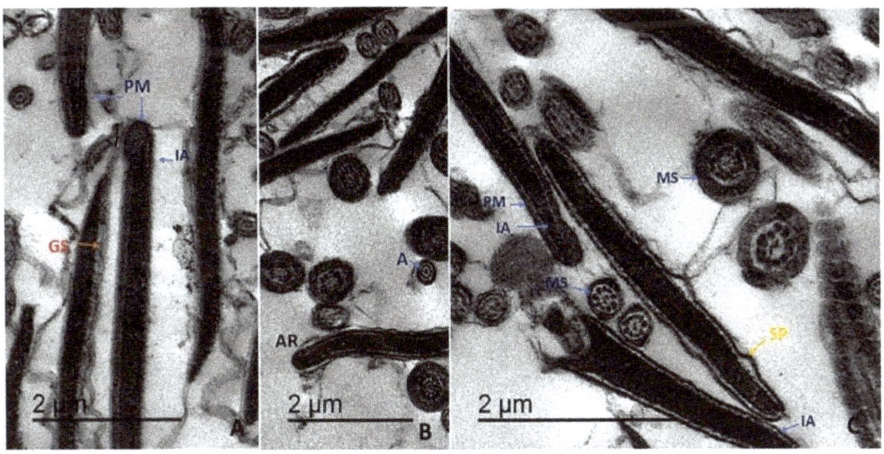

Figure 1. *Cont.*

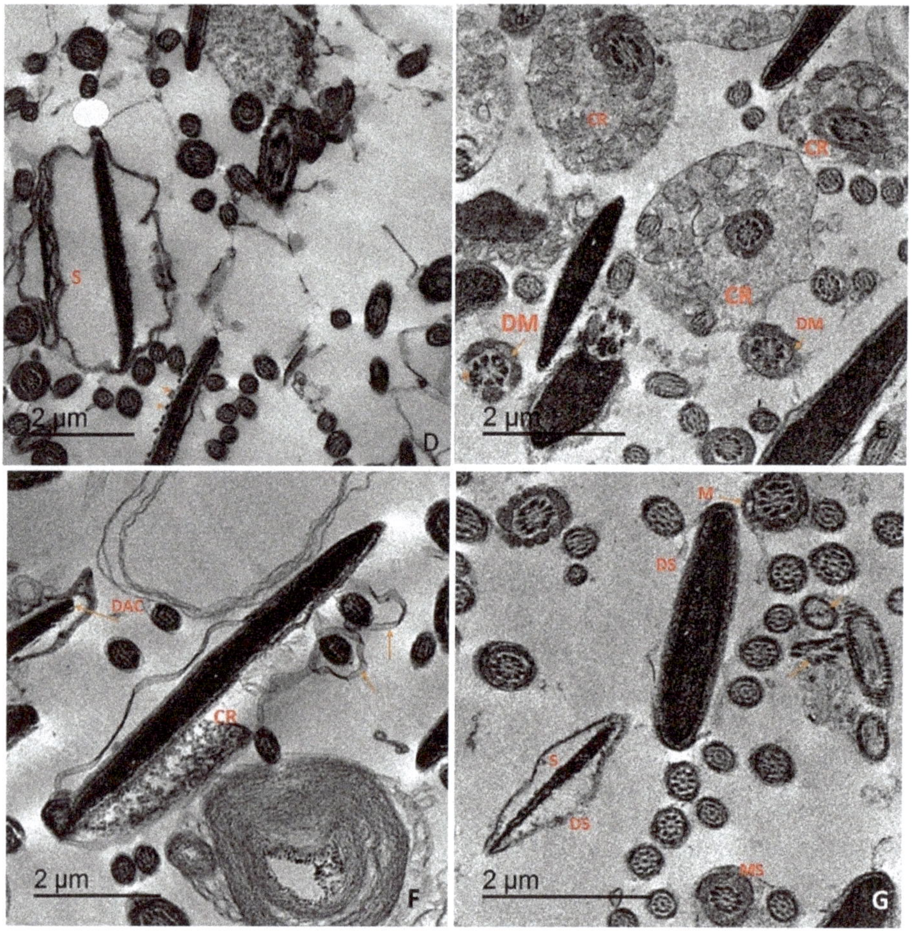

Figure 1. Addition of *T. fischeri* leaf extract to semen extender during cryopreservation significantly improved post-thaw sperm ultrastructure. (**A–C**) Intact sperm with structurally intact acrosomes (IA) completely enclosed by contiguous plasma membranes (PM). The PM appears continuous alongside the nucleus in longitudinal sections and around the mitochondrial sheath (MS) in cross-sections. Few sperm cells exhibited a separated plasma membrane (SP) or diffusion of ground substance (GS) under a detached PM. (**D**) Damaged sperm showing a swollen 'wavy' PM leaving a large space around the nucleus (S). (**E**) Cross-sections of the tail region showing swollen PM containing cytoplasmic residue (CR) and mitochondrial sheathes enveloping damaged mitochondria (DM). (**F**) Damaged acrosomal cap (DAC), dilated PM in the tail region (Arrows), and cytoplasmic residue (CR). (**G**) Discontinuous (DS) PM in longitudinal and cross-sections of different sperm regions.

3.2.4. Effects on Oxidative Biomarkers and Enzyme Activity

Addition of TFLE also dose-dependently enhanced insignificantly post-thaw TAC and reduced significantly H_2O_2 ($p < 0.05$, Table 7). Alternatively, TFLE had no significant effects on LDH, AST, and ALT activities as shown in Table 7.

Table 7. Effect of *T. fischeri* leaf extract supplementation on seminal antioxidant capacity and enzymatic activities (means ± SE, $n = 3$).

Concentration (mg/mL)	TAC (Mm/L)	H_2O_2 (nm/L)	LDH (U/mL)	AST (U/L)	ALT (U/L)
Control	0.57 ± 0.02	1.3 ± 0.09 [a]	91.7 ± 8.67	63.3 ± 6.36	14.0 ± 2.00
Extract 125 μg/mL	0.62 ± 0.03	0.9 ± 0.04 [b]	112.4 ± 13.25	60.0 ± 10.58	13.7 ± 0.88
Extract 250 μg/mL	0.67 ± 0.01	0.7 ± 0.09 [b]	105.2 ± 21.64	60.0 ± 4.00	14.0 ± 2.00
Extract 375 μg/mL	0.69 ± 0.05	0.8 ± 0.05 [b]	96.2 ± 15.91	60.0 ± 4.00	12.7 ± 0.67

[a] and [b]: Values in the same column with different superscripts are significantly different at $p < 0.05$. TAC = total antioxidant capacity, H_2O_2 = hydrogen peroxide, LDH = lactic dehydrogenase. AST = aspartate transaminase, ALT = alanine transaminase.

4. Discussion

Supplementation of goat semen extender with 375 μg/mL TFLE improved the functional and ultrastructural characteristics of cryopreserved sperm by maintaining antioxidant capacity, thereby preventing membrane injury and reducing apoptosis.

Semen cryopreserved in semen extender alone (control group) demonstrated the lowest proportion of viable sperm compared to samples preserved with extender containing varying concentrations of the leaf extract. Freezing and thawing can damage cellular membranes, reducing sperm number and quality as shown by flow cytometry and TEM. Our results are in accord with previous studies demonstrating the harmful effects of thawing on sperm structure and function. Oxidative stress during cryopreservation reduces the reproductive potential of semen by impairing sperm motility, reducing mitochondrial activity, damaging DNA, and activating apoptotic pathways [8,9,29]. Thus, supplementation of semen extender with antioxidants prior to cryopreservation is recommended to facilitate efficient goat breeding [30]. Here, we demonstrate improved cryopreservation using a natural plant extract containing multiple bioactive agents with known beneficial effects against cellular stress.

Freezing and thawing induced apoptosis as evidenced by Annexin staining, which reveals the translocation of phosphatidylserine from the inner to the outer PM layer. Further, some sperm cells became necrotic during cryopreservation as evidenced by PI staining [31]. Both apoptosis and necrosis are associated with loss of PM integrity, which is necessary to maintain sperm function within the female reproductive tract [32]. During cryopreservation, rearrangement of membrane lipids alters fluidity and increases susceptibility to disruption, which then induces further cellular damage and ultimately death [33]. These pathological effects are manifested by changes in sperm morphology during the freezing-thawing process [34]. High concentrations of polyunsaturated fatty acids (PUSFAs) such as arachidonic and docosahexaenoic acids in the PM increase the vulnerability to ROS-induced peroxidative damage and membrane dysfunction [35]. Further, oxidative injury may be spread throughout the spermatozoa population by a subset of cells overproducing ROS [36], leading to generally reduced mitochondrial metabolic activity, motility, and viability [37]. Maintenance of cell membrane integrity and mitochondrial function under oxidative stress are thus essential for successful fertilization using cryopreserved semen [33].

Following cryopreservation, damage to the PM and acrosomal cap was predominantly observed in the head region, in accordance with previous observations of human sperm [28]. Membrane swelling is most probably caused by changes in the extracellular osmotic pressure during freezing and thawing, causing cells to accumulate or lose water. The sperm PM is known to mediate the exchange of sodium, potassium [38,39], and calcium [40], and these ion fluxes regulate motility and mitochondrial function as well as osmotic balance. An intact PM is also necessary for fusion with the outer acrosomal membrane and induction of the acrosome reaction [41]. Acrosomal integrity is also essential for fertilization as this organelle contains hydrolytic enzymes such as hyaluronidase, acrosin, and esterases required for lysis of the zona pellucida and penetration of the oocyte corona radiata [42]. Freezing and thawing significantly increased the number of sperm cells with atypical acrosomal structure, which has previously been attributed to degeneration and apoptosis [43].

The addition of TFLE to sperm extender dose-dependently increased the TAC of goat semen and reduced the concentration of H_2O_2, a major ROS generator. On the other hand, TFLE had little effect on the activities of LDH, AST, and ALT. The dose-dependent increase in TAC was strongly associated with the progressive rise in sperm cell viability and the decreases in apoptosis, necrosis, and structural abnormalities. Consistent with these findings, Salimi, et al. [44] reported positive correlations between TAC and both sperm motility and normal sperm morphology, while Pahune, et al. [45] observed positive correlations between TAC and multiple seminogram parameters including sperm concentration, sperm motility, and normal sperm morphology. Collectively, these findings suggest that an imbalance between TAC and ROS production is a major contributor to impaired sperm function following cryopreservation [46].

Mitochondrial enzymatic activities in human spermatozoa are strongly correlated with motility [47]. Aspartate transaminase (AST) and ALT are essential for metabolic processes that provide energy for sperm survival, motility, and fertility [48], and so are good indicators of sperm membrane stability and semen quality [49]. An increase in spermatozoa damage within the liquid storage medium results in an elevated concentration of transaminase enzymes [50]. Indeed, AST and ALT activities were slightly higher in control samples than samples containing 375 µg/mL TFLE, although the difference did not reach statistical significance.

TFLE contains secondary metabolites such as malic acid, quercetin, and kaempferol that may contribute to these improved functional and structural characteristics. Indeed, malic acid decreases the accumulation of ROS and enhances the glutathione cycle by regulating various endogenous antioxidant pathways [51,52], while the flavonoids can directly scavenge ROS, thereby resisting oxidative damage during cryopreservation [53]. Finally, kaempferol is a flavonoid compound with potent activity against inflammation caused by oxidative stress [54].

The effect of TFLE on semen cryopreservation was stronger compared with other plant extracts such as *Albizia harveyi* leaf extract in bull [8], *Entada abyssinica* bark extract in ram [9], and nanoformulations of mint, thyme, and curcumin in goat [10]. All these extracts enhanced semen preservability and sperm characteristics after freezing and thawing.

TFLE is rich in phenolic and flavonoid compounds, which have antioxidant properties. This is evident in the antioxidant activity of the extract via DPPH, FRAP, and TPC assays or by reducing the concentration of hydrogen peroxide in the semen extender after thawing. This can be explained by the ability of polyphenolic compounds to scavenge reactive oxygen species such as superoxide anion radicals and hydroxyl radicals, thus interrupting free radical chain reaction [55]. Therefore, we expect the same effect in preserving the semen in other species.

5. Conclusions

The addition of 375 µg/mL TFLE to Tris-soybean lecithin extender significantly improved the cryopreservation of goat semen as evidenced by a greater proportion of cells retaining robust motility, viability (low apoptosis rate), and normal ultrastructure after thawing. These benefits were associated with elevation of semen antioxidant capacity. The efficacy of the extract in artificial insemination needs to be studied in more detail.

Supplementary Materials: The following are available online at https://www.mdpi.com/article/10.3390/ani11102840/s1, Detailed methods of the LC-MS analysis and the in vitro assays (DPPH, FRAP and TPC) are included in the Supplementary File.

Author Contributions: Conceptualization, W.A.K., S.A.H. and M.S.; methodology, W.A.K., M.S., S.A.H., M.A.E.H. and A.I.Y.; software, W.A.K., M.S. and S.A.H.; formal analysis, W.A.K. and M.S.; investigation, W.A.K., M.S., S.A.H. and M.A.E.H.; data curation, W.A.K., M.S., S.A.H., O.M.S. and M.A.E.H.; writing—original draft preparation, W.A.K., M.S., S.A.H., M.A.E.H. and A.I.Y.; writing—review and editing, W.A.K., M.S., O.M.S., M.W. and S.A.H.; visualization, W.A.K. and M.S.; supervision, W.A.K., M.S. and M.W. All authors have read and agreed to the published version of the manuscript.

Funding: This research received no external funding. APC was paid by UM6P.

Institutional Review Board Statement: The animal study was reviewed and approved by the Ethics Committee of Mansoura University.

Informed Consent Statement: Not applicable.

Data Availability Statement: The data that support the findings of this study are available from the corresponding author, W.A.K., upon reasonable request.

Conflicts of Interest: The authors declare no conflict of interest.

References

1. Holt, W. Alternative strategies for the long-term preservation of spermatozoa. *Reprod. Fertil. Dev.* **1997**, *9*, 309–320. [CrossRef]
2. Agarwal, A.; Virk, G.; Ong, C.; Du Plessis, S.S. Effect of oxidative stress on male reproduction. *World J. Men's Health* **2014**, *32*, 1–17. [CrossRef] [PubMed]
3. Valko, M.; Leibfritz, D.; Moncol, J.; Cronin, M.T.; Mazur, M.; Telser, J. Free radicals and antioxidants in normal physiological functions and human disease. *Int. J. Biochem. Cell Biol.* **2007**, *39*, 44–84. [CrossRef]
4. De Lamirande, E.; Jiang, H.; Zini, A.; Kodama, H.; Gagnon, C. Reactive oxygen species and sperm physiology. *Rev. Reprod.* **1997**, *2*, 48–54. [CrossRef] [PubMed]
5. Maia da Silva, M.; Bicudo, S.D.; Sicherle, C.C.; Rodello, L.; Gallego, I.C.S. Lipid peroxidation and generation of hydrogen peroxide in frozen-thawed ram semen cryopreserved in extenders with antioxidants. *Anim. Reprod. Sci.* **2010**, *122*, 118–123. [CrossRef]
6. Ziaullah, M.; Ijaz, A.; Aleem, M.; Mahmood, A.; Rehman, H.; Bhatti, S.; Farooq, U.; Sohail, M. Optimal inclusion level of butylated hydroxytoluene in semen extender improves the quality of post-thawed canine sperm. *Czech J. Anim. Sci.* **2012**, *57*, 377–381. [CrossRef]
7. Van Wyk, B.; Wink, M. *Medicinal Plants of the World*; CABI: Wallingford, UK, 2017.
8. Sobeh, M.; Hassan, S.; El Raey, M.; Khalil, W.; Hassan, M.; Wink, M. Polyphenolics from *Albizia harveyi* exhibit antioxidant activities and counteract oxidative damage and ultra-structural changes of cryopreserved bull semen. *Molecules* **2017**, *22*, 1993. [CrossRef]
9. Sobeh, M.; Hassan, S.A.; Hassan, M.A.; Khalil, W.A.; Abdelfattah, M.A.; Wink, M.; Yasri, A. A polyphenol-rich extract from Entada abyssinica reduces oxidative damage in cryopreserved ram semen. *Front. Vet. Sci.* **2020**, *7*, 1–9. [CrossRef] [PubMed]
10. Ismail, A.A.; Abdel-Khalek, A.-K.E.; Khalil, W.A.; Yousif, A.I.; Saadeldin, I.M.; Abomughaid, M.M.; El-Harairy, M.A. Effects of mint, thyme, and curcumin extract nanoformulations on the sperm quality, apoptosis, chromatin decondensation, enzyme activity, and oxidative status of cryopreserved goat semen. *Cryobiology* **2020**, *97*, 144–152. [CrossRef]
11. Mabberley, D.J. *Mabberley's Plant.-Book: A Portable Dictionary of Plants, Their Classification and Uses*; Cambridge University Press: Cambridge, UK, 2017.
12. Mokoka, T.A.; Xolani, P.K.; Zimmermann, S.; Hata, Y.; Adams, M.; Kaiser, M.; Moodley, N.; Maharaj, V.; Koorbanally, N.A.; Hamburger, M. Antiprotozoal screening of 60 south african plants, and the identification of the antitrypanosomal germacranolides schkuhrin i and ii. *Planta Med.* **2013**, *79*, 1380–1384. [CrossRef]
13. Augustino, S.; Hall, J.B.; Makonda, F.B.; Ishengoma, R.C. Medicinal resources of the miombo woodlands of urumwa, tanzania: Plants and its uses. *J. Med. Plants Res.* **2011**, *5*, 6352–6372.
14. Sobeh, M.; Mahmoud, M.F.; Sabry, O.M.; Adel, R.; Dmirieh, M.; El-Shazly, A.M.; Wink, M. Hplc-pda-ms/ms characterization of bioactive secondary metabolites from *Turraea fischeri* bark extract and its antioxidant and hepatoprotective activities in vivo. *Molecules* **2017**, *22*, 2089. [CrossRef] [PubMed]
15. Yuan, C.-M.; Tang, G.-H.; Zhang, Y.; Wang, X.-Y.; Cao, M.-M.; Guo, F.; Li, Y.; Di, Y.-T.; Li, S.-L.; Hua, H.-M. Bioactive limonoid and triterpenoid constituents of *Turraea pubescens*. *J. Nat. Prod.* **2013**, *76*, 1166–1174. [CrossRef]
16. Nibret, E.; Ashour, M.L.; Rubanza, C.D.; Wink, M. Screening of some tanzanian medicinal plants for their trypanocidal and cytotoxic activities. *Phytother. Res.* **2010**, *24*, 945–947. [CrossRef] [PubMed]
17. Moskovtsev, S.; Librach, C. Methods of sperm vitality assessment. In *Spermatogenesis*; Springer: Berlin/Heidelberg, Germany, 2013; pp. 13–19.
18. Aamdal, J.; Andersen, K.; Fougner, J. Insemination with Frozen Semen in the Blue Fox. In Proceedings of the 7th International Congress on Animal Reproduction and Artificial Insemination, Munich, Germany, 1972; pp. 1713–1716.

19. Neild, D.; Chaves, G.; Flores, M.; Mora, N.; Beconi, M.; Agüero, A. Hypoosmotic test in equine spermatozoa. *Theriogenology* **1999**, *51*, 721–727. [CrossRef]
20. Koracevic, D.; Koracevic, G.; Djordjevic, V.; Andrejevic, S.; Cosic, V. Colorimetric method for determination of total antioxidant capacity. *J. Clin. Pathol.* **2001**, *54*, 356–361. [CrossRef]
21. Aebi, H. Catalase in vitro. In *Methods Enzymology*; Elsevier: Amsterdam, The Netherlands, 1984.
22. Bais, R.; Philcox, M. IFCC methods for the measurement of catalytic concentration of enzymes. Part 8. IFCC method for lactate dehydrogenase (L-lactate: NAD+ oxidoreductase, EC 1.1. 1.27). *J. Anal. Methods Chem.* **1994**, *16*, 167–182. [CrossRef] [PubMed]
23. Reitman, S.; Frankel, S. A colorimetric method for the determination of serum glutamic oxalacetic and glutamic pyruvic transaminases. *Am. J. Clin. Pathol.* **1957**, *28*, 56–63. [CrossRef]
24. Masters, A.; Harrison, P. Platelet counting with the bd accuritm c6 flow cytometer. *Platelets* **2014**, *25*, 175–180. [CrossRef]
25. Oliveira, L.Z.; de Lima, V.F.H.; Levenhagen, M.A.; Dos Santos, R.M.; Assumpção, T.I.; Jacomini, J.O.; de Andrade, A.F.; de Arruda, R.P.; Beletti, M.E. Transmission electron microscopy for characterization of acrosomal damage after percoll gradient centrifugation of cryopreserved bovine spermatozoa. *J. Vet. Sci.* **2011**, *12*, 267–272. [CrossRef]
26. SAS. *Statistical Analysis System. Stat. User's Guid. Release 9.1.3*; Sas Institute: Cary, NC, USA, 2007.
27. Duncan, D.B. Multiple range and multiple f tests. *Biometrics* **1955**, *11*, 1–42. [CrossRef]
28. Zhu, W.-J.; Liu, X.-G. Cryodamage to plasma membrane integrity in head and tail regions of human sperm. *Asian J. Androl.* **2000**, *2*, 135–138.
29. Sarıözkan, S.; Bucak, M.N.; Tuncer, P.B.; Ulutaş, P.A.; Bilgen, A. The influence of cysteine and taurine on microscopic–oxidative stress parameters and fertilizing ability of bull semen following cryopreservation. *Cryobiology* **2009**, *58*, 134–138. [CrossRef]
30. Bucak, M.N.; Sarıözkan, S.; Tuncer, P.B.; Sakin, F.; Ateşşahin, A.; Kulaksız, R.; Çevik, M. The effect of antioxidants on post-thawed angora goat (capra hircus ancryrensis) sperm parameters, lipid peroxidation and antioxidant activities. *Small Rumin. Res.* **2010**, *89*, 24–30. [CrossRef]
31. Anzar, M.; He, L.; Buhr, M.M.; Kroetsch, T.G.; Pauls, K.P. Sperm apoptosis in fresh and cryopreserved bull semen detected by flow cytometry and its relationship with fertility. *Biol. Reprod.* **2002**, *66*, 354–360. [CrossRef] [PubMed]
32. Emamverdi, M.; Zhandi, M.; Zare Shahneh, A.; Sharafi, M.; Akbari-Sharif, A. Optimization of ram semen cryopreservation using a chemically defined soybean lecithin-based extender. *Reprod. Domest. Anim.* **2013**, *48*, 899–904. [CrossRef] [PubMed]
33. Holt, W.; North, R. Effects of temperature and restoration of osmotic equilibrium during thawing on the induction of plasma membrane damage in cryopreserved ram spermatozoa. *Biol. Reprod.* **1994**, *51*, 414–424. [CrossRef] [PubMed]
34. Nur, Z.; Zik, B.; Ustuner, B.; Sagirkaya, H.; Ozguden, C. Effects of different cryoprotective agents on ram sperm morphology and dnaintegrity. *Theriogenology* **2010**, *73*, 1267–1275. [CrossRef]
35. Nair, S.J.; Brar, A.; Ahuja, C.; Sangha, S.; Chaudhary, K. A comparative study on lipid peroxidation, activities of antioxidant enzymes and viability of cattle and buffalo bull spermatozoa during storage at refrigeration temperature. *Anim. Reprod. Sci.* **2006**, *96*, 21–29. [CrossRef] [PubMed]
36. Pagl, R.; Aurich, J.E.; Müller-Schlösser, F.; Kankofer, M.; Aurich, C. Comparison of an extender containing defined milk protein fractions with a skim milk-based extender for storage of equine semen at 5 °C. *Theriogenology* **2006**, *66*, 1115–1122. [CrossRef]
37. Ball, B.A.; Vo, A.T.; Baumber, J. Generation of reactive oxygen species by equine spermatozoa. *Am. J. Vet. Res.* **2001**, *62*, 508–515. [CrossRef]
38. Darszon, A.; Trevino, C.; Wood, C.; Galindo, B.; Rodriguez-Miranda, E.; Acevedo, J.; Hernandez-González, E.; Beltrán, C.; Martínez-López, P.; Nishigaki, T. Ion channels in sperm motility and capacitation. *Soc. Reprod. Fertil. Suppl.* **2007**, *65*, 229–244.
39. Peralta-Arias, R.D.; Vívenes, C.Y.; Camejo, M.I.; Piñero, S.; Proverbio, T.; Martínez, E.; Marín, R.; Proverbio, F. Atpases, ion exchangers and human sperm motility. *Reproduction* **2015**, *149*, 475–484. [CrossRef]
40. Triphan, J.; Aumüller, G.; Brandenburger, T.; Wilhelm, B. Localization and regulation of plasma membrane Ca^{2+}-ATPase in bovine spermatozoa. *Eur. J. Cell Biol.* **2007**, *86*, 265–273. [CrossRef]
41. Tomes, C.N. Molecular mechanisms of membrane fusion during acrosomal exocytosis. *Soc. Reprod. Fertil. Suppl.* **2007**, *65*, 275–291.
42. Rovan, E. Biochemistry of spermatozoa. In *Veterinary Medicine Andrology, Physiology and Pathology of Reproduction in Male Animals*; Springer: Berlin/Heidelberg, Germany, 2001; Volume 22, pp. 23–54.
43. Cross, N.L.; Meizel, S. Methods for evaluating the acrosomal status of mammalian sperm. *Biol. Reprod.* **1989**, *41*, 635–641. [CrossRef] [PubMed]
44. Salimi, S.; Fazeli, F.; Khosravi, P.; Nabizadeh, S. Association of seminal plasma total antioxidant capacity and malondialdehyde levels with sperm parameters in infertile men with varicocele. *Avicenna J. Med. Biochem.* **2016**, *4*, 1–6. [CrossRef]
45. Pahune, P.P.; Choudhari, A.R.; Muley, P.A. The total antioxidant power of semen and its correlation with the fertility potential of human male subjects. *J. Clin. Diagn. Res.* **2013**, *7*, 991–995. [CrossRef] [PubMed]
46. Koca, Y.; Özdal, Ö.; Celik, M.; Ünal, S.; Balaban, N. Antioxidant activity of seminal plasma in fertile and infertile men. *Arch. Androl.* **2003**, *49*, 355–359. [CrossRef]
47. Ruiz-Pesini, E.; Alvarez, E.; Enríquez, J.A.; López-Pérez, M.J. Association between seminal plasma carnitine and sperm mitochondrial enzymatic activities. *Int. J. Androl.* **2001**, *24*, 335–340. [CrossRef]
48. Perumal, P.; Vupru, K.; Khate, K. Effect of addition of melatonin on the liquid storage (5 °C) of mithun (*bos frontalis*) semen. *Int. J. Zool.* **2013**, *2013*, 642632. [CrossRef]

49. Gündoğan, M. Short term preservation of ram semen with different extenders. *Kafkas Üniversitesi Veteriner Fakültesi Dergisi* **2009**, *15*, 429–435.
50. Gündoğan, M. Some reproductive parameters and seminal plasma constituents in relation to season in akkaraman and awassi rams. *Turk. J. Vet. Anim. Sci.* **2006**, *30*, 95–100.
51. Lee, Y.-P.; Kim, S.-H.; Bang, J.-W.; Lee, H.-S.; Kwak, S.-S.; Kwon, S.-Y. Enhanced tolerance to oxidative stress in transgenic tobacco plants expressing three antioxidant enzymes in chloroplasts. *Plant. Cell Rep.* **2007**, *26*, 591–598. [CrossRef] [PubMed]
52. Gill, S.S.; Tuteja, N. Reactive oxygen species and antioxidant machinery in abiotic stress tolerance in crop plants. *Plant. Physiol. Biochem.* **2010**, *48*, 909–930. [CrossRef]
53. Xu, D.; Hu, M.-J.; Wang, Y.-Q.; Cui, Y.-L. Antioxidant activities of quercetin and its complexes for medicinal application. *Molecules* **2019**, *24*, 1123. [CrossRef]
54. Xiao, H.-B.; Lu, X.-Y.; Sun, Z.-L.; Zhang, H.-B. Kaempferol regulates opn–cd44 pathway to inhibit the atherogenesis of apolipoprotein e deficient mice. *Toxicol. Appl. Pharmacol.* **2011**, *257*, 405–411. [CrossRef]
55. Hashem, N.M.; Gonzalez-Bulnes, A.; Simal-Gandara, J. Polyphenols in farm animals: Source of reproductive gain or waste? *Antioxidants* **2020**, *9*, 1023. [CrossRef]

animals

MDPI

Article

Correlations between Antioxidant and Biochemical Parameters of Blood Serum of Duroc Breed Pigs

Sergei Yu. Zaitsev *, Anna A. Belous, Oksana A. Voronina, Roman A. Rykov and Nadezhda V. Bogolyubova

Federal Research Center for Animal Husbandry Named after Academy Member L.K. Ernst, Dubrovitsy 60, 142132 Moscow, Russia; belousa663@gmail.com (A.A.B.); voroninaok-senia@inbox.ru (O.A.V.); brukw@bk.ru (R.A.R.); 652202@mail.ru (N.V.B.)

* Correspondence: s.y.zaitsev@mail.ru; Tel.: +74-95-6765-1363

Simple Summary: Human nutrition is currently one of the most important factors that determine health, performance, duration and quality of life. The increasing demand for high-quality livestock products requires scientists and practitioners to develop an advanced complex approach to assessing the composition of animal meat and methods of its regulation. One of these products is pork, the value of which lies primarily in the lipid and protein content, which are necessary for human nutrition. The significance of this paper is also determined by the high popularity of pork in Russia and in a number of other countries worldwide.

Abstract: Correlations between the major biochemical (BC) and antioxidant (TAWSA) parameters of pigs' blood are necessary to study in order to assess physiological–biochemical status (PhBS), animal health, production, etc. Blood samples were obtained from Duroc breed boars (n = 77), divided into groups 1 (n = 25), 2 (n = 40) and 3 (n = 12), which were fattened for 65, 72 and 100 days, respectively. Significant positive and negative correlations were found between TAWSA and BC parameters of pigs' blood for group 3: very high in the case of total protein (TP) (−0.75) and aspartate aminotransferase (AST) (−0.79); high in the case of cholesterol (−0.72), glucose (0.66), alkaline phosphatase (0.66), calcium ions (−0.60) and globulins (0.53); moderate in the case of albumins (−0.36), triglycerides (−0.35), magnesium (−0.32) and phosphorus (−0.27). The same was found for group 2: high in the case of TP (0.51); moderate in the case of globulins (0.48), cholesterol (0.33) and phosphates (0.25). The only moderate correlation was found for group 1: magnesium (−0.48), glucose (0.36) and calcium (−0.25). This tendency indicated the stabilization of pig PhBS during growth and fattening, which can be useful for understanding the PhBS and antioxidant features of pigs, the factors of their nutrition, maintenance, etc.

Keywords: Duroc breed boars; antioxidants; biochemistry; blood parameters; feeding time; correlation coefficients

Citation: Zaitsev, S.Y.; Belous, A.A.; Voronina, O.A.; Rykov, R.A.; Bogolyubova, N.V. Correlations between Antioxidant and Biochemical Parameters of Blood Serum of Duroc Breed Pigs. *Animals* **2021**, *11*, 2400. https://doi.org/10.3390/ani11082400

Academic Editors: Youssef A. Atti, Woo Kyun Kim, Nesrein Hashem and Maria de Olivera

Received: 27 July 2021
Accepted: 10 August 2021
Published: 13 August 2021

Publisher's Note: MDPI stays neutral with regard to jurisdictional claims in published maps and institutional affiliations.

1. Introduction

It is known that animal products are important sources of high-quality proteins, fats, vitamins, minerals and macro- and microelements in human nutrition [1]. In this regard, some approaches have been developed to improve the introduction of animal breeding [2–4] in order to ensure a higher production efficiency and meat quality (including pork), which is especially important for Russia [2,4–6]. Duroc is one of the most popular breeds of pig, possessing numerous positive characteristics in growth rate; total body weight; chemical composition of meat and fat, i.e., high intramuscular fat content in meat [6–8].

To assess the physiological–biochemical status (PhBS) and animal health, safety and quality of meat products [8–10], a complex of biochemical and hematological, antioxidant and "zootechnical" indicators must be studied.

To date, the PhBS of pigs, taking into account the changes in the content of antioxidants in the blood and other tissues of fattening animals, has not been sufficiently studied (both in the Russian Federation and in Asian countries) even for purebred animals. The assessment of PhBS of breeding pigs in Russia is mainly based on the study of changes in biochemical parameters in connection with feeding conditions [11–13], genotype [14,15], sex [16], age [17] and others (only some Russian works is cited as references in this part of the paper). For example, there was an interesting study on selective biochemical and hematological parameters of blood, as well as indicators of mineral metabolism in pigs of the Landrace and Kemerovo breeds, fed at large industrial enterprises of the "Chistogorsk" and "Altaymyasoprom" complexes in Russia [18]. These indicators are close to those obtained in our study and are within physiological norms [18].

On the other hand, some hematological and biochemical parameters of blood were studied in novel hybrids of boars (Large White Landrace) [19]. In industrial combinations (three breeds), where the meat breeds Pietrain and Duroc were used, the content of hemoglobin and erythrocytes was higher as compared to purebred animals [19]. From a practical point of view, data on the relationship of blood biochemical parameters with fattening and meat qualities [20–22] are especially important. There are a number of dissertations and articles on the assessment of PhBS in pigs of the Russian population based on the biochemical parameters of blood and meat of animals [4,9,17,23,24], which we will discuss in detail in the part 4 of our article. The least studied issue in this area is the study of the total amount of water-soluble antioxidants (TAWSA) in the blood serum of fattening pigs. In Russia in this area, only the groups from our Federal Research Center work in collaboration with physiologists and livestock specialists [24–26]. At the same time, individual indicators of antioxidant protection (for example, a concentration of TBA-active products) have been studied more fully [26,27] than TAWSA, but do not provide a complete knowledge of the antioxidant status of the pigs' organisms.

Some authors [8–10,28] believe that the assessment of PhBS through the specified set of indicators is very important, and "the average population values of biochemical parameters are the starting point for the subsequent monitoring of breeding populations" [28]. For example, using these data, it is possible to analyze "in which direction the biochemical status of the population changes when pigs are selected for various characteristics of productivity" and to carry out the ecological monitoring of populations of certain breeds of pigs "in zones with different anthropogenic pressures" [8–10,28]. In addition, the obtained data may be the "average population norm" for healthy animals [28].

The studies carried out to assess the PhBS and health of animals of the Duroc breed, which showed a comparison of hematological and biochemical parameters (more than 40 parameters) between purebred and crossbred offspring [29]. The data obtained are consistent with previous studies [30] on crossbred and purebred offspring, in which there were small differences in some biochemical parameters of pigs at 15 and 27 weeks [30], including creatine, alkaline phosphatase, phosphorus and calcium [29], and were "corrected" with the further growth of animals [29,30].

For a correct assessment of PhBS, it is necessary to study the possibilities of protecting the body from reactive oxygen species (ROS) [31–33]. ROS are formed during many metabolic processes in human and animal cells, capable of oxidizing biologically active compounds (BAC), and damaging the membranes and cells of the body [31–33]. The body's antioxidant defense system is designed to maintain the balance of bioactive substances (lipids, peptides, vitamins and other compounds) in organs and tissues of humans and animals, protecting them from ROS [34–36]. Antioxidant activity is a valuable source of information about the state of health and the level of stress resistance in humans and productive animals in industrial conditions [34–36], which led to a huge variety of methods for its study [33,36]. All these methods are based on a model reaction (most often oxidation) of an individual compound proceeded by a radical mechanism [33–36]. Among these, one of the relatively simple and reliable methods is the electrochemical method based on the amperometric detection of the oxidation reaction signal [33–36].

Duroc is one of the most popular breeds in Russia and Asian countries (including China), which is used not only for meat production, but also for reproduction in pig breeding. It is important to highlight that pork is one of the key directions in Russian livestock production.

In connection with the above-stated relevance, the aim of the work was to study the biochemical and antioxidant parameters of the blood of Duroc breed boars of the Russian population and to identify the most significant correlations between these parameters. These data will be used by the Ministry of Agriculture of the Russian Federation and are especially important in order to establish the range of particular biochemical references (standard norms) for the Duroc breed of pigs in the Russian population.

2. Materials and Methods

2.1. *The Samples of the Pig Blood Serum*

The studies were carried out on the basis of a selection-hybrid center on boars of the Duroc breed (n = 77), from which blood samples were taken from the ear vein during setting and withdrawing from fattening. These pigs were divided onto groups 1 (n = 25), 2 (n = 40) and 3 (n = 12) that were fattened for 65, 72 and 100 days, respectively. The age of boars' setting at the feeding stations was 75 $\pm$ 1 days, with an initial live weight of approximately 35.0 $\pm$ 0.5 kg. The times of boars' withdrawing were from 130 to 175 days, the final live weights were from 92 to 122 kg. Only clinically healthy animals that were periodically examined by veterinary specialists participated in the studies.

The experimental protocols concerning these animals were approved by the Bioethical Committee of the Federal Research Center for Animal Husbandry named after Academy Member L.K. Ernst. All experiments and conditions (animal care, feeding, biological material sampling, etc.) are fulfilled in accordance with the applicable regulations (internationally recognized guidelines and local acts).

2.2. *Measurements of the Biochemical Parameters of Pig Blood Serum Samples*

The biochemical parameters of animal blood serum samples were determined using a "ChemWell" automatic biochemical analyzer (Awareness Technology, Palm City, FL, USA). The open system of this analyzer allows the use of any method or reagent. Therefore, in this study, we used the reagents "Analyticon Biotechnologies AG" (Lichtenfels, Germany) and "Spinreact" (Carretera Santa Coloma, Spain). All reactions were performed in standard "microwell" plates. Pre-dilution, mixing, incubation, rinsing and measurement of the samples were controlled automatic. The following biochemical indicators were determined [37–39]: the concentration of total protein (TP)—by the biuret method; albumin (A)—by the colorimetric method with bromcresol green; urea—by enzymatic colorimetric analysis (Berthelot method); creatinine—by the kinetic Yaffe method; glucose—by the enzymatic glucose oxidase method; cholesterol (Chol) and triglycerides (TG) by the enzyme-colorimetric method; bilirubin (quantification by the Walters and Gerarde method); calcium (Ca)—by the O-cresolphthalein complexon method; phosphorus (P), magnesium (Mg) and iron (Fe)—by the colorimetric method; alanine aminotransferase (ALT) activity—by the UV-kinetic method; aspartate aminotransferase (AST) activity by—the UV-kinetic method; alkaline phosphatase (ALP) activity—by the kinetic method. The following ratios and indicators were determined by calculation: A/G, Ca/P, ALT/AST and the concentration of globulins (G) [37–39]. The results of these measurements were statistically processed using the MS Excel program.

2.3. *Measurements of the Total Amount of Water-Soluble Antioxidants of Pig Blood Serum Samples*

The amperometric method [33–36,39] was used to study the total amount of water-soluble antioxidants (TAWSA). The measurements were carried out on a "Tsvet-Yauza 01-AA" device [39]. The TAWSA values were determined by measuring the strength of the electric current arising during the oxidation of molecules on the surface of the working electrode at a potential of ~500 mV. TAWSA was measured in equivalent to gallic acid as in reference [35]. For this, the "working solutions" were prepared from a gallic acid solution (100 mg/dm^3) for calibration with a mass concentration of 0.2, 0.5, 1.0 and 4.0 mg/dm^3. An amount of 2.2 mmol/dm^3 phosphoric acid solution was used as an "eluent" [35,39]. The results of measuring the total antioxidant activity of the samples were statistically processed using the MS Excel program.

When analyzing the studied indicators, the methods of variation statistics were used to calculate the mean and standard error (M $\pm$ m) and the coefficient of variation (Cv,%), as well as performing correlation analysis using the STATISTICA 10 program (StatSoft, Moscow, Russia).

The obtained datasets of the antioxidant and biochemical parameters boars of the Duroc breed (n = 77) are available online on the website of the L.K. Ernst Federal Research Center for Animal Husbandry (https://www.vij.ru/institut/struktura-organizatsii/nauchnye-podrazdeleniya/52-gruppa-analiticheskoj-biohimii accessed on 1 June 2021).

3. Results

The blood of animals has a complicated biochemical composition: proteins (including albumins, globulins and enzymes), lipids (triglycerides, cholesterol, etc.), minerals (calcium, magnesium, phosphate ions, etc.), etc. From the point of view of biological chemistry, blood serum is a multiphase colloidal system in which the main phase is aqueous moderate, and one of the important integral characteristics is the total amount of water-soluble antioxidants (TAWSA). The authors determined the main biochemical and antioxidant parameters of the blood serum of the Duroc breed of pigs, divided onto groups 1 (n = 25), 2 (n = 40) and 3 (n = 12), which were fattened for 65, 72 and 100 days, respectively (Tables 1–3).

Table 1. Biochemical and antioxidant parameters of the blood serum of the Duroc breed pigs (n = 25) in group 1 (65 days of fattening).

Parameters [1]	Mean Value [1]	Standard Deviation [2]	Error [1]	Coefficient of Variation [3]
Total protein, g/L	75.29	4.44	0.91	5.89
Albumin, g/L	41.60	3.56	0.73	8.56
Globulin, g/L	33.70	5.24	1.07	15.55
A/G	1.27	0.24	0.05	19.06
Urea, mM/L	7.60	1.63	0.33	21.41
Creatinine, μM/L	112.88	21.46	4.38	19.00
Glucose, mM/L	5.33	0.55	0.11	10.27
Triglycerides, mM/L	0.28	0.02	0.01	8.06
Cholesterol, mM/L	2.12	0.25	0.05	11.67
ALT, IU/L	25.03	8.48	1.73	33.88
AST, IU/L	29.60	13.85	2.82	46.79
De Ritis coefficient AST/ALT	1.20	0.54	0.11	45.00
Alk. phosphatase, IU/L	187.91	45.77	9.3	24.36
Ca, mM/L	2.74	0.22	0.05	8.08
P, mM/L	2.96	0.38	0.08	12.88
Mg, mM/L	1.28	0.17	0.04	13.42
TAWSA, mg/L	9.77	3.01	0.61	30.81

[1] same unit as mentioned in the "Parameters" (i.e., g/L, mM/L, μM/L, etc.); [2] rel. units; [3] %.

Table 2. Biochemical and antioxidant parameters of the blood serum of the Duroc breed pigs (n = 40) in group 2 (72 days of fattening).

Parameters [1]	Mean Value [1]	Standard Deviation [2]	Error [1]	Coefficient of Variation [3]
Total protein, g/L	76.76	8.30	1.31	10.81
Albumin, g/L	41.93	4.59	0.73	10.90
Globulin, g/L	34.80	6.84	1.08	19.66
A/G	1.26	0.30	0.05	23.81

Table 2. *Cont.*

Parameters [1]	Mean Value [1]	Standard Deviation [2]	Error [1]	Coefficient of Variation [3]
Urea, mM/L	7.20	0.99	0.16	13.75
Creatinine, µM/L	97.08	20.91	3.31	21.54
Glucose, mM/L	3.88	0.99	0.16	25.52
Triglycerides, mM/L	0.28	0.02	0.01	7.14
Cholesterol, mM/L	2.50	0.47	0.07	18.80
ALT, IU/L	30.39	6.56	1.04	21.59
AST, IU/L	24.04	11.49	1.82	47.80
De Ritis coefficient AST/ALT	0.81	0.34	0.05	41.98
Alk. phosphatase, IU/L	172.17	52.00	8.22	30.20
Ca, mM/L	2.48	0.23	0.04	9.27
P, mM/L	3.48	0.36	0.06	10.34
Mg, mM/L	1.15	0.14	0.02	12.14
TAWSA, mg/L	7.50	1.77	0.28	23.60

[1] same unit as mentioned in the "Parameters" (i.e., g/L, mM/L, µM/L, etc.); [2] rel. units; [3] %.

Table 3. Biochemical and antioxidant parameters of the blood serum of the Duroc breed pigs (n = 12) in group 3 (100 days of fattening).

Parameters [1]	Mean Value [1]	Standard Deviation [2]	Error [1]	Coefficient of Variation [3]
Total protein, g/L	69.95	4.03	1.52	5.76
Albumin, g/L	36.66	2.13	0.99	5.81
Globulin, g/L	33.20	4.25	1.11	12.76
A/G	1.12	0.21	0.09	18.75
Urea, mM/L	7.96	1.21	0.26	15.20
Creatinine, µM/L	104.14	15.64	0.93	15.02
Glucose, mM/L	3.96	0.76	0.14	19.19
Triglycerides, mM/L	0.93	0.09	0.10	9.68
Cholesterol, mM/L	2.21	0.21	0.96	9.50
ALT, IU/L	29.31	5.64	0.86	19.24
AST, IU/L	29.61	8.73	0.96	29.48
De Ritis coefficient AST/ALT	1.05	0.35	0.49	33.33
Alk. phosphatase, IU/L	118.36	39.54	1.58	33.41
Ca, mM/L	2.37	0.35	0.92	14.77
P, mM/L	3.64	0.27	0.09	7.42
Mg, mM/L	0.97	0.06	0.08	6.19
TAWSA, mg/L	19.16	2.44	1.07	12.73

[1] same unit as mentioned in the "Parameters" (i.e., g/L, mM/L, µM/L, etc.); [2] rel. units; [3] %.

By measuring the biochemical and antioxidant (TAWSA) blood parameters of the Duroc breed pigs (n = 77), the most significant differences between these parameters in connection with the days of animal fattening were revealed. It was shown that all biochemical and antioxidant parameters of the blood of pigs of both groups were within the physiological norms for this animal species. The main indicators of protein metabolism in the blood were fairly constant for groups 1 and 2, but changed significantly for group 3. Thus, the parameters of total protein (TP) and albumins (A) changed by less than 2% for groups 1 and 2, but significantly decreased (−7.1% and −11.9% for TP and A, respectively) for group 3 as compared to group 1 (Table 1). The content of globulins (G) varied in the range of +3.3% for group 2 up to −1.2% for group 3 versus group 1, respectively (Table 1).

It seemed logical that the A/G ratio changed by less than 1% for group 2 and more than 11.8% for group 3 (as compared to group 1); however, there was a significant decrease in the albumin values (−11.9%) for group 3 as compared with group 1, respectively, and insignificant changes in the globulin content for all the groups studied (Tables 1–3).

Some of these biochemical parameters differed slightly in a number of values: urea—by −5.3% and +4.7%, and creatinine—by −14.0% and −7.8%, for groups 2 and 3 compared to group 1, respectively (Tables 1–3). This is normal for a young growing organism and indicates an improvement in protein metabolism in a number of animals [37–39]. Similar values of the total protein content in the blood serum of piglets were also noted by other authors [9,10,37–39], and it was indicated that this parameter "differed by variability" [28].

On the other hand, a significant (more than −27.2% and −25.7%) decrease in pigs' blood glucose values in groups 2 and 3 as compared to group 1 is surprising (Tables 1–3). Generally, with the increase in the age of pigs, there is a slight increase in the values of glucose in their blood. Therefore, in [28], it was found that the glucose content in piglets (at the age of 20–60 days) was at the level of 5.2–5.3 mmol/L, although the author pointed out the highest coefficient of variation (Cv = 42.2%) among all other biochemical parameters. In our studies, the coefficient of variation in the case of glucose ranged from 4% to 10%, which was typical for most other biochemical parameters.

It is interesting that for groups 1 and 2 (fed for 65 days and 72 days), such an important biochemical parameter of lipid metabolism as the content of triglycerides in the blood practically did not change, while the content of TG for group 3 animals (fed for 100 days) increased by 232%, i.e., almost 2.3 times (Tables 1–3).

The cholesterol content increased by 18.0% and 4.2% for groups 1 and 2 (Tables 1–3), respectively, which is indirect evidence of changes in lipid metabolism in a number of animals [39–41]. Some authors [9,10,39–41] note that at a young age, piglets of the Large White breed have the highest serum cholesterol content (for example, 4.21 ± 0.90 mmol/L at the age of 2–3 weeks [28], whereas by the age of two months, its content in piglets decreases to 2.70 ± 0.58 mmol/L and remains practically at the same level as in adult sows (2.67 ± 0.75 mmol/L) [28]. Of course, the coefficient of variation of this feature, according to the same authors [28], is a fairly large value of the order of 21–28%. These authors explain the tendency to a high cholesterol content in piglets by "more active metabolic processes in their body, including glycolysis" [28].

The following changes in enzyme activity were observed: ALT—by 21.0% and 17.1% (for groups 2 and 3) and AST—by −19% and approximately 0% (for groups 2 and 3), which did not go beyond the physiological norms. Since for such enzymes, only changes in values from 30% and higher are significant, only the change in the "de Ritis coefficient" should be considered, which decreased by more than −32% for the 2nd group of pigs as compared to the 1st and changed in the ALP activity by −37.0% for the 3rd group of pigs as compared to the 1st, respectively. Changes in ALP activity by −8.4% for the 2nd group of pigs as compared to the 1st are not significant (Tables 1–3).

The calcium content in the blood serum fell by −9.5% and −13.5%, whereas magnesium content sometimes increased by 10.2%, then fell by −24.2%, and the phosphorus content always increased by 17.6% and by 23.0% for the 2nd and 3rd groups of pigs as compared to the 1st one, respectively (Tables 1–3). These changes had a positive effect on the ratio of calcium to phosphorus in the blood serum of Duroc pigs.

Finally, the total amount of water-soluble antioxidants decreased by more than −23% in the 2nd group of pigs, and then increased by 96.1% in the 3rd group compared to the 1st group, respectively (Tables 1–3). This is directly related to the changes in a number of basic biochemical parameters of the blood of pigs of the Duroc breed, for example, albumin and TG listed above (Tables 1–3).

The experimental conclusions are summarized later. It should be noted that an increase in the duration of feeding (from 65 to 72 and 100 days) led to a tendency for a significant decrease in the coefficients of variation (Cv) for most biochemical parameters. This indicates the stabilization of the physiological and biochemical status of the growing organism in the 2nd and 3rd pig groups in comparison with the 1st one (Tables 1–3).

4. Discussion

4.1. The Relationship between the Biochemical Parameters of the Blood Serum of Boars of the Duroc Breed

In recent years, lipid peroxidation has become the subject of extensive research in terms of mechanisms, dynamics, product analysis, disease involvement, inhibition and biological signaling. Some types of antioxidants with different functions inhibit lipid peroxidation and the harmful effects caused by lipid peroxidation products. Much attention has recently been paid to the biological role of lipid peroxidation products, but it is topical to study the relationship between biochemical parameters and indicators of antioxidant protection [38,42].

The relationships of biochemical parameters in group 1 are presented in Table S1 (65 days of fattening, n = 25). The presence of 5 very strong, 5 strong (i.e., subtotal—10 significant) and 36 moderate correlations (i.e., subtotal—46 meaningful) from the 136 total independent correlations was found. In particular, there were only 2 very strong, 1 strong and 16 moderate correlations between the 4 protein indicators in group 1 (i.e., correlations—8 for TP, 6 for albumins and 5 for globulins) correlations from the 58 total independent correlations. There was only 1 very strong correlation between enzymes and 2 strong correlations between enzymes and inorganic ions, as well as 2 strong correlations between cholesterol and inorganic ions. There were only 2 very strong correlations between inorganic ions, as well as numerous moderate correlations in the case of glucose, triglycerides, cholesterol, enzymes, magnesium ions, phosphates, inorganic ions and the Ca/P ratio.

The relationships of biochemical parameters in group 2 are presented in Table S2 (72 days of fattening, n = 40). The presence of 4 very strong, 11 strong (i.e., subtotal 15 significant) and 37 moderate (i.e., subtotal 52 meaningful) correlations from the 136 total independent correlations was found. In particular, there were only 3 very strong, 7 strong and 16 moderate correlations between the 4 protein indicators in group 2 (i.e., correlations—9 for TP, 7 for albumins, 5 for globulins and 3 for the A/G ratio) from the 58 total independent correlations. There was only 1 very strong correlation between enzymes and 3 strong correlations between inorganic ions, as well as 1 strong correlation between urea and Mg^{2+} ions. There were numerous moderate correlations in the case of glucose, triglycerides, cholesterol, enzymes and magnesium ions with Ca/P ratio.

The relationships of biochemical parameters in group 3 are presented in Table S3 (100 days of fattening, n = 12). The presence of 12 very strong, 28 strong (i.e., subtotal 40 significant) and 47 moderate correlations (i.e., subtotal 87 meaningful) from the 136 total independent correlations was found. In particular, there were 6 very strong, 9 strong and 20 moderate correlations between the 4 protein indicators in group 3 (i.e., correlations—11 for TP, 9 for albumins, 10 for globulins and 5 for the A/G ratio) correlations from the 58 total independent correlations. There were 6 very strong, 19 strong and numerous moderate correlations in the case of all other organic compounds and inorganic ions, as well as their ratios.

Thus, animal group 3 was preferential for blood biochemistry correlations as compared to groups 1 or 2. There were 2.5–3.0 times more very strong correlations, 2.5–5.6 times more strong correlations and approximately 1.3 times more moderate correlations in the case of animal group 3 (in total) as compared to the groups 1 or 2. Moreover, there were 2.0–3.0 times more very strong correlations, 1.3–9.0 times more strong correlations and approximately 1.25 times more moderate correlations between the 4 protein indicators in the case of animal group 3 (in total) as compared to the groups 1 or 2. It is important to highlight that there were no meaningful correlations of the total protein indicator and its fractions with the A/G ratio for group 3 as compared to groups 1 or 2.

It is well known [42–46] that the PhBS of animals is initially determined by the multilevel and complex interaction of the systems of the animal body. Therefore, a comparative analysis of the biochemical parameters of blood is especially important. For example, according to Molyanova G.V. [17], the total protein content in pigs' blood at the age of 120 days is 61.12 or 62.02 g/L and at the age of 180 days—63.02 or 72.05 g/L (for Duroc breed pigs in Samar region) in the cold or warm weather periods, respectively. According to Giro T.M. et al. [44], the total protein content in pigs' blood at the age of 120 days is approximately 74.6 g/L; albumins—38.9 g/L; globulins—35.5 g/L (for Duroc breed pigs in Saratov region). In the work of Nikolaev D.V. et al. [45], the level of total protein in pigs' blood at the age of 180 days is 78.5 g/L; albumins—33.7 g/L; globulins—44.8 g/L (for Duroc breed pigs in the Volgograd region). We compared the data of 11 biochemistry indicators (total protein, albumin, globulins, A/G, glucose, triglycerides, ALT, AST, ALP, Ca and P) from all of these studies, which have different values between them, with our results (for Duroc breed pigs in the Voronezh region), but within the range of general data of blood biochemistry for healthy pigs [38]. This is why a detailed study on the major biochemical parameters of the blood serum of such pigs is especially important in order to establish the range of particular biochemical norms for the Duroc breed pigs in the Russian population.

The data of biochemical analysis, obtained in the work of Gu T. [46], did not show significant differences in more than 40 parameters in purebred Duroc pigs, their clones and offspring (including two-breed hybrids in various variations). According to Gu T. et al. [46], on the 112th day, the level of total protein in the blood of Duroc pigs was 76.38 g/L; albumin—40.08 g/L; globulins—36.30 g/L; glucose—3.35 mmol/L; calcium—2.65 mmol/L. However, in the work of Gu T. [46], we did not find data characterizing the state of the antioxidant system in pigs. At the same time, the activity of enzymes such as ALT, AST and ALP is more variable and can vary significantly. Thus, in our study, ALT and AST activity was observed up to 30 U/L for all studied groups of animals (from 140 days of fattening to 175 days), while in [19], the ALT activity was 91.66 U/L and AST activity was 89.63 U/L, which is within the normal range (the norms for ALT 22–98 IU/L and AST 13–95 IU/L according to Gusev I.V. [39]). The data of biochemical analysis, which we obtained in the study of animals of the Duroc breed, are in good agreement with the data obtained in the above works [44–46], and there are also relatively close values in the levels of total protein, glucose and calcium content in [19] for three-breed hybrids. As a literature search showed, most of the works are focused on biochemical parameters, while scarce attention is paid to the study of antioxidant systems in a comprehensive assessment of health status. Of course, the rate of metabolic processes in the body directly affects the formation of free radicals and their neutralization by both enzymatic and low molecular weight antioxidants. The assessment of these processes is extremely important, as confirmed by the work of Kotenkova E.A. [47], which studied the antioxidant potential of the pig spleen, heart and aorta extracts (by determining their total antioxidant capacity after slaughter). The highest total antioxidant capacity was observed in the spleen extract. However, no attention was paid to blood in the context of this work, which could be interesting for the development of a strategy to increase the antioxidant activity of food products.

4.2. The Relationship of TAWSA with Biochemical Parameters of the Blood Serum of Boars of the Duro Breed

For the first time, the calculation of phenotypic correlations of biochemical and antioxidant parameters of the blood of pigs of the Duroc breed was carried out (Table 4).

Table 4. Correlations of antioxidant parameters (TAWSA) with biochemical parameters of the blood serum of the Duroc breed pigs in groups 1, 2 and 3 (65, 72 and 100 days of fattening, respectively).

Parameters	Group 1, n = 25 (65 Days of Fattening).	Group 2, n = 40 (72 Days of Fattening).	Group 3, n = 12 (100 Days of Fattening).
Total protein, g/L	−0.15	0.51	−0.75
Albumin, g/L	0.00	0.21	−0.36
Globulin, g/L	−0.12	0.48	−0.53
Urea, mM/L	0.23	0.20	0.26
Creatinine, μM/L	0.23	0.20	−0.18
Glucose, mM/L	0.36	−0.09	0.66
Triglycerides, mM/L	0.24	0.09	−0.35
Cholesterol, mM/L	0.21	0.33	−0.72
ALT, IU/L	0.20	0.21	0.13
AST, IU/L	−0.16	0.16	−0.79
Alk. phosphatase, IU/L	−0.03	−0.19	0.66
Ca, mM/L	−0.25	−0.05	−0.60
P, mM/L	−0.12	0.25	−0.27
Mg, mM/L	−0.48	0.37	−0.32

As shown in our previous works on the study of correlations between TAWSA and biochemical parameters of sheep blood [38], positive or negative values of correlations are not as important as their absolute values, i.e., whether these correlations are very strong (0.75–1.0), strong (0.50–0.74), moderate (0.25–0.49) or weak (0.01–0.24) [38]. In the latter case (weak correlations), it makes no sense to discuss their direction, i.e., whether they are positive or negative [38]. Here, we focused on describing very strong, strong and moderate correlations (in this sequence mainly) between the studied parameters.

Significant positive and negative correlations were found between TAWSA and the following biochemical parameters of pigs' blood for group 3: very high in the case of TP (−0.75) and AST (−0.79); high in the case of cholesterol (−0.72), glucose (0.66), alkaline phosphatase (0.66), calcium ions (−0.60) and globulins (0.53); and moderate in the case of albumin (−0.36), triglycerides (−0.35), magnesium ions (−0.32) and phosphorus (−0.27) (Table 4).

Significant positive correlations were found between the following biochemical and antioxidant parameters of pigs' blood for group 2 (Table 4): high in the case of TP (0.51) and moderate in the case of globulins (0.48), cholesterol (0.33) and phosphates (0.25).

The correlation between TAWSA and biochemical parameters was the most significant only in the case of magnesium ions (−0.48), glucose (0.36) and calcium ions (−0.25) for group 1 (Table 4), i.e., only moderate correlations were found.

Significantly higher values of the correlation coefficients between TAWSA and biochemical parameters (Table 4) and a large number of significant correlations were obtained for group 3 (11 in total, including 7 strong and very strong) as compared to groups 2 (4 in total, including 1 strong) and 1 (3 in total, only moderate). Thus, a clear tendency indicated the stabilization of the PhBS of animals during their growth and fattening. The same dependences concerning correlations of TAWSA and the biochemical parameters of pigs' blood were found in our ongoing research of hybrid animals (Duroc, Landras and Large White pigs of the Russian population).

It is important to highlight that the pronounced correlations between antioxidant and biochemical blood parameters were found earlier in our previous research in hybrid sheep breeds [38] and only in a short presentation concerning different sheep breeds [43]. For example, the author of [43] found the relationship of lipid peroxidation (LPO) parameters with some hematological and biochemical blood parameters of the following breeds: Texel x Manych Merino, Grozny Small-haired Merino and Soviet Merino. At a high level of lipid peroxidation (5.2 mmol/L), an inverse correlation was established among hemoglobin, total protein and the gamma-globulin fraction of protein for crossbred animals [43]. The high ($r = -0.6$–0.7) and very high ($r = -0.9$) negative correlations were revealed between LPO and total proteins (or albumins) for Soviet merino at the same antioxidant background (MDA = 5.0 mmol/L) [43]. Significant correlations were also established for the Grozny Yark breed ($r = 0.8$–0.7 for hemoglobin and albumin and $r = -0.9$ for globulins) [43]. All of these data indicated a particular balance between oxidative and reduction processes in the body of these sheep breeds, as considered by the authors of [34,35,38,43].

5. Conclusions

The presented data on the biochemical and antioxidant parameters of the blood serum of the Duroc breed pigs and their specific correlations were obtained for the first time. The tendency towards the optimization of blood biochemical parameters in adult animals can be useful for understanding the features of PhBS and the antioxidant status of pigs. The revealed tendencies can be explained only by a closer (than previously thought) relationship between the biochemical and antioxidant functionality of the pig body. In our opinion, the rate of ROS neutralization can be related to the rate of reaching a live weight of 100 kg, which is an important economic feature. This can be explained by the fact that the metabolic energy with an increase in the fattening period will be directed mainly to gaining muscle mass, but not "to fight free radicals" (ROS). The authors assume that the revealed tendency of changes in the biochemical and antioxidant parameters of

Duroc breed pigs (at a longer feeding duration) will be monitored in our further ongoing experiments with hybrid animals. These data will be used by the Ministry of Agriculture of the Russian Federation and are especially important in order to establish the range of particular biochemical references (standard norms) for the Duroc breed pigs of the Russian population.

Supplementary Materials: The following are available online at https://www.mdpi.com/article/10.3390/ani11082400/s1, Table S1: Correlations of biochemical parameters* of the blood serum of the Duroc breed pigs in group 1 (65 days of fattening), Table S2. Correlations of biochemical parameters of the blood serum of the Duroc breed pigs in group 2 (72 days of fattening), Table S3. Correlations of biochemical parameters of the blood serum of the Duroc breed pigs in group 3 (100 days of fattening).

Author Contributions: Conceptualization, S.Y.Z.; methodology, A.A.B., O.A.V. and R.A.R.; validation, N.V.B. and O.A.V.; formal analysis, A.A.B.; investigation, A.A.B., R.A.R. and O.A.V.; resources, S.Y.Z. and N.V.B.; data curation, A.A.B. and O.A.V.; writing—original draft preparation, S.Y.Z. and N.V.B.; writing—review and editing, S.Y.Z. and A.A.B.; supervision, S.Y.Z. and N.V.B.; project administration, S.Y.Z. and N.V.B.; funding acquisition, S.Y.Z. All authors have read and agreed to the published version of the manuscript.

Funding: The methodological topics of this research (parts 1 and 2) were supported by the Ministry of Science and Higher Education of the Russian Federation (registration number of the topic of the state task: 121052600314–1); parts 3, 4 and 5 were supported by a grant from the Russian Science Foundation (project no. 20-16-00032).

Institutional Review Board Statement: The study was conducted according to the guidelines of the Declaration of Helsinki, internationally recognized guidelines (concerning experiments with animals), and approved by the Ethics Committee of L.K. Ernst Federal Research Center for Animal Husbandry (protocol code: 2021–2303; date of approval: 23 March 2021).

Data Availability Statement: Data supporting reported results can be found at: https://www.vij.ru/goszadanie-i-proekty/proekty/proekty-rnf/749-proekt-20-16-00032-rnf (accessed on 1 June 2021).

Acknowledgments: The authors are grateful to T.V. Karpushkina and A.A. Savina for technical assistance.

Conflicts of Interest: The authors declare no conflict of interest.

References

1. Ernst, L.K.; Zinovieva, N.A. *Biological Problems of Animal Husbandry in the XXI Century*; Russian Academy of Sciences: Moscow, Russia, 2008; pp. 225–501.
2. Belous, A.A.; Trebunskikh, E.A.; Kostyunina, O.V.; Sermyagin, A.A.; Zinovieva, N.A. Evaluation of signs of feed conversion and feeding behavior of Duroc boars using automatic feeding stations. *Achiev. Sci. Technol. Agro-Ind. Complex* **2019**, *33*, 63–67.
3. Colpoys, J.D.; Johnson, A.K.; Gabler, N.K. Daily feeding regimen impacts pig growth and behavior. *Physiol. Behav.* **2016**, *159*, 27–32. [CrossRef]
4. Neupokoeva, A.S. Productive and Biological Qualities of Pigs of Canadian Selection in the Conditions of the Trans-Urals. Ph.D. Thesis, Kurgan State Agricultural Academy Named after T.S. Maltseva, Kurgan, Russia, 2019.
5. Aaslyng, M.D.; Oksama, M.; Olsen, E.V.; Bejerholm, C.; Baltzer, M.; Andersen, G.; Bredie, W.L.P.; Byrne, D.V.; Gabrielsen, G. The impact of sensory quality of pork on consumer preference. *Meat Sci.* **2007**, *76*, 61–73. [CrossRef] [PubMed]
6. Nevrkla, P.; Kapelański, W.; Václavková, W.E.; Hadaš, Z.; Cebulska, A. Meat quality and fatty acid profile of pork and backfat from an indigenous breed and a commercial hybrid of pigs. *Ann. Anim. Sci.* **2017**, *17*, 1215–1227. [CrossRef]
7. Velichko, V.A.; Patieva, A.M. The influence of the genotype on the nutritional value of pig meat. *Proc. Kuban State Agrar. Univ.* **2011**, *31*, 254–258.
8. Burmistrov, V.; Pustovit, I. Physico-chemical composition of muscle and adipose tissue in pigs of different genotypes. *Pig Breed.* **2005**, *2*, 14–16.
9. Dementyeva, T.A. Characteristics of the Productivity of Pigs by Biochemical and Cytochemical Tests in Purebred Breeding and Crossbreeding. Ph.D. Thesis, Novosibirsk State University, Novosibirsk, Russia, 1998.
10. Pond, W.G.; Houpt, K.A. *The Biology of the Pig, Transl. from English, V.V. Popova*; Kolos: Moscow, Russia, 1983; pp. 230–334.
11. Shalak, M.V.; Portnoy, A.I.; Katushenok, N.N. Aromatic additive in the diet of fattening pigs and its effect on morphological and biochemical blood parameters. *Sci. Notes Educ. Inst. Vitebsk Order Badge Honor State Acad. Vet. Med.* **2010**, *2*, 323–326.
12. Yunusova, O.Y. The influence of the feed additive "kostovit-forte" on the morphological and biochemical parameters of the blood of pigs on fattening. *Perm Agrar. Bull.* **2017**, *18*, 144–148.

13. Serdyukova, Y.A.; Zlepkin, V.A. The effect of feed additives on morphological and biochemical blood parameters of fattened pigs. *Bull. Agro-Ind. Complex Stavropol* **2016**, *23*, 112–115.
14. Kosilov, V.I.; Perevoiko, Z.A. Biochemical parameters of blood serum of young pigs of large white breed of different genotypes. *Proc. Orenbg. State Agrar. Univ.* **2015**, *53*, 194–196.
15. Garskaya, N.A.; Peretyatko, L.G. Some biochemical blood parameters and their variability in boars of the Poltava meat breed depending on the genotype. *Genet. Anim. Breed.* **2019**, *3*, 63–67. [CrossRef]
16. Perevoiko, Z.A.; Kosilov, V.I. Basic biochemical blood parameters of boars and sows of large white breed. *Proc. Orenbg. State Agrar. Univ.* **2014**, *49*, 196–199.
17. Molyanova, G.V. Formation of the Physiological Immune Status Pigs with Age Depending on Heliogeophysical and Climatic Factors the Middle Volga Region and It's Correction with Thymosin-α1. Ph.D. Thesis, Federal State Budgetary Institution of Higher Professional Education "Moscow State Academy of Veterinary Medicine and Biotechnology named after K. I. Scriabin", Moscow, Russia, 2011.
18. Sebezhko, O.I.; Korotkevich, O.S.; Konovalova, T.V.; Biryulya, I.K.; Petukhov, V.L.; Kamaldinov, E.V.; Osadchuk, L.V. Biochemical, hematological and mineral parameters in pigs of two breeds reared in large industrial complexes of Western Siberia. In Proceedings of the 3rd International Symposium for Agriculture and Food–ISAF 2017, Ohrid, Republic of Macedonia, 18–20 October 2017; p. 100.
19. Chernev, I.F. Influence of boor genotype on hematological and biochemical indicators of blood in hybrid pigs. *Anim. Breed. Genet.* **2021**, *61*, 186–191. [CrossRef]
20. Shakhbazova, O.P. Biochemical blood parameters and their relationship with fattening and meat qualities in pigs of different genotypes. *Vet. Pathol.* **2011**, *36*, 100–103.
21. Maksimov, G.V.; Polozyuk, O.N. The relationship of meat productivity with biochemical parameters of blood of pigs of different genotypes according to the RYR-1 gene. *Bull. Don State Agrar. Univ.* **2011**, *2*, 18–20.
22. Halak, V.I.; Chernyavsky, S.E.; Ilchenko, M.A.; Petulko, P.V.; Gorchanok, A.V. Biochemical parameters of blood serum and their relationship with fattening and meat qualities of young pigs of different genotypes according to snp c. 1426 g>a of the melanocortin 4 receptor gene (MC4R). In *Biotechnology: Achievements and Development Prospects. Collection of Materials of the IV International Scientific and Practical Conference*; Shebeko, K.K., Volkova, E.M., Kruchinsky, N.G., Nikandrov, V.N., Cheshchevik, V.T., Eds.; Ministry of Education of the Republic of Belarus, Polessky State University: Pinsk, Belarus, 2019; pp. 38–41.
23. Kireev, I.V.; Orobets, V.A.; Sevostyanova, O.I.; Shakhova, V.N.; Agarkov, A.V. Prospects of using antioxidant drugs for the treatment and prevention diseases of farm animals. *Res. J. Pharm. Biol. Chem. Sci.* **2018**, *9*, 2031–2036.
24. Magomedaliev, I.M.; Nekrasov, R.V.; Chabaev, M.G.; Dzhavakhiya, V.V.; Glagoleva, E.V.; Kartashov, M.I.; Durnikin, D.A.; Matsyura, A.V. Use of different concentrations of Enzymesporin probiotic in feeding of growing young pigs. *Ukr. J. Ecol.* **2019**, *9*, 704–708. [CrossRef]
25. Fomichev, Y.P.; Bogolyubova, N.V.; Nekrasov, R.V.; Chabaev, M.G.; Rykov, R.A.; Semenova, A.A. Physiological and biochemical effects of two feed antioxidants in modeling technological stress in pigs. (Sus scrofa domesticus Erxleben, 1777). *Biol. Agric.* **2020**, *55*, 750–769.
26. Nikanova, L.A.; Fomichev, Y.P. Bioprotective effect of dihydroquercetin and arabinogalactan in weakening the influence of extreme environmental factors on the body of pigs. *Collect. Sci. Pap. North Cauc. Res. Inst. Anim. Husb.* **2016**, *5*, 89–96.
27. Lyubin, N.A.; Stetsenko, I.I.; Lyubina, E.N. Functional state of the system of antioxidant protection and free radical oxidation in pigs depending on the use of various forms of vitamin A and beta-carotene. *Bull. Ulyanovsk State Agric. Acad.* **2013**, *21*, 54–59.
28. Tyan, E.A. Biochemical status of pigs of large white breed of Western Siberia. *Successes Mod. Nat. Sci.* **2004**, *6*, 21–24.
29. Maja, W.I.; Øyvind, N.; Gjerlaug-Enger, E.; Eli, G.; Marcos, S.L.; Theo, M. Effects of heterozygosity on performance of purebred and crossbred pigs. *Genet. Sel. Evol.* **2019**, *51*, 1–13.
30. Maselyne, J.; Saeys, W.; Van Nuffel, A. Review: Quantifying animal feeding behavior with a focus on pigs. *Physiol. Behav.* **2015**, *138*, 37–51. [CrossRef] [PubMed]
31. Halliwell, B.; John, M.C. *Free Radicals in Biology and Medicine*, 5th ed.; Oxford University Press: Oxford, UK, 2015; pp. 750–905.
32. Bohn, T. Carotenoids and Markers of Oxidative Stress in Human Observational Studies and Intervention Trials: Implications for Chronic Diseases. *Antioxidants* **2019**, *8*, 179. [CrossRef]
33. Yashin, Y.I.; Ryzhnev, V.Y.; Yashin, A.Y.; Chernousov, N.I. *Natural Antioxidants. Content in food Products and Their Impact on Human Health and Aging*; TransLit: Moscow, Russia, 2009; pp. 109–212.
34. Savina, A.A.; Voronina, O.A.; Bogolyubov, N.V.; Zaitsev, S.Y. Amperometric detection of antioxidant activity of model and biological fluids. *Bull. Mosc. Univ. Chem.* **2020**, *58*, 97–103.
35. Voronina, O.A.; Savina, A.A.; Bogolyubov, N.V.; Zaitsev, S.Y. The total amount of water-soluble antioxidants in the blood serum of productive animals. Veterinary Medicine. *Anim. Sci. Biotechnol.* **2019**, *12*, 75–78.
36. Yashin, A.Y. Injector-Flow System with Amperometric Detector for Selective Determination of Antioxidants in Meal Additives and Drinks. *Russ. Chem. J.* **2008**, *2*, 130–135.
37. Zaitsev, S.Y.; Bogolyubov, N.V.; Zhang, X.; Brenig, B. Biochemical parameters, dynamic tensiometry and circulating nucleic acids for cattle blood analysis. *PeerJ* **2020**, *8*, 8997. [CrossRef] [PubMed]
38. Zaitsev, S.Y.; Savina, A.A.; Volnin, A.A.; Voronina, O.A.; Bogolyubova, N.V. Comparative Study of the Water-Soluble Antioxidants in Fodder Additives and Sheep Blood Serum by Amperometric and Biochemical Methods. *Animals* **2020**, *10*, 1186. [CrossRef]

39. Gusev, I.V.; Bogolyubova, N.V.; Rykov, R.A.; Levin, G.N. *Control of the Biochemical Status of Pigs and Cows*; L.K. Ernst Federal Research Center for Animal Husbandry: Moscow, Russia, 2019; pp. 3–40.
40. Romanov, V.N.; Bogolyubova, N.V.; Mishurov, A.V. Multifunctional feed additive for increasing the adaptive capabilities and productivity growth of ruminants. *Anim. Sci.* **2020**, *8*, 21–24.
41. Bogolyubova, N.V.; Rykov, R.A.; Romanov, V.N. Assessment of the metabolic status of the body of bulls when using the energy-vitamin-mineral complex in nutrition. *Anim. Husb.* **2018**, *5*, 2–4.
42. Niki, E.; Yoshida, Y.; Saito, Y.; Noguchi, N. Lipid peroxidation: Mechanisms, inhibition, and biological effects. *Biochem. Biophys. Res. Commun.* **2005**, *338*, 668–676. [CrossRef]
43. Dyakova, S.P. The relationship of lipid peroxidation with some hematological and biochemical parameters of blood is bright in different breeds. In Proceedings of the International Scientific-Practical Conference "Free Radicals, Antioxidants and Animal Health", Voronezh, Russia, 21–23 September 2004; pp. 31–36.
44. Giro, N.M.; Egorova, Z.G.; Vornikova, D.V.; Zakharova, N.B. Biochemical parameters of the blood serum of pigs and pork quality. *Meat Ind.* **2013**, *3*, 58–59.
45. Nikolaev, D.V.; Kukushkina, I.Y.; Randelin, D.A. Morphological and biochemical properties of the blood of pigs of Canadian selection. *Bull. Altai State Agrar. Univ.* **2011**, *86*, 62–64.
46. Gu, T.; Shi, J.; Luo, L.; Li, Z.; Yang, J.; Cai, G.; Zheng, E.; Hong, L.; Wu, Z. Study on hematological and biochemical characters of cloned Duroc pigs and their progeny. *Animals* **2019**, *9*, 912. [CrossRef] [PubMed]
47. Kotenkova, E.A.; Kupaeva, N.V. Assessment of the antioxidant potential of some by-products of pig slaughter. *Food Ind.* **2020**, *7*, 34–40.

animals

MDPI

Article

Bayesian Analysis of the Effects of Olive Oil-Derived Antioxidants on Cryopreserved Buck Sperm Parameters

Ander Arando Arbulu [1,2], Francisco Javier Navas González [1,3,*], Alejandra Bermúdez-Oria [4], Juan Vicente Delgado Bermejo [1], África Fernández-Prior [4], Antonio González Ariza [1], Jose Manuel León Jurado [5] and Carlos Carmelo Pérez-Marín [6]

1 Department of Genetics, Faculty of Veterinary Medicine, University of Cordoba, 14014 Cordoba, Spain; anderarando@hotmail.com (A.A.A.); juanviagr218@gmail.co (J.V.D.B.); angoarvet@outlook.es (A.G.A.)
2 Animal Breeding Consulting, S.L., University of Cordoba, 14014 Cordoba, Spain
3 Institute of Agricultural Research and Training (IFAPA), Alameda del Obispo, 14004 Cordoba, Spain
4 Instituto de la Grasa, Consejo Superior de Investigaciones Científicas (CSIC), 41013 Sevilla, Spain; aleberori@ig.csic.es (A.B.-O.); mafprior@ig.csic.es (Á.F.-P.)
5 Centro Agropecuario Provincial de la Diputación de Cordoba, 14014 Cordoba, Spain; jomalejur@yahoo.es
6 Department of Animal Medicine and Surgery, Faculty of Veterinary Medicine, University of Cordoba, 14014 Cordoba, Spain; pv2pemac@uco.es
* Correspondence: fjng87@hotmail.com; Tel.: +34-638-535-046 (ext. 621262)

Citation: Arando Arbulu, A.; Navas González, F.J.; Bermúdez-Oria, A.; Delgado Bermejo, J.V.; Fernández-Prior, Á.; González Ariza, A.; León Jurado, J.M.; Pérez-Marín, C.C. Bayesian Analysis of the Effects of Olive Oil-Derived Antioxidants on Cryopreserved Buck Sperm Parameters. *Animals* **2021**, *11*, 2032. https://doi.org/10.3390/ani11072032

Academic Editors: Woo Kyun Kim, Youssef A. Attia, Maria de Olivera, Nesrein Hashem and Manuel Álvarez-Rodríguez

Received: 21 May 2021
Accepted: 6 July 2021
Published: 7 July 2021

Simple Summary: The use of olive oil by-products for caprine sperm cryopreservation offers an interesting opportunity to improve post-thawed sperm quality, as antioxidants such as hydroxytyrosol (HT) and 3,4-dihydroxyphenylglycol (DHPG) could reduce lipid peroxidation. Therefore, this study provides evidence of the positive effect of the addition of HT, DHPG, or the mixture of both antioxidants in cryopreserved buck sperm. In addition, the application of Bayesian statistics for data analysis may enable quantifying the dimensionality of the real effect of antioxidants on sperm.

Abstract: The present study evaluates the effect of olive oil-derived antioxidants, hydroxytyrosol (HT) and 3,4-dihydroxyphenylglycol (DHPG), on cryopreserved caprine sperm using Bayesian inference of ANOVA. For this proposal, sperm was collected, pooled and diluted in freezing media supplemented with different concentrations of HT, DHPG and the mixture (MIX) of both antioxidants. Sperm motility, viability, acrosome integrity, mitochondrial status, and lipid peroxidation (LPO) were assessed in fresh and frozen-thawed sperm samples. The results provided evidence that HT at low concentrations improves sperm motility and viability, and reduces the LPO. Contrastingly, DHPG and MIX exert a positive effect by reducing LPO values as concentration increases. Additionally, mitochondrial potential was reduced when samples were supplemented with HT at low concentrations and mixture of both antioxidants. Conclusively, the addition of olive oil-derived antioxidants (HT at 10 μg/mL and DHPG at 30 μg/mL) implements a protective effect in cryopreserved buck sperm. Bayesian analysis alternatives offer new possibilities to determine the repercussion of antioxidants on sperm, both quantitatively and qualitatively.

Keywords: phenolic antioxidant; olive oil; caprine; spermatozoa; Bayesian inference

1. Introduction

For the goat industry, the combination of artificial insemination and sperm cryopreservation is an optimal manner in which to speed up genetic improvement while reducing the incidence of sexually transmitted diseases. However, sperm cryopreservation procedures are associated with irreversible damage in sperm cells, which compromises sperm fertility due to of cold shock, osmotic stress, and intracellular ice crystals formation, among others [1]. One of the reasons of the vulnerability of goat spermatozoa to freezing-thawing procedures is the composition of their plasma membrane, which contains large amounts of polyunsaturated fatty acids [2]. As a consequence, such sperm turns highly susceptible to membrane peroxidation derived from the lipid oxidation of membrane by reactive oxygen species (ROS) [3]. The presence of prooxidant molecules, such as free radicals and ROS, is strongly related to metabolic stress and spermatozoa damage during cryopreservation [4].

Nevertheless, ROS have a variable effect on spermatozoa and their impact hinges on the nature and concentration of these substances. When in physiological concentrations, a promoting effect has been reported on capacitation, acrosomal reaction, and sperm zona pellucida interactions [5]. By contrast, high concentrations of ROS are related to spermatozoa normal function inhibition, thus the reduction of sperm viability due to the oxidative stress (OE) and the subsequent peroxidation of polyunsaturated fatty acids in their membranes [6].

Under normal conditions, spermatozoa have endogenous mechanisms to deal with OE. In the sperm plasma, enzymatic and non-enzymatic antioxidants are present, and they participate in the balance mechanism to prevent OE [7]. By contrast, when spermatozoa are exposed to stress conditions, endogenous antioxidants cannot counteract the excess of free radicals; thus, the addition of exogenous antioxidants may be essential to preserve the quality of these cells. In this sense, several studies have evaluated the use of exogenous antioxidants in goat sperm [3,8–11]. In this framework, a large number of natural compounds has been tested in cell cultures assessing the antioxidant, anti-inflammatory or chelating properties of antioxidants, in recent years, e.g., mentha [12], *Feijoa sellowiana* [13], grapes [14], or olives [15].

Olive fruits (*Olea europea*), olive oil, and its derivates present a large amount of phenolic components, for which remarkable antioxidant properties have been reported [16], with corroborated advantages for human health [17]. Two of the most important phenolic compounds present in olive fruit are hydroxytyrosol (3,4 dyhydroxyphenylethanol, HT) and 3,4-dihydroxyphenylglycol (DHPG), which are isolated from the alperujo olive pulp (semi-solid waste generated in the two-phase system used in olive oil extraction). HT is a simple phenol with significant antioxidant properties [16], which reduces the oxidation of low-density lipoproteins, protect against H_2O_2 cytotoxicity and minimize lactate dehydrogenase activity [18–20]. Regarding DHPG, powerful antioxidant and potential anti-inflammatory effects have been reported which even compare to those reported for vitamin E [21].

The use of HT and DHPH is widespread in human health studies. By contrast, there are few studies on the effects derived from the addition of these antioxidants to sperm dilution media in animals. Contextually, HT-supplemented sperm extender has previously been evaluated in studies conducted on rats [22], humans [23], and rams [15,24,25]. However, the effect of DHPG supplementation has only been reported in ram sperm [15,25]. Taking into account the properties of both antioxidants, the present study hypothesizes that extenders supplemented with these compounds might counteract the sperm damage inflicted by the cryopreservation process. Thus, the aim of the present study was to evaluate the effect of freezing extenders supplemented by different concentrations of HT, DHPG, and the mixture of both substances on the post-thawed sperm quality of goat semen. The effects of the increasing concentrations on each sperm parameter pair correlation were studied.

2. Materials and Methods

2.1. Chemicals

HT and DHPG, stock solution 76.9 mM and 14.7 mM, respectively (Figure 1), were extracted and purified from olive by-products (alperujo) following the processes described by Fernandez-Bolaños et al. [26] and Fernández-Bolaños Guzmán et al. [27]. A commercial TRIS-based extender (Biladyl, Minitube Iberica, Tarragona, Spain) was used to centrifuge and freeze sperm. LIVE/DEAD® sperm viability kit, composed by SYBR-14 and propidium iodide (PI), Mitotracker Red CMXRos and C_{11}-BODIPY$^{581/591}$ were purchased from Molecular Probes Europe (Leiden, The Netherlands). Peanut agglutinin conjugated with fluorescein isothiocyanate (PNA-FITC) was obtained in Sigma-Aldrich (St. Louis, MO, USA).

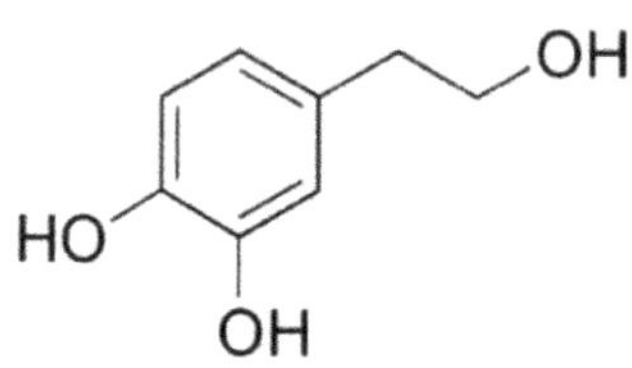

Hydroxytyrosol (HT)

Molecular formula: $C_8H_{10}O_3$
Molar Mas: 154.16 g/mol

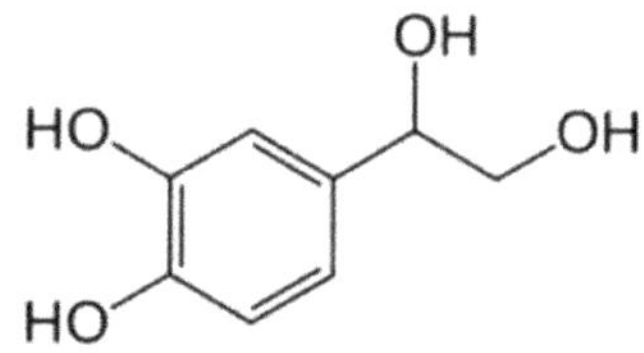

3,4-dihydroxyphenylglycol (DHPG)

Molecular formula: $C_8H_{10}O_4$
Molar Mas: 170.16 g/mol

Figure 1. Chemical structure of olive oil-derived antioxidants used in the present study.

2.2. *Animals and Semen Collection*

Semen was collected from six Murciano-Granadina breed bucks (4–5 years old). The animals involved in the study were located at the Centro Agropecuario Provincial de Córdoba (Córdoba, Spain) and were managed following the prescriptions and regulations of the European Union (2010/63/EU) in its transposition to Spanish law (RD 53/2013). A total of 12 ejaculates per animal (72 ejaculates in total) were collected with an artificial vagina twice a week during the non-breeding season. Previously, semen had been collected for one month in order to ensure the renewal of the epididymal reserves.

After collection, the ejaculates were placed in a water bath at 37 °C during evaluation and they were assessed to determine volume by graduated tubes, sperm concentration by photometer (Accurread, IMV technologies, France) and mass motility, by placing 5 μL of raw semen on a preheated slide (37 °C) and observed in the optical microscope (40× magnification; Olympus, Tokyo, Japan). Sperm mass motility was scored subjectively from 0 (no motile spermatozoa) to 5 (numerous rapid waves) as described by Evans and Maxwell [28] and Lopes et al. [29]. The inclusive criteria for ejaculates to be considered in the study were: volume $\geq$ 0.5 mL, concentration $\geq 3000 \times 10^6$ spz/mL and masal motility $\geq$ 4 (ejaculates with more than 70% total motility).

2.3. *Experimental Design*

As schematized in Figure 2, every sperm collection day, sperm samples were splitted into 13 different aliquots and diluted with extenders containing different (or null) concentrations of HT, DHPG, or a mixture of both antioxidants (MIX), to obtain the final concentrations as follows: Control (without antioxidant); HT1 (10 μg/mL); HT2 (30 μg/mL); HT3 (50 μg/mL) and HT4 (70 μg/mL); DHPG1 (10 μg/mL); DHPG2 (30 μg/mL); DHPG3 (50 μg/mL); DHPG4 (70 μg/mL); MIX1 (5 μg/mL HT + 5 μg/mL DHPG); MIX2 (15 μg/mL HT + 15 μg/mL DHPG); MIX3 (25 μg/mL HT + 25 μg/mL DHPG); MIX4 (35 μg/mL HT + 35 μg/mL DHPG). In order to maintain the aforementioned final antioxidant concentrations, the fact that the extender used (Biladyl) requires a two-step process was considered. Therefore, the same antioxidant concentration was added to both fractions used (FAey and FBey), as described below.

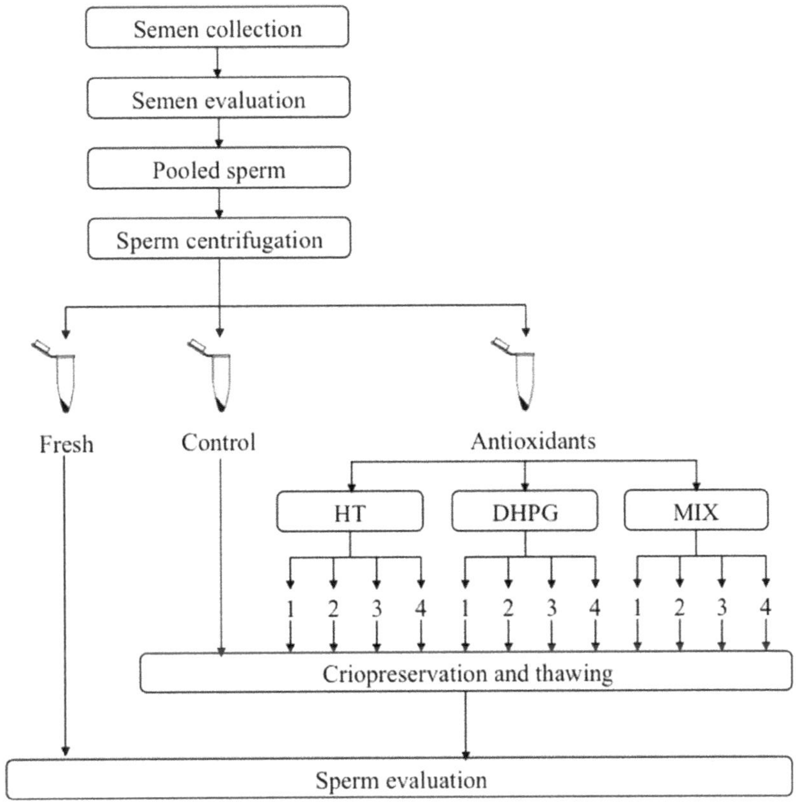

Figure 2. Schematic overview of the experimental design.

Sperm motility, viability, acrosome integrity, mitochondrial membrane potential (HMMP), and membrane lipid peroxidation (LPO) were assessed in frozen-thawed sperm samples. Twelve repetitions of the experiment were performed.

2.4. Semen Dilution and Freezing

After collection, ejaculates were diluted at 1:2 with TRIS-based extender (FA of Biladyl without egg yolk) for individual evaluation and, if inclusion criteria were reached, these were pooled and diluted to reach a dilution of 1:10. To remove seminal plasma, pooled samples were centrifuged at 600× *g* for 15 min. The supernatant was removed and the pellet was resuspended, adding a volume of fraction A of Biladyl containing egg yolk (FAey). Then, the pooled sample was split into 13 different aliquots; 12 samples were prepared by adding FAey supplemented with the previously described HT, DHPG and MIX antioxidant concentrations. A control group (no antioxidant) was also prepared. The samples were then immediately placed in a programmable freezer (cell incubator SH-020S, Welson, Korea) to reach 5 °C and maintained for two hours at 5 °C.

Then, each aliquot was diluted with Tris-egg yolk-glycerol extender (FBey) supplemented with HT, DHPG and/or MIX, obtaining a final volume of 1000 μL per sample (with a concentration of 400 × 10^6 spz/mL). The samples were loaded into 0.25 mL straws (100 × 10^6 spz/straws) and maintained for two hours at 5 °C.

Three straws per sample were frozen using liquid nitrogen vapors. Straws were horizontally placed in racks 4 cm above the liquid nitrogen level for 10 min and then plunged in liquid nitrogen pending analysis. For thawing, the samples were immersed in a water bath at 37 °C for 30 s.

2.5. Sperm Quality Assessment

2.5.1. Motility

ISAS software v.1.2 (Integrated Semen Analyser System, Proiser, Valencia, Spain) equipped with an HS640C video camera was used to assess sperm motility. Sperm samples were diluted in FAey at a final concentration of 25 $\times$ 10^6 spz/mL and, after 10 min incubation, 5 µL of each diluted sample was evaluated using a slide and covered (22 $\times$ 22 mm). Four fields and a minimum of 500 spermatozoa were randomly captured at 10$\times$ magnification using a UB203i phase contrast microscope (Chongqing UOP Photoelectric Technology Co., Ltd, Beibei District, Chongqing, China). A total of 25 images per second were acquired, selecting particles with an area of between 10 and 70 µm^2 and categorized as motile when VAP >10 µm/s, and linearly motile when they deviated >75% from a straight line. The analyses provided information about total motility (TM, %) and progressive motility (PM, %), curvilinear velocity (VCL, µm/s), straight line velocity (VSL, µm/s), average path velocity (VAP, µm/s), straightness (STR, %), linearity (LIN, %), wobble (WOB, %), amplitude of lateral head displacement (ALH, µm), and beat/cross frequency (BCF, Hz).

2.5.2. Flow Cytometer

The recommendations of the International Society for Advancement of Cytometry were followed to perform flow cytometric analyses [30] using a FACScalibur flow cytometer (BD Biosciences, San Jose, CA, USA) equipped with a 488 nm argon blue laser. Sheath flow rate was set at 12.0 $\pm$ 3 µL/min in all analyses (LOW mode). Green fluorescence from SYBR-14, PNA-FITC and C_{11}-BODIPY$^{581/591}$ was read with an FL1 photodetector (530/30 band-pass filter). Red fluorescence PI and Mitotracker Red CMXRos was read with an FL2 photodetector (585/42 nm bandpass filter). Approximately 10,000 events of a gated population were counted per sample.

Forward scatter (FSC) and side scatter (SSC) were recorded in a linear mode (in FSC vs. SSC dot plots). Data were acquired as FSC files using BD Cell Quest Pro v. 6.0, (Becton Dickinson Immunochemistry, San Jose, CA, USA). FlowJo® Version 7.6.2 software (FlowJo™, Ashland, OR, USA) was used to analyze the acquired data, using dot plots with the relative cell size (FSC), the internal complexity (SSC) and the specific fluorescence intensity for each probe.

2.5.3. Viability

A LIVE/DEAD® sperm viability kit was used to evaluate sperm viability and the evaluation was conducted along the lines recommended by Arando et al. [31]. A total of 100 µL of sperm was diluted with 150 µL of cytometer buffer to reach a final concentration ~4 $\times$ 10^6 spz/mL. Then, 2.5 µL SYBR-14 (2 µM) and 5 µL PI (480 µM) were added and incubated in darkness conditions for 15 min. After incubation, the proportion of live/dead sperm cells was measured. Spermatozoa emitting in the green wavelength were deemed to be spermatozoa with intact plasma membranes and the results were reported as the percentage of spermatozoa with intact plasma membrane. Unstained and single-stained samples were used for calibrating the FSC gain, FL-1 and FL-2 PMT voltages and for compensation of SYBR-14 spill over into the PI channel (9.8%). Non-sperm particles (also called "alien events") were located in the SYBR-14$^-$/PI$^-$ quadrant and they did not contain DNA. Spermatozoa with intact plasma membrane were located in the SYBR-14$^+$/PI$^-$ quadrant.

2.5.4. Acrosome Integrity

Acrosome integrity was assessed using the combination of PNA-FITC and PI. One-hundred microliters of sperm (containing around 40 $\times$ 10^6 spz) was incubated in darkness for 5 min with 5 µL of PNA-FITC stock solution (100 µg/mL in DMSO) and 5 µL PI

(480 μM). After incubation, 400 μL of cytometer buffer was added and samples were analyzed. PI^- and $PNA\text{-}FITC^-$ cells were categorized as sperm with intact acrosome and plasma membrane. Unstained and single-stained samples were used for setting the FSC gain, FL-1 and FL-2 PMT voltages and for compensation of PNA-FITC spill over into the PI channel (9.8%). The percentages of alien particles (f) determined by SYBR-14/PI staining were used to correct the percentages of non-stained spermatozoa (q1) in each sample in order to obtain the corrected percentage of non-stained spermatozoa (q1′), in accordance with Petrunkina & Harrison [32]:

$$q1' = [(q1 - f)/(100 - f)] \times 100. \quad (1)$$

2.5.5. Mitochondrial Membrane Potential

The combination of Mitotracker Red CMXRos and SYBR-14 was used to estimate mitochondrial membrane potential and it was assessed using a modified protocol [33,34]. A volume of 50 μL of sperm (containing 20×10^6 spz) was mixed with 350 μL of cytometer buffer and immediately loaded with 2 μL SYBR-14 (2 μM) and 2 μL Mitotracker Red CMXRos (20 μM) was added. Sperm doses were incubated for 10 min at 37 °C in the dark and only sperm with high mitochondrial potential (HMMP) were reported. Unstained and single-stained samples were used for setting the FSC gain, FL-1 and FL-2 PMT voltages. Data were not compensated.

2.5.6. Lipid Peroxidation

Lipid peroxidation (LPO) was estimated using C_{11}-BODIPY$^{581/591}$ (Molecular Probes Europe, Leiden, The Netherlands) using a modified protocol [35]. A volume of 100 μL of diluted sperm (containing around 2×10^6 spz) was mixed with 1 μL C_{11}-BODIPY$^{581/591}$ (0.2 mM) and incubated at 37 °C for 30 min. After incubation, 1 mL of PBS was added for centrifugation at 600× *g* for 8 min. The pellet was resuspended with 100 μL of PBS prior to assessment. Spermatozoa with LPO emitted light in the green wavelength and were deemed to be BODIPY-positive cells. Unstained and single-stained samples were used for setting the FSC gain, FL-1 and FL-2 PMT voltages. Data were not compensated.

2.6. Data Analysis

Bayesian inference for ANOVA was run to test for statistical differences in the mean across antioxidants (HT, DHPG and MIX) at different concentrations (Control, HT1, HT2, HT3, HT4, DHPG1, DHPG2, DHPG3, DHPG4, MIX1, MIX2, MIX3 and MIX4) on buck sperm parameters (Table 1).

The Bayes factor (BF) quantifies the strength of the evidence of null and alternative hypotheses and is used instead of frequentist *p* values when Bayesian approaches are applied to issue conclusions. As BF increases, the degree at which evidence favors the alternative hypothesis compared to the null hypothesis increases as well. In this context, Cleophas and Zwinderman [36], suggested a method to extrapolate between the Bayes factor used in Bayesian approaches and *p* values from frequentist approaches to favor the interpretability of results.

Sample descriptive posterior statistics are modeled from the means and variances of the measured unpaired groups and are provided as sources of variation, while the prior element was modeled as an uninformative prior using the Jeffreys–Zellener–Siow (JZS) method or, equivalently, from the computation of a reference prior based on a gamma distribution with a standard error of 1. As suggested by Martins-Bessa et al. [37], the 95% credibility interval shows that there is a 95% probability that these regression coefficients (posterior distribution mean value for each covariate and factor) in the population lie within the corresponding credibility intervals. When 0 is not contained in the credibility interval, a significant effect for such factor is detected. Integral calculation of factors is required for BF accuracy purposes. Therefore, afterwards, it can be used as a precise statistical index to measure the amount of support in favor of either H_1 (the difference between the unpaired means is larger than zero) or H_0 (the difference between the unpaired means is not larger than zero). Contextually, Bayesian approaches provide a better perspective of the structure model of the H_1 and H_0. Therefore, the maximal likelihoods of likelihood distributions are not always identical to the mean effect of traditional tests, which specifically fits the context of biological inferences, given biological likelihoods may better respond biological questions than numerical means of non-representative subgroups do.

Table 1. Descriptive statistics of sperm viability, acrosome integrity, HMMP, and LPO parameters in fresh and frozen-thawed buck sperm. Data are expressed as Mean ± SD.

	Viability (%)	Acrosome Integrity (%)	HMMP (%)	LPO (%)
FRESH	84.4 ± 4.5	72.0 ± 6.4	79.8 ± 7.1	1.9 ± 0.8
CONTROL	43.9 ± 7.2	43.0 ± 7.7	40.3 ± 6.6	2.1 ± 0.5
HT1	50.2 ± 6.5	40.2 ± 9.8	32.8 ± 4.4	1.7 ± 0.6
HT2	44.5 ± 5.9	37.7 ± 5.5	32.6 ± 7.0	2.2 ± 0.4
HT3	44.5 ± 8.8	42.2 ± 11.4	38.2 ± 5.6	2.5 ± 0.7
HT4	43.1 ± 9.0	42.3 ± 6.9	40.3 ± 7.1	2.2 ± 0.8
DHPG1	49.9 ± 14.2	47.5 ± 11.3	40.8 ± 9.0	2.2 ± 1.3
DHPG2	43.8 ± 6.7	41.7 ± 6.4	35.8 ± 8.2	1.6 ± 0.5
DHPG3	43.6 ± 7.0	38.8 ± 6.2	36.8 ± 9.1	1.8 ± 0.9
DHPG4	47.0 ± 8.5	35.0 ± 8.3	40.0 ± 6.0	1.3 ± 0.7
MIX1	43.4 ± 7.9	39.0 ± 5.2	35.8 ± 7.3	2.9 ± 1.5
MIX2	44.2 ± 8.6	39.9 ± 7.9	36.5 ± 7.3	1.6 ± 0.7
MIX3	42.6 ± 13.3	41.7 ± 12.7	38.0 ± 3.1	1.6 ± 0.6
MIX4	43.2 ± 8.5	38.8 ± 8.9	34.0 ± 8.1	1.6 ± 0.7

HMMP: mitochondrial potential; LPO: lipid peroxidation; HT: hydroxytyrosol; DHPG: 3,4-dihydroxyphenylglycol; CONTROL (without antioxidant); HT1 (10 μg/mL); HT2 (30 μg/mL); HT3 (50 μg/mL) and HT4 (70 μg/mL); DHPG1 (10 μg/mL); DHPG2 (30 μg/mL); DHPG3 (50 μg/mL); DHPG4 (70 μg/mL); MIX1 (5 μg/mL HT + 5 μg/mL DHPG); MIX2 (15 μg/mL HT + 15 μg/mL DHPG); MIX3 (25 μg/mL HT + 25 μg/mL DHPG); MIX4 (35 μg/mL HT + 35 μg/mL DHPG).

IBM SPSS Statistics Algorithms version 25.0 by IBM Corp. [38] suggests Bayesian inference of ANOVA is approached as a special case of the Bayesian general multiple linear regression model. The algorithms used by SPSS to perform Bayesian Inference on Analysis of Variance (ANOVA) in this study are described in IBM SPSS Statistics Algorithms version 25.0 by IBM Corp. [38]. The tolerance value for the numerical methods and the number of method iterations were set as a default by SPSS v25.0 (IBM Corp., Armonk, NY, USA) [39].

The around 0 symmetric JZS prior was used as it is appropriate for Bayesian inference of ANOVA, provided positive and negative values of the slope parameters a priori have the same probability of occurring [40]. Furthermore, its scale-invariant properties, permits comparing parameters measured in different units, as it occurs in the present study. Bayesian inference for ANOVA was performed using the Bayesian Package of SPSS v25.0 (IBM Corp., Armonk, NY, USA) [39].

3. Results

3.1. Prior Descriptive Statistics

The descriptive statistics of sperm viability, acrosome integrity, HMMP, and LPO parameters in fresh and frozen-thawed buck sperm are shown in the Tables 1 and 2.

3.2. Bayesian Inference of Olive Oil Derived Antioxidant Effect

Table 3 and Supplementary Table S1 report the outputs from Bayesian ANOVA analysis and present posterior distribution statistics for sperm parameters across antioxidants and concentrations. Regarding to motility parameters, an evident/significant ($p < 0.05$) increase in PM was observed when HT2 treatment when compared to control treatment. However, at higher HT concentrations (70 μg/mL) an evident/significant ($p < 0.05$) decrease in PM occurred.

Table 2. Descriptive statistics of sperm motility and kinematic parameters in fresh and frozen-thawed buck sperm.

	MT (%)	PM (%)	VLC (µm/s)	VSL (µm/s)	VAP (%)	LIN (%)	STR (%)	WOB (%)	ALH (µm)	BFC (Hz)
FRESH	91.6 ± 3.4	38.9 ± 4.8	105.4 ± 9.4	36.9 ± 4.9	64.3 ± 4.3	35.2 ± 4.7	57.4 ± 5.7	61.1 ± 3.3	3.3 ± 0.4	10.1 ± 1.1
CONTROL	60.2 ± 9.4	31.6 ± 6.5	82.5 ± 14.3	39.7 ± 8.9	54.5 ± 11.7	47.7 ± 3.8	72.7 ± 2.2	65.7 ± 4.0	3.0 ± 0.3	11.0 ± 0.6
HT1	55.7 ± 6.4	30 ± 8.5	78.1 ± 10.0	37.6 ± 7.8	51.2 ± 9.0	48.0 ± 7.1	73.1 ± 4.7	65.4 ± 6.6	2.8 ± 0.3	11.4 ± 1.0
HT2	60.0 ± 8.9	35.2 ± 5.3	79.7 ± 6.8	38.9 ± 4.9	52.3 ± 5.2	49.0 ± 6.7	74.3 ± 4.2	65.8 ± 6.3	2.9 ± 0.4	11.2 ± 1.0
HT3	63.7 ± 8.1	33.9 ± 5.8	77.0 ± 10.2	36.1 ± 6.4	49.6 ± 9.1	46.9 ± 5.6	73.0 ± 3.2	64.2 ± 6.6	2.9 ± 0.3	11.1 ± 1.2
HT4	54.5 ± 7.9	27.7 ± 8.2	85.7 ± 9.2	36.2 ± 4.6	52.9 ± 6.8	42.2 ± 2.9	68.5 ± 3.1	61.6 ± 2.6	3.0 ± 0.2	11.9 ± 0.2
DHPG1	58.9 ± 9.8	32.8 ± 8	83.7 ± 10.8	37.6 ± 3.3	52.8 ± 5.4	45.3 ± 3.9	71.4 ± 2.3	63.3 ± 3.8	3.1 ± 0.3	12.1 ± 0.9
DHPG2	55.5 ± 9.8	31.8 ± 4.7	85.0 ± 10.0	39.8 ± 5.6	55.1 ± 8.3	46.8 ± 3.2	72.5 ± 2.3	64.6 ± 3.7	2.9 ± 0.2	11.8 ± 0.4
DHPG3	61.6 ± 9.1	32.4 ± 7.9	78.5 ± 12.4	35.2 ± 5.1	49.0 ± 6.2	45.2 ± 4.9	71.9 ± 4.4	62.7 ± 3.4	3.0 ± 0.3	11.6 ± 1.0
DHPG4	59.7 ± 9.8	33.9 ± 4.4	84.9 ± 8.5	37.9 ± 3.6	52.5 ± 4.7	44.7 ± 3.1	72.2 ± 2.7	62.0 ± 4.0	3.1 ± 0.3	12.2 ± 1.1
MIX1	59.8 ± 9.4	30.7 ± 8.3	81.0 ± 10.7	36.3 ± 6.0	50.7 ± 6.4	44.9 ± 4.3	71.4 ± 4.1	62.8 ± 2.9	3.0 ± 0.2	11.4 ± 1.3
MIX2	60.0 ± 6.6	31.5 ± 7.8	80.1 ± 6.6	35.3 ± 4.0	49.6 ± 3.5	44.1 ± 2.8	71.1 ± 3.5	62.0 ± 1.6	3.1 ± 0.2	11.4 ± 0.9
MIX3	57.4 ± 9.0	32.3 ± 9.1	76.8 ± 5.5	34.4 ± 3.1	47.2 ± 4.2	44.8 ± 3.5	72.9 ± 5.0	61.4 ± 2.0	3.0 ± 0.2	11.7 ± 1.0
MIX4	58.1 ± 8.1	31.4 ± 3.6	77.3 ± 5.5	34.5 ± 2.6	47.5 ± 3.9	44.6 ± 2.0	72.7 ± 1.9	61.4 ± 2.8	3.1 ± 0.1	11.9 ± 0.7

TM: Total motility; PM: progressive motility, VCL: curvilinear velocity; VSL:, straight line velocity; VAP: average path velocity; STR: straightness; LIN: linearity; WOB: wobble; ALH: amplitude of lateral head displacement; BCF: beat/cross frequency; HT: hydroxytyrosol; DHPG: 3,4-dihydroxyphenylglycol; CONTROL (without antioxidant); HT1 (10 µg/mL); HT2 (30 µg/mL); HT3 (50 µg/mL) and HT4 (70 µg/mL); DHPG1 (10 µg/mL); DHPG2 (30 µg/mL); DHPG3 (50 µg/mL); DHPG4 (70 µg/mL); MIX1 (5 µg/mL HT + 5 µg/mL DHPG); MIX2 (15 µg/mL HT + 15 µg/mL DHPG); MIX3 (25 µg/mL HT + 25 µg/mL DHPG); MIX4 (35 µg/mL HT + 35 µg/mL DHPG).

Table 3. Summary of the outputs of Bayesian inference ANOVA to detect differences in the mean of sperm parameters across the different concentrations of HT, DHPG, and MIX antioxidants.

	BG Sum of Squares	BG df	BG Mean Square	WG Sum of Squares	WG df	WG Mean Square	F	*p* Value	Bayes Factor
Viability (%)	9146.527	12	703.579	5301.865	70	75.741	9.289	0.001	14,617,034.109
Acrosome integrity (%)	6166.972	12	474.382	5044.732	70	72.068	6.582	0.001	11,102.852
HMMP (%)	10,748.051	12	826.773	3465.358	70	49.505	16.701	0.001	24,196,033,653,804.700
LPO (%)	13.991	12	1.076	48.149	70	0.688	1.565	0.117	0.000
TM (%)	6445.744	12	495.826	4985.240	70	71.218	6.962	0.001	33,431.056
PM (%)	530.067	12	40.774	3293.809	70	47.054	0.867	0.591	0.000
VCL (µm/s)	4134.502	12	318.039	6422.702	70	91.753	3.466	0.001	0.252
VSL (µm/s)	247.939	12	19.072	1998.821	70	28.555	0.668	0.787	0.000
VAP (µm/s)	1402.105	12	107.854	3178.839	70	45.412	2.375	0.011	0.002
LIN (%)	869.610	12	66.893	1361.286	70	19.447	3.440	0.001	0.227
STR (%)	1344.017	12	103.386	960.978	70	13.728	7.531	0.001	163,617.988
WOB (%)	222.771	12	17.136	1195.420	70	17.077	1.003	0.458	0.000
ALH (µm)	0.842	12	0.065	4.724	70	0.067	0.960	0.499	0.000
BCF (Hz)	23.095	12	1.777	60.923	70	0.870	2.041	0.029	0.000

HMMP: mitochondrial potential; LPO: lipid peroxidation; BG: Between groups; df: degrees of freedom; WG: Within groups; TM: Total motility; PM: progressive motility, VCL: curvilinear velocity; VSL: straight line velocity; VAP: average path velocity; STR: straightness; LIN: linearity; WOB: wobble; ALH: amplitude of lateral head displacement; BCF: beat/cross frequency.

When velocity parameters (VCL, VSL, and VAP) were evaluated, the mixture of both antioxidants produced a dose-dependent decrease in VSL and VAP values, with this being evidently/significantly ($p < 0.05$) lower when high concentrations were used. However, the rest of the studied kinetic parameters were not evidently/significantly affected ($p > 0.05$) by the addition of antioxidants, except for LIN when high concentrations of HT were added (70 μg/mL).

Samples supplemented with low concentrations of HT and DHPG (10 μg/mL) reported evident/significant ($p < 0.05$) sperm viability increases in comparison to Control. This positive evident/significant ($p < 0.05$) effect was also observed for acrosome integrity in samples supplemented with low DHPG concentrations (10 μg/mL). By contrast, higher DHPG concentrations offered evidently lower acrosome integrity. Similarly, a negative effect was observed when high concentrations (70 μg/mL) of the mixture of both antioxidants were used.

Mitochondrial potential evidently/significantly decreased ($p < 0.05$) in comparison to when the Control treatment was considered in samples supplemented with HT at low concentrations (10 and 30 μg/mL). Similarly, the addition of a mixture of both antioxidants evidently/significantly ($p < 0.05$) reduced the mitochondrial potential in frozen-thawed spermatozoa. Regarding to LPO values, the addition of 10 μg/mL of HT reported an evident/significant ($p < 0.05$) protective effect, reducing its value in comparison with Control treatments. However, when the dose increased, an opposite evident/significant ($p < 0.05$) effect was observed. By contrast, DHPG and MIX provided a dose-dependent better protection, diminishing LPO values as antioxidant concentration increases.

4. Discussion

Sperm cryopreservation offers goat breeders several benefits over fresh sperm storage. However, recent studies suggested that ROS concentration increases considerably during cryopreservation, disrupting sperm functions and subsequent fertilization [41]. Sometimes the endogenous antioxidant capacity of sperm cells is compromised due to the proliferation of ROS, producing an imbalance that promotes oxidative stress and consequently LPO, which affects membrane structure and distorts its functions such as membrane fluidity, membrane enzymes, ion gradients, receptor transduction, and transport processes [42]. In this context, the addition of antioxidants to the semen extender seems to have the potential to mitigate the negative impact of oxidative stress, as antioxidants capture free radicals and conclude the chain reaction, maintaining a redox state and offsetting their capacity to reduce molecular oxygen [43].

The addition of natural or synthetic antioxidants to the cryopreservation medium in goat sperm has attracted the attention of researchers as an alternative to diminish the negative effect of oxidative stress produced by ROS and to improve post-thawed sperm quality [11,44,45]. However, it is not easy to prove the exact nature of the action of antioxidants on sperm quality, and the degree to which interactions with other factors such as the species, the extender and the type or the concentration of antioxidant may be involved [46]. There is a knowledge gap on whether antioxidants are absorbed unchanged or metabolized into completely different compounds. Furthermore, the efficacy of the common antioxidants, such as vitamins C and E, selenium, and herbal supplements to reduce pathological ROS has not yet been determined [47].

In this context, HT is soluble in both lipid and water solutions, and therefore soluble in all phases of the heterogeneous system studied in the present research. The concentration of HT in biological systems is very similar in both aqueous and lipid areas [48]. As far as DHPG is concerned, this component has only recently been isolated and there is relatively little information regarding the way it behaves. On the basis of its chemical structure, which is very similar to HT, it should be broadly analogous to HT, and it is also soluble in both lipid and water solutions.

To the present authors' knowledge, this is the first study in which HT and DHPG were tested as antioxidants for caprine sperm cryopreservation. Biological activity and risk/benefit of polyphenolic compounds are dependent on their diversity, dual-effects, biological activity, and source [49]. In this regard, the effect of olive oil and olive oil-derived antioxidants on sperm quality has previously been investigated in other species. In rabbits, olive oil administered at 7% v/w for 16 weeks succeeded in recovering the loss of volume, count, motility, and normal spermatozoa in males exposed to a hypercholesterolemic diet [50]. Similarly, the oral administration of olive oil in healthy rats at 0.4 mL daily for six weeks improved the sperm parameters [51]. Banihani [52] concluded that the addition of olive oil preserves semen quality by enhancing the gonadal function, reducing oxidative injury and lipid peroxidation, and promoting nitric oxide signaling.

Results derived from the present study reported the fact that in samples supplemented with 30 μg/mL of HT an increase in progressive motility of 11% was reported in comparison to Control treatment which agrees the results reported by Hamden et al. [22], who supplemented rat sperm with 50 μg/mL HT and those by Krishnappa et al. [24] who reported an improvement in total motility when 80 mM HT was added to ovine sperm. These findings suggest that the presence of HT could mitigate ROS concentration, preventing the negative impact of moderately elevated ROS concentrations on the sperm movement, mostly via depletion of intracellular ATP and the successive reduction in the phosphorylation of axonemal proteins [53]. However, in the present study, PM considerably decreased when high concentrations (70 μg/mL) of HT were added. One of the reasons for this negative effect associated to the increase of the dose would be the extender acidification, as previously described by Ibrahim et al. [54], who supplemented goat sperm with alpha lipoic acid. By contrast, no effect was observed in the present study for TM and PM when DHPG and MIX were added, in line with previous studies carried out on sheep using the same antioxidants and concentrations [15,25] and in incubated human sperm [23] after HT supplementation.

In regard to kinematic parameters, no effect was observed when HT and DHPG were independently added. However, the mixture of both antioxidants induced a decrease of VSL and VAP in dose-dependent manner, being ~13% when high concentrations were used. This similar trend was previously described by Arando, et al. [15] in liquid ram sperm stored at 5 °C, suggesting that high concentrations of these antioxidants could be deleterious for spermatozoa. Contextually, broad evidence suggests higher concentrations may not necessarily translate into better quality, but indeed may be detrimental [55]. In agreement with the present study, the use of other antioxidants such as arbutin, butylated hydroxyanisole, rosemary or lycopene, have reported a positive effect on goat sperm motility [11,44,45,56].

A high amount of PUFA in sperm membranes could interact with ROS, affecting membrane fluidity, facilitating Ca^{2+} influx, and provoking membrane protein reorganization and the destabilization of the plasma membrane [57]. Based on the current results, the addition of HT at low concentrations (10 μg/mL) may improve membrane integrity, showing an increase of 14% compared to Control treatment, as reported by Hamden et al. [22] when HT was added. Similarly, some authors reported an increase of membrane integrity when other antioxidants were added in extender medium [8,56,58,59].

Acrosome is a specialized sperm structure comprising membranes and proteins which makes it highly susceptible to ROS-derived damage [60]. In this context, a positive effect was observed in acrosome integrity when samples were supplemented with low DHPG concentrations (10 μg/mL). By contrast, as DHPG concentration increases or when high concentrations of mixture (70 μg/mL) were used, acrosome integrity decreased, as previously described by Arando et al. [25]. On the other hand, Hashem et al. [61] reported higher acrosome integrity values when oleic acid was added to ram sperm. Similarly, recent studies in buck sperm showed that the addition of different antioxidants, as vitamin C or lycopene, could mitigate the acrosome damage [11,62].

Mitochondria are involved in the generation of ROS in spermatozoa through the pathway of nicotinamide adenine dinucleotide-dependent oxide reductase reactions, which directly affects their normal functions [63]. Endogenous antioxidant supplementation has bene hypothesized to reduce oxidative stress and, as a consequence, to maintain post-thawed mitochondrial potential. In this sense, the present study reveals that mitochondrial membrane potential is not improved in freezing-thawing sperm samples supplemented with HT, DHPG, and/or MIX, in comparison to the Control treatment, as reported by Zanganeh et al. [45] and Arando et al. [25].

However, in a recent study using mitochondria-targeted antioxidants, the authors noted a slight significant mitochondrial potential increase [64]. The olive-oil antioxidants tested in the present study did not offer any advantage for the mitochondrial activity after cryopreservation. However, further studies should be conducted to elucidate why the intense impact of freezing-thawing process on the mitochondrial activity in buck sperm cannot be counteract by the mentioned antioxidants. By contrast, other authors observed a significant increase in mitochondrial potential when cysteine, coenzyme $Q_{10,}$ lycopene or alpha-lipoic acid were used [10,11,65].

As far as LPO is concerned, the present study showed that the addition of 10 µg/mL of HT may present a protective function against lipid peroxidation, reducing it by 19% in comparison to Control group. In reference to oil-derived antioxidants, a recent study also reported that the addition of HT and DHPG improved LPO values in ram sperm [25]. However, the opposite effect was observed as dose increased. DHPG and the mixture of both antioxidants provided better protection properties, diminishing LPO values around 20%, when sperm samples were supplemented with high doses of antioxidants.

5. Conclusions

HT could mitigate ROS concentration, preventing their negative impact on the sperm movement. Dose-dependent extender acidification produces a considerable reduction of PM. The mixture of both antioxidants induced a decrease of VSL and VAP in dose dependent manner, being approximately 13% when high concentrations were used. Higher concentrations may not necessarily lead to better results but may be detrimental for sperm quality. The addition of HT at low concentrations (10 µg/mL) may improve membrane integrity. Acrosome integrity improves when supplementing with low DHPG concentrations (10 µg/mL). By contrast, as DHPG concentration increases or when high concentrations of mixture (70 µg/mL) were used, acrosome integrity decreased. Mitochondrial membrane potential is not improved in freezing-thawing sperm supplemented with HT, DHPG and/or MIX. Olive-oil antioxidants did not benefit mitochondrial activity after cryopreservation. The addition of 10 µg/mL of HT may present a protective function against lipid peroxidation, reducing it by 19%. However, the opposite effect was observed as dose increased. DHPG and the mixture of both antioxidants provided better protection properties, diminishing LPO values around 20%, when sperm was supplemented with high doses of antioxidants. Therefore, in light of the obtained results the addition of HT at 10 µg/mL and DHPG at 30 µg/mL were the most suitable treatments since they improved post-thawing sperm quality.

Supplementary Materials: The following are available online at https://www.mdpi.com/article/10.3390/ani11072032/s1, Table S1. Summary of the posterior distribution of Bayesian statistics and the percentage of improvement (+) or detriment (−) on the quality of buck sperm parameters ascribed to the addition of olive oil-derived antioxidants.

Author Contributions: Conceptualization, F.J.N.G., J.M.L.J. and C.P.-M.; Data curation, A.A.A., F.J.N.G., A.G.A., J.M.L.J. and C.P.-M.; Formal analysis, A.A.A., F.J.N.G., A.G.A. and J.M.L.J.; Funding acquisition, Á.F.-P. and J.M.L.J.; Investigation, A.A.A., A.G.A., J.M.L.J. and C.P.-M.; Methodology, A.A.A., F.J.N.G., Á.F.-P., A.G.A. and C.P.-M.; Project administration, Alejandra Bermúdez-Oria and C.P.-M.; Resources, F.J.N.G., A.B.-O., J.V.D.B. and Á.F.-P.; Software, A.A.A., F.J.N.G. and C.P.-M.; Supervision, F.J.N.G., A.B.-O., J.V.D.B., Á.F.-P. and C.P.-M.; Validation, F.J.N.G.; Visualization, F.J.N.G., J.V.D.B., Á.F.-P. and C.P.-M.; Writing—original draft, A.A.A., F.J.N.G., A.G.A. and J.M.L.J.; Writing—review and editing, A.A.A., F.J.N.G., A.B.-O., J.V.D.B., Á.F.-P. and C.P.-M. All authors have read and agreed to the published version of the manuscript.

Funding: This research was funded by the Torres Quevedo Programme from the Ministry of Science and Innovation of Spain, grant number PTQ2019-010670, granted to Ander Arando Arbulu.

Institutional Review Board Statement: The study was conducted according to the guidelines of the Declaration of Helsinki, and approved by the Institutional Review Board (or Ethics Committee) of the University of Córdoba (n. 2018PI/29).

Informed Consent Statement: Not applicable.

Data Availability Statement: Data will be made accessible from corresponding authors upon reasonable request.

Acknowledgments: The authors would like to acknowledge the Agricultural and Livestock Center of the Córdoba Provincial Council (Diputación de Cordoba, Spain) for their help with the supply of semen samples.

Conflicts of Interest: The authors declare no conflict of interest.

References

1. Yoon, S.J.; Rahman, M.S.; Kwon, W.S.; Park, Y.J.; Pang, M.G. Addition of Cryoprotectant Significantly Alters the Epididymal Sperm Proteome. *PLoS ONE* **2016**, *11*, e0152690.
2. Kargar, R.; Forouzanfar, M.; Ghalamkari, G.; Nasr Esfahani, M.H. Dietary flax seed oil and/or Vitamin E improve sperm parameters of cloned goats following freezing-thawing. *Cryobiology* **2017**, *74*, 110–114. [CrossRef] [PubMed]
3. Bucak, M.N.; Sarıözkan, S.; Tuncer, P.B.; Ulutaş, P.A.; Akçadağ, H.İ. Effect of antioxidants on microscopic semen parameters, lipid peroxidation and antioxidant activities in Angora goat semen following cryopreservation. *Small Rumin. Res.* **2009**, *81*, 90–95. [CrossRef]
4. Mata-Campuzano, M.; Álvarez-Rodríguez, M.; Álvarez, M.; Tamayo-Canul, J.; Anel, L.; de Paz, P.; Martínez-Pastor, F. Post-thawing quality and incubation resilience of cryopreserved ram spermatozoa are affected by antioxidant supplementation and choice of extender. *Theriogenology* **2015**, *83*, 520–528. [CrossRef] [PubMed]
5. De Lamirande, E.; Gagnon, C. Impact of reactive oxygen species on spermatozoa: A balancing act between beneficial and detrimental effects. *Hum. Reprod.* **1995**, *10* (Suppl. 1), 15–21. [CrossRef]
6. Aziz, N.; Saleh, R.A.; Sharma, R.K.; Lewis-Jones, I.; Esfandiari, N.; Thomas, A.J.; Agarwal, A. Novel association between sperm reactive oxygen species production, sperm morphological defects, and the sperm deformity index. *Fertil. Steril.* **2004**, *81*, 349–354. [CrossRef] [PubMed]
7. Agarwal, A.; Saleh, R.A. Role of oxidants in male infertility: Rationale, significance, and treatment. *Urol. Clin. N. Am.* **2002**, *29*, 817–828. [CrossRef]
8. Shafiei, M.; Forouzanfar, M.; Hosseini, S.M.; Nasr Esfahani, M.H. The effect of superoxide dismutase mimetic and catalase on the quality of postthawed goat semen. *Theriogenology* **2015**, *83*, 1321–1327. [CrossRef] [PubMed]
9. Alcay, S.; Gokce, E.; Toker, M.B.; Onder, N.T.; Ustuner, B.; Uzabacı, E.; Gul, Z.; Cavus, S. Freeze-dried egg yolk based extenders containing various antioxidants improve post-thawing quality and incubation resilience of goat spermatozoa. *Cryobiology* **2016**, *72*, 269–273. [CrossRef]
10. Yousefian, I.; Emamverdi, M.; Karamzadeh-Dehaghani, A.; Sabzian-Melei, R.; Zhandi, M.; Zare-Shahneh, A. Attenuation of cryopreservation-induced oxidative stress by antioxidant: Impact of Coenzyme Q10 on the quality of post-thawed buck spermatozoa. *Cryobiology* **2018**, *81*, 88–93. [CrossRef] [PubMed]
11. Ren, F.; Feng, T.; Dai, G.; Wang, Y.; Zhu, H.; Hu, J. Lycopene and alpha-lipoic acid improve semen antioxidant enzymes activity and cashmere goat sperm function after cryopreservation. *Cryobiology* **2018**, *84*, 27–32. [CrossRef]
12. Haydari, M.; Maresca, V.; Rigano, D.; Taleei, A.; Shahnejat-Bushehri, A.A.; Hadian, J.; Sorbo, S.; Guida, M.; Manna, C.; Piscopo, M.; et al. Salicylic Acid and Melatonin Alleviate the Effects of Heat Stress on Essential Oil Composition and Antioxidant Enzyme Activity in Mentha × piperita and *Mentha arvensis* L. *Antioxidants* **2019**, *8*, 547. [CrossRef]
13. Tortora, F.; Notariale, R.; Maresca, V.; Good, K.V. Phenol-Rich Feijoa sellowiana (Pineapple Guava) Extracts Protect Human Red Blood Cells from Mercury-Induced Cellular Toxicity. *Antioxidants* **2019**, *8*, 220. [CrossRef]
14. Boo, Y.C. Human Skin Lightening Efficacy of Resveratrol and Its Analogs: From in Vitro Studies to Cosmetic Applications. *Antioxidants* **2019**, *8*, 332. [CrossRef]
15. Arando, A.; Delgado, J.V.; Bermúdez-Oria, A.; León, J.M.; Fernández-Prior, Á.; Nogales, S.; Pérez-Marín, C.C. Effect of olive-derived antioxidants (3,4-dihydroxyphenylethanol and 3,4 dihydroxyphenylglycol) on sperm motility and fertility in liquid ram sperm stored at 15 °C or 5 °C. *Reprod. Domest. Anim.* **2020**, *55*, 325–332. [CrossRef] [PubMed]
16. Fernandez-Bolanos, J.G.; Lopez, O.; Fernandez-Bolanos, J.; Rodriguez-Gutierrez, G. Hydroxytyrosol and derivatives: Isolation, synthesis, and biological properties. *Curr. Org. Chem.* **2008**, *12*, 442–463. [CrossRef]
17. Bermúdez-Oria, A.; Rodríguez-Gutiérrez, G.; Rubio-Senent, F.; Lama-Muñoz, A.; Fernández-Bolaños, J. Complexation of hydroxytyrosol and 3,4-dihydroxyphenylglycol with pectin and their potential use for colon targeting. *Carbohydr. Polym.* **2017**, *163*, 292–300. [CrossRef] [PubMed]

18. Choucair, F.; Saliba, E.; Jaoude, I.A.; Hazzouri, M. Antioxidants modulation of sperm genome and epigenome damage: Fact or fad? Converging evidence from animal and human studies. *Middle East Fertil Soc. J.* **2018**, *23*, 85–90. [CrossRef]
19. Roche, M.; Dufour, C.; Loonis, M.; Reist, M.; Carrupt, P.-A.; Dangles, O. Olive phenols efficiently inhibit the oxidation of serum albumin-bound linoleic acid and butyrylcholine esterase. *Biochim. Biophys. Acta Gen. Subj.* **2009**, *1790*, 240–248. [CrossRef] [PubMed]
20. Hashimoto, T.; Ibi, M.; Matsuno, K.; Nakashima, S.; Tanigawa, T.; Yoshikawa, T.; Yabe-Nishimura, C. An endogenous metabolite of dopamine, 3, 4-dihydroxyphenylethanol, acts as a unique cytoprotective agent against oxidative stress-induced injury. *Free Radic. Biol. Med.* **2004**, *36*, 555–564. [CrossRef]
21. Rodríguez, G.; Rodríguez, R.; Fernández-Bolaños, J.; Guillén, R.; Jiménez, A. Antioxidant activity of effluents during the purification of hydroxytyrosol and 3, 4-dihydroxyphenyl glycol from olive oil waste. *Eur. Food Res. Technol.* **2007**, *224*, 733–741. [CrossRef]
22. Hamden, K.; Allouche, N.; Damak, M.; Carreau, S.; Elfeki, A. Potential protective effect of hydroxytyrosol against glucose-induced toxicity in spermatozoa in vitro. *Asia Life Sci.* **2010**, *19*, 333–344.
23. Kedechi, S.; Zribi, N.; Louati, N.; Menif, H.; Sellami, A.; Lassoued, S.; Ben Mansour, R.; Keskes, L.; Rebai, T.; Chakroun, N. Antioxidant effect of hydroxytyrosol on human sperm quality during in vitro incubation. *Andrologia* **2017**, *49*, e12595. [CrossRef] [PubMed]
24. Krishnappa, B.; Srivastava, S.; Kumar, D.; Ghosh, S.; Kalyan, D.; Paul, R.; Bahire, S.; Naqvi, S. Effect of hydroxytyrosol on sperm post-thaw motion and velocity of cryopreserved ram semen. *Indian J. Small Rumin.* **2018**, *24*, 75–79. [CrossRef]
25. Arando, A.; Delgado, J.; Fernández-Prior, A.; León, J.; Bermúdez-Oria, A.; Nogales, S.; Pérez-Marín, C. Effect of different olive oil-derived antioxidants (hydroxytyrosol and 3, 4-dihydroxyphenylglycol) on the quality of frozen-thawed ram sperm. *Cryobiology* **2019**, *86*, 33–39. [CrossRef] [PubMed]
26. Fernández-Bolaños, J.; Rodríguez, G.; Rodríguez, R.; Heredia, A.; Guillén, R.; Jiménez, A. Production in large quantities of highly purified hydroxytyrosol from liquid− solid waste of two-phase olive oil processing or "Alperujo". *J. Agric. Food Chem.* **2002**, *50*, 6804–6811. [CrossRef]
27. Fernández-Bolaños Guzmán, J.; Rodríguez-Gutiérrez, G.; Lama Muñoz, A.; Fernández-Bolaños Guzmán, J.M.; Maya, I.; López López, Ó.; Marset Castro, A. Method for obtaining hydroxytyrosol extract, mixture of hydroxytyrosol and 3, 4-dihydroxyphenylglycol extract, and hydroxytyrosyl acetate extract, from by-products of the olive tree, and the purification thereof. International Patent Application No. PCT/ES2012/070491. U.S. Patent Application No. WO 2013/007850, 3 July, 2013.
28. Evans, G.; Maxwell, W.C. *Salamons' Artificial Insemination of Sheep and Goats*; Butterworths: Sydney, Australia, 1987.
29. Lopes, S.A.F.; Rosa, H.J.D.; Chaveiro, A.; da Silva, F.M. Influence of Different Freezing Curves on the Acrosome Integrity of Male Goat Sperm Cells. *Am. J. Anim. Vet. Sci.* **2021**, *16*, 56–61. [CrossRef]
30. Lee, J.A.; Spidlen, J.; Boyce, K.; Cai, J.; Crosbie, N.; Dalphin, M.; Furlong, J.; Gasparetto, M.; Goldberg, M.; Goralczyk, E.M.; et al. MIFlowCyt: The minimum information about a flow cytometry experiment. *Cytom. Part A* **2008**, *73A*, 926–930. [CrossRef]
31. Arando, A.; Gonzalez, A.; Delgado, J.; Arrebola, F.; Perez-Marín, C. Storage temperature and sucrose concentrations affect ram sperm quality after vitrification. *Anim. Reprod. Sci.* **2017**, *181*, 175–185. [CrossRef]
32. Petrunkina, A.; Waberski, D.; Bollwein, H.; Sieme, H. Identifying non-sperm particles during flow cytometric physiological assessment: A simple approach. *Theriogenology* **2010**, *73*, 995–1000. [CrossRef]
33. Santiani, A.; Ugarelli, A.; Evangelista-Vargas, S. Characterization of functional variables in epididymal alpaca (Vicugna pacos) sperm using imaging flow cytometry. *Anim. Reprod. Sci.* **2016**, *173*, 49–55. [CrossRef]
34. Hallap, T.; Nagy, S.; Jaakma, Ü.; Johannisson, A.; Rodriguez-Martinez, H. Mitochondrial activity of frozen-thawed spermatozoa assessed by MitoTracker Deep Red 633. *Theriogenology* **2005**, *63*, 2311–2322. [CrossRef]
35. Thuwanut, P.; Axner, E.; Johanisson, A.; Chatdarong, K. Detection of Lipid Peroxidation Reaction in Frozen-Thawed Epididymal Cat Spermatozoa Using BODIPY581/591 C11. *Reprod. Domest. Anim.* **2009**, *44*, 373–376. [CrossRef]
36. Cleophas, T.J.; Zwinderman, A.H. Bayesian Analysis of Variance (Anova). In *Modern Bayesian Statistics in Clinical Research*; Cleophas, T.J., Zwinderman, A.H., Eds.; Springer International Publishing: Cham, Switzerland, 2018; pp. 83–89.
37. Martins-Bessa, A.; Quaresma, M.; Leiva, B.; Calado, A.; Navas González, F.J. Bayesian Linear Regression Modelling for Sperm Quality Parameters Using Age, Body Weight, Testicular Morphometry, and Combined Biometric Indices in Donkeys. *Animals* **2021**, *11*, 176. [CrossRef]
38. IBM Corp. *IBM SPSS Statistics Algorithms*; IBM Corp: Armonk, NY, USA, 2017; p. 110.
39. IBM Corp. *IBM SPSS Statistics for Windows, 25.0*; IBM Corp: Armonk, NY, USA, 2017.
40. Bayarri, M.J.; Berger, J.O.; Forte, A.; García-Donato, G. Criteria for Bayesian model choice with application to variable selection. *Ann. Stat.* **2012**, *40*, 1550–1577. [CrossRef]
41. Seifi-Jamadi, A.; Ahmad, E.; Ansari, M.; Kohram, H. Antioxidant effect of quercetin in an extender containing DMA or glycerol on freezing capacity of goat semen. *Cryobiology* **2017**, *75*, 15–20. [CrossRef] [PubMed]
42. Sikka, S.C.; Rajasekaran, M.; Hellstrom, W.J. Role of oxidative stress and antioxidants in male infertility. *J. Androl.* **1995**, *16*, 464–468.
43. Flora, S.J. Structural, chemical and biological aspects of antioxidants for strategies against metal and metalloid exposure. *Oxid. Med. Cell. Longev.* **2009**, *2*, 191–206. [CrossRef] [PubMed]

44. Aboagla, E.M.-E.; Maeda, T. Arbutin's suppression of cryodamage in goat sperm and its mechanism of cryoprotection. *Theriogenology* **2011**, *76*, 538–546. [CrossRef]
45. Zanganeh, Z.; Zhandi, M.; Zare-Shahneh, A.; Najafi, A.; Nabi, M.M.; Mohammadi-Sangcheshmeh, A. Does rosemary aqueous extract improve buck semen cryopreservation? *Small Rumin. Res.* **2013**, *114*, 120–125. [CrossRef]
46. Silva, E.C.B.; Cajueiro, J.F.P.; Silva, S.V.; Soares, P.C.; Guerra, M.M.P. Effect of antioxidants resveratrol and quercetin on in vitro evaluation of frozen ram sperm. *Theriogenology* **2012**, *77*, 1722–1726. [CrossRef]
47. Salehi, B.; Martorell, M.; Arbiser, J.L.; Sureda, A.; Martins, N.; Maurya, P.K.; Sharifi-Rad, M.; Kumar, P.; Sharifi-Rad, J. Antioxidants: Positive or Negative Actors? *Biomolecules* **2018**, *8*, 124. [CrossRef] [PubMed]
48. Rietjens, S.J.; Bast, A.; Haenen, G.R. New insights into controversies on the antioxidant potential of the olive oil antioxidant hydroxytyrosol. *J. Agric. Food Chem.* **2007**, *55*, 7609–7614. [CrossRef] [PubMed]
49. Hashem, N.M.; Gonzalez-Bulnes, A.; Simal-Gandara, J. Polyphenols in Farm Animals: Source of Reproductive Gain or Waste? *Antioxidants* **2020**, *9*, 1023. [CrossRef] [PubMed]
50. Lancellotti, T.E.S.; Boarelli, P.V.; Romero, A.A.; Funes, A.K.; Cid-Barria, M.; Cabrillana, M.E.; Monclus, M.A.; Simón, L.; Vicenti, A.E.; Fornes, M.W. Semen quality and sperm function loss by hypercholesterolemic diet was recovered by addition of olive oil to diet in rabbit. *PLoS ONE* **2013**, *8*, e52386. [CrossRef] [PubMed]
51. Mansour, S.W.; Sangi, S.; Harsha, S.; Khaleel, M.A.; Ibrahim, A. Sensibility of male rats fertility against olive oil, Nigella sativa oil and pomegranate extract. *Asian Pac. J. Trop. Biomed.* **2013**, *3*, 563–568. [CrossRef]
52. Banihani, S.A. Semen quality as affected by olive oil. *Int. J. Food Prop.* **2017**, *20*, 1901–1906. [CrossRef]
53. Agarwal, A.; Gupta, S.; Sharma, R.K. Role of oxidative stress in female reproduction. *Reprod. Biol. Endocrin.* **2005**, *3*, 28. [CrossRef]
54. Ibrahim, S.F.; Osman, K.; Das, S.; Othman, A.M.; Majid, N.A.; Rahman, M.P.A. A study of the antioxidant effect of alpha lipoic acids on sperm quality. *Clinics* **2008**, *63*, 545–550. [CrossRef]
55. Halliwell, B. The antioxidant paradox: Less paradoxical now? *Br. J. Clin. Pharmacol.* **2013**, *75*, 637–644. [CrossRef]
56. Rahmatzadeh, M.; Kohram, H.; Zare Shahneh, A.; Seifi-Jamadi, A.; Ahmad, E. Antioxidative effect of BHA in soya bean lecithin-based extender containing glycerol or DMSO on freezing capacity of goat semen. *Reprod. Domest. Anim.* **2017**, *52*, 985–991. [CrossRef]
57. Ahmad, E.; Aksoy, M.; Serin, İ.; Küçük, N.; Ceylan, A.; Uçan, U. Cholesterol-loaded cyclodextrin pretreatment of ram spermatozoa protects structural integrity of plasma membrane during osmotic challenge and reduces their ability to undergo acrosome reaction in vitro. *Small Rumin. Res.* **2013**, *115*, 77–81. [CrossRef]
58. Daramola, J.O.; Adekunle, E.O.; Oke, O.E.; Onagbesan, O.M.; Iyasere, O.S.; Williams, T.J.; James, I.J.; Oyewusi, I.K.; Oyewusi, J.A. Effects of pyridoxine supplementation or in combination with other antioxidants on motility, in vitro capacitation and acrosome reaction of goat buck spermatozoa during cryopreservation. *Small Rumin. Res.* **2015**, *131*, 113–117. [CrossRef]
59. Falchi, L.; Pau, S.; Pivato, I.; Bogliolo, L.; Zedda, M.T. Resveratrol supplementation and cryopreservation of buck semen. *Cryobiology* **2020**, *95*, 60–67. [CrossRef] [PubMed]
60. Huang, T., Jr.; Yanagimachi, R. Inner acrosomal membrane of mammalian spermatozoa: Its properties and possible functions in fertilization. *Am. J. Anat.* **1985**, *174*, 249–268. [CrossRef]
61. Zadeh Hashem, E.; Haddad, R.; Eslami, M. Evaluation of ram semen enrichment with oleic acid on different spermatozoa parameters during low temperature liquid storage. *Small Rumin. Res.* **2017**, *150*, 30–39. [CrossRef]
62. Gangwar, C.; Kharche, S.D.; Ranjan, R.; Kumar, S.; Goel, A.K.; Jindal, S.K.; Agarwal, S.K. Effect of vitamin C supplementation on freezability of Barbari buck semen. *Small Rumin. Res.* **2015**, *129*, 104–107. [CrossRef]
63. Agarwal, A.; Virk, G.; Ong, C.; du Plessis, S.S. Effect of oxidative stress on male reproduction. *World J. Mens Health* **2014**, *32*, 1–17. [CrossRef]
64. Lu, X.; Zhang, Y.; Bai, H.; Liu, J.; Li, J.; Wu, B. Mitochondria-targeted antioxidant MitoTEMPO improves the post-thaw sperm quality. *Cryobiology* **2018**, *80*, 26–29. [CrossRef]
65. Sharafi, M.; Zhandi, M.; Akbari Sharif, A. Supplementation of soybean lecithin-based semen extender by antioxidants: Complementary flowcytometric study on post-thawed ram spermatozoa. *Cell Tissue Bank.* **2015**, *16*, 261–269. [CrossRef]

animals

Article

Antioxidant and Antimicrobial Activity of *Cleome droserifolia* (Forssk.) Del. and Its Biological Effects on Redox Status, Immunity, and Gut Microflora

Nesrein M. Hashem [1,*] and **Mohamed G. Shehata** [2]

1 Department of Animal and Fish Production, Faculty of Agriculture, Alexandria University, Alexandria 21545, Egypt
2 Department of Food Technology, Arid Lands Cultivation Research Institute, City of Scientific Research and Technological Applications (SRTACITY), New Borg El Arab, Alexandria 21545, Egypt; gamalsng@gmail.com
* Correspondence: nesreen.hashem@alexu.edu.eg

Simple Summary: The antioxidant, antimicrobial, and immunomodulatory activities of the *Cleome droserifolia* (Forssk.) Del. (Cd) shrub were investigated considering the biological activity of its phytogenic compounds. Cd shrub encompasses several phenolic compounds, mainly phenolic acids, such as benzoic acid. The methanolic extract of Cd exhibited strong in vitro antioxidant and antimicrobial activities. Anin vivo study using rabbits as an animal model confirmed the ability of a powder of Cd aerial parts to improve humoral and innate immunity, as well as gastrointestinal microbiota homeostasis. In conclusion, Cd shrub represents a novel source of secondary active metabolites that can be employed as antibiotic alternative in the livestock production field and/or in human pharmaceutical applications.

Citation: Hashem, N.M.; Shehata, M.G. Antioxidant and Antimicrobial Activity of *Cleome droserifolia* (Forssk.) Del. and Its Biological Effects on Redox Status, Immunity, and Gut Microflora. *Animals* **2021**, *11*, 1929. https://doi.org/10.3390/ani11071929

Academic Editors: Cesare Castellini and Ilias Giannenas

Received: 17 May 2021
Accepted: 25 June 2021
Published: 28 June 2021

Abstract: This study aimed to investigate the antioxidant, antimicrobial, and immunomodulatory activities of a *Cleome droserifolia* (Forssk.) Del. (Cd) shoot methanolic extracts considering the biological activity of its phytogenic compounds. For this purpose, the Cd phenolic compounds were detected, and an in vitro evaluation of the antioxidant and antimicrobial activities of the Cd extract was performed. For a biological evaluation, 30 v-line rabbits were randomly distributed into three groups with treatments including: a basal diet without Cd shoots powder supplement (C group) or supplemented with 1.25- (Cdl group) or 2.5 (Cdh group)-mg Cd/kg dry matter (DM). The Cd extract showed a linear scavenging activity for 2,2-diphenyl-1-picrylhydrazyl and 2,2′-azino-bis (3-ethylbenzothiazoline-6-sulfonic acid), with the maximal activity observed at a concentration of 1 mg/mL. A total of 16 phenolic compounds were identified by reverse-phase high-performance liquid chromatography (RP-HPLC) in the Cd methanolic extract, among which benzoic acid, rutin, ellagic acid, *naringenin*, and o-coumaric acid were the major compounds. The methanolic extract of Cd showed inhibitory actions against microbial pathogen species. The in vivo study showed that the two concentrations of Cd significantly improved the redox status of the blood plasma and lysozyme activity. Treatment with Cdh significantly decreased the levels of interleukin-β1 in the blood plasma compared with the control. Moreover, the two concentrations of Cd significantly increased the counts of intestinal and cecal yeast and *Lactobacillus* species and decreased the *Salmonella* and *Coliform* species compared with the control. The aerial parts of the Cd shrub had strong antioxidant, antimicrobial, and immunomodulatory activities, which can improve the overall health status and seem to be related to its impressive range of biologically active phenolic compounds.

Keywords: phytogenic; phenols; antioxidant; antimicrobial activity; immunity

1. Introduction

Medicinal plants can serve as a natural source of therapeutic drugs, nutraceuticals/food supplements, and feed additives that can be safely used to improve human and animal health. The interest in exploring plants as a new source of different drugs, specifically antimicrobials, has increased in recent decades as an attempt to fight multidrug-resistant bacteria [1,2]. Among the medicinal plants, the *Cleome* genus is one of the largest genera belonging to the family Cleomaceae. This genus encompasses about 180–200 species that are geographically distributed in Egypt, Libya, Palestine, Syria, and other arid and semi-arid regions [3]. Moreover, they are perennial, low, and aromatic cushion-like shrubs with a length of 25–60 cm that exhibit intricately branched stems and broad oval-shaped, three-nerved leaves with swollen glandular hairs [3,4]. The shrubs that belong to this genus have medicinal and ecological importance. *Cleome* genus shrubs are well-known in folk medicine for treating stomachache, skin allergies, and open wounds, as well as for exhibiting anticancer and hepatoprotective properties [5–7]. In addition, *Cleome* genus shrubs have shown strong antidiabetic properties; the aqueous extract of *Cleome* has been found to contain a very high percentage of flavonols that showed 63.3% activity, similar to that of the metformin synthetic drug [8]. *Cleome* genus shrubs have antioxidant, antiparasitic, and antimicrobial activities [6]. These biological effects are related to the vast array of secondary metabolites that occur naturally in *Cleome* genus shrubs. Several terpenes, flavonoids, glucosinolates, anthocyanin alkaloids, and polyphenols have been isolated from *Cleome* genus shrubs [3]. Given these biological activities of *Cleome* genus shrubs, additional studies are required to explore the active secondary metabolites of these shrubs and their eligibility to innovate natural feed and food supplements that could be applied for improving animal and human health. Therefore, this study was devoted to the evaluation of the antioxidant, antimicrobial, and immunomodulatory properties of *Cleome droserifolia* (Forssk.) Del., the most famous species among the *Cleome* genus in Egypt, considering phenolic compounds as active secondary metabolites.

2. Materials and Methods

2.1. *In Vitro Evaluation of* Cleome droserifolia *(Forssk.) Del.*

2.1.1. Plant Source and Extraction

Shoots of Cleome droserifolia (Forssk.) Del. (Cd) were collected at "Megerah" Valley, Dahab, the Eastern Desert, South Sina, Egypt during the month of November 2018. The plant was authenticated by the Plant Protection and Biomolecular Diagnosis Department, STR-City, New Borg El-Arab, Egypt. The collected shoots were pooled, dried at 40 °C for 72 h, and milled through a 0.25-mm screen to obtain a fine powder. Representative samples of dried Cd powder were extracted and pooled for evaluating the phenolic content, antioxidant activity, and antimicrobial activity of the plant. Briefly, each 100 g of dried Cd powder was extracted in 1000 mL of hydro-methanolic solution (700-mL methanol and 300-mL water; 70%) at 40 °C for 72 h. The extract was filtered through Whatman No. 1 filter paper (Whatman No. 1, Camlab, Cambridge, UK). The collected filtrate was evaporated at 45 °C to complete dryness, and the residues were then stored at −20 °C for further analyses.

2.1.2. Determination of Total Phenolic and Total Flavonoid Content

The concentrations of total phenols (TPC) and total flavonoids (TFC) of the Cd methanolic extract were colorimetrically (T80 UV/Vis spectrometer PG Instruments LDT, Leicestershire, UK) determined using the Folin–Ciocalteu and aluminum trichloride ($AlCl_3$) methods, respectively. Gallic acid (GA) and catechol (CAT) were used as a standard for TPC and TFC measurements, respectively. Results were expressed as mg of GA equivalent/g of the DM extract (mg GA/g DM) and mg of CAT equivalent/g of the DM extract (mg CAT/g DM), as in reference [9].

2.1.3. Determination of Polyphenol Content

All analytical chemicals were gradient grade for the HPLC analysis. All chemicals and standards were purchased from Sigma-Aldrich® (Merck KGaA, Darmstadt, Germany). The polyphenol profile of the Cd extract was assessed using reverse-phase high-performance liquid chromatography (RP-HPLC) in an apparatus coupled with a variable wavelength detector (VWD; Agilent1260 infinity HPLC Series, Agilent, Santa Clara, CA, USA) at a wavelength of 284 nm and fitted with a C18 column (a Kinetex®5lJm EVO C18, 106 × 4.6 mm, Phenomenex, Torrance, CA, USA) that was maintained at 35 °C [10,11]. The flow rate of the binary elution phase (A: 0.1% trifluoroacetic acid in water and B: 50% acetonitrile, 49.8% water, and 0.2% trifluoroacetic acid) was kept at 1.0 mL/min using a ternary linear elution gradient (A: 0.2% phosphoric acid, B: 100% methanol, and C: 100% acetonitrile). The measured values were expressed as µg/g of dry weight (µg/g DM) of Cd.

2.1.4. Determination of Antioxidant Activity

The antioxidant activity of the Cd methanolic extract was assessed via a radical scavenging assay using 2,2-diphenyl-1-picrylhydrazyl (DPPH) and 2,2′-azino-bis (3-ethylbenzo thiazoline-6-sulfonic acid) (ABTS)-based methods [12]. The antiradical activity of the Cd methanolic extract was determined based on its ability to scavenge the DPPH free radical. In brief, a mixture of 500 µL of the extract at various concentrations with 375 µL of ethanol and 125 µL of DPPH solution (0.02% in ethanol) was prepared. A control containing 875 µL of ethanol and 125 µL of DPPH solution was also prepared. After incubation for 60 min in the dark, the absorbance at 517 nm was measured. The antiradical activity was determined using the following formula: inhibition activity (%) of the DPPH radical = (absorbance (Abs) of the control − Abs of the sample/Abs of the sample) × 100. To determine the scavenging activity of the ABTS radical, two stock solutions were prepared as follows: 7-mM ABTS and 2.4-mM potassium persulfate. The working solution was then prepared by mixing the two stock solutions in equal quantities and allowing them to react for 12 h at room temperature in the dark. The solution was then diluted by mixing 1 mL of the ABTS solution with 60 mL of methanol to obtain an absorbance of 0.802 ± 0.005 units at 734 nm, as assessed using a spectrophotometer (T80 UV/Vis spectrometer PG Instruments LDT, Leicestershire, UK). One milliliter of the Cd methanolic extract was allowed to react with 1 mL of the ABTS solution, and the absorbance was set at 734 nm after 7 min using a spectrophotometer. The antiradical activity was determined using the following formula: inhibition activity of ABTS (%) = (absorbance (Abs) of the control–Abs of the sample/Abs of the control) × 100. The DPPH and ABTS scavenging activities of the Cd extract were compared with the scavenging activity of ascorbic acid.

2.1.5. In Vitro Antimicrobial Activity of *Cleome droserifolia* (Forssk.) Del.

The agar well diffusion method was used to determine the diameters of the inhibition zones of the Cd methanolic extract against five pathogenic strains, including *Staphylococcus aureus* NCTC 10788, *Salmonella senftenberg* ATCC 8400, *Escherichia coli* BA 12296B, *Candida albicans* ATCC MYA-2876, and *Listeria monocytogenes* ATCC 19116. Tests were performed in triplicate, and the results are presented as the mean ± standard error of the mean (SE) [13].

2.2. *In Vivo Evaluation of* Cleome droserifolia *(Forssk.) Del.*

2.2.1. Experimental Design

Thirty v-line (a maternal line selected for high litter size at weaning) male rabbits (70 days of age) weighing 1224.0 ± 19.91 g at allocation were individually placed in galvanized wire cages (40 × 50 × 35 cm^3) and housed at the rabbitry of the Laboratory of Rabbit Physiology Research, Agricultural Experimental Station, Faculty of Agriculture, Alexandria University, Alexandria, Egypt. Rabbits were kept under similar management and hygiene conditions. Rabbits were equally allocated into three groups and received the same standard diet supplemented with 0 (C group), 1.25 (Cdl group), or 2.5 (Cdh group) g of Cd shoots powder/kg of DM diet for 4 consecutive weeks (day 0: first day of the treatment and day 30: last day of the treatment). Rabbits were fed on a pellet diet containing (g/kg): 300 alfalfa hay, 230 wheat bran, 180 soybean, 180 barley, 60 yellow maize, 30 molasses, 10 NaCl 10, and $CaCo_3$ (18.90% CP and 10.25-MJ/kg digestible energy), covering their daily nutritional requirements according to NRC (1977) [14]. The values of

TPC and TFC of the standard diet were 12.81 ± 0.83-mg GA/g DM and 4.16 ± 0.22-mg CAT/g DM, respectively. The half-maximal inhibitory concentration (IC_{50}) of the standard diet was 1730 ± 3.12 µg/mL for DPPH and 1492 ± 4.21 µg/mL for ABTS.

Weight, feed consumption, and rectal temperature were recorded weekly for each rabbit. Fecal score was also recorded twice a week for each rabbit and assigned one of the following scores: 1, normal; 2, soft; 3, mixed soft and liquid; and 4, completely liquid [15].

2.2.2. Blood Plasma Hemato-Biochemical Attributes, Redox Status Indicator, and Immunological Variables

Blood samples were collected from the marginal ear vein of rabbit (n = 6/group) on experimental days 0 and 30. Each blood sample was divided into two subplots: the first subplot (whole blood) was used to assess the hematological and innate immune variables, and the second subplot was centrifuged at 2000× g for 20 min at 4 °C to obtain plasma samples. The plasma samples were stored at −20 °C pending analyses. The counts of red and white blood cells and the packed corpuscular volume were determined. A differential white blood cell count test was also performed to identify the percentage of specific white blood cells [1]. The concentrations of hemoglobin were assessed colorimetrically using commercial kits (Biosystems S.A., Costa Brava, Barcelona, Spain). Phagocytic activity (PA) and the phagocytic index (PI) were determined. A sample of whole blood was mixed (1:1) with *Staphylococcus albus* (1.0×10^5 cells/mL) in phosphate-buffered solution (PBS; pH = 7.2) and incubated for 30 min at 37 °C. A drop of the mixture was transferred to a slide, and a smear was prepared. After drying, the smear was fixed with methanol for 30 min, then processed using Levowitz-Weber staining for 2 min and washed three times with distilled water [16]. Phagocytic cells and engulfed bacteria were counted on a light microscope at a magnification of 100×, and the PA and PI were calculated as follows: PA = percentage of phagocytic cells containing bacterial cells and PI = number of bacterial phagocytosed cells/number of phagocytic cells [16].

The plasma lysozyme activity (LA) was determined using lyophilized *Micrococcus lysodekticus* as the substrate in PBS (pH = 6.4). A plasma sample of 50 µL was mixed with 3 mL of bacterial suspension. The absorbance of the mixture was measured at 570 nm twice at time 0 (directly after plasma addition; A0) and again after incubation of the mixture for 30 min (A30) at 37 °C. The LA was calculated using the following formula: LA = (A0 − A30)/A30 [17].

Blood plasma metabolites, including total protein, albumin, and glucose, were determined using commercial kits obtained from Biodiagnostics (Giza, Egypt). The linearity of the methods was up to 10.0 g/dL, 7.0 g/dL, and 500 mg/dL for the total protein content, albumin content, and glucose, respectively. The total antioxidant capacity and malondialdehyde concentration in the plasma were also determined as indicators of the antioxidant and redox status of plasma using commercial kits (Biodiagnostics, Giza, Egypt), according to the instructions of the manufacturer. The linearities of the methods were up to 120 U/mL, 1000 mg/dL, and 2 mM, respectively. The enzyme-linked immunosorbent assay (ELISA) technique was applied to assess the concentrations of immunoglobulin G (IgG), immunoglobulin E (IgE), and immunoglobulin A (IgA) (IBL America Immuno-Biological Laboratories, Inc., Spring Lake Park, MN, USA). According to the manufacturer's instructions, the sensitivity and specificity of the assays exceeded 96%. Interleukin-1β (IL-1β) was determined in the blood plasma samples (Cat. No. MBS262525, MyBioSource, Inc., San Diego, CA, USA). The lower limit of detection was 5 pg/mL, and the intra- and inter-assay precisions were ≥8% and ≥12%, respectively.

2.2.3. Intestinal and Cecal Microflora Composition

At the end of the experiment (day 30), six rabbits were randomly chosen from each group and were slaughtered [1]. The intestine and cecum were ligated with light twine before separating the cecum from the small intestine. The first part of small intestinal tract and the last part of the cecum were removed and stored in sterile bags at −4 °C. For bacterial enumeration, the intestinal and cecal contents were separately diluted 10-fold (i.e.,

10% w/v) with sterile ice-cold anoxic PBS (0.1 M; pH 7.0) and subsequently homogenized for 3 min in a stomacher. Each homogenate was serially diluted from 10–1 to 10–7. Dilutions were subsequently plated in duplicate on selective agar media for target bacterial groups, and the enumeration results were expressed as colony-forming units (cfu) log 10/g. In particular, Sabouraud Dextrose Agar for yeast counts; de Man, Rogosa and Sharpe (MRS) agar for LAB counts; MacConkey agar media for *coliform* counts; and *Salmonella* and *Shigella* agar plates for *Salmonella* counts were used. Plates were then incubated at 37 °C for 24 to 72 h [18].

The procedures and methods performed to evaluate Cd's in vitro and in vivo biological activities are shown in Figure 1.

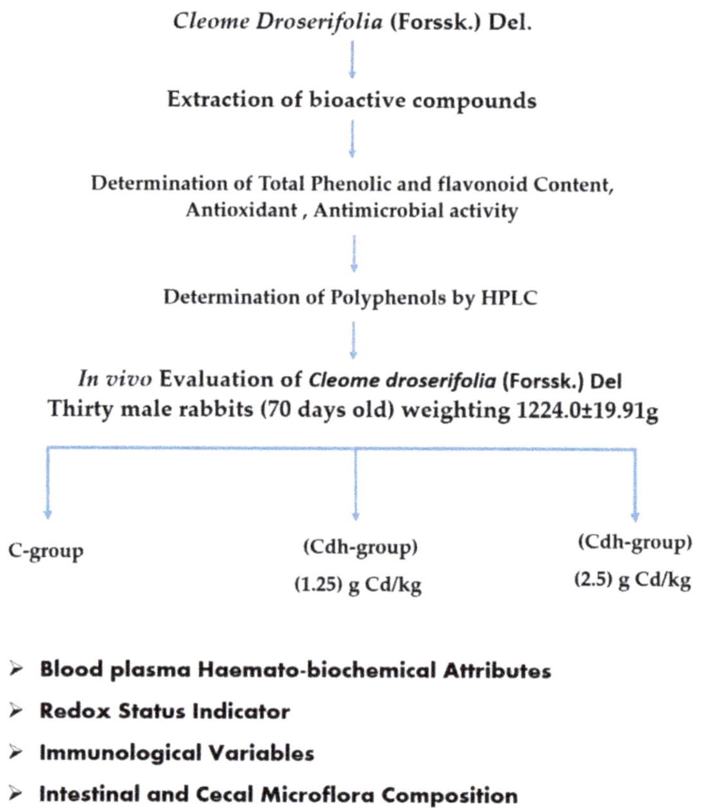

Figure 1. Flow chart of in vitro and in vivo evaluations of *Cleome droserifolia* (Forssk.) Del.

2.3. Statistical Analysis

Statistical Analysis System [19] software was used for analyzing all results. Body weight; feed intake; fecal score; rectal temperature; and hematological, biochemical, and immunological variables were analyzed by the Generalized Linear Model (GLM) method using the following model: yij = μ + Ti + eij, in which yij is the observed value of the dependent variable, μ is the overall mean, Ti is the fixed effect of the *i*th treatment, and eij is the residual error. Comparisons between treatment means were performed using Duncan's multiple range test. All results were expressed as the mean ± SE. Significance was set at $p \leq 0.05$.

3. Results

3.1. RP-HPLC Assessment of Total Phenol and Flavonoid Contents and Phenolic Compound Profile

The values of TPC and TFC of the Cd methanolic extract were 32.55- ± 2.26-mg GA/g DM and 12.78- ±1.86-mg CAT/g DM, respectively (Table 1). The phenolic profile of the Cd methanolic extract detected by RP-HPLC is shown in Table 1. These results revealed that, among the 16 phenolic compounds identified here, the most abundant phenolic compounds, ranging between 1460.62 and 7657.15 μg/g DM, were benzoic acid, rutin, ellagic acid, naringenin, and o-coumaric acid. The second-most abundant phenolic compounds, ranging between 432.14 and 264.06 μg/g DM, were rosmarinic acid, p-hydroxybenzoic acid, resveratrol, kaempferol, quercetin, and ferulic acid. The third-most abundant phenolic compounds were caffeic acid, p-coumaric acid, chlorogenic acid, catechin, syringic acid, and catechin, which were detected in low quantities, ranging between 10.43 and 59.59 μg/g DM.

Table 1. Contents of the total phenolic, total flavonoid, and individual phenolic compounds (as detected by reverse-phase high-performance liquid chromatography; RP-HPLC) in the *Cleome droserifolia* (Forssk.) Del. methanolic extract (Cd extract).

Analysis	Content
Total phenols (mean ± SE, mg GA equivalent/g DM)	32.55 ± 0.23
Total flavonoids (mean ± SE, mg CAT equivalent/g DM)	12.78 ± 0.19
Individual detected phenolic compounds (μg/g DM)	
Benzoic acid	7657.15
Rutin	2987.63
Ellagic acid	1641.98
Naringenin	1516.25
o-Coumaric acid	1460.62
Rosmarinicacid	955.27
p-Hydroxybenzoic acid	924.57
Resveratrol	895.77
Kaempferol	778.80
Quercetin	432.14
Ferulic acid	264.06
Caffeic acid	59.59
p-Coumaric acid	39.55
Chlorogenic acid	29.33
Syringic acid	19.29
Catechin	10.43

GA = gallic acid, CAT = catechol, and DM = dry matter.

3.2. Antioxidant Activity of the Cd Extract

The antiradical capacity (scavenging activity) of the Cd methanolic extract, as determined by the DPPH and ABTS colorimetric tests, is shown in Table 2. The percent inhibition values of the Cd extract were not much greater than those of the standard antioxidant (ascorbic acid). The Cd extract showed a linear increase in the DPPH and ABTS radical scavenging activities with increasing concentrations, reaching 66.09% ± 1.92% and 81.14% ± 1.26% scavenging activity for DPPH and ABTS, respectively, at concentrations of 1000 μg/mL vs. 87.52% ± 0.62% and 92.44% ± 0.14% for ascorbic acid. The half-maximal inhibitory concentration (IC_{50}) of the Cd extract was 470.27 ± 2.24 μg/mL for DPPH and 387.53 ± 3.11 μg/mL for ABTS vs. 16.62 ± 0.91 μg/mL and 14.03 ± 0.67 μg/mL for ascorbic acid, respectively.

Table 2. Antioxidant activity of the *Cleome droserifolia* (Forssk.) Del. methanolic extract (Cd extract), as assessed by the 2,2-diphenyl-1-picrylhydrazyl (DPPH) and 2,2′-azino-bis (3-ethylbenzothiazoline-6-sulfonic acid (ABTS) tests.

Antioxidant Concentration (μg/mL)	DPPH Scavenging Activity, %		ABTS Scavenging Activity, %	
	Cd Extract	Ascorbic Acid	Cd Extract	Ascorbic Acid
7.81	12.87± 0.91 [b]	62.66 ± 0.12 [a]	23.16 ± 0.76 [b]	64.58 ± 0.24 [a]
15.6	16.03 ± 0.84 [b]	75.68 ± 0.45 [a]	25.09 ± 0.84 [b]	76.10 ± 0.92 [a]
31.25	26.88 ± 0.86 [b]	77.6 ± 0.86 [a]	34.16 ± 0.92 [b]	80.21 ± 1.04 [a]
62.5	31.45 ± 1.12 [b]	79.11 ± 1.14 [a]	43.79 ± 1.16 [b]	82.30 ± 0.86 [a]
125	34.56 ± 1.24 [b]	80.20 ± 0.88 [a]	49.28 ± 1.13 [b]	85.12 ± 0.45 [a]
250	46.87 ± 1.16 [b]	83.2 ± 0.62 [a]	54.03 ± 0.76 [b]	88.07 ± 0.93 [a]
500	53.16 ± 0.85 [b]	85.4 ± 0.56 [a]	64.51 ± 0.85 [b]	89.02 ± 0.88 [a]
1000	66.09 ± 1.92 [b]	87.52 ± 0.62 [a]	81.14 ± 1.26 [b]	92.44 ± 0.14 [a]
	Half-maximal inhibitory concentration			
(IC_{50}) (μg/mL)	470.27 ± 2.24 [a]	16.62 ± 0.91 [b]	387.53 ± 3.11 [a]	14.03 ± 0.67 [b]

The mean values indicated in the same rows within variable with different superscripts (a and b) were significantly different ($p < 0.05$).

3.3. In Vitro Antimicrobial Activity

The methanolic extract of Cd exhibited striking inhibitory actions against *Staphylococcus aureus* NCTC 10788, *Salmonella senftenberg* ATCC 8400, *Escherichia coli* BA 12296, and *Candida albicans* ATCC MAY-2876 (Table 3 and Figure 2). Conversely, the Cd extract was inactive against *Lissteria monocytogenes* ATCC 19116 (Table 3 and Figure 2).

Table 3. In vitro antimicrobial activity of *Cleome droserifolia* (Forssk.) Del. against pathogenic microorganisms.

Pathogens Microorganisms	Inhibition Zone (mm)
Staphylococcus aureus NCTC 10788	15.63 ± 1.30 [a]
Salmonella senftenberg ATCC 8400	12.70 ± 0.81 [a]
Escherichia coli BA 12296	8.06 ± 1.72 [b]
Candida albicans ATCC MYA-2876	7.16 ± 2.92 [b]
Listeria monocytogenes ATCC 19116	NI

NI, no inhibitory action.

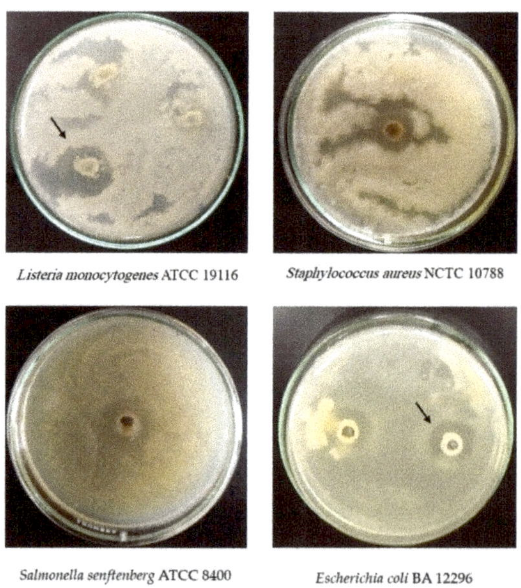

Figure 2. Antimicrobial activity of *Cleome droserifolia* (Forssk.) Del. extract against pathogenic microorganisms.

3.4. Effect of Treatment on Weight, Feed Intake, and Health Indicators in Rabbits

The treatments with different concentrations of Cd (0, 1.25, or 2.5 g/kg of DM diet) did not affect the overall mean body weight and feed intake of rabbits during the 30-day experimental period (Table 4). The treatment with Cdh tended ($p < 0.085$) to decrease the fecal score compared with the other treatments (Table 4). Compared with the control, the two concentrations of Cd decreased significantly the overall mean rectal temperature (Table 4).

Table 4. Body weight, feed intake, fecal score, and rectal temperature of rabbits treated with different concentrations of *Cleome droserifolia* (Forssk.) Del. (Cd) (0: C, Cdl: 1.25 g/kg of DM diet, or Cdh: 2.5 g/kg of DM diet).

Treatment	Variable (Mean± Standard Error of the Mean, n = 10/Treatment)			
	Body Weight, g	**Feed Intake, g/day**	**Fecal Score**	**Rectal Temperature, °C**
C	1454 ± 37.01	99.47 ± 17.36	1.19 ± 0.083	39.07 ± 0.112 [a]
Cdl	1413 ± 33.75	100.06 ± 15.88	1.21 ± 0.073	38.80 ± 0.091 [b]
Cdh	1393 ± 37.82	99.14 ± 17.31	1.13 ± 0.063	38.74 ± 0.123 [b]
p-Value	0.478	0.968	0.085	0.007

The mean values indicated in the same columns with different superscripts (a and b) were significantly different ($p < 0.05$).

3.5. Effect of Treatment on Hemato-Chemistry and Redox Status

The hematological attributes, blood plasma metabolites, and antioxidant activity of rabbits treated with different concentrations of Cd (0, 1.25, or 2.5 g/kg of DM diet) are shown in Table 5. No differences were observed for any of the variables at day 0, confirming the homogeneity of the experimental groups before the beginning of the treatment. At day 30 (the end of the experimental period), the treatment had not affected the hematological attributes or blood plasma metabolites. However, both concentrations of Cd significantly increased the levels of the total antioxidant activity and significantly decreased the levels of malondialdehyde in the blood plasma.

Table 5. Hematological attributes, blood plasma metabolites, and antioxidant activity of rabbits treated with different concentrations of *Cleome droserifolia* (Forssk.) Del. (Cd) (0: C, Cdl: 1.25 g/kg of DM diet, or Cdh: 2.5 g/kg of DM diet).

Treatment	Variable (Mean± Standard Error of the Mean, n = 6)							
	Red Blood Cell Count (10^6/mL)	**Packed Cell Volume(%)**	**Hemoglobin, g/dL**	**Total Protein, g/dL**	**Albumin, g/dL**	**Glucose, mg/dL**	**Total Antioxidant Capacity, Mm/L**	**Malondialdehyde, nmol/mL**
At day 0								
C	6.31 ± 1.01	32.67 ± 2.45	10.16 ± 0.32	6.34 ± 0.15	4.37 ± 0.11	93.61 ± 2.27	492.75 ± 0.53	5.23 ± 0.43
Cdl	5.85 ± 1.09	33.05 ± 3.45	10.79 ± 0.58	6.72 ± 0.19	4.01 ± 0.07	91.51 ± 1.15	425.40 ± 1.64	4.92 ± 0.06
Cdh	6.36 ± 0.97	34.45 ± 3.47	10.58 ± 0.79	6.65 ± 0.27	4.08 ± 0.12	93.31 ± 1.24	430.23 ± 1.59	5.20 ± 0.13
p-Value	0.764	0.947	0.764	0.742	0.641	0.369	0.4752	0.379
At day 30								
C	5.94 ± 1.21	31.35 ± 3.72	10.91 ± 0.67	6.44 ± 0.23	4.37 ± 0.24	91.92 ± 1.01	440.40 ± 0.30 [b]	4.19 ± 0.25 [a]
Cdl	5.85 ± 0.98	30.37 ± 1.99	10.91 ± 0.37	6.28 ± 0.24	4.59 ± 0.06	91.56 ± 1.62	444.09 ± 0.95 [a]	3.83 ± 0.04 [b]
Cdh	5.61 ± 1.23	30.01± 2.01	9.86± 0.34	6.14 ± 0.11	4.38 ± 0.15	92.41 ± 0.97	443.37 ± 0.92 [a]	3.73 ± 0.04 [b]
p-Value	0.967	0.641	0.143	0.281	0.287	0.258	0.034	0.002

Mean values indicated in the same columns with different superscripts (a and b) were significantly different ($p < 0.05$).

3.6. Effect of Treatment on Immune Indicators

3.6.1. Innate Immune System

The innate immune indicators of rabbits treated with different concentrations of Cd (0, 1.25, or 2.5 g/kg of DM diet) are shown in Table 6. No differences were observed for any of the variables at day 0, confirming the homogeneity of the experimental groups before the beginning of the treatment. At day 30 (the end of the experimental period), the treatment had not affected the white blood cell count/differential count, PI, or PA. The treatment with Cdh significantly increased the blood plasma lysozyme activity compared with the C and Cdl treatments. Moreover, the treatment with Cdh significantly decreased the levels of interleukin-β1 in the blood plasma compared with the C treatment, whereas Cdl yielded an intermediate value.

Table 6. Innate immune indicators of rabbits treated with different concentrations of *Cleome droserifolia* (Forssk.) Del. (Cd) (0: C, Cdl: 1.25 g/kg of DM diet, or Cdh: 2.5 g/kg of DM diet).

Treatment	Variable (Mean± Standard Error of the Mean, n = 6)								
	White Blood Cells, 10^3/mm^3	**Lymphocytes, %**	**Neutrocytes, %**	**Echinocytes, %**	**Monocytes, %**	**Phagocytic Index**	**Phagocytic Activity, %**	**Lysozyme Activity, U/mL**	**Interleukin-β1, pg/mL**
At day 0									
C	7.29 ± 1.26	39.90 ±1.28	38.85 ± 2.33	12.49 ± 0.78	13.18 ± 2.00	1.94 ± 0.27	24.90 ± 1.24	0.113 ± 0.37	16.91 ± 0.34
Cdl	6.47 ± 0.88	38.88 ± 1.91	33.75 ± 5.29	10.79 ± 1.18	11.65 ± 2.61	2.04 ± 0.13	19.39 ± 0.80	0.092 ± 0.01	15.21 ± 0.72
Cdh	6.33 ± 1.42	38.71 ± 1.99	37.59 ± 3.32	10.12 ± 1.09	12.91 ± 0.69	1.96 ± 0.41	20.95 ± 0.12	0.101 ± 0.01	15.74 ± 0.82
p-Value	0.560	0.240	0.338	0.327	0.679	0.804	0.258		0.175
At day 30									
C	6.49 ± 0.84	39.56 ± 1.32	32.69 ± 1.35	11.74 ± 0.52	13.22 ± 1.20	2.10 ± 0.35	20.56 ± 1.63	0.104 ± 0.02 ^{b}	18.66 ± 0.22 ^{a}
Cdl	6.33 ± 0.56	42.01 ± 1.68	33.65 ± 3.05	10.22 ± 0.65	11.28 ± 1.53	2.19 ± 0.54	20.63 ± 1.01	0.106 ± 0.12 ^{b}	17.01 ± 0.81 ^{ab}
Cdh	6.02 ± 1.40	44.52 ± 1.21	37.63 ± 1.92	11.51 ± 0.89	10.97 ± 1.37	2.49 ± 0.24	21.2 ± 2.01	0.142 ± 0.01 ^{a}	15.25 ± 0.92 ^{b}
p-Value	0.449	0.123	0.236	0.531	0.195	0.446	0.561	0.046	0.001

Mean values indicated in the same columns with different superscripts (a and b) were significantly different ($p < 0.05$).

3.6.2. Humoral Immune System

The humoral immune indicators of rabbits treated with different concentrations of Cd (0, 1.25, or 2.5 g/kg of DM diet) are shown in Table 7. No differences were observed for any of the variables at day 0, confirming the homogeneity of the experimental groups before the beginning of the treatment. At day 30 (the end of the experimental period), the two concentrations of Cd had significantly increased the levels of IgG in the blood plasma compared with the control. Conversely, the treatments did not affect the levels of IgA and IgE in the blood plasma.

Table 7. Humoral immune indicators (immunoglobulins (Igs)) of rabbits treated with different concentrations of *Cleome droserifolia* (Forssk.) Del. (Cd) (0: C, Cdl: 1.25 g/kg of DM diet, or Cdh: 2.5 g/kg of DM diet).

Treatment	Variable (Mean± Standard Error of the Mean, n = 6)		
	IgG, mg/dL	**IgA, mg/dL**	**IgE, mg/dL**
At day 0		-	
C	981.32 ± 6.65	84.77 ± 2.68	7.73 ± 1.35
Cdl	989.90 ± 10.41	85.79 ± 4.82	6.69 ± 0.67
Cdh	985.74 ± 8.83	88.47 ± 3.45	7.99 ± 0.49
p-Value	0.516	0.329	0.1602
At day 30		-	
C	974.57 ± 3.84 ^{b}	91.78 ± 2.39	7.99 ± 0.78
Cdl	987.91 ± 6.01 ^{a}	93.86 ± 2.78	6.88 ± 0.38
Cdh	982.99 ± 7.48 ^{a}	93.06 ± 4.26	7.73 ± 0.28
p-Value	0.016	0.647	0.359

Mean values indicated in the same columns with different superscripts (a and b) were significantly different ($p < 0.05$). IgG, immunoglobulin G; IgE, immunoglobulin E; and IgA, immunoglobulin A.

3.7. Intestinal and Cecal Microflora Composition

The gastrointestinal (small intestine and cecum) microflora composition of rabbits treated with different concentrations of Cd (0, 1.25, or 2.5 g/kg of the DM diet) is shown in Table 8. At day 30 (the end of the experimental period), the counts of intestine and cecum *Salmonella* and *Coliform* species were significantly reduced in the Cd-treated groups compared with the control group. The two concentrations of Cd significantly increased the counts of intestinal and cecal yeast and Lactobacillus species compared with the control.

Table 8. Small intestinal and cecal microflora composition of rabbits treated with different concentrations of *Cleome droserifolia* (Forssk.) Del. (Cd) (0: C, Cdl: 1.25 g/kg of DM diet, or Cdh: 2.5 g/kg of DM diet).

Treatment	Variable (Mean± Standard Error of the Mean, $n = 6$)			
	Yeast	Lactobacillus	*Salmonella*	*Coliform*
	Intestinal microflora (log cfu/g)			
C	4.83 ± 0.65 [b]	6.80 ± 0.91 [a]	5.96 ± 0.55 [a]	6.30 ± 0.70 [a]
Cdl	7.60 ± 0.52 [a]	8.10 ± 0.94 [a]	3.10 ± 0.65 [b]	4.86 ± 0.77 [a]
Cdh	8.06 ± 0.66 [a]	8.06 ± 0.70 [a]	3.13 ± 0.85 [b]	4.83 ± 0.85 [a]
	Cecal microflora (log cfu/g)			
C	3.56 ± 0.81 [b]	5.40 ± 0.55 [a]	7.63 ± 0.86 [a]	8.13 ± 0.61 [a]
Cdl	5.60 ± 1.13 [a]	6.57 ± 1.70 [a]	5.50 ± 0.45 [b]	6.06 ± 1.30 [b]
Cdh	5.27 ± 0.83 [ab]	6.93 ± 1.53 [a]	5.34 ± 0.67 [b]	6.20 ± 0.79 [b]

Mean values indicated in the same rows with different superscripts (a and b) were significantly different ($p < 0.05$).

4. Discussion

The literature on natural antioxidant utilization as stabilizing and protecting agents for biological macromolecular components is vast. This covers many aspects of their activity against, mainly, reactive oxygen species (ROS), whereas other aspects are less-known or totally ignored [1,12,20]. Despite their remarkable potential for commercial exploitation, species in the *Cleome* genus have attracted interest, and they are currently used as folk medicine for treating stomachaches, cancer, and liver disorders [7,21]. In this study, the values of TPC and TFC obtained for a Cd methanolic extract were close and/or higher than those documented in the literature. For example, the value of TPC obtained in our study was higher than that reported by Aicha et al. [22] in the leaves of Algerian *Cleome L.* varieties (TPC = 35.17-mg GA equivalent/g extract and TFC = 11.35-mg rutin equivalent equivalent/g extract). In another study [21], the TPC was 2.38 mg GA equivalent/g of dried plant extract. It is well-known that the *Cleome* species are an excellent source of phenolic compounds; however, the variations of the TPC and TFC values among the studied samples could be related to many factors, such as the nature of the agro-ecological zones (soil and fertilizers), plant parameters (plant parts and growth phases), extraction method, and assay procedures [23,24].

The antiradical scavenging activity (antioxidant potential) of the Cd methanolic extract was assessed using two colorimetric tests (DPPH and ABTS). Both tests confirmed the strong antioxidant activity of the Cd methanolic extract. These findings are consistent with those of previous studies (DPPH method: reference [25] and ABTS method: reference [26]). In our study, the IC_{50} value of the Cd methanolic extract was less than that of ascorbic acid by about 29-fold. In another study, this difference was only two-fold when the IC_{50} of the extract was compared with that of CAT, which was used as the standard antioxidant [25,27]. In general, these differences in IC_{50} values can be mainly ascribed to variations in the selection of endpoints, the expression of results even within the same method, and the standard antioxidant used. Therefore, comparisons between the values quantified by different laboratories can be quite difficult [28,29].

The strong activity of the Cd extract observed here may be attributed to the presence of phenolic and flavonoid compounds, which are known for their antioxidant activity. Phenolic compounds—in particular, flavonoids and phenolic acids—are able to directly scavenge ROS, such as superoxide anion radicals (O_{2-}) and hydroxyl radicals ($OH-$). They are also able to enhance the expression and activity of antioxidant enzymes via different pathways, such as the nuclear factor erythroid 2-related factor 2 signaling pathway [23,30].

The RP-HPLC analysis performed in our study confirmed the presence of an array of phenolic compounds in the Cd methanolic extract, i.e., phenolic acids, both hydroxybenzoic acid and hydroxycinnamic acid derivatives, and flavonoids. These findings are in line with those obtained by El-Askary et al. (2019) [4], who detected 20 different phenolic compounds in a water extract of Cd, in which phenolic acids (caffeoyl and feruloylquinic acid derivatives) were the major components. Previous studies have reported the presence of numerous active secondary metabolites in *Cleome* species, including phenolic compounds, terpenes, glucosinolates, tannins, and steroids, with different biological activities [3,8,21]. The results of the in vitro antimicrobial activity obtained in our study confirmed the remarkable antimicrobial activity of the Cd methanolic extract against the *Staphylococcus aureus* NCTC 10788, *Salmonella senftenberg* ATCC 8400, *Escherichia coli* BA 12296 *B*, and *Candida albicans* ATCC MYA-2876 pathogen species. The antibacterial activity of the secondary metabolites of the *Cleome* species against both Gram-positive and Gram-negative bacteria has been reported in previous studies [12,31,32].

In this study, the effects of the inclusion of a powder of Cd shoots, to assess its active secondary metabolites, on the health of animals were evaluated using rabbits as a model. Overall, no negative effects of the inclusion of the powder of Cd shoots in the diets of rabbits were observed regarding the hematological parameters, protein and glucose metabolism, and feed intakes. Moreover, all of these variables were in the normal physiological range reported for rabbits. Blood variables can be analyzed to indicate the animal health status and to aid in detecting different nutritional, environmental, or physical stresses [1]. Moreover, linking in vivo results with the antioxidant properties and in vitro antimicrobial activities of the Cd methanolic extract supports the biological activity of the secondary metabolites detected in the Cd methanolic extract. Rabbits that were fed Cd-containing diets had a better redox status and intestinal and cecal microbial homeostasis (lower pathogenic microbes and higher beneficial microbes) than those that were fed the control diet. The antimicrobial activity of the Cd methanolic extract against pathogen species may be related to the presence of many phenolic compounds with antioxidant and antimicrobial activities. Interestingly, the major phenolic compound detected in the Cd methanolic extract was benzoic acid. This phenolic acid and its derivatives can exert antioxidant effects against various types of ROS by reducing their overproduction [33]. These components also possess antibacterial and antifungal proprieties by inhibiting the microbial active uptake of several essential amino acids [34]. Based on such properties, Cd supplements could be used as an adjuvant in treating many oxidative stress-induced diseases without any detected harm. Moreover, the naringenin, rutin, o-coumaric acid, and ellagic acid components detected in the Cd methanolic extract exhibited strong antioxidant and antimicrobial activities [12,31,32,35–38]. According to the results obtained for the small intestinal and cecal microflora composition, we suggest that the phenolic compounds of Cd can inhibit the growth of pathogenic bacteria (*Salmonella* and *Coliform* species) while stimulating the growth of beneficial microbes (yeast and Lactobacillus species) among the intestinal and cecal microbiota in rabbits, thus optimizing the intestinal microbiota ecosystem. Such enhancements in the intestinal microbiota ecosystem can improve the immune status and digestive health of rabbits.

It is worth noting that the inclusion of a powder of Cd shoots in the diet of rabbits yielded several immunomodulatory effects. These effects occurred mainly through the improvement of the innate immune system, the increase in lysozyme activity, the decrease in the production of the proinflammatory cytokine IL-β1, the improvement of the humoral immune system, and the increase in IgG levels. The immunomodulatory effects of the phenolic compounds of the Cd extract, such as rutin, quercetin, kaempferol, and phenolic acids, have been reported in several studies and depend on many factors, such as the bioavailability and chemical structure of the component [39]. Interestingly, many of the phenolic compounds detected in the Cd methanolic extract are known for their bioavailability because of their increased intestinal absorbance ability. For example, Manach et al. [40] suggested that GA and isoflavones, catechins, flavanones, and quercetin glucosides are among the most well-absorbed phenolic compounds, whereas the least well-absorbed compounds are proanthocyanidins and anthocyanins. In this context, ellagic acid, which was detected in abundance in the Cd methanolic extract, has been found to significantly increase the serum IgM and IgG levels, whereas both IgA and IgE remain unchanged [41]. This phenolic acid also exerted an inhibitory effect on IL-1b secretion in ex vivo and in vivo experiments [41]. In another study, the IgG response was increased after a treatment with a pomegranate extract rich in polyphenols (16.9% GA equivalent/day in calves) [42]. As observed here, Cd-treated rabbits had lower rectal temperatures and gastrointestinal microflora homeostasis, which could be attributed to the enhancement of the immune system function. The levels of the inflammatory factors can be increased as a result of a pathogenic infection and are often associated with elevated body temperatures (rectal temperatures) [43]. In our study, the inclusion of Cd shoots powder in the diets of rabbits decreased the numbers of intestinal and cecal pathogenic bacteria (Salmonella and Coliform). This finding might have explained the decrease in rectal temperature in Cd-treated rabbits. Moreover, the improved lysozyme activity may contribute to the elimination of pathogens because of its enzymatic degradative potential [17]. Finally, increased IgG levels can improve the health of animals in the long term, as these antibodies are responsible for long-term immunological memory [39].

5. Conclusions

The results of the present study indicate the impressive range of active phenolic compounds of the shoots of Cd shrub with a multifunctional biological activity. The Cd shrub exhibited strong antioxidant and antimicrobial activities, which were confirmed in vitro and in vivo in our study. These results suggest the possibility of using Cd as an antimicrobial and antioxidant agent. Moreover, this shrub has positive immunomodulatory effects. According to our results, the positive effects of Cd shoots powder on the health status of rabbits can be obtained in vivo at a level of 1.25-mg Cd/kg DM diet. Prospective studies are needed to discover the bioactive natural components of the Cd extract and their specific biological activities.

Author Contributions: Conceptualization, N.M.H.; methodology, N.M.H. and M.G.S.; investigation, N.M.H. and M.G.S.; resources, N.M.H. and M.G.S.; data curation, N.M.H. and M.G.S.; writing—original draft preparation, N.M.H. and M.G.S.; writing—review and editing, N.M.H.; and visualization, N.M.H. and M.G.S. All authors have read and agreed to the published version of the manuscript.

Funding: This research received no external funding.

Institutional Review Board Statement: The study was conducted according to the guidelines of the Declaration of Pharmaceutical & Fermentation Industries Development Center and approved by the Institutional Animal Care and Use Committee (IAUCU) of SRTA-city (protocol code #31-1Z-0521).

Data Availability Statement: Data of this study are confidential.

Conflicts of Interest: The authors declare no conflict of interest.

References

1. Hashem, N.; Soltan, Y.; El-Desoky, N.; Morsy, A.; Sallam, S. Effects of Moringaoleifera extracts and monensin on performance of growing rabbits. *Livest. Sci.* **2019**, *228*, 136–143. [CrossRef]
2. Awad, O.M.E.; El Sohaimy, S.; Ghareeb, D.A.; Aboulenein, A.; Saleh, S.R.; El-Aziz, N.M.A. Phytochemical analysis and toxicity assessment of artichoke by-product extract. *Pak. J. Biol. Sci.* **2019**, *23*, 81–91. [CrossRef] [PubMed]
3. Moustafa, A.; Sarah, R.; Qiqa, S.; Mansour, S.; Alotaibi, M. *Cleome droserifolia*: An Egyptian natural heritage facing extinction. *Asian J. Plant Sci. Res.* **2019**, *9*, 14–21.
4. El-Askary, H.; Handoussa, H.; Badria, F.; El-Khatib, A.H.; Alsayari, A.; Linscheid, M.W.; Motaal, A.A. Characterization of hepatoprotective metabolites from *Artemisia annua* and *Cleome droserifolia* using HPLC/PDA/ESI/MS–MS. *Rev. Bras.Farm.* **2019**, *29*, 213–220. [CrossRef]
5. Abdel-Kader, M.S.; Alqasoumi, S.I.; Al-Taweel, A.M. Hepatoprotective constituents from Cleome droserifolia. *Chem. Pharm. Bull.* **2009**, *57*, 620–624. [CrossRef]
6. Ezzat, S.M.; Motaal, A.A. Isolation of new cytotoxic metabolites from Cleome droserifolia growing in Egypt. *Z. FürNat. C* **2012**, *67*, 266–274. [CrossRef]
7. Maksoud, H.A.; Zaid, O.A.A.; Elharrif, M.G.; Omnia, M.; Alaa, E. Selenium *Cleome droserifolia* nanoparticles (Se-CNPs) and it's ameliorative effects in experimentally induced diabetes mellitus. *Clin. Nutr. ESPEN* **2020**, *40*, 383–391. [CrossRef] [PubMed]
8. Motaal, A.; Ezzat, S.; El-askary, H. Antihyperglycemic Activity and standardization of the bioactive extract of *Ceome droserifolia* growing in Egypt. *Pharmacogn. J.* **2014**, *6*, 15–21. [CrossRef]

9. Shehata, M.G.; Ahmad, F.T.; Badr, A.N.; Masry, S.H.; El-Sohaimy, S.A. Chemical analysis, antioxidant, cytotoxic and antimicrobial properties of propolis from different geographic regions. *Ann. Agric. Sci.* **2020**, *65*, 209–217. [CrossRef]
10. Ma, X.; Laaksonen, O.; Zheng, J.; Yang, W.; Trépanier, M.; Kallio, H.; Yang, B. Flavonol glycosides in berries of two major subspecies of sea buckthorn (*Hippophaë rhamnoides* L.) and influence of growth sites. *Food Chem.* **2016**, *200*, 189–198. [CrossRef] [PubMed]
11. Teleszko, M.; Wojdyło, A.; Rudzińska, M.; Oszmiański, J.; Golis, T. Analysis of lipophilic and hydrophilic bioactive compounds content in sea buckthorn (*Hippophaë rhamnoides* L.) Berries. *J. Agric. Food Chem.* **2015**, *63*, 4120–4129. [CrossRef]
12. Shehata, M.G.; Abu-Serie, M.M.; El-Aziz, N.M.A.; El-Sohaimy, S.A. Nutritional, phytochemical, and in vitro anticancer potential of sugar apple (Annona squamosa) fruits. *Sci. Rep.* **2021**, *11*, 1–13. [CrossRef]
13. Panda, S.K.; Mohanta, Y.K.; Padhi, L.; Park, Y.-H.; Mohanta, T.K.; Bae, H. Large scale screening of ethnomedicinal plants for identification of potential antibacterial compounds. *Molecules* **2016**, *21*, 293. [CrossRef]
14. National Research Council (NRC). *Nutrient Requirements of Rabbits*, 2nd ed.; National Academy Press: Washington, DC, USA, 1977; p. 24.
15. Phuoc, T.L.; Jamikorn, U. Effects of probiotic supplement (Bacillus subtilis and Lactobacillus acidophilus) on feed efficiency, growth performance, and microbial population of weaning rabbits. *Asian-Australas. J. Anim. Sci.* **2016**, *30*, 198–205. [CrossRef] [PubMed]
16. Calder, P.C. Immunological parameters: What do they mean? *J. Nutr.* **2007**, *137*, 773S–780S. [CrossRef] [PubMed]
17. Ishii, S.; Matsuura, A.; Itakura, E. Identification of a factor controlling lysosomal homeostasis using a novel lysosomal trafficking probe. *Sci. Rep.* **2019**, *9*, 1–12. [CrossRef]
18. Pourakbari, M.; Seidavi, A.; Asadpour, L.; Martínez, A. Probiotic level effects on growth performance, carcass traits, blood parameters, cecal microbiota, and immune response of broilers. *An. Acad. Bras. Ciênc.* **2016**, *88*, 1011–1021. [CrossRef]
19. Lovie, P. Book Review: Everitt, B.S.; Dunn, G. Applied Multivariate Data Analysis. London: Edward Arnold. 316pp. £29.95 PB. ISBN 0 340 54529 1. *Stat. Methods Med. Res.* **1993**, *2*, 114–115. [CrossRef]
20. Badr, A.N.; Abdel-Razek, A.G.; Youssef, M.M.; Shehata, M.G.; Hassanein, M.M.; Amra, H.A. Natural antioxidants: Preservation roles and mycotoxicological safety of food. *Egypt. J. Chem.* **2020**, *64*, 285–298. [CrossRef]
21. Yasmin, F.; Yusoff, N.A.; Hossain, A. Phytochemical screening and pharmacological activities of the ethanolic stem extract of *Cleome gynandra*. *Pak. J. Nutr.* **2020**, *19*, 153–159. [CrossRef]
22. Aicha, M.; Nadia, Z.; Sihem, H.; Abdelmalik, B. Antioxidant activity and phenolic compounds contents of spider flower (*Cleome Arabica* ssp. *Arabica*), a well acclimated species in the Algerian desert areas. *Eur. Sci. J. ESJ* **2017**, *13*, 13. [CrossRef]
23. Hashem, N.M.; Gonzalez-Bulnes, A.; Simal-Gandara, J. Polyphenols in farm animals: Source of reproductive gain or waste? *Antioxidants* **2020**, *9*, 1023. [CrossRef]
24. Badr, A.N.; Gromadzka, K.; Shehata, M.G.; Stuper-Szablewska, K.; Drzewiecka, K.; Abdel-Razek, A.G.; Youssef, M.M. Encapsulated bioactive ingredients of grape by-products applicate in fresh-cut fruit and juices diminished the ochratoxins. *J. Food Process. Preserv.* **2021**, *45*. [CrossRef]
25. El-Gawad, A.M.A.; El-Amier, Y.A.; Bonanomi, G. Essential oil composition, antioxidant and allelopathic activities *of Cleome droserifolia* (Forssk.) Delile. *Chem. Biodivers.* **2018**, *15*, e1800392. [CrossRef] [PubMed]
26. Badr, A.; Ali, H.; Abdel-Razek, A.; Shehata, M.; Albaridi, N. Bioactive components of pomegranate oil and their influence on mycotoxin secretion. *Toxins* **2020**, *12*, 748. [CrossRef]
27. Ismael, N.M.M.; Shehata, M.G. Improvement of lipid profile and antioxidant of hyperlipidemic albino rats by functional Plantago psyllium Cake. *Curr. Res. Nutr. Food Sci. J.* **2020**, *8*, 424–437. [CrossRef]
28. Harzallah, A.; Bhouri, A.M.; Amri, Z.; Soltana, H.; Hammami, M. Phytochemical content and antioxidant activity of different fruit parts juices of three figs (*Ficuscarica L.*) varieties grown in Tunisia. *Ind. Crop. Prod.* **2016**, *83*, 255–267. [CrossRef]
29. Shehata, M.G.; Darwish, A.; El Sohaimy, S. Physicochemical, structural and functional properties of water-soluble polysaccharides extracted from Egyptian agricultural by-products. *Ann. Agric. Sci.* **2020**, *65*, 21–27. [CrossRef]
30. Hashem, N.; Hassanein, E.; Simal-Gandara, J. Improving reproductive performance and health of mammals using honeybee products. *Antioxidants* **2021**, *10*, 336. [CrossRef]
31. Muhaidat, R.; Al-Qudah, M.A.; Samir, O.; Jacob, J.H.; Hussein, E.; Al-Tarawneh, I.N.; Bsoul, E.; Orabi, S.T.A. Phytochemical investigation and in vitro antibacterial activity of essential oils from *Cleome droserifolia* (Forssk.) Delile and *C. trinervia* Fresen. (Cleomaceae). *S. Afr. J. Bot.* **2015**, *99*, 21–28. [CrossRef]
32. Al-Mugdadi, S.F.H.; Al-Sudani, B.; Mohsin, R.A.; Mjali, A.J. Anticarcinogenic and antimicrobial activity effects of the ellagic acid extract. *Int. J. Res. Pharm. Sci.* **2019**, *10*, 1172–1180. [CrossRef]
33. Velika, B.; Kron, I. Antioxidant properties of benzoic acid derivatives against superoxide radical. *Free Radic. Antioxid.* **2012**, *2*, 62–67. [CrossRef]
34. Koul, B.; Chase, N. Moringa oleifera Lam: Panacea to several maladies. *J. Chem. Pharm. Res.* **2015**, *7*, 687–707.
35. Park, E.-S.; Moon, W.-S.; Song, M.-J.; Kim, M.-N.; Chung, K.-H.; Yoon, J.-S. Antimicrobial activity of phenol and benzoic acid derivatives. *Int. Biodeterior. Biodegrad.* **2001**, *47*, 209–214. [CrossRef]
36. Arima, H.; Ashida, H.; Danno, G.-I. Rutin-Enhanced antibacterial activities of flavonoids against *Bacillus cereus* and *Salmonella enteritidis*. *Biosci. Biotechnol. Biochem.* **2002**, *66*, 1009–1014. [CrossRef]

37. Céliz, G.; Daz, M.; Audisio, M. Antibacterial activity of naringin derivatives against pathogenic strains. *J. Appl. Microbiol.* **2011**, *111*, 731–738. [CrossRef]
38. Bassolé, I.H.N.; Juliani, H.R. Essential oils in combination and their antimicrobial properties. *Molecules* **2012**, *17*, 3989–4006. [CrossRef] [PubMed]
39. Chen, L.; Wei, Y.; Zhao, S.; Zhang, M.; Yan, X.; Gao, X.; Li, J.; Gao, Y.; Zhang, A.; Gao, Y. Antitumor and immunomodulatory activities of total flavonoids extract from persimmon leaves in H22 liver tumor-bearing mice. *Sci. Rep.* **2018**, *8*, 10523. [CrossRef]
40. Manach, C.; Williamson, G.; Morand, C.; Scalbert, A.; Rémésy, C. Bioavailability and bioefficacy of polyphenols in humans. I. Review of 97 bioavailability studies. *Am. J. Clin. Nutr.* **2005**, *81*, 230S–242S. [CrossRef] [PubMed]
41. Allam, G.; Abuelsaad, A.; Alblihed, M.A.; Alsulaimani, A.A. Ellagic acid reduces murine schistosomiasis mansoni immunopathology via up-regulation of IL-10 and down-modulation of pro-inflammatory cytokines production. *Immunopharmacol. Immunotoxicol.* **2016**, *38*, 286–297. [CrossRef]
42. Oliveira, R.; Narciso, C.; Bisinotto, R.; Perdomo, M.; Ballou, M.; Dreher, M.; Santos, J. Effects of feeding polyphenols from pomegranate extract on health, growth, nutrient digestion, and immunocompetence of calves. *J. Dairy Sci.* **2010**, *93*, 4280–4291. [CrossRef] [PubMed]
43. Tesch, T.; Bannert, E.; Kluess, J.; Frahm, J.; Hüther, L.; Kersten, S.; Breves, G.; Renner, L.; Kahlert, S.; Rothkötter, H.-J.; et al. Relationships between body temperatures and inflammation indicators under physiological and pathophysiological conditions in pigs exposed to systemic lipopolysaccharide and dietary deoxynivalenol. *J. Anim. Physiol. Anim. Nutr.* **2017**, *102*, 241–251. [CrossRef]

animals

MDPI

Article

Alleviating Effects of Vitamins C and E Supplementation on Oxidative Stress, Hematobiochemical, and Histopathological Alterations Caused by Copper Toxicity in Broiler Chickens

Mohamed A. Hashem [1], Sahar S. Abd El Hamied [2], Eman M. A. Ahmed [2], Shimaa A. Amer [3,*] and Aziza M. Hassan [4]

1 Clinical Pathology Department, Faculty of Veterinary Medicine, Zagazig University, Zagazig 44511, Egypt; mhashem.vet@gmail.com

2 Animal Health Institute, Zagazig Branch, Zagazig 44511, Egypt; sahar_elmasry91@yahoo.com (S.S.A.E.H.); emanalilogy@gmail.com (E.M.A.A.)

3 Department of Nutrition & Clinical Nutrition, Faculty of Veterinary Medicine, Zagazig University, Zagazig 44511, Egypt

4 Department of Biotechnology, College of Science, Taif University, P.O. Box 11099, Taif 21944, Saudi Arabia; a.hasn@tu.edu.sa

* Correspondence: shimaa.amer@zu.edu.eg

Simple Summary: Excessive copper in diets is associated with numerous disadvantageous impacts on poultry. The current study evaluated the efficacy of vitamin C and vitamin E in mitigating oxidative stress, hematobiochemical, and histopathological changes in the kidney induced by copper sulfate ($CuSO_4$) toxicity in broiler chickens. The birds were assigned to five experimental groups: 1st group—basal diet with no additives (control group), 2nd group—basal diet complemented with $CuSO_4$ (300 mg/kg diet), 3rd group—basal diet with $CuSO_4$ (300 mg/kg diet) + vitamin C (250 mg/kg diet), 4th group—basal diet with $CuSO_4$ (300 mg/kg diet) + vitamin E (250 mg/kg diet), and 5th group—basal diet with $CuSO_4$ (300 mg/kg diet) + vitamin C (250 mg/kg diet) + vitamin E (250 mg/kg diet). The current study's findings showed the possible preventive impacts of dietary antioxidants on hematobiochemical alterations, oxidative stress, and kidney damage induced by $CuSO_4$ toxicity.

Abstract: The current investigation evaluated the alleviating effects of vitamin C and vitamin E on oxidative stress, hematobiochemical, and histopathological changes in the kidney induced by copper sulfate ($CuSO_4$) toxicity in chickens. Two hundred and fifty-one-day-old male broiler chicks were randomly allotted into five experimental groups (five replicates/group, ten chicks/replicate): 1st group—basal diet with no additives (control group), 2nd group—basal diet complemented with $CuSO_4$ (300 mg/kg diet), 3rd group—basal diet with $CuSO_4$ (300 mg/kg diet) + vitamin C (250 mg/kg diet), 4th group—basal diet with $CuSO_4$ (300 mg/kg diet) + vitamin E (250 mg/kg diet), and 5th group—basal diet with $CuSO_4$ (300 mg/kg diet) + vitamin C (250 mg/kg diet) + vitamin E (250 mg/kg diet) for a 42 day feeding period. The results showed a significant reduction in red blood cells (RBCs), hemoglobin (Hb) concentration, and hematocrit values as well as total leukocyte counts (WBCs), lymphocyte, heterophil, and monocyte counts in the $CuSO_4$-intoxicated birds ($2.42 \times 10^6/\mu L$, 9.54 g/dL, 26.02%, $15.80 \times 10^3/\mu L$, $7.86 \times 10^3/\mu L$, $5.26 \times 10^3/\mu L$, and $1.18 \times 10^3/\mu L$, respectively, at the 6th week) compared to ($2.79 \times 10^6/\mu L$, 10.98 g/dL, 28.46%, $21.07 \times 10^3/\mu L$, $10.84 \times 10^3/\mu L$, $7.12 \times 10^3/\mu L$, and $1.60 \times 10^3/\mu L$, respectively) in the control group. Moreover, $CuSO_4$-intoxicated birds showed hypoglycemia with a rise in serum uric acid and creatinine levels (122.68, 5.18, and 0.78 mg/dL at the 6th week) compared to (159.46, 4.41, and 0.61 mg/dL) in the control group. The $CuSO_4$ toxicity in birds induced oxidative stress, indicated by a high serum malondialdehyde level (MDA) and diminished activity of the antioxidant enzymes (glutathione peroxidase (GSH-Px) and superoxide dismutase (SOD)) (2.01 nmol/mL, 37.66 U/mL, and 2.91 U/mL, respectively, at the 6th week) compared to (1.34 nmol/mL, 57.00 U/mL, 4.99 U/mL, respectively) in the control group. High doses of Cu exposure caused severe microscopic alterations in kidney architecture. The addition of vitamins C and E, singularly or in combination, displayed a beneficial

Citation: Hashem, M.A.; Abd El Hamied, S.S.; Ahmed, E.M.A.; Amer, S.A.; Hassan, A.M. Alleviating Effects of Vitamins C and E Supplementation on Oxidative Stress, Hematobiochemical, and Histopathological Alterations Caused by Copper Toxicity in Broiler Chickens. *Animals* **2021**, *11*, 1739. https://doi.org/10.3390/ani11061739

Academic Editors: Youssef A. Attia, Woo Kyun Kim, Nesrein Hashem and Maria de Olivera

Received: 6 May 2021
Accepted: 7 June 2021
Published: 10 June 2021

effect in alleviating these harmful effects of Cu toxicity. These findings showed the possible mitigating impacts of dietary antioxidants on the hematobiochemical alterations, oxidative stress, and kidney damage induced by $CuSO_4$ toxicity.

Keywords: copper toxicity; vitamin C; vitamin E; hematological parameters; oxidative stress; histopathology

1. Introduction

Copper is a vital micromineral in living animals' diets. It is essential for cellular metabolism and enzyme activity like Cu–Zn superoxide dismutase, tyrosinase, lysyl oxidase, and cytochrome C oxidase [1,2], which are engaged in a range of vital processes necessary for growth and maturation [2]. Supplementing diets with organic and inorganic Cu sources has beneficial effects on poultry production [3–6]. Attia et al. [7] reported that supplementing inorganic Cu (8 mg/kg) is better for growth of male White Pekin ducks from 1–56 d of age than inorganic Cu. Moreover, Cu supplementation increased plasma Cu and cholesterol and decreased Zn levels. They also reported improved liver Cu concentration and Cu excretion and retention by organic Cu supplementation compared to inorganic form. Copper supplementation (10 mg/kg) was sufficient for productive and reproductive performance and egg quality of laying hens [6]. Copper sulfate is the most common form used as a feed additive in poultry and livestock feed [8]. Supplementation with up to 250 ppm Cu has been proven for growth stimulation [1]. However, supplementation in excess amounts lowers growth, feed intake, and the feed conversion ratio in broiler chickens [9,10]. Several studies have demonstrated the toxic impacts of copper on broiler chickens [9–11]. The addition of 325 ppm copper to poultry diets initiates growth retardation and muscle atrophy [12]. The increased intake of copper in the feed results in morphological alterations in the visceral organs [6,13,14]. In our previous study, we reported alterations in the liver tissues in Cu-intoxicated chickens represented by hyperplastic and necrotic biliary epithelium with various degenerative and necrotic changes at the third week. Also, cholestasis, necrotic bile duct epithelia, in addition to lymphocytic portal aggregation and fibroblast proliferation, were encountered at the 6th week [15]. Chronic exposure also leads to hemolytic anemia and affects the central nervous system [14,16]. Exposure to excessive levels of Cu can lead to oxidative stress in broiler chickens [17,18], decrease the activities of copper–zinc superoxide dismutase (SOD), glutathione peroxidase (GSH-Px), and enhance the malondialdehyde (MDA) contents in ducklings [19].

When copper is found in supply more than cell's demands, it stimulates the creation of free radicals and the immediate oxidation of fats, proteins, and DNA [20]. Numerous tools have been suggested to justify copper-induced cytotoxicity [2]. The base for these concepts is that free copper ions behave as potent catalysts for creating reactive oxygen species (ROS) [2,21]. Cupric and cuprous copper ions can perform an essential part in redox reactions. Such as, cupric ions (Cu^{2+}) can be reduced to cuprous ($Cu($^{+}$ in the existence of superoxide ($O2^{-}$), which can catalyze the production of reactive hydroxyl radicals (OH) from hydrogen peroxide (H_2O_2) breakdown through the Haber–Weiss reaction [22]. Hydroxyl radicals are the most potent oxidative radicals potentially present in living systems and may stimulate lipid peroxidation causing tissue injury [23]. These outcomes can be diminished by antioxidant protection systems such as catalase, glutathione peroxidase, superoxide dismutase, and vitamins C and E [21,24].

The dietary addition of various antioxidants such as vitamin C, vitamin E, and selenium was efficiently practiced, mitigating oxidative stress in vivo and in animal products [25–29]. Vitamin E is a coating antioxidant that protects the intracellular structures of live organisms. It acts through mitigating the toxic effects of free radicals and reactive oxygen species that motivate oxidation of phospholipids and sulphydryl groups, leading to an impaired cell membrane structure [30]. Likewise, vitamin C is the most essential water-soluble antioxidant, as it protects biofilms from lipid peroxidation by removing peroxyl radicals in the aqueous stage before the oxidation process starts [30]. It also works to replenish the reduced vitamin E. Vitamin C cannot directly eliminate the lipophilic radicals formed in the membranes, but it reduces the number of tocopheroxyl radicals that adhere to the membrane during the transition of the lipophilic to the aqueous phase [31]. Numerous investigations disclosed good performance with supplementing of vitamins (C and E) in broiler chick diets [12,26,27,32] or fish [25,28,33,34]. Recently, different reports have been reported that vitamins E and C, due to the fact of their role as antioxidants, can protect against toxic infections from xenobiotics and those from minerals, too [35]. Both vitamins can function synergistically to inhibit the adverse outcomes of copper toxicity [12]. In our previous study, it was reported that a combination of vitamin C and E can mitigate the histopathological and DNA changes in the liver of $CuSO_4$-intoxicated birds [15]. The present study aimed to evaluate the efficacy of single and combined supplementation of vitamin C and vitamin E in mitigating oxidative stress, erythrogram, leukogram changes, and histopathological alterations in the kidney induced by copper sulfate ($CuSO_4$) toxicity in broiler chickens.

2. Material and Methods

2.1. Experimental Birds, Design

This study was conducted in a poultry research unit in the faculty of veterinary medicine, Zagazig University, Egypt. The ethics of the experimental protocol were approved by the Institutional Animal Care and Use Committee of Zagazig University, Egypt (ZU-IACUC/2020). All animal experiments were performed following the recommendations described in "The Guide for the Care and Use of Laboratory Animals in Scientific Investigations".

Two hundred and fifty-one-day-old male broiler chickens (COBB-500) were attained from Al-Kahira Poultry Company, 10th of Ramadan City, Sharkia Governorate, Egypt. The experiment lasted for 42 days with good ventilation. Birds were bred in an open, well-ventilated house with sawdust. The chicks were stocked in pens with 10 birds each. A pen is considered a replicate. The room temperature was thermostatically controlled and regulated by two heaters. The room temperature during the first week was set at 34 °C and gradually decreased by 3 °C every week until it reached 24 °C. The lighting program for the first week was 24 h a day and then changed to 16 h of light and 8 h of darkness over 7–42 days. Freshwater and feed were accessible for ad libitum consumption throughout the experiment. The chicks were given a starter diet from day one until the 10th day of age, a grower diet (11th–22nd day), followed by a finisher diet until 42 days of age. The ingredients and chemical composition of the diets were formulated as defined in the COBB-500 Broiler Handbook [36] (Table 1). All birds were vaccinated at 7 and 14 days old against Newcastle disease and at 11 and 22 days old for Gumboro disease [37]. Birds were monitored for any disease challenge or mortalities.

The chicks were randomly allotted into five experimental groups (five replicates/group, ten chicks/replicate): the 1st group: basal diet with no additives (control group), the 2nd group: basal diet supplemented with $CuSO_4$ (300 mg/kg diet), the 3rd group: basal diet with $CuSO_4$ (300 mg/kg diet) + vitamin C (250 mg/kg diet), the 4th group: basal diet with $CuSO_4$ (300 mg/kg diet) + vitamin E (250 mg/kg diet), and the 5th group: basal diet with $CuSO_4$ (300 mg/kg diet) + vitamin C (250 mg/kg diet) + vitamin E (250 mg/kg diet) for a six-week feeding period. Copper sulfate ($CuSO_4 \cdot 5H_2O$, El-Gomhoria Industry, Zagazig, Egypt), vitamin C (L-Ascorbic acid, phosphate, ROVIMIX® STAY-C®35, DSM, Heerlen, Holland), and vitamin E (DL-α-tocopherol acetate, Pharco Pharmaceutical Industries, Zagazig, Egypt). The experimental diets were stored in a cool and well-ventilated place until used in the experiment. The toxic dose of Cu used in this investigation was determined according to Cinar et al. [12]. In contrast, the vitamins C and E doses were used as described by Sahin et al. [38].

Table 1. Proximate and chemical composition of the basal diets (%).

Ingredients	Starter Stage (1–10 Day)	Grower Stage (11–22 Day)	Finisher Stage (23–42 Day)
Soybean meal, 48%	34.65	28.1	24.9
Corn gluten, 60%	1.5	3	3
Yellow corn	58.1	62.1	63.6
Wheat bran	-	1.10	1.80
Soy oil	2.00	2.00	3.26
Calcium carbonate	1.00	1.00	1.00
Calcium dibasic phosphate	1.80	1.70	1.50
Premix *	0.300	0.300	0.300
Common salt	0.300	0.300	0.300
DL-Methionine, 98%	0.180	0.140	0.110
Lysine, Hcl, 78%	0.160	0.160	0.130
Anti-mycotoxin	0.100	0.100	0.100
	Proximate composition (%)		
ME, Kcal/Kg	3047.53	3090.13	3178.59
Crude protein	22.14	20.40	19.07
Crude fiber	2.60	2.61	2.63
Fat	4.50	4.61	5.87
Available P	0.49	0.46	0.42
Calcium	0.96	0.93	0.87
Lysine	1.38	1.21	1.09
Methionine	0.56	0.49	0.47

* Premix per kg of diet: vitamin D3, 200 IU; vitamin A, 1500 IU; vitamin E, 10 mg; vitamin K3, 0.5 mg; thiamine, 1.8 mg; riboflavin, 3.6 mg; pantothenic acid, 10 mg; folic acid, 0.55 mg; pyridoxine, 3.5 mg; niacin, 35 mg; cobalamin, 0.01 mg; biotin, 0.15 mg; Cu, 8 mg; Fe, 80 mg; Zn, 40 mg; Mn, 60 mg; Se, 0.15 mg; I, 0.35 mg. ME: metabolizable energy; P: phosphorus.

2.2. Determination of Hematological and Biochemical Parameters

On the termination of the 3rd and 6th week, blood samples (two aliquots) were attained from the wing vein of (two birds/replicate, 10 birds/group). The first aliquot of blood was placed in tubes containing dipotassium salt of EDTA as an anticoagulant for hematological analysis by Hemascreen 18 Automatic Cell Counter (Hospitex Diagnostics, Sesto Fiorentino, Italy), including red blood cells (RBCs), hemoglobin (Hb), hematocrit (HCT), mean corpuscular volume (MCV), mean corpuscular hemoglobin (MCH), mean corpuscular hemoglobin concentration (MCHC), total leukocyte counts (WBCs), and differential leukocyte counts [18].

The second aliquot of blood was collected without anticoagulant, left to clot at room temperature, and centrifuged for 15 min at 1500 rpm for serum separation, which was stored at −20 °C in deep freezing until biochemical analysis. An automatic biochemical analyzer (Robotnik Prietest ECO-India) was used to measure the level of glucose, creatinine, and uric acid in the blood by following the described methods [39–41]. An UV-Vis Spectrophotometer (OPTIMA, PHOTOMECH. 301-D+, Japan) was used to estimate oxidative stress and antioxidants markers, including serum malondialdehyde (MDA), glutathione peroxidase (GSH-Px), and superoxide dismutase (SOD) according to the methods described in [42–44], respectively.

2.3. Histopathological Investigations

Kidney samples were harvested and fixed in 10% neutral formalin, fixative, dehydrated, and embedded in paraffin. Five-micron-thick paraffin slices were stained with hematoxylin and eosin (H&E) [45] and inspected microscopically.

2.4. Statistical Analysis

Data were analyzed with one-way analysis of variance (ANOVA) using the GLM procedure in SPSS (SPSS Inc., Chicago, IL, USA), after the Shapiro–Wilk's test was used to verify the normality and Levene's test was used to verify the homogeneity of the variance components among experimental treatments. Tukey's test was used to compare the differences between the means at a 5% probability. Variations in the data were expressed as the mean ± SD, and the significance level was set at $p < 0.05$.

3. Results

3.1. Hematological Parameters

As shown in Table 2, a significant decrease in RBCs, Hb concentration, and hematocrit (HCT) values were detected in the $CuSO_4$-intoxicated birds compared with the CON group at the 3rd and 6th weeks ($p < 0.05$). Meanwhile, the MCV, MCH, and MCHC values displayed a non-significant change at the 3rd week indicating normocytic normochromic anemia, but at the 6th week only MCHC significantly decreased leading to normocytic hypochromic anemia. Compared with the $CuSO_4$-intoxicated birds, dietary co-supplementation with vitamin C and E, singularly or in combination, led to amelioration in all changes in the erythrogram that returned towards the normal values of control all over the experimental period, especially in the $CuSO_4$ + vitamin C + vitamin E group.

Table 2. Impact of single or combined supplementation of vitamin C and E on the erythrogram of $CuSO_4$-intoxicated broiler chickens (mean ± SD, $n = 5$).

	Parameters	CON	$CuSO_4$	$CuSO_4$ + vitamin C	$CuSO_4$ + vitamin E	$CuSO_4$ + vitamin C + vitamin E	*p*-Value
At the 3rd week	RBCs ($\times 10^6/\mu L$)	2.15 ± 0.068 ^{a}	1.85 ± 0.094 ^{b}	2.08 ± 0.058 ^{a}	2.09 ± 0.066 ^{a}	2.11 ± 0.052 ^{a}	0.00
	Hb (g/dL)	9.73 ± 0.228 ^{a}	8.05 ± 0.545 ^{c}	9.00 ± 0.158 ^{b}	9.15 ± 0.145 ^{b}	9.41 ± 0.303 ^{a,b}	0.00
	HCT (%)	25.1 ± 0.47 ^{a}	21.5 ± 1.32 ^{b}	23.9 ± 0.54 ^{a}	24.1 ± 0.66 ^{a}	24.5 ± 0.61 ^{a}	0.00
	MCV (fL)	116 ± 1.8	116 ± 2.5	115 ± 1.8	1151 ± 1.4	116 ± 1.0	0.59
	MCH (Pg)	45.1 ± 1.83	43.4 ± 1.77	43.1 ± 1.06	43.8 ± 1.13	44.5 ± 0.48	0.12
	MCHC (%)	38.8 ± 1.10	37.3 ± 0.97	37.6 ± 0.71	38.0 ± 0.82	38.3 ± 0.73	0.05
At the 6th week	RBCs ($\times 10^6/\mu L$)	2.79 ± 0.124 ^{a}	2.42 ± 0.045 ^{d}	2.55 ± 0.049 ^{c,d}	2.60 ± 0.088 ^{b,c}	2.70 ± 0.077 ^{a,b}	0.00
	Hb (g/dL)	10.98 ± 0.184 ^{a}	9.54 ± 0.383 ^{c}	9.93 ± 0.109 ^{b,c}	10.18 ± 0.268 ^{b}	10.68 ± 0.226 ^{a}	0.00
	HCT (%)	28.4 ± 0.56 ^{a}	26.0 ± 0.73 ^{c}	27.1 ± 0.51 ^{b}	27.5 ± 0.60 ^{a,b}	28.1 ± 0.84 ^{a,b}	0.00
	MCV (fL)	101 ± 1.6	107 ± 1.6	106 ± 1.7	106 ± 1.7	103 ± 1.0	0.51
	MCH (Pg)	40.1 ± 2.55	38.7 ± 1.11	38.6 ± 0.84	39.1 ± 0.89	39.5 ± 0.94	0.37
	MCHC (%)	38.2 ± 0.51 ^{a}	36.1 ± 0.69 ^{c}	36.5 ± 0.53 ^{c}	37.0 ± 0.51 ^{b,c}	37.9 ± 1.04 ^{a,b}	0.00

^{a,b,c,d} Means carrying different superscripts are significantly different at $p < 0.05$. RBCs: Red blood cells, Hb: hemoglobin, HCT: hematocrit, MCV: mean corpuscular volume, MCH: mean corpuscular hemoglobin, MCHC: mean corpuscular hemoglobin concentration.

The WBCs, lymphocyte, and monocyte counts were significantly diminished in the $CuSO_4$-intoxicated birds relative to the control group ($p < 0.05$). However, heterophil, eosinophil, and basophil counts were displayed non-statistical variations at the end of the 3rd week ($p < 0.05$). The same picture in the leukogram was observed at the termination of the 6th week, but the heterophil count significantly dropped ($p < 0.05$). Compared with the Cu-intoxicated group, supplementation of vitamin C and vitamin E to $CuSO_4$-intoxicated birds improved the leukocyte picture in all groups, especially the $CuSO_4$ + vitamin C + vitamin E group in which the WBCs, lymphocyte, heterophil, and monocyte counts were significantly increased and returned to the values of the control group (Table 3).

3.2. Serum Biochemical Parameters

A significant decrease in serum glucose was found in Cu-intoxicated birds at the 3rd and 6th weeks. The uric acid and creatinine levels were not significantly different in all experimental groups at the 3rd week ($p > 0.05$). At the sixth week, the uric acid and creatinine levels in the copper intoxicated birds increased compared to the control group ($p < 0.05$). Vitamins-treated groups displayed favorable effects in serum glucose, creatinine, and uric acid levels that reverted close to the control values at the termination of the 6th week (Table 4).

Table 3. Effect of single or combined supplementation of vitamin C and E on the leukogram of $CuSO_4$-intoxicated broiler chickens (mean ± SD, n = 5).

	Parameters	CON	$CuSO_4$	$CuSO_4$ + Vitamin C	$CuSO_4$ + Vitamin E	$CuSO_4$ + Vitamin C + Vitamin E	p-Value
At the 3rd week	WBCs ($\times 10^3/\mu L$)	20.4 ± 1.33 [a]	17.2 ± 1.21 [c]	18.0 ± 0.88 [b,c]	18.7 ± 0.77 [a,b,c]	19.7 ± 1.33 [a,b]	0.001
	Heterophils ($\times 10^3/\mu L$)	6.79 ± 0.459	6.23 ± 0.393	6.30 ± 0.312	6.56 ± 0.379	6.70 ± 0.459	0.09
	Lymphocytes ($\times 10^3/\mu L$)	10.54 ± 0.425 [a]	8.32 ± 0.927 [c]	9.03 ± 0.654 [b,c]	9.41 ± 0.685 [a,b,c]	0.15 ± 0.758 [a,b]	0.00
	Monocytes ($\times 10^3/\mu L$)	1.59 ± 0.157 [a]	1.16 ± 0.340 [b]	1.27 ± 0.071 [a,b]	1.30 ± 0.084 [a,b]	1.44 ± 0.105 [a,b]	0.007
	Eosinophils ($\times 10^3/\mu L$)	0.980 ± 0.0604	1.03 ± 0.288	0.970 ± 0.0484	0.982 ± 0.1042	0.992 ± 0.1308	0.95
	Basophils ($\times 10^3/\mu L$)	0.506 ± 0.0493	0.468 ± 0.0497	0.472 ± 0.0396	0.472 ± 0.0438	0.486 ± 0.0461	0.57
At the 6th week	WBCs ($\times 10^3/\mu L$)	21.1 ± 1.90 [a]	15.8 ± 1.68 [c]	16.8 ± 0.87 [c]	17.9 ± 0.58 [b,c]	19.9 ± 1.25 [a,b]	0.00
	Heterophils ($\times 10^3/\mu L$)	7.12 ± 0.674 [a]	5.26 ± 1.088 [b]	5.99 ± 0.321 [a,b]	6.14 ± 0.571 [a,b]	6.75 ± 0.447 [a]	0.001
	Lymphocytes ($\times 10^3/\mu L$)	10.84 ± 0.993 [a]	7.86 ± 0.616 [b]	8.16 ± 0.532 [b]	9.01 ± 0.564 [b]	10.20 ± 0.700 [a]	0.00
	Monocytes ($\times 10^3/\mu L$)	1.60 ± 0.150 [a]	1.18 ± 0.340 [b]	1.29 ± 0.071 [a,b]	1.34 ± 0.084 [a,b]	1.45 ± 0.102 [a,b]	0.008
	Eosinophils ($\times 10^3/\mu L$)	1.00 ± 0.090	1.05 ± 0.388	0.980 ± 0.0484	0.992 ± 0.1042	1.08 ± 0.130	0.83
	Basophils ($\times 10^3/\mu L$)	0.518 ± 0.0476 [b]	0.450 ± 0.0509	0.462 ± 0.0396	0.472 ± 0.0438	0.506 ± 0.0461	0.07

[a,b,c] Means carrying different superscripts are significantly different at $p < 0.05$.

3.3. Oxidative Stress and Antioxidant Status

As evidenced in Table 5, $CuSO_4$ toxicity in birds induced oxidative stress as indicated statistically by increased serum MDA levels and diminished SOD activity at the end of the 3rd and 6th week and decreased activity of GSH-Px at the 6th week compared with the control group ($p < 0.05$). However, the administration of vitamins induced a beneficial effect in the picture of the oxidative stress by significantly decreasing serum MDA levels and increasing the activity of GSH-Px and SOD compared with the Cu-intoxicated group.

Table 4. Effect of single or combined supplementation of vitamin C and E on the serum glucose and renal biomarkers of CuSO4-intoxicated broiler chickens (Mean ± SD, n = 5).

	Parameters	CON	$CuSO_4$	$CuSO_4$ + Vitamin C	$CuSO_4$ + Vitamin E	$CuSO_4$ + Vitamin C + Vitamin E	p Value
At the 3rd week	Glucose (mg/dL)	157 ± 10.01 [a]	117 ± 7.20 [b]	127 ± 8.00 [a,b]	136 ± 7.5 [a,b]	150 ± 9.7 [a]	0.009
	Uric acid (mg/dL)	3.29 ± 0.553	3.70 ± 0.266	3.19 ± 0.158	3.39 ± 0.503	3.14 ± 0.479	0.25
	Creatinine (mg/dL)	0.508 ± 0.0679	0.618 ± 0.1160	0.566 ± 0.0585	0.532 ± 0.0438	0.528 ± 0.0759	0.13
At the 6th week	Glucose (mg/dL)	159 ± 17.9 [a]	122 ± 9.8 [b]	135 ± 13.6 [a,b]	143 ± 11.3 [a,b]	152 ± 14.0 [a]	0.003
	Uric acid (mg/dL)	4.41 ± 0.570 [b]	5.18 ± 0.362 [a]	4.85 ± 0.352 [a,b]	4.83 ± 0.651 [a,b]	4.54 ± 0.405 [b]	0.008
	Creatinine (mg/dL)	0.614 ± 0.0665 [b]	0.784 ± 0.0971 [a]	0.688 ± 0.0664 [a,b]	0.624 ± 0.0646 [b]	0.586 ± 0.1059 [b]	0.009

[a,b] Means carrying different superscripts are significantly different at $p < 0.05$.

Table 5. Impact of single or combined supplementation of vitamin C and E on the serum oxidative stress indicator and antioxidant status of CuSO4-intoxicated broiler chickens (mean ± SD, n = 5).

	Parameters	CON	$CuSO_4$	$CuSO_4$ + Vitamin C	$CuSO_4$ + Vitamin E	$CuSO_4$ + Vitamin C + Vitamin E	p-Value
At the 3rd week	MDA (nmol/mL)	1.28 ± 0.177^{c}	1.78 ± 0.094^{a}	$1.64 \pm 0.134^{a,b}$	$1.53 \pm 0.182^{a,b,c}$	$1.44 \pm 0.131^{b,c}$	0.00
	GSH-Px (U/mL)	37.3 ± 5.71	30.3 ± 3.50	31.5 ± 3.13	34.0 ± 8.00	36.1 ± 6.38	0.07
	SOD (U/mL)	4.04 ± 0.114^{a}	2.48 ± 0.211^{c}	2.95 ± 0.124^{b}	3.08 ± 0.174^{b}	3.10 ± 0.234^{b}	0.00
At the 6th week	MDA (nmol/mL)	1.34 ± 0.157^{c}	2.01 ± 0.222^{a}	$1.84 \pm 0.152^{a,b}$	$1.74 \pm 0.147^{a,b}$	1.65 ± 0.249^{b}	0.00
	GSH-Px (U/mL)	57.0 ± 6.70^{a}	37.6 ± 3.60^{c}	43.6 ± 3.57^{b}	46.3 ± 11.23^{b}	$50.4 \pm 4.97^{a,b}$	0.00
	SOD (U/mL)	4.99 ± 0.234^{a}	2.91 ± 0.177^{d}	$3.24 \pm 0.350^{c,d}$	$3.61 \pm 0.165^{b,c}$	3.85 ± 0.350^{b}	0.00

[a,b,c,d] Means carrying different superscripts are significantly different at $p < 0.05$. MDA: Malondialdehyde, SOD: superoxide dismutase, GSH-Px: glutathione peroxidase.

3.4. Histopathological Findings

The kidney of control chickens exhibited normal histological structures at the 3rd and 6th week (Figure 1a,b). In contrast, the kidney of $CuSO_4$-intoxicated birds revealed nephrotoxicity represented by extensive necrosis of the renal tubular epithelial and hypercellularity of glomeruli at the 3rd week (Figure 2a). Moreover, extensive hemorrhage focally replaced renal parenchyma with congested blood vessels, and capillaries were detected, besides lymphocytic infiltrations in some glomeruli and hyaline cast inside tubular lumina. At the termination of the 6th week, renal damage became more extensive and characterized by pronounced oncotic necrosis of the renal tubules and glomeruli (Figure 3a), in addition to extensive hemorrhage and hemolysis were seen. Kidney of $CuSO_4$ + vitamin C showed moderate nephrotic changes involving most renal tubular epithelia with few extravasated erythrocytes at the termination of 3rd week. In addition, most glomeruli restored their normal picture, and a few had necrosed glomerular tuft (Figure 2b). At the 6th week post-supplementation, moderate renal lesion represented by degeneration or necrosis of some tubular epithelia with hypercellularity of the glomeruli were encountered, with dilatation and hyperemia in the blood vessels (Figure 3b). Kidney of $CuSO_4$ + vitamin E displayed various mild degenerative changes, mainly cloudy swelling or hydropic degeneration of tubular epithelia with partial intravascular hemolysis at the termination of 3rd week (Figure 2c). A few regenerative attempts in the tubular epithelia were detected. At the end of the 6th week, mild interstitial lymphocytic aggregation could be seen, with improvements of lesions in most renal parenchyma and the regeneration of histomorphology of all nephron segments (Figure 3c). Kidney of chickens supplemented with $CuSO_4$ and both vitamins showed a reduction in renal toxicity. A few tubules still suffered from nephritic changes, while great regenerative attempts were encountered in the adjacent tubular epithelia at the end of 3rd week (Figure 2d). After six weeks of the experiment, the renal parenchyma had an intense reduction in lesions of copper toxicity. All segments of nephrons restored their normal histomorphologic picture with extreme regenerative attempts in the tubules and interstitium (Figure 3d).

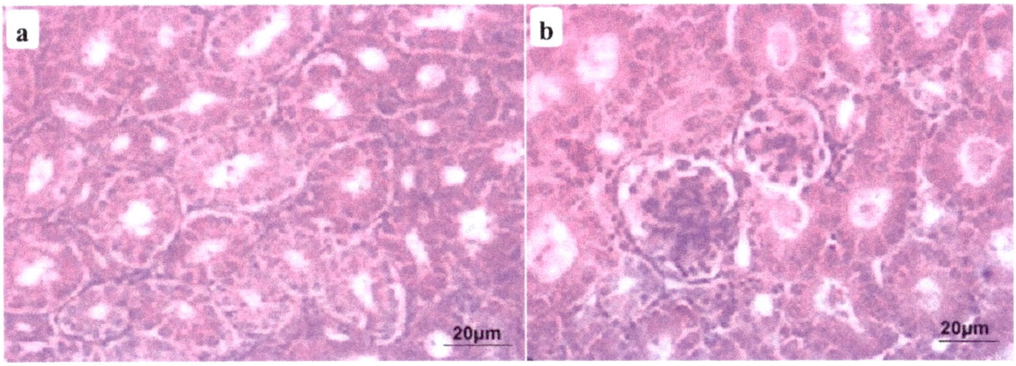

Figure 1. Photomicrographs of kidney sections from the control group at the 3rd week (**a**) and 6th week (**b**) showed normal histological structures.

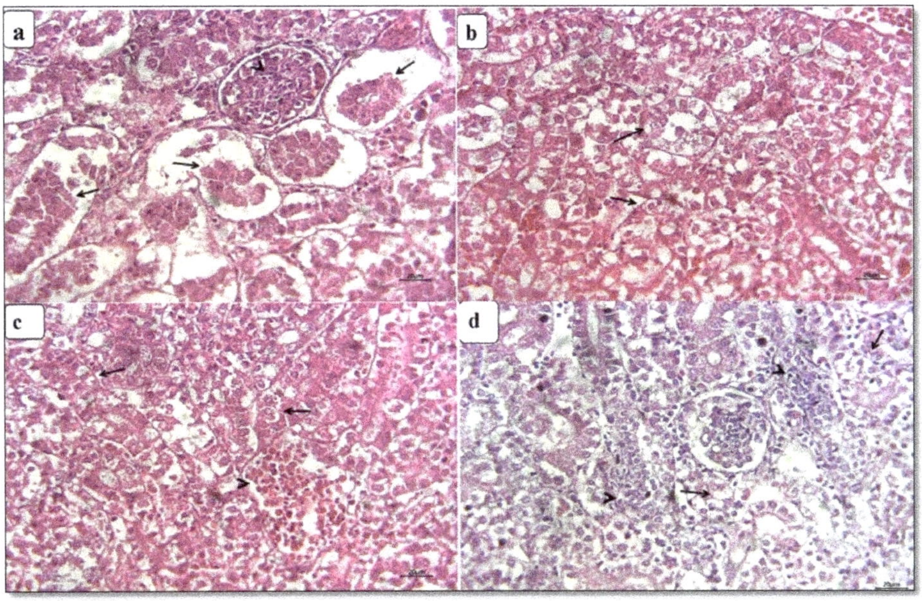

Figure 2. Photomicrographs of kidney sections from $CuSO_4$-intoxicated and vitamin-treated groups at the 3rd week: (**a**) the kidney of $CuSO_4$-intoxicated chickens shows extensive necrotic tubular epithelia (arrow) and hypercellularity of glomeruli (arrowhead); (**b**) the kidney of $CuSO_4$ + vitamin C group shows moderate nephritic changes in renal tubular epithelia (arrow); (**c**) the kidney of the $CuSO_4$ + vitamin E group shows various degenerative changes in some tubular epithelia (arrow) and partial intravascular hemolysis (arrowhead); (**d**) the kidney of $CuSO_4$ + vitamin C + vitamin E group shows regenerative attempts from tubular epithelia (arrowhead) adjacent to nephritic tubules (arrow).

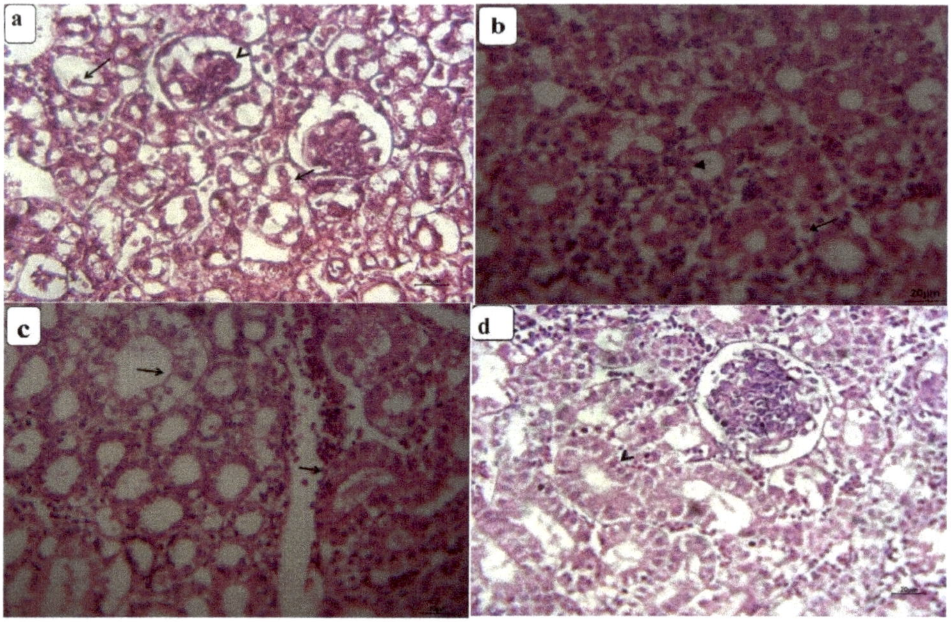

Figure 3. Photomicrographs of kidney sections from $CuSO_4$-intoxicated and vitamin-treated groups at the 6th week: (**a**) the kidney of $CuSO_4$-intoxicated chickens show pronounced oncotic necrosis of renal tubules (arrow) and glomeruli (arrowhead); (**b**) the kidney of the $CuSO_4$ + vitamin C group show degeneration and necrosis of some tubular epithelia (arrow); (**c**) the kidney of the $CuSO_4$ + vitamin E group shows normal renal parenchyma (arrowhead) with mild interstitial lymphocytic aggregation (arrow); (**d**) the kidney of the $CuSO_4$ + vitamin C + vitamin E group shows normal nephron segments and intense regenerative attempts in the tubules (arrow).

4. Discussion

Copper is a vital element for animals, but dietary inclusion at high doses or over a long period can harm performance and increased lipid peroxidation [46]. Along with this experiment, no clinical signs were observed in all groups of birds, except the $CuSO_4$–intoxicated group, which exhibited mild diarrhea, anorexia, and weight reduction with no mortalities. These results coordinated with other studies [47,48].

The erythrogram results in this study showed a significant reduction in RBC count, Hb concentration, and hematocrit value, whereas there were no significant changes in MCV, MCH, and MCHC values in copper-intoxicated birds at the 3rd and 6th weeks. The MCHC value significantly decreased only at the 6th week, indicating normocytic hypochromic anemia. Similar findings were reported in laying hens [46], turkeys [49], and Wister albino rats [50]. This anemia may be due to the excess copper bound to mucosal ferritin, interfering with the re-utilization of iron from ferritin in reticuloendothelial cells; thus, copper acted as a competitive inhibitor of iron, leading to iron-deficient anemia [51]. The decrease in Hb concentration could be due to the interaction of copper with copper-containing enzyme cytochrome oxidase, which is involved in heme synthesis by reducing Fe^{+3} to Fe^{+2} [52]. A recent investigation stated that intravascular hemolysis is a typical sign of copper toxicity resulting in a reduced Hb level [53]. Supplementation with vitamin C and vitamin E showed an improvement in the hematological values. The enhancing action of vitamins may be due to the fact of their roles as antioxidants, protecting the cell membranes of red blood cells against oxidative damages induced by heavy metal toxicity in chicks [54–56].

Regarding the results of the leukogram, the $CuSO_4$-intoxicated birds showed leukopenia, lymphopenia, heteropenia, and monocytopenia. These results may be attributed to general injury produced by the toxic dose of $CuSO_4$ on the hematopoietic stem cells in the bone marrow and other erythropoietic organs [57]. Another study informed that leukopenia might be due to the fact of increased leukocyte mobilization to protect the body in copper-damaged tissue [58]. Moreover, lymphocytopenia could be attributed to the weakening of the immune system caused by Cu or by oxidative stress release due to the fact of Cu poisoning [59,60]. Another experiment stated that excess dietary supplementation levels with nano-Cu led to the reduction of WBCs, Hb, hematocrit, and RBCs in chickens [61]. Dietary inclusion of vitamins C and E, singularly or in combination with Cu, resulted in a correction and improvement in the values of TLC, lymphocytes, heterophils, and monocytes. The associations between metal toxicity and vitamins' protective effects have been formerly reported [54–56]. Both vitamins (i.e., C and E) activate the phagocyte population and immunostimulants or significantly protect WBC against hydrogen peroxide by scavenging [62,63].

Concerning the biochemical results in the current inquiry, significant hypoglycemia was recorded in the blood of $CuSO_4$-intoxicated birds. This finding was attributed to a diminution in feed intake and agrees with previously obtained results [10,49,56]. Moreover, hypoglycemia indicates a depletion of energy resources (glycogen) and, consequently, deterioration of the organism's state [64]. Renal function tests in the current work revealed an increase in serum uric acid and creatinine level in the blood of $CuSO_4$-intoxicated birds at the sixth week. These elevations indicate a renal impairment (nephrotoxicity) as a result of the alterations in the tubular reabsorption threshold, renal blood flow, and glomerular infiltration rate [65], which suggests that the kidney cannot excrete these products due to the fact of impaired kidney function [66]. Other studies reported that high creatinine level happens due to the fact of severe muscle or renal damage [67], or it indicates signs of renal failure [68]. In birds, uric acid is the main final product of nitrogen catabolism [69]. Raised serum uric acid levels are related to the toxic effects from Cu and kidney failure due to the impact of copper metabolites [53,56,67]. Elevation in the concentration of uric acid in the blood might be confirmed lipid peroxidation caused by Cu toxicity, causing a disturbance in kidney excretion function [12,70]. Another study stated significantly higher uric acid concentrations in the serum of broilers given high dietary copper levels [71]. These findings were confirmed by the histopathological findings in the kidney, which showed extensive necrosis of the renal tubular epithelial, hypercellularity and lymphocytic infiltrations of glomeruli, extensive hemorrhages, hemolysis, and hyaline cast inside some tubular lumina. Some studies conducted in numerous animal models have shown that Cu toxicity causes severe pathological findings in kidneys [46,53,72,73]. In addition, Rasool et al. [74] revealed degenerative and necrotic changes in the kidneys of birds fed CuSO4. Addition of vitamins C and E with copper reduced serum creatinine and uric acid levels compared with the copper-intoxicated group. This might be because of the nephroprotective effect of these vitamins due to the fact of their antioxidant effect. Several investigations have revealed that vitamins C and E act synergistically [30]. This was confirmed pathologically, where the kidney displayed moderate nephritic changes with moderate renal lesions in the $CuSO_4$ + vitamin C group, mild degenerative changes with a few regenerative attempts in the $CuSO_4$ + vitamin E group, great regenerative attempts in all nephron segments, and the interstitium restored its normal histomorphologic picture in the $CuSO_4$ + vitamin C + vitamin E group.

Concerning oxidant and antioxidant biomarkers in the current study, Cu toxicity induced lipid peroxidation, as evidenced by an increased serum MDA concentration, with diminished GSH-Px and SOD activities. Oxidative damage is one of the hallmarks of copper toxicity [75]. Previous studies disclosed that copper toxicity increased MDA production and decreased antioxidant activity in broiler blood [61,76,77]. Increased MDA in copper-treated broilers has been associated with excessive production of ROS or ROS accretion due to the ineffectiveness of the antioxidant system after long-term exposure [78]. The declined SOD and GSH-Px activity could be explained because of the efficiency of antioxidant enzymes in the detoxification of the lipid peroxidation products due to the high concentration of copper in the blood and liver [11]. The fall in the levels of these enzymes will consequently upsurge ROS [79,80]. Supplementation of vitamin C or E alone or in combination with Cu-intoxicated birds revealed enhancement in oxidative stress and antioxidant markers. This was demonstrated by a significantly reduced MDA level with a significant rise in GSH-Px and SOD activities compared to the Cu-intoxicated group. This improvement was more pronounced in broilers given a combination of vitamins ($CuSO_4$ + vitamin C + vitamin E group) in comparison to other supplemented groups. However, the protective effect of vitamins for copper-induced oxidative stress was not entirely similar to that of the MDA level, and the antioxidant enzyme activity did not return to the normal control values. Alleviating the effects of both vitamins may increase membranes' ROS scavenging and their consequent reduction into hydroperoxides, restoring antioxidant enzyme activities [24]. Another recent study reported that lipid peroxidation and hydrogen peroxide production levels were significantly reduced in $CuSO_4$-fed fish supplemented with vitamin E [33]. Other investigators have been informed on the ameliorating effects of vitamin C or E in metal-induced toxicity [12,81,82].

5. Conclusions

It could be concluded that co-supplementation of vitamins C and E, singularly or in combination, in chickens with $CuSO_4$-induced toxicity displayed an enhancement in the hematological and biochemical parameters. Moreover, their supplementation mitigated oxidative stress as well as the histopathological alterations in the kidney. The combination of vitamins C and E showed more beneficial effects due to the fact of their synergistic activity in normalizing the levels of most assessed parameters.

Author Contributions: M.A.H.: conceptualization, methodology, visualization, writing—review and editing. S.S.A.E.H.: conceptualization, methodology, resources, software, formal analysis, investigation, data curation, visualization, writing—original draft. E.M.A.A.: conceptualization, methodology, visualization, writing—review and editing. S.A.A.: conceptualization, methodology, writing—review and editing. A.M.H.: conceptualization, funding acquisition. All authors have read and agreed to the published version of the manuscript.

Funding: The work was funded by Taif University Researchers Supporting Project (TURSP-2020/76), Taif University, Taif, Saudi Arabia.

Institutional Review Board Statement: The ethics of the experimental protocol were approved by the Institutional Animal Care and Use Committee of Zagazig University, Egypt (ZU-IACUC/2020). All animal experiments were performed following the recommendations described in "The Guide for the Care and Use of Laboratory Animals in Scientific Investigations".

Informed Consent Statement: Not applicable.

Data Availability Statement: Data sharing not applicable.

Acknowledgments: This work was carried out using the facilities and materials at the Taif University Researchers Supporting Project (TURSP-2020/76), Taif University, Taif, Saudi Arabia. The authors would like to express gratitude to Abd-ELmonem A. Ali at Zagazig University, for his valuable help in histopathological studies.

Conflicts of Interest: The authors declare no conflict of interest.

References

1. Gupta, R.C. *Veterinary Toxicology: Basic and Clinical Principles*; Academic Press: Cambridge, MA, USA, 2012.
2. Gaetke, L.M.; Chow, C.K. Copper toxicity, oxidative stress, and antioxidant nutrients. *Toxicology* **2003**, *189*, 147–163. [CrossRef]
3. Das, T.; Mondal, M.; Biswas, P.; Bairagi, B.; Samanta, C. Influence of level of dietary inorganic and organic copper and energy level on the performance and nutrient utilization of broiler chickens. *Asian Australas. J. Anim. Sci.* **2010**, *23*, 82–89. [CrossRef]
4. Samanta, B.; Biswas, A.; Ghosh, P. Effects of dietary copper supplementation on production performance and plasma biochemical parameters in broiler chickens. *Br. Poult. Sci.* **2011**, *52*, 573–577. [CrossRef]
5. Samanta, B.; Ghosh, P.; Biswas, A.; Das, S. The effects of copper supplementation on the performance and hematological parameters of broiler chickens. *Asian Australas. J. Anim. Sci.* **2011**, *24*, 1001–1006. [CrossRef]

6. Attia, Y.; Abdalah, A.; Zeweil, H.; Bovera, F.; El-Din, A.T.; Araft, M. Effect of inorganic or organic copper additions on reproductive performance, lipid metabolism and morphology of organs of dual-purpose breeding hens. *Arch. Geflugelk.* **2011**, *3*, 169–178.
7. Attia, Y.; Qota, E.; Zeweil, H.; Bovera, F.; Abd Al-Hamid, A.; Sahledom, M. Effect of different dietary concentrations of inorganic and organic copper on growth performance and lipid metabolism of White Pekin male ducks. *Br. Poult. Sci.* **2012**, *53*, 77–88. [CrossRef] [PubMed]
8. Wen, A.; Dai, S.; Wu, X.; Cai, Z. Copper bioavailability, mineral utilization, and lipid metabolism in broilers. *Czech J. Anim. Sci.* **2019**, *64*, 483–490. [CrossRef]
9. Funk, M.; Baker, D. Toxicity and tissue accumulation of copper in chicks fed casein and soy-based diets. *J. Anim. Sci.* **1991**, *69*, 4505–4511. [CrossRef]
10. Rahman, Z.; Besbasi, F.; Afan, A.; Bengali, E.; Zendah, M.; Hilmy, M.; Mukhtar, M.; Jaspal, S.; Aslam, N. Effects of copper supplement on haematological profiles and broiler meat composition. *Int. J. Agric. Biol.* **2001**, *3*, 203–205.
11. Yigit, A.; Cinar, M.; Yildirim, E. The effects of levamisole on oxidative stress induced by copper intoxication in broilers. *N. Z. Vet. J.* **2012**, *60*, 273–277. [CrossRef]
12. Cinar, M.; Yildirim, E.; Yigit, A.A.; Yalcinkaya, I.; Duru, O.; Kisa, U.; Atmaca, N. Effects of dietary supplementation with vitamin C and vitamin E and their combination on growth performance, some biochemical parameters, and oxidative stress induced by copper toxicity in broilers. *Biol. Trace Elem. Res.* **2014**, *158*, 186–196. [CrossRef]
13. Wideman, R., Jr.; Kirby, Y.K.; Barton, T.; Clark, D.; Bayyari, G.; Huff, W.; Moore, P., Jr.; Dunn, P. Excess dietary copper triggers enlargement of the proventriculus in broilers. *J. Appl. Poult. Res.* **1996**, *5*, 219–230. [CrossRef]
14. Hardej, D.; Trombetta, L.D. *Clinical Toxicology Principles and Mechanism*; CRC Press: Boca Raton, FL, USA, 2004.
15. Hashem, M.; Abd El Hamied, S.; Ahmed, E. *Mitigation Effects of Vitamin C and E Against Copper Toxicity in Broiler Chickens, with Reference to Biochemical, Genotoxicity and Pathological Studies*; Research Square: Durham, NC, USA, 2021. [CrossRef]
16. Tanner, M.; Portmann, B.; Mowat, A.; Williams, R.; Pandit, A.; Mills, C.; Bremner, I. Increased hepatic copper concentration in Indian childhood cirrhosis. *Lancet* **1979**, *313*, 1203–1205. [CrossRef]
17. Min, L.; Cui, W.; Peng, X.; Bai, C.; Cui, H. Effect of dietary high copper on the antioxidase activities of brain tissue in chickens. *Chin. J. Anim. Vet. Sci.* **2010**, *41*, 220–223.
18. Yang, F.; Liao, J.; Pei, R.; Yu, W.; Han, Q.; Li, Y.; Guo, J.; Hu, L.; Pan, J.; Tang, Z. Autophagy attenuates copper-induced mitochondrial dysfunction by regulating oxidative stress in chicken hepatocytes. *Chemosphere* **2018**, *204*, 36–43. [CrossRef]
19. Zhao, L.; Yang, F.; Peng, X.; Deng, J.; Cui, H. Effects of high copper on the production of the hydroxy radical and nitrogen monoxide in liver of ducklings. *Chin. Vet. Sci.* **2008**, *38*, 787–790.
20. Valko, M.; Morris, H.; Cronin, M. Metals, toxicity and oxidative stress. *Curr. Med. Chem.* **2005**, *12*, 1161–1208. [CrossRef] [PubMed]
21. Akkuş, İ. Serbest radikaller ve fizyopatolojik etkileri. *Mimoza Yayınları Konya* **1995**, *1*, 57–63.
22. Bremner, I. Manifestations of copper excess. *Am. J. Clin. Nutr.* **1998**, *67*, 1069S–1073S. [CrossRef] [PubMed]
23. Halliwell, B.; Chirico, S. Lipid peroxidation: Its mechanism, measurement, and significance. *Am. J. Clin. Nutr.* **1993**, *57*, 715S–725S. [CrossRef]
24. Cinar, M.; Yigit, A.; Eraslan, G. Effects of vitamin C or vitamin E supplementation on cadmium induced oxidative stress and anaemia in broilers. *Rev. Méd. Vétér.* **2010**, *161*, 449–454.
25. Ibrahim, R.E.; Amer, S.A.; Farroh, K.Y.; Al-Gabri, N.A.; Ahmed, A.I.; El-Araby, D.A.; Ahmed, S.A. The effects of chitosan-vitamin C nanocomposite supplementation on the growth performance, antioxidant status, immune response, and disease resistance of Nile tilapia (*Oreochromis niloticus*) fingerlings. *Aquaculture* **2021**, *534*, 736269. [CrossRef]
26. Amer, S.A.; Mohamed, W.A.; Gharib, H.S.; Al-Gabri, N.A.; Gouda, A.; Elabbasy, M.T.; Abd El-Rahman, G.I.; Omar, A.E. Changes in the growth, ileal digestibility, intestinal histology, behavior, fatty acid composition of the breast muscles, and blood biochemical parameters of broiler chickens by dietary inclusion of safflower oil and vitamin C. *BMC Vet. Res.* **2021**, *17*, 68. [CrossRef] [PubMed]
27. Gouda, A.; Amer, S.A.; Gabr, S.; Tolba, S.A. Effect of dietary supplemental ascorbic acid and folic acid on the growth performance, redox status, and immune status of broiler chickens under heat stress. *Trop. Anim. Health Prod.* **2020**, *52*, 2987–2996. [CrossRef] [PubMed]
28. Ibrahim, R.E.; Ahmed, S.A.; Amer, S.A.; Al-Gabri, N.A.; Ahmed, A.I.; Abdel-Warith, A.-W.A.; Younis, E.-S.M.; Metwally, A.E. Influence of vitamin C feed supplementation on the growth, antioxidant activity, immune status, tissue histomorphology, and disease resistance in Nile tilapia, *Oreochromis niloticus*. *Aquac. Rep.* **2020**, *18*, 100545. [CrossRef]
29. Leskovec, J.; Levart, A.; Svete, A.N.; Perić, L.; Stojčić, M.Đ.; Žikić, D.; Salobir, J.; Rezar, V. Effects of supplementation with α-tocopherol, ascorbic acid, selenium, or their combination in linseed oil-enriched diets on the oxidative status in broilers. *Poult. Sci.* **2018**, *97*, 1641–1650. [CrossRef] [PubMed]
30. McDowell, L.R. *Vitamins in Animal and Human Nutrition*; John Wiley & Sons: Hoboken, NJ, USA, 2008.
31. Niki, E. Interaction of ascorbate and α-tocopherol. *Ann. N. Y. Acad. Sci.* **1987**, *498*, 186–199. [CrossRef]
32. Zhu, Y.; Li, S.; Sun, Q.; Yang, X. Effect of in ovo feeding of vitamin C on antioxidation and immune function of broiler chickens. *Animal* **2019**, *13*, 1927–1933. [CrossRef]
33. Azeez, O.; Braimah, S. Mitigating Effect of Vitamin-E on Copper Sulphate-Induced Toxicity in African Catfish (Clarias gariepinus). *Eur. J. Med. Res.* **2020**, *2*, 4. [CrossRef]

34. Ahmed, S.A.; Ibrahim, R.E.; Farroh, K.Y.; Moustafa, A.A.; Al-Gabri, N.A.; Alkafafy, M.; Amer, S.A. Chitosan vitamin E nanocomposite ameliorates the growth, redox, and immune status of Nile tilapia (*Oreochromis niloticus*) reared under different stocking densities. *Aquaculture* **2021**, *541*, 736804. [CrossRef]
35. Ajuwon, O.R.; Idowu, O. Vitamin C attenuates copper-induced oxidative damage in broiler chickens. *Afr. J. Biotechnol.* **2010**, *9*, 7525–7530.
36. Vantress, C. Cobb Broiler Management Guide. Cobb Vantress. Siloam Springs USA. 2012. Available online: https://www.cobb-vantress.com/assets/5c7576a214/Broiler-guide-R1.pdf (accessed on 27 May 2021).
37. Giambrone, J.; Clay, R.P. Vaccination of day-old broiler chicks against Newcastle disease and infectious bursal disease using commercial live and/or inactivated vaccines. *Avian Dis.* **1986**, *30*, 557–561. [CrossRef]
38. Sahin, K.; Sahin, N.; Yaralioglu, S. Effects of vitamin C and vitamin E on lipid peroxidation, blood serum metabolites, and mineral concentrations of laying hens reared at high ambient temperature. *Biol. Trace Elem. Res.* **2002**, *85*, 35–45. [CrossRef]
39. Trinder, P. Determination of blood glucose using an oxidase-peroxidase system with a non-carcinogenic chromogen. *J. Clin. Pathol.* **1969**, *22*, 158–161. [CrossRef] [PubMed]
40. Henry, R. Determination of serum creatinine. In *Clinical Chemistry: Principles and Techniques*, 2nd ed.; Harper and Row Publishers: New York, NY, USA, 1974.
41. Sanders, G.; Pasman, A.; Hoek, F. Determination of uric acid with uricase and peroxidase. *Clin. Chim. Acta* **1980**, *101*, 299–303. [CrossRef]
42. Ohkawa, H.; Ohishi, N.; Yagi, K. Assay for lipid peroxides in animal tissues by thiobarbituric acid reaction. *Anal. Biochem.* **1979**, *95*, 351–358. [CrossRef]
43. Paglia, D.E.; Valentine, W.N. Studies on the quantitative and qualitative characterization of erythrocyte glutathione peroxidase. *J. Lab. Clin. Med.* **1967**, *70*, 158–169.
44. Martin, J.P., Jr.; Dailey, M.; Sugarman, E. Negative and positive assays of superoxide dismutase based on hematoxylin autoxidation. *Arch. Biochem. Biophys.* **1987**, *255*, 329–336. [CrossRef]
45. Suvarna, S.; Layton, C.; Bancroft, J. The hematoxylins and eosin. In *Bancroft's Theory and Practice of Histological Techniques*, 7th ed.; Churchill Livingstone: London, UK, 2013; pp. 172–186.
46. Zhou, Q.; Zhu, J.; Liu, B.; Qiu, J.; Lu, X.; Curtin, B.; Ji, F.; Yu, D. Effects of High-Dose of Copper Amino Acid Complex on Laying Performance, Hematological and Biochemical Parameters, Organ Index, and Histopathology in Laying Hens. *Biol. Trace Elem. Res.* **2020**, 1–8. [CrossRef]
47. Luo, X.; Ji, F.; Lin, Y.; Steward, F.; Lu, L.; Liu, B.; Yu, S. Effects of dietary supplementation with copper sulfate or tribasic copper chloride on broiler performance, relative copper bioavailability, and oxidation stability of vitamin E in feed. *Poult. Sci.* **2005**, *84*, 888–893. [CrossRef] [PubMed]
48. Wang, Z.; Cerrate, S.; Coto, C.; Yan, F.; Waldroup, P. Evaluation of Mintrex copper as a source of copper in broiler diets. *Int. J. Poult. Sci.* **2007**, *6*, 308–313. [CrossRef]
49. Makarski, B.; Gortat, M.; Lechowski, J.; Zukiewicz-Sobczak, W.; Sobczak, P.; Zawislak, K. Impact of copper (Cu) at the dose of 50 mg on haematological and biochemical blood parameters in turkeys, and level of Cu accumulation in the selected tissues as a source of information on product safety for consumers. *Ann. Agric. Environ. Med.* **2014**, *21*, 567–570. [CrossRef] [PubMed]
50. Ozcelik, D.; Toplan, S.; Ozdemir, S.; Akyolcu, M. Effects of excessive copper intake on hematological and hemorheological parameters. *Biol. Trace Elem. Res.* **2002**, *89*, 35–42. [CrossRef]
51. Chan, W.; Rennert, O. The role of copper in iron metabolism. *Ann. Clin. Lab. Sci.* **1980**, *10*, 338–344. [PubMed]
52. Selim, N.; Youssef, S.; Abdel-Salam, A.; Nada, S.A. Evaluations of some natural antioxidant sources in broiler diets: 1-effect on growth, physiological and immunological performance of broiler chicks. *Int. J. Poult. Sci.* **2013**, *12*, 561–571. [CrossRef]
53. Baruah, S.; Goswami, S.; Kalita, D. Haematobiochemical and Pathological Alterations of Chronic Copper Toxicity in Ducks. *J. Anim. Res.* **2018**, *8*, 283–287.
54. Kalavathi, S.; Kumar, A.A.; Reddy, A.G.; Srilatha, C.; Reddy, A.R. *Sodium Arsenite Toxicity in Broiler Chicks and its Amelioration: Haemato-Biochemical and Pathological Studies*; AGRIS: Chatham, ON, Canada, 2011.
55. Kandpal, V.; Kumar, D.; Bisht, R. Protective effect of vitamin E on haematological parameters in chronic toxicity of hexavalent chromium in laboratory chicks. *J. Drug Deliv. Ther.* **2019**, *9*, 388–392.
56. Kumari, R.R.; Kumar, P.; Mondal, T.K. Effect of vitamin E and selenium on haematological parameters in sub-acute toxicity of hexavalent chromium in broiler chick. *Natl. J. Physiol. Pharm. Pharmacol.* **2013**, *3*, 158–161. [CrossRef]
57. Oliveira, D.C.; Nogueira-Pedro, A.; Santos, E.W.; Hastreiter, A.; Silva, G.B.; Borelli, P.; Fock, R.A. A review of select minerals influencing the haematopoietic process. *Nutr. Res. Rev.* **2018**, *31*, 267–280. [CrossRef]
58. Mazon, A.; Monteiro, E.; Pinheiro, G.; Fernadez, M. Hematological and physiological changes induced by short-term exposure to copper in the freshwater fish, *Prochilodus scrofa*. *Braz. J. Biol.* **2002**, *62*, 621–631. [CrossRef]
59. Guo, C.-H.; Liu, P.-J.; Hsia, S.; Chuang, C.-J.; Chen, P.-C. Role of certain trace minerals in oxidative stress, inflammation, CD4/CD8 lymphocyte ratios and lung function in asthmatic patients. *Ann. Clin. Biochem.* **2011**, *48*, 344–351. [CrossRef]
60. El-Kazaz, S.E.; Hafez, M.H. Evaluation of copper nanoparticles and copper sulfate effect on immune status, behavior, and productive performance of broilers. *J. Adv. Vet. Anim. Res.* **2020**, *7*, 16. [CrossRef] [PubMed]

61. Ognik, K.; Sembratowicz, I.; Cholewińska, E.; Jankowski, J.; Kozłowski, K.; Juśkiewicz, J.; Zduńczyk, Z. The effect of administration of copper nanoparticles to chickens in their drinking water on the immune and antioxidant status of the blood. *Anim. Sci. J.* **2018**, *89*, 579–588. [CrossRef] [PubMed]
62. Eo, J.; Lee, K.-J. Effect of dietary ascorbic acid on growth and non-specific immune responses of tiger puffer, *Takifugu rubripes*. *Fish Shellfish Immunol.* **2008**, *25*, 611–616. [CrossRef] [PubMed]
63. Carr, A.; Maggini, S. Vitamin C and Immune Function. *Nutrients* **2017**, *9*, 1211. [CrossRef] [PubMed]
64. Afaghi, A.; Zare, S.; Heidari, R.; Asadpoor, Y.; Viayeh, R.M. Effects of copper sulfate (CuSO4) on the levels of glucose and cortisol in common carp, *Cyprinus carpio*. *Pak. J. Biol. Sci. PJBS* **2007**, *10*, 1655–1660. [CrossRef] [PubMed]
65. Wang, T.; Long, X.; Cheng, Y.; Liu, Z.; Yan, S. The potential toxicity of copper nanoparticles and copper sulphate on juvenile *Epinephelus coioides*. *Aquat. Toxicol.* **2014**, *152*, 96–104. [CrossRef]
66. Hassan, S.; Shaikh, M.U.; Ali, N.; Riaz, M. Copper sulphate toxicity in a young male complicated by methemoglobinemia, rhabdomyolysis and renal failure. *J. Coll. Phys. Surg. Pak.* **2010**, *20*, 490–491.
67. Lierz, M. Avian renal disease: Pathogenesis, diagnosis, and therapy. *Vet. Clin. North Am. Exot. Anim. Pract.* **2003**, *6*, 29–55. [CrossRef]
68. Khan, A.; Sharaf, R.; Khan, M.; Saleemi, M.K.; Mahmood, F. Arsenic toxicity in broiler chicks and its alleviation with ascorbic acid: A toxico-patho-biochemical study. *Int. J. Agric. Biol.* **2013**, *15*, 6.
69. Hosseintabar, B.; Dadashbeiki, M.; Bouyeh, M.; Seidavi, A.; van den Hoven, R.; Gamboa, S. Effect of different levels of L-carnitine and lysine-methionine on broiler blood parameters. *Rev. MVZ Córdoba* **2015**, *20*, 4698–4708. [CrossRef]
70. Torki, M.; Habibian, M.; Rostami, T.; Moradi, A. *Effects of High Dietary Levels of Selenium and Copper on Growth Performance, Selected Blood Biochemical Parameters and Antibody Production Against Newcastle Disease Vaccine Virus in Broiler Chickens*; Scientific Information Database: Tehran, Iran, 2015.
71. Apsîte, M.; Berzina, N.; Basova, N. Effects of high but non-toxic dietary intake of selenium and copper on indices of the antioxidant defence system and on accumulation of trace elements in chicks. *Proc. Latvian Acad. Sci.* **2012**, 117. [CrossRef]
72. Wang, Y.; Zhao, H.; Shao, Y.; Liu, J.; Li, J.; Xing, M. Copper or/and arsenic induce oxidative stress-cascaded, nuclear factor kappa B-dependent inflammation and immune imbalance, trigging heat shock response in the kidney of chicken. *Oncotarget* **2017**, *8*, 98103. [CrossRef] [PubMed]
73. Iqbal, R.; Malik, F.; Aziz, T.; Sarfraz, I.; Ahmed, Z.; Shafqat, S. The study of histopathological changes upon exposure to vinegerized copper sulphate in liver and kidney of broiler chick. *Middle East J. Sci. Res.* **2012**, *12*, 36–41.
74. Rasool, A.; Javed, M.T.; Akhtar, M.; Bhatti, S.S.; Shahzad, M.; Hussain, R. Effects of urea and copper sulphate on some serum biochemical and meat parameters in broiler chicken. *Pak. Vet. J.* **2012**, *33*, 27–31.
75. Tchounwou, P.B.; Newsome, C.; Williams, J.; Glass, K. Copper-induced cytotoxicity and transcriptional activation of stress genes in human liver carcinoma (HepG2) cells. In Proceedings of the International Symposium on Metal Ions in Biology and Medicine (Les Ions Metalliques en Biologie et en Medecine, Symposium International Sur Les Ions Metalliques), Lisbon, Portugal, 21–24 May 2008; p. 285.
76. Oguz, E.O.; Yuksel, H.; Enli, Y.; Tufan, A.C.; Turgut, G. The effects of copper sulfate on liver histology and biochemical parameters of term ross broiler chicks. *Biol. Trace Elem. Res.* **2010**, *133*, 335–341. [CrossRef]
77. Oğuz, E.; Enli, Y.; Tufan, A.; Turgut, G. Toxic effects of copper sulfate on the brains of term Hubbard broiler chicks: A stereological and biochemical study. *Biotech. Histochem.* **2014**, *89*, 23–28. [CrossRef]
78. Ajuwon, O.; Idowu, O.; Afolabi, S.; Kehinde, B.; Oguntola, O.; Olatunbosun, K. The effects of dietary copper supplementation on oxidative and antioxidant systems in broiler chickens. *Arch. Zootec.* **2011**, *60*, 275–282. [CrossRef]
79. Gupta, R.S.; Gupta, E.S.; Dhakal, B.K.; Thakur, A.R.; Ahnn, J. Vitamin C and vitamin E protect the rat testes from cadmium-induced reactive oxygen species. *Mol. Cells* **2004**, *17*, 132–139.
80. Liu, H.; Guo, H.; Jian, Z.; Cui, H.; Fang, J.; Zuo, Z.; Deng, J.; Li, Y.; Wang, X.; Zhao, L. Copper induces oxidative stress and apoptosis in the mouse liver. *Oxidative Med. Cell. Longev.* **2020**, *2020*. [CrossRef] [PubMed]
81. Ademuyiwa, O.; Ugbaja, R.; Ojo, D.; Owoigbe, A.; Adeokun, S. Reversal of aminolevulinic acid dehydratase (ALAD) inhibition and reduction of erythrocyte protoporphyrin levels by vitamin C in occupational lead exposure in Abeokuta, Nigeria. *Environ. Toxicol. Pharmacol.* **2005**, *20*, 404–411. [CrossRef] [PubMed]
82. Mattie, M.D.; Freedman, J.H. Protective effects of aspirin and vitamin E (α-tocopherol) against copper-and cadmium-induced toxicity. *Biochem. Biophys. Res. Commun.* **2001**, *285*, 921–925. [CrossRef] [PubMed]

animals

Article

Effects of Activated Charcoal on Growth, Immunity, Oxidative Stress Markers, and Physiological Responses of Nile Tilapia Exposed to Sub-Lethal Imidacloprid Toxicity

Samah A. A. Abd El-hameed [1], Samar S. Negm [2], Nahla E. M. Ismael [2], Mohammed A. E. Naiel [3], Mohamed Mohamed Soliman [4], Mustafa Shukry [5] and Hany M. R. Abdel-Latif [6,*]

1 Fish Health and Management Department, Central Laboratory for Aquaculture Research (CLAR), Agriculture Research Center, Abbassa, Abu Hammad, Sharkia 44661, Egypt; samahbasha222@yahoo.com
2 Fish Biology and Ecology Department, Central Laboratory for Aquaculture Research (CLAR), Agriculture Research Center, Abbassa, Abu Hammad, Sharkia 44661, Egypt; sss.negm1984@gmail.com (S.S.N.); nahla_aquaecology@yahoo.com (N.E.M.I.)
3 Department of Animal Production, Faculty of Agriculture, Zagazig University, Zagazig 44511, Egypt; mohammednaiel.1984@gmail.com
4 Clinical Laboratory Sciences Department, Turabah University College, Taif University, P.O. Box 11099, Taif 21944, Saudi Arabia; mmsoliman@tu.edu.sa
5 Department of Physiology, Faculty of Veterinary Medicine, Kafrelsheikh University, Kafrelsheikh 33516, Egypt; mostafa.ataa@vet.kfs.edu.eg
6 Department of Poultry and Fish Diseases, Faculty of Veterinary Medicine, Alexandria University, Alexandria 22758, Egypt
* Correspondence: hmhany@alexu.edu.eg

Citation: Abd El-hameed, S.A.A.; Negm, S.S.; Ismael, N.E.M.; Naiel, M.A.E.; Soliman, M.M.; Shukry, M.; Abdel-Latif, H.M.R. Effects of Activated Charcoal on Growth, Immunity, Oxidative Stress Markers, and Physiological Responses of Nile Tilapia Exposed to Sub-Lethal Imidacloprid Toxicity. *Animals* **2021**, *11*, 1357. https://doi.org/10.3390/ani11051357

Academic Editors: Clive J. C. Phillips, Woo Kyun Kim, Youssef A. Attia, Maria de Olivera and Nesrein Hashem

Received: 22 March 2021
Accepted: 29 April 2021
Published: 11 May 2021

Simple Summary: Finding a suitable feed supplement is important for maintaining fish health and sustainability of the aquaculture industry. From these supplements, research studies have shown that activated charcoal (AC) has been extensively used for veterinary and aquaculture objectives as a "Universal Antidote" against several toxicants and aquatic pollutants. Therefore, the mitigating roles of dietary supplementation with different AC levels on physiological responses of Nile tilapia exposed to sub-lethal imidacloprid (IMID) toxicity were evaluated. The findings of this study revealed that dietary supplementation with 14.30 g AC/kg diet positively modulated the toxic impacts of IMID-intoxicated fish.

Abstract: The existing study was designed to assess the influences of dietary activated charcoal (AC) on the growth performance, immune responses, antioxidative status, and its mitigating roles against the physiological responses of Nile tilapia exposed a sub-lethal dose of a neonicotinoid agriculture pesticide, namely, as imidacloprid (IMID). Nile tilapia juveniles were fed on diets supplemented with graded AC levels as 0 (control), 5, 10, 15, and 20 g/kg diet for eight weeks. Growth, hemato-biochemical indices, and antioxidant and immune responses of fish in all groups were evaluated at the end of the feeding experiment. Afterward, fish in all experimental groups were subjected to a sub-lethal dose of IMID (0.0109 μg/L) for two weeks. Then, fish mortalities, stress indicators, and IMID residual levels in liver and flesh were examined. Results of the feeding experiment showed that total feed intake, weight gain, final body weights, and feed efficiency ratio were significantly increased in all AC groups compared with the control group. The survival rate was 100% in all experimental groups. No statistical differences were observed in the hematological picture of all experimental groups except the lymphocyte count, which was significantly increased in all AC groups compared to the control group. Total protein, albumin, globulin, nitric oxide levels, lysozyme, and respiratory burst activities were significantly increased in all AC groups. Serum alanine transaminase, aspartate transaminase, alkaline phosphatase activities, and malondialdehyde (MDA) levels were significantly decreased in all AC groups compared with the AC0 group. After exposure to a sub-lethal dose of IMID, survival rates were significantly elevated, and IMID residual levels in liver and flesh were significantly decreased in all AC groups than in the control group. Moreover, second-order polynomial regression showed that dietary supplementation with 14.30 g AC/kg diet resulted in the lowest blood glucose and serum MDA levels. Conclusively, we suggest dietary supplementation with 14.30 g AC/kg diet to modulate physiological responses of Nile tilapia to sub-lethal IMID toxicity.

Keywords: *Oreochromis niloticus*; antioxidant; hematology; nitric oxide; lysozyme

1. Introduction

Activated charcoal (AC) is an odorless, tasteless, and very fine black powder that acts as an "adsorbent " for toxicants, gases, poisons, and several impurities [1]. It has been widely applied for medicinal, veterinary, and aquatic medical purposes as a "universal antidotal treatment" for several poisons and environmental toxicants [2]. The mechanisms of actions of AC showed its potential in vitro affinity in the adsorption and elimination of several toxicants such as aflatoxins [3] and pesticide tissue residues [4]. The detoxifying properties of AC have been attributed to its physical and chemical properties, such as its pore size, surface area, and adsorption capability [5]. In terrestrial animals, dietary supplementation with AC has been used to absorb ammonia and nitrogen, improve the functions of the gastrointestinal tract (GIT), and eliminate the impurities and poisonous materials from the GIT [6–8].

Studies on Nile tilapia (*Oreochromis niloticus*) showed that optimal dietary levels of AC could improve the health status, fillet composition [2], growth performance, and intestinal histomorphology [9], boost the antioxidant capacity, and reduce the tissue bioaccumulation after environmental heavy metal exposure [10]. Moreover, reports showed that AC could enhance the growth of African catfish (*Clarias gariepinus*) [11], decrease heavy metal bioaccumulation in tissues of big sturgeon (*Huso huso*) [12], and improve the intestinal histomorphology of giant trevally (*Caranx ignobilis*) [13,14].

Other AC forms, for instance, dietary supplementation with bamboo charcoal, considerably enhanced the growth of Tiger puffer fish (*Takifugu rubripes*) [15] and Japanese flounder (*Paralichthys olivaceus*) [16] and reduced the percentage of nitrogen and ammonia in rearing water of Striped catfish (*Pangasianodon hypophthalmus*) [17]. Other studies showed that dietary supplementation with charcoal and wood vinegar mixture improved the body composition analysis of *P. olivaceus* [18]. Moreover, commercial wood charcoal could reduce the environmental load in the rearing water of red tilapia (*Oreochromis* sp.) [19] and enhance the water quality of gilthead seabream (*Sparus aurata*) [20].

Imidacloprid (IMID) (as a neonicotinoid pesticide) has been extensively used for insect and pest control, especially those affecting crops [21]. Although Tišler et al. [22] showed that IMID was steady in the water and did not quickly undergo biodegradation in the ecosystems, the unsafe and unhygienic disposal of IMID will subsequently provoke serious toxic impacts on the exposed organisms [23]. In Nile tilapia, previous studies reported that IMID exposure induced genotoxicity [24], hematological changes [25], histopathological alterations [26,27], oxidative stress, and growth depression [28]. Furthermore, IMID induced neurotoxicity in zebrafish (*Danio rerio*) and rainbow trout (*Oncorhynchus mykiss*) [29,30], oxidative stress injury and genotoxicity in Streaked prochilod (*Prochilodus lineatus*) [31], spinal cord malformations of the Japanese rice fish (*Oryzias latipes*) [32], and genotoxicity and immunotoxicity in Chinese rare minnows (*Gobiocypris rarus*) [33].

Our previously published studies reported the potential efficacy of dietary supplementation with vitamin C, chitosan nanoparticles (ChNPs) [28], and, recently, clinoptilolite and ChNPs in attenuation of sub-acute IMID toxicity in Nile tilapia [34]. The present study aimed to evaluate dietary supplementation effects with AC on growth, hemato-biochemical indices, antioxidant capacity, immunological assays, and modulation of sub-acute IMID toxicity in the exposed Nile tilapia juveniles.

2. Materials and Methods

2.1. Fish and Rearing Conditions

Two hundred healthy Nile tilapia (*Oreochromis niloticus*) juveniles were procured from the governmental hatchery (Central Laboratory for Aquaculture Research (CLAR), Abbassa, Egypt). Laboratory experiments were achieved at the Fish Wet Laboratory (Department of Fish Biology and Ecology, CLAR, Egypt).

Fish were maintained in the indoor fiberglass tanks for two weeks to be acclimated to the laboratory conditions with 12 Light photoperiod. Each tank was equipped with dechlorinated, fresh tap water supplied with continuous compressed air (through air-stones by using air pumps). Before starting the experiments, fish were fed daily up to apparent satiation on a commercially purchased basal diet (30% crude protein) (Aller-Aqua Co., Egypt). During the acclimation period and thereafter, diets were offered to the fish according to their live weight (3% of their live body weight).

2.2. Maintenance of Water Quality Parameters

For maintaining healthy water, one-third of the water (per each aquarium) was siphoned (each two days) to eliminate feces and get rid of the uneaten food particles. The physical and chemical features of the rearing water were biweekly examined throughout the whole experimental period. The mean values of water parameters were maintained as pH (7.50–8.50), un-ionized ammonia (0.03 $\pm$ 0.01 mg/L), dissolved oxygen (7.50 $\pm$ 0.05 mg/L), nitrite (0.013 $\pm$ 0.01 mg/L), and water temperature (27.5 $\pm$ 0.5 °C) (these levels were within the suitable ranges necessitated for Nile tilapia).

2.3. Experiment I: Feeding Trial

2.3.1. Experimental Diets

The basal commercial diet was obtained to meet the appropriate nutritional requirements for Nile tilapia juveniles. Feed ingredients and chemical composition of the basal diet (%) (on an air-dry basis) (Table 1) were previously published in our study by Abdelghany et al. [35].

Table 1. Feed ingredients and proximate chemical composition of the commercially purchased basal diet (%) (on air-dry basis) (previously published in Abdelghany et al. [35]).

Feed Ingredients	(%)
Fish meal	15
Yellow corn	32
Soybean meal (44%)	20
Corn gluten meal (60%)	14
Wheat bran	13
Vegetable oil	4
Vitamin premix [1]	1
Mineral premix [2]	1
Total	100
Proximate chemical composition (g/kg) as fed basis	
Crude protein (CP) ($n \times 6.25$)	311.5
Crude lipids (CL)	75.6
Ash	40.8
Crude fiber (CF)	53.7
Nitrogen free extract (NFE) [3]	518.4
Gross energy (MJ per 100 g) [4]	19.17

[1] Composition (per kg): Manganese (53 g), Zinc (40 g), Iron (20 g), Copper (2.7 g), Iodine (0.34 g), Selenium (70 mg), Cobalt (70 mg), and Calcium carbonate (as carrier) up to 1 kg. [2] Composition (per kg): Vitamin A (8,000,000 IU), Vitamin C (500 mg), Vitamin D3 (2,000,000 IU), Vitamin E (7000 mg), Vitamin K3 (1500 mg), Vitamin B1 (700 mg), Vitamin B2 (3500 mg), Vitamin B6 (1000 mg), Vitamin B12 (7 mg), Biotin (50 mg), folic acid (700 mg), Nicotinic acid (20,000 mg), and Pantothenic acid (7000 mg). [3] NFE = 100 − (CP + CL+ Ash + CF). [4] Calculated as 23.4 kJ g^{-1}, 39.2 kJ g^{-1}, and 17.2 kJ g^{-1} for protein, lipids, and carbohydrates, respectively.

Activated charcoal (AC) powder (Sigma-Aldrich, St. Louis, MO, USA) (#161551) (CAS Number 7440–44-0) as decolorizing carbonaceous material with high purity, a molecular weight of 12.01 g mol^{-1}, and particle size of –100 (mesh) was used in the current study.

Different graded levels of AC were used and were mixed with the basal diet, whereas five experimental diets were formulated containing 0, 5, 10, 15, and 20 g/kg diet [10]. Diet ingredients were finely ground, and each dose level of AC was then suspended in 100 mL water per kg diet and thoroughly mixed with the other diet ingredients for 40 min using a blender. The mixture was then pelleted using a grinder with a 1-mm diameter paste extruder. All diets were left to dry and then packed into plastic bags and refrigerated at −4 °C until further use.

2.3.2. Experimental Design

Fish (with initial body weight = 33.06 ± 0.74 g) were allocated into five groups known as AC 0 (control), AC 5, AC 10, AC 15, and AC 20. Each group was composed of four replicates (each replicate contained 10 fish per 100-L aquarium) (0.75 × 0.50 × 0.50 m). Fish in each group were fed the corresponding diets for 8 weeks.

2.3.3. Growth Performance

At the end of the feeding experiment (8 weeks), fish were assembled, counted, and group weighed. Fish mortality was documented daily, and dead fish were daily collected. Equations used to evaluate the growth and feed utilization parameters were previously illustrated in Abdel-Latif et al. [36] and Mohammadi et al. [37].

$$\text{Weight gain (WG) (g)} = \text{FBW} - \text{IBW} \quad (1)$$

$$\text{Specific growth rate (SGR, \%/day)} = 100\ [\text{Ln FBW (g)} - \text{Ln IBW (g)}]/\text{T} \quad (2)$$

where IBW is initial body weight, FBW is final body weight, and T is the rearing period.

Total feed intake (TFI) (g feed/fish) is the summation of the amounts of diets (g) fed to fish in each group throughout the experiment/fish number,

$$\text{Feed conversion ratio (FCR)} = \text{TFI (g)/WG (g)} \quad (3)$$

$$\text{Feed efficiency ratio (FER)} = \text{WG (g)/TFI (g)} \quad (4)$$

$$\text{Survival rate (SR) (\%)} = 100\ \text{(number of fish at the end of the experiment/number at the start).} \quad (5)$$

2.3.4. Blood Sampling and Serum Separation

By the end of the feeding trial, fish had fasted for 24 h before blood sampling. Eight fish were sampled from each group (n = 8) and anesthetized using MS-222 (Argent Chemical Laboratories, Redmond, WA, USA) (200 mg/L). Blood was sampled from the caudal veins and divided into two parts (one part was mixed with anticoagulant into Eppendorf tubes for hematological parameters and the other part was left at room temperature for collection of fish serum). The serum was separated into centrifuge tubes by centrifugation (3000× g for 15 min) and stored at −20 °C until being used.

2.3.5. Hematological Indices

Red blood cells' (RBCs) and white blood cells' (WBCs) counts were measured by using a hemocytometer [38]. Hemoglobin (Hb) values were measured according to Collier [39]. Hematocrit (HTC) and mean corpuscular hemoglobin concentration (MCHC) values were evaluated as defined by Wintrobe's method [40]. Mean corpuscular volume (MCV) was assessed by an automated Coulter LH 750 hematology analyzer (Beckman Coulter, Fullerton, CA, USA) [41]. Differential leucocytic counts were calculated according to Klontz's method [42].

2.3.6. Serum Biochemical Measurements

Serum biochemical indices were measured using colorimetric methods using commercial, fish-specific diagnostic kits (Bio-diagnostic Co. for Modern Laboratory Chemicals, Giza, Egypt). The blood protein profile, including serum total protein (TP) and albumin (ALB) values, were assessed as illustrated by Henry [43] and Wotton and Freeman [44], respectively. Globulin (GLO) values were evaluated from the differences between TP and ALB levels. Serum transaminases such as aspartate aminotransferase (AST) and alanine aminotransferase (ALT) activities in fish sera were assessed according to Reitman and Frankel [45]. Serum alkaline phosphatase (ALP) activity was evaluated by the kinetic assay method [46]. Blood glucose (GLU) concentration was analyzed using specific diagnostic kits (Glu L 1000, PLIVA-Lachema Diagnostika, Brno, Czech Republic) [47].

2.3.7. Antioxidant and Immunological Assays

Nitric oxide (NO) was assessed spectrophotometrically (according to a protocol supported by the manufacturer) by the commercial diagnostic kits (BioChain Institute Inc., Newark, CA, USA). Serum malondialdehyde (MDA) levels (as a marker of lipid peroxidation (LPO)) were calorimetrically assessed by using a commercial diagnostic kit (Lipid peroxide (LPO), OXIS International Inc., Portland, OR, USA) [48]. Respiratory burst activity of the whole blood sample was assessed by nitro blue tetrazolium (NBT) dye [49]. Lysozyme activity (LYZ) was evaluated by using turbidity measurement, described by Siwicki and Studnicka [50] and Ellis [51].

2.4. Experiment II: Modulation of Sub-Lethal Imidacloprid Toxicity

2.4.1. Experimental Design

The remaining fish in all experimental groups (32 fish per group) continued feeding on the corresponding diets containing the graded AC levels and then exposed to a sub-lethal dose of imidacloprid (IMID) (Imidacloprid 35% SC) (Tagros Chemicals India Ltd., Chennai, India) for an additional 2 weeks. The selected sub-lethal dose of IMID was one-tenth of the previously calculated 96h LC50 = 0.0109 μg/L. The 96h LC50 of IMID in Nile tilapia was calculated in our study as 0.109 μg/L [28]. To maintain the needed IMID dose, water (in each aquarium) was daily substituted, and the calculated IMID dose was then admixed into a little amount of water before being added to the aquarium water. Dead fish were removed daily and recorded to estimate the mortality rate (MR) (%) and SR (%). Relative percent of survival (RPS) (%) was calculated according to this equation. RPS (%) = 100 × (1 − % of mortality in experimental/% of mortality in control).

2.4.2. Sampling

Serum and tissue (liver and flesh) samples were assembled from eight fish per group (n = 8) at the end of the sub-acute toxicity test.

2.4.3. Determination of Serum Stress Biomarkers

Serum MDA levels and blood glucose levels as stress biomarkers were evaluated according to the previously described methods (please see Sections 2.3.6 and 2.3.7.

2.4.4. Determination of Imidacloprid (IMID) Residues

One gram from the fish tissues (either from liver or flesh from the dorsal muscles) was pooled from each fish (n = 12) from each experimental group and then mixed with 5 mL acetonitrile for 4 min.

The tissue homogenate was processed, and IMID residues were determined using high-performance liquid chromatography (HPLC), as in methods previously clarified in Dewangan et al. [52] and Ismael et al. [34].

2.5. Statistical Analytics

One-way ANOVA was used to assess the effect of AC after 8 weeks of the feeding experiment. Second-order polynomial regression analysis was done to estimate the optimum dietary AC level for the lowest blood glucose and MDA levels of fish in all experimental groups and exposed to sub-lethal IMID dose for 2 weeks. Differences between experimental groups were clarified using Duncan's multiple range test as a post hoc test, and $p < 0.05$ was determined as statistically significant. The analyzed data are represented as the mean ± standard error (S.E.). Data analyses were performed using SPSS program version 22 (SPSS, v 22.0; SPSS Inc., Chicago, IL, USA) and GraphPad Prism Software 5.0 (GraphPad Software, Inc., San Diego, CA, USA).

3. Results

3.1. Results of Experiment I

3.1.1. Growth Performance

Table 2 shows the growth indices and survival rate (SR) (%) of Nile tilapia fed diets supplemented with different AC levels for 8 weeks. There were no significant differences in the initial body weight (IBW) of fish in all groups ($p > 0.05$). TFI was significantly increased ($p < 0.05$) in all AC groups compared to the control group. FBW values were significantly increased in AC5, AC10, and AC15 over AC 0 values. Moreover, WG values were statistically the highest in the AC5 group among all groups. FER values were statistically the highest values in the AC10 group among all groups. FCR was significantly decreased ($p < 0.05$) in all AC groups compared to the control group. SR (%) was 100% in all experimental groups with no recorded mortalities throughout the whole experimental period. This finding suggests that dietary AC supplementation does not induce toxic effects on the treated fish.

Table 2. Growth parameters and survival rate (%) of Nile tilapia fed diets supplemented with graded, activated charcoal (AC) levels for 8 weeks.

Parameters	Experimental Groups					p Value
	AC 0	AC 5	AC 10	AC 15	AC 20	
IBW (g)	33.66 ± 0.05	33.22 ± 0.09	33.17 ± 0.06	33.14 ± 0.12	33.15 ± 0.32	0.356
FBW (g)	51.05 ± 0.08 c	55.35 ± 0.33 a	54.75 ± 0.25 a	54.44 ± 0.26 a	53.82 ± 0.61 ab	0.002
WG (g)	18.39 ± 0.98 b	22.13 ± 0.34 a	21.58 ± 0.24 ab	21.30 ± 0.76 ab	20.67 ± 0.63 ab	0.029
TFI (g)	30.33 ± 0.13 c	31.32 ± 0.92 b	33.19 ± 0.52 a	33.17 ± 0.02 a	31.29 ± 0.32 b	<0.001
FER (g/g)	0.55 ± 0.12 b	0.66 ± 0.02 ab	0.71 ± 0.02 a	0.68 ± 0.10 ab	0.66 ± 0.05 ab	0.003
FCR (g/g)	1.80 ± 0.25 a	1.50 ± 0.85 bc	1.41 ± 0.28 d	1.47 ± 0.37 c	1.51 ± 0.92 bc	0.011
SR (%)	100	100	100	100	100	0.476

IBW: initial body weight, FBW: final body weight, WG: weight gain, TFI: total feed intake, FER: feed efficiency ratio, FCR: feed conversion ratio, SR: survival rate. Values in the same row showing different letters are statistically significantly different ($p < 0.05$). Data are presented as the mean ± S.E.M.

3.1.2. Hemato-Biochemical Measurements

There were no significant differences ($p > 0.05$; Table 3) in the hematological picture of fish in all experimental groups, except the lymphocyte counts were significantly increased in all AC-supplemented groups compared with AC 0 group. Moreover, there was a significant increase ($p < 0.05$) in TP, ALB, and GLO values (Table 4) and a significant decrease ($p < 0.05$) in ALP, AST, and ALT activities in all AC groups (Table 4).

Table 3. Hematological profile of Nile tilapia fed diets supplemented with graded, activated charcoal (AC) levels for 8 weeks.

Parameters	Experimental Groups					*p* Value
	AC 0	AC 5	AC 10	AC 15	AC 20	
Erythrocyte constituents						
RBCs ($10^6 \times$ µL)	2.22 ± 0.19	2.53 ± 0.36	2.80 ± 0.44	2.61 ± 0.07	2.66 ± 0.39	0.593
Hb (g/dL)	8.80 ± 0.57	9.38 ± 0.79	10.20 ± 0.23	9.74 ± 0.95	9.67 ± 0.93	0.596
PCV (mg/L)	29.5 ± 1.44	32.1 ± 2.57	34.5 ± 1.44	32.6 ± 0.57	33.7 ± 3.33	0.568
MCV (fL)	134.6 ± 4.23	129.0 ± 2.33	123.4 ± 1.27	125.8 ± 1.75	128.7 ± 3.43	0.585
MCHC (g/dL)	29.75 ± 0.52	29.20 ± 0.21	29.63 ± 0.57	29.87 ± 0.41	28.72 ± 0.80	0.328
Leucocyte constituents						
WBCs ($10^6 \times$ µL)	3.40 ± 0.17	3.68 ± 0.23	3.85 ± 0.87	3.64 ± 0.05	4.03 ± 0.03	0.068
Lymphocyte ($10^6 \times$ µL)	1.34 ± 0.79 b	1.71 ± 0.72ab	1.83 ± 0.11ab	1.65 ± 0.38ab	2.01 ± 0.01a	0.002
Heterophils ($10^6 \times$ µL)	1.15 ± 0.03	1.11 ± 0.31	1.20 ± 0.32	1.12 ± 0.10	1.22 ± 0.33	0.129
Eosinophils (%)	8.53 ± 0.09	7.07 ± 0.06	6.49 ± 0.29	7.69 ± 0.12	5.46 ± 0.23	0.199
Monocytes (%)	18.53 ± 0.02	16.30 ± 0.27	14.99 ± 0.17	16.48 ± 0.09	14.39 ± 0.43	0.588

RBCs: red blood cells, Hb: hemoglobin, PCV: packed cell volume, MCV: mean corpuscular volume, MCHC: mean corpuscular hemoglobin concentration, WBCs: white blood cells. Values in the same row showing different letters are statistically significantly different ($p < 0.05$). Data are presented as the mean ± S.E.M.

Table 4. Serum biochemical indices of Nile tilapia fed diets supplemented with graded, activated charcoal (AC) levels for 8 weeks.

Parameters	Experimental Groups					*p* Value
	AC 0	AC 5	AC 10	AC 15	AC 20	
Blood protein profile						
TP (g/dL)	3.03 ± 0.32 d	3.41 ± 0.26 c	5.01 ± 0.36 b	5.98 ± 0.19 a	5.74 ± 0.23 ab	<0.001
ALB (g/dL)	1.03 ± 0.05 d	1.77 ± 0.81 c	2.79 ± 0.28 b	3.93 ± 0.88 a	3.07 ± 0.66 ab	<0.001
GLO (g/dL)	1.34 ± 0.55 d	1.90 ±0.07c	2.23 ± 0.38 bc	2.06 ± 0.13 c	2.67 ± 0.29 a	<0.001
Liver function enzymes						
ALP (U/L)	13.82 ± 0.73 a	9.98 ± 1.05 b	7.11 ± 1.16 c	8.55 ± 0.49 b	7.28 ± 0.71 c	<0.001
AST (IU/L)	78.5 ± 1.75 a	36.5 ± 1.75 b	28.0 ± 1.69 b	31.0 ± 1.44 b	30.5 ± 1.02 b	<0.001
ALT (IU/L)	100.5 ± 1.56 a	90.6 ± 1.83 b	91.0 ± 1.97 b	66.5 ± 1.89 c	70.5 ± 1.88 bc	0.035

TP: total protein, ALB: albumin, GLO: globulin, ALP: alkaline phosphatase, AST: aspartate transaminase, ALT: alanine transaminase. Values in the same row showing different letters are statistically significantly different ($p < 0.05$). Data are presented as the mean ± S.E.M.

3.1.3. Antioxidation and Immune Responses

Serum LYZ activities (Figure 1A) and NBT values (Figure 1B) were significantly elevated ($p < 0.05$) in all AC groups compared to the control group. On the other hand, serum MDA levels (Figure 2A) were significantly lowered ($p < 0.05$) in all AC groups compared to the control group. However, NO levels (Figure 2B) were statistically increased in AC groups compared to the control.

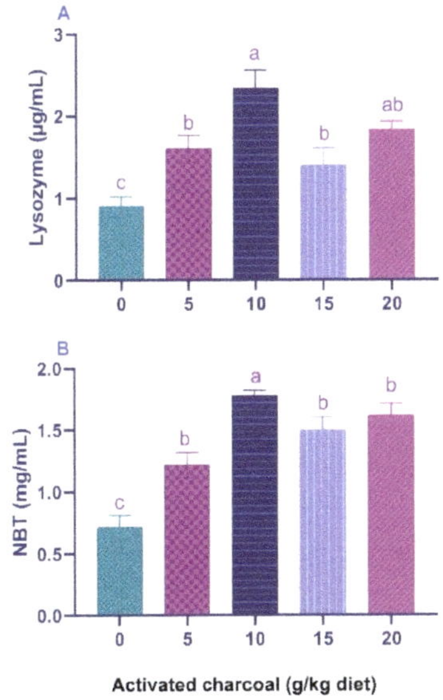

Figure 1. Serum lysozyme (LYZ) activity (**A**) and Nitro blue tetrazolium (NBT) levels (**B**) of Nile tilapia fed diets supplemented with graded, activated charcoal (AC) levels for 8 weeks. (a, b, c) indicate significant differences between groups.

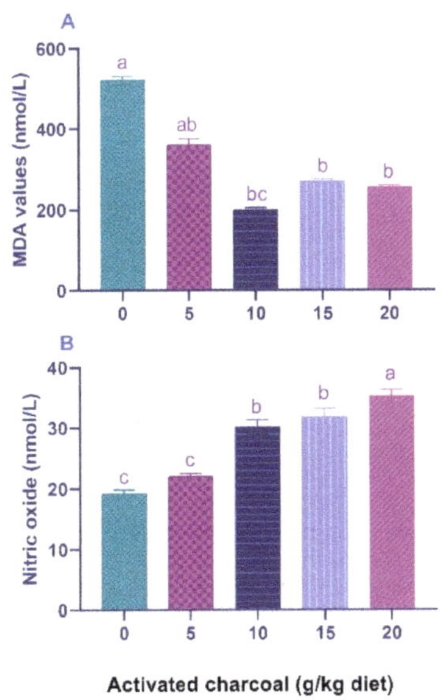

Figure 2. Serum malondialdehyde (MDA) (**A**) and nitric oxide (NO) levels (**B**) of Nile tilapia fed diets supplemented with graded, activated charcoal (AC) levels for 8 weeks. (a, b, c) indicate significant differences between groups.

3.2. Results of Experiment II: Responses to Sub-Lethal Imidacloprid (IMID) Toxicity

SR (%) and RPS (%) (Table 5) were significantly increased ($p < 0.05$), and IMID residual levels in the flesh (Figure 3A) and liver (Figure 3B) were significantly decreased ($p < 0.05$)

in all AC groups compared with the AC 0 group after exposure to sub-lethal IMID toxicity for 2 weeks.

Table 5. Relative percentage survival and mortality rate (%) of Nile tilapia fed diets supplemented with graded, activated charcoal (AC) levels for 8 weeks and then exposed to a sub-lethal level of imidacloprid (IMID) for 2 weeks (Experiment II).

Table	Total No.	Dead Fish	SR (%)	MR (%)	RPS (%)
AC 0	32	6	81.25	18.75	-
AC 5	32	5	84.40	15.60	16.66
AC 10	32	2	93.75	6.25	66.66
AC 15	32	2	93.75	6.25	66.66
AC 20	32	2	93.75	6.25	66.66

SR: survival rate (%), MR: mortality rate, RPS: relative percentage survival.

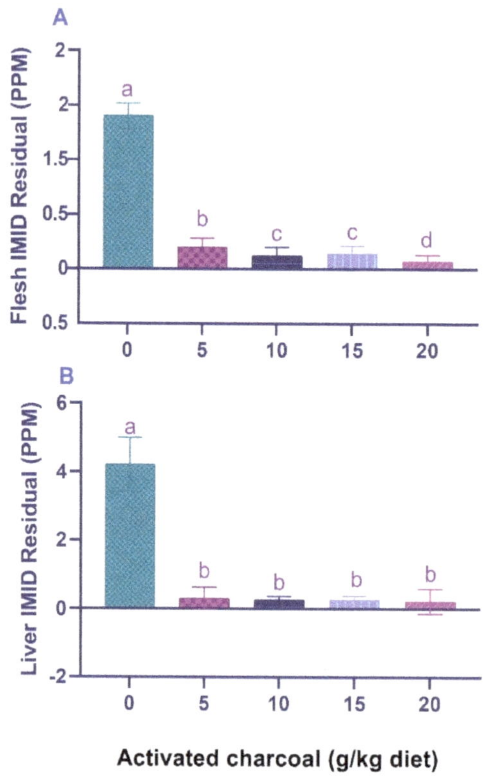

Figure 3. Residual imidacloprid (IMID) levels in the flesh (**A**) and liver (**B**) of Nile tilapia fed on diets supplemented with graded, activated charcoal (AC) levels for 2 months and then exposed to a sub-lethal IMID level for 2 weeks. (a, b, c, d) indicate significant differences between groups.

There was a significant decrease ($p < 0.05$) in blood glucose (Figure 4) and serum MDA levels (Figure 5) in all AC groups compared to the AC 0 group after exposure to sub-lethal IMID toxicity. Second-order polynomial regression analysis showed that the lowest glucose and MDA levels (Figures 4 and 5) were found at dietary supplementation with 14.30 g/kg diet after exposure to a sub-lethal IMID dose. This result suggests that dietary supplementation with 14.30 g/kg diet in the formulated feeds of Nile tilapia could mitigate the stressful effects of sub-lethal IMID toxicity.

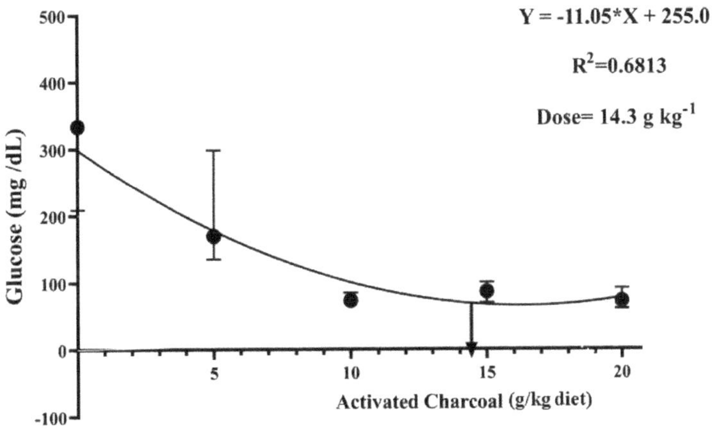

Figure 4. Second-order polynomial regression equation between blood glucose levels of Nile tilapia fed different dietary activated charcoal (AC) levels for 8 weeks and then exposed to a sub-lethal imidacloprid (IMID) level for 2 weeks. Values expressed as means ± S.E.M.

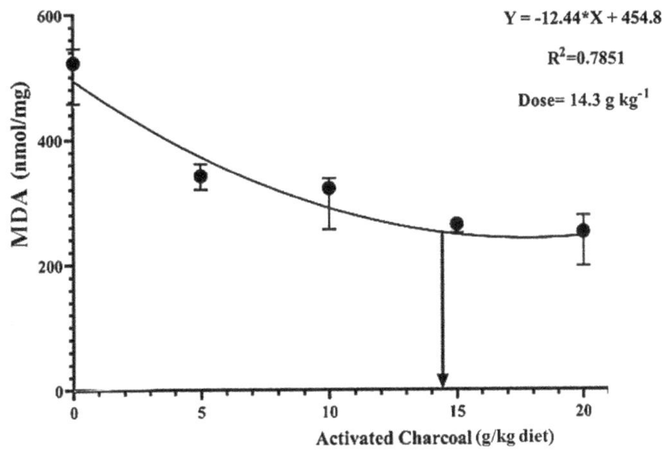

Figure 5. Second-order polynomial regression equation between malondialdehyde (MDA) levels of Nile tilapia fed different dietary activated charcoal (AC) levels for 8 weeks and then exposed to a sub-lethal imidacloprid (IMID) level for 2 weeks. Values expressed as means ± S.E.M.

4. Discussion

4.1. Experiment I-Feeding Trial

4.1.1. Growth Performance

The present study reported significant enhancement of growth indices of Nile tilapia fed on diets supplemented with graded AC levels for 8 weeks. Pirarat et al. [9] illustrated that Nile tilapia fed on 2% AC-supplemented diet for 4 weeks showed noticeably improved growth performance and intestinal histomorphological criteria. Abdel-Tawwab et al. [10] illustrated that a 7.0 g/kg diet considerably enhanced the TFI and growth indices of Nile tilapia. Michael et al. [19] showed that 3% commercial wood charcoal improved the growth performance, nutrient utilization parameters, and proximate chemical composition of red tilapia juveniles. Dietary supplementation with a diet of 5 g of bamboo charcoal kg-1 noticeably enhanced the growth of flounder juveniles [16]. Moreover, dietary supplementation with charcoal and wood vinegar mixture at 5 and 10 g/kg diet for 8 weeks considerably increased the FER and WG of flounder [18]. However, Boonanuntanasarn et al. [2] reported no statistically significant differences in the growth of Nile tilapia fed diets supplemented with graded AC levels for 4 weeks. These inconsistencies may be associated with several factors, including different AC (source, composition, and supplementation levels), experiment (design, rearing, and period), and fish differences (species, IBW, and feeding habits).

The improvement of the growth performance of Nile tilapia in the current study may be explained by several factors such as (1) the adsorptive ability of AC, which helps in the elimination of the impurities and gases from the intestinal tract, which will, in turn, improve the digestion of diets [6,7], (2) dietary supplementation with AC, which could improve the absorptive functions of the intestinal villi, which will consequently increase the feed utilization [8], and (3) improving the intestinal histomorphometric parameters such as the intestinal villi height of Nile tilapia [2,9], and giant trevally [13,14]. The increase of the intestinal villus height will increase the surface area of absorption, contributing positively to the absorption of nutrients [53].

4.1.2. Hemato-Biochemical Indices

Hematological profile could be considered important physiological bioindicators for evaluating the overall performances and health status of fish [54]. In the present study, there were no statistical differences in the RBC, PCV, Hb, and WBC count of Nile tilapia in all experimental groups. Our results were consistent with those reported by Boonanuntanasarn et al. [2], who reported no significant differences in RBCs' count and Hb and hematocrit values of Nile tilapia fed AC-supplemented diets. Moreover, similar findings were reported in African catfish fed on AC-based diets [11]. In a similar sense, there were no differences in Hb and hematocrit values in Tiger puffer fish fed diets supplemented with bamboo charcoal [15]. These findings suggest that dietary AC levels did not negatively affect the health status of the treated fish.

On the other hand, a significant increase in TP, ALB, and GLO values was recorded in all AC-supplemented groups in the current study. Samadaii and Bahrekazemi [12] illustrated that a 15 g/kg diet significantly decreased the TP and ALB values of big sturgeon (*Huso huso*).

Transaminases are biomarkers of fish liver functions [36,55]. The increase of these enzymes in fish plasma is regarded as an indicator of liver damage after exposure to aquatic toxicants [56–58]. The results reported a significant decrease of ALP, AST, and ALT activities observed in all AC-supplemented groups, and these findings suggest a healthy status of the liver of the treated fish. Yoo et al. [18] found that dietary charcoal and vinegar mixture noticeably decreased ALT and AST activities in Olive flounder. A similar decrease in ALT and AST activities were also noticed in Nile tilapia fed AC-supplemented diets [10]. Moreover, dietary supplementation with 15 g/kg diet considerably decreases ALT and AST activities in big sturgeon [12]. Contrarily, there were no significant changes in ALT and AST activities in Nile tilapia fed AC-supplemented diets [2]. These differences may be attributed to different AC sources, supplementation levels, experimental design, rearing conditions, and study period.

4.1.3. Immunity and Antioxidant Biomarkers

LYZ is an important enzyme in the non-specific immune responses of fish, required for breaking the cell walls of G+ve and G−ve bacteria [59]. Respiratory burst activity of fish phagocytes is associated with the attack of the challenged pathogens during the process of phagocytosis [60]. NO increases the ability of macrophages to engulf and destroy the challenged foreign pathogens [61]. In the current study, there were significant increments in serum LYZ, respiratory burst activities, and NO values in AC-supplemented groups, which indicate an enhancement of immune responses of the treated fish.

MDA levels are indicators of LPO, which occurs during the oxidative damage of host tissues due to the overproduction of reactive oxygen species [62,63]. A decrease in serum MDA levels in all AC groups suggests a decrease in LPO. Our results were inconsistent with those reported by Abdel-Tawwab et al. [10], who found no significant changes in the MDA levels of Nile tilapia fed AC-supplemented diets.

4.2. Experiment II-Responses to Sub-Lethal Imidacloprid (IMID) Toxicity

Reports showed the toxicological influences of IMID in different fish species [23–28]. Herein, it was found that after exposure to a sub-lethal IMID dose for 2 weeks, there were significantly increased SR% and decreased blood glucose, MDA levels, and IMID residual levels in the flesh and liver in all AC groups compared with the control group. These findings indicate that dietary AC can protect Nile tilapia against sub-lethal IMID toxicity and reduce its bioaccumulation in fish body. Moreover, dietary supplementation with 14.30 g AC kg^{-1} diet was the ideal dose to counteract the stress effects of IMID. In the same sense, Abdel-Tawwab et al. [10] found a significant decrease of MDA and glucose levels of Nile tilapia fed AC-supplemented diets after environmental heavy metals' exposure. Moreover, a significant decrease in heavy metals' bioaccumulation was also observed in tissues of big Sturgeon fed AC-supplemented diets compared to those fed the control diet [12]. Furthermore, Naiel et al. [28] demonstrated a significant decrease of IMID residual levels in flesh of Nile tilapia previously fed on diets supplemented with both ChNPs and vitamin C. Our recently published study showed that dietary supplementation with clinoptilolite and/or ChNPs significantly decreased the mortality rates and IMID residual levels in flesh of Nile tilapia [34].

These findings could be attributed to the characteristics of AC as a universal antidotal treatment of several toxicants and pollutants. The possible mechanisms of dietary AC could be related to several factors, including (1) its chelating properties, (2) increased elimination rate, and (3) increased adsorption to toxic elements, which will subsequently decrease the absorption of toxicants [2].

5. Conclusions

The findings of the present study indicated that dietary AC could enhance the growth performance and improve serum biochemical measurements, antioxidant capacity, and non-specific immunity of Nile tilapia juveniles. Moreover, a 14.30 g/kg diet could be recommended as an ideal dietary supplementation dose to mitigate the stressful effects of sub-lethal IMID toxicity in Nile tilapia.

Author Contributions: H.M.R.A.-L. and S.A.A.A.E.-h., methodology; M.M.S., M.S., S.S.N. and N.E.M.I., software; M.A.E.N., M.M.S. and H.M.R.A.-L., validation; H.M.R.A.-L. and S.A.A.A.E.-h., formal analysis; M.M.S., M.S., S.S.N., N.E.M.I. and H.M.R.A.-L., writing—original draft preparation; M.A.E.N., M.M.S. and H.M.R.A.-L., visualization; H.M.R.A.-L., M.A.E.N. and M.S., supervision. All authors have read and agreed to the published version of the manuscript.

Funding: This study was endorsed by Taif University Researchers Supporting Project (TURSP-2020/09), Taif University, Taif, Saudi Arabia.

Institutional Review Board Statement: Experimental techniques in the present investigation were accomplished in accordance with the local ethics of the Experimental Animal Care Committee and permitted by the Governmental Institutional Ethics Committee, Department of Animal Production, Faculty of Agriculture, Zagazig University, Egypt (Approval No. Zu-IACUC/2/F/99/2018).

Informed Consent Statement: Not applicable.

Data Availability Statement: Available from the corresponding author on call.

Acknowledgments: We deeply thank Taif University for the financial assistance for the Taif University Researchers Supporting Project (TURSP-2020/09), Taif University, Taif, Saudi Arabia.

Conflicts of Interest: The authors declare no conflict of interest. The funders had no role in the design of the study; in the collection, analyses, or interpretation of data; in the writing of the manuscript, or in the decision to publish the results.

References

1. Osol, A. *Remington's Pharmaceutical Sciences*, 15th ed.; Mack Publishing Co.: Easton, PA, USA, 1975.
2. Boonanuntanasarn, S.; Khaomek, P.; Pitaksong, T.; Hua, Y. The effects of the supplementation of activated charcoal on the growth, health status and fillet composition-odor of Nile tilapia (*Oreochromis niloticus*) before harvesting. *Aquac. Int.* **2014**, *22*, 1417–1436. [CrossRef]
3. Galvano, F.; Pietri, A.; Fallico, B.; Bertuzzi, T.; ScirÈ, S.; Galvano, M.; Maggiore, R. Activated Carbons: In Vitro Affinity for Aflatoxin B1 and Relation of Adsorption Ability to Physicochemical Parameters. *J. Food Prot.* **1996**, *59*, 545–550. [CrossRef]
4. Wilson, L.L.; Kurtz, D.A.; Rugh, M.C.; Chase, L.E.; Zieglek, J.H.; Varela-Alvarez, H.; Borger, M.L. Effects of Feeding Activated Carbon on Growth Rate and Pesticide Concentrations in Adipose Tissues of Steers Fed Apple Waste. *J. Anim. Sci.* **1971**, *33*, 1361–1364. [CrossRef] [PubMed]
5. McFarland, A.K.; Chyka, P.A. Selection of Activated Charcoal Products for the Treatment of Poisonings. *Ann. Pharmacother.* **1993**, *27*, 358–361. [CrossRef]
6. Kutlu, H.R.; Ünsal, I.; Görgülü, M. Effects of providing dietary wood (oak) charcoal to broiler chicks and laying hens. *Anim. Feed Sci. Technol.* **2001**, *90*, 213–226. [CrossRef]
7. Samanya, M.; Yamauchi, K.E. Morphological Changes of the Intestinal Villi in Chickens Fed the Dietary Charcoal Powder Including Wood Vinegar Compounds. *J. Poult. Sci.* **2001**, *38*, 289–301. [CrossRef]
8. Mekbungwan, A.; Yamauchi, K.; Sakaida, T. Intestinal Villus Histological Alterations in Piglets fed Dietary Charcoal Powder Including Wood Vinegar Compound Liquid. *Anat. Histol. Embryol.* **2004**, *33*, 11–16. [CrossRef]
9. Pirarat, N.; Boonananthanasarn, S.; Krongpong, L.; Katagiri, T.; Maita, M. Effect of activated charcoal-supplemented diet on growth performance and intestinal morphology of Nile tilapia (*Oreochromis niloticus*). *Thai J. Vet. Med.* **2015**, *45*, 113–119.
10. Abdel-Tawwab, M.; El-Sayed, G.O.; Shady, S.H. Effect of dietary active charcoal supplementation on growth performance, biochemical and antioxidant responses, and resistance of Nile tilapia, *Oreochromis niloticus* (L.) to environmental heavy metals exposure. *Aquaculture* **2017**, *479*, 17–24. [CrossRef]
11. Lawal, M.O.; Aderolu, A.Z.; Aarode, O.O.; Seriki, B.M.; Alonge, T.O. Evaluation of soybean meal replacement with sesame seed meal using activated charcoal as an additive in the diet of African catfish juveniles, *Clarias gariepinus*. *Int. J. Aquat. Biol.* **2016**, *4*, 43–50.
12. Samadaii, S.; Bahrekazemi, M. The effect of diets containing different levels of active charcoal on growth performance, body composition, haematological parameters and possibility of heavy metals detoxification in big sturgeon (*Huso huso*). *Aquac. Res.* **2020**, *51*, 91–101. [CrossRef]
13. Firdus, F.; Samadi, S.; Muhammadar, A.A.; Sarong, M.A.; Muchlisin, Z.A.; Sari, W.; Batubara, A.S. Gut and intestinal biometrics of the giant trevally, *Caranx ignobilis*, fed an experimental diet with difference sources of activated charcoal. *F1000Research* **2020**, *9*, 444. [CrossRef] [PubMed]
14. Firdus, F.; Samadi, S.; Muhammadar, A.A.; Sarong, M.A.; Muchlisin, Z.A.; Sari, W.; Mellisa, S.; Satria, S.; Boihaqi, B.; Batubara, A.S. Supplementation of rice husk activated charcoal in feed and its effects on growth and histology of the stomach and intestines from giant trevally, *Caranx ignobilis*. *F1000Research* **2020**, *9*, 1274. [CrossRef]
15. Thu, M.; Koshio, S.; Ishikawa, M.; Yokoyama, S. Effects of dietary bamboo charcoal on growth parameters, apparent digestibility and ammonia nitrogen excretion of tiger puffer fish, *Takifugu rubripes*. *Aquac. Sci.* **2009**, *57*, 53–60.
16. Thu, M.; Koshio, S.; Ishikawa, M.; Yokoyama, S. Effects of supplementation of dietary bamboo charcoal on growth performance and body composition of juvenile Japanese flounder, *Paralichthys olivaceus*. *J. World Aquac. Soc.* **2010**, *41*, 255–262. [CrossRef]
17. Jahan, R.; Quaiyum, M.A.; Jahan, N.; Akhter, T.; Islam, M.S. Dietary added bamboo charcoal can evoke *Pangasianodon* growth and can reduce ammonia from culture medium. *Int. J. Fish. Aquac.* **2014**, *6*, 87–93.
18. Yoo, J.H.; Ji, S.C.; Jeong, G.S. Effect of dietary charcoal and wood vinegar mixture (CV82) on body composition of Olive flounder *Paralichthys alivaceus*. *J. World Aquac. Soc.* **2005**, *36*, 203–208. [CrossRef]
19. Michael, F.R.; Saleh, N.E.; Shalaby, S.M.; Sakr, E.M.; Abd-El-Khalek, D.E.; Abd Elmonem, A.I. Effect of different dietary levels of commercial wood charcoal on growth, body composition and environmental loading of red tilapia hybrid. *Aquac. Nutr.* **2017**, *23*, 210–216. [CrossRef]
20. Michael, F.R.; Helal, A.M. Rule of dietary activated wood charcoal on the growth and biochemical composition of gilthead Seabream (*Sparus aurata*) reared under different stocking densities. *Life Sci. J.* **2018**, *15*, 79–86.
21. Tomizawa, M.; Casida, J.E. Neonicotinoid insecticide toxicology: Mechanisms of selective action. *Annu. Rev. Pharmacol. Toxicol.* **2005**, *45*, 247–268. [CrossRef]
22. Tišler, T.; Jemec, A.; Mozetič, B.; Trebše, P. Hazard identification of imidacloprid to aquatic environment. *Chemosphere* **2009**, *76*, 907–914. [CrossRef] [PubMed]
23. Naiel, M.A.E.; Shehata, A.M.; Negm, S.S.; Abd El-Hack, M.E.; Amer, M.S.; Khafaga, A.F.; Bin-Jumah, M.; Allam, A.A. The new aspects of using some safe feed additives on alleviated imidacloprid toxicity in farmed fish: A review. *Rev. Aquac.* **2020**, *12*, 2250–2267. [CrossRef]

24. Ansoar-Rodríguez, Y.; Christofoletti, C.A.; Marcato, A.C.; Correia, J.E.; Bueno, O.C.; Malaspina, O.; Fontanetti, C.S. Genotoxic potential of the insecticide imidacloprid in a non-target organism (*Oreochromis niloticus*-Pisces). *J. Environ. Prot.* **2015**, *6*, 1360. [CrossRef]
25. Américo-Pinheiro, J.H.P.; da Cruz, C.; Aguiar, M.M.; Torres, N.H.; Ferreira, L.F.R.; Machado-Neto, J.G. Sublethal effects of Imidacloprid in hematological parameters of Tilapia (*Oreochromis niloticus*). *Water Air Soil Pollut.* **2019**, *230*, 193. [CrossRef]
26. Ansoar-Rodríguez, Y.; Christofoletti, C.A.; Correia, J.E.; de Souza, R.B.; Moreira-de-Sousa, C.; Marcato, A.C.d.C.; Bueno, O.C.; Malaspina, O.; Silva-Zacarin, E.C.M.; Fontanetti, C.S. Liver alterations in *Oreochromis niloticus* (Pisces) induced by insecticide imidacloprid: Histopathology and heat shock protein in situ localization. *J. Environ. Sci. Health Part B* **2016**, *51*, 881–887. [CrossRef]
27. Günal, A.Ç.; Erkmen, B.; Paçal, E.; Arslan, P.; Yildirim, Z.; Erkoç, F. Sub-lethal Effects of Imidacloprid on Nile tilapia (*Oreochromis niloticus*). *Water Air Soil Pollut.* **2020**, *231*, 4. [CrossRef]
28. Naiel, M.A.E.; Ismael, N.E.M.; Abd El-hameed, S.A.A.; Amer, M.S. The antioxidative and immunity roles of chitosan nanoparticle and vitamin C-supplemented diets against imidacloprid toxicity on *Oreochromis niloticus*. *Aquaculture* **2020**, *523*, 735219. [CrossRef]
29. Crosby, E.B.; Bailey, J.M.; Oliveri, A.N.; Levin, E.D. Neurobehavioral impairments caused by developmental imidacloprid exposure in zebrafish. *Neurotox. Teratol.* **2015**, *49*, 81–90. [CrossRef] [PubMed]
30. Topal, A.; Alak, G.; Ozkaraca, M.; Yeltekin, A.C.; Comaklı, S.; Acıl, G.; Kokturk, M.; Atamanalp, M. Neurotoxic responses in brain tissues of rainbow trout exposed to imidacloprid pesticide: Assessment of 8-hydroxy-2-deoxyguanosine activity, oxidative stress and acetylcholinesterase activity. *Chemosphere* **2017**, *175*, 186–191. [CrossRef] [PubMed]
31. Vieira, C.E.D.; Pérez, M.R.; Acayaba, R.D.A.; Raimundo, C.C.M.; dos Reis Martinez, C.B. DNA damage and oxidative stress induced by imidacloprid exposure in different tissues of the Neotropical fish *Prochilodus lineatus*. *Chemosphere* **2018**, *195*, 125–134. [CrossRef]
32. Sánchez-Bayo, F.; Goka, K. Unexpected effects of zinc pyrithione and imidacloprid on Japanese medaka fish (*Oryzias latipes*). *Aquat. Toxicol.* **2005**, *74*, 285–293. [CrossRef] [PubMed]
33. Hong, X.; Zhao, X.; Tian, X.; Li, J.; Zha, J. Changes of hematological and biochemical parameters revealed genotoxicity and immunotoxicity of neonicotinoids on Chinese rare minnows (*Gobiocypris rarus*). *Environ. Pollut.* **2018**, *233*, 862–871. [CrossRef] [PubMed]
34. Ismael, N.E.M.; Abd El-hameed, S.A.A.; Salama, A.M.; Naiel, M.A.E.; Abdel-Latif, H.M.R. The effects of dietary clinoptilolite and chitosan nanoparticles on growth, body composition, haemato-biochemical parameters, immune responses, and antioxidative status of Nile tilapia exposed to imidacloprid. *Environ. Sci. Pollut. Res.* **2021**. [CrossRef] [PubMed]
35. Abdelghany, M.F.; El-Sawy, H.B.; Abd El-hameed, S.A.A.; Khames, M.K.; Abdel-Latif, H.M.R.; Naiel, M.A.E. Effects of dietary *Nannochloropsis oculata* on growth performance, serum biochemical parameters, immune responses, and resistance against *Aeromonas veronii* challenge in Nile tilapia (*Oreochromis niloticus*). *Fish Shellfish Immunol.* **2020**, *107*, 277–288. [CrossRef] [PubMed]
36. Abdel-Latif, H.M.R.; Abdel-Tawwab, M.; Khafaga, A.F.; Dawood, M.A.O. Dietary oregano essential oil improved the growth performance via enhancing the intestinal morphometry and hepato-renal functions of common carp (*Cyprinus carpio* L.) fingerlings. *Aquaculture* **2020**, *526*, 735432. [CrossRef]
37. Mohammadi, G.; Rafiee, G.; El Basuini, M.F.; Abdel-Latif, H.M.R.; Dawood, M.A.O. The growth performance, antioxidant capacity, immunological responses, and the resistance against *Aeromonas hydrophila* in Nile tilapia (*Oreochromis niloticus*) fed *Pistacia vera* hulls derived polysaccharide. *Fish Shellfish Immunol.* **2020**, *106*, 36–43. [CrossRef]
38. Brown, B.A. Routine hematology procedures. In *Hematology: Principles and Procedures*; Brown, B.A., Ed.; Leo and Fabiger: Philadelphia, PA, USA, 1988; pp. 7–122.
39. Collier, H.B. Standardization of blood haemoglobin determinations. *Can. Med. Assoc. J.* **1944**, *50*, 550–552. [PubMed]
40. Wintrobe, M.M. Variations in the size and hemoglobin content of erythrocytes in the blood of various vertebrates. *Folia Haematol.* **1934**, *51*, 32–49.
41. Dacie, J.V.; Lewis, S.M. *Practical Haematology*, 6th ed.; Churchill Livingstone: London, UK, 1984.
42. Klontz, G. *Fish hematology. Techniques in Fish Immunology*; Stolen, J.S., Fletcher, T.C., Rowley, A.F., Kelikoff, T.C., Kaatari, S.L., Smith, S.A., Eds.; SOS Publications: Fair Haven, NJ, USA, 1994; Volume 3.
43. Henry, R. Colorimetric Determination of Total Protein. In *Clinical Chemistry*; Harper and Row Publ.: New York, NY, USA, 1964.
44. Wotton, I.D.; Freeman, H. *Microanalysis in Medical Biochemistry*; Churchill: New York, NY, USA, 1982.
45. 4Reitman, S.; Frankel, S. A colorimetric method for the determination of serum glutamic oxalacetic and glutamic pyruvic transaminases. *Am. J. Clin. Pathol.* **1957**, *28*, 56–63. [CrossRef]
46. Tietz, N.; Burtis, C.; Duncan, P.; Ervin, K.; Petitclerc, C.; Rinker, A.; Shuey, D.; Zygowicz, E. A reference method for measurement of alkaline phosphatase activity in human serum. *Clin. Chem.* **1983**, *29*, 751–761. [CrossRef]
47. Bartoňková, J.; Hyršl, P.; Vojtek, L. Glucose determination in fish plasma by two different moderate methods. *Acta Vet. Brno* **2017**, *85*, 349–353. [CrossRef]
48. Uchiyama, M.; Mihara, M. Determination of malonaldehyde precursor in tissues by thiobarbituric acid test. *Anal. Biochem.* **1978**, *86*, 271–278. [CrossRef]
49. Anderson, D.; Siwicki, A. *Basic Haematology and Serology for Fish Health Programs*; Fish Health Section, Asian Fisheries Society: Manila, Phillipines, 1995.
50. Siwicki, A.; Studnicka, M. The phagocytic ability of neutrophils and serum lysozyme activity in experimentally infected carp, *Cyprinus carpio* L. *J. Fish Biol.* **1987**, *31*, 57–60. [CrossRef]

51. Ellis, A.E. Lysozyme assays. *Tech. Fish Immunol.* **1990**, *1*, 101–103.
52. Dewangan, G.; Mishra, A.; Mandal, T.K. A simple RP-HPLC method for determining imidacloprid residues in goat tissues. *Biomed. Res.* **2016**, *27*, 1.
53. Dawood, M.A.; Gewaily, M.S.; Soliman, A.A.; Shukry, M.; Amer, A.A.; Younis, E.M.; Abdel-Warith, A.-W.A.; Van Doan, H.; Saad, A.H.; Aboubakr, M. Marine-Derived Chitosan Nanoparticles Improved the Intestinal Histo-Morphometrical Features in Association with the Health and Immune Response of Grey Mullet (*Liza ramada*). *Mar. Drugs* **2020**, *18*, 611. [CrossRef] [PubMed]
54. Dawood, M.A.O.; Metwally, A.E.-S.; El-Sharawy, M.E.; Ghozlan, A.M.; Abdel-Latif, H.M.R.; Van Doan, H.; Ali, M.A.M. The influences of ferulic acid on the growth performance, haemato-immunological responses, and immune-related genes of Nile tilapia (*Oreochromis niloticus*) exposed to heat stress. *Aquaculture* **2020**, *525*, 735320. [CrossRef]
55. Abdel-Wahab, M.; Taha, N.; Lebda, M.; Elfeky, M.; Abdel-Latif, H.M.R. Effects of bovine lactoferrin and chitosan nanoparticles on serum biochemical indices, antioxidative enzymes, transcriptomic responses, and resistance of Nile tilapia against *Aeromonas hydrophila*. *Fish Shellfish Immunol.* **2021**. [CrossRef] [PubMed]
56. Abdel-Latif, H.; Abou Khashaba, A. Subchronic toxicity of Nile tilapia with different exposure routes to *Microcystis aeruginosa*: Histopathology, liver functions, and oxidative stress biomarkers. *Vet. World* **2017**, *10*, 955. [CrossRef]
57. El Euony, O.I.; Elblehi, S.S.; Abdel-Latif, H.M.; Abdel-Daim, M.M.; El-Sayed, Y.S. Modulatory role of dietary *Thymus vulgaris* essential oil and *Bacillus subtilis* against thiamethoxam-induced hepatorenal damage, oxidative stress, and immunotoxicity in African catfish (*Clarias garipenus*). *Environ. Sci. Pollut. Res.* **2020**, *27*, 23108–23128. [CrossRef] [PubMed]
58. Khafaga, A.F.; Naiel, M.A.E.; Dawood, M.A.O.; Abdel-Latif, H.M.R. Dietary *Origanum vulgare* essential oil attenuates cypermethrin-induced biochemical changes, oxidative stress, histopathological alterations, apoptosis, and reduces DNA damage in *Common carp* (*Cyprinus carpio*). *Aquat. Toxicol.* **2020**, *228*, 105624. [CrossRef] [PubMed]
59. Alexander, J.B.; Ingram, G.A. Noncellular nonspecific defence mechanisms of fish. *Annu. Rev. Fish Dis.* **1992**, *2*, 249–279. [CrossRef]
60. Secombes, C.J. Isolation of salmonid macrophages and analysis of their killing activity. *Tech. Fish Immunol.* **1990**, *1*, 137–154.
61. Villamil, L.; Tafalla, C.; Figueras, A.; Novoa, B. Evaluation of Immunomodulatory Effects of Lactic Acid Bacteria in Turbot (*Scophthalmus maximus*). *Clin. Diagn. Lab. Immunol.* **2002**, *9*, 1318. [CrossRef]
62. Awed, E.M.; Sadek, K.M.; Soliman, M.K.; Khalil, R.H.; Younis, E.M.; Abdel-Warith, A.-W.A.; Van Doan, H.; Dawood, M.A.; Abdel-Latif, H.M. *Spirulina platensis* Alleviated the Oxidative Damage in the Gills, Liver, and Kidney Organs of Nile Tilapia Intoxicated with Sodium Sulphate. *Animals* **2020**, *10*, 2423. [CrossRef] [PubMed]
63. Hoseinifar, S.H.; Yousefi, S.; Van Doan, H.; Ashouri, G.; Gioacchini, G.; Maradonna, F.; Carnevali, O. Oxidative stress and antioxidant defense in fish: The implications of probiotic, prebiotic, and synbiotics. *Rev. Fish. Sci. Aquac.* **2020**, *29*, 198–217. [CrossRef]

animals

MDPI

Article

Ameliorative Effects of Boswellic Acid on Fipronil-Induced Toxicity: Antioxidant State, Apoptotic Markers, and Testicular Steroidogenic Expression in Male Rats

Hossam G. Tohamy [1], Sara E. El-Kazaz [2], Saqer S. Alotaibi [3], Hawary S. Ibrahiem [4], Mustafa Shukry [5,*] and Mahmoud A. O. Dawood [6,*]

1 Department of Pathology, Faculty of Veterinary Medicine, Alexandria University, Alexandria 22758, Egypt; hossam.gafar@yahoo.com
2 Animals and Poultry Behavior and Management, Department of Animal Husbandry and Animal Wealth Development, Faculty of Veterinary Medicine, Alexandria University, Alexandria 22758, Egypt; sara@agr.kfs.edu.eg
3 Department of Biotechnology, College of Science, Taif University, P.O. Box 11099, Taif 21944, Saudi Arabia; saqer@tu.edu.sa
4 Department of Veterinary Pharmacology, Faculty of Veterinary Medicine, Alexandria University, Alexandria 22758, Egypt; hawary@agr.kfs.edu.eg
5 Department of Physiology, Faculty of Veterinary Medicine, Kafrelsheikh University, Kafrelsheikh 33516, Egypt
6 Department of Animal Production, Faculty of Agriculture, Kafrelsheikh University, Kafrelsheikh 33516, Egypt
* Correspondence: mostafa.ataa@vet.kfs.edu.eg (M.S.); Mahmouddawood@agr.kfs.edu.eg (M.A.O.D.)

Citation: Tohamy, H.G.; El-Kazaz, S.E.; Alotaibi, S.S.; Ibrahiem, H.S.; Shukry, M.; Dawood, M.A.O. Ameliorative Effects of Boswellic Acid on Fipronil-Induced Toxicity: Antioxidant State, Apoptotic Markers, and Testicular Steroidogenic Expression in Male Rats. *Animals* **2021**, *11*, 1302. https://doi.org/10.3390/ani11051302

Academic Editors: Woo Kyun Kim, Youssef A. Attia, Maria de Olivera and Nesrein Hashem

Received: 29 March 2021
Accepted: 26 April 2021
Published: 30 April 2021

Simple Summary: Fipronil (FPN) is an insecticide that can be used in insect control in various cereal crops in agriculture, veterinary activities, and public health management. Boswellic acid (BA) is a pentacyclic triterpene, which is a compound isolated from Boswellia serrata gum resin. This study was designed to determine BA's potential protective impact against oxidative and testicular damage caused by FPN insecticide poisoning on the male rat model. BA significantly improved the reproductive parameters assessed, such as the number of pregnant females, index of pregnancy and the number of litters, weights of the reproductive organ, sperm cell quality, morphological alterations of testes, epididymis, and sex glands by accessory caused by FPN oxidative stress, as well as the improvement of steroidogenesis, antioxidants, and antiapoptotic marker.

Abstract: The study investigated the ability of boswellic acid (BA) to alleviate the testicular and oxidative injury FPN insecticide intoxication in the male rat model. Rats were randomly assigned to six equivalent groups (six rats each) as the following: control rats orally administered with 2 mL physiological saline/kg of body weight (bwt); boswellic acid (BA1) rats orally administered 250 mg BA/kg bwt; boswellic acid (BA2) rats orally administered 500 mg BA/kg bwt; fipronil (FPN) rats orally administered 20 mg FPN/kg bwt; (FPN + BA1) rats orally administered 20 mg FPN/kg bwt plus 250 mg BA/kg bwt, and (FPN + BA2) rats orally administered 20 mg FPN/kg bwt plus 500 mg BA/kg bwt. After 60 days, semen viability percentage and live spermatozoa percentage were decreased, and a considerably increased abnormality of the sperm cells in FPN-administered rats improved substantially with the co-administration of BA. BA had refinement of the histological architecture of testes and sexual glands. Quantitative analysis recorded a noticeable decline in the nuclear cell-proliferating antigen (PCNA) percentage area. FPN triggered cell damage, which was suggested by elevated malondialdehyde and interleukin 6, tumor necrosis factors alpha, and decreased glutathione level. Proapoptotic factor overexpression is mediated by FPN administration, while it decreased the antiapoptotic protein expression. Similarly, BA has shown significant upregulation in steroidogenic and fertility-related gene expression concerning the FPN group. Pathophysiological damages induced by FPN could be alleviated by BA's antioxidant ability and antiapoptotic factor alongside the upregulation of steroidogenic and fertility-related genes and regimented the detrimental effects of FPN on antioxidant and pro-inflammatory biomarkers.

Keywords: fipronil; boswellic acid; PCNA; semen; fertility-related markers

1. Introduction

Fipronil (FPN) is an N-phenyl-pyrazole insecticide with a wide range and can be used in insect control in various cereal crops in agriculture, veterinary activities, and public health management [1]. FPN is an insect neurotoxin agent, and the critical mechanism is active when the GABA-regulated chloride channels are blocked, causing depression and death in the central nervous system [2]. FPN insecticides are the only organic toxicants to be applied to the ecosystem in a targeted manner to improve food safety by battling pests and regulating disease vectors [3]. Insecticide toxicity is well known to have multiple consequences; creating oxidative injury is a high concept due to reactive oxygen species [4]. Extensive use of these insecticides in agriculture and residential settings causes chronic neurological syndromes, teratogenicity, male reproductive failure, fetal growth retardation, embryo fetotoxicity, and genotoxicity [5]. Beyond liver toxicity, FPN also had reproductive effects, as studies found that the application of FPN influences fertility [6]. The excess output of reactive oxygen species leads to oxidative stress that decreases sperm fertility [7]. The level of serum hepatic enzymes and renal function biomarkers (creatinine and urea), cholesterol, and lactate dehydrogenase increased markedly in FPN intoxication; however, overall serum protein, albumin, and triglycerides decreased significantly, as well as a major increase in malondialdehyde and nitric oxide levels with a significant reduction in glutathione (GSH), glutathione peroxidase (GPx), superoxide dismutase (SOD), and catalase (CAT) [8]. FPN induced deterioration within the seminiferous tubules and apoptosis in the epididymides. Upregulated *interleukin-1β*, *nitric oxide synthase 2*, *caspase-3* (*Casp3*) as well as downregulated *Burkitt-cell lymphomas*, inhibin *B proteins*, and *androgen receptor* mRNA expressions *Casp3*, *nitric oxide synthase*, *ionized calcium-binding adapter molecule 1*, and *IL-1β* immunoreactions were increased. There was also a reduction of *proliferating cell nuclear antigen (PCNA)*, *mouse vasa homolog* (*MVH*), and *SOX9* protein reactions [9].

Many studies report the effect of fipronil on the reproductive system on different animals in which Ohi et al. [10] reported that when fipronil was topically utilized to rats (single dose) at different concentrations (70, 140, and 280 mg/kg), it altered the cyclicity of female rats and had harmful reproductive effects in female rats. Mazzo et al. [6] reported that male rats that received fipronil 5 mg/kg for 14 days had decreased sperm production, reduced epidydimal sperm count, a reduction in GSH, and an increase in the concentration of malondialdehyde. In addition, Eisa et al. [5] reported that rats treated with different doses of fipronil 1/10 LD50 (2.1 mg/kg bwt) and 1/30 LD50 (0.7 mg/kg bwt) at the 6th to 15th days of pregnancy lead to teratogenic and embryotoxic effects. De Barros et al. [11] reported that pregnant rats exposed (via gavage) to fipronil (0.03, 0.3, or 3 mg/kg) from gestational day 15 until postnatal day 7 had infertility. In addition, Kitulagodage et al. [12] breeding female zebra finches orally dosed with single sublethal levels of fipronil (1, 5, and 10 mg/kg body weight) had a decrease in hatchability percentage.

Medicinal herbs are used extensively because of their antimicrobial, antioxidant, and less toxic effects than chemical substances. Boswellic acid (BA) is a pentacyclic triterpene, which is a compound isolated from Boswellia serrata gum resin. It is potent against several inflammatory diseases, including cancer, arthritis, ulcerative colitis, respiratory inflammatory disorders, brain tumor, fertility, and memory [13]. This gum-resin is applied in conventional Chinese medicine to remedy many aspects of well-being [14]. Many in vitro trials indicated that BA hinders the synthesis of the pro-inflammatory enzyme such as *5-lipoxygenase (5-LO)*, *cyclooxygenase-1 (COX-1)*, *human elastase of the leucocytes* (*HLE*), *cathepsin G* (*cat G*), and *microsomal prostaglandin E* (*mPGES-1*), together with *nuclear factor kappa B* (*NF-μB*) inhibition and various cytokines such as *TNFα*, *IL-1β*, and *IL-6*, respectively [15]. Many clinical reports have also illustrated BA's possible value as an anti-inflammatory agent [16–18] and antioxidant activity [19]. Therefore, this study was designed to determine

BA's potential protective impact against oxidative and testicular damage caused by FPN insecticide poisoning on the male rat model.

2. Materials and Methods

2.1. Animals and Management

Thirty-six healthy male albino rats (with a weight of 150 ± 10 g, ten weeks of age) were purchased from the Medical Research Institute, Alexandria University, Egypt. Rats were kept in separate plastic cages under unique conditions (23 ± 2 °C, 55% RH, and 12-h light/dark cycle) and had regular food and water ad libitum. The rats were acclimatized 14 days before the commencement of the experiment to restore normal behavior and growth. During the investigation, they were held under the same hygiene and environmental conditions. The research protocol was accepted by the Institutional Animal Care and Use Committee, Faculty of Veterinary Medicine, Alexandria University, and it was precisely designed under the consideration of animal welfare (AU013202062958).

2.2. Chemicals and Reagents

Fipronil (FPN) was purchased as a commercial product (Rado-X 80%WG, Cruz Agro Development and Investment, Ltd., Shanghai, China), and boswellic acid (BA) has also been procured as hard gelatin capsules (Boswellia serrata dry powder extract 500 mg) (Atos Pharma Phyto Pharmaceuticals Company, Cairo, Egypt). Biochemical kits have been purchased from Biodiagnostics Co. (Cairo, Egypt).

2.3. Experimental Design

Rats were randomly assigned to six equivalent groups (six rats/each) as follows: Group 1 (control) rats were orally administered with two ml physiological saline/kg body weight (bwt) (a vehicle for other drugs); Group 2 (BA1) rats were orally administered 250 mg BA/kg bwt; Group 3 (BA2) rats were orally administered 500 mg BA/kg bwt according to Nusier et al. [20]. The utilized doses of boswellic acid in the current study were tested previously as showed by Sami et al. [21] and Nusier et al. [20] as well as Barakat et al. [22] and Tawfik et al. [23]. Al-Yahya et al. [24] reported that the boswellic acid is safe up to 1000 mg/kg in rats. Still, this dose is relatively high, considering the amount of extract consumed by humans. So, in this study, we try to investigate the effect of boswellic acid in ameliorating the toxic impact of fipronil in a dose-response manner. Group 4 (FPN) rats were orally administered 20 mg FPN/kg bwt, which corresponded to one-fifth of the LD50 [25]; Group 5 (FPN + BA1) rats were orally administered 20 mg FPN/kg bwt plus 250 mg BA/kg bwt, and Group 6 (FPN + BA2) rats were orally administered 20 mg FPN/kg bwt plus 500 mg BA/kg bwt. All treatments were given by stomach tube once daily/five times weekly for 60 days. The co-treatment group with BA was administered an hour before FPN administration. The FPN concentrations were determined by using FPN commercial formulation depending on the percentage of the active ingredient. Concentrations of FPN and BA were freshly made, and body weights were checked weekly throughout the experiment. Ten days after the last dose, the rats have been euthanized and the sample was collected. The experimental designs are shown in Figure 1.

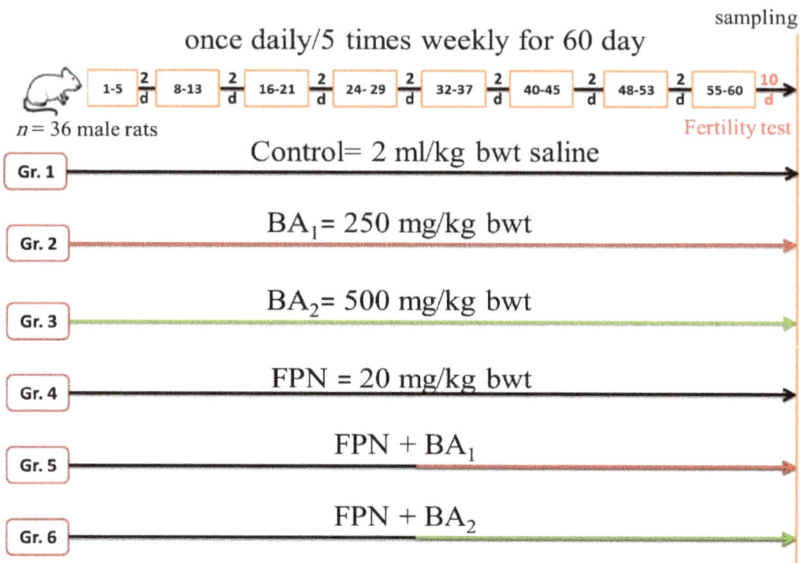

Figure 1. The experimental design. FBN, Fipronil. BA1, Boswellic Acid 250 mg/kg bwt. BA2, Boswellic Acid 500 mg/kg bwt. Gr, Group.

2.4. Fertility Test

Adult male rats were introduced to porous untreated females in the ratio 1:2. Animals were left together for ten days, during which two estrous cycles should have elapsed. Female rats were injected intraperitoneally with LUTALYSE® (dinoprost tromethamine) 0.1 mg/100 g bwt twice per day in the morning and late afternoon to synchronize estrous. Then, vaginal smear was performed to check for the presence of estrous. Females in estrous were introduced to each male present in the plastic cage individually. Vaginal smears were collected daily and examined under a microscope, and every positive female was followed up until parturition [26]. Day zero of pregnancy was considered the day of vaginal sperm detection. The number of positive sperm females, number of females pregnant, pregnancy index (number of pregnant females/numbers of positive sperm females), and number of cohabitation litters were reported.

2.5. Reproductive Organs Weights

Testes, epididymis, seminal vesicles, and prostate glands were scrutinized out and weighed from each rat. The dissected organ weight was calculated as the index weight (IW) = (organ weight (gm)/body weight (gm)) 100, as reported by Matousek [27].

2.6. Sperm Morphology

Sperm concentration was assayed microscopically using a hemacytometer following Yokoi et al. [28]. Cauda epididymis was diluted after mincing in 5 mL of saline, and then, the supernatant was diluted in an alkaline aqueous solution. Progressive motility, sperm, and live sperm have been evaluated (Sönmez et al. [29]). A dye exclusion approach was used to perform the viability test. For sperm abnormality evaluation, each of the epididymis contents were mixed with an eosin–nigrosin stain drop and the thin blood film scattered over slides, with a random examination of three hundred spermatozoa per slide [30].

2.7. Serum Testosterone Concentration Assessment

Immediately after blood was collected, sodium pentobarbital anesthesia was then left to clot. The sera were extracted at 500 RPM for 30 min by centrifugation and preserved for subsequent use at −20 °C. The blood sera utilized for testosterone measurement, according to Demetrious [31], using rats highly sensitive ELISA Kits (Immunometrics Ltd., London, UK).

2.8. Assays for Oxidative Stress Markers

The samples were washed three times in cold normal physiological saline solution (PBS, 0.9% NaCl). Then, the tissues were homogenized in ice-cold Tris-HCl buffer solution within a homogenizer for 2 min at 12,000× *g*. The homogenate was centrifuged at 20,000× *g* (4 °C) for 30 min, and supernatant was obtained. The levels of malondialdehyde (MDA) were tested in the homogenate. For a further extraction procedure, the supernatant was extracted in ethanol/chloroform mixture (5/3 *v*/*v*). After a second centrifugation at 3500× *g* for 20 min, the clear upper layer was taken and used for glutathione (GSH) activity determination according to Parlaktas et al. [32]. The principle of Colorimetric Evaluation was based on the response inhomogeneous form of a molecule of MDA with two thiobarbituric acid molecules resulting in a rose-colored complex with an absorbance assessed at 532 nm [33]. The GSH assay depended on reducing GSH to yield a colored complex of 5.5′-dithiobis (2-nitrobenzoic acid); its absorption was read by 405 nm within 15 min [34].

2.9. Testicular Pro-Inflammatory Cytokines Biomarkers

A system of quantitative sandwich enzyme immunoassay with Rat High-Sensitivity ELISA kits (Sigma-Aldrich, St. Louis, MO, USA) has been tested for testicular necrosis factor-alpha (TNF-α) and interleukin-6 (IL-6) testicular homogenates.

2.10. Gene Expression Analysis

The whole RNA of approximately 100 mg testicular tissue was obtained with (Invitrogen, Life Technologies, Carlsbad, CA, USA) TRIzol reagent and Nanodrop for quantification. For DNA synthesis, RNA samples of 1.8 or more A260/A280 were used using a cDNA synthesis kit (Fermentas, Waltham, MA, USA). The SYBR Green Master Mix and the primers (GAPDH) of the household gene were indicated in Table 1 added to amplify cDNA. Data on amplification were analyzed using $2^{-\Delta\Delta T}$ methods [35].

Table 1. Primers for gene expression by RT-PCR.

Gene	Direction	Primer Sequence	Accession Number
Bax	Sense	GGCGAATTGGCGATGAACTG	NM_017059.2
	Antisense	ATGGTTCTGATCAGCTCGGG	
Bcl-2	Sense	GATTGTGGCCTTCTTTGAGT	NM_016993.1
	Antisense	ATAGTTCCACAAAGGCATCC	
HSP70	Sense	TCAGAGCTGCTATGTCGCTG	NM_153629.1
	Antisense	GCAGCGGTCGCTATACTCAT	
CYP17A1	Sense	ACTGAGGGTATCGTGGATGC	NM_012753.2
	Antisense	TCGAACTTCTCCCTGCACTT	
StAR	Sense	CTGCTAGACCAGCCCATGGAC	NM_031558.3
	Antisense	TGATTTCCTTGACATTTGGGTTCC	
KISS1	Sense	TGCTGCTTCTCCTCTGTGTGG	NM_181692.1
	Antisense	ATTAACGAGTTCCTGGGGTCC	

Table 1. *Cont.*

Gene	Direction	Primer Sequence	Accession Number
Cyp11a1	Sense	AGGTGTAGCTCAGGACTT	J05156
	Antisense	AGGAGGCTATAAAGGACACC	
3β-HSD	Sense	CCCATACAGCAAAAGGATGG	M38178
	Antisense	GCCGCAAGTATCATGACAGA	
Cyp19	Sense	GCTTCTCATCGCAGAGTATCCGG	M33986
	Antisense	CAAGGGTAAATTCATTGGGCTTGG	
GAPDH	Sense	TCAAGAAGGTGGTGAAGCAG	NM_017008.4
	Antisense	AGGTGGAAGAATGGGAGTTG	

Bax, Bcl-2-associated X protein. *Bcl-2,* B-cell lymphoma 2. *CYP17A1,* cytochrome P450 17A1. *GAPDH,* glyceraldehyde-3-phosphate dehydrogenase. *HSP70,* heat shock protein 70. *KISS1,* kisspeptin. *StAR,* steroidogenic acute regulatory protein. *Cyp11a1,* cholesterol side-chain cleavage enzyme mRNA (P450SCC). *3β-HSD,* 3-beta-hydroxysteroid dehydrogenase/delta-5-delta-4 isomerase type I; *Cyp19,* cytochrome P450 aromatase.

2.11. Morphopathological Studies

Each rat's right testis has been separated at the end of the experiment and fixed quickly in Davidson's modified solution. Epididymis, prostate, and seminal vesicles were easily fixed for at least 24 h in 10% neutral formalin buffered. The paraffin-embedding technique was used to process the samples and to cut them into four-five μm thick. The sections were deparaffinized with xylene, stained with hematoxylin–eosin (HE), and then analyzed by light microscopy and recorded with a digital camera [36]. The testis, epididymis, prostate, and seminal vesicles damage was evaluated using a semiquantitative scoring assay, in which five random fields were examined from each section. The severity of lesions was scored and graded as follows: (−) absence of the lesion = 0%, (+) mild = 0–25%, (++) moderate = 25–50%, and (+++) severe = 50–100% of the examined tissue sections.

2.12. Immunohistochemistry Analysis

The Davidson-fixed rat testes use the complicated immunohistochemical process of avidin–biotin–peroxidase (Elite-ABC; Vector Laboratory, Burlingam, CA, USA) against proliferating the nuclear cell antigen (1:100 dilution; Dako Japan Co., Ltd., Tokyo, Japan) [37] was applied to positively charged paraffin tissue slides. Sections were deparaffinized, rehydrated, soaked in PBS (3–5 min), and cleaned. Then, there was a 30-min quenching of peroxidase activity utilizing 0.3% hydrogen peroxide in methyl alcohol. Samples were washed and incubated subsequently in PBS at 25 ± 1 °C and a blocking solution for ten min. Sections were incubated for 30–60 min and the primary antibody was placed in a moist chamber after rinsing with PBS (0.9% NaCl), and then it was rinsed with PBS again. Samples have been set at room temperature for 10 min with streptavidin–peroxidase and flushed with the PBS. A complex antibody–peroxidase was developed for 2–5 min using diaminobenzidine chromogen at 18–24 °C. The sections were finally cleaned up with the PBS, dehydrated, and mounted with hematoxylin by Mayer. The primary antibody had been removed or substituted by the iso-type-matched mouse IgG2a for negative controls. Positive staining microscopically is described through the visual detection of brown color. Images of 10 different fields were analyzed at a magnification of (400×) Image J. software program (ImageJ Version 1.47, National Institutes of Health, Bethesda, MD, USA for Positive Brown Immunostaining Cells estimates.

2.13. Data Analysis

Data have been represented as mean ± SEM. One-way variance analysis (ANOVA) was performed, which was accompanied by Duncan's group differences detection study. The *p*-value of less than 0.05 was significantly different for all statistical analyses using "version 24" of SPSS/PC+.

3. Results

3.1. Fertility Test

The impact of FBN and BA on fertility tests are presented in Table 2. The number of positive sperm females was the lowest in FBN-administered rats. Conversely, almost all females recorded positive sperm in all other groups, and even the positive sperm females in the FBN group were not low pregnant. Moreover, the number of pregnant females and pregnancy index (%) was highest in the control and BA groups. Simultaneously, it decreased in both groups given FBN and BA (250 mg and 500 mg) and the lowest was recorded in the FBN group. The number of litters was negatively affected by FBN. Still, BA decreases this adverse effect, while the control group and BA groups (250 mg and 500 mg) were the highest in the number of litters.

Table 2. The effect of fipronil (FPN) and co-treatment with boswellic acid (BA) on fertility index of rat.

Group	Control	BA_1	BA_2	FBN	FBN + BA_1	FBN + BA_2
No. of females	12	12	12	12	12	12
No. of positive sperm female	10/12 (83.3%)	11/12 (91.7%)	11/12 (91.7%)	4/12 (33.3%)	9/12 (75%)	10/12 (83.3%)
No. of pregnant female	9/12 (75%)	10/12 (83.3%)	11/12 (91.7%)	3/12 (25%)	7/12 (58.3%)	8/12 (66.7%)
Pregnancy index (%)	90	90.9	100	75	77.8	80
No. of litters	8.52 ± 1.43^{c}	10.36 ± 2.01^{b}	12.86 ± 2.85^{a}	3.58 ± 0.98^{e}	5.63 ± 1.56^{de}	6.09 ± 1.74^{d}

The data shown are the mean and standard deviation. $^{a–e}$ Means bearing different superscript letters within the same row are significantly different ($p < 0.05$). Group 1 (control); Group 2 (BA1) rats were orally administered 250 mg BA/kg bwt; Group 3 (BA2) rats were orally administered 500 mg BA/kg bwt; Group 4 (FPN) rats were orally administered 20 mg FPN/kg bwt; Group 5 (FPN + BA1) rats were orally administered 20 mg FPN/kg bwt plus 250 mg BA/kg bwt, and Group 6 (FPN + BA2) rats were orally administered 20 mg FPN/kg bwt plus 500 mg BA/kg bwt.

3.2. Reproductive Organs Weights

The weight of the testes index, epididymis, and accessory sex organs in the FPN-administered group decreased dramatically. This relative index weight was retained in the group's control values and concomitantly allocated with BA and FPN (Table 3).

Table 3. The effect of fipronil (FPN) and co-treatment with boswellic acid (BA) on the index weight of rat reproductive organs.

Groups/Parameters	Control	BA_1	BA_2	FBN	FBN + BA_1	FBN + BA_2
I.W. of testes	1.67 ± 0.05^{a}	1.63 ± 0.04^{a}	1.64 ± 0.06^{a}	1.17 ± 0.04^{b}	1.57 ± 0.07^{a}	1.58 ± 0.07^{a}
I.W. of epididymis	0.82 ± 0.01^{a}	0.80 ± 0.02^{a}	0.83 ± 0.02^{a}	0.65 ± 0.02^{b}	0.79 ± 0.01^{a}	0.78 ± 0.02^{a}
I.W. of accessory gland	0.94 ± 0.03^{a}	0.95 ± 0.04^{a}	0.96 ± 0.02^{a}	0.77 ± 0.02^{b}	0.90 ± 0.02^{a}	0.91 ± 0.03^{a}

All values are expressed as mean ± S.E. $^{a–b}$ Mean values with different letters at the same raw are significantly different at $p \leq 0.05$ (ANOVA) with Duncan's multiple range test. Group 1 (control); Group 2 (BA1) rats were orally administered 250 mg BA/kg bwt; Group 3 (BA2) rats were orally administered 500 mg BA/kg bwt; Group 4 (FPN) rats were orally administered 20 mg FPN/kg bwt; Group 5 (FPN + BA1) rats were orally administered 20 mg FPN/kg bwt plus 250 mg BA/kg bwt, and Group 6 (FPN + BA2) rats were orally administered 20 mg FPN/kg bwt plus 500 mg BA/kg bwt.

3.3. Sperm Morphology

Sperm cell concentration, as well as motility percentage, and live sperm cell percentage were substantially ($p \leq 0.05$) decreased, while the sperm cell abnormality percentage was considerably ($p \leq 0.05$) improved in the FPN-administered rats concerning to the control rats (Table 4). These parameters stayed as in control, one in the groups concomitantly given with BA and FPN. Sperm abnormalities appear in the form of a bent/amorphous head and coiled/short tail. These abnormalities increased in FBN administration rats and decreased in co-administration rats.

Table 4. The effect of fipronil (FPN) and co-treatment with boswellic acid (BA) on semen analysis.

Groups/Parameters		Control	BA_1	BA_2	FBN	FBN + BA_1	FBN + BA_2
Sperm cell count ($\times 10^6$/mL)		150.40 ± 1.51 [a]	149.50 ± 2.42 [a]	152.50 ± 1.50 [a]	112.20 ± 1.49 [b]	145.00 ± 1.71 [a]	146.40 ± 1.51 [a]
Sperm motility %		90.00 ± 1.22 [a]	90.00 ± 3.74 [a]	91.00 ± 3.67 [a]	74.00 ± 2.92 [b]	88.00 ± 1.87 [a]	89.00 ± 1.87 [a]
Live spermatozoa %		92.00 ± 2.22 [a]	91.00 ± 1.74 [a]	90.00 ± 2.67 [a]	82.00 ± 1.92 [b]	90.00 ± 2.02 [a]	91.00 ± 1.92 [a]
Abnormality %		8.20 ± 0.51 [b]	8.00 ± 0.58 [b]	8.33 ± 0.33 [b]	14.30 ± 0.71 [a]	8.50 ± 1.00 [b]	8.63 ± 0.88 [b]
Abnormalities							
1	Bent head	1.60 ± 0.24 [b]	1.67 ± 0.33 [b]	2.00 ± 0.58 [a]	2.80 ± 0.65 [a]	2.33 ± 0.33 [a]	2.17 ± 0.33 [a]
2	Amorphous head	1.80 ± 0.51 [b]	2.00 ± 0.58 [b]	1.33 ± 0.33 [d]	4.00 ± 0.71 [a]	1.67 ± 0.33 [c]	2.00 ± 0.03 [b]
3	Coiled tail	1.80 ± 0.58 [c]	2.33 ± 0.33 [b]	2.38 ± 0.88 [b]	3.25 ± 0.25 [a]	2.17 ± 0.88 [b]	2.35 ± 0.88 [b]
4	Short tail	3.00 ± 0.55 [b]	2.00 ± 0.58 [c]	2.67 ± 0.33 [c]	4.25 ± 0.85 [a]	2.33 ± 0.88 [c]	2.13 ± 0.33 [c]

All values are expressed as mean ± S.E. [a–d] Mean values with different letters at the same raw are significantly different at $p \leq 0.05$ (ANOVA) with Duncan's multiple range test. Group 1 (control); Group 2 (BA1) rats were orally administered 250 mg BA/kg bwt; Group 3 (BA2) rats were orally administered 500 mg BA/kg bwt; Group 4 (FPN) rats were orally administered 20 mg FPN/kg bwt; Group 5 (FPN + BA1) rats were orally administered 20 mg FPN/kg bwt plus 250 mg BA/kg bwt, and Group 6 (FPN + BA2) rats were orally administered 20 mg FPN/kg bwt plus 500 mg BA/kg bwt.

3.4. Serum Testosterone, Testicular Antioxidant, and Pro-Inflammatory Cytokines

Blood serum testosterone and testicular levels of GSH in the FBN-treated group were appreciably ($p \leq 0.05$) decreased compared with control and two levels of BA-administered rats. Concomitant administration of BA with FPN significantly increased testosterone and GSH levels but still dropped compared with the control rats for GSH. FPN-applied rats increased substantially ($p \leq 0.05$) MDA, interleukin-6, and TNF-α concentrations compared with the control and two levels of BA groups of rats (Table 5).

Table 5. The effect of fipronil (FPN) and co-treatment with boswellic acid (BA) on serum testosterone, antioxidant, and pro-inflammatory cytokines.

Groups/Parameters	Control	BA_1	BA_2	FBN	FBN + BA_1	FBN + BA_2
Testosterone (ng/mL)	2.43 ± 0.042 [a]	2.31 ± 0.022 [a]	2.29 ± 0.039[a]	1.63 ± 0.023 [b]	2.34 ± 0.032 [a]	2.22 ± 0.025 [a]
MDA (nmol/mg protein)	47.50 ± 0.76 [c]	48.00 ± 0.58 [c]	48.17 ± 0.60 [c]	71.83 ± 0.95 [a]	55.83 ± 0.60 [b]	56.50 ± 0.76 [b]
GSH (mmol/mg protein)	42.33 ± 0.67 [a]	43.50 ± 0.76 [a]	43.83 ± 0.60 [a]	19.17 ± 0.60 [d]	30.67 ± 0.49 [c]	33.17 ± 0.60 [b]
IL-6 (pg/mL)	102.00 ± 0.97 [c]	103.50 ± 1.15 [c]	103.33 ± 0.88 [c]	203.00 ± 0.97 [a]	138.63 ± 1.70 [b]	139.17 ± 0.60 [b]
TNF-α (pg/mL)	76.80 ± 0. 60 [c]	77.50 ± 0.76 [c]	78.50 ± 0.60 [c]	151.33 ± 1.05 [a]	99.50 ± 0.76 [b]	100.50 ± 0.76 [b]

All values are expressed as mean ± SEM. [a–c] Mean values of different letters within the same row are significantly different ($p < 0.05$, ANOVA with Duncan's multiple range test). Group 1 (control); Group 2 (BA1) rats were orally administered 250 mg BA/kg bwt; Group 3 (BA2) rats were orally administered 500 mg BA/kg bwt; Group 4 (FPN) rats were orally administered 20 mg FPN/kg bwt; Group 5 (FPN + BA1) rats were orally administered 20 mg FPN/kg bwt plus 250 mg BA/kg bwt, and Group 6 (FPN + BA2) rats were orally administered 20 mg FPN/kg bwt plus 500 mg BA/kg bwt.

3.5. Gene Expression

FPN-administered rats showed significant downregulation in steroidogenic and fertility-related gene expression, including *CYP17A1* and cytochrome *P450 17A1 KISS1, kisspeptin, STAR, Cyp11a1,* cholesterol side-chain cleavage enzyme mRNA (*P450SCC*) *Hsd3b1,* 3-beta-hydroxysteroid dehydrogenase/delta-5-delta-4 isomerase type I, *Cyp19,* and cytochrome P450 aromatase, concerning other administered groups and the control one as shown in Figure 2. Interestingly, co-administration of BA with FPN showed significant upregulation and stabilization of the steroidogenic gene expression, as presented in Figure 2. In addition, the FPN-administered groups showed a considerable increase in Bax, gene expression, and heat shock protein with significant downregulation to Bcl-2 gene expression concerning other administered groups in which the BA groups showed significant normalization of these genes concerning the control group. The kisspeptin gene showed no substantial changes between all treated groups.

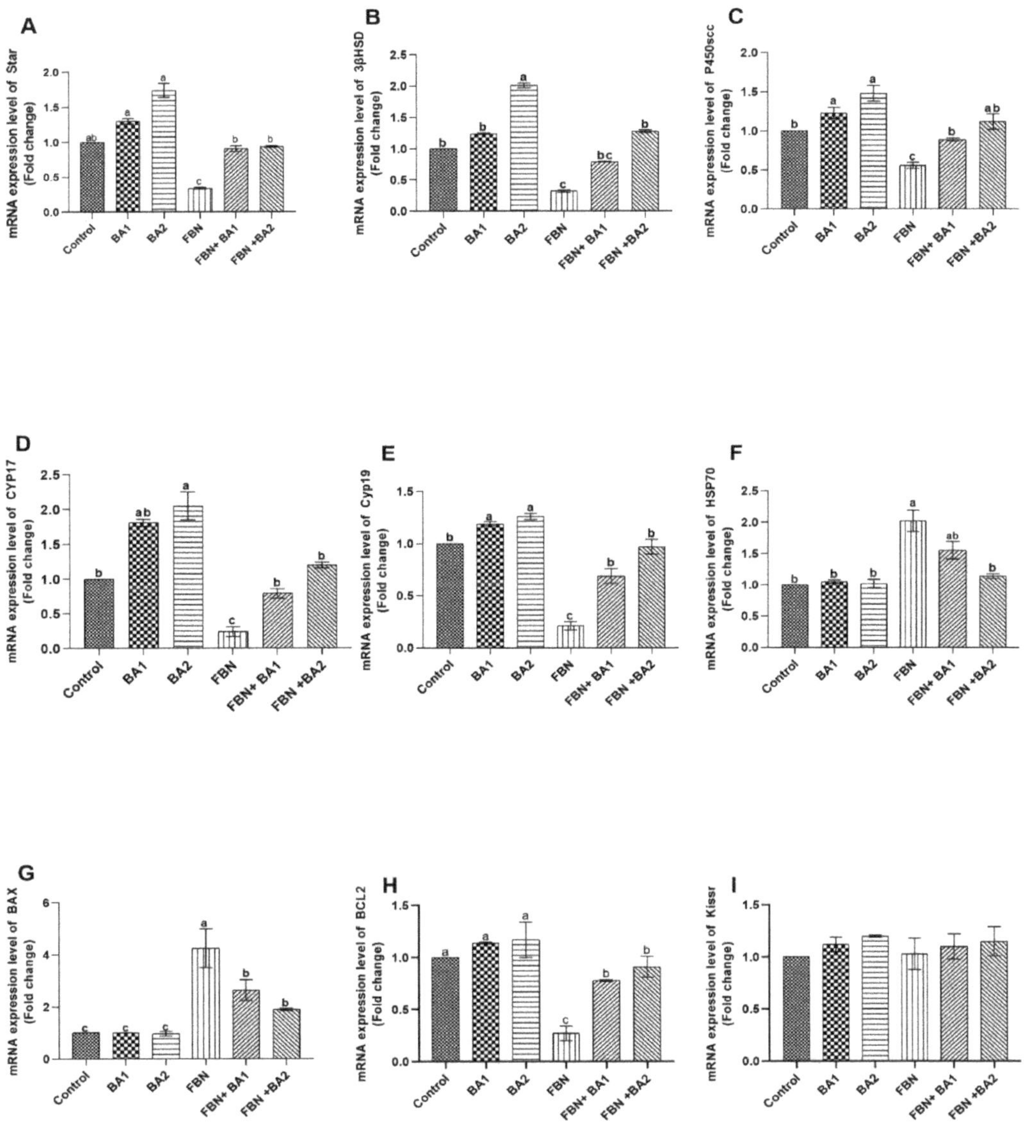

Figure 2. The effect of fipronil (FPN) and co-treatment with boswellic acid (BA) on gene expression of steroidogenic markers genes, apoptotic genes, and heat shock protein gene. STAR (**A**), 3β-HSD (**B**), P450SCC (**C**), CYP17A1 (**D**), Cyp19 (**E**), HSP70 (**F**), Bax (**G**), Bcl-2 (**H**), and KISS1 (**I**). All values are expressed as mean ± SEM. (**a**–**c**) Mean values of different letters within the same row are significantly different ($p < 0.05$, ANOVA with Duncan's multiple range test).

3.6. Histopathological Findings

There were no histological architecture differences among the control and two levels of BA (250–500 mg/kg) groups in the examined tissues. The incidence and severity of histopathological findings in the examined testis, epididymis, prostate gland, and seminal vesicle of Fipronil and co-treatment with boswellic acid are summarized in Table 6.

Table 6. Incidence and severity of histopathological findings in the examined tissue of fipronil and co-treatment with boswellic acid.

Group/Lesion	Control Rats				FPN-Rats				FPN + BA_1 Rats				FPN + BA_2 Rats			
	−	+	++	+++	−	+	++	+++	−	+	++	+++	−	+	++	+++
a—Seminiferous tubules																
sloughing of the germinal epithelium	6	0	0	0	0	1	2	3	5	1	0	0	5	1	0	0
necrosis of tubular epithelium	6	0	0	0	0	2	3	1	6	0	0	0	6	0	0	0
giant cell formations	6	0	0	0	1	1	2	2	6	0	0	0	6	0	0	0
hyalinization of the luminal contents	6	0	0	0	1	2	2	1	6	0	0	0	6	0	0	0
shrunken, buckled and disorganized	5	1	0	0	0	2	3	1	5	1	0	0	5	1	0	0
Atrophied tubules	6	0	0	0	1	1	2	2	6	0	0	0	6	0	0	0
b—Interstitial tissue																
inflammatory cell infiltration	6	0	0	0	2	3	1	0	6	0	0	0	6	0	0	0
hyperplasia endocrine cells	6	0	0	0	2	3	1	0	6	0	0	0	6	0	0	0
congestion of blood vessels	5	1	0	0	0	1	3	2	5	1	0	0	5	1	0	0
c—Epididymis																
sloughing of germinal epithelial cells	6	0	0	0	1	3	2	0	6	0	0	0	6	0	0	0
vacuolation of germinal epithelial cells	6	0	0	0	2	3	1	0	6	0	0	0	6	0	0	0
interstitial congestion of blood vessel	4	2	0	0	1	1	3	1	5	1	0	0	5	1	0	0
interstitial inflammatory cell infiltrations	6	0	0	0	1	1	2	2	6	0	0	0	6	0	0	0
perivascular inflammatory cell infiltrations	6	0	0	0	1	2	2	1	6	0	0	0	6	0	0	0
sperm density	6	0	0	0	0	2	2	2	6	0	0	0	6	0	0	0
d—Prostate gland																
interstitial congestion	4	2	0	0	0	1	3	2	5	1	0	0	5	1	0	0
perivascular inflammatory cell infiltration	6	0	0	0	1	1	2	2	6	0	0	0	6	0	0	0
low luminal secretions	6	0	0	0	1	1	1	3	6	0	0	0	6	0	0	0
desquamation of glandular epithelium	5	1	0	0	1	1	2	2	6	0	0	0	6	0	0	0
interstitial leukocytes infiltration	6	0	0	0	1	1	1	3	6	0	0	0	6	0	0	0
necrosis of glandular acini	5	1	0	0	1	1	2	2	6	0	0	0	6	0	0	0
e—Seminal vesicle																
leukocytes infiltration	6	0	0	0	1	1	1	3	6	0	0	0	6	0	0	0
congestion of blood vessel	5	1	0	0	0	2	2	2	5	1	0	0	5	1	0	0
necrotic tubuloalveolar glandular epithelial cells	6	0	0	0	1	2	2	1	6	0	0	0	6	0	0	0
low luminal secretions	6	0	0	0	2	2	2	0	6	0	0	0	6	0	0	0

Incidence is the number of rats with lesions per total examined. Severity of lesions was scored and graded by determining the percentage of tissue involvement. Lesion scoring: (−) absence of the lesion = 0%, (+) mild = 0–25%, (++) moderate = 25–50%, and (+++) severe = 50–100% of the examined tissue sections. FPN: fipronil; BA: boswellic acid.

3.7. Testicular Tissue

The control rats' testicular tissue showed regular, uniform, well-organized seminiferous tubules and normal interstitial connective tissue with entire spermatogenesis (Figure 3a). Testes of the FPN-administered group exhibited exfoliation of the germinal epithelium in the lumen of seminiferous tubules (Figure 3b) as well as necrosis of tubular epithelium (Figure 3c) with multinucleated giant cell formations in some seminiferous tubule's lumen (Figure 3d). Moreover, some tubules showed reduced germinal cells and coagulative necrosis with luminal content hyalinization, in addition to the degenerative changes of most of the seminiferous tubules as diminished or collapsed. Incoherent seminiferous tubules (Figure 3e) were noticed. Furthermore, interstitial connective tissue exhibited edema defined by faint eosinophilic substances, mild inflammatory cell infiltration, and mild interstitial endocrine cell hyperplasia with interstitial vessel obstruction (Figure 3f).

Finally, it atrophied some seminiferous tubules (Figure 3g), which was characterized by tiny seminiferous tubules, and thickened basement membrane and marked reduced numbers of germinal cells. Conversely, rats that received FPN plus BA (250 and 500 mg/kg) showed major enhancement of most seminiferous tubules' spermatogenesis by the inclusion of elongated spermatids and spermatozoa (Figure 3h,i).

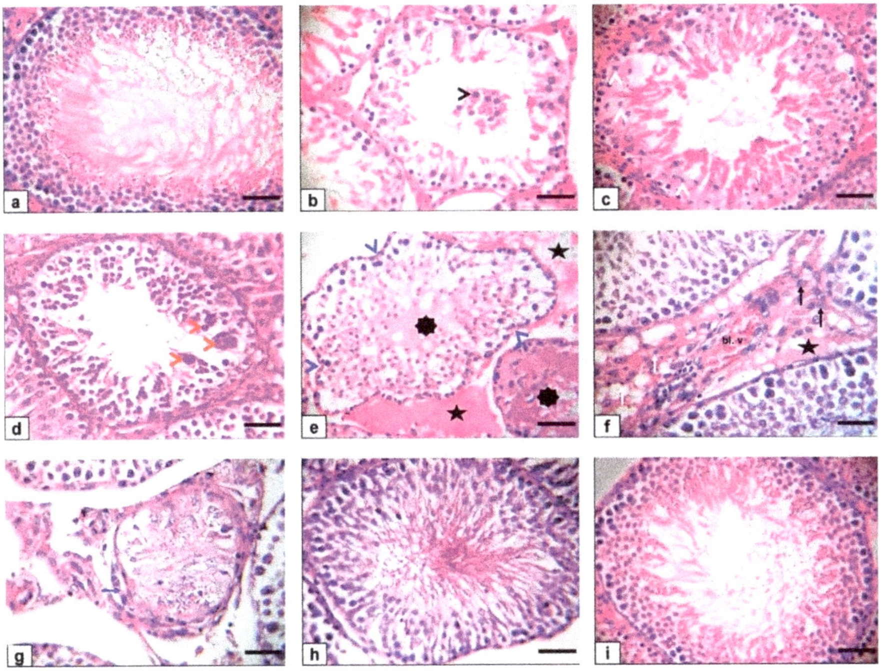

Figure 3. Photomicrograph of rat testes stained with HE. Bar = 50 µm. (**a**) Normal testes histoarchitecture of control rats (**b–g**) FPN-administered rats showing sloughing of the germinal epithelium in the lumen of seminiferous tubules (black arrowhead), fragmentation or necrosis of tubular epithelium (white arrowheads) with giant cell formations in the lumen of seminiferous tubules (red arrowheads), depletion of germinal cells and hyalinization of the luminal contents (asterisks) besides shrunken, buckled, and disorganized seminiferous tubules (blue arrowheads) as well as interstitial edema (stars), mild inflammatory cell infiltration (white arrows), mild hyperplasia of interstitial endocrine cells (black arrows) with congestion of the interstitial blood vessels (Bl. v), and finally atrophy of some seminiferous tubules (blue arrow). (**h**) FPN + BA_1 rats showing the normal histoarchitecture of seminiferous tubules with the presence of spermatids and spermatozoa in their lumen. (**i**) FPN + BA_2 rats showing normal histoarchitecture of seminiferous tubules.

3.8. Epididymis

In the control group, the caput and cauda epididymis exhibited ordinary sperm intensity histological architecture. The caput and cauda epididymis of FPN-administered rats showed histopathological alterations as a shedding of germinal cells in their lumen, vacuolation of few germinal epithelial cells, interstitial blood vessels congestion with interstitial and perivascular inflammatory cell permeations beside low or/and free sperm density, and luminal content hyalinization of some ductal epididymis. Epididymal ductules of FPN plus BA showed typical histological architecture with marked enhancement in sperm density (Figure 4).

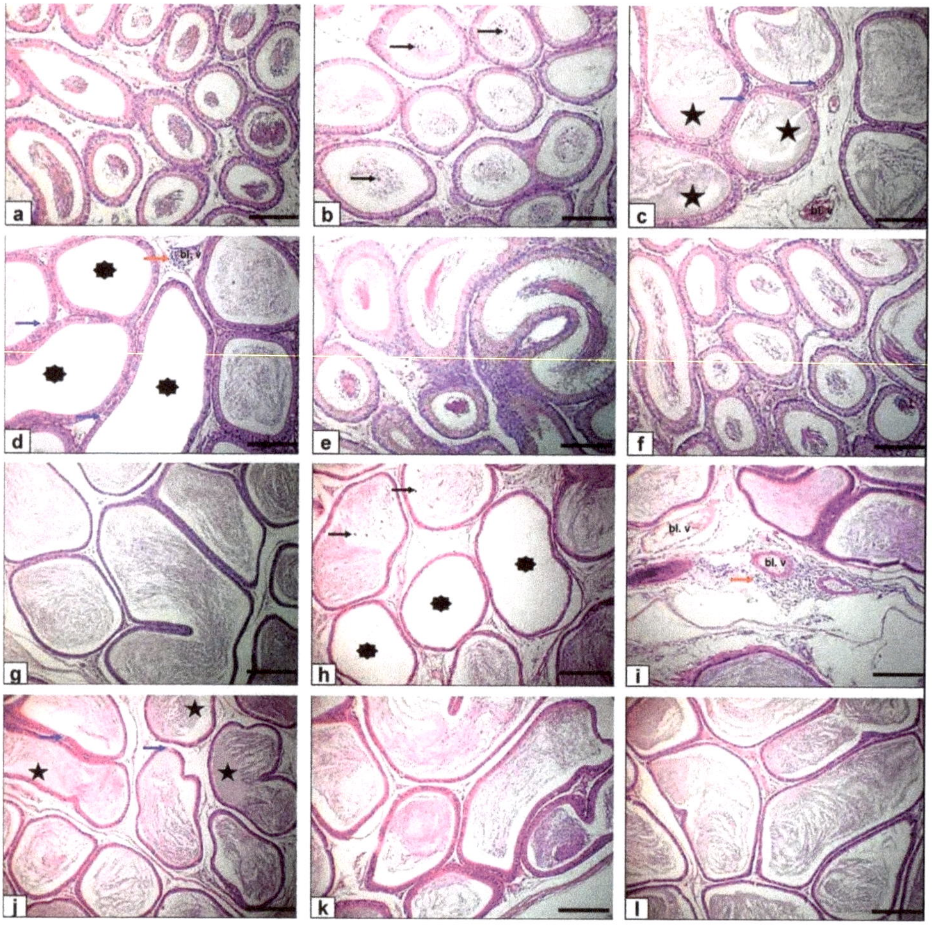

Figure 4. Photomicrograph of rat epididymis stained with HE. Bar = 100 µm. (**a–f**) Caput epididymis (**g–l**) cauda epididymis. (**a**,**g**) Normal histological structure with normal sperm density of control rats. (**b–d**,**h–j**) FPN-treated rats showing sloughing of some germinal epithelial cells (black arrows), vacuolation of few germinal epithelial cells (blue arrows), interstitial congestion of blood vessel (Bl. v) with interstitial and perivascular inflammatory cell infiltrations (red arrow) in addition to low or/and free (asterisks) sperm density with hyalinization of the luminal contents (stars) of some ductal epididymis. (**e**,**k**) FPN + BA_1 rats showing normal structure with the marked enhancement of sperm density. (**f**,**l**) FPN + BA_2 rats showing normal structural integrity and sperm density.

3.9. Prostate Gland

The control group's prostate revealed the typical histological arrangement of the glandular epithelium and normal luminal secretions (Figure 5a). FPN-administered rats showed interstitial congestion and perivascular inflammatory cell infiltration with low luminal secretions (Figure 5b) in addition to glandular epithelial desquamation (Figure 5c) and severe white blood cells infiltration with necrosis of glandular acini (Figure 5d). The prostate gland of FPN plus BA showed normal histoarchitecture of glandular epithelium and moderate luminal secretions (Figure 5e,f).

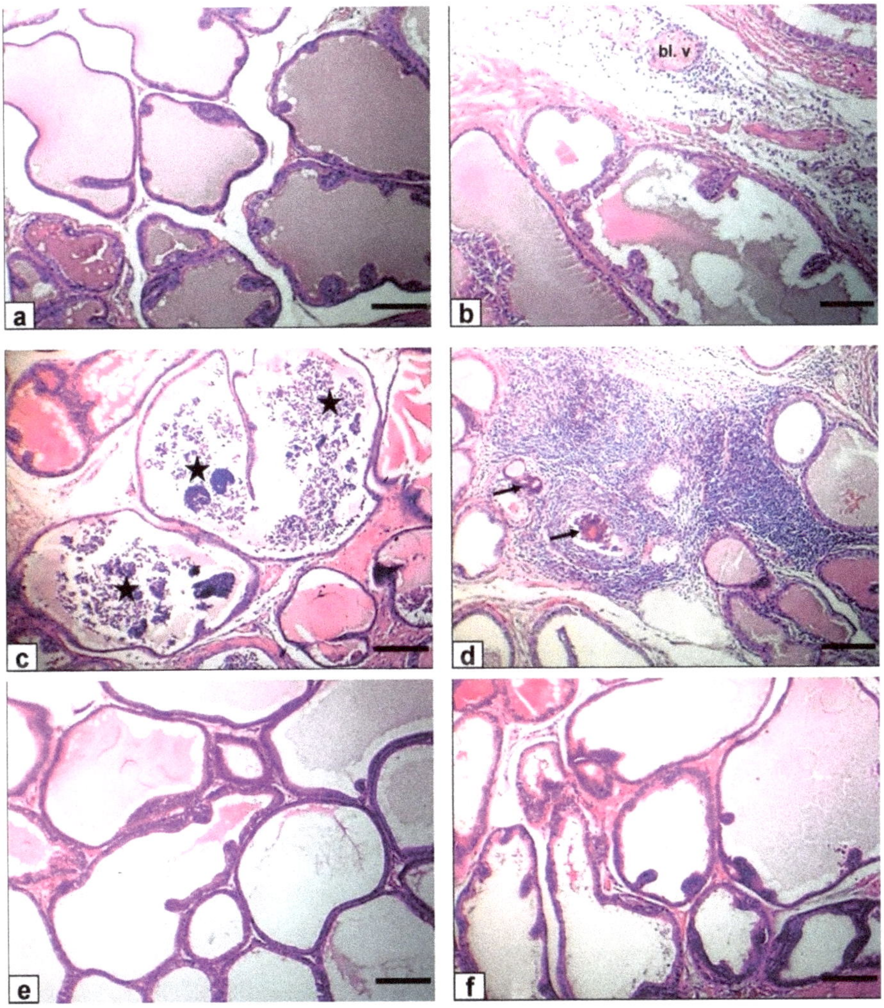

Figure 5. Photomicrograph of rat prostate stained with HE. Bar = 100 µm). (**a**) A prostate of control rats with the normal histological structure of the glandular epithelium and normal luminal secretions. (**b–d**) FPN-treated rats showing interstitial congestion (Bl. v) and perivascular inflammatory cell infiltration (white arrow) with low luminal secretions beside the desquamation of glandular epithelium (stars) and severe leukocytes in the interstitium with necrosis of glandular acini (black arrows). (**e**) FPN + BA_1 rats showing normal histoarchitecture of glandular epithelium and moderate luminal secretions. (**f**) FPN + BA_2 rats showing nearly normal histoarchitecture and moderate luminal secretions.

3.10. Seminal Vesicle

Our study showed that the control rats' seminal vesicle had a standard structure and normal luminal secretions (Figure 6a). FPN-administered rats exhibited seminal vesiculitis expressed in the serosa tunic and muscular tunica by white blood cell infiltrations, mainly neutrophils, plasma cells, and lymphocytes, besides congestion of blood vessels desquamation of specific necrotic epithelial glandular cells with low luminal secretions (Figure 6b). The seminal vesicle of FPN plus BA showed nearly normal histoarchitecture (Figure 6c,d).

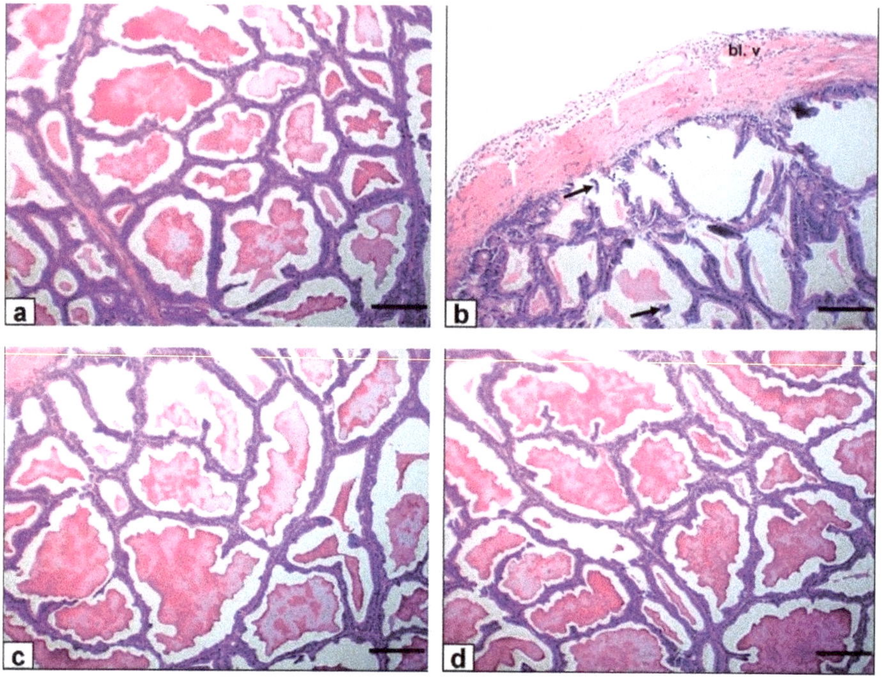

Figure 6. Photomicrograph of rat seminal vesicle stained with HE. Bar = 100 µm). (**a**) The seminal vesicle of the control rats with normal structure and normal luminal secretions. (**b**) FPN-treated rats showing neutrophils, plasma cells, and lymphocytes infiltration in the tunica serosa and the tunica muscularis (white arrows) beside the congestion of blood vessel (Bl. v) and desquamation of some necrotic tubuloalveolar glandular epithelial cells (black arrows) with low luminal secretions. (**c**) FPN + BA_1 rats showing nearly normal histoarchitecture. (**d**) FPN + BA_2 rats showing nearly normal histoarchitecture.

3.11. Immunohistochemistry and Quantitative Analysis

The proliferating cell nuclear antigen (PCNA) is useful for assessing germ-cell kinetics, especially for pathological diagnosis of germinal arrest, difficult to differentiate by the HE staining technique. The control rat's seminiferous tubules exhibited positive brown nuclei of spermatogonia and spermatocytes PCNA immunoreactions (Figure 7a). FPN-administered rats showed that most seminiferous tubules have negative immune-stained spermatogonia and spermatocyte nuclei (Figure 7b). In comparison, other seminiferous tubules have few positive brown PCNA nuclei immunoreactions of the spermatogenic cells and negative spermatocytes (Figure 7c). The seminiferous tubules of FPN plus BA showed positive brown PCNA immunoreactions in the spermatogonia and spermatocytes nuclei (Figure 7d,e). The quantitative analysis indicated a marked decline in the area percentage of PCNA immunopositive cells in FPN-administered rats concerning the control one. Co-administration with BA showed a decrease in the area percentage of PCNA immunopositive cells (Figure 7f).

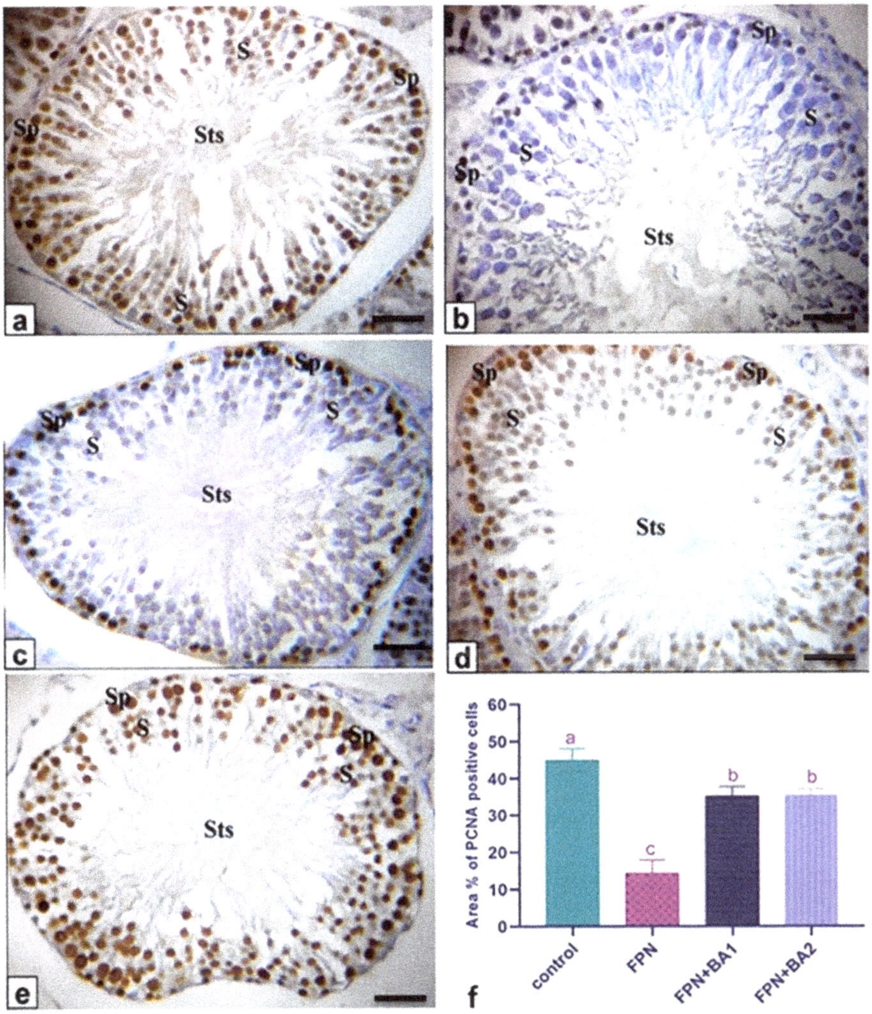

Figure 7. Photomicrograph of rat seminiferous tubules (Sts) showing positive brown proliferating cell nuclear antigen (PCNA) immunostaining. Bar = 50 μm. (**a**) Control rat showing positive brown PCNA immunoreactions in the nuclei of spermatogonia (Sp) and spermatocytes (S). (**b**,**c**) FPN-treated rats showing that most of the seminiferous tubules have negative immunostained in the nuclei of spermatogonia and spermatocytes. In contrast, other seminiferous tubules have few PCNA immunoreactions in the nuclei of the spermatogenic cells and negative spermatocytes. (**d**) FPN + BA_1 rats were showing positive brown PCNA immunoreactions in the nuclei of spermatogonia and spermatocytes (**b**) FPN + BA_2 rats showing positive brown PCNA immunoreactions in the nuclei of spermatogonia and spermatocytes. (**f**) Area percentage of PCNA positive brown-stained cells. All the values were expressed as mean ± SEM. Different small letters indicate significant at $p < 0.0001$. (**a**–**c**) Mean values with different letters within columns are significantly different at $p \leq 0.05$ (ANOVA) with Duncan's multiple range test.

4. Discussion

Chemical insecticides are widely utilized worldwide in the agriculture sector and for other purposes [38]. Food residues, contaminated tap water, occupational exposure, repellence, household use, and application against fleas and ticks are various sources for endangering insecticides for animals and people [39,40]. FPN is an insecticide with phenylpyrazole in chemical form. It is a common insecticide used both in agriculture and in domestic pest management [25]. However, few studies are evaluating its consequence on the fertility of males and reproductive efficiency. Therefore, this experiment was conducted to determine the effects of FPN on fertility test, the weight of the male reproductive organ, seminal study, serum testosterone level, oxidative status, cytokines level, some gene expression, and histopathology and to evaluate the ameliorative effect of BA. In our study, FPN has a distinct adverse effect on fertility tests by decreasing pregnant females, pregnancy index, and many litters.

Moreover, it gave rise to a noteworthy decrease in the reproductive organs' weights due to the decline in serum testosterone concentration, sperm quantity, sperm progressive motility, and live sperm cell percentage. It considerably boosted sperm cell abnormalities percentage in the FPN-administered group, which many attribute to its hazard impact. Consequently, the findings obtained indicate that FPN decreases sperm cell quality, leading to male rats' infertility. The pathway of insecticide toxicity on the testicular tissue can be correlated with the activation of oxidative injuries. These results agree with Mazzo, Balieira, Bizerra, and Mingatto [6], who reported that FPN-induced harmful impacts on sperm quality. Sperm motility is impaired by FPN's long-term exposure [11]. In addition, the overproduction of reactive oxygen species (ROS) exceeds the cellular capability, leads to oxidative damage, and reduces sperm viability and fertility [1]. The sperm contains a large proportion of polyunsaturated fatty acids and is highly susceptible to harm caused by excessive oxidative damage and peroxidation to its plasma membrane, leading to a loss of motility and decreased number [41,42]. FPN-administered rats showed a remarkable increase in MDA, which considers the consequence of lipid peroxidation and lipid degradation triggering radicals and reduction of GSH levels, which reverberate the degree of oxidative harm. These results are following the work of Mossa et al. [43] that demonstrated decreased concentrations of GSH in the kidneys and liver of rats handled with FPN (10 mg/kg bwt) [44]. The reduction in the content of GSH in the kidney and brain of FPN-treated mice (5 and 10 mg/kg bwt) in the same FPN-treated line (5 mg/kg bwt) resulted in a decrease in the concentration of GSH in the testis [6]. These results are probably attributed to Fipronil's oxidative anxiety, including reduced GSH levels, and antioxidant activity and consequent lipoperoxidation [45]. FPN-administered rats showed an overexpression of TNF-α and IL-6 pro-inflammatory cytokines. TNF-α is a major inflammatory and immune response cytokine [46], and IL-6 is a cytokine pleiotropic rendered by macrophages of tissue and monocytes [47].

The utilized doses of boswellic acid in the current study were tested previously as reported by Sami et al. [21] in which they found that boswellic acids can ameliorate doxorubicin-induced nephrotoxicity in mice, they used different doses of boswellic acid (125 mg/kg), (250 mg/kg), and (500 mg/kg), and they found that the effect of the high dose of BAs (500 mg/kg) was different (more ameliorative) from that observed with the lowest dose (125 mg/kg). In the same line, Nusier et al. [20] studied the effect of two different doses of boswellic acid: 250 and 500 mg/kg. They reported variation in their impact on the reproductive system of the rat. Barakat et al. [22,23] investigated the protective effect of boswellic acid in different doses 250 and 500 in doxorubicin-induced hepatic damaged. The anti-aggregatory effect of boswellic acid in high-fat fed rats found a variation in the boswellic acid in a dose-dependant manner. Al-Yahya et al. [24] reported that the boswellic acid is safe up to 1000 mg/kg in rats. Still, this dose is relatively high considering the amount of extract consumed by humans. So, in this study, we try to investigate the ameliorative effect of boswellic acid on the toxic impact of fipronil in a dose–response manner.

The impact of FBN and BA on fertility tests is presented in Table 2. Almost all females recorded positive sperm in all other groups other than the FBN group; moreover, the pregnancy index (%) was highest in the control and BA groups. Still, BA decreases this adverse effect, while the control group and BA groups (250 mg and 500 mg) were the highest in the number of litters. The obtained results were inconsistent with [20]. They reported that oral administration of boswellic acid increased the fertility in rats and the number of implantations as well as increased spermatogenesis due to its antioxidant activity. In the same line, our work was in harmony with [20], in which they reported non-significant improvement in the fertility parameters in BA (500 mg/kg) dose concerning (250 mg/kg). The concomitant administration of both doses of BA with FPN significantly increases testosterone and GSH with a significant decrease in MDA, interleukin-6, and TNF-α concentrations concerning the FBN-treated group, as shown in Table 5. with non-significant improvement in the BA higher dose concerning the lower dose of BA. The obtained result was in the same line with Sami et al. [21] in which they reported the nephroprotective effect of BA with a non-significant improvement of the higher dose of BA (500 mg/kg) due to the antioxidant and antiapoptotic effect of boswellic acid.

The results showed that FPN-administered rats showed a significant downregulation in steroidogenic and fertility-related gene expression, including *CYP17A1* and *cytochrome P450 17A1, KISS1, kisspeptin, STAR, Cyp11a1,* cholesterol side-chain cleavage enzyme mRNA (*P450SCC*) *Hsd3b1*, 3-beta-hydroxysteroid dehydrogenase/delta-5-delta-4 isomerase type I, *Cyp19*, and cytochrome P450 aromatase compared with other administered groups and the control one. However, co-administration of BA (both doses) with FPN showed significant upregulation and stabilization of the steroidogenic gene expression. Steroid hormones are produced from cholesterol through various steroid cytochrome P450 hydroxylases-induced reactions [48,49]. Cholesterol transfer from the external to the mitochondria's inner membrane by steroidogenic acute regulating protein (StAR) involves a rate-restricted steroidogenesis phase [50]. Then, steroidogenesis begins with the transition of P450 (P450scc/CYP11A1/Cyp11a1) cholesterol to pregnenolone, which is an essential molecule in developing the body steroid hormones [48].

Light microscope examination of FPN-administered rats testes showed degenerated and exfoliated germ cells in the seminiferous tubular lumen, which was attributed by the main effect of the cell-to-cell junction of Sertoli and germ cells or microtubular deterioration that lead to Sertoli cell damage [51,52]. In the current work, fragmentation or necrosis of tubular epithelium with giant cell formations in the seminiferous tubular lumen, germ cell loss, and luminous hyalinization dwindled, collapsed, and incoherent seminiferous tubules were reported. FPN caused interstitial edema because of increased vascular permeability and mononuclear cell infiltration that secrete cytokines, causing congested interstitial blood vessels. Another important finding in atrophied seminiferous tubules was noticed in our study due to the cytotoxic effect of FPN. The epididymal lesions were sloughing off some germinal epithelial in the lumen of some ducts of the epididymis, which indicate testicular malfunction [53,54] and interstitial congestion of blood vessel with perivascular inflammatory cell infiltrations; besides, most epididymal ducts seemed to have no or low sperm quantities in their lumen, which reflected the cessation of spermatogenesis. The prostates and seminal vesicle lesions experienced necrosis and desquamation of some glandular epithelial cells with low luminal discharges and severe interstitial leukocytes infiltration [53]. These alterations may be due to reduced testosterone, which needs differentiation, development, and the preservation of epithelial cells of accessory sex glands [55]. To estimate germ cell kinetics and an indication of DNA synthesis deterioration, PCNA is considered a valuable molecular marker [56]. Similarly, it measures the cell proliferation and spermatogenic role of studies in male infertility. In spermatogonia and the proliferating control rat spermatocytes, many positive brown nuclear reactions have been found, while a reduced countenance and area percentage of PCNA antibody was detected in FPN-administered rats compared with the control group. There were no histological architecture differences among the control and two levels of BA (250–500 mg/kg) groups in the examined tissues. Conversely, rats that received FPN plus BA (250 and 500 mg/kg) showed significant enhancement of most seminiferous tubules' spermatogenesis by including elongated spermatids and spermatozoa. These results were inconsistent with [21] in which they reported that the most convenient restoration of these parameters was achieved in the doxorubicin + BA 500 mg/kg group. Supporting our result, data obtained by Kruger et al. [57] have reported antioxidants and chelate metals of BAs in oxidative injury pathways.

The co-administration of BA significantly improved the evaluated parameters such as the number of pregnant rats, pregnancy index, and the number of litters, reproductive organs weights, normalized testosterone levels, and sperm quality, which were attributed to the increasing number of spermatocytes and normal spermatogenesis as found by Nusier et al. [20], which reported the promising effect of boswellia on adult male rat fertility due to its antioxidant activity [19]. BA treatment (250 and 500 mg/kg) normalizes the apoptotic effect of FPN, which is cleared by our result, in which FPN significantly upregulates Bax, heat shock protein, and gene expression with significant downregulation to Bcl-2. These results show inconsistency with [21]; they revealed the antioxidant and antiapoptotic effects of BA. BA's prospective role in the scavenging of FPN ROS was linked in histological architecture to testes, the epididymis, and sexual accessory glands. Upon our result, the molecular mechanism of FBN that induced infertility was summarized in the downregulation in steroidogenic and fertility-related gene expression, including *CYP17A1* and *cytochrome P450 17A1*, *KISS1*, *kisspeptin*, *STAR*, *Cyp11a1*, cholesterol side-chain cleavage enzyme mRNA (*P450SCC*) *Hsd3b1*, 3-beta-hydroxysteroid dehydrogenase/delta-5-delta-4 isomerase type I, *Cyp19*, and cytochrome P450 aromatase. In addition, FPN-treated rats showing few PCNA immunoreactions in the nuclei of the spermatogenic cells and negative spermatocytes FPN caused interstitial edema because of increased vascular permeability and mononuclear cell infiltration that secrete cytokines *IL-6* and *TNF-α*, causing congested interstitial blood vessels. Another important finding regarding atrophied seminiferous tubules was noticed in our study due to the cytotoxic effect of FPN blood serum testosterone, and testicular levels of GSH in FBN-treated group were appreciably ($p \leq 0.05$) decreased.

Many studies are concerned with the significant mechanism driving the protective actions in which boswellic acid-treated rats showed reduced serum expression of *TNF-α* and *IL-6* and hepatic iNOS. At the cellular level, also, boswellic acids increased the expression in the white adipose tissue of thermogenesis associated mitochondrial uncoupling protein-1 and carnitine palmitoyl transferase-1 [58].

Gayathri et al. [59] showed that boswellic acid has anti-inflammatory effects in human peripheral mononuclear blood cells and mouse macrophages by inhibiting *tumor necrosis factor-α* (*TNF-α*), *IL-1β*, *nitric oxide*, and mitogenic protein kinases. Several clinical trials highlighted boswellic acid as a potentially effective anti-inflammatory drug [60]. In addition, BA extract inhibited the *TGF-ß*-induced fibrosis ($p = 0.01$) and 5-lipooxygenase activity levels that prevent fibrosis, as mentioned by Ali and Mansour [61]. In the same line, Sharma, et al. [62] reported the anti-inflammatory role of the boswellic acid through the inflammatory mediators *TNF-α* and *IL-6*. The inhibition of *NF-κB* activity by the boswellic acid family may be considered an alternative therapy for chronic inflammatory disorders [63].

5. Conclusions

BA significantly improved the reproductive parameters assessed, such as the number of pregnant females, index of pregnancy and the number of litters, weights of the reproductive organ, sperm cell quality, morphological alterations of testes, epididymis, and sex glands by accessory caused by FPN oxidative stress, as well as the improvement of steroidogenesis, antioxidants, and antiapoptotic marker. This is the first report to address the defensive function of BA in male rats against reproductive FPN lesions.

Fipronil (FPN) is a commonly used phenylpyrazole pesticide used to manage insects and remove fleas, ticks, and other parasites. While FPN poses health risks, it is frequently encountered in everyday life. Taken together, FPN can restrain various sperm functions directly and indirectly. Thus, FPN can adversely affect male fertility, which leads to infertility. We suggest that using FPN as a pesticide demands the attention of reproductive toxicity from these findings.

Author Contributions: Conceptualization, H.G.T., S.E.E.-K., H.S.I., M.S. and M.A.O.D. Data curation, M.S.; Formal analysis, H.G.T., S.E.E.-K. and H.S.I.; Funding acquisition, S.S.A.; Investigation, H.G.T. and M.A.O.D.; Methodology, H.G.T., S.E.E.-K., H.S.I. and M.S.; Project administration, H.G.T., M.S. and M.A.O.D.; Resources, S.S.A. and M.A.O.D.; Writing—original draft, M.S. and M.A.O.D.; Writing—review and editing, S.S.A. All authors have read and agreed to the published version of the manuscript.

Funding: This work was funded by Taif University Researchers Supporting Project number (TURSP-2020/38), Taif University, Taif, Saudi Arabia.

Institutional Review Board Statement: The research protocol was accepted by the Institutional Animal Care and Use Committee, Faculty of Veterinary Medicine, Alexandria University, and it was precisely designed under the consideration of animal welfare (AU013202062958).

Informed Consent Statement: Not applicable.

Data Availability Statement: Not applicable.

Acknowledgments: The authors extend their appreciation to Taif University Researchers Supporting Project number (TURSP-2020/38), Taif University, Taif, Saudi Arabia.

Conflicts of Interest: The authors declare no conflict of interest.

References

1. Soll, M.D.; Pate, J.; Baker, L.A. Topical Compositions Comprising Fipronil and Permethrin and Methods of Use. U.S. Patent 9,949,953, 24 April 2018.
2. Khalaf, A.; Galal, M.K.; Ibrahim, M.A.; Abd Allah, A.; Afify, M.M.; Refaat, R. The Terminalia laxiflora modulates the neurotoxicity induced by fipronil in male albino rats. *Biosci. Rep.* **2019**, 39. [CrossRef]
3. Pavlidi, N.; Vontas, J.; Van Leeuwen, T. The role of glutathione S-transferases (GSTs) in insecticide resistance in crop pests and disease vectors. *Curr. Opin. Insect Sci.* **2018**, *27*, 97–102. [CrossRef] [PubMed]
4. Kalyanaraman, B.; Hardy, M.; Zielonka, J. A critical review of methodologies to detect reactive oxygen and nitrogen species stimulated by NADPH oxidase enzymes: Implications in pesticide toxicity. *Curr. Pharmacol. Rep.* **2016**, *2*, 193–201. [CrossRef] [PubMed]
5. Eisa, A.A.; Abo-Elghar, G.; Ammar, I.; Metwally, H.G.; Arafa, S.S. Embryotoxicity and teratogenicity of fipronil in rats (Rattus norvegicus). *Zagazig J. Agric. Res.* **2017**, *44*, 1851–1861.
6. Mazzo, M.; Balieira, K.V.B.; Bizerra, P.F.V.; Mingatto, F.E. Fipronil-induced decrease in the epididymal sperm count: Oxidative effect and protection by vitamin E. *Animal Reprod. AR* **2018**, *15*, 1223–1230. [CrossRef]
7. Agarwal, A.; Sengupta, P. Oxidative Stress and its Association with Male Infertility. In *Male Infertility*; Springer: Berlin, Germany, 2020; pp. 57–68.
8. Al Basher, G.; Abdel-Daim, M.M.; Almeer, R.; Ibrahim, K.A.; Hamza, R.Z.; Bungau, S.; Aleya, L. Synergistic antioxidant effects of resveratrol and curcumin against fipronil-triggered oxidative damage in male albino rats. *Environ. Sci. Poll. Res.* **2020**, *27*, 6505–6514. [CrossRef]
9. Saleh, H.; Nassar, A.M.; Noreldin, A.E.; Samak, D.; Elshony, N.; Wasef, L.; Elewa, Y.H.; Hassan, S.; Saati, A.A.; Hetta, H.F. Chemo-protective potential of cerium oxide nanoparticles against fipronil-induced oxidative stress, apoptosis, inflammation and reproductive dysfunction in male white albino rats. *Molecules* **2020**, *25*, 3479. [CrossRef] [PubMed]
10. Ohi, M.; Dalsenter, P.R.; Andrade, A.J.; Nascimento, A.J. Reproductive adverse effects of fipronil in Wistar rats. *Toxicol. Lett.* **2004**, *146*, 121–127. [CrossRef]
11. De Barros, A.L.; Bae, J.H.; Borges, C.S.; Rosa, J.L.; Cavariani, M.M.; Silva, P.V.; Pinheiro, P.F.F.; Anselmo-Franci, J.A.; Arena, A.C. Perinatal exposure to insecticide fipronil: Effects on the reproductive system in male rats. *Reprod. Fertil. Dev.* **2017**, *29*, 1130–1143. [CrossRef]
12. Kitulagodage, M.; Buttemer, W.A.; Astheimer, L.B. Adverse effects of fipronil on avian reproduction and development: Maternal transfer of fipronil to eggs in zebra finch Taeniopygia guttata and in ovo exposure in chickens Gallus domesticus. *Ecotoxicology* **2011**, *20*, 653–660. [CrossRef]
13. Ammon, H. Boswellic acids and their role in chronic inflammatory diseases. In *Anti-inflammatory Nutraceuticals and Chronic Diseases*; Springer: Berlin, Germany, 2016; pp. 291–327.
14. Mehrzadi, S.; Tavakolifar, B.; Huseini, H.F.; Mosavat, S.H.; Heydari, M. The effects of Boswellia serrata gum resin on the blood glucose and lipid profile of diabetic patients: A double-blind randomized placebo-controlled clinical trial. *J. Evid. Based Integr. Med.* **2018**, *23*, 2515690X18772728. [CrossRef] [PubMed]
15. Ahmed, M.A.; Ahmed, A.A.; El Morsy, E.M. Acetyl-11-keto-β-boswellic acid prevents testicular torsion/detorsion injury in rats by modulating 5-LOX/LTB4 and p38-MAPK/JNK/Bax/Caspase-3 pathways. *Life Sci.* **2020**, *260*, 118472. [CrossRef]
16. Du, Z.; Liu, Z.; Ning, Z.; Liu, Y.; Song, Z.; Wang, C.; Lu, A. Prospects of boswellic acids as potential pharmaceutics. *Planta Med.* **2015**, *81*, 259–271. [CrossRef] [PubMed]

17. Iram, F.; Khan, S.A.; Husain, A. Phytochemistry and potential therapeutic actions of Boswellic acids: A mini-review. *Asian Pac. J. Trop. Biomed.* **2017**, *7*, 513–523. [CrossRef]
18. Al-Harrasi, A.; Csuk, R.; Khan, A.; Hussain, J. Distribution of the anti-inflammatory and anti-depressant compounds: Incensole and incensole acetate in genus Boswellia. *Phytochemistry* **2019**, *161*, 28–40. [CrossRef] [PubMed]
19. Ebrahimpour, S.; Fazeli, M.; Mehri, S.; Taherianfard, M.; Hosseinzadeh, H. Boswellic acid improves cognitive function in a rat model through its antioxidant activity:-neuroprotective effect of boswellic acid. *J. Pharmacopunct.* **2017**, *20*, 10.
20. Nusier, M.K.; Bataineh, H.N.; Bataineh, Z.M.; Daradka, H.M. Effect of frankincense (*Boswellia thurifera*) on reproductive system in adult male rat. *J. Health Sci.* **2007**, *53*, 365–370. [CrossRef]
21. Sami, M.M.; Ali, E.A.; Galhom, R.A.; Youssef, A.M.; Mohammad, H.M. Boswellic acids ameliorate doxorubicin-induced nephrotoxicity in mice: A focus on antioxidant and antiapoptotic effects. *Egyp. J. Basic Appl. Sci.* **2019**, *6*, 10–24. [CrossRef]
22. Barakat, B.M.; Ahmed, H.I.; Bahr, H.I.; Elbahaie, A.M. Protective effect of boswellic acids against doxorubicin-induced hepatotoxicity: Impact on Nrf2/HO-1 defense pathway. *Oxid. Med. Cell. Longev.* **2018**, *2018*. [CrossRef]
23. Tawfik, M.K. Anti-aggregatory effect of boswellic acid in high-fat fed rats: Involvement of redox and inflammatory cascades. *Arch. Med. Sci. AMS* **2016**, *12*, 1354. [CrossRef] [PubMed]
24. Al-Yahya, A.A.; Asad, M.; Sadaby, A.; Alhussaini, M.S. Repeat oral dose safety study of standardized methanolic extract of Boswellia sacra oleo gum resin in rats. *Saudi J. Biol. Sci.* **2020**, *27*, 117–123. [CrossRef]
25. Tingle, C.C.; Rother, J.A.; Dewhurst, C.F.; Lauer, S.; King, W.J. Fipronil: Environmental fate, ecotoxicology, and human health concerns. In *Reviews of Environmental Contamination and Toxicology*; Springer: Berlin, Germany, 2003; pp. 1–66.
26. Bronson, F. The reproductive ecology of the house mouse. *Q. Rev. Biol.* **1979**, *54*, 265–299. [CrossRef] [PubMed]
27. Matousek, J. Effects on spermatogenesis in guinea-pigs, rabbits and sheep after their immunization with sexual organ fluids of bulls. *Reproduction* **1969**, *19*, 63–72. [CrossRef] [PubMed]
28. Yokoi, K.; Uthus, E.O.; Nielsen, F.H. Nickel deficiency diminishes sperm quantity and movement in rats. *Biol. Trace Elem. Res.* **2003**, *93*, 141–153. [CrossRef]
29. Sönmez, M.; Türk, G.; Yüce, A. The effect of ascorbic acid supplementation on sperm quality, lipid peroxidation and testosterone levels of male Wistar rats. *Theriogenology* **2005**, *63*, 2063–2072. [CrossRef] [PubMed]
30. Bearden, H.J.; Fuquay, J.W. Applied Animal Reproduction. Reston Publishing Company, Inc.: Reston, VA, USA, 1980.
31. Demetrious, J. Testosterone in methods. In *Clinical Chemistry Tech AG*, 2nd ed.; CVMOS Co.: Washington, DC, USA, 1987; p. 268.
32. Parlaktas, B.; Atilgan, D.; Gencten, Y.; Akbas, A.; Markoc, F.; Erdemir, F.; Ozyurt, H.; Uluocak, N. The effects of carvedilol on ischemia-reperfusion injury in the rat testis. *Int. Braz. Jurol.* **2014**, *40*, 109–117. [CrossRef]
33. Uchiyama, M.; Mihara, M. Determination of malonaldehyde precursor in tissues by thiobarbituric acid test. *Anal. Biochem.* **1978**, *86*, 271–278. [CrossRef]
34. Adams, J.; Lauterburg, B.; Mitchell, J. Plasma glutathione and glutathione disulfide in the rat: Regulation and response to oxidative stress. *J. Pharmacol. Exp. Ther.* **1983**, *227*, 749–754.
35. Livak, K.J.; Schmittgen, T.D. Analysis of relative gene expression data using real-time quantitative PCR and the $2-\Delta\Delta CT$ method. *Methods* **2001**, *25*, 402–408. [CrossRef]
36. Bancroft, J.; Layton, C. *The hematoxylins and eosin. Bancroft's Theory and Practice of Histological Techniques*; Elsevier: Amsterdam, The Netherlands, 2013.
37. Shivji, M.K.; Kenny, M.K.; Wood, R.D. Proliferating cell nuclear antigen is required for DNA excision repair. *Cell* **1992**, *69*, 367–374. [CrossRef]
38. Dawood, M.A.O.; Abdel-Razik, N.I.; Gewaily, M.S.; Sewilam, H.; Paray, B.A.; Soliman, A.A.; Abdelhiee, E.Y.; Aboubakr, M.; Van Doan, H.; El-Sabagh, M.; et al. β-Glucan improved the immunity, hepato-renal, and histopathology disorders induced by chlorpyrifos in Nile tilapia. *Aquac. Rep.* **2020**, *18*, 100549. [CrossRef]
39. Lipton, E. EPA Chief, Rejecting Agency's Science, Chooses not to Ban Insecticide. 2018. Available online: https://www.nytimes.com/2017/03/29/us/politics/epa-insecticide-chlorpyrifos.html (accessed on 28 February 2021).
40. Dawood, M.A.O.; El-Shamaa, I.S.; Abdel-Razik, N.I.; Elkomy, A.H.; Gewaily, M.S.; Abdo, S.E.; Soliman, A.A.; Paray, B.A.; Abdelkhalek, N. The effect of mannanoligosaccharide on the growth performance, histopathology, and the expression of immune and antioxidative related genes in Nile tilapia reared under chlorpyrifos ambient toxicity. *Fish Shellfish Immunol.* **2020**, *103*, 421–429. [CrossRef]
41. Nowicka-Bauer, K.; Nixon, B. Molecular changes induced by oxidative stress that impair human sperm motility. *Antioxidants* **2020**, *9*, 134. [CrossRef]
42. Mahmoud, S.; Saad, M.; Farrag, F.; Abo Ghanima, M.M.; Dawood, M.A.; Abdel-Daim, M.M.; Alkahtani, S.H.; Shukry, M. Promoting effect of L. tyrosine supplement on New Zealand rabbit bucks' performance and reproduction through upregulation of steroidogenic markers. *Front. Vet. Sci.* **2020**, *7*, 605. [CrossRef]
43. Mossa, A.-T.H.; Swelam, E.S.; Mohafrash, S.M. Sub-chronic exposure to fipronil induced oxidative stress, biochemical and histopathological changes in the liver and kidney of male albino rats. *Toxicol. Rep.* **2015**, *2*, 775–784. [CrossRef]
44. Badgujar, P.C.; Chandratre, G.A.; Pawar, N.N.; Telang, A.; Kurade, N. Fipronil induced oxidative stress involves alterations in SOD 1 and catalase gene expression in male mice liver: Protection by vitamins E and C. *Environ. Toxicol.* **2016**, *31*, 1147–1158. [CrossRef]

45. Bevilaqua, F.; Sachett, A.; Chitolina, R.; Garbinato, C.; Gasparetto, H.; Marcon, M.; Mocelin, R.; Dallegrave, E.; Conterato, G.; Piato, A. A mixture of fipronil and fungicides induces alterations on behavioral and oxidative stress parameters in zebrafish. *Ecotoxicology* **2020**, *29*, 140–147. [CrossRef]
46. Lebda, M.A.; Sadek, K.M.; Abouzed, T.K.; Tohamy, H.G.; El-Sayed, Y.S. Melatonin mitigates thioacetamide-induced hepatic fibrosis via antioxidant activity and modulation of proinflammatory cytokines and fibrogenic genes. *Life Sci.* **2018**, *192*, 136–143. [CrossRef]
47. Palin, K.; Bluthé, R.-M.; McCusker, R.H.; Levade, T.; Moos, F.; Dantzer, R.; Kelley, K.W. The type 1 TNF receptor and its associated adapter protein, FAN, are required for TNFα-induced sickness behavior. *Psychopharmacology* **2009**, *201*, 549–556. [CrossRef] [PubMed]
48. Miller, W.L.; Auchus, R.J. The molecular biology, biochemistry, and physiology of human steroidogenesis and its disorders. *Endocr. Rev.* **2011**, *32*, 81–151. [CrossRef] [PubMed]
49. Miller, W.L. Steroidogenesis: Unanswered questions. *Trends Endocrinol. Metab.* **2017**, *28*, 771–793. [CrossRef]
50. Stocco, D.M.; Zhao, A.H.; Tu, L.N.; Morohaku, K.; Selvaraj, V. A brief history of the search for the protein (s) involved in the acute regulation of steroidogenesis. *Mol. Cell. Endocrinol.* **2017**, *441*, 7–16. [CrossRef]
51. Creasy, D.M. Pathogenesis of male reproductive toxicity. *Toxicol. Pathol.* **2001**, *29*, 64–76. [CrossRef] [PubMed]
52. Kumar, S.G.; Narayana, K.; Bairy, K.; D'Souza, U.J.; Samuel, V.P.; Gopalakrishna, K. Dacarbazine induces genotoxic and cytotoxic germ cell damage with concomitant decrease in testosterone and increase in lactate dehydrogenase concentration in the testis. *Mutat. Res. Genet. Toxicol. Environ. Mutagenesis* **2006**, *607*, 240–252. [CrossRef]
53. Tohamy, H.G.; El-Karim, D.R.G.; El-Sayed, Y.S. Attenuation potentials of royal jelly against hydroxyurea-induced infertility through inhibiting oxidation and release of pro-inflammatory cytokines in male rats. *Environ. Sci. Poll. Res.* **2019**, *26*, 21524–21534. [CrossRef] [PubMed]
54. Narayana, K.; Prashanthi, N.; Nayanatara, A.; Kumar, S.G.; Kumar, H.H.C.; Bairy, K.; D'Souza, U.J. A broad-spectrum organophosphate pesticide O, O-dimethyl O-4-nitrophenyl phosphorothioate (methyl parathion) adversely affects the structure and function of male accessory reproductive organs in the rat. *Environ. Toxicol. Pharmacol.* **2006**, *22*, 315–324. [CrossRef]
55. Ono, Y.; Suzuki, K.; Kashiwagi, B.; Shibata, Y.; Ito, K.; Fukabori, Y.; Yamanaka, H. Role of androgen on blood flow and capillary structure in rat seminal vesicles. *Tohoku J. Exp. Med.* **2004**, *202*, 193–201. [CrossRef]
56. Zeng, L.; Kong, X.-T.; Su, J.-W.; Xia, T.-L.; Na, Y.-Q.; Guo, Y.-L. Evaluation of germ-cell kinetics in infertile patients with proliferating cell nuclear antigen proliferating index. *Asian J. Androl.* **2001**, *3*, 63–66.
57. Krueger, P.; Daneshfar, R.; Eckert, G.P.; Klein, J.; Volmer, D.A.; Bahr, U.; Müller, W.E.; Karas, M.; Schubert-Zsilavecz, M.; Abdel-Tawab, M. Metabolism of boswellic acids in vitro and in vivo. *Drug Metab. Dispos.* **2008**, *36*, 1135–1142. [CrossRef] [PubMed]
58. Zaitone, S.A.; Barakat, B.M.; Bilasy, S.E.; Fawzy, M.S.; Abdelaziz, E.Z.; Farag, N.E. Protective effect of boswellic acids versus pioglitazone in a rat model of diet-induced non-alcoholic fatty liver disease: Influence on insulin resistance and energy expenditure. *Naunyn Schmiedeberg arch. Pharmacol.* **2015**, *388*, 587–600. [CrossRef]
59. Gayathri, B.; Manjula, N.; Vinaykumar, K.; Lakshmi, B.; Balakrishnan, A. Pure compound from Boswellia serrata extract exhibits anti-inflammatory property in human PBMCs and mouse macrophages through inhibition of TNFα, IL-1β, NO and MAP kinases. *Int. Immunopharmacol.* **2007**, *7*, 473–482. [CrossRef]
60. Kimmatkar, N.; Thawani, V.; Hingorani, L.; Khiyani, R. Efficacy and tolerability of Boswellia serrata extract in treatment of osteoarthritis of knee–a randomized double blind placebo controlled trial. *Phytomedicine* **2003**, *10*, 3–7. [CrossRef]
61. Ali, E.N.; Mansour, S.Z. Boswellic acids extract attenuates pulmonary fibrosis induced by bleomycin and oxidative stress from gamma irradiation in rats. *Chin. Med.* **2011**, *6*, 1–14. [CrossRef] [PubMed]
62. Sharma, S.; Gupta, S.; Khajuria, V.; Bhagat, A.; Ahmed, Z.; Shah, B.A. Analogues of boswellic acids as inhibitors of pro-inflammatory cytokines TNF-α and IL-6. *Bioorg. Med. Chem. Lett.* **2016**, *26*, 695–698. [CrossRef] [PubMed]
63. Cuaz-Pérolin, C.; Billiet, L.; Baugé, E.; Copin, C.; Scott-Algara, D.; Genze, F.; Büchele, B.; Syrovets, T.; Simmet, T.; Rouis, M. Antiinflammatory and antiatherogenic effects of the NF-κB inhibitor acetyl-11-keto-β-boswellic acid in LPS-challenged ApoE−/− mice. *Arterioscler. Thromb. Vasc. Biol.* **2008**, *28*, 272–277. [CrossRef]

animals

MDPI

Article

Blue-Green Algae (*Spirulina platensis*) Alleviates the Negative Impact of Heat Stress on Broiler Production Performance and Redox Status

Eman S. Moustafa [1,*], Walaa F. Alsanie [2], Ahmed Gaber [3], Nancy N. Kamel [4], Abdulaziz A. Alaqil [5] and Ahmed O. Abbas [1,5]

1 Department of Animal Production, Faculty of Agriculture, Cairo University, Giza 12613, Egypt; aabbas@kfu.edu.sa
2 Department of Clinical Laboratories Sciences, College of Applied Medical Sciences, Taif University, P.O. Box 11099, Taif 21944, Saudi Arabia; w.alsanie@tu.edu.sa
3 Department of Biology, College of Science, Taif University, P.O. Box 11099, Taif 21944, Saudi Arabia; a.gaber@tu.edu.sa
4 Department of Animal Production, National Research Center, El Buhouth St., Dokki, Giza, Cairo P.O. Box 12622, Egypt; nancy.nk3@gmail.com
5 Department of Animal and Fish Production, College of Agricultural and Food Sciences, King Faisal University, P.O. Box 420, Al-Ahsa 31982, Saudi Arabia; aalaqil@kfu.edu.sa
* Correspondence: emansm23@agr.cu.edu.eg

Citation: Moustafa, E.S.; Alsanie, W.F.; Gaber, A.; Kamel, N.N.; Alaqil, A.A.; Abbas, A.O. Blue-Green Algae (*Spirulina platensis*) Alleviates the Negative Impact of Heat Stress on Broiler Production Performance and Redox Status. *Animals* **2021**, *11*, 1243. https://doi.org/10.3390/ani11051243

Academic Editors: Woo Kyun Kim, Youssef A. Attia, Nesrein Hashem and Maria de Olivera

Received: 23 February 2021
Accepted: 23 April 2021
Published: 26 April 2021

Simple Summary: Heat stress is the leading cause of poor broiler productivity in tropical and subtropical countries. To face such stress, natural antioxidant feed additives are attracting interest due to their high effectiveness and safety. Dietary algae *Spirulina platensis* have received much attention in the last decade due to its high protein content. The effectiveness of (*Spirulina platensis*) as a feed additive to alleviate the negative impacts of heat stress on production performance was investigated. Under heat stress conditions, *Spirulina* supplementation improved broiler productivity and was able to bring back redox balance. It can be inferred that *Spirulina* can be used as a natural antioxidant supplementation to heat-stressed broilers for improving the production performance and modulating serum metabolites to bring them to the normal values.

Abstract: The modern broiler industry faces huge challenges to keep high production quality and quantity, especially under environmental heat stress conditions. The negative effect of heat stress on broiler productivity is mediated by oxidative stress induction. The blue-green alga (*Spirulina platensis*) has many applications in poultry nutrition with the high levels of bioactive antioxidant compounds, which can alleviate the oxidative stress damage induced by high ambient temperature. The current study was designed to investigate the effects of dietary *Spirulina* inclusion at different levels on growth performance, redox status, carcass traits, meat quality, blood hematology, and metabolites profile of broilers subjected to cyclic heat stress. A total of 300 one-day-old Cobb-500 broiler chicks were recruited. Starting from day 21 to 42 of age, birds were randomly divided into five treatment groups with 6 replicates × 10 birds per group, where the first one was provided with the basal diet and reared under normal thermal conditions (23 ± 1 °C) to serve as a negative control. Meanwhile, the other four groups were exposed to cyclic heat stress (34 ± 1 °C for 8 h per day) and were fed a basal diet supplemented with *Spirulina* at a concentration of 0, 0.5, 1 or 1.5%. *Spirulina* supplementation to heat-stressed broilers was able to alleviate the negative impacts of heat stress on the final average daily gain, body weight and feed conversion ratio, with the best impact observed among the chickens fed 1% *Spirulina*. Hematological results indicate increasing hemoglobin and hematocrit levels with *Spirulina* supplementation compared to the non-supplemented stressed group. Further, *Spirulina* supplementation significantly influenced blood lipid metabolites marked by reduced serum cholesterol and low-density lipoprotein (LDL), and increased high-density lipoprotein (HDL) levels. The lipid peroxidation level was reduced ($p < 0.05$), while the antioxidant enzyme activity was increased with *Spirulina* supplementation to the heat-stressed group. *Spirulina* supplementation at 0.5 or 1% improved carcass dressing, breast and leg percentages. It can be concluded that dietary

Spirulina supplementation at 0.5 or 1% to broiler reared under heat stress conditions can effectively improve broiler production performance and balance the redox status.

Keywords: cyclic heat stress; *Spirulina platensis*; redox status; serum metabolites; blood hematology; meat quality; broiler

1. Introduction

One of the challenges facing the modern broiler chicken industry is the high ambient temperatures, especially in regions depending on the open production system [1]. Modern, rapidly growing broiler genotypes are more sensitive to heat stress due to higher metabolic activity, generating more body heat [2]. Lin et al. [3] reported that the optimal production temperature for growing broilers ranged from 18 to 22 °C, while heat stress may occur at an average of 30 °C. Heat stress reported having a negative influence on broiler chicken digestion, nutrient absorption, carcass characteristics, development of immune organs, immune response and survival [4–8]. Broilers exposed to acute heat stress at the market age have been reported to have lower growth performance accompanied by serum metabolites disorder and imbalance redox status [9]. Furthermore, meat quality was reported to be impaired in broilers subjected to chronic heat stress [10,11]. Decreased ultimate pH and increased lightness (L*), cooking loss and shear force are the commonly reported negative effects of heat stress on broiler meat quality [11]. Eventually, heat stress threatens the profitability of poultry in many countries of the world, particularly during the summer months [12,13]. Therefore, there is a continuous endeavor for functional food materials that can be safely utilized to enhance the health and wellbeing of birds subjected to heat stress conditions.

Heat stress is known to break the body redox balance, resulting in the generation of reactive oxygen species (ROS), which induces oxidative damage, and subsequently affects nutrient absorption and metabolism negatively [14]. To overcome the detrimental effects of heat stress and the subsequent oxidative stress induction on poultry production, many intervention strategies have been suggested but with variable or inconsistent outcomes [5]. The fundamental ground and ultimate goal for choosing a mitigation strategy confronting heat stress are to reduce oxidative stress and relieve its tissue-damage impact. Numerous additives have been proposed to improve the performance of birds suffering from heat stress. Mineral and vitamins [15,16], organic acids [17,18], phytogenics [19,20], probiotics [21] and prebiotics [22] are some of the presented additives used to mitigate the negative impact of heat stress on poultry performance. Furthermore, phytochemicals with antioxidant activity, such as polyphenols, were presented as a potentially effective feed additive to confront heat stress in poultry [14].

Spirulina platensis is a blue-green alga rich in protein content, vitamins, minerals and phytopigments [23]. It can be produced in marine or freshwater aquatic systems, with marginal land requirement [24], which reduces the conflict of using limited farmland in cultivating animal feed. *Spirulina* is currently presented as an effective alternative dietary source to substitute the costly supplies of fishmeal and fish oil required in the poultry diet [25,26]. Tavernari et al. [27] demonstrated the potentials of using *Spirulina* as an alternative dietary ingredient for formulating diets that require higher amounts of essential amino acid and metabolizable energy. It can replace up to 15% of broiler diets as a partial substitute of traditional protein sources without negatively affecting production performance or meat quality [28,29]. Moreover, *Spirulina* contains bioactive compounds (e.g., gamma-linoleic acid, phycocyanins, phenolic acids, β-carotene and chlorophyll) [30]. Agustini et al. [31] found that dried *Spirulina* contains high bioactive compounds (i.e., phenolic, flavonoid, saponin, triterpenoid and steroid), which contribute to its high antioxidant activity. Furthermore, Park et al. [32] reported that applying the *Spirulina* addition to broiler chicken diets could enhance growth efficiency, nutrients digestibility, increasing antioxidant enzyme activity, modulating cecal microflora and reducing excreta noxious gas emission. Dietary supplementation of *Spirulina* to chickens reared under chronic heat stress conditions may mitigate the adverse effects of such stress on the growth performance and the immune status of both local chicken strains [33] as well as commercial strains [34,35]. Hence, the present study was designed to investigate the effect of dietary inclusion of *Spirulina platensis* to heat-stressed broiler diets on growth performance, carcass traits, meat quality, hematological and serum biochemical profile.

2. Materials and Methods

2.1. Experimental Material

The *Spirulina platensis* microalgae were obtained from a commercial supplier (Inner Mongolia Rejuve Biotech. Co., Ltd., Ordos, China) in the form of freeze-dried powder. According to the manufacturer, the nutrients composition of *Spirulina* powder was 5.6, 56.4, 7.2, 0.02 and 7.5% for moisture, protein, ether extract, fiber and ash, respectively.

2.2. Experimental Design

The ethical committee of King Faisal University, Al-Ahsa, Saudi Arabia, approved the experimental procedures. It is committed to favoring animal rights and reducing the discomfort, pain, and misery of the birds. A total of 300 unsexed one-day-old Cobb-500 broiler chicks with an average body weight of 45 $\pm$ 1.3 g were obtained from Al Watania Poultry Co., kept under normal controlled conditions and nourished with corn–soybean meal basal diet ad libitum with continuous access to fresh water. The basal diet was formulated to meet the nutritional requirements according to the NRC [36] and the management guide of Cobb-500 broiler and offered in a mash form. Starting from the 22nd day and until the 42nd day of the breeding, the chicks were randomly distributed (considering the equal ratio of males to females) into five treatments with 6 replicates (10 chickens each). Each replicate of birds was reared in a floor pen (0.90 lengths $\times$ 0.90 widths $\times$ 0.38 m height) and relative humidity of 50 $\pm$ 5%. Birds were fed ad libitum. In a temperature-controlled room, the broilers in the thermoneutral group (TN) were raised at 23 $\pm$ 1 °C and 50 $\pm$ 5% relative humidity receiving a corn–soybean meal basal diet (Table 1). In another temperature-controlled room, the heat-stressed groups (HS) broilers were exposed to cyclic heat stress at 34 $\pm$ 1 °C for 8 h per day (from 9:00 a.m. to 5:00 p.m.) and fed the basal diet mixed with either 0, 0.5, 1, or 1.5% of *Spirulina platensis* (SP) powder, respectively. After the high heatwave, the excess heat was removed by increasing ventilation and evaporation cooling after the heat challenge. Birds were kept on an illumination program of 23 L: 1D.

2.3. Production Performance Parameters

Feed consumption was recorded daily, and the birds were weighed by replication after fasting for 12 h. Feed conversion ratio and average daily gain were calculated at the end of the experiment. The mortality rate of birds was recorded in all the treatments, and the European broiler index (EBI) was estimated following Islam et al. [37] formula; European broiler index (EBI) = daily gain $\times$ survival rate/10 $\times$ feed conversion ratio.

2.4. Blood Hematological, Biochemical and Redox Profile Analysis

At the end of the trial, 12 blood samples per treatment (2 birds per replicate) were obtained from the brachial vein into heparinized and non-heparinized tubes. The serum was separated by centrifugation at 1500$\times$ g for 10 min at 4 °C and stored at -18 °C until further analysis. Serum cholesterol, triglyceride, high-density lipoprotein (HDL), creatinine and urea were determined using commercial diagnostic kits according to the manufacturer guidelines (Diamond Diagnostics Company, Cairo, Egypt). Low-density lipoprotein (LDL) was determined according to the formula of Friedewald et al. [38] (LDL = total cholesterol (TC) minus high-density lipoprotein (HDL) minus triglycerides (TGs)/5 in (mg/dL)). The activities (U/L) of alanine aminotransferase (ALT) and aspartate aminotransferase (AST) enzymes were measured using commercial kits (BioDiagnostic, Giza, Egypt) according to the method described by Reitman and Frankel [39]. Serum glutathione reduced concentration (GSH), superoxide dismutase (SOD) activity and total antioxidant capacity (TAC) were measured using commercial kits (Nanjing Jianheng Bioengineering Institute, Nanjing, Jiangsu, China). The serum malondialdehyde (MDA) concentration was determined following the thiobarbituric acid reaction method using a commercial colorimetric assay kit (Nanjing Jiancheng Bioengineering Institute, Nanjing, Jiangsu, China).

Table 1. Ingredients and nutrient composition used of the experimental diets as fed basis (days 22 to 42).

Ingredients	g/kg as Fed
Corn	626
Gluten meal	20.0
Soybean meal, 48% CP	292
Soya oil	25.0
Di-calcium phosphate	16.5
Limestone	7.00
Salt	4.50
Vitamin–mineral premix *	5.00
L-threonine	0.50
DL-methionine	0.80
L-lysine	1.70
Choline chloride	0.20
3-Phytase	0.80
Total	1000
Nutrient content	
Chemical analysis	
Metabolizable energy (kcal/kg)	3150
Crude protein, g/kg	202
Crude fat, g/kg	58.8
Ash, g/kg	5.63
Calculated analysis	
Calcium, g/kg	8.48
Available phosphorus, g/kg	4.21
DL-methionine, g/kg	5.68
L-lysine, g/kg	11.00
Sodium, g/kg	1.40

* Premix provided the following per kg of diet: vitamin A, 1500 IU; vitamin D3, 200 IU; vitamin E, 10 mg; vitamin K3, 0.5 mg; thiamine, 1.8 mg; riboflavin, 3.6 mg; pantothenic acid, 10 mg; folic acid, 0.55 mg; pyridoxine, 3.5 mg; niacin, 35 mg; cobalamin, 0.01 mg; biotin, 0.15 mg; Fe, 80 mg; Cu, 8 mg; Mn, 60 mg; Zn, 40 mg; I, 0.35 mg; Se, 0.15 mg.

Total red blood cells (RBCs) were counted by a Bright-Line™ hemocytometer (American Optical, Buffalo, NY, USA), using a light microscope at 1900× magnification. Before counting, blood samples were diluted 200 times with physiological saline. The hemoglobin concentration was calculated using the cyanomethemoglobin method [40]. For the hematocrit calculation, Wintrobe hematocrit tubes were used, blood samples were centrifuged at 1900× *g* for 20 min and 4 °C, then the hematocrit values were determined on the graduated scale by reading the packed cell volume. The values of MCV, MCH and MCHC% were calculated using the following formulas: The average volume (size) of RBC (MCV, mm^3) = (hematocrit%/RBC] × 10. The average weight of hemoglobin in RBC (MCH, pg) = (hemoglobin concentration (g/dL)/RBC) × 10. The average concentration of hemoglobin in the RBC (MCHC, %) = (hemoglobin (g/dL)/ hematocrit%) × 100.

2.5. Carcass Characteristics

Twelve chicks were taken randomly from each treatment at the end of the trial, two birds per replicate, weighed individually and slaughtered. At 54 °C for 2 min, the chicks were scalded and then de-feathered, and eventually, their heads were cut. Breast muscles and abdominal fat were immediately separated from the hot carcass. The internal organs from the beginning of the esophagus to the end of the outlet were separated in a detailed anatomical fashion [41]. Using a sensitive balance of 0.1 g, the intestines were cleaned and weighed, and the data were expressed as a percentage of carcass weight. The dressing percentage was calculated by dividing the clean carcass weight by the living body weight. The breast muscle, thigh muscle, abdominal fat, gizzard, liver, heart and small intestine weights were estimated as a percentage of carcass weight.

2.6. Meat Quality Measurements

The meat color of breast muscle samples at 45 min postmortem were determined using a color reader (Minolta CR-10, Konica Minolta, Tokyo, Japan) to measure the meat color applying the CIELAB method (L* = lightness; a* = redness; b * = yellowness). At a depth of 1 cm, a pH meter (HI9125, HANNA Instruments, Venice, Italy) was used to assess the pH value at 24 h postmortem. Cooking loss and shear strength of samples were assessed at 24 h postmortem as previously described by Lu et al. [42]. Briefly, after slaughter, muscle samples were weighed, hung in a sealed plastic bag at 4 °C for 24 h and weighed again. The drip loss was expressed as the percentage of weight loss during storage. The samples were eventually put into a lined plastic bag and heated for 20 min in a water bath to reach an interior temperature of 75 °C. The samples were weighed after cooling to room temperature, and cooking loss was expressed as the percentage of loss of weight after cooking. A C-LM3 texture analyzer (Northeast Agricultural University, Harbin, China) was then used to assess shear intensity.

2.7. Statistical Analyses

Data for all variables excluding the normal temperature treatment (thermoneutral) were analyzed using one-way analysis of variance (ANOVA) using the general linear model (GLM) procedure of SAS/STAT® 9.3 software (Copyright© 2021, SAS Institute Inc., Cary, NC, USA). A single degree of linear freedom contrast was used to evaluate the effect of thermoneutral versus cyclic heat stress for birds on the basal diet without *Spirulina*. The results obtained were expressed as mean of standard error (SEM), and significant differences between treatment means were determined using Tukey's test. Each floor pen was considered as an experimental unit for growth performance measurements. Meanwhile, for other measurements, each bird was considered as an experimental unit. Significant values were determined at $p < 0.05$.

3. Results

3.1. Growth Performance

The effects of the different dietary supplementation levels of *Spirulina platensis* on the growth performance of Cobb-500 broiler chickens subjected to cyclic heat stress are presented in Table 2. Growth performance was negatively influenced by heat stress exposure. Feed intake was significantly reduced ($p < 0.037$) by 12% in the HS group compared to the thermoneutral group with no effect of *Spirulina* supplementation. The average daily gain and final body weight were significantly reduced ($p < 0.009$ and $p = 0.012$, respectively) in HSgroup compared to the thermoneutral. Under heat stress conditions, dietary *Spirulina* supplementation at the level of 1% significantly increased both average daily gain and final body weight by 27 and 19%, respectively, compared to the HS group. The other *Spirulina* supplemented groups, HS 0.5 and HS 1.5%, showed an increase ($p = 0.031$) in the average daily gain by 13 and 14% and higher ($p = 0.034$) final body weight by 16 and 15%, respectively, compared to HS group. Moreover, the feed conversion ratio was improved ($p = 0.023$) with *Spirulina* supplementation at different levels than HSgroup. European broiler index (EBI) is a function of daily gain, survival rate and feed conversion ratio. EBI was significantly reduced by heat stress exposure. Nevertheless, *Spirulina* supplementation significantly elevates EBI values with the best production efficiency achieved with 1% and 1.5%, followed by 0.5% supplementation level.

Table 2. Effect of dietary *Spirulina platensis* supplementation at different levels on growth performance of broiler chickens subjected to cyclic heat stress starting from 22 to 42 days of age.

Traits	TN	HS	HS Added SP, %			NT vs. HS	Effect of SP under Heat Stress	
			0.5%	1%	1.5%	*p*-Value	SEM	*p*-Value
IBW, g	660	640	655	650	644	0.560	96.75	0.124
FBW, g	2325	1860 ^{b}	2150 ^{ab}	2210 ^{a}	2140 ^{ab}	0.012	53.25	0.034
ADG, g/d	76.33	54.43 ^{c}	61.57 ^{b}	69.20 ^{a}	62.05 ^{b}	0.009	4.86	0.031
FI, g/d	156.58	138.25	139.29	144.45	134.58	0.037	10.49	0.087
FCR	1.87	2.26 ^{a}	2.03 ^{b}	1.90 ^{c}	1.94 ^{c}	0.002	0.083	0.023
EBI	394.1	227.3 ^{c}	297.2 ^{b}	382.7 ^{a}	337.2 ^{a}	0.007	49.83	0.013

Means within a row with different superscripts significantly differ ($p < 0.05$). SEM, Standard error of the mean, TN, fed the basal diet and reared under thermoneutral condition; HS, cyclic heat stress exposure; SP, *Spirulina platensis*; IBW, initial body weight; FBW, final body weight; ADG, average daily gain; FI, feed intake; FCR, feed conversion ratio; EBI, European broiler index; SEM, Standard error of the mean.

3.2. Hematological Parameters

The hematological profile of broiler chickens exposed to cyclic heat stress and supplemented with different levels of *Spirulina* is illustrated in Table 3. Chronic heat stress reduced ($p = 0.047$) the total number of broiler red blood cells (RBC), moreover, it showed significant negative effect ($p = 0.041$) on hemoglobin (HGB) and hematocrit (HCT) values with subsequent reduction ($p = 0.019$) of mean cell volume (MCV) and low ($p = 0.033$) mean cell hemoglobin (MCH) compared to the thermoneutral group. *Spirulina* supplementation to heat-stressed chickens, irrespective of the addition level, was able to improve) RBC, HGB and HCT values compared to the HS group. Meanwhile, MCV and MCH showed significant elevation in HS 0.5 and 1% *Spirulina* supplementation compared to the non-supplemented HS group.

Table 3. Effect of dietary *Spirulina platensis* supplementation at different levels on hematological parameters of broiler chickens subjected to cyclic heat stress starting from 22 to 42 days of age.

Parameters	TN	HS	HS Added SP, %			NT vs. HS	Effect of SP under Heat Stress	
			0.5%	1%	1.5%	*p*-Value	SEM	*p*-Value
RBC, $10^6/mm^3$	2.25	2.14 ^{c}	2.35 ^{a}	2.27 ^{b}	2.25 ^{b}	0.047	0.031	0.032
HGB, g/dL	9.7	9.1 ^{c}	9.6 ^{b}	10.4 ^{a}	10.2 ^{a}	0.041	0.029	0.033
HCT, %	33.56	28.53 ^{b}	31.82 ^{a}	32.77 ^{a}	30.31 ^{a}	0.013	0.232	0.025
MCV, μm^3/RBC	205.14	191.30 ^{b}	215.51 ^{a}	207.72 ^{a}	200.87 ^{b}	0.019	0.563	0.047
MCH, pg	64.2	62.6 ^{b}	66.1 ^{a}	65.0 ^{a}	62.97 ^{b}	0.033	0.121	0.036
MCHC, %	32.69	33.30	30.90	31.28	31.83	0.109	0.301	0.110

Means within the same row with different superscripts significantly differ ($p < 0.05$). SEM, Standard error of the mean, TN fed the basal diet and reared under thermoneutral condition; HS, cyclic heat stress exposure; SP, *Spirulina platensis*; RBC, red blood cell; HGB, hemoglobin; HCT, hematocrit; MCV, mean cell volume MCH, mean cell hemoglobin and MCHC: mean cell hemoglobin concentration.

3.3. Serum Biochemical Parameters

Broiler blood serum biochemical parameters of the different experimental groups are shown in Table 4. The results demonstrated the presence of significant changes in blood metabolites related to heat stress exposure. Results related to lipid metabolism showed significant increase in serum cholesterol ($p = 0.015$), LDL ($p = 0.011$) and triglycerides ($p = 0.023$) levels and a decrease in HDL ($p = 0.003$) level in HS group compared to the thermoneutral group. Serum concentrations of creatinine and urea, as well as the activity of AST, were significantly elevated in the HS group compared to the thermoneutral group. Results demonstrated that *Spirulina* supplementation generally revoked the negative impact of heat stress on serum lipid metabolites, liver function and kidney function. *Spirulina* supplementation was able to significantly decrease serum cholesterol, LDL, triglycerides, creatinine, urea concentrations and AST activity while increase HDL level compared to the non-supplemented HS group.

Table 4. Effect of dietary *Spirulina platensis* supplementation at different levels on blood serum profile of broiler chickens subjected to cyclic heat stress starting from 22 to 42 days of age.

Parameters	TN	HS	HS Added SP, %			NT vs. HS	Effect of SP under Heat Stress	
			0.5%	1%	1.5%	*p*-Value	SEM	*p*-Value
CHOL, mg/dL	191.3	236.5 ^{a}	201.3 ^{b}	190.5 ^{b}	186.6 ^{b}	0.015	4.123	0.022
LDL, mg/dL	91.7	138.6 ^{a}	100.2 ^{b}	85.2 ^{c}	78.5 ^{c}	0.011	3.204	0.031
HDL, mg/dL	64.4	61.1 ^{c}	65.9 ^{b}	70.3 ^{a}	73.1 ^{a}	0.003	1.378	0.034
TG, mg/dL	176	184 ^{a}	176 ^{b}	175 ^{b}	175 ^{b}	0.023	1.123	0.028
AST, U/L	62.4	65.7 ^{a}	62.3 ^{b}	61.6 ^{b}	62.2 ^{b}	0.031	0.543	0.045
ALT, U/L	69.70	70.68	71.09	73.97	72.55	0.071	1.137	0.091
Creatinine, mg/dL	0.28	0.46 ^{a}	0.28 ^{b}	0.25 ^{b}	0.27 ^{b}	0.002	0.021	0.018
urea, mg/dL	5.35	6.3 ^{a}	4.82 ^{b}	4.17 ^{c}	5.22 ^{b}	0.019	2.125	0.023

Means within the same row with different superscripts significantly differ ($p < 0.05$). SEM, Standard error of the mean, TN fed the basal diet and reared under thermoneutral condition; HS, cyclic heat stress exposure; SP, *Spirulina platensis*; CHOL, total cholesterol; LDL, low-density lipoprotein cholesterol; HDL, high-density lipoprotein cholesterol; TG, triglycerides.

3.4. Blood Redox Status

The redox status of broiler chicken reared under cyclic heat stress with different levels of *Spirulina* supplementation is presented in Table 5. Heat stress disturbed the redox status of broiler chickens. The lipid peroxidation increased significantly ($p = 0.021$) with heat stress exposure, as indicated by the high levels of MDA. The activity of serum antioxidant SOD enzyme and GSH concentration were significantly reduced in the HS group by 1.5 and 1.6-fold compared to the thermoneutral group. Consequently, the total antioxidant capacity showed a 33% reduction ($p = 0.040$) in the cyclic-heat-stress-exposed group with 0% *Spirulina* supplementation compared to the thermoneutral group. Moreover, *Spirulina* supplementation, at all experimented levels, modulated the negative impact of heat stress on broiler redox status by reducing MDA concentration while increasing SOD and TAC activities and GSH concentration.

Table 5. Effect of dietary *Spirulina platensis* supplementation at different levels on blood redox profile of broiler chickens subjected to cyclic heat stress starting from 22 to 42 days of age.

Parameters	TN	HS	HS Added SP, %			NT vs. HS	Effect of SP under Heat Stress	
			0.5%	1%	1.5%	*p*-Value	SEM	*p*-Value
MDA, nmol/mL	1.9	3.5 ^{a}	3.1 ^{b}	2.7 ^{c}	2.8 ^{c}	0.021	0.141	0.009
SOD, U/mL	5.2	3.4 ^{b}	6.3 ^{a}	6.1 ^{a}	5.8 ^{a}	0.019	0.451	0.004
GSH, µmol/L	37.2	23.3 ^{c}	29.7 ^{b}	34.6 ^{a}	32.1 ^{a}	0.029	2.570	0.009
TAC, U/mL	8.4	5.6 ^{c}	7.2 ^{b}	8.1 ^{a}	8.6 ^{a}	0.040	0.248	0.011

Means within the same row with different superscripts significantly differ ($p < 0.05$). SEM, Standard error of the mean, TN fed the basal diet and reared under thermoneutral condition; HS, cyclic heat stress exposure; SP, *Spirulina platensis*; MDA, malondialdehyde; SOD, superoxide dismutase; GSH, glutathione reduced; TAC, total antioxidant capacity.

3.5. Carcass Characteristics

Carcass characteristics were calculated and presented in (Table 6). Dressing percentage, as well as breast and leg percentages, decreased significantly with heat stress exposure. No effect was found on the liver, heart and gizzard proportions among the different experimental groups. Meanwhile, carcass fat percentage was significantly lower ($p = 0.11$) in the HS group compared to the thermoneutral group. Under heat stress, dietary *Spirulina* supplementations at 1% improved carcass dressing percentage ($p = 0.020$) and breast ($p = 0.032$) proportion. Furthermore, the intestine percentage was higher ($p = 0.05$) by about 18% in *Spirulina* supplemented groups compared to the non-supplemented HS group. These findings indicate that *Spirulina* supplementation has a positive effect on carcass characteristics in broilers reared under heat stress, with the best outcome observed among chickens fed at 0.5 and 1% supplemented levels.

Table 6. Effect of dietary *Spirulina platensis* supplementation at different levels on carcass characteristics of broiler chickens subjected to cyclic heat stress starting from 22 to 42 days of age.

Traits (%)	TN	HS	HS Added SP, %			NT vs. HS	Effect of SP under Heat Stress	
			0.5%	1%	1.5%	*p*-Value	SEM	*p*-Value
Dressing	69.01	67.83 ^{b}	68.50 ^{a}	68.99 ^{a}	68.00 ^{b}	0.030	0.195	0.020
Breast	36.71	35.63 ^{c}	37.94 ^{ab}	38.51 ^{a}	36.99 ^{b}	0.020	0.120	0.032
Leg	30.21	29.31	29.57	29.85	28.38	0.070	0.125	0.091
Fat	2.48	1.95 ^{b}	2.45 ^{a}	2.39 ^{a}	2.40 ^{a}	0.011	0.088	0.042
Liver	2.85	2.89	2.90	2.91	2.89	0.083	0.225	0.083
Heart	0.83	0.83	0.84	0.83	0.82	0.231	0.021	0.112
Gizzard	3.71	3.51	3.62	3.72	3.63	0.124	0.141	0.117
Intestine	6.53	6.31 ^{b}	7.70 ^{a}	7.74 ^{a}	7.70 ^{a}	0.128	0.148	0.031

Means within the same row with different superscripts significantly differ ($p < 0.05$). SEM, Standard error of the mean, TN fed the basal diet and reared under thermoneutral condition; HS, cyclic heat stress exposure; SP, *Spirulina platensis*.

3.6. Meat Quality Measurements

Meat is the end product of the broiler industry. Thus its quality is crucial for the breeders' final profitability. In the present study, meat quality parameters showed no changes across the HS group compared to the thermoneutral group (Table 7). Nevertheless, *Spirulina* supplementation, at all levels, increased (p = 0.31) the meat yellowness (b*) compared to the HS group.

Table 7. Effect of dietary *Spirulina platensis* supplementation at different levels on meat quality of broiler chickens subjected to cyclic heat stress starting from 22 to 42 days of age.

Traits	TN	HS	HS Added SP, %			NT vs. HS	Effect of SP under Heat Stress	
			0.5%	1%	1.5%	*p*-Value	SEM	*p*-Value
pH 24 h	5.78	5.76	5.80	5.79	5.74	0.070	0.161	0.119
Lightness (L^*)	49.3	48.5	47.4	46.8	47.1	0.170	1.318	0.074
Yellowness (b^*)	4.37	4.40 ^{c}	10.68 ^{b}	11.59 ^{a}	12.24 ^{a}	0.290	0.228	0.031
Redness (a^*)	4.52	4.49	5.32	5.48	5.36	0.120	0.509	0.101
Cooking loss,%	14.6	12.5	13.5	12.3	12.6	0.100	0.657	0.121
Shear force, kg	1.64	1.59	1.61	1.55	1.58	0.140	0.217	0.108

Means within the same row with different superscripts significantly differ ($p < 0.05$). SEM, Standard error of the mean, TN fed the basal diet and reared under thermoneutral condition; HS, cyclic heat stress exposure; SP, *Spirulina platensis*.

4. Discussion

The results of the current study demonstrated the presence of significant negative effects of heat stress on broiler growth performance. Akbarian et al. [43] reviewed the side effects of heat stress exposure on poultry to be numerous and significantly influence both wellbeing and productivity of the birds. Reduction of voluntary feed intake, decreased energy availability, alteration of various nutrients digestibility and metabolism, as well as the disintegration of intestinal epithelium structure and function, are some of the direct negative impacts of heat stress exposure on poultry behavior and physiology [42]. Low feed intake and body weight gain, as well as increased feed conversion ratio and mortality, were reported in chickens subjected to cyclic heat stress from day 22 to 35 of age [10]. The negative effects of heat stress on broiler growth performance were suggested to be mediated by changes in the intestinal morphology and permeability rather than the alteration in feed intake [44,45]. Liu et al. [46] reported that chronic cyclic heat stress-induced intestinal damage and altered cecal microflora profile via oxidative stress induction. Acute heat stress was also reported to cause serious damages in chickens' small intestine and liver [47]. Adding *Spirulina* to heat-stressed chickens significantly improved growth performance [35]. Although *Spirulina* supplementation in the current study did not increase feed intake, the supplementation improved feed conversion and subsequently the final bodyweight of the heat-stressed chickens. Hajati and Zaghari [48] reported a significant increase in the final body weight and the European production efficiency factor of Japanese quails supplemented with 5 g/kg *Spirulina*. A significant increase in growth rate and improvement in feed conversion ratio was reported in broiler-fed-*Spirulina*-enclosed diet with 5 or 10 g/kg feed [49]. Moreover, *Spirulina* supplementation at 0.25, 0.5, 0.75, or 1.0% linearly improved body weight gain, feed conversion ratio and European production index [32]. The positive impact of *Spirulina* supplementation on broiler performance can partially be justified by the high apparent metabolizable energy and amino acid digestibility of *Spirulina* supplemented diet [27], especially with the negative impact of heat stress on intestinal morphology and feed consumption. Moreover, *Spirulina* addition may beneficially alter intestinal microbial population with a reported increase in *Lactobacillus* sp. and a decrease in *E. coli* population [49], as well as improving gut morphology with higher villi length and increasing goblet cells [50]. The significant increase in relative intestinal weight observed in the present study with *Spirulina* addition could be involved in the adaptation to low feed intake, which subsequently plays a role in compensating the reduction in feed efficiency of heat-stressed birds [51].

Blood biochemical profiles can reflect different physiological changes in birds (e.g., species, age, season, nutrition, and physiological condition) [52]. In the present study, heat stress exposure elevates broiler serum cholesterol, LDL and triglycerides levels while reducing HDL levels. These results can be partially justified by activating the hypothalamic–pituitary–adrenal axis and the release of glucocorticoids, which mediate such serum biochemical changes to confront the excessive heat load [11]. Moreover, heat stress mediates multiple pathways involving glucose, amino acid and lipid metabolism [53] and is reported to increase liver triglyceride synthesis [54] and serum triglyceride and cholesterol levels [55]. Guo et al. [56] reported alteration of serum lipid metabolism in heat-stressed-indigenous slow-growing chickens. Plasma triglyceride and liver enzyme concentrations increased significantly in broilers exposed to chronic heat stress for 14 days [57]. *Spirulina* supplementation at 0.5 to 2% reported decreasing serum cholesterol, LDL, total lipids and triglycerides of broilers subjected to heat stress [34,35]. Moreover, adding *Spirulina* to broiler diet at 5 or 10 g/kg feed was reported to increase hemoglobin level with no effect on total RBC count [49]. The low level of hemoglobin observed in heat-stressed chicken can be justified by the negative effect of heat exposure on ion absorption (i.e., iron), leading to a reduction in hemoglobin formation [58]. Further, the rich minerals content [59] as well as the modulation of intestinal integrity and permeability [60] that *Spirulina* supplementation presents directly modified the blood hemoglobin level. From these findings, it can be concluded that *Spirulina* supplementation to heat-stressed broilers helps recover the normal blood metabolite and hematological profile.

The negative impacts of heat stress on bird physiology and behavior are generally mediated by the induction of oxidative stress and redox imbalance [5,43,46]. The current results demonstrated that chronic heat stress negatively disturbed the redox balance with increasing MDA levels and decreasing GSH concentration, the activity of SOD and the total antioxidant capacity. However, *Spirulina* supplementation, due to its bioactive antioxidant compounds [31], was able to bring back redox balance. Dietary *Spirulina* at 0.25, 0.5, 0.75, or 1.0% showed linear increase of broiler serum antioxidant enzymes (SOD and glutathione peroxidase) [32]. Furthermore, under chronic heat stress conditions, *Spirulina* supplementation to broiler chicken significantly elevates SOD and glutathione peroxidase while decreasing MDA levels [34,35]. It has been reported that *Spirulina* supplementation effectively activates antioxidant enzymes while reduces MDA production in rats subjected to oxidative stress induced by heavy metal [61]. They justified these effects to the bioactive compounds found in *Spirulina*, such as α-tocopherol, ascorbic acid, β-carotene and selenium [61].

Carcass composition and meat quality are largely affected by exposure to high environmental temperature conditions [10,11]. Lu et al. [62] investigated the effect of chronic heat stress on broiler breast muscle quality and reported a reduction in pH $_{45\ min}$, increased lightness, drip loss and intramuscular fat deposition. They further explained such changes by the negative impact of heat stress on the mitochondrial function that causes a decrease in fat and glucose aerobic metabolism and increases in glycolysis and fat deposition. Under our study condition, the lack of differences among groups in meat quality traits can be justified by birds adaptation to the cyclic heat stress imposed, especially under the relieving procedures that were practiced after the heat stress period (i.e., offering cool water and increasing ventilation to remove the excess heat). Chronic heat stress was reported to reduce the proportion of broiler breast muscle and increase the proportion of thigh muscle as well as increase fat deposition [11]. However, the present results showed that *Spirulina* supplementation to heat-stressed broilers at 1% improved carcass composition. Dressing percentage was improved by *Spirulina* addition at 2 g/kg feed with no effect on abdominal fat pad [50]. Hajati and Zaghari [48] supplemented *Spirulina* at the levels of 2.5 or 5 g/kg diet and reported increases in the relative weight of the breast in quail reared under normal environmental conditions. *Spirulina* supplementation increased the yellowness (b*) of meat, which can be attributed to its high carotenoid content [61,63]. Pestana et al. [64] reported higher yellowness values and total carotenoids in breast and thigh meats of chicken fed 15% *Spirulina*. The positive influence of *Spirulina* on carcass composition and meat quality can be justified by the improvement in energy partitioning in favor of muscle development. Furthermore, *Spirulina* has a positive impact on high nutrient content leading to improvement in feed efficiency and nutrient conversion to lean meat.

5. Conclusions

The present study demonstrated the negative impact of chronic cyclic heat stress on broiler growth performance, carcass composition and blood hematological and metabolite profiles. The negative impacts of heat stress on the studied broiler physiological aspects can be justified by the induction of oxidative stress. Dietary *Spirulina* supplementation to heat-stressed chicken positively affects different production performance aspects as well as hematological and biochemical parameters. Furthermore, *Spirulina* addition at 0.5 and 1% to broiler feed can be used to alleviate the negative impacts of heat stress on broiler carcass composition without any negative impact on meat quality. In addition, the blood biochemical and hematological profile of broiler exposed to chronic heat stress can be enhanced by *Spirulina* addition. Furthermore, *Spirulina* supplementation can balance the redox status of broilers reared under heat stress conditions and hence improve productivity.

Author Contributions: Conceptualization, E.S.M. and A.O.A.; data curation, E.S.M.; formal analysis, A.A.A.; funding acquisition, A.G.; investigation, E.S.M.; methodology, A.O.A.; project administration, W.F.A.; resources, W.F.A.; software, E.S.M.; supervision, A.O.A.; validation, W.F.A., A.G., N.N.K. and A.O.A.; visualization, N.N.K.; writing—original draft, E.S.M.; writing—review and editing, A.O.A. All authors have read and agreed to the published version of the manuscript.

Funding: This research was funded by Taif University Researchers Supporting Project number (TURSP-2020/39), Taif University, Taif, Saudi Arabia.

Institutional Review Board Statement: The study was conducted according to the guidelines for ethical conduct for the use of animals in research and approved by the King Faisal University, Scientific Research Deanship Review Board (protocol code 200693 obtained on June 2020).

Data Availability Statement: All data sets obtained and evaluated during the current study are available upon appropriate request from the corresponding author. All data sets obtained and evaluated during the current study are available upon appropriate request from the corresponding author.

Acknowledgments: The authors appreciated Taif University, Taif, Saudi Arabia, for funding the present study.

Conflicts of Interest: The authors declare no conflict of interest.

References

1. He, S.P.; Arowolo, M.A.; Medrano, R.F.; Li, S.; Yu, Q.F.; Chen, J.Y.; He, J.H. Impact of heat stress and nutritional interventions on poultry production. *World's Poult. Sci. J.* **2018**, *74*, 647–664. [CrossRef]
2. He, S.; Yin, Q.; Xiong, Y.; Liu, D.; Hu, H. Effects of dietary fumaric acid on the growth performance, immune response, relative weight and antioxidant status of immune organs in broilers exposed to chronic heat stress. *Czech J. Anim. Sci.* **2020**, *65*, 104–113. [CrossRef]
3. Lin, H.; Decuypere, E.; Buyse, J. Acute heat stress induces oxidative stress in broiler chickens. *Comp. Biochem. Physiol. Part A Mol. Integr. Physiol.* **2006**, *144*, 11–17. [CrossRef] [PubMed]
4. Habibu, B.; Dzenda, T.; Ayo, J.O.; Yaqub, L.S.; Kawu, M.U. Haematological changes and plasma fluid dynamics in livestock during thermal stress, and response to mitigative measures. *Livest. Sci.* **2018**, *214*, 189–201. [CrossRef]
5. Lara, L.J.; Rostagno, M.H. Impact of heat stress on poultry production. *Animals* **2013**, *3*, 356–369. [CrossRef]
6. Niu, Z.Y.; Liu, F.Z.; Yan, Q.L.; Li, W.C. Effects of different levels of vitamin E on growth performance and immune responses of broilers under heat stress. *Poult. Sci.* **2009**, *88*, 2101–2107. [CrossRef]
7. Hirakawa, R.; Nurjanah, S.; Furukawa, K.; Murai, A.; Kikusato, M.; Nochi, T.; Toyomizu, M. Heat Stress Causes Immune Abnormalities via Massive Damage to Effect Proliferation and Differentiation of Lymphocytes in Broiler Chickens. *Front. Vet. Sci.* **2020**, *7*, 46. [CrossRef]
8. Zhang, J.F.; Hu, Z.P.; Lu, C.H.; Yang, M.X.; Zhang, L.L.; Wang, T. Dietary curcumin supplementation protects against heat-stress-impaired growth performance of broilers possibly through a mitochondrial pathway. *J. Anim. Sci.* **2015**, *93*, 1656–1665. [CrossRef]
9. He, S.J.; Yin, Q.R.; Xiong, Y.J.; Li, J.; Liu, D.Y. Characterization of heat stress affecting the growth performance, blood biochemical profile, and redox status in male and female broilers at market age. *Trop. Anim. Health Prod.* **2020**, *52*, 3833–3841. [CrossRef]
10. Awad, E.A.; Najaa, M.; Zulaikha, Z.A.; Zulkifli, I.; Soleimani, A.F. Effects of heat stress on growth performance, selected physiological and immunological parameters, caecalmicroflora, and meat quality in two broiler strains. *Asian Australas. J. Anim. Sci.* **2020**, *33*, 778–787. [CrossRef]
11. Gonzalez-Rivas, P.A.; Chauhan, S.S.; Ha, M.; Fegan, N.; Dunshea, F.R.; Warner, R.D. Effects of heat stress on animal physiology, metabolism, and meat quality: A review. *Meat Sci.* **2020**, *162*, 108025. [CrossRef]
12. Attia, Y.A.; Hassan, R.A.; Tag El-Din, A.E.; Abou-Shehema, B.M. Effect of ascorbic acid or increasing metabolizable energy level with or without supplementation of some essential amino acids on productive and physiological traits of slow-growing chicks exposed to chronic heat stress. *J. Anim. Physiol. Anim. Nutr.* **2011**, *95*, 744–755. [CrossRef] [PubMed]
13. Hajati, H.; Hassanabadi, A.; Golian, A.; Nassiri-Moghaddam, H.; Nassiri, M.R. The effect of grape seed extract and vitamin C feed supplementation on some blood parameters and HSP70 gene expression of broiler chickens suffering from chronic heat stress. *Ital. J. Anim. Sci.* **2015**, *14*, 3273. [CrossRef]
14. Hu, R.; He, Y.; Arowolo, M.A.; Wu, S.; He, J. Polyphenols as potential attenuators of heat stress in poultry production. *Antioxidants* **2019**, *8*, 67. [CrossRef] [PubMed]
15. Del Barrio, A.S.; Mansilla, W.D.; Navarro-Villa, A.; Mica, J.H.; Smeets, J.H.; den Hartog, L.A.; García-Ruiz, A.I. Effect of mineral and vitamin C mix on growth performance and blood corticosterone concentrations in heat-stressed broilers. *J. Appl. Poult. Res.* **2020**, *29*, 23–33. [CrossRef]
16. Shakeri, M.; Oskoueian, E.; Le, H.H.; Shakeri, M. Strategies to combat heat stress in broiler chickens: Unveiling the roles of selenium, vitamin E and vitamin C. *Vet. Sci.* **2020**, *7*, 71. [CrossRef]
17. Ding, J.; He, S.; Xiong, Y.; Liu, D.; Dai, S.; Hu, H. Effects of dietary supplementation of fumaric acid on growth performance, blood hematological and biochemical profile of broiler chickens exposed to chronic heat stress. *Braz. J. Poult. Sci.* **2020**, *22*. [CrossRef]
18. Gouda, A.; Amer, S.A.; Gabr, S.; Tolba, S.A. Effect of dietary supplemental ascorbic acid and folic acid on the growth performance, redox status, and immune status of broiler chickens under heat stress. *Trop. Anim. Health Prod.* **2020**, *52*, 2987–2996. [CrossRef]
19. Al-Sagan, A.A.; Khalil, S.; Hussein, E.O.S.; Attia, A.Y. Effects of Fennel Seed Powder Supplementation on Growth Performance, Carcass Characteristics, Meat Quality, and Economic Efficiency of Broilers under Thermoneutral and Chronic Heat Stress Conditions. *Animals* **2020**, *10*, 206. [CrossRef]
20. Wen, C.; Liu, Y.; Ye, Y.; Tao, Z.; Cheng, Z.; Wang, T.; Zhou, Y. Effects of gingerols-rich extract of ginger on growth performance, serum metabolites, meat quality and antioxidant activity of heat-stressed broilers. *J. Therm. Biol.* **2020**, *89*, 102544. [CrossRef]
21. Sugiharto, S.; Yudiarti, T.; Isroli, I.; Widiastuti, E.; Putra, F.D. Effect of dietary supplementation with Rhizopusoryzae or Chrysoniliacrassa on growth performance, blood profile, intestinal microbial population, and carcass traits in broilers exposed to heat stress. *Arch. Anim. Breed.* **2017**, *60*, 347–356. [CrossRef]
22. Houshmand, M.; Azhar, K.; Zulkifli, I.; Bejo, M.H.; Kamyab, A. Effects of prebiotic, protein level, and stocking density on performance, immunity, and stress indicators of broilers. *Poult. Sci.* **2012**, *91*, 393–401. [CrossRef]
23. Farag, M.R.; Alagawany, M.; Abd El-Hack, M.E.; Dhama, K. Nutritional and healthical aspects of *Spirulina* (Arthrospira) for poultry, animals and human. *Int. J. Pharmacol.* **2016**, *12*, 36–51. [CrossRef]
24. Taelman, S.E.; De Meester, S.; Van Dijk, W.; da Silva, V.; Dewulf, J. Environmental sustainability analysis of a protein-rich livestock feed ingredient in The Netherlands: Microalgae production versus soybean import. *Resour. Conserv. Recycl.* **2015**, *101*, 61–72. [CrossRef]

25. Chen, J.; Wang, Y.; Benemann, J.R.; Zhang, X.; Hu, H.; Qin, S. Microalgal industry in China: Challenges and prospects. *J. Appl. Phycol.* **2016**, *28*, 715–725. [CrossRef]
26. Altmann, B.A.; Neumann, C.; Velten, S.; Liebert, F.; Mörlein, D. Meat quality derived from high inclusion of a micro-alga or insect meal as an alternative protein source in poultry diets: A pilot study. *Foods* **2018**, *7*, 34. [CrossRef]
27. Tavernari, F.D.C.; Roza, L.F.; Surek, D.; Sordi, C.; Silva, M.; Albino, L.F.T.; Migliorini, M.J.; Paiano, D.; Boiago, M.M. Apparent metabolisable energy and amino acid digestibility of microalgae *Spirulinaplatensis* as an ingredient in broiler chicken diets. *Br. Poult. Sci.* **2018**, *59*, 562–567. [CrossRef]
28. Altmann, B.A.; Wigger, R.; Ciulu, M.; Mörlein, D. The effect of insect or microalga alternative protein feeds on broiler meat quality. *J. Sci. Food Agric.* **2020**, *100*, 4292–4302. [CrossRef]
29. Spolaore, P.; Joannis-Cassan, C.; Duran, E.; Isambert, A. Commercial applications of microalgae. *J. Biosci. Bioeng.* **2006**, *101*, 87–96. [CrossRef]
30. Mariey, B.; Samak, H.; Ibrahem, M.A. Effect of using *Spirulinaplatensis* algae as a feed additive for poultry diets: 1-Productive and reproductive performances of local laying hens. *Egypt. Poult. Sci. J.* **2012**, *32*, 201–215. [CrossRef]
31. Agustini, T.W.; Suzery, M.; Sutrisnanto, D.; Ma'ruf, W.F.; Hadiyanto. Comparative study of bioactive substances extracted from fresh and dried *Spirulina* sp. *Procedia Environ. Sci.* **2015**, *23*, 282–289. [CrossRef]
32. Park, J.H.; Lee, S.I.; Kim, I.H. Effect of dietary *Spirulina (Arthrospira) platensis* on the growth performance, antioxidant enzyme activity, nutrient digestibility, cecalmicroflora, excreta noxious gas emission, and breast meat quality of broiler chickens. *Poult. Sci.* **2018**, *97*, 2451–2459. [CrossRef] [PubMed]
33. Zeweil, H.S.; Im, A.; Zahran, S.; Ahmed, M.H.; Haiam, M.; As, A. Effect of *Spirulinaplatensis* as dietary supplement on some biological traits for chickens under heat stress condition. *Asian J. Biomed. Pharm. Sci.* **2016**, *6*, 8–12. [CrossRef]
34. Mirzaie, S.; Zirak-Khattab, F.; Hosseini, S.A.; Donyaei-Darian, H. Effects of dietary Spirulina on antioxidant status, lipid profile, immune response and performance characteristics of broiler chickens reared under high ambient temperature. *Asian Australas. J. Anim. Sci.* **2018**, *31*, 556–563. [CrossRef] [PubMed]
35. Abdel-Moneim, A.M.E.; Shehata, A.M.; Mohamed, N.G.; Elbaz, A.M.; Ibrahim, N.S. Synergistic effect of *Spirulina platensis* and selenium nanoparticles on growth performance, serum metabolites, immune responses, and antioxidant capacity of heat-stressed broiler chickens. *Biol. Trace Elem. Res.* **2021**, 1–12. [CrossRef]
36. Council, N.R. *Nutrient Requirements of Poultry: Ninth Revised Edition, 1994;* The National Academies Press: Washington, DC, USA, 1994; p. 176. [CrossRef]
37. Islam, K.M.; Schaeublin, H.; Wenk, C.; Wanner, M.; Liesegang, A. Effect of dietary citric acid on the performance and mineral metabolism of broiler. *J. Anim. Physiol. Anim. Nutr.* **2012**, *96*, 808–817. [CrossRef]
38. Friedewald, W.T.; Levy, R.I.; Fredrickson, D.S. Estimation of the concentration of low-density lipoprotein cholesterol in plasma, without use of the preparative ultracentrifuge. *Clin. Chem.* **1972**, *18*, 499–502. [CrossRef]
39. Reitman, S.; Frankel, S. A colorimetric method for the determination of serum glutamic oxalacetic and glutamic pyruvic transaminases. *Am. J. Clin. Pathol.* **1957**, *28*, 56–63. [CrossRef] [PubMed]
40. Eilers, R.J. Notification of final adoption of an international method and standard solution for hemoglobinometry specifications for preparation of standard solution. *Am. J. Clin. Pathol.* **1967**, *47*, 212–214. [CrossRef]
41. Fletcher, D.L. Slaughter technology. *Poult. Sci.* **1999**, *78*, 277–281. [CrossRef] [PubMed]
42. Lu, Q.; Wen, J.; Zhang, H. Effect of chronic heat exposure on fat deposition and meat quality in two genetic types of chicken. *Poult. Sci.* **2007**, *86*, 1059–1064. [CrossRef] [PubMed]
43. Akbarian, A.; Michiels, J.; Degroote, J.; Majdeddin, M.; Golian, A.; De Smet, S. Association between heat stress and oxidative stress in poultry; mitochondrial dysfunction and dietary interventions with phytochemicals. *J. Anim. Sci. Biotechnol.* **2016**, *7*, 37. [CrossRef]
44. Nanto-Hara, F.; Kikusato, M.; Ohwada, S.; Toyomizu, M. Heat stress directly affects intestinal integrity in broiler chickens. *J. Poult. Sci.* **2020**, *57*, 284–290. [CrossRef]
45. Tabler, T.W.; Greene, E.S.; Orlowski, S.K.; Hiltz, J.Z.; Anthony, N.B.; Dridi, S. Intestinal barrier integrity in heat-stressed modern broilers and their ancestor wild jungle fowl. *Front. Vet. Sci.* **2020**, *7*, 249. [CrossRef] [PubMed]
46. Liu, G.H.; Zhu, H.B.; Ma, T.H.; Yan, Z.Y.; Zhang, Y.Y.; Geng, Y.Y.; Zhu, Y.; Shi, Y.X. Effect of chronic cyclic heat stress on the intestinal morphology, oxidative status and cecal bacterial communities in broilers. *J. Therm. Biol.* **2020**, *91*. [CrossRef] [PubMed]
47. Siddiqui, S.H.; Kang, D.; Park, J.; Choi, H.W.; Shim, K. Acute heat stress induces the differential expression of heat shock proteins in different sections of the small intestine of chickens based on exposure duration. *Animals* **2020**, *10*, 1234. [CrossRef]
48. Hajati, H.; Zaghari, M. Effects of *Spirulinaplatensis* on growth performance, carcass characteristics, egg traits and immunity response of Japanese quails. *Iran. J. Appl. Anim. Sci.* **2019**, *9*, 347–357.
49. Alwaleed, E.A.; El-Sheekh, M.; Abdel-Daim, M.M.; Saber, H. Effects of *Spirulinaplatensis* and Amphora coffeaeformis as dietary supplements on blood biochemical parameters, intestinal microbial population, and productive performance in broiler chickens. *Environ. Sci. Pollut. Res.* **2020**, *28*, 1801–1811. [CrossRef]
50. Khan, S.; Mobashar, M.; Mahsood, F.K.; Javaid, S.; Abdel-Wareth, A.A.; Ammanullah, H.; Mahmood, A. Spirulinainclusion levels in a broiler ration: Evaluation of growth performance, gut integrity, and immunity. *Trop. Anim. Health Prod.* **2020**, *52*, 3233–3240. [CrossRef] [PubMed]

51. De Verdal, H.; Mignon-Grasteau, S.; Jeulin, C.; Le Bihan-Duval, E.; Leconte, M.; Mallet, S.; Martin, C.; Narcy, A. Digestive tract measurements and histological adaptation in broiler lines divergently selected for digestive efficiency. *Poult. Sci.* **2010**, *89*, 1955–1961. [CrossRef] [PubMed]
52. Hrabčáková, P.; Voslářová, E.; Bedáňová, I.; Pištěková, V.; Chloupek, J.; Večerek, V. Haematological and biochemical parameters during the laying period in common Pheasant hens housed in enhanced cages. *Sci. World J.* **2014**, *2014*, 364602. [CrossRef] [PubMed]
53. Jastrebski, S.F.; Lamont, S.J.; Schmidt, C.J. Chicken hepatic response to chronic heat stress using integrated transcriptome and metabolome analysis. *PLoS ONE* **2017**, *12*, e0181900. [CrossRef]
54. Lu, Z.; He, X.F.; Ma, B.B.; Zhang, L.; Li, J.L.; Jiang, Y.; Zhou, G.H.; Gao, F. Increased fat synthesis and limited apolipoprotein B cause lipid accumulation in the liver of broiler chickens exposed to chronic heat stress. *Poult. Sci.* **2019**, *98*, 3695–3704. [CrossRef] [PubMed]
55. Erol, H.; Imik, H.; Gumus, R.; Halici, M. The effects of different amount of protein and vitamin E supplementation in rations on lipid and antioxidant metabolism of broilers exposed to heat stress. *Braz. J. Poult. Sci.* **2017**, *19*, 289–296. [CrossRef]
56. Guo, Y.; Balasubramanian, B.; Zhao, Z.H.; Liu, W.C. Heat stress alters serum lipid metabolism of Chinese indigenous broiler chickens-a lipidomics study. *Environ. Sci. Pollut. Res.* **2020**, *28*, 10707–10717. [CrossRef]
57. Lu, Z.; He, X.; Ma, B.; Zhang, L.; Li, J.; Jiang, Y.; Zhou, G.; Gao, F. Dietary taurine supplementation decreases fat synthesis by suppressing the liver X receptor α pathway and alleviates lipid accumulation in the liver of chronic heat-stressed broilers. *J. Sci. Food Agric.* **2019**, *99*, 5631–5637. [CrossRef] [PubMed]
58. Wei, C.; Zhou, J.; Huang, X.; Li, M. Effects of psychological stress on serum iron and erythropoiesis. *Int. J. Hematol.* **2008**, *88*, 52–56. [CrossRef]
59. Seghiri, R.; Kharbach, M.; Essamri, A. Functional composition, nutritional properties, and biological activities of Moroccan Spirulina microalga. *J. Food Qual.* **2019**, *2019*, 3707219. [CrossRef]
60. Yu, T.; Wang, Y.; Chen, X.; Xiong, W.; Tang, Y.; Lin, L. *Spirulinaplatensis* alleviates chronic inflammation with modulation of gut microbiota and intestinal permeability in rats fed a high-fat diet. *J. Cell. Mol. Med.* **2020**, *24*, 8603–8613. [CrossRef]
61. El-Desoky, G.E.; Bashandy, S.A.; Alhazza, I.M.; Al-Othman, Z.A.; Aboul-Soud, M.A.; Yusuf, K. Improvement of mercuric chloride-induced testis injuries and sperm quality deteriorations by *Spirulinaplatensis* in rats. *PLoS ONE* **2013**, *8*, e59177. [CrossRef]
62. Lu, Z.; He, X.; Ma, B.; Zhang, L.; Li, J.; Jiang, Y.; Zhou, G.; Gao, F. Chronic heat stress impairs the quality of breast-muscle meat in broilers by affecting redox status and energy-substance metabolism. *J. Agric. Food Chem.* **2017**, *65*, 11251–11258. [CrossRef] [PubMed]
63. Anderson, D.W.; Tang, C.-S.; Ross, E. The xanthophylls of *Spirulina* and their effect on egg yolk pigmentation. *Poult. Sci.* **1991**, *70*, 115–119. [CrossRef]
64. Pestana, J.M.; Puerta, B.; Santos, H.; Madeira, M.S.; Alfaia, C.M.; Lopes, P.A.; Pinto, R.M.A.; Lemos, J.P.C.; Fontes, C.; Lordelo, M.M.; et al. Impact of dietary incorporation of Spirulina (*Arthrospiraplatensis*) and exogenous enzymes on broiler performance, carcass traits, and meat quality. *Poult. Sci.* **2020**, *99*, 2519–2532. [CrossRef] [PubMed]

animals

MDPI

Article

Potential Application of Cornelian Cherry Extract on Broiler Chickens: Growth, Expression of Antioxidant Biomarker and Glucose Transport Genes, and Oxidative Stability of Frozen Meat

Doaa Ibrahim [1,*], Amira Moustafa [2], Aya Sh. Metwally [3], Mohamed A. Nassan [4], Karima Abdallah [5], Fatma Eldemery [6], Vincenzo Tufarelli [7], Vito Laudadio [7] and Asmaa T. Y. Kishawy [1,*]

1 Department of Nutrition and Clinical Nutrition, Faculty of Veterinary Medicine, Zagazig University, Zagazig 44519, Egypt
2 Department of Physiology, Faculty of Veterinary Medicine, Zagazig University, Zagazig 44519, Egypt; amiramostafa@zu.edu.eg
3 Department of Pharmacology, Faculty of Veterinary Medicine, Aswan University, Aswan 81511, Egypt; ayashawky@aswu.edu.eg
4 Department of Clinical Laboratory Sciences, Turabah University College, Taif University, P.O. Box 11099, Taif 21944, Saudi Arabia; m.nassan@tu.edu.sa
5 Department of Food Control, Faculty of Veterinary Medicine, Zagazig University, Zagazig 44519, Egypt; karimaeissa1989@yahoo.com
6 Department of Hygiene and Zoonoses, Faculty of Veterinary Medicine, Mansoura University, Mansoura 35516, Egypt; Fatmaelbaz@mans.edu.eg
7 Department of DETO, Section of Veterinary Science and Animal Production, University of Bari, Strada Provinciale per Casamassima km 3, 70010 Valenzano, Italy; vincenzo.tufarelli@uniba.it (V.T.); vito.laudadio@uniba.it (V.L.)
* Correspondence: doibrahim@vet.zu.edu.eg (D.I.); No.dispair2000@gmail.com (A.T.Y.K.)

Citation: Ibrahim, D.; Moustafa, A.; Metwally, A.S.; Nassan, M.A.; Abdallah, K.; Eldemery, F.; Tufarelli, V.; Laudadio, V.; Kishawy, A.T.Y. Potential Application of Cornelian Cherry Extract on Broiler Chickens: Growth, Expression of Antioxidant Biomarker and Glucose Transport Genes, and Oxidative Stability of Frozen Meat. *Animals* **2021**, *11*, 1038. https://doi.org/10.3390/ani11041038

Academic Editors: Woo Kyun Kim, Youssef A. Attia, Nesrein Hashem and Maria de Olivera

Received: 23 February 2021
Accepted: 5 April 2021
Published: 7 April 2021

Simple Summary: Supplementation of the poultry diet with plant extracts rich in polyphenolic compounds could improve the performance of animals as well as the oxidative stability of their derived meat. The present study evaluated the efficacy of cornelian cherry extract (CCE) on the expression of genes controlling glucose transporters and different assays regulating the oxidative stability of frozen, stored meat over a long period of time (90 days of storage). The results indicated that the addition of 200 mg/kg of CCE to the diet could improve the growth rate and antioxidant status of broiler chickens and thus increase their productivity. Moreover, polyphenolic compounds rich in CCE can act as antioxidant agents to increase the shelf-life extension of frozen, stored poultry meat. Finally, supplementation with CCE could increase the total concentration of phenolic compounds in poultry meat offered to human consumers.

Abstract: The use of natural plant extracts in poultry feed could improve their productivity as well as the oxidative stability of stored derived meat. The roles of cornelian cherry extract (CCE) in growth, cecal microbes, and meat antioxidative markers of broiler chickens were evaluated. A total of 500 Ross 308 broiler chicks were fed diets supplemented with CCE (0, 50, 100, 200, 400 mg/kg of diet) for 38 days. The highest levels of weight gain and feed utilization were observed in a group fed 200 mg/kg of CCE. Maximum upregulation of glucose transporters—1 and 2 and sodium-dependent glucose transporter genes—were found in the group fed 200 mg/kg of CCE. *Lactobacilli* and *Bifidobacterium* colonization increased as the CCE levels increased. The greatest upregulation of antioxidant genes (glutathione peroxidase, catalase, and superoxide dismutase) in breast meat was observed in groups fed CCE (200 and 400 mg/kg). Dietary CCE significantly delayed the lipid oxidation of breast meat compared with that of the control group. The total phenolic content, 2,2-Diphenyl-1-Picrihydrzyl (DPPH) radical scavenging activity and reducing power in meat improved with higher levels of CCE. Dietary CCE improved the growth, performance of broilers, and meat antioxidant stability after 90 days of storage.

Keywords: cornelian cherry extract; glucose transporter; gene expression; oxidative biomarker; chicken

1. Introduction

The intention for the widespread use of phytonutrients in the poultry industry is ultimately associated with the growing discouragement of the use of antibiotics in feed. Recently, natural, active, plant-derived compounds have been gaining great importance, because of their ability to enhance poultry growth performance by improving nutrient digestibility, increasing the concentration of nutrient transporters, sustaining a healthy gut environment, and improving the quality of their products [1]. A phytogenic diet has also been reported to produce changes in the cell membrane permeability, resulting in a higher absorption rate of micronutrients from the small intestine [2]. Additionally, herbal and medicinal plant additives might have the capacity to control intestinal pathogenic bacteria and improve the beneficial intestinal microbiota [3–5] due to their antimicrobial, fungicidal, antiviral, anticoccidial, and antioxidant properties [6,7]. Furthermore, plant extracts that are rich in polyphenols can be effective for preserving meat and their products against oxidative deterioration, pathogen growth, and bacterial spoilage [8].

On the other hand, modern large-scale broiler production prompts stressful conditions such as high-stocking density, heat stress, immunological challenges, handling, feed quality, and transportation [9]. These stressors can enhance reactive oxygen species (ROS) production and interrupt the balance between the antioxidant defense systems and oxidation inside a bird's body, causing oxidative stress [10]. The harmful effect of oxidative stress can be reduced by the dietary inclusion of antioxidants [5]. The use of natural plant-derived compounds rich in polyphenolic compounds can improve the antioxidative status of the living birds and increase the oxidative stability of their derived meat [11].

Among these natural compounds is cornelian cherry (*Cornus mas*) extract (CCE), which is composed of several active compounds, including five anthocyanins: delphinidin 3-galactoside, cyanidin 3-rhamnosylgalactoside, cyanidin 3-galactoside, pelargonidin 3-rhamnosylgalactoside, and pelargonidin 3-galactoside [12]. Ursolic acid is an important constituent of CCE [13–16] possessing antioxidant and antibacterial properties. Additionally, it contains vitamin C, trace minerals, organic acids, pectintriterpenoids, iridoids, pectins, and tannins that range within the safe standards levels of food [17–19]. It is rich in flavonoids such as quercetin 3-O-rhamnoside, quercetin 3-O-rutinoside, and quercetin 3-O-glucuronide [20] and phenolic compounds such as caffeic acid, caffeoylquinic acids, p-coumaric acid, and ellagic acid [21]. Furthermore, CCE has been shown to have antimicrobial [22] anti-inflammatory, and antioxidant activity [23] as well as a hypoglycemic effect [24]. Moreover, the high content of iridoids, such as cyclopentanopyran, found in CCE provides pharmacological properties such as antibiotic and anti-inflammatory effects [25].

Poultry meat is highly susceptible to quality deterioration by lipid oxidation during storage, leading to a decrease in nutritive value and the production of a high content of lipid oxidation products [26]. The oxidative stability of poultry meat is influenced by birds' diets and dietary inclusion of CCE with an abundant amount of polyphenolic bioactive compounds that have been demonstrated to scavenge free radicals and chelate metal ions, helping to increase the oxidation resistance of meat. Moreover, the application of CCE to broiler breast meat was associated with a lower thiobarbituric acid reactive substances (TBARS) value and an increased meat shelf life [15].

Digested carbohydrates, protein, and lipids are transferred into the body by certain transporter proteins located in the brush border enterocytes of the small intestine [27,28]. These include GLUT1 and GLUT2, which are responsible for monosaccharide transportation (glucose, galactose, fructose, and mannose) across the intestinal membrane [29]. The greater expression of transporter-encoding genes leads to a higher flood of nutrients into the intestinal cells and, subsequently, into the blood [30].

Cornelian cherry extract can play an important role in chickens' intestinal health and meat quality owing to its active principle content. Thus, this work investigated different mechanisms and provides new data about the effects of cornelian cherry extract on growth performance, glucose transporters, gut microbes, and meat oxidative stability in broiler chickens.

2. Materials and Methods

The management practices and procedures followed animal welfare, ethical norms, and guidelines of the Institutional Animal Care and Use Committee of the Faculty of Veterinary Medicine at Zagazig University.

2.1. Birds, Diets, and Experimental Design

A total of 500 male Ross broiler chicks (ROSS 308), on the day of hatching (initial body weight 45.8 ± 1), were purchased from a commercial hatchery. Chicks were weighed and randomly divided into five treatment groups with 10 replicate pens containing 10 birds each. The study was organized at the Faculty of Veterinary Medicine, Zagazig University, Egypt. The experimental protocol was accepted by the ethics committee of the Institutional Animal Care and Use Committee of Zagazig University, Egypt. All animal experiments were done according to the recommendations described in "The Guide for the Care and Use of Laboratory Animals in scientific investigations" to ensure their welfare, maintain their rights, and cause minimal stress. All chicks were housed in the same environmental and sanitary conditions all over the experimental period. Birds were raised in floor pens with wood shavings (bird density: 10 broilers/m^2) in an environmentally controlled room. The photoperiod in all experimental pens was maintained at 23 L:1 D h for the first 3 days, followed by 20 L:4 D h until the end of the experiment. The relative humidity ranged from 65 to 75% throughout the trial. During the 1st week, the room temperature was initially set at 33 °C and then gradually decreased until the final temperature of 23 °C was reached. The control starter (d 1–10 d) and grower-finisher (d 11–38) diets were formulated to cover the nutrient requirements of Ross broilers according to the nutritional specifications of ROSS [31]. All birds were allowed access to water and feed ad libitum. The birds were offered a basal diet supplemented with (0 (control), 50, 100, 200, and 400 mg/kg diet of cornelian cherry extract (CCE). The quantities of feed ingredients and the chemical composition of the control diet are listed in Table 1. The proximate analysis of the feed ingredients was done according to the standard procedures of the Association of Official Agricultural Chemists [32]. Asiatic cornelian cherry extract was obtained from Shaanxi Sinuote Biotech Co. Ltd. China. The extract was collected by water alcohol extraction with an extraction ratio of 10:1. The HPLC analysis of the extract based on the manufacturing company was 100 g of cornelian cherry extract containing 203 parts per thousand (PPT) iridoids (consisting of 85% loganic acid), 2.8 PPT ellagic acid, 8.9 PPT anthocyanins, and 4.1 PPT flavonols such as quercetin 3-glucuronide, kaempferol galactoside, and kaempferol glucoside.

2.2. Growth Parameters and Digestibility Trial

The body weight and feed intake (FI) were estimated during the starter period (d 1–10) and grower–finisher period (d 11–38) to calculate the body weight gain, FI, and feed conversion ratio for the whole experimental period (d 1–38). The apparent nutrient digestibility was determined with titanium oxide. At 38 days of age, titanium oxide was added to experimental diets at a rate of 5 g/kg diet. The excreta from each replicate pen was collected every 8 h for five days and analyzed for dry matter, crude protein, ether extract, and crude fiber according to the Association of Official Agricultural Chemists [32]. The titanium oxide content in the diets and excreta was analyzed spectrophotometrically after acid digestion in accordance with the method presented by Short et al. [33] The apparent digestibility coefficient of nutrients was calculated in accordance with the equation presented by McDonald [34].

$$\text{Apparent nutrient digestibility} = 100 - [100 \times (\text{Indicator content (diet)}/\text{Indicator content (feces)} \times \text{Nutrient content (feces)}/\text{Nutrient content (diet)}] \quad (1)$$

Table 1. Ingredients and chemical composition of the basal diet (as dry matter).

Ingredients, g/kg	Starter	Grower–Finisher
Yellow corn grain	57.40	60.1
Soybean meal, 47.5%	34.66	29.00
Corn gluten, 60%	3.00	4.00
Soybean oil	1.10	3.00
Calcium carbonate	1.00	1.00
Dicalcium phosphate	1.80	1.90
Common salt	0.30	0.30
Premix *	0.30	0.30
DL- Methionine, 98%	0.18	0.14
Lysine, HCl, 78%	0.16	0.16
Anti-mycotoxin	0.10	0.10
	Analyzed Chemical Composition	
ME, Kcal/Kg **	3004	3158
CP %	23.01	21.10
EE %	3.63	5.55
CF %	2.66	2.53
Ca %	0.97	0.98
Available P %	0.47	0.47
Lysine %	1.37	1.22
Methionine %	0.56	0.51

* Supplied per kg of diet: Vitamin A, 12 000 IU; Vitamin D3, 2200 IU; Vitamin E, 26 IU; Vitamin K3, 6.25 mg; Vitamin B1, 3.75 mg; Vitamin B2, 6.6 mg; Vitamin B6, 1.5 g; Pantothenic acid, 18.8 mg; Vitamin B12, 0.31 mg; Niacin, 30 mg; Folic acid, 1.25 mg; Biotin, 0.6 mg; Fe, 50 mg; Mn, 60 mg; Cu, 6 mg; I, 1 mg; Co, 1 mg; Se, 0.20 mg; Zn, 50 mg; Choline chloride, 500 mg; ** ME calculated according to National Research Council 1994; ** ME, metabolic energy; CP, crude protein; EE, ether extract; CF, crude fiber; Ca, calcium; P, Phosphorus.

2.3. Sample Collection and Analytical Procedures

At the end of the experimental period, randomly selected birds were weighed and slaughtered.

For serum biochemical measurements, 3 mL blood samples were collected from each bird and then centrifuged for 15 min at 2000 rpm. Clear serum samples were kept at −20 °C until further biochemical analysis.

For meat chemical composition analysis, meat samples were collected from the breast and thigh and then stored at −20 °C until chemical analysis.

For the meat antioxidant analysis, breast meat samples were frozen immediately until the total phenolic content (TPC) and thiobarbituric acid reactive substance (TBARS) content were analyzed and the 2,2-Diphenyl-1-picrihydrzyl (DPPH) assay and Ferric reducing antioxidant power (FRAP) were conducted at 7 and 90 days of storage at −20 °C.

For the molecular analysis, breast meat samples were collected and stored at −80 °C until the analysis of antioxidant-related genes. The small intestine (jejunal part) was separated and the digesta was squeezed out from it and rinsed 3 times in PBS (NaH_2PO_4, 1.47 mmol/L; Na_2HPO_4, 8.09 mmol/L; and NaCl, 145 mmol/L). One square centimeter of the distal jejunum immediately before the Meckel's diverticulum was dissected and kept in Trizol reagent at −80 °C until the analysis of nutrient transporter encoding genes.

2.4. Serum Biochemical Analysis

Serum aspartate aminotransferase (AST), and alanine aminotransferase (ALT), total cholesterol (TC), triglycerides (TGs), high-density lipoprotein cholesterol (HDL-C), low-density lipoprotein cholesterol (LDL-C), very-low-density lipoprotein cholesterol (VLDL-C), creatinine, uric acid, total protein, globulin, and albumin concentrations were determined using commercial diagnostic kits (Spinreact Co., Santa Coloma, Spain).

2.5. Chemical Copmostion of Meat

The dry matter, crude protein, fat and ash contents of breast and thigh meat were analyzed according to the Association of Official Analytical Chemists (AOAC) [32].

2.6. Antioxidant Potential of Broiler Meat

Breast meat samples (5 g) were mixed with phosphate buffer (20 mL; pH 7.4) and glycerol (20 mL; 20%) and homogenized and filtered to ensure they were free from connective tissues.

2.6.1. Total Phenolic Content (TPC)

The total phenolic content of breast meat was measured in accordance with [35]. Briefly, a previously prepared meat sample (100 μL), distilled water, 2.5 mL of 95% ethanol (500 μL), and 50% Folin–Ciocalteu reagent (250 μL) were mixed. This mixture settled for 5 min, and then 5% Na_2CO_3 (500 μL) was added. The mixture was rotated in a vortex meter and left in a dark place for 1 h. The absorbance of samples was measured at 725 nm via a spectrophotometer. The quantity of TPC in meat was measured as Gallic acid equivalents (milligrams of gallic acid per 100 g of meat sample).

2.6.2. 2,2-Diphenyl-1-picrihydrzyl (DPPH) Assay

The DPPH scavenging capacity of the meat sample was calculated as described by [36]. The DPPH A solution (0.25 mM) was formed in methanol. Each sample (100 mL) was mixed with 100 μL of DPPH solution and maintained for 30 min at 25 °C in a dark place, and the sample absorbance was read at 517 nm by a spectrophotometer. The scavenging activity percentage of DPPH in the meat was determined with the following equation:

$$\text{Scavenging activity of DPPH (\%)} = 1 - [A_1 - A_2] \times 100 \quad (2)$$

Blank absorbance
A_1 = Sample absorbance; A_2 = Blank absorbance

2.6.3. Ferric Reducing Antioxidant Power (FRAP)

FRAP in meat was detected according to Oyaizu [37]. A 200-μL homogenized meat sample was mixed with sodium phosphate buffer (500 μL). After that, the prepared solution was maintained in a water bath for 20 min at 50 °C and then centrifuged for 10 min with trichloroacetic acid (2.5 mL) and ferric chloride solution (100 μL). The spectrophotometric absorbance of the samples was measured at 700 nm. The FRAP was estimated as μmol/Fe^2+/g meat.

2.7. TBARS Assay

Lipid oxidation in breast meat was evaluated based on the malondialdehyde (MDA) content, as the MDA concentration in breast meat was determined as previously described by [38]. Briefly, perchloric acid (27 mL) was added to 5 g breast meat samples and then homogenized and filtered. Supernatant samples were mixed with thiobarbituric acid (2 mL) and incubated for 20 min in a water bath (100 °C). Consequently, direct cooling and centrifugation were done for 15 min, and the absorbance was measured by a spectrophotometer at 532 nm. The values are expressed as milligrams of malondialdehyde per kilogram of meat.

2.8. Real-Time PCR to Assess Nutrient Transporter Encoding Genes

Total mRNA was extracted from jejunum and breast meat samples (n = 10 per treatment) using Trizol reagent (TaKaRa Biotechnology Co. Ltd., Dalian, Liaoning, China). The isolated RNA was treated with the RNeasy Mini Kit; Qiagen, Cat. No. 74104 according to the manufacturer's guidelines. The quantity and purity of the total RNA were determined by a NanoDrop ND-8000 spectrophotometer (Thermo Fisher Scientific, Waltham, USA). Complementary DNA (cDNA) was obtained by reverse-transcription of isolated RNA samples using RevertAidTM H Minus kits (Fermentas Life Science, Pittsburgh, PA, USA). One microliter of this cDNA was mixed with 2× maxima SYBR Green PCR mix (12.5 μL) and RNase free water (10.5 μL), and then, 0.5 μL of each forward and reverse primer for the selected genes was added. The primers' sequences of genes encoding the glucose transporter and antioxidant enzymes are described in Table 2. Glyceraldehyde-3-phosphate dehydrogenase (GAPDH) was selected as a reference gene.

Table 2. Primer sequences and target genes used for Q-PCR reactions.

Gene	Gene Full Name	Primer Sequence (5′–3′)	Reference No
Glucose transporters			
GLUT1	Glucose transporter 1	F-TCCTCCTGATCAACCGCAAT R-TGTGCCCCGGAGCTTCT	NM_205209.1
GLUT2	Glucose transporter 2	F-TGATCGTGGCACTGATGGTT R-CCACCAGGAAGAC↓GGAGATA	NM_207178.1
SGLT-1	Sodium-dependent glucose transporter	F-TGCCGGAGTATCTGAGGAAG R-CCCCATGGCCAACTGTATAA	XM_015275173.2
Antioxidant related genes			
GPX1	Glutathione peroxidase	F- GCTGTTCGCCTTCCTGAGAG R- GTTCCAGGAGACGTCGTTGC	NM_001277853.1
SOD1	Superoxide dismutase	F- CACTGCATCATTGGCCGTACCA R- GCTTGCACACGGAAGAGCAAGT	NM_205064.1
CAT	Catalase	F- TGGCGGTAGGAGTCTGGTCT R- GTCCCGTCCGTCAGCCATTT	NM_001031215.1
House keeping			
GAPDH	Glyceraldahyde-3-phosphate dehydrogenase	F-GGTGGTGCTAAGCGTGTTA R-CCCTCCACAATGCCAA	NM205518
TBP	TATA-binding protein	F: GTCCACGGTGAATCTTGGTT R: GCGCAGTAGTACGTGGTTCTC	Acc:8484

2.9. Microbiological Assay

Spread plate counting method for different microbes: One gram of cecal content was mixed with 9 mL phosphate-buffered saline and vortexed for 1 min. Samples were serially diluted in sterile diluents (0.5 g/kg peptone water in distilled water). De Man, Rogosa, and Sharpe (MRS, CM1153, Oxoid, Basingstoke Hampshire, UK) agar medium was used for the enumeration of *Lactobacilli* (MRS, CM1153, Oxoid), Bifidus selective agar was utilized to determine the *Bifidobacteria* content (BSM-Agar, 88517, Sigma, St. Louis, MO, USA), and violet-red bile glucose agar (VRBG, CM485, Oxoid) was used to determine the *Escherichia coli* content. After incubation under appropriate conditions for each group of bacteria (72 h at 37 °C under anaerobic conditions for *lactobacilli* and *Bifidobacteria* and 48 h at 39 °C under aerobic conditions for *E. coli*), colonies were counted on the plates, and the microbial population was expressed as $\log_{10}$ CFU/g of cecal content.

2.10. Statistical Analysis

The data analysis was conducted using the general linear model (GLM) procedure of Statistical Package for the Social Sciences, software (SPSS), after confirming the homogeneity among experimental groups using the Levene test and the normality using the Shapiro–Wilk test. Tukey's test was used to test for significant differences between the mean values. All results are expressed as the standard error of the mean (SEM), and the statistical significance was set at $p < 0.05$. Cecal colony-forming unit (CFU) data were converted to $\log_{10}$ CFU numbers before analysis. The fold change was measured by the following equation: (B–A)/A where the lowest value is A and the highest value is B. Relative fold changes in the expression of target genes were calculated by the $2^{-\Delta\Delta}$Ct method using the GAPDH gene as an internal control gene to normalize the target gene expression levels [39].

3. Results

3.1. Growth Performance and Nutrient Digestibility

The growth performance parameters of the total growing period and nutrient digestibility are shown in Table 3. The body weight gain (BWG) was significantly greater ($p < 0.05$) in groups fed 100, 200, and 400 mg/kg of CCE when compared with the control group. Moreover, the highest BWG was observed in the group fed 200 mg/kg of CCE (increased by 9% in comparison with the control group). The feed conversion ratio (FCR) was improved in all groups fed CCE supplemented diets at different levels, and the biggest improvement in FCR was detected in the group fed diets supplemented with 100 or 200 mg/kg CCE. Concerning the nutrient digestibility, the groups fed 100 or 200 mg/kg CCE showed higher dry matter (DM) contents and CP digestibility levels in comparison with diets containing other levels of CCE, while the control group showed the lowest level of DM digestibility. No significant difference in the digestibility of CF was observed among the different groups.

Table 3. Effect of different levels of cornelian cherry extract (CCE) on growth performance and nutrient digestibility of broiler chickens.

	CCE (mg/kg Diet)						
Parameters	**0**	**50**	**100**	**200**	**400**	***p*-Value**	**SEM**
Total growing period							
BW (g/bird)	2423 ^{c}	2439 ^{b,c}	2508 ^{b}	2649 ^{a}	2503 ^{b}	<0.001	14.10
BWG (g/bird)	2377 ^{c}	2393 ^{b,c}	2462 ^{b}	2603 ^{a}	2457 ^{b}	<0.001	14.12
FI (g/bird)	4331 ^{b}	4219 ^{c}	4224 ^{c}	4503 ^{a}	4400 ^{b}	<0.001	17.53
FCR	1.82 ^{a}	1.76 ^{a,b,c}	1.72 ^{c}	1.73 ^{b,c}	1.79 ^{a,b}	<0.001	0.01
Digestibility %							
Dry matter	69.37 ^{d}	71.87 ^{b,c}	73.33 ^{a,b}	74.80 ^{a}	71.54 ^{c}	<0.05	0.50
Crude protein	62.45 ^{c}	63.63 ^{b,c}	65.92 ^{a,b}	69.22 ^{a}	63.69 ^{b}	<0.001	0.40
Crude fiber	27.30	27.37	27.95	29.47	29.18	<0.20	0.60

BW (body weight); BWG = body weight gain; FI = feed intake; FCR = feed conversion ratio; Number of birds/replicates = 10; $^{a–d}$ Means within a row carrying different superscript letters denote significant differences ($p < 0.05$).

3.2. Serum Biochemical Parameters

Data regarding the impact of CCE on serum biochemical parameters after 38 days are shown in Table 4. The levels of serum AST, ALT, creatinine, uric acid, total protein, albumin, globulin, TGs, HDL-C, and VLDL-C were not affected by dietary CCE ($p > 0.05$). However, total cholesterol and LDL-C concentrations were significantly reduced ($p < 0.05$) in the group supplemented with 400 mg/kg of CCE in comparison with the other groups.

Table 4. Effect of different levels of cornelian cherry extract (CCE) on serum biochemical parameters of broiler chickens.

	CCE (mg/kg Diet)						
Parameter	**0**	**50**	**100**	**200**	**400**	***p*-Value**	**SEM**
ALT, U/L	20.82	20.78	20.98	20.60	19.58	0.18	0.40
AST, U/L	55.16	55.40	54.00	54.64	53.58	0.06	0.48
Uric acid, mg/dL	5.70	5.96	5.74	6.04	5.91	0.55	0.06
Creatinine, mg/dL	0.96	1.02	0.98	1.1	0.94	0.69	0.01
Total Cholesterol, mg/dL	109.16 ^{a}	109.30 ^{a}	107.76 ^{a}	109.34 ^{a}	100.78 ^{b}	<0.001	3.62
TGs, mg/dL	61.18	60.78	61.24	60.32	60.28	0.34	0.38
HDL-C, mg/dL	43.34	44.58	44.48	44.78	43.90	0.17	0.36
LDL-C, mg/dL	53.58 ^{a}	52.56 ^{a}	51.03 ^{a}	52.49 ^{a}	44.82 ^{b}	0.02	4.20
VLDL-C, mg/dL	12.23	12.16	12.24	12.06	12.00	0.34	0.02
Total protein (g/dL)	4.43	4.40	4.41	4.47	4.44	0.96	0.03
Albumin (g/dL)	2.28	2.25	2.32	2.25	2.25	0.97	0.03
Globulin (g/dL)	2.15	2.16	2.09	2.21	2.19	0.96	0.05

^{a,b} Means with different superscripts within the same row differ significantly ($p < 0.05$); ALT: Alanine aminotransferase, AST: Aspartate aminotransferase, TGs: Triglycerides, HDL-C: high- density lipoprotein cholesterol, LDL-C: low-density lipoprotein cholesterol, VLDL-C: very-low density lipoprotein cholesterol.

3.3. Chemical Composition of Meat:

The dry matter, crude protein, fat, and ash contents both breast and thigh meat were not affected ($p > 0.05$) by dietary CCE, as shown in Table 5.

Table 5. Effect of different levels of cornelian cherry extract (CCE) on breast and thigh muscle chemical analysis of broiler chickens % on wet basis.

	CCE (mg/kg Diet)						
Parameters	**0**	**50**	**100**	**200**	**400**	***p*-Value**	**SEM**
	Breast Muscle Analysis % of Wet Basis						
DM	25.76	25.26	25.61	24.73	25.35	0.614	0.21
CP	23.61	23.04	24.19	23.24	23.92	0.322	0.19
EE	4.52	4.52	4.63	4.64	4.56	0.949	0.06
Ash	1.28	1.20	1.27	1.22	1.25	0.864	0.02
	Thigh Muscles Analysis% of Wet Basis						
DM	28.42	28.34	27.21	26.52	28.36	0.273	0.34
CP	22.06	20.66	20.38	21.95	21.74	0.051	0.24
EE	6.94	7.21	6.62	6.31	6.91	0.087	0.11
Ash	1.18	1.23	1.29	1.11	1.24	0.399	0.03

Dry matter: DM, crude protein: CP, ether extract: EE.

3.4. Gene Expression of Glucose Transporter

The mRNA expression levels of the jejunal nutrient transport genes (*GLUT1, GLUT2*, and SGLT-1) are presented in Figure 1. The expression levels of glucose transporter genes (*GLUT-1, GLUT-2*, and *SGLT-1*) were significantly upregulated in groups fed CCE, and the highest level was found in the group supplemented with 200 mg/kg of CCE ($p < 0.05$) with fold changes of 2.55, 2.32, and 2.36, respectively. The aforementioned upregulation was declined in the group supplemented with 400 mg/kg of CCE. Of note, GLUT1 expression in the group supplemented with 50 mg/kg of CCE showed no significant difference as compared with the control group. Moreover, *SGLT-1* expression was significantly upregulated in all groups supplemented with CCE, and the maximum level of upregulation was found in the group supplemented with 200 mg/kg of CCE.

3.5. Gut Microbiota

The gut microbiota data are presented in Table 6. The mean cecal populations of beneficial lactobacilli, bifidobacteria significantly increased after dietary supplementation with 100, 200, or 400 mg/kg of CCE when compared with the control group ($p < 0.05$). The population of *E. coli* in cecal samples significantly decreased ($p < 0.05$) as the CCE level increased, and the lowest reduction of *E. coli* was observed with diets supplemented with 200 or 400 mg/kg of CCE.

Table 6. Effect of different levels of *Cornelian cherry extract* (CCE) on cecal microorganisms (Log_{10} cfu/g fresh digesta) of broiler chickens at slaughter.

	CCE (mg/kg Diet)						
	0	**50**	**100**	**200**	**400**	***p*-Value**	**SEM**
Bifidobacterium	6.17 ^{b}	6.57 ^{b}	7.70 ^{a}	7.90 ^{a}	8.17 ^{a}	<0.01	0.20
Lactobacillus	6.50 ^{d}	6.70 ^{d}	7.73 ^{c}	8.13 ^{b}	8.87 ^{a}	<0.001	0.10
Escherichia coli	8.23 ^{a}	7.9 ^{ab}	7.23 ^{b}	6.47 ^{c}	6.27 ^{c}	<0.008	0.16

Number of birds/replicate = 10. $^{a-d}$ Means within a row carrying different superscript letters denote significant differences ($p < 0.05$).

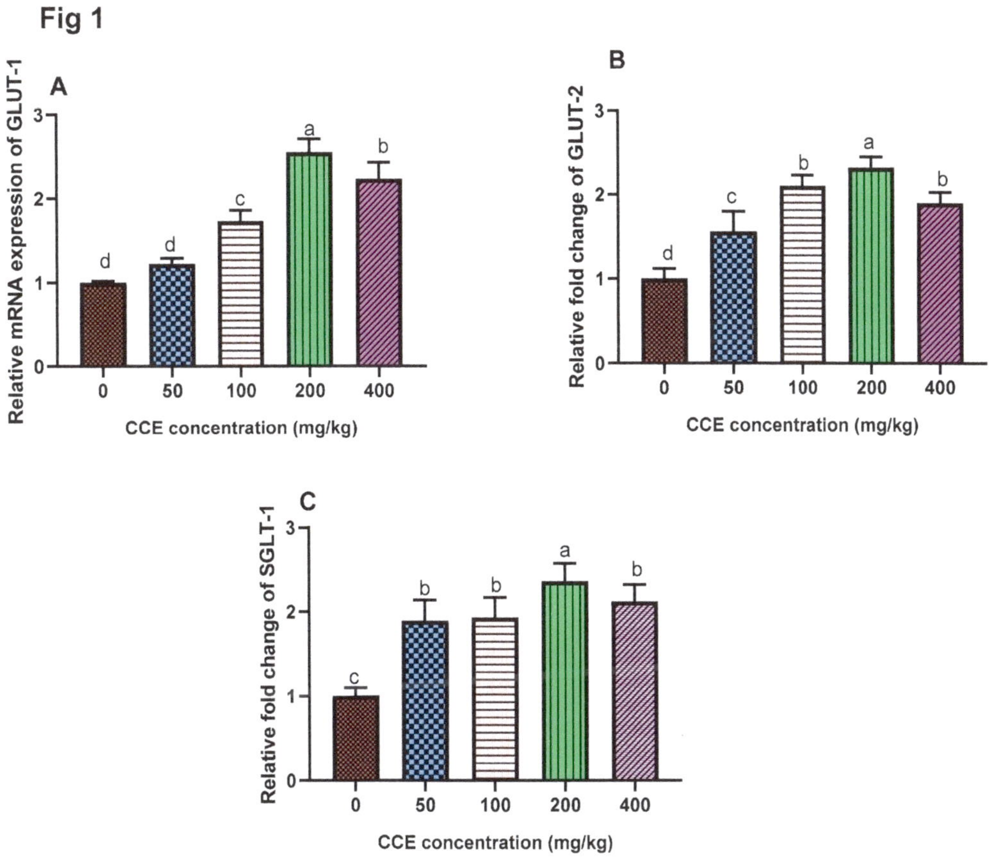

Figure 1. Effect of different levels of cornelian cherry extract (CCE) on glucose transporters genes expression (glucose transporter-1 (**A**) *GLUT-1*, glucose transporter-2 (**B**) *GLUT-2*, Sodium-dependent glucose transporter (**C**) *SGLT-1*. $^{a-d}$ Means within the same column carrying different superscripts are significantly different at $p < 0.05$. Values are means ± standard error. Number of birds/replicates = 10

3.6. Antioxidant-Related Genes

The expression patterns of selected antioxidant-related genes (glutathione peroxidase, *GPx*, superoxide dismutase, *SOD* and catalase, *CAT*) are presented in Figure 2. The highest expression levels of *GPX* were observed in groups supplemented with 400 mg/kg of CCE followed by the groups fed 200 mg/kg of CCE then 50 and 100 mg/kg of CCE. The mRNA expression of *GPx* was significantly upregulated ($p < 0.05$) in the groups fed 200 or 400 mg/kg of CCE compared with groups fed 50 or 100 mg/kg of CCE when compared with the control group. The mRNA expression level of *SOD* was significantly increased as the CCE level increased when compared with the control. The highest expression levels of catalase were observed in groups supplemented with 200 or 400 mg/kg of CCE followed by the groups fed 50 or 100 mg/kg of CCE when compared with the control group.

Fig 2

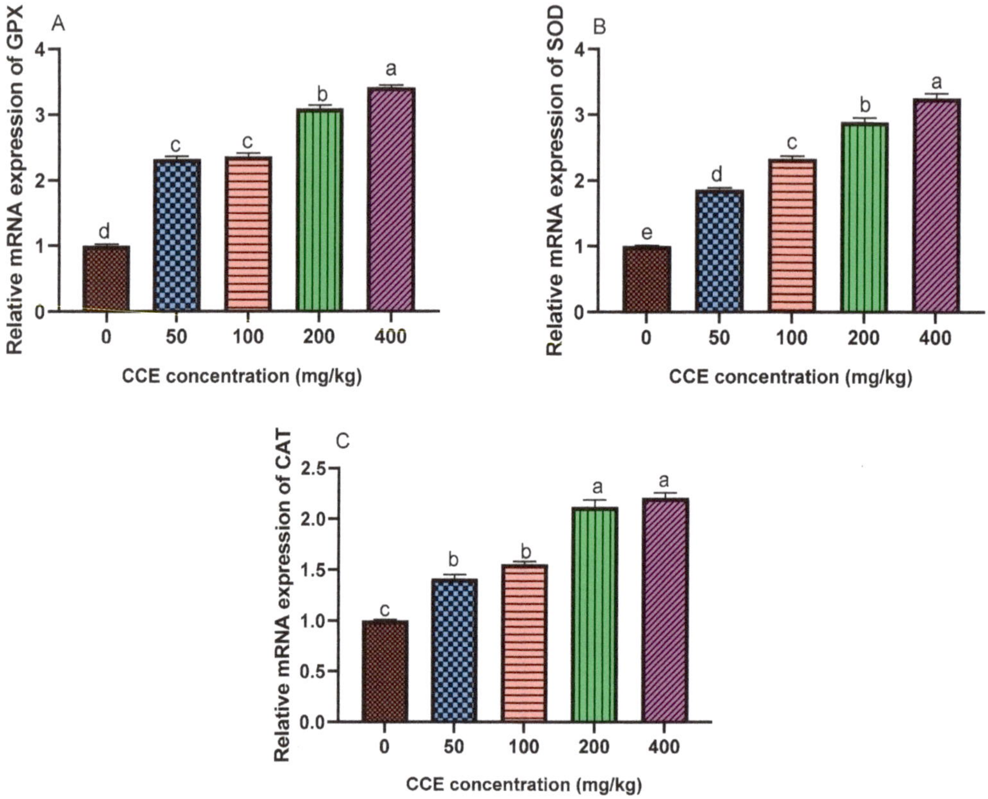

Figure 2. Effect of different levels of cornelian cherry extract (*CCE*) on (glutathione peroxidase, (*GPx*) (**A**), superoxide dismutase, (*SOD*) (**B**) and catalase, (*CAT*) (**C**). [a–e] Means with a row carrying different superscripts are significantly different at $p < 0.05$. Values are means ± standard error. Number of birds/replicates = 10.

3.7. *Antioxidant Potential of Breast Meat*

Data from the analysis of the total phenolic content (TPC), 2,2-diphenyl-1-picrihydrzyl (DPPH) assay, and ferric reducing antioxidant power (FRAP) are presented in Table 5.

3.7.1. Total Phenolic Content (TPC) in Breast Meat

After a short storage period (7 days), the TPC content of breast meat was significantly increased with an increasing level of dietary CCE. After a long storage period (at day 90), the highest values of TPC were observed in the meat of bird groups fed 200 or 400 mg/kg of CCE.

3.7.2. The Free (DPPH) Radical Scavenging Activity

The DPPH activity significantly increased ($p < 0.05$) in breast meat from groups fed dietary CCE in a dose-dependent manner, even after 90 days of storage.

3.7.3. FRAP Reducing Activity

The capacity of the breast myofibrillar protein to reduce Fe^{3+} to Fe^{2+} was greater in the meat of groups fed an increased level of dietary CCE. This capacity was increased after 90 days of storage by 1.1 and 1.7 times, respectively, in the meat from groups fed 200 and 400 mg/kg of CCE when compared to the control group.

3.8. Lipid Peroxidation

Lipid peroxidation, determined as the concentration of MDA in breast meat after 7 and 90 days of storage, is presented in Table 7. At day 7 (short term storage), the MDA level was significantly reduced ($p < 0.05$) in all groups supplemented with CCE, and the lowest levels were observed in the groups fed 200 and 400 mg/kg of CCE when compared with the control group. At day 90 (long-term storage), all groups supplemented with CCE, except the group fed 50 mg/kg, showed reduced levels of MDA when compared with the control group. The greatest reduction in MDA was observed in the group fed 400 mg/kg of CCE. Moreover, the meat MDA content in the group supplemented with 400 mg/kg of CCE was decreased by up to 59.5% and 68.6% after 7 and 90 days of storage, respectively, when compared with the control ($p < 0.05$).

Table 7. Effect of different levels of cornelian cherry extract (CCE) on breast meat total phenolic content oxidative stability and lipid peroxidation during freezing storage at −20 °C.

	CCE (mg/kg Diet)						
Parameters	0	50	100	200	400	*p*-Value	SEM
TPC at d 7 of storage	68.71 ^{e}	111.06 ^{d}	124.06 ^{c}	131.68 ^{b}	143.72 ^{a}	4.91	<0.001
TPC at d 90 of storage	49.77 ^{d}	96.71 ^{c}	109.51 ^{b}	121.95 ^{a}	128.02 ^{a}	3.27	<0.001
DPPH assay at d 7 of storage	86.99 ^{e}	118.20 ^{d}	133.67 ^{c}	148.58 ^{b}	155.88 ^{a}	4.45	<0.001
DPPH at d 90 of storage	77.58 ^{e}	106.93 ^{d}	124.61 ^{c}	134.58 ^{b}	144.03 ^{a}	3.89	<0.001
FRAP assay at d 7 of storage	229.44 ^{e}	435.81 ^{d}	447.99 ^{c}	514.62 ^{b}	618.37 ^{a}	8.09	<0.001
FRAP assay at d 90 of storage	175.44 ^{d}	292.34 ^{c}	301.89 ^{c}	369.88 ^{b}	474.55 ^{a}	7.51	<0.001
MDA content at d 7 of storage	0.47 ^{a}	0.26 ^{b}	0.22 ^{c}	0.15 ^{d}	0.19 ^{d}	0.12	<0.03
MDA content at d 90 of storage	0.67 ^{a}	0.46 ^{a}	0.30 ^{b}	0.28 ^{b}	0.21 ^{c}	0.03	<0.008

TPC = Total phenolic contents; DPPH = 2,2-Diphenyl-1-picrihydrzyl; FRAP = Ferric reducing antioxidant power; MDA = malondialdehyde; Number of birds/replicate = 10; $^{a–e}$ Means within a row carrying different superscript letters denote significant differences ($p < 0.05$).

4. Discussion

Natural plant extracts have been shown to improve the performance of birds by augmenting nutrient utilization and bacterial modulation in the gastrointestinal tract as well as improving the meat quality and oxidative stability. Among these natural products, polyphenols gained growing interest due to their numerous functional properties. Cornelian cherry extract (CCE) has been reported to have an abundance of polyphenolic compounds. The current study demonstrated that CCE could be used to improve the growth rate of broilers by controlling nutrient utilization and absorption and the gut microbiota, and increasing the shelf-life storage of poultry meat by boosting the oxidative stability of meat. In the present study, supplementation of CCE has been shown to improve the feed efficiency of birds by lowering the overall FCR by nearly 5% in the group supplemented with 100 and 200 g/kg CCE. Herein, CCE supplementation has been shown to improve the feed efficiency of birds by lowering the overall FCR by nearly 5% in groups supplemented with 100 or 200 g/kg CCE. The present results are consistent with the results of other researchers [40], who reported that the dietary inclusion of different phenolic compounds for broilers had a positive impact on growth performance parameters. Plant-rich phenolics could promote the growth performance of broilers [40,41] through their potential to improve the antioxidant status of the gut [42]. Furthermore, the application of green tea extract, which is rich in catechins, at concentrations of 100 or 200 mg/kg, in feed has been found to boost the growth performance of broiler chickens [43,44]. Additionally, Herrero-Encinas et al. [45] showed that dietary supplementation with olive extract rich in polyphenolic compounds significantly improved the weight gain and feed conversion ratio of broilers. Similarly, the body weight gain of broiler chickens improved after feeding with grape seed proanthocyanidin extract [46]. Additionally, a sugarcane-served polyphenol mix had a positive effect on the growth performance of broilers and modulated the negative effect of heat exposure [47]. Moreover, the consumption of berries, which have numerous bioactive phenolic compounds, was shown to upregulate the expression of growth-related genes such as insulin-like growth factor binding proteins [48]. Furthermore, certain phytogenic derived agents could improve gastrointestinal barrier function and nutrient absorption [49,50]. Sarker et al. [51] showed that feeding with *Cornus officinalis* had no adverse effect on the growth rate in broilers.

On the other hand, the concentrations of serum AST and ALT indirectly reveal the liver health status and increases in their levels are considered markers of liver damage [52]. In addition, the function of kidney can be estimated via the decrease or increase in of urea and creatinine serum levels. Herein, serum concentration of AST, ALT, uric acid, and creatinine were not affected by dietary CCE and were within normal range, which indicates healthy liver and kidney functions in both control and CCE supplemented groups. Moreover, the current study revealed that higher levels of dietary CCE (400 mg/kg) significantly reduced total cholesterol and LDL-C levels in serum. Similarly, Zhang et al. [53] specified that the feeding of broiler chickens with Chinese bayberry leaves, which are rich in phenolic compounds, significantly reduced the serum cholesterol concentration. The presence of a higher concentration of proanthocyanidins in sorghum, which has antioxidative properties, was also reported to be associated with cholesterol-lowering [54]. Furthermore, the higher concentration of anthocyanins in cornelian cherry powder had a hypercholesterolemic effect in rats via augmenting peroxisome proliferator-activated receptor (PPARα) protein expression and controlling reactive oxygen species (ROS) production and, subsequently, the inflammatory process [55]. Additionally, serum protein and globulin concentrations were not significantly different between treatments and were within the normal range [56]. Similarly, supplementation of broiler chickens with polyphenol extract did not affect their serum protein and globulin concentrations [57]. In addition, supplementation with CCE had no significant effects on the chemical composition (DM, CP, EE and Ash) of meat; these results are in accordance with Gopi et al. [40].

Additionally, the enhanced growth performance of the broilers could have resulted from increasing levels of probiotic bacteria such as *lactobacilli* and *Bifidobacteria*. Moreover, the population of these beneficial bacteria increased in groups supplemented with higher levels of CCE. These positive effects could be related to the role of phenolic-rich cornelian cherry extract on the intestinal microflora, leading to an increase in the concentration of beneficial bacteria (probiotic effect) or inhibiting the growth of pathogenic species (antimicrobial effect) [58]. In agreement with our results, *Bifidobacterium* and *Lactobacillus* have been shown to be the most widely used probiotic bacteria, exerting health-promoting properties, such as the maintenance of gut barrier function [59]. Additionally, an increase in probiotic bacteria, such as *Lactobacillus,* is accompanied by a decrease in the concentration of pathogenic *E. coli,* in accordance with authors [60,61] who stated that *Lactobacillus* can quantitatively inhibit the adherence of pathogenic *E. coli.* The potential effect of phenolic compounds in CCE on the gut microbiota may result from modulation of the bacterial population by acting as prebiotics and enriching the beneficial bacteria [62]. The antimicrobial properties of polyphenols are of primary significance and inhibit biofilm formation in the gut by suppressing harmful bacteria [63]. In vitro testing of the antimicrobial activity of cornelian cherry extract demonstrated its inhibitory effects against *Staphylococcus* aureus and *Escherichia coli* [64]. Caffeic acid, present in CCE, has been described as a potential inhibitor of the growth of *E. coli* and *Clostridium* [65].

Furthermore, flavonoids can change the microbiota ecosystem through their bacteriostatic or bactericidal properties [66]. In addition, blueberry flavonoids can inhibit the activity of *E. coli,* which reduces the integrity of the intestinal barrier as a key mechanism of its pathogenesis [67]. Moreover, feeding with dietary polyphenol-rich grapes was shown to significantly increase the Lactobacillus population in the ceca of broiler chickens [4].

In the small intestine, glucose transporter 1 (GLUT1), recognized as part of solute carrier family 2 (SLC2A1), facilitates glucose transport across the apical surface of the enterocytes, whereas glucose and fructose transport across the basal side of the enterocytes and into the blood circulation is facilitated by GLUT2 [68]. In the current study, the inclusion of CCE in the broiler diet upregulated the expression of glucose nutrient transporters such as GLUT1, GLUT2, and SGLT-1, which are responsible for the transportation of fructose, galactose, mannose, and glucosamine. This can be attributed to the presence of anthocyanins in CCE that are characterized by α-glucosidase inhibitor activity and the capacity to combine with and activate peroxisome proliferator-activated receptor gamma (PPARγ) [69]. The activation of PPARγ accelerates lipid metabolism and glucose uptake by increasing the actions of insulin in glucose utilization in animals [70]. The enhanced expression of GLUT2 can regulate the digestion and absorption of poultry by modulating food consumption by controlling the feedback signal to the brain [71]. Additionally, cornelian cherry extract administration can initiate the uptake of glucose by tissues such as muscle and participate in increasing muscle mass in poultry [72,73]. Moreover, the upregulation of GLUT1 and GLUT2 can enhance their absorptive functions and increase the final body weight of broiler chickens via increasing nutrient transporter expression in the small intestine [74,75]. Additionally, plant-rich phenolics could promote the growth performance of broilers [41] through their potential to improve the antioxidant status of the gut [42]. Antioxidant capacity is a critical factor in poultry health that affects meat quality after slaughter. The radical scavenging ability is linked to the rich polyphenolic compound composition [76]. Animals have developed effective methods of protection against oxidative stress. SOD can eliminate superoxide anion free radicals, and GSH-PX and catalase can catalyze hydrogen peroxide decomposition [77]. Lipid peroxidation is produced by high levels of free radicals, and it leads to an increase in the content of MDA, the end product of lipid oxidation [78]. Levels of thiobarbituric acid reactive substance (MDA) are biomarkers for assessing the lipid peroxidation degree [79]. To the best of our knowledge, there are no data available on the impact of CCE supplementation on the expression of genes encoding the antioxidant enzymes in broilers. In the current study, the expression of antioxidant-related genes (SOD, GPX, and catalase) was upregulated in breast meat by increasing the CCE level. Chickens' breast meat that was not supplemented with CCE had elevated MDA levels and, as a result, lowered oxidative stability, compared with groups supplemented with CCE. Thus, dietary supplementation with CCE had a postmortem effect of decreasing the rate of lipid oxidation by decreasing the MDA content in breast meat, even after 90 days of storage. Similarly, the total antioxidant capacity and phenolic content in the broilers' breast meat improved following dietary supplementation with pomegranate peel extract [80]. Additionally, previous studies showed that lipid oxidation in chicken meat was reduced by feeding with dietary antioxidants, such as plant extracts rich in phenolic compounds [81,82]. A similar positive effect of dietary phenolic compounds on TBARS in breast meat was detected [83].

The higher antioxidant capacity of CCE may be related to the higher polyphenol and flavonoid contents [84], which augment the ability to scavenge radicals. Parallel results support the idea that lipid oxidation could be prevented by fortifying flavonoid antioxidants in animal feed [85]. Furthermore, supplementation of the broiler diet with blackcurrant-extract-rich-polyphenolic compounds enhanced the oxidative stability of their meat after 90 days of frozen storage [86].

Additionally, the reduced TBARS value was useful for lengthening the shelf time of meat products and improving the meat quality [11]. Additionally, the increased total phenolic content in the breast meat of broilers fed increased levels of CCE compared with broilers fed the control diet indicated a higher total antioxidant capacity [87], as these phenolic compounds are able to scavenger free radicals [88]. Besides this, supplementation with dietary CCE improved the DPPH scavenging activity of breast meat, especially at higher levels, even after 90 days of storage. The higher scavenging activity of DPPH indicated an increased antioxidant content in broiler meat, which has the potential to provide one proton to produce a stable DPPH2 compound, thus scavenging the free radicals [89]. Additionally, Cerit et al. [90] reported that cornelian cherry fruits have greater DPPH radical scavenging and ferric-reducing activity.

Moreover, the highest FRAP values were measured in meat enriched with higher levels of supplemental CCE. Faiz et al. [91] showed decreased TBARS values and better activity against 2,2-diphenyl-1-picrylhydrazyl (DPPH) free radicals, associated with a dose-dependent increase in the total phenolic compound concentration detected in chickens' meat after they received different levels of citrus waste. Furthermore, Jang et al. [92] stated that the oxidative stability of the breast meat of chickens fed a diet enriched with Coptis chinensis extract was mainly attributed to the higher concentration of polyphenolic compounds. The antioxidant content present in broiler meat tended to convert ferric ion (Fe^{3+}) to ferrous ion (Fe^{2+}) by providing one electron. Higher metal-chelating potential after supplementation with CCE protected tissues from damage resulting from oxidation. Similarly, a higher metal ion reducing capacity was observed in meat enriched with natural antioxidants in contrast with a control treatment group [93]. Similarly, cornelian cherry extract can reduce 20.41 μmol of Fe^{2+}/g of solution [13].

5. Conclusions

Supplementation with cornelian cherry extract, which is rich in antioxidants, improved the growth performance of broiler chickens via several mechanisms, such as increasing the favorable probiotic populations and lowering the concentration of harmful bacteria, such as *E. coli* spp. The modulation of genes expression responsible for glucose absorption and antioxidant enzymes indicates that CCE can play an effective role in the previously mentioned molecular mechanisms. CCE can scavenge free radicals, thereby improving the antioxidant capacity and lipid peroxidation of poultry meat without affecting its chemical composition. The results demonstrate that the application of dietary CCE (200 mg/kg) is recommended in chickens' diets to boost their growth performance, health, and meat shelf stability during long periods of frozen storage.

Author Contributions: Conceptualization, D.I., A.T.Y.K., A.M., A.S.M., K.A., F.E., V.T. and V.L.; methodology, D.I., A.T.Y.K., A.M., and K.A.; software, D.I., A.T.Y.K., A.S.M. A.M., M.A.N., F.E., K.A., V.T., and V.L.; validation, D.I., A.T.Y.K., A.M., and K.A., F.E.; formal analysis, D.I., A.T.Y.K., F.E., A.M., K.A., V.T. and V.L.; Investigation, D.I., A.T.Y.K., A.M., K.A., A.S.M., M.A.N., F.E., V.T., and V.L.; resources, D.I., A.T.Y.K., A.M., K.A., M.A.N., and F.E.; Data curation, D.I., A.T.Y.K., A.M., A.S.M., K.A., V.T., V.L., and F.E.; Writing—original draft preparation, D.I., A.T.Y.K., A.M., K.A., A.S.M., M.A.N., and F.E.; Writing—review and editing, D.I., A.T.Y.K., A.M., F.E., K.A., A.S.M., V.T., and V.L.; Visualization, D.I., A.T.Y.K., A.M., A.S.M., and K.A.; Supervision, D.I., A.T.Y.K., M.A.N., F.E., A.M., and K.A.; Funding acquisition, D.I., A.T.Y.K., A.M., M.A.N., F.E., A.S.M., and K.A. All authors have read and agreed to the published version of the manuscript.

Funding: This research received no external funding.

Institutional Review Board Statement: The management practices and procedures followed the ethical norms and guidelines of the Institutional Animal Care and Use Committee of the Faculty of Veterinary Medicine at Zagazig University (ZU-IACUC/2020).

Data Availability Statement: The data source was sent to a section of nonpublished material.

Acknowledgments: The authors would like to acknowledge and thank Taif University research supporting project number (TURSP-2020/71), Taif University, Taif, Saudi Arabia.

Conflicts of Interest: The authors declare no conflict of interest.

References

1. Kurekci, C.; Al Jassim, R.; Hassan, E.; Bishop-Hurley, S.L.; Padmanabha, J.; McSweeney, C.S. Effects of feeding plant-derived agents on the colonization of *Campylobacter jejuni* in broiler chickens. *Poult. Sci.* **2014**, *93*, 2337–2346. [CrossRef] [PubMed]
2. Prakash, U.N.; Srinivasan, K. Beneficial influence of dietary spices on the ultrastructure and fluidity of the intestinal brush border in rats. *Br. J. Nutr.* **2010**, *104*, 31–39. [CrossRef] [PubMed]
3. Tiihonen, K.; Kettunen, H.; Bento, M.H.L.; Saarinen, M.; Lahtinen, S.; Ouwehand, A.; Schulze, H.; Rautonen, N. The effect of feeding essential oils on broiler performance and gut microbiota. *Br. Poult. Sci.* **2010**, *51*, 381–392. [CrossRef] [PubMed]

4. Viveros, A.; Chamorro, S.; Pizarro, M.; Arija, I.; Centeno, C.; Brenes, A. Effects of dietary polyphenol-rich grape products on intestinal microflora and gut morphology in broiler chicks. *Poult. Sci.* **2011**, *90*, 566–578. [CrossRef]
5. Liu, H.; Liu, Y.; Hu, L.; Suo, Y.; Zhang, L.; Jin, F.; Feng, X.; Teng, N.; Li, Y. Effects of dietary supplementation of quercetin on performance, egg quality, cecal microflora populations, and antioxidant status in laying hens. *Poult. Sci.* **2014**, *93*, 347–353. [CrossRef]
6. Applegate, T.; Klose, V.; Steiner, T.; Ganner, A.; Schatzmayr, G. Probiotics and phytogenics for poultry: Myth or reality? *J. Appl. Poult. Res.* **2010**, *19*, 194–210. [CrossRef]
7. Kumar, P.; Patra, A. Beneficial uses of black cumin (*Nigella sativa* L.) seeds as a feed additive in poultry nutrition. *Worlds Poult. Sci. J.* **2017**, *73*, 872–885. [CrossRef]
8. Papuc, C.; Goran, G.V.; Predescu, C.N.; Nicorescu, V.; Stefan, G. Plant polyphenols as antioxidant and antibacterial agents for shelf-life extension of meat and meat products: Classification, structures, sources, and action mechanisms. *Compr. Rev. Food Sci. Food Saf.* **2017**, *16*, 1243–1268. [CrossRef]
9. Mehaisen, G.M.; Eshak, M.G.; Elkaiaty, A.M.; Atta, A.-R.M.; Mashaly, M.M.; Abass, A.O. Comprehensive growth performance, immune function, plasma biochemistry, gene expressions and cell death morphology responses to a daily corticosterone injection course in broiler chickens. *PLoS ONE* **2017**, *12*, e0172684. [CrossRef]
10. Akbarian, A.; Michiels, J.; Degroote, J.; Majdeddin, M.; Golian, A.; De Smet, S. Association between heat stress and oxidative stress in poultry; mitochondrial dysfunction and dietary interventions with phytochemicals. *J. Anim. Sci. Biotechnol.* **2016**, *7*, 37. [CrossRef]
11. Jiang, J.; Xiong, Y.L. Natural antioxidants as food and feed additives to promote health benefits and quality of meat products: A review. *Meat Sci.* **2016**, *120*, 107–117. [CrossRef]
12. Tural, S.; Koca, I. Physico-chemical and antioxidant properties of cornelian cherry fruits (*Cornus mas* L.) grown in Turkey. *Sci. Hortic.* **2008**, *116*, 362–366. [CrossRef]
13. De Biaggi, M.; Donno, D.; Mellano, M.G.; Riondato, I.; Rakotoniaina, E.N.; Beccaro, G.L. *Cornus mas* (L.) fruit as a potential source of natural health-promoting compounds: Physico-chemical characterisation of bioactive components. *Plant Foods Hum. Nutr.* **2018**, *73*, 89–94. [CrossRef]
14. Do Nascimento, P.G.; Lemos, T.L.; Bizerra, A.; Arriaga, Â.; Ferreira, D.A.; Santiago, G.M.; Braz-Filho, R.; Costa, J.G.M. Antibacterial and antioxidant activities of ursolic acid and derivatives. *Molecules* **2014**, *19*, 1317–1327. [CrossRef]
15. Sarker, M.; Sharmin, M.; Huque, K.; Yang, C. Effect of Medicinal Plants and Probiotics on Thiobarbituric acid value in broiler chicken meat. In Proceedings of the 58th International Congress of Meat Science and Technology, Montreal, QC, Canada, 12–17 August 2012.
16. Quah, Y.; Lee, S.-J.; Lee, E.-B.; Birhanu, B.T.; Ali, M.; Abbas, M.A.; Boby, N.; Im, Z.-E.; Park, S.-C. *Cornus officinalis* Ethanolic Extract with Potential Anti-Allergic, Anti-Inflammatory, and Antioxidant Activities. *Nutrients* **2020**, *12*, 3317. [CrossRef]
17. Deng, S.; West, B.J.; Jensen, C.J. UPLC-TOF-MS characterization and identification of bioactive iridoids in *Cornus mas* fruit. *J. Anal. Methods Chem.* **2013**, *2013*. [CrossRef]
18. Krośniak, M.; Gąstoł, M.; Szałkowski, M.; Zagrodzki, P.; Derwisz, M. Cornelian cherry (*Cornus mas* L.) juices as a source of minerals in human diet. *J. Toxicol. Env. Health A* **2010**, *73*, 1155–1158. [CrossRef]
19. Dinda, B.; Kyriakopoulos, A.M.; Dinda, S.; Zoumpourlis, V.; Thomaidis, N.S.; Velegraki, A.; Markopoulos, C.; Dinda, M. *Cornus mas* L. (cornelian cherry), an important European and Asian traditional food and medicine: Ethnomedicine, phytochemistry and pharmacology for its commercial utilization in drug industry. *J. Ethnopharmacol.* **2016**, *193*, 670–690. [CrossRef]
20. Pawlowska, A.M.; Camangi, F.; Braca, A. Quali-quantitative analysis of flavonoids of *Cornus mas* L. (Cornaceae) fruits. *Food Chem.* **2010**, *119*, 1257–1261. [CrossRef]
21. Dzydzan, O.; Bila, I.; Kucharska, A.Z.; Brodyak, I.; Sybirna, N. Antidiabetic effects of extracts of red and yellow fruits of cornelian cherries (*Cornus mas* L.) on rats with streptozotocin-induced diabetes mellitus. *Food Funct.* **2019**, *10*, 6459–6472. [CrossRef]
22. Krisch, J. Effect of fruit juices and pomace extracts on the growth of Gram-positive and Gram-negative bacteria. *Acta Biol. Szeged* **2008**, *52*, 267–270.
23. Ersoy, N.; Bagci, Y.; Gok, V. Antioxidant properties of 12 cornelian cherry fruit types (*Cornus mas* L.) selected from Turkey. *Sci. Res. Essays* **2011**, *6*, 98–102.
24. Narimani-Rad, M.; Zendehdel, M.; Abbasi, M.M.; Abdollahi, B.; Lotfi, A. Cornelian cherry (*Cornus mas* L.) extract affects glycemic status in Wistar rats. *Bull. Env. Pharm. Life Sci.* **2013**, *2*, 48–50.
25. Waksmundzka-Hajnos, M.; Sherma, J. *High Performance Liquid Chromatography in Phytochemical Analysis*; CRC Press: Boca Raton, FL, USA, 2010.
26. Aziza, A.; Quezada, N.; Cherian, G. Antioxidative effect of dietary Camelina meal in fresh, stored, or cooked broiler chicken meat. *Poult. Sci.* **2010**, *89*, 2711–2718. [CrossRef] [PubMed]
27. Fotiadis, D.; Kanai, Y.; Palacín, M. The SLC3 and SLC7 families of amino acid transporters. *Mol. Asp. Med.* **2013**, *34*, 139–158. [CrossRef] [PubMed]
28. Bröer, S. Amino acid transport across mammalian intestinal and renal epithelia. *Physiol. Rev.* **2008**, *88*, 249–286. [CrossRef] [PubMed]
29. Mueckler, M.; Thorens, B. The SLC2 (GLUT) family of membrane transporters. *Mol. Asp. Med.* **2013**, *34*, 121–138. [CrossRef] [PubMed]

30. Ruhnke, I.; Röhe, I.; Goodarzi Boroojeni, F.; Knorr, F.; Mader, A.; Hafeez, A.; Zentek, J. Feed supplemented with organic acids does not affect starch digestibility, nor intestinal absorptive or secretory function in broiler chickens. *J. Anim. Physiol. Anim. Nutr.* **2015**, *99*, 29–35. [CrossRef]
31. Aviagen, W. Ross 308: Broiler Nutrition Specification. Available online: https://tmea.aviagen.com/assets/Tech_Center/Ross_Broiler/RossBroilerNutritionSpecs2019-EN.pdf (accessed on 30 September 2019).
32. AOAC. *Official Methods of Analysis of AOAC International*; Association of Official Analytical Chemists: Washington, DC, USA, 2012.
33. Short, F.; Gorton, P.; Wiseman, J.; Boorman, K. Determination of titanium dioxide added as an inert marker in chicken digestibility studies. *Anim. Feed Sci. Technol.* **1996**, *59*, 215–221. [CrossRef]
34. McDonald, P. *Animal Nutrition*; Pearson Education: London, UK, 2002.
35. Senevirathne, M.; Kim, S.-H.; Siriwardhana, N.; Ha, J.-H.; Lee, K.-W.; Jeon, Y.-J. Antioxidant potential of *Ecklonia cava* on reactive oxygen species scavenging, metal chelating, reducing power and lipid peroxidation inhibition. *Food Sci. Technol. Int.* **2006**, *12*, 27–38. [CrossRef]
36. Hwang, K.-E.; Choi, Y.-S.; Choi, S.-M.; Kim, H.-W.; Choi, J.-H.; Lee, M.-A.; Kim, C.-J. Antioxidant action of ganghwayakssuk (*Artemisia princeps* Pamp.) in combination with ascorbic acid to increase the shelf life in raw and deep fried chicken nuggets. *Meat Sci.* **2013**, *95*, 593–602. [CrossRef]
37. Oyaızu, M. Studies on product of browning reaction prepared from glucose amine. *Jpn. J. Nutr.* **1986**, *44*, 307–315. [CrossRef]
38. Zeb, A.; Ullah, F. A simple spectrophotometric method for the determination of thiobarbituric acid reactive substances in fried fast foods. *J. Anal. Methods Chem.* **2016**. [CrossRef]
39. Livak, K.J.; Schmittgen, T.D. Analysis of relative gene expression data using real-time quantitative PCR and the $2-\Delta\Delta CT$ method. *Methods* **2001**, *25*, 402–408. [CrossRef]
40. Starčević, K.; Krstulović, L.; Brozić, D.; Maurić, M.; Stojević, Z.; Mikulec, Ž.; Bajić, M.; Mašek, T. Production performance, meat composition and oxidative susceptibility in broiler chicken fed with different phenolic compounds. *J. Sci. Food Agric.* **2015**, *95*, 1172–1178. [CrossRef]
41. Luo, J.; Song, J.; Liu, L.; Xue, B.; Tian, G.; Yang, Y. Effect of epigallocatechin gallate on growth performance and serum biochemical metabolites in heat-stressed broilers. *Poult. Sci.* **2018**, *97*, 599–606. [CrossRef]
42. Johannah, N.; Joseph, A.; Maliakel, B.; Krishnakumar, I. Dietary addition of a standardized extract of turmeric (TurmaFEED TM) improves growth performance and carcass quality of broilers. *J. Anim. Sci. Technol.* **2018**, *60*, 8.
43. El-Hack, A.; Mohamed, E.; Alagawany, M.; Abdel-Moneim, A.-M.E.; Mohammed, N.G.; Khafaga, A.F.; Bin-Jumah, M.; Othman, S.I.; Allam, A.A.; Elnesr, S.S. Cinnamon (*Cinnamomum zeylanicum*) Oil as a Potential Alternative to Antibiotics in Poultry. *Antibiotics* **2020**, *9*, 210. [CrossRef]
44. Erener, G.; Ocak, N.; Altop, A.; Cankaya, S.; Aksoy, H.M.; Ozturk, E. Growth performance, meat quality and caecal coliform bacteria count of broiler chicks fed diet with green tea extract. *Asian-Australas. J. Anim. Sci.* **2011**, *24*, 1128–1135. [CrossRef]
45. Herrero-Encinas, J.; Blanch, M.; Pastor, J.; Mereu, A.; Ipharraguerre, I.; Menoyo, D. Effects of a bioactive olive pomace extract from *Olea europaea* on growth performance, gut function, and intestinal microbiota in broiler chickens. *Poult. Sci.* **2020**, *99*, 2–10. [CrossRef]
46. Wang, M.; Suo, X.; Gu, J.; Zhang, W.; Fang, Q.; Wang, X. Influence of grape seed proanthocyanidin extract in broiler chickens: Effect on chicken coccidiosis and antioxidant status. *Poult. Sci.* **2008**, *87*, 2273–2280. [CrossRef] [PubMed]
47. Shakeri, M.; Cottrell, J.J.; Wilkinson, S.; Le, H.H.; Suleria, H.A.; Warner, R.D.; Dunshea, F.R. A Dietary Sugarcane-Derived Polyphenol Mix Reduces the Negative Effects of Cyclic Heat Exposure on Growth Performance, Blood Gas Status, and Meat Quality in Broiler Chickens. *Animals* **2020**, *10*, 1158. [CrossRef] [PubMed]
48. Johnson, M.H.; De Mejia, E.G. Phenolic compounds from fermented berry beverages modulated gene and protein expression to increase insulin secretion from pancreatic β-cells in vitro. *J. Agric. Food Chem.* **2016**, *64*, 2569–2581. [CrossRef] [PubMed]
49. Patra, A.K.; Amasheh, S.; Aschenbach, J.R. Modulation of gastrointestinal barrier and nutrient transport function in farm animals by natural plant bioactive compounds–A comprehensive review. *Crit. Rev. Food Sci. Nutr.* **2019**, *59*, 3237–3266. [CrossRef]
50. Ibrahim, D.; Sewid, A.H.; Arisha, A.H.; Abd El-Fattah, A.H.; Abdelaziz, A.M.; Al-Jabr, O.A.; Kishawy, A.T. Influence of Glycyrrhiza glabra Extract on Growth, Gene Expression of Gut Integrity, and Campylobacter jejuni Colonization in Broiler Chickens. *Front. Vet. Sci.* **2020**, *7*, 2063. [CrossRef]
51. Sarker, M.S.K.; Kim, G.M.; Sharmin, F.; Yang, C.J. Effects of medicinal plants, *Alisma canaliculatum*, *Laminaria japonica* and *Cornus officinalis*, treated with probiotics on growth performance, meat composition and internal organ development of broiler chicken. *Asian J. Med. Biol. Res.* **2016**, *2*, 696–702. [CrossRef]
52. Son, H.-K.; Kang, S.-T.; Lee, J.-J. Effects of *Peucedanum japonicum* Thunb. on lipid metabolism and antioxidative activities in rats fed a high-fat/high-cholesterol diet. *J. Korean Soc. Food Sci. Nutr.* **2014**, *43*, 641–649. [CrossRef]
53. Zhang, Y.; Chen, S.; Wei, C.; Chen, J.; Ye, X. Proanthocyanidins from Chinese bayberry (*Myrica rubra* Sieb. et Zucc.) leaves regulate lipid metabolism and glucose consumption by activating AMPK pathway in HepG2 cells. *J. Funct. Foods* **2017**, *29*, 217–225. [CrossRef]
54. Dykes, L.; Rooney, L.W. Sorghum and millet phenols and antioxidants. *J. Cereal Sci.* **2006**, *44*, 236–251. [CrossRef]

55. Sozański, T.; Kucharska, A.; Szumny, A.; Magdalan, J.; Bielska, K.; Merwid-Ląd, A.; Woźniak, A.; Dzimira, S.; Piórecki, N.; Trocha, M. The protective effect of the *Cornus mas* fruits (cornelian cherry) on hypertriglyceridemia and atherosclerosis through PPARα activation in hypercholesterolemic rabbits. *Phytomedicine* **2014**, *21*, 1774–1784. [CrossRef]
56. Allison, R.W. Laboratory evaluation of plasma and serum proteins. In *Thrall MA. Veterinary Hematology and Clinical Chemistry*, 2nd ed.; Blackwell Publishing: London, UK, 2012; pp. 460–475.
57. Gopi, M.; Dutta, N.; Pattanaik, A.K.; Jadhav, S.E.; Madhupriya, V.; Tyagi, P.K.; Mohan, J. Effect of polyphenol extract on performance, serum biochemistry, skin pigmentation and carcass characteristics in broiler chickens fed with different cereal sources under hot-humid conditions. *Saudi J. Biol. Sci.* **2020**, *27*, 2719–2726. [CrossRef]
58. Williamson, G.; Kay, C.D.; Crozier, A. The bioavailability, transport, and bioactivity of dietary flavonoids: A review from a historical perspective. *Compr. Rev. Food Sci. Food Saf.* **2018**, *17*, 1054–1112. [CrossRef]
59. Bermudez-Brito, M.; Muñoz-Quezada, S.; Gomez-Llorente, C.; Matencio, E.; Bernal, M.J.; Romero, F.; Gil, A. Human intestinal dendritic cells decrease cytokine release against Salmonella infection in the presence of *Lactobacillus paracasei* upon TLR activation. *PLoS ONE* **2012**, *7*, e43197. [CrossRef]
60. Mack, D.R.; Michail, S.; Wei, S.; McDougall, L.; Hollingsworth, M.A. Probiotics inhibit enteropathogenic *E. coli* adherence in vitro by inducing intestinal mucin gene expression. *Am. J. Physiol. Gastrointest. Liver Physiol.* **1999**, *276*, G941–G950. [CrossRef]
61. Ibrahim, D.; Abdelfattah-Hassan, A.; Arisha, A.H.; Abd El-Aziz, R.M.; Sherief, W.R.; Adil, S.H.; El Sayed, R.; Metwally, A.E. Impact of feeding anaerobically fermented feed supplemented with acidifiers on its quality and growth performance, intestinal villi and enteric pathogens of mulard ducks. *Livest. Sci.* **2020**, *242*, 104299. [CrossRef]
62. Abbas, M.; Saeed, F.; Anjum, F.M.; Afzaal, M.; Tufail, T.; Bashir, M.S.; Ishtiaq, A.; Hussain, S.; Suleria, H.A.R. Natural polyphenols: An overview. *Int. J. Food Prop.* **2017**, *20*, 1689–1699. [CrossRef]
63. Espín, J.C.; González-Sarrías, A.; Tomás-Barberán, F.A. The gut microbiota: A key factor in the therapeutic effects of (poly)phenols. *Biochem. Pharm.* **2017**, *139*, 82–93. [CrossRef]
64. Savaş, E.; Tavşanlı, H.; Çatalkaya, G.; Çapanoğlu, E.; Tamer, C.E. The antimicrobial and antioxidant properties of garagurt: Traditional Cornelian cherry (*Cornus mas*) marmalade. *Qual. Assur. Saf. Crop. Foods* **2020**, *12*, 12–23. [CrossRef]
65. Cueva, C.; Moreno-Arribas, M.V.; Martín-Álvarez, P.J.; Bills, G.; Vicente, M.F.; Basilio, A.; Rivas, C.L.; Requena, T.; Rodríguez, J.M.; Bartolomé, B. Antimicrobial activity of phenolic acids against commensal, probiotic and pathogenic bacteria. *Res. Microbiol.* **2010**, *161*, 372–382. [CrossRef]
66. Etxeberria, U.; Fernández-Quintela, A.; Milagro, F.I.; Aguirre, L.; Martínez, J.A.; Portillo, M.P. Impact of polyphenols and polyphenol-rich dietary sources on gut microbiota composition. *J. Agric. Food Chem.* **2013**, *61*, 9517–9533. [CrossRef]
67. Polewski, M.A.; Esquivel-Alvarado, D.; Wedde, N.S.; Kruger, C.G.; Reed, J.D. Isolation and Characterization of Blueberry Polyphenolic Components and Their Effects on Gut Barrier Dysfunction. *J. Agric. Food Chem.* **2019**, *68*, 2940–2947. [CrossRef] [PubMed]
68. Williams, A.R.; Hansen, T.V.; Krych, L.; Ahmad, H.F.B.; Nielsen, D.S.; Skovgaard, K.; Thamsborg, S.M. Dietary cinnamaldehyde enhances acquisition of specific antibodies following helminth infection in pigs. *Vet. Immunol. Immunopathol.* **2017**, *189*, 43–52. [CrossRef] [PubMed]
69. Ruhnke, I.; Röhe, I.; Krämer, C.; Goodarzi Boroojeni, F.; Knorr, F.; Mader, A.; Schulze, E.; Hafeez, A.; Neumann, K.; Löwe, R. The effects of particle size, milling method, and thermal treatment of feed on performance, apparent ileal digestibility, and pH of the digesta in laying hens. *Poult. Sci.* **2015**, *94*, 692–699. [CrossRef] [PubMed]
70. Leonardini, A.; Laviola, L.; Perrini, S.; Natalicchio, A.; Giorgino, F. Cross-talk between PPAR and insulin signaling and modulation of insulin sensitivity. *PPAR Res.* **2009**, *2009*. [CrossRef]
71. Brubaker, P.L. The glucagon-like peptides: Pleiotropic regulators of nutrient homeostasis. *Ann. N. Y. Acad. Sci.* **2006**, *1070*, 10–26. [CrossRef]
72. Guo, X.; Sun, W.; Luo, G.; Wu, L.; Xu, G.; Hou, D.; Hou, Y.; Guo, X.; Mu, X.; Qin, L. Panax notoginseng saponins alleviate skeletal muscle insulin resistance by regulating the IRS 1–PI 3K–AKT signaling pathway and GLUT 4 expression. *FEBS Open Bio* **2019**, *9*, 1008–1019. [CrossRef]
73. Changxing, L.; Chenling, M.; Alagawany, M.; Jianhua, L.; Dongfang, D.; Gaichao, W.; Wenyin, Z.; Syed, S.; Arain, M.; Saeed, M. Health benefits and potential applications of anthocyanins in poultry feed industry. *Worlds Poult. Sci. J.* **2018**, *74*, 251–264. [CrossRef]
74. Hu, X.; Guo, Y.; Huang, B.; Bun, S.; Zhang, L.; Li, J.; Liu, D.; Long, F.; Yang, X.; Jiao, P. The effect of glucagon-like peptide 2 injection on performance, small intestinal morphology, and nutrient transporter expression of stressed broiler chickens. *Poult. Sci.* **2010**, *89*, 1967–1974. [CrossRef]
75. Al-Khalaifah, H.S.; Shahin, S.E.; Omar, A.E.; Mohammed, H.A.; Mahmoud, H.I.; Ibrahim, D. Effects of graded levels of microbial fermented or enzymatically treated dried brewer's grains on growth, digestive and nutrient transporter genes expression and cost effectiveness in broiler chickens. *BMC Vet. Res.* **2020**, *16*, 1–15. [CrossRef]
76. Czerwińska, M.E.; Melzig, M.F. *Cornus mas* and *Cornus officinalis*—Analogies and differences of two medicinal plants traditionally used. *Front. Pharm.* **2018**, *9*, 894. [CrossRef]
77. Crespo, I.; García-Mediavilla, M.V.; Almar, M.; González, P.; Tuñón, M.J.; Sánchez-Campos, S.; González-Gallego, J. Differential effects of dietary flavonoids on reactive oxygen and nitrogen species generation and changes in antioxidant enzyme expression induced by proinflammatory cytokines in Chang Liver cells. *Food Chem. Toxicol.* **2008**, *46*, 1555–1569. [CrossRef]

78. Kim, J.E.; Clark, R.M.; Park, Y.; Lee, J.; Fernandez, M.L. Erratum: Lutein decreases oxidative stress and inflammation in liver and eyes of guinea pigs fed a hypercholesterolemic diet. *Nutr. Res. Pr.* **2013**, *7*, 146. [CrossRef]
79. Kotunia, A.; Wolinski, J.; Laubitz, D.; Jurkowska, M.; Rome, V.; Guilloteau, P.; Zabielski, R. Effect of sodium butyrate on the small intestine. *J. Physiol. Pharm.* **2004**, *55*, 59–68.
80. Saleh, H.; Golian, A.; Kermanshahi, H.; Mirakzehi, M.T. Effects of dietary α-tocopherol acetate, pomegranate peel, and pomegranate peel extract on phenolic content, fatty acid composition, and meat quality of broiler chickens. *J. Appl. Anim. Res.* **2017**, *45*, 629–636. [CrossRef]
81. Hu, Z.; Wang, T.; Ahmad, H.; Zhang, J.; Zhang, L.; Zhong, X. Effects of different formulations of α-tocopherol acetate (vitamin E) on growth performance, meat quality and antioxidant capacity in broiler chickens. *Br. Poult. Sci.* **2015**, *56*, 687–695. [CrossRef]
82. Akbarian, A.; Michiels, J.; Golian, A.; Buyse, J.; Wang, Y.; De Smet, S. Gene expression of heat shock protein 70 and antioxidant enzymes, oxidative status, and meat quality of cyclically heat challenged finishing broilers fed *Oreganum compactum* and *Curcuma xanthorrhiza* essential oils. *Poult. Sci.* **2014**, *93*, 1930–1941. [CrossRef]
83. Jung, S.; Choe, J.H.; Kim, B.; Yun, H.; Kruk, Z.A.; Jo, C. Effect of dietary mixture of gallic acid and linoleic acid on antioxidative potential and quality of breast meat from broilers. *Meat Sci.* **2010**, *86*, 520–526. [CrossRef]
84. Moldovan, B.; David, L. Influence of temperature and preserving agents on the stability of cornelian cherries anthocyanins. *Molecules* **2014**, *19*, 8177–8188. [CrossRef]
85. Zhong, R.; Tan, C.; Han, X.; Tang, S.; Tan, Z.; Zeng, B. Effect of dietary tea catechins supplementation in goats on the quality of meat kept under refrigeration. *Small Rumin. Res.* **2009**, *87*, 122–125. [CrossRef]
86. Sierżant, K.; Korzeniowska, M.; Król, B.; Orda, J.; Wojdyło, A. Oxidative stability of the meat of broilers fed diets supplemented with various levels of Blackcurrant extract (*Ribes nigrum* L.) during different time period. *J. Chem.* **2018**, *2018*. [CrossRef]
87. Moldovan, B.; Filip, A.; Clichici, S.; Suharoschi, R.; Bolfa, P.; David, L. Antioxidant activity of Cornelian cherry (*Cornus mas* L.) fruits extract and the in vivo evaluation of its anti-inflammatory effects. *J. Funct. Foods* **2016**, *26*, 77–87. [CrossRef]
88. Dave, R. In vitro models for antioxidant activity evaluation and some medicinal plants possessing antioxidant properties: An overview. *Afr. J. Microbiol. Res.* **2009**, *3*, 981–996.
89. Ibrahim, D.; Kishawy, A.T.; Khater, S.I.; Hamed Arisha, A.; Mohammed, H.A.; Abdelaziz, A.S.; El-Rahman, A.; Ghada, I.; Elabbasy, M.T. Effect of dietary modulation of selenium form and level on performance, tissue retention, quality of frozen stored meat and gene expression of antioxidant status in Ross broiler chickens. *Animals* **2019**, *9*, 342. [CrossRef] [PubMed]
90. Cerit, İ.; Şenkaya, S.; Tulukoğlu, B.; Kurtuluş, M.; Seçilmişoğlu, Ü.R.; Demirkol, O. Enrichment of functional properties of white chocolates with cornelian cherry, spinach and pollen powders. *Gida J. Food* **2016**, *41*, 311–316. [CrossRef]
91. Faiz, F.; Khan, M.I.; Butt, M.S.; Nawaz, H. Enhancement of broiler meat oxidative stability through dietary supplementation of citrus processing waste. *Pak. J. Agric. Sci.* **2017**, *54*, 893–898.
92. Jang, A.; Liu, X.-D.; Shin, M.-H.; Lee, B.-D.; Lee, S.-K.; Lee, J.-H.; Jo, C. Antioxidative potential of raw breast meat from broiler chicks fed a dietary medicinal herb extract mix. *Poult. Sci.* **2008**, *87*, 2382–2389. [CrossRef]
93. Banerjee, R.; Verma, A.K.; Das, A.K.; Rajkumar, V.; Shewalkar, A.; Narkhede, H. Antioxidant effects of broccoli powder extract in goat meat nuggets. *Meat Sci.* **2012**, *91*, 179–184. [CrossRef]

animals

MDPI

Article

A Multi-Biomarker Approach in European Sea Bass Exposed to Dynamic Temperature Changes under Dietary Supplementation with *Origanum vulgare* Essential Oil

Francesca Rita Dinardo, Aristide Maggiolino, Elisabetta Casalino *, Michele Deflorio and Gerardo Centoducati

Department of Veterinary Medicine, University of Bari Aldo Moro, Casamassima km 3, 70010 Valenzano, Italy; francesca.dinardo@uniba.it (F.R.D.); aristide.maggiolino@uniba.it (A.M.); michele.deflorio@uniba.it (M.D.); gerardo.centoducati@uniba.it (G.C.)

* Correspondence: elisabetta.casalino@uniba.it; Tel.: +39-080-544-3864

Simple Summary: Temperature fluctuations may induce metabolic and physiological imbalances over marine organisms, involving reproduction, growth, immune response, osmoregulatory capacity, and antioxidant defenses. It is of great importance to find tools, including nutritional interventions on farms, able to reduce such imbalances and the consequent stress for animals. In light of this perspective, we investigated the correlations between temperature and metabolic performance in sea bass fed on diet containing oregano essential oil. Under the condition of our study, thermal changes affected the levels of several biomarkers (e.g., triglycerides and cholesterol) highlighting an attempt to provide for additional energy, to counterbalance the oxidative damage, and to maintain cell homeostasis. On the other side, the activity of antioxidant enzymes, TBARS levels, and the energetic balance seemed to benefit from the intake of oregano essential oil under exposure to thermal changes.

Abstract: A feeding trial for 150 days was carried out to evaluate the cross-effects between oregano essential oil (EO) dietary supplementation and dynamic temperature change in sea bass. Under exposure to rising temperature (13–25 °C), fish were fed with a control diet (CD) and two experimental diets supplemented with 100 (D100) and 200 ppm (D200) of EO. Feed inclusion of EO promoted the activity of antioxidant enzymes in sea bass exposed to increasing temperature. Consistently with the temperature rise, TBARS concentrations increased in CD and D200 groups, whereas were almost stable in D100. Trend of blood glucose in fish fed on CD was likely affected by glycogenolysis and gluconeogenesis. Similarly, the depletion of triglycerides and cholesterol in fish fed on CD likely supported the energy cost of gluconeogenesis. On the other hand, the reduction of glucose, triglycerides, and cholesterol in D100 and D200 was mainly attributable to the hypoglycemic and hypolipidemic effects of EO. The higher levels of serum protein observed in D100 and D200 groups were also associated to a reduced thermal stress compared to CD. EO dietary supplementation may be a promising strategy to alleviate the negative effects of temperature shift on sea bass physiological and oxidative state.

Keywords: oxidative stress; biochemical parameters; oregano essential oil; temperature; sea bass

Citation: Dinardo, F.R.; Maggiolino, A.; Casalino, E.; Deflorio, M.; Centoducati, G. A Multi-Biomarker Approach in European Sea Bass Exposed to Dynamic Temperature Changes under Dietary Supplementation with *Origanum vulgare* Essential Oil. *Animals* **2021**, *11*, 982. https://doi.org/10.3390/ani11040982

Academic Editor: Woo Kyun Kim

Received: 8 March 2021
Accepted: 23 March 2021
Published: 1 April 2021

1. Introduction

Anthropogenic impacts, including the emission of carbon dioxide, the overuse of pesticides and the discharge of industrial and household chemical wastes have been considered the main causes of environmental temperature global change. Most aquatic communities are particularly vulnerable to the thermal fluctuations, which may affect ecological aspects and anatomical (e.g., body insulation and locomotor system) and physiological functions of aquatic organisms (e.g., blood–water countercurrent respiratory system, metabolic rate, and osmoregulation) [1–3]. Several studies have previously described the strong relationship between metabolic processes and temperature, whose variations may slow down or accelerate enzyme-catalyzed reactions [4,5]. Low and high temperature extremes decrease the secretion and activity of digestive enzymes, leading to minor feed intake and growth efficiency. Alternatively, temperatures close to the optimum level promote food digestion by increasing the metabolic rate or reducing the intestinal transit time [6]. Continuous exposure to thermic variations may counter sex ratios and reproduction status, making fish sterile or sexually incompetent [7]. Temperature is the major driver of embryogenesis and gonadal differentiation processes [8,9]. For these reasons, the environmental temperature has been defined as the main abiotic regulatory factor of basic physiological processes involved in animal growth, reproduction, and welfare [10]. In order to adapt its metabolic rate to thermal variations, aquatic fauna has developed a temperature range with maximum and minimum tolerance limits. Physiological alterations impair the cellular redox balance with the consequent formation of reactive oxygen species (ROS), which may damage vital biomolecules such as DNA, protein, and lipid [11]. For these reasons, antioxidant enzymes (e.g., superoxide dismutase and catalase) are used as biomarkers of fish health status due to their pivotal role in cellular defense systems against temperature-induced oxidative stress [12,13].

Plant-enriched diets may represent an effective strategy to increase fish immunity and disease resistance in aquaculture. Several studies have revealed the immunostimulant, antioxidant, and antimicrobial potential of plant-based supplementation in fish diet [14–16]. The use of plant supplements can also reduce the mortality rates and improve growth and feed assimilation, contributing to a better optimization of aquaculture resources. Whether medicinal plants can be administered to fish in several ways (e.g., intramuscular and intraperitoneal injection, immersion, or baths), oral administration seems to be the most suitable for aquaculture. Plants can be administered as a whole plant or parts (leaf, root, seed, fruit) and can either be used fresh or as powder, plant-extract, or essential oils (EO) [17,18]. Recently, much attention has been given to EO-enriched diets in order to preserve the fish healthiness and to minimize the farming cost, as well as to improve the fillet nutritional quality [16]. Blood parameters act as insightful markers of physiological perturbations depending on extrinsic (e.g., temperature, season, dissolved oxygen, water quality, stocking density, photoperiod) and intrinsic factors (e.g., age, nutritional state, species) [19]. Thus, previous studies investigated the levels of several blood parameters after EO administration, showing an enhanced health status of treated fish [20–22]. Carvacrol is the main component of *Origanum vulgare* (L.) EO and has a broad spectrum of biological activity [23]. Consequently, carvacrol-rich oregano oils have been applied in farmed fish diet as growth- and health-promoter. Anyway, the specific mechanisms behind the observed physiological effects of *Origanum vulgare* (L.) EO or carvacrol are still poorly described, as well as more research is needed to establish the most suitable preparations and the most appropriate treatment strategies. Adequate dosing is crucial to obtain the desired effects, since most of the plant bioactive molecules may result toxic or antinutritional at high levels. Treatment duration is another important parameter, since it directly affects the treatment effectiveness [17,18].

Due to its wide thermal tolerance range, large acclimation capacity, and low generation times, sea bass (*Dicentrarchus labrax* L.) has acquired a considerable commercial importance [5]. Anyway, the climate change effects are predicted to represent great ecological challenges for this species [24,25]. Sea bass grows best at temperature ranging from 22 to 24 °C, depending on the geographic region. In the wild, sea bass moves between deeper and shallow water, and seasonally between the open sea and lagoons/estuaries. Farmed sea bass is cultured in cage systems along the coast; therefore, it cannot alter their position to cope with daily and/or seasonally temperature changes and it has to face with continuous metabolic adjustments [26]. European sea bass may represent a good biological model for evaluating the impacts of temperature fluctuations in aquatic ecosystem [27]. In our study, we tested the effects of oregano EO-enriched diets on the physiological and oxidative state of sea bass exposed to temperature shifts. Aiming to provide a scientific base for effective aquaculture practices, we collected a broad range of fish responses under different doses of EO and during a long treatment period.

2. Materials and Methods

2.1. Animals and Experimental Design

European sea bass (*Dicentrarchus labrax* (Linnaeus, 1758)) were obtained from the commercial fish farm "Ittica Caldoli" (San Nazario, Lesina, FG, Italy). All fish were visually healthy and showed no clinical signs of abnormalities or infestation. Prior to the trial, 420 fish (12.48 ± 0.7 g body weight and 15.0 ± 0.90 cm total body length) were acclimated to 13 °C for 15 days. During this period, they were fed twice a day until apparent satiation with a commercial food pellet provided by Veronesi Mangimi A.I.A. S.p.A (Verona, Italy). The nutritional composition of commercial feed is reported in Table 1.

Table 1. Nutritional composition of commercial feed [1].

Ingredients	Composition (%)
Crude protein	42.00
Crude fat	18.00
Crude fiber	3.20
Ash	9.00
Total carbohydrates	18.80
Mineral supplement	1.40
Vitamin C (mg/Kg)	160.00
Vitamin E (mg/Kg)	160.00
Gross energy (MJ/Kg)	18.44

[1] The feed was supplied by Veronesi Mangimi A.I.A. S.p.A-Italy.

Then, specimens were randomly distributed in 12 cylindrical fiberglass 2000-L tanks (4 treatments × 3 replications; n = 35 individuals per tank). Nine tanks were exposed to same temperature conditions because they were inter-connected and linked to a re-circulating system, with a water flow of 7200 L/h (about eight total volume renewal per day) and equipped with mechanical and UV filters, a skimmer, a 3000 L biological filter and a 3000 L/h heat/cool pump. Ten percent of the water volume was renewed with reconstituted water every week. These tanks were placed in an air-conditioned room to support the increase in water temperature and to avoid heat loss during the experimental trial. There was constant aeration of the water with supplemental oxygen to keep dissolved oxygen values within the optimal range. Fish were progressively exposed to five experimental temperatures: 15, 18, 21, 23, and 25 °C, which reflected the natural water temperature range occurring in the southern Mediterranean region from mid-winter to mid-summer (https://www.seatemperature.org, accessed on 4 November 2019) (Figure 1). The water temperature was monthly increased regularly (2–3 °C month^{-1}) according to the seasonal trend. Four days before starting each exposure phase, water temperatures were progressively increased by 0.50–0.75 °C. After this period, the required experimental temperatures were kept constant for the remaining 26 days. The other three tanks were used as control at constant temperature (data not shown). These ones were placed in a different air-conditioned room and water temperature was constantly kept through a heat/cool pump at 13 °C during all the experimental trial (Figure 1). The tanks were linked to re-circulating system with a water flow of 2400 L/h (about eight total volume renewals per day) and equipped with mechanical and UV filters, a skimmer, a 1000 L biological filter and a 1000 L/h heat/cool pump. In control tanks, constant aeration of the water was sufficient to maintain optimal dissolved oxygen values. All tanks were maintained at a 14:10 L:D light–dark regime. During the experimental trial, the water quality parameters (e.g., temperature, dissolved oxygen, pH, total ammonia, nitrite, and nitrate) were monitored daily. Data about water temperature, dissolved oxygen, pH and salinity were collected by means of a tester HI-9829 (Hanna Instruments, Padova, Italy) whereas total ammonia, nitrite, and nitrate were measured with colorimetric kit (Testlab Marin, JBL). The mean temperature values during each exposure phase were 13.05 ± 0.03 °C; 15.35 ± 0.01 °C; 17.90 ± 0.08 °C; 21.20 ± 0.01 °C; 23.15 ± 0.03; and 25.07 ± 0.21 °C. These range values were marked as groups 13, 15, 18, 21, 23, and 25 °C. During the 150-days experimental period, specimens were maintained under the following conditions: 7.4 ± 0.5 mg/L of dissolved oxygen, 7.5 ± 0.1 of pH and 30 ‰ ± 2 of salinity. Ammonia (NH_4^+), nitrite (NO_2^-), and nitrate (NO_3^-) concentrations were kept below 0.05 mg/L, 0.20 mg/L, and 2.0 mg/L, respectively.

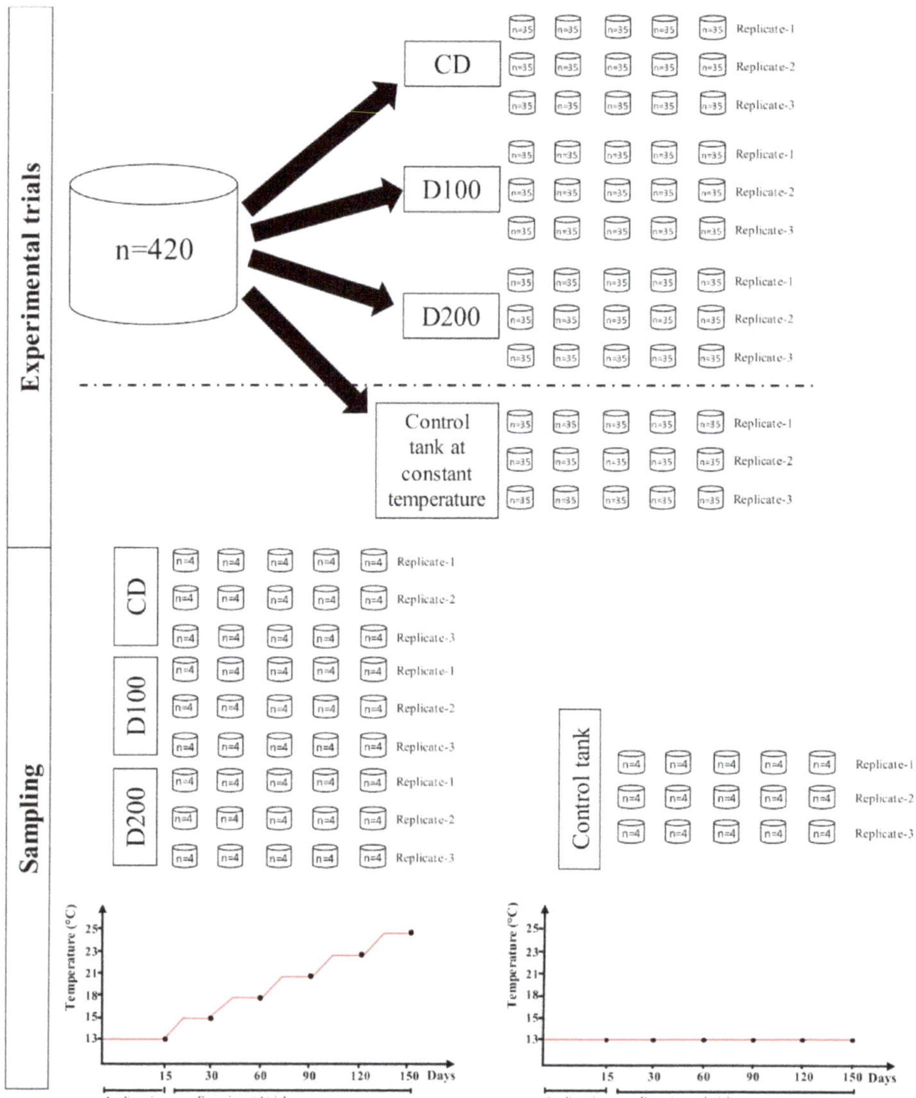

Figure 1. Schematic design of experimental trials and sampling protocol.

2.2. Experimental Diets

Fish were fed on three experimental diets: a control diet (Basic 3 commercial food pellet) and two diets which were supplemented with different concentrations of oregano (*Origanum vulgare* L., 1753) essential oil (EO). The EO used in this study was obtained from Farmalabor S.r.l. (Canosa di Puglia, Italy). The chemical composition provided by the manufacturer is reported in Table 2. Supplemented diets were prepared according to the protocol described by Dairiki et al. [28] and Dinardo et al. [29]. Briefly, oregano EO was dissolved in grain alcohol to prepare EO suspensions at different concentrations. Commercial pellets were ground and the resultant powder was mixed with the EO suspensions to reach the final concentrations of 100 (D100) and 200 ppm (D200) of EO [30]. In the control diet (CD), the same amount of pure grain alcohol was added to the feed, without EO supplementation. The mixtures were homogenized, pelleted, left to dry for 24 h at 25 °C, and stored at −18 °C until feeding. Each diet was tested in triplicate (three tanks per treatment). The fish were fed twice a day for 150 days until apparent satiety.

Table 2. Chemical composition of *Origanum vulgare* L. essential oil [1].

Compounds	Concentrations
Carvacrol	60–80%
Hydrocarbons	15%
Citral	2.5–8.0%
Beta-caryophyllene	0.5%
Geraniol	0.2%
Limonene	0.3%
Linalool	0.3%
Eugenol	0.1%
Arsenic	<1 mg/kg
Lead	<1 mg/kg
Mercury	<1 mg/kg
Cadmium	<1 mg/kg
Total heavy metals	<10 mg/kg

[1] The essential oil of *Origanum vulgare* L. was obtained by Farmalabor S.R.L.-Italy.

Animal management and sampling was carried out aiming at minimizing stress and health risks. The experiments were performed in accordance with the Italian guidelines for animal care (DL 26/14) and the European Communities Council Directive (2010/63/UE), and approved by the General Directorate of Animal Health and Veterinary Drugs of Ministry of Health, with authorization no. 444/2019-PR on 12 June 2019.

2.3. Blood Sampling

Every 30 days, twelve fish from each treatment (4 fish per tank) were randomly sampled, anesthetized with fish clove oil at a dose of 30 mg/L and soaked in ice-slurry to achieve death by hypothermia [31,32]. Blood samples (ca. 1 mL) were drawn from the caudal vein, using a 1-mL syringe, collected in plastic tubes and allowed to clot at room temperature. Subsequently, serum was separated by centrifugation at 3000 rpm for 5 min, stored at −80 °C and analyzed one week later.

2.4. Fish Performance

During and at the end of the feeding trial, the fish were weighed (g/fish) and the specific growth rate (SGR) was calculated as follow: SGR = 100 * [Ln (final body weight)—Ln (initial body weight)]/days of feeding trial.

2.5. Oxidative Stress Parameters

The thiobarbituric acid-reactive substances (TBARS) assay was performed in serum to quantify the peroxidative damage to lipids that occurs with free radical generation [33]. Free radical damage to lipids result in the production of malonaldehyde (MDA), which reacts with thiobarbituric acid (TBA) under conditions of high temperature and acidity generating a chromogen that can be measured spectrophotometrically at 535 nm. TBARS levels were reported as nmol MDA/mL.

Serum superoxide dismutase (SOD, EC1.15.1.1) activity was carried out as described by Misra [34]. The enzymatic activity was based on the 50% inhibition rate of epinephrine auto-oxidation at 480 nm. SOD activity was expressed as U/mL. Serum catalase (CAT, EC 1.11.1.6) activity was evaluated by following the decrease in absorbance of H_2O_2 at 240 nm [35]. One unit of enzyme activity was defined as the amount of enzyme required to degrade 1 μmol of H_2O_2 in 60 s. CAT activity was expressed as U/mL. Each sample analysis was performed in triplicate.

2.6. Serum Biochemical Analysis

Bradford assay [36] was carried out to quantify the total protein levels in each sample, using bovine serum albumin as standard. The protein concentrations were expressed as g/dL. Cholesterol, triglycerides, and glucose were measured using commercial colorimetric kits following manufacturer instructions (FAR S.r.l., Pescantina, VR, Italy), and concentrations were reported as mg/dL. Each sample analysis was performed in triplicate.

2.7. Statistical Analysis

Treatments were performed in triplicate. Results were reported as means ± standard deviations. Fish were used as statistical units (n = 12) after verifying the absence of a tank effect through a three-way nested analysis of variance (ANOVA), with temperature and feeding treatment as fixed factors and the tank as aleatory factor. Growth performances data were submitted to one-way ANOVA. A two-way ANOVA was used to analyze the effect of temperature and feeding treatments on oxidative stress biomarkers and biochemical parameters. ANOVA analyses were followed by the Tukey post hoc tests with significance level of 5%. In addition, oxidative stress biomarkers and biochemical parameters were subjected to principal components analysis (PCA) and statistical differences were evaluated using two-way PERMANOVA analyses. PERMANOVA test was performed with 999 permutations, with Euclidean distances as the distance measure and obtaining p-values from permutations. Data were analyzed using Statistica 13.0 (Statsoft Inc., Tulsa, OK, USA) and PAST 4.05 (University of Oslo, Oslo, Norway).

3. Results

3.1. Growth Parameters

After 150 days, fish fed on diet containing 100 ppm (D100) oregano EO showed a significantly ($p < 0.05$) higher final body weight compared both to control (CD) and 200 ppm EO diet (D200) (Table 3 and Supplementary Table S1). Similarly, the highest ($p < 0.05$) specific growth rate (SGR) value was found in D100. Weight values recorded at each sampling point were shown in Table S2 of the Supplementary Files.

Table 3. Growth performances of sea bass fed on control diet (CD) and experimental diets with different concentration of oregano essential oil: 100 ppm (D100) and 200 ppm (D200)

Growth Parameters	CD	D100	D200	p-Value
Initial body weight (g)	12.48 ± 0.70	12.48 ± 0.70	12.48 ± 0.70	n.s.
Final body weight (g)	125.75 ± 2.91 [b]	142.52 ± 2.11 [a]	108.82 ± 2.16 [c]	<0.001
Specific growth rate (%/d)	1.51 ± 0.03 [b]	1.60 ± 0.04 [a]	1.41 ± 0.04 [c]	<0.001

Values are reported as means ± standard deviations. Values followed by different letters ([a–c]) in the same row are significantly different by Tukey post hoc tests ($p < 0.05$). p-values from one-way analysis are also provided. n.s. = not significant.

3.2. Oxidative Stress Biomarkers

According to the temperature rise from 13 to 25 °C, TBARS concentrations increased significantly in CD and D200 groups (plus 60 and 74%, respectively) (Figure 2 and

Supplementary Table S1). Excepted at 25 °C, no significant differences ($p > 0.05$) were observed between CD and D200 during the experimental trials. Conversely, in fish fed on D100, TBARS levels ($p < 0.05$) were stable between 13 and 15 °C, then increased rising the temperature from 15 to 21 °C (plus 10%). At 23 and 25 °C TBARS levels in D100 dropped again to the values found at 13 and 15 °C.

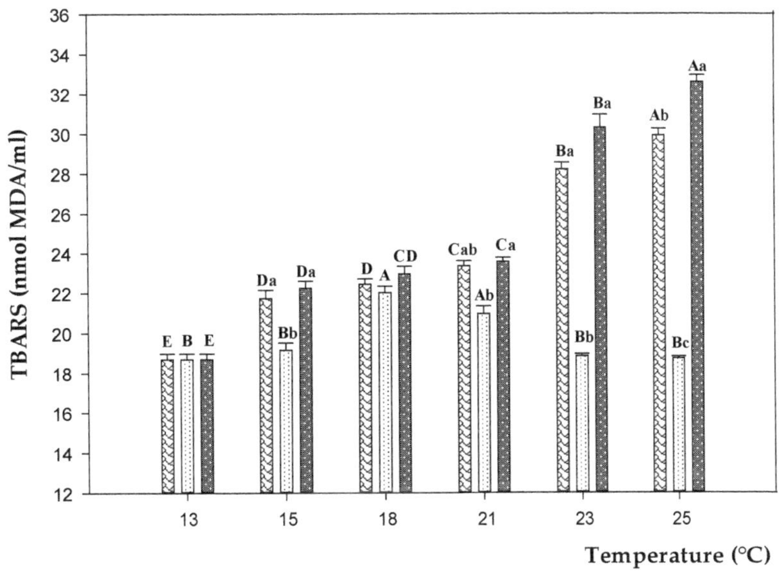

Figure 2. TBARS levels in sea bass fed on control diet (CD, ▨) and experimental diets with different concentration of oregano essential oil: 100 ppm (D100, ▭) and 200 ppm (D200, ▩) and, exposed to different water temperature (13, 15, 18, 21, 23, and 25 °C). Data are reported as means ± standard deviations (at each temperature value, $n = 12$ per treatment). Different capital letters (A–E) indicate significant differences ($p < 0.05$) among water temperature within the same diet. Different lowercase letters (a–c) indicate significant differences ($p < 0.05$) among diets within the same water temperature.

Superoxide dismutase (SOD) and catalase (CAT) activities were significantly ($p < 0.05$) affected by temperature and feeding treatments (Figure 3 and Supplementary Table S1). SOD activity increased significantly ($p < 0.05$) in both control and experimental groups rising the temperature from 13 to 18 °C (plus 23–35%), with the highest ($p < 0.05$) levels observed in fish fed on D200 diet (Figure 3a). Switching from 18 to 25 °C, enzyme levels showed different trends according to diet. SOD activity significantly ($p < 0.05$) dropped in fish fed on CD (minus 34%), whereas it remained almost stable in D100 groups. A slight but significant ($p < 0.05$) decrease of SOD activity was observed in D200 groups switching from 18 to 25 °C (minus 10%). The values of CAT activities were consistent with SOD activities trend (Figure 3b). Switching from 13 to 18 °C, the activity levels increased significantly ($p < 0.05$) in fish fed on CD and D200 diets (plus 116% and 151%, respectively), and then continually decreased (minus 43% and 30%, respectively). Overall, D200 groups showed higher ($p < 0.05$) activity than CD. In fish fed on D100, CAT activity increased ($p < 0.05$) at 15 °C and 18 °C, and then remained almost stable during the experimental trials.

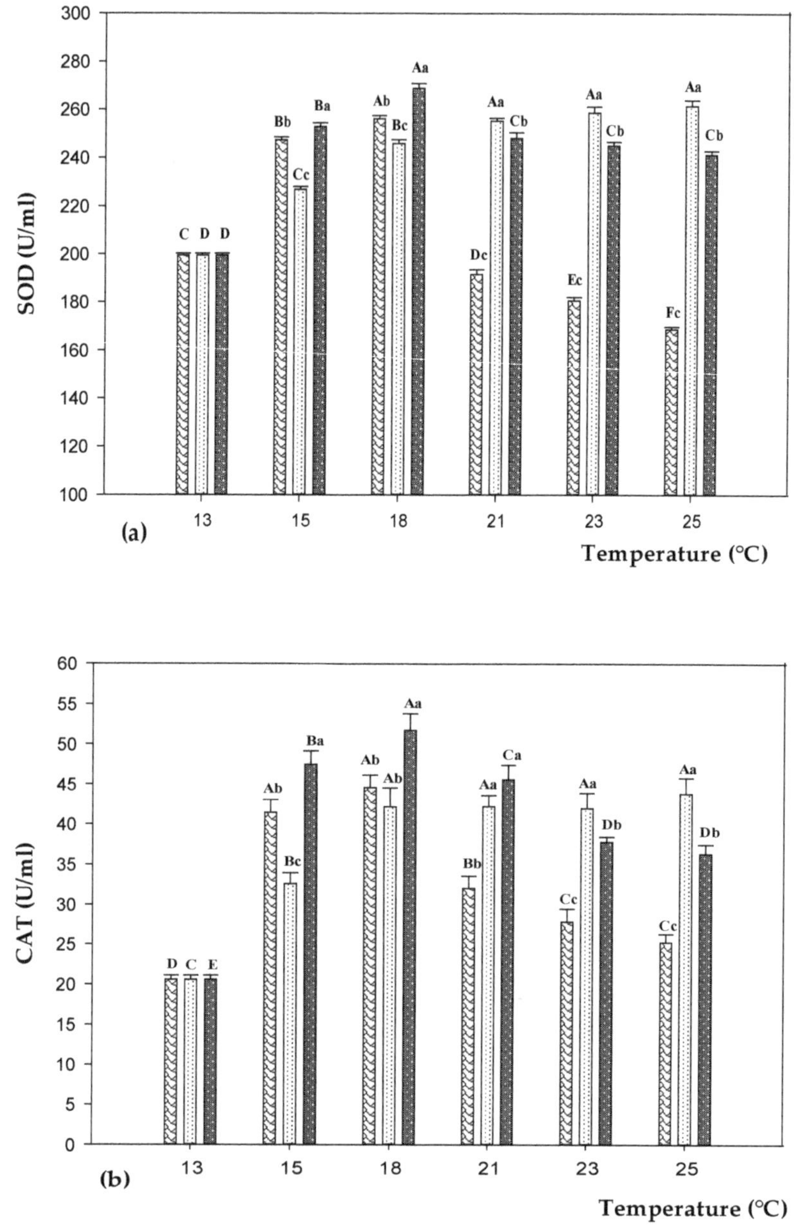

Figure 3. SOD (**a**) and CAT (**b**) levels in sea bass fed on control diet (CD, ▨) and experimental diets with different concentration of oregano essential oil: 100 ppm (D100, ⬚) and 200 ppm (D200, ■) and, exposed to different water temperature (13, 15, 18, 21, 23, and 25 °C). Data are reported as means ± standard deviations (at each temperature value, n = 12 per treatment). Different capital letters (A–F) indicate significant differences ($p < 0.05$) among water temperature within the same diet. Different lowercase letters (a–c) indicate significant differences ($p < 0.05$) among diets within the same water temperature.

3.3. Serum Biochemical Parameters

The total protein concentration fluctuated during the experimental trials (Figure 4 and Supplementary Table S1). According to the temperature rise, the protein levels followed an increasing ($p < 0.05$) trend until 21 °C (plus 45–60%) under all experimental conditions. Compared to CD, higher ($p < 0.05$) values were found at 21 °C in fish fed on D200 diet and, especially, D100. Temperature shifts from 21 to 25 °C caused a significant decrease ($p < 0.05$) of total protein in fish fed on CD (minus 20%), and to a lesser extent D100 and D200 groups (minus 8 and 4%, respectively). No significant differences ($p > 0.05$) were observed for the protein levels between fish fed on 100 and 200 ppm diets at 25 °C.

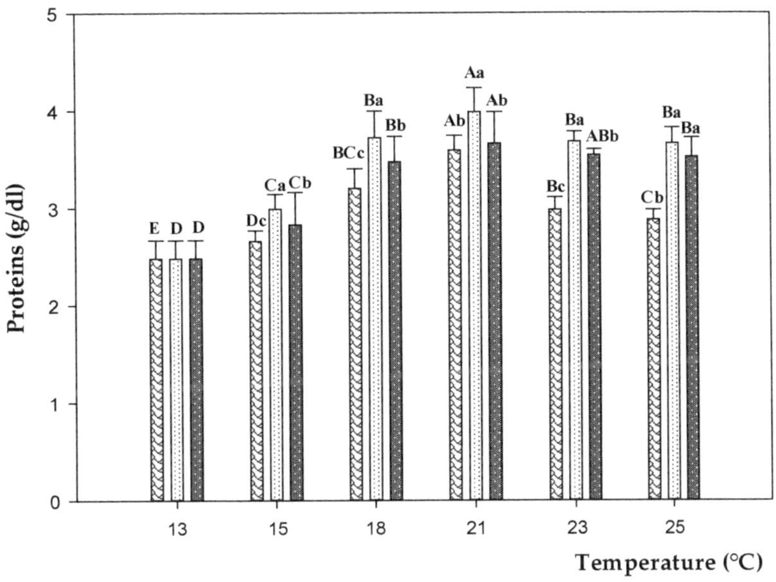

Figure 4. Protein levels in sea bass fed on control diet (CD, ▨) and experimental diets with different concentration of oregano essential oil: 100 ppm (D100, ▭) and 200 ppm (D200, ▩) and, exposed to different water temperature (13, 15, 18, 21, 23, and 25 °C). Data are reported as means ± standard deviations (at each temperature value, $n = 12$ per treatment). Different capital letters (A–E) indicate significant differences ($p < 0.05$) among water temperature within the same diet. Different lowercase letters (a–c) indicate significant differences ($p < 0.05$) among diets within the same water temperature.

Serum triglycerides levels also showed a temperature-related trend (Figure 5a and Supplementary Table S1). The triglycerides content decreased switching from 13 to 21 °C (minus 53–59%) under all experimental conditions. In particular, a dramatic drop was observed between 15 and 18 °C in D100 and D200 groups (minus 32 and 29%, respectively). Switching from 21 to 23 °C, the trend was inverted (plus 55–69%). No changes ($p > 0.05$) were observed between 23 and 25 °C. Overall, the lowest ($p < 0.05$) triglycerides levels were always found in D100 and D200 groups. A similar pattern was observed for cholesterol levels (Figure 5b and Supplementary Table S1). Switching from 13 to 21 °C, the cholesterol content constantly went down (minus 25–50%) ($p < 0.05$) under all experimental conditions. Within the range 15–21 °C, the lowest ($p < 0.05$) values were recorded in fish fed on both D100 and D200. At 23 °C the cholesterol levels significantly ($p < 0.05$) increased in all groups (plus 13–32%) with respect to 21 °C. Switching from 23 to 25 °C a slight increase ($p < 0.05$) was found (plus 7%) only in fish fed on D200, whereas no changes ($p > 0.05$) were observed in CD and D100. Within the range 23–25 °C, the lowest ($p < 0.05$) values were recorded in fish fed on D100.

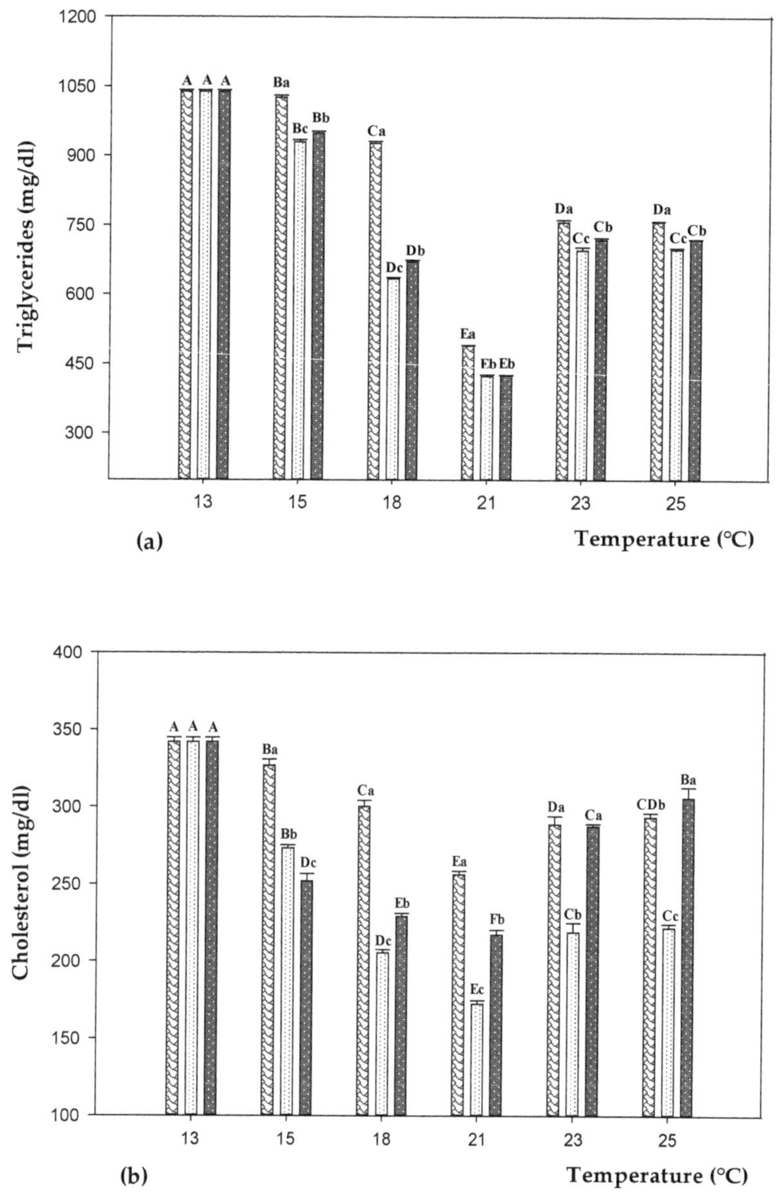

Figure 5. Triglycerides (**a**) and cholesterol (**b**) levels in sea bass fed on control diet (CD, ▨) and experimental diets with different concentration of oregano essential oil: 100 ppm (D100, ▭) and 200 ppm (D200, ■) and, exposed to different water temperature (13, 15, 18, 21, 23, and 25 °C). Data are reported as means ± standard deviations (at each temperature value, n = 12 per treatment). Different capital letters (A–F) indicate significant differences ($p < 0.05$) among water temperature within the same diet. Different lowercase letters (a–c) indicate significant differences ($p < 0.05$) among diets within the same water temperature.

During the experimental trials, the glucose levels were affected both by temperature and feeding treatments. The glucose content in fish fed on CD increased switching from 13 to 18 °C (plus 9%), significantly decreased at 21 °C (minus 34%), and increased ($p < 0.05$) again with the temperature rise to 25 °C (plus 50%) (Figure 6 and Supplementary Table S1). Conversely, in D100 and D200 the glucose concentration significantly went down ($p < 0.05$) switching from 13 to 18 °C (minus 24 and 12%, respectively), and increased ($p < 0.05$) again rising the temperature to 25 °C (plus 34 and 20%, respectively). The lowest glucose levels were observed in D100 groups, excepted at 21 °C, when the lowest level was in fish fed on CD.

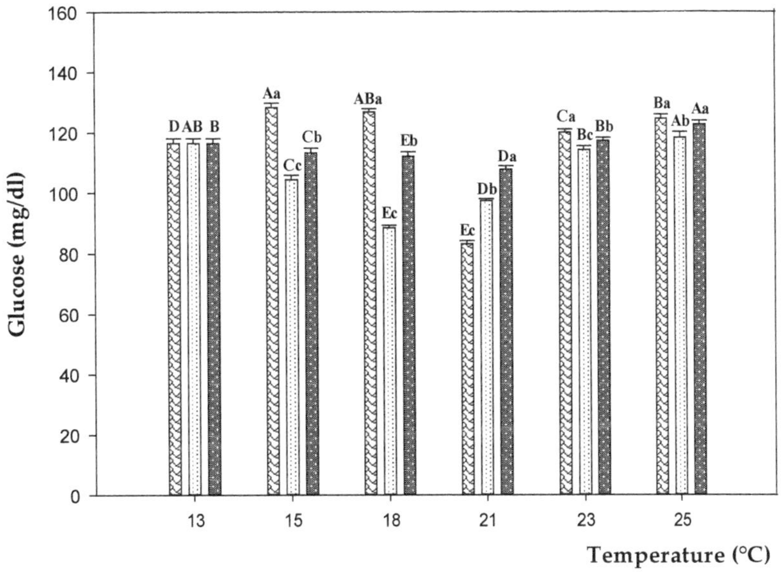

Figure 6. Glucose levels in sea bass fed on control diet (CD, ▨) and experimental diets with different concentration of oregano essential oil: 100 ppm (D100, ▭) and 200 ppm (D200, ▩) and, exposed to different water temperature (13, 15, 18, 21, 23, and 25 °C). Data are reported as means ± standard deviations (at each temperature value, $n = 12$ per treatment). Different capital letters (A–E) indicate significant differences ($p < 0.05$) among water temperature within the same diet. Different lowercase letters (a–c) indicate significant differences ($p < 0.05$) among diets within the same water temperature.

3.4. Multivariate Analysis

The principal component analysis (PCA) biplot was applied to oxidative stress biomarkers and biochemical parameters (Figure 7). Permanova analyses indicated the significance of temperature (p = 0. 001), diet (p = 0. 001) and their interaction ($p = 0.001$), giving importance to PCA analysis. The two components (factor 1 and factor 2) explained ca. 76% of total variance. PCA analysis showed that control and experimental groups exposed at low temperatures (13, 15, and 18 °C) were well separated on the plane from the same groups exposed to high temperatures (21, 23, and 25 °C), with some exceptions. Fish fed on CD and D200 diets and exposed to 21–25 °C were scattered on the upper part of the plane and shared the highest TBARS levels. On the contrary, D100 groups exposed to the same temperature were scattered on the right zone of the plane, and were mainly distinguished by the improved antioxidant enzyme activity and high serum protein levels.

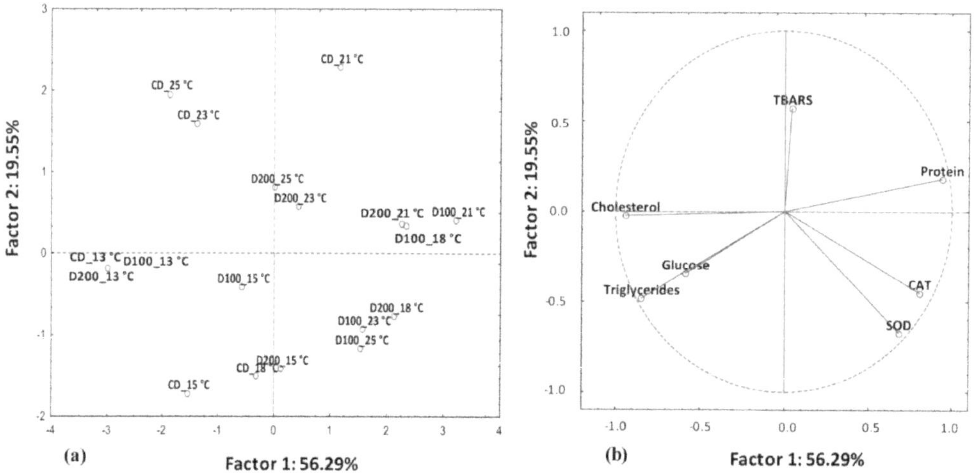

Figure 7. Score (**a**) and loading (**b**) plots of first and second principal components after principal component analysis performed on oxidative stress biomarkers and serum biochemical parameters in sea bass fed on control diet (CD) and experimental diets with different concentration of oregano essential oil (100 ppm (D100) and 200 ppm (D200)) and, exposed to different water temperatures (13, 15, 18, 21, 23, and 25 °C).

4. Discussion

Temperature fluctuations may induce metabolic and physiological imbalances over marine organisms. Aiming to find nutritional interventions on farms, able to reduce such imbalances, we investigated the correlations between temperature and metabolic performances in sea bass fed on oregano EO. In the present study, apart from the water temperature shifts, the other environmental conditions were held constant, including photoperiod and water quality [9].

It is known that one of the first signs of stress in fish undergoing temperature changes is the alteration of the redox state [37]. TBARS are good indicators of induced oxidative damage in cells. By raising temperature, our results showed an increased serum level of TBARS in fish fed on CD, especially switching from 21 to 25 °C when the activity of antioxidant enzymes (SOD and CAT) dropped. Results are in line with previous studies describing the oxidative stress in sea bass due to stressful temperatures [38]. This was likely due to the inability of the antioxidant enzyme machinery to compensate for ROS-generating stressful conditions [39,40]. Feed inclusion of EO reduced TBARS levels and promoted the activity of antioxidant enzymes. In particular, the protective effect against oxidative damages occurred with supplementation of 100 ppm diet, whereas it was negligible with 200 ppm. The interactive effect of EO and stressful temperatures has not been studied before and makes it difficult to compare our findings with others. High doses of EO were previously reported to be inefficacy or deleterious in sea bass dietary supplementation by Dinardo et al. [29]. On the other hand, high temperatures may also result in higher toxicity of chemicals, by affecting the uptake and detoxification mechanisms, the metabolic rates, and the enzymatic activities [24,41]. Dietary supplementation with 100 ppm EO boosted the antioxidative status of sea bass through a considerable elevation of serum SOD and CAT. The same effect was not observed with D200, especially under the warmer temperature, likely due to the inability of the antioxidant enzyme machinery to compensate for the presence of both stressors (high temperature and high EO dose exposure). This resulted in a diminished protective action against oxidative stress and, ultimately, an increased lipid peroxidation [39].

The antioxidant properties of EO have been widely established, and attributed to the presence of phenols such as carvacrol, having a hydroxyl group in the phenolic ring lending a radical scavenging or metal chelating activity [42,43]. Within this frame, we speculated that the antioxidant constituents of EO counteracted the oxidative stress induced by temperature increase. We also hypothesized a beneficial effect of EO going beyond the inherent antioxidant activity of carvacrol. As previously reported, the antioxidant enzyme machinery can be impaired when excessive oxidative damage occurs and substrate is accumulated (negative feedback) [39,40]. Supporting our thesis, carvacrol administration was previously shown to recover the activities of CAT and SOD and to mitigate the lipid peroxidation in mice [44].

Serum glucose level is another index of thermal stress, and supply of glucose in bloodstream allows to cope with high metabolic needs in stressed organisms [45,46]. In the present report, fish fed on CD showed an increase of serum glucose content upon exposure to 15 and 18 °C. As primary response to cold stress, the stimulation of glycogenolysis by catecholamines promotes the breakdown of hepatic glycogen and the release of glucose into the blood [47,48]. The same hyperglycemic responses were reported in many species, such as sea bream, Nile tilapia, silver catfish, milkfish, and grass carp [27,49–51]. Trend of blood glucose in control group at temperatures between 21 and 25 °C could be a consequence of depletion of hepatic glycogen stores and the subsequent activation of gluconeogenesis. The stimulation of glycogenolysis and gluconeogenesis in sea bass subjected to thermal stress has been previously reported in several studies [9,38,52]. Islam et al. [38] detected low amount of blood glucose in fish reared both at low and high temperature extremes (8 and 32 °C). Samaras et al. [9] found lower levels of circulating glucose in sea bass exposed to a temperature of 15 °C with respect to 25 °C. A direct comparison of results from different studies is not always possible due to differences in experimental design and techniques employed, anyway most of the authors agreed that thermal stress set higher energy demands, resulting in high glucose consumption rate and stimulation of glycogenolysis and gluconeogenesis [9,38]. A similar pattern was reported in fish undergoing starvation and crowding stress or captivity [52–54]. On the contrary, the intake of diets supplemented with oregano EO led to reduced glucose levels. Several authors associated the reduction in blood glucose to the hypoglycemic effects of carvacrol [55–58]. By improving insulin sensibility and promoting intracellular glucose uptake, carvacrol treatment may also prevent ROS production and oxidative damage [59]. To some extent, the energy stores depletion during stress exposure [26] could explain the poor growth rates observed with CD compared to D100. On the other hand, the same beneficial effect of EO on growth performances was not observed with D200, likely due to the high EO dose exposure, which appeared to have a deleterious impact [29].

A reduction of triglycerides in fish subjected to temperature shifts was previously reported by other authors [60]. The decrease observed in fish fed on CD was likely due to the depletion at liver level of triglycerides and cholesterol to support the energy cost of gluconeogenesis [53]. On the other hand, the sharp reduction of triglycerides in D100 and D200 was mainly attributable to the hypolipidemic effect of carvacrol rather than to the consumption of triglycerides [61]. The lower cholesterol content in fish fed on experimental diets may be also ascribed to the suppression of 3-hydroxy-3- methylglutaryl coenzyme A reductase (HMG-CoA), a key regulatory enzyme in cholesterol synthesis [62]. Indeed, Kim et al. [63] demonstrated that carvacrol lowers hepatic cholesterol through the downregulation of genes involved in lipogenesis. Our results are in line with the study of Hong et al. [64] which revealed a significant reduction of serum cholesterol when broiler chickens fed on essential oils containing carvacrol as major component.

Serum proteins are key indicator of vital functions, such as humoral defense, coagulation, metabolite transport, and homeostasis [65]. Their levels may be influenced by water quality and seasonal changes or by endogenous factors (e.g., hemodilution and reproductive cycle) [66,67]. In the present study, serum proteins increased according to the temperature rise, reaching the highest peak when exposed to 21 °C. Further switching the temperature up to 25 °C, serum protein decreased, in particular in fish fed on CD. Our results for serum protein showed similarity with the study of Islam et al. [26], which found a decreasing trend in sea bass exposed to thermal stress. In response to the environmental fluctuations, proteins are released into circulation where they are catabolized in order to produce ATP, to support gluconeogenesis and to maintain the physiological homeostasis [68]. The activation of protein catabolism exposes fish to immune dysfunction [69], skeletal muscle degradation [70], growth restriction, and makes fish susceptible to death [71]. On the other hand, the elevated protein levels observed in fish fed on experimental diets may be attributed to EO supplementation, likely due to activation of regulatory and metabolic pathways to protect proteins from degradation. Many studies reported a high protein level and the consequent immuno-stimulating effect in rainbow trout, carp, or catfish treated with carvacrol [72–75].

Principal component analysis (PCA) based on oxidative stress biomarkers and serum biochemical parameters showed a clear separation of control and experimental groups as a function of temperature changes. However, diet supplementation with 100 ppm of EO seemed to mitigate the effects of high temperatures.

5. Conclusions

Summing up, we investigated the physiological responses triggered by temperature changes in European sea bass, and the effectiveness of dietary EO supplementation in counteracting the thermal stress. Under exposure to thermal shift, EO affected fish growth and metabolic biomarkers in a dose-depending manner. The dosage of the EO is crucial to obtain the desired effects and thus deserves to be appropriately investigated. The addition of 100 ppm oregano EO improved growth performances, restored the antioxidant enzyme machinery (SOD and CAT), and activated the non-specific immune system by increasing the serum protein level. On the other hand, a higher dose (200 ppm) of EO was less effective in counteracting the thermal stress and was detrimental to the fish growth. Our findings are crucial for promoting the economic sustainability of dietary supplementation with EO, as low-dose treatments are cheaper than higher dosages. The nutritional strategy we proposed is easily transposable into the field and can benefit different aquaculture sectors, from small-scale fish farmers to intensive productions. The role of EO in regulating the antioxidant enzyme machinery and the non-specific immune system suggests a versatility of application. Likely, EO administration might potentially be effective as preventive treatment and as an alternative to antibiotics in several cultured species, and as sustainable treatment for diseases and stress management in farms of high-income countries. Further investigations should be taken on target tissues to understand biological mechanisms ameliorating fish conditions under temperature changes.

Supplementary Materials: The following are available online at https://www.mdpi.com/article/10.3390/ani11040982/s1, Table S1: Results of two-way ANOVA evaluating the effects of temperature and feeding treatments on weight values and serum SOD and CAT activities and TBARS, proteins, triglycerides, cholesterol and glucose levels; Table S2: Weight values of sea bass fed on control and experimental diets and exposed to different temperatures.

Author Contributions: Conceptualization, E.C. and G.C.; Methodology, E.C.; Validation, E.C. and F.R.D.; Formal analysis, E.C., F.R.D. and M.D.; Investigation, E.C., F.R.D. and M.D.; Resources, E.C. and G.C.; Data curation, E.C. and F.R.D.; Writing—original draft preparation, E.C.; Writing—review and editing, E.C. and G.C.; Supervision, E.C., A.M. and G.C.; Project administration, G.C.; Funding acquisition, G.C. All authors have read and agreed to the published version of the manuscript.

Funding: This research was supported by grants from Italian Ministry of Education, University and Research (project no. PON02 00186 2937475).

Institutional Review Board Statement: The study was conducted according to the guidelines of the Declaration of Helsinki, and approved by the General Directorate of Animal Health and Veterinary Drugs of Ministry of Health (authorization no. 444/2019-PR on 12 June 2019).

Data Availability Statement: Not applicable.

Acknowledgments: The authors are grateful to G. Calzaretti, S. Giacummo, and F. Giannico for their technical support in laboratory activities.

Conflicts of Interest: The authors declare no conflict of interest.

References

1. Makrinos, D.L.; Bowden, T.J. Natural environmental impacts on teleost immune function. *Fish Shellfish Immunol.* **2016**, *53*, 50–57. [CrossRef]
2. Boltana, S.; Sanhueza, N.; Aguilar, A.; Gallardo-Escarate, C.; Arriagada, G.; Valdes, J.A.; Soto, D.; Quinones, R.A. Influences of thermal environment on fish growth. *Ecol. Evol.* **2017**, *7*, 6814–6825. [CrossRef] [PubMed]
3. Navarro, J.M.; Paschke, K.; Ortiz, A.; Vargas-Chacoff, L.; Pardo, L.M.; Valdivia, N. The Antarctic fish *Harpagifer antarcticus* under current temperatures and salinities and future scenarios of climate change. *Prog. Oceanogr.* **2019**, *174*, 37–43. [CrossRef]
4. Mazumder, S.K.; De, M.; Mazlan, A.G.; Zaidi, C.C.; Rahim, S.M.; Simon, K.D. Impact of global climate change on fish growth, digestion and physiological status: Developing a hypothesis for cause and effect relationships. *J. Water Clim. Chang.* **2015**, *6*, 200–226. [CrossRef]
5. Almeida, J.R.; Gravato, C.; Guilhermino, L. Effects of temperature in juvenile seabass (*Dicentrarchus labrax* L.) biomarker responses and behaviour: Implications for environmental monitoring. *Estuaries Coasts* **2015**, *38*, 45–55. [CrossRef]
6. Mazumder, S.K.; Das, S.K.; Rahim, S.M.; Ghaffar, M.A. Temperature and diet effect on the pepsin enzyme activities, digestive somatic index and relative gut length of Malabar blood snapper (*Lutjanus malabaricus* Bloch & Schneider, 1801). *Aquac. Rep.* **2018**, *9*, 1–9. [CrossRef]
7. Majhi, S.K.; Das, S.K. Thermal tolerance, oxygen consumption and stress response in *Danio dangila* and *Brachydanio rerio* (Hamilton, 1822) acclimated to four temperatures. *Turk. J. Fish Aquat. Sci.* **2013**, *13*, 359–365. [CrossRef]
8. Arfuso, F.; Guerrera, M.C.; Fortino, G.; Fazio, F.; Santulli, A.; Piccione, G. Water temperature influences growth and gonad differentiation in European sea bass (*Dicentrarchus labrax* L. 1758). *Theriology* **2017**, *88*, 145–151. [CrossRef] [PubMed]
9. Samaras, A.; Papandroulakis, N.; Lika, K.; Pavlidis, M. Water temperature modifies the acute stress response of European sea bass, *Dicentrarchus labrax* L. (1758). *J. Ther. Biol.* **2018**, *78*, 84–91. [CrossRef] [PubMed]
10. Bowyer, J.N.; Booth, M.A.; Qin, J.G.; D'Antignana, T.; Thomson, M.J.S.; Stone, D.A.J. Temperature and dissolved oxygen influence growth and digestive enzyme activities of yellowtail kingfish *Seriola lalandi* (Valenciennes, 1833). *Aquac. Res.* **2014**, *45*, 2010–2020. [CrossRef]
11. Liu, M.P.; Li, J.; Gao, B.Q.; Chen, P. Physiological responses of swimming crab *Portunus trituberculatus* under cold acclimation: Antioxidant defense and heat shock proteins. *Aquaculture* **2014**, *434*, 11–17. [CrossRef]
12. Madeira, A.; Camps, M.; Zorzano, A.; Moura, T.F.; Soveral, G. Biophysical assessment of human aquaporin-7 as a water and glycerol channel in 3T3-L1 adipocytes. *PLoS ONE* **2013**, *8*, e83442. [CrossRef] [PubMed]
13. Fazio, F.; Piccione, G.; Saoca, C.; Caputo, A.R.; Cecchini, S. Assessment of oxidative stress in Flathead mullet (*Mugil cephalus*) and Gilthead sea bream (*Sparus aurata*). *Vet. Med.* **2015**, *60*, 12. [CrossRef]
14. El-Hack, M.E.A.; Alagawany, M.; Saeed, M.; Arif, M.; Arain, M.A.; Bhutto, Z.A.; Fazlani, S.A. Effect of gradual substitution of soyabean meal by *Nigella Sativa* meal on growth performance, carcass traits and blood lipid profile of growing Japanese quail. *J. Anim. Feed Sci.* **2016**, *25*, 244–249. [CrossRef]
15. Diler, O.; Gormez, O.; Diler, I.; Metin, S. Effect of oregano (*Origanum onites* L.) essential oil on growth, lysozyme and antioxidant activity and resistance against *Lactococcus garvieae* in rainbow trout, *Oncorhynchus mykiss* (*Walbaum*). *Aquac. Nutr.* **2017**, *23*, 844–851. [CrossRef]
16. Sutili, F.J.; Gatlin III, D.M.; Heinzmann, B.M.; Baldisserotto, B. Plant essential oils as fish diet additives: Benefits on fish health and stability in feed. *Rev. Aquac.* **2018**, *10*, 716–726. [CrossRef]
17. Reverter, M.; Tapissier-Bontemps, N.; Sasal, P.; Saulnier, D. Use of medicinal plants in aquaculture. In *Diagnosis and Control of Diseases of Fish and Shellfish*, 1st ed.; Austin, B., Newaj-Fyzul, A., Eds.; John Wiley & Sons Ltd.: Chichester, UK, 2017; Volume 1, pp. 223–261.
18. Reverter, M.; Tapissier-Bontemps, N.; Sarter, S.; Sasal, P.; Caruso, D. Moving towards more sustainable aquaculture practices: A meta-analysis on the potential of plant-enriched diets to improve fish growth, immunity and disease resistance. *Rev. Aquac.* **2021**, *13*, 537–555. [CrossRef]
19. Ahmed, I.; Reshi, Q.M.; Fazio, F. The influence of the endogenous and exogenous factors on hematological parameters in different fish species: A review. *Aquac. Int.* **2020**, *28*, 869–899. [CrossRef]
20. Zargar, A.; Rahimi-Afzal, Z.; Soltani, E.; Mirghaed, A.T.; Ebrahimzadeh-Mousavi, H.A.; Soltani, M.; Yuosefi, P. Growth performance, immune response and disease resistance of rainbow trout (*Oncorhynchus mykiss*) fed *Thymus vulgaris* essential oils. *Aquac. Res.* **2019**, *50*, 3097–3106. [CrossRef]
21. Parrino, V.; Kesbiç, O.S.; Acar, Ü.; Fazio, F. Hot pepper (*Capsicum* sp.) oil and its effects on growth performance and blood parameters in rainbow trout (*Oncorhynchus mykiss*). *Nat. Prod. Res.* **2019**, *34*, 3226–3230. [CrossRef]
22. Acar, Ü.; Parrino, V.; Kesbiç, O.S.; Lo Paro, G.; Saoca, C.; Abbate, F.; Yilmaz, S.; Fazio, F. Effects of different levels of pomegranate seed oil on some blood parameters and disease resistance against *Yersinia ruckeri* in rainbow trout. *Front. Physiol.* **2018**, *9*, 596. [CrossRef]

23. Dundar, E.; Olgun, E.G.; Isiksoy, S.; Kurkcuoglu, M.; Baser, C.K.H.; Bal, C. The Effects of intra-rectal and intra-peritoneal application of *Origanum onites* L. essential oil on 2, 4, 6-trinitro- benzenesulfonic acid-induced colitis in the rat. *Exp. Toxicol. Pathol.* **2008**, *59*, 399–408. [CrossRef] [PubMed]
24. Marques, A.; Nunes, M.L.; Moore, S.K.; Strom, M.S. Climate change and seafood safety: Human health implications. *Food Res. Int.* **2010**, *43*, 1766–1779. [CrossRef]
25. Rosa, R.; Marques, A.; Nunes, M.L. Impact of climate change in Mediterranean aquaculture. *Rev. Aquac.* **2012**, *4*, 163–177. [CrossRef]
26. Islam, M.D.J.; Slater, M.S.; Bogner, M.; Zeytin, S.; Kunzmanna, A. Extreme ambient temperature effects in European seabass, *Dicentrarchus labrax*: Growth performance and hemato-biochemical parameters. *Aquaculture* **2020**, *522*, 735093. [CrossRef]
27. Fazio, F.; Ferrantelli, V.; Piccione, G.; Saoca, C.; Levanti, M.; Mucciardi, M. Biochemical and hematological parameters in European sea bass (*Dicentrarchus labrax* Linnaeus, 1758) and Gilthead sea bream (*Sparus aurata* Linnaeus, 1758) in relation to temperature. *Vet. Arh.* **2018**, *88*, 397–411. [CrossRef]
28. Dairiki, J.K.; Majolo, C.; Chagas, E.C.; Chaves, F.C.M.; Oliveira, M.R.; Morais, I.S. Procedimento para inclusão de óleos essenciais em rações para peixes. *Circ. Tec. Embrapa Manaus* **2013**, *42*, 1–8.
29. Dinardo, F.R.; Deflorio, M.; Casalino, E.; Crescenzo, G.; Centoducati, G. Effect of feed supplementation with *Origanum vulgare* L. essential oil on sea bass (*Dicentrarchus labrax*): A preliminary framework on metabolic status and growth performances. *Aquac. Rep.* **2020**, *18*, 100511. [CrossRef]
30. Arciuli, M.; Fiocco, D.; Fontana, S.; Arena, M.P.; Frassanito, M.A.; Gallone, A. Administration of a polyphenol-enriched feed to farmed sea bass (*Dicentrarchus labrax* L.): Kidney melanomacrophages response. *Fish Shellfish Immunol.* **2017**, *68*, 404–410. [CrossRef]
31. Blessing, J.J.; Marshall, J.C.; Balcombe, S.R. Humane killing of fishes for scientific research: A comparison of two methods. *J. Fish Biol.* **2010**, *76*, 2571–2577. [CrossRef] [PubMed]
32. Mylonas, C.C.; Cardinaletti, G.; Sigelaki, I.; Polzonetti-Magni, A. Comparative efficacy of clove oil and 2-phenoxyethanol as anesthetics in the aquaculture of European sea bass (*Dicentrarchus labrax*) and gilthead sea bream (*Sparus aurata*) at different temperatures. *Aquaculture* **2005**, *246*, 467–481. [CrossRef]
33. Buege, J.A.; Aust, S.D. Microsomal lipid peroxidation. *Methods Enzymol.* **1978**, *52*, 302–310. [CrossRef]
34. Misra, H.P. Adrenochrome assay. In *Handbook of Methods for Oxygen Radical Research*, 1st ed.; Greenwald, R.A., Ed.; CRC Press: Boca Raton, FL, USA, 1985; pp. 237–241. ISBN 9781315893822.
35. Clairborne, A. Catalase activity. In *Handbook of Methods for Oxygen Radical Research*, 1st ed.; Greenwald, R.A., Ed.; CRC Press: Boca Raton, FL, USA, 1985; pp. 283–284. ISBN 9781315893822.
36. Bradford, M.M. A rapid and sensitive method for the quantification of microgram quantities of protein utilizing the principle of protein-dye binding. *Anal. Biochem.* **1976**, *72*, 248–254. [CrossRef]
37. Mauvault, A.L.; Barbosa, V.; Alves, R.; Custodio, A.; Anacleto, P.; Repolho, T.; Ferreira, P.P.; Rosa, R.; Marques, A.; Diniz, M. Ecophysiological responses of juvenile seabass (*Dicentrarchus labrax*) exposed to increased temperature and dietary methylmercury. *Sci. Total Environ.* **2017**, *586*, 551–558. [CrossRef] [PubMed]
38. Islam, M.J.; Kunzmann, A.; Bögner, M.; Meyer, A.; Thiele, R.; Slater, M.J. Metabolic and molecular stress responses of European seabass, *Dicentrarchus labrax* at low and high temperature extremes. *Ecol. Indic.* **2020**, *112*, 106118. [CrossRef]
39. Carvalho, C.S.; Bernusso, V.A.; Araújo, H.S.S.; Espíndola, E.L.G.; Fernandes, M.N. Biomarker responses as indication of contaminant effects in *Oreochromis niloticus*. *Chemosphere* **2012**, *89*, 60–69. [CrossRef] [PubMed]
40. Karadag, H.; Firat, O. Use of oxidative stress biomarkers in *Cyprinus carpio* L. for the evaluation of water pollution in Ataturk Dam Lake (Adiyaman, Turkey). *Bull. Environ. Contam. Toxicol.* **2014**, *92*, 289–293. [CrossRef]
41. Noyes, P.D.; McElwee, M.K.; Miller, H.D.; Clark, B.W.; Van Tiem, L.A.; Walcott, K.C.; Erwin, K.N.; Levin, E.D. The toxicology of climate change: Environmental contaminants in a warming world. *Environ. Int.* **2009**, *35*, 971–986. [CrossRef]
42. Jnaid, Y.; Yacoub, R.; Al-Biski, F. Antioxidant and antimicrobial activities of *Origanum vulgare* essential oil. *Int. Food Res. J.* **2016**, *23*, 1706–1710.
43. Maqsood, S.; Benjakul, S.; Shahidi, F. Emerging Role of Phenolic Compounds as Natural Food Additives in Fish and Fish Products. *Crit. Rev. Food Sci. Nutr.* **2013**, *53*, 162–179. [CrossRef]
44. El-Sayed, E.M.; Abd-Allah, A.R.; Mansour, A.M.; El-Arabey, A.A. Thymol and carvacrol prevent cisplatin-induced nephrotoxicity by abrogation of oxidative stress, inflammation, and apoptosis in rats. *J. Biochem. Mol. Toxicol.* **2015**, *29*, 165–172. [CrossRef] [PubMed]
45. Roychowdhury, P.; Aftabuddin, M.; Pati, M.K. Thermal stress altered growth performance and metabolism and induced anaemia and liver disorder in *Labeo rohita*. *Aquac. Res.* **2020**, *51*, 1406–1414. [CrossRef]
46. Dallarés, S.; Dourado, P.; Sanahuja, I.; Solovyev, M.; Gisbert, E.; Montemurro, N.; Torreblanca, A.; Blàzquez, M.; Solé, M. Multibiomarker approach to fipronil exposure in the fish *Dicentrarchus labrax* under two temperature regimes. *Aquat. Toxicol.* **2020**, *219*, 105378. [CrossRef]
47. Cho, H.C.; Kim, J.E.; Kim, H.B.; Baek, H.J. Effects of water temperature change on the hematological responses and plasma cortisol levels in growing of Red spotted grouper *Epinephelus akaara*. *Dev. Reprod.* **2015**, *19*, 19–24. [CrossRef] [PubMed]

48. Tort, L.; Rotllant, J.; Liarte, C.; Acerete, L.; Hernández, A.; Ceulemans, S.; Coutteau, P.; Padros, F. Effects of temperature decrease on feeding rates, immune indicators and histopathological changes of gilthead sea bream *Sparus aurata* fed with an experimental diet. *Aquaculture* **2004**, *229*, 55–65. [CrossRef]
49. Panase, P.; Saenphet, S.; Saenphet, K. Biochemical and physiological responses of Nile tilapia *Oreochromis niloticus* Lin subjected to cold shock of water temperature. *Aquac. Rep.* **2018**, *11*, 17–23. [CrossRef]
50. Lermen, C.L.; Lappe, R.; Crestani, M.; Vieira, V.P.; Gioda, C.R.; Schetinger, M.R.C.; Baldisserotto, B.; Moraes, G.; Morsch, V.M. Effect of different temperature regimes on metabolic and blood parameters of silver catfish *Rhamdia quelen*. *Aquaculture* **2004**, *239*, 497–507. [CrossRef]
51. Kuo, C.M.; Hsieh, S.L. Comparisons of physiological and biochemical responses between milkfish (*Chanos chanos*) and grass carp (*Ctenopharyngodon idella*) to cold shock. *Aquaculture* **2006**, *251*, 525–536. [CrossRef]
52. Kavadias, S.; Castritsi-Catharios, J.; Dessypris, A. Annual cycles of growth rate, feeding rate, food conversion, plasma glucose and plasma lipids in a population of European sea bass (*Dicentrarchus labrax* L.) farmed in floating marine cages. *J. Appl. Ichthyol.* **2003**, *19*, 29–34. [CrossRef]
53. Echevarría, G.; Martínez-Bebiá, M.; Zamora, S. Evolution of Biometric Indices and Plasma Metabolites During Prolonged Starvation in European Sea Bass (*Dicentrarchus labrax* L.). *Comp. Biochem. Physio.* **1997**, *118*, 111–123. [CrossRef]
54. Ortuno, J.O.; Esteban, M.A.; Meseguer, J. Effects of short-term crowding stress on the gilthead seabream (*Sparus aurata* L.) innate immune response. *Fish Shellfish Immunol.* **2001**, *11*, 187–197. [CrossRef]
55. Deng, J.; BI, B.; Kang, B.; Kong, L.; Wang, Q.; Zhang, X. Improving the growth performance and cholesterol metabolism of rainbow trout (*Oncorhynchus mykiss*) fed soyabean meal- based diets using dietary cholesterol supplementation. *Br. J. Nutr.* **2013**, *110*, 29–39. [CrossRef] [PubMed]
56. Alagawany, M.; Abd El-Hack, M.E.; Farag, M.R.; Shaheen, H.M.; Abdel-Latif, M.A.; Noreldin, A.E.; Khafaga, A.F. The applications of *Origanum vulgare* and its derivatives in human, ruminant and fish nutrition—A review. *Ann. Anim. Sci.* **2020**, *20*, 389–407. [CrossRef]
57. Saravanan, S.; Pari, L. Role of thymol on hyperglycaemia and hyperlipidemia in High fat diet-induced type 2 diabetic C57BL/6J mice. *Eur. J. Pharm.* **2015**, *761*, 279–287. [CrossRef]
58. Jing, L.; Zhang, Y.; Shengjie, F.; Gu, M.; Guan, Y.; Lu, X.; Huang, C.; Zhou, Z. Preventive and ameliorating effects of citrus D-limonene on dyslipidemia and hyperglycemia in mice with high-fat diet-induced obesity. *Eur. J. Pharm.* **2013**, *715*, 46–55. [CrossRef]
59. Li, Y.; Mai, Y.; Qiu, X.; Xiaoqing Chen, X.; Li, C.; Wenchang Yuan, W.; Hou, N. Effect of long-term treatment of Carvacrol on glucose metabolism in Streptozotocin- induced diabetic mice. *BMC Complement. Med. Ther.* **2020**, *20*, 142. [CrossRef]
60. Shim, K.S.; Hwang, K.T.; Sonand, M.W.; Park, G.H. Lipid Metabolism and peroxidation in broiler chicks under chronic heat stress. *Asian Australas. J. Ani. Sci.* **2006**, *19*, 1206–1211. [CrossRef]
61. Ahmed, L.A.; Hassan, W.A.; Fawzy, H.M.; Moawad, H. Combination of carvacrol with simvastatin improves the lipid-lowering efficacy and alleviates simvastatin side effects. *J. Biochem. Mol. Toxicol.* **2017**, *31*, e21981. [CrossRef]
62. Lee, K.W.; Everts, H.; Beynen, A.C. Essential oils in broiler nutrition. *Int. J. Poult. Sci.* **2004**, *3*, 738–752.
63. Kim, J.H.; Park, H.J.; Kim, K.W.; Hwang, I.K.; Kim, D.H.; Oh, C.W.; Lee, J.S.; Kang, J.C. Growth performance, oxidative stress, and non-specific immune responses in juvenile sablefish, *Anoplopoma fimbria*, by changes of water temperature and salinity. *Fish Physiol. Biochem.* **2017**, *43*, 1421–1431. [CrossRef]
64. Hong, J.C.; Steiner, T.; Aufy, A.; Lien, T.-F. Effects of supplemental essential oil on growth performance, lipid metabolites and immunity, intestinal characteristics, microbiota and carcass traits in broilers. *Livest. Sci.* **2012**, *144*, 253–262. [CrossRef]
65. Jenkins, F.; Smith, J.; Bajanna, B.; Shameem, U.; Umadevi, K.; Sandhya, V.; Madhavi, R. Effect of sub-lethal concentrations of endosulfan on hematological and serum biochemical parameters in the carp *Cyprinus carpio*. *Bull. Environ. Contam. Toxicol.* **2003**, *70*, 993–997. [CrossRef]
66. Sala-Rabanal, M.; Sánchez, J.; Ibarz, A.; Fernández-Borràs, J.; Blasco, J.; Gallardo, M.A. Effects of low temperatures and fasting on hematology and plasma composition of gilthead sea bream (*Sparus aurata*). *Fish Physiol. Biochem.* **2003**, *29*, 105–115. [CrossRef]
67. Fazio, F.; Marafioti, S.; Torre, A.; Sanfilippo, M.; Panzera, M.; Faggio, C. Haematological and serum protein profiles of *Mugil cephalus*: Effect of two different habitat. *Ichthyol. Res.* **2013**, *60*, 36–42. [CrossRef]
68. Das, P.C.; Ayyappan, S.; Jena, J.K. Haematological changes in the three Indian major carps, *Catla catla* (Hamilton), *Labeo rohita* (Hamilton) and *Cirrhinus mrigala* (Hamilton) exposed to acidic and alkaline water pH. *Aquaculture* **2006**, *256*, 80–87. [CrossRef]
69. Aluru, N.; Vijayan, M.M. Stress transcriptomics in fish: A role for genomic cortisol signaling. *Gen. Comp. Endocrinol.* **2009**, *164*, 142–150. [CrossRef] [PubMed]
70. Nemova, N.N.; Lysenko, L.A.; Kantserova, N.P. Degradation of skeletal muscle protein during growth and development of salmonid fish. *Russ. J. Dev. Biol.* **2016**, *47*, 161–172. [CrossRef]
71. Pérez-Sánchez, J.; Terova, G.; Simó-Mirabet, P.; Rimoldi, S.; Folkedal, O.; Calduch-Giner, J.A.; Olsen, R.E.; Sitjà-Bobadilla, A. Skin mucus of Gilthead Sea bream (*Sparus aurata* L.). protein mapping and regulation in chronically stressed fish. *Front. Physiol.* **2017**, *8*, 34. [CrossRef] [PubMed]
72. Yilmaz, E.; Ergün, S.; Yilmaz, S. Influence of carvacrol on the growth performance, hematological, non-specific immune and serum biochemistry parameters in rainbow trout (*Oncorhynchus mykiss*). *Food Nutr. Sci.* **2015**, *6*, 523–531. [CrossRef]

73. Pourmoghim, H.; Haghighi, M.; Rohani, M.S. Effect of dietary inclusion of *Origanum vulgare* extract on nonspecific immune responses and hematological parameters of rainbow trout (*Oncorhynchus mykiss*). *Bull. Environ. Pharmacol. Life Sci.* **2015**, *4*, 33–39.
74. Abdel-Latif, H.M.; Abdel-Tawwab, M.; Khafaga, A.F.; Dawood, M.A. Dietary oregano essential oil improved the growth performance via enhancing the intestinal morphometry and hepato-renal functions of common carp (*Cyprinus carpio* L.) fingerlings. *Aquaculture* **2020**, *526*, 735432. [CrossRef]
75. Chelladurai, G.; Veni, T.; Mohanraj, J.; Nagarajan, R. Effect of Herbal Extracts Supplemented diets on Non Specific Immunity and Resistance to *Aeromonas hydrophila* in Indian cat fish (*Mystus montanus*). *J. Zool. Biosci. Res.* **2014**, *1*, 10–14.

animals

MDPI

Review

Roles of Nitrocompounds in Inhibition of Foodborne Bacteria, Parasites, and Methane Production in Economic Animals

Po-Yun Teng and **Woo Kyun Kim** *

Department of Poultry Science, University of Georgia, Athens, GA 30605, USA; pyteng@uga.edu
* Correspondence: wkkim@uga.edu

Simple Summary: Supplementation of nitrocompounds in animal diets has been studied to investigate their effects on economic animals. It has been known that nitrocompounds are capable of inhibiting pathogens, parasites, methane and ammonia production. The toxicity, metabolism, and mechanisms of actions have been discussed in the review to conclude the advantages and disadvantages of application of nitrocompounds in animal production.

Abstract: Nitrocompounds are derivatives of hydrocarbons, alcohols, fatty acids, and esters, consisting one or more nitro functional groups. Either natural sources of nitrocompounds or synthetic chemicals have been applied in animal diets to investigate their effects on economic animals, since conjugates of 3-nitropropanol and 3-nitropropionic acid were isolated from *Astragalus oblongifolius*. In this review, emphasis will be placed on nitrocompounds' antimicrobial activity, toxicity, metabolisms and mechanisms of actions. Nitrocompounds can be metabolized by ruminal microbials, such as *Denitrobacterium detoxificans*, or alcohol dehydrogenase in the liver. Moreover, it has been found that nitrocompounds are capable of inhibiting pathogens, parasites, methane and ammonia production; however, overdose of nitrocompounds could cause methemoglobinemia or interfere with energy production in mitochondria by inhibiting succinate dehydrogenase.

Keywords: nitrocompounds; nitropropanol; nitroethanol

Citation: Teng, P.-Y.; Kim, W.K. Roles of Nitrocompounds in Inhibition of Foodborne Bacteria, Parasites, and Methane Production in Economic Animals. *Animals* **2021**, *11*, 923. https://doi.org/10.3390/ani11040923

Academic Editor: Sven Dänicke

Received: 4 February 2021
Accepted: 21 March 2021
Published: 24 March 2021

1. Introduction

Nitrocompounds are derivatives of hydrocarbons, alcohols, fatty acids, and esters which contain one or more nitro functional groups (-NO2) [1]. Short chain aliphatic nitrocompounds, such as unitary alphatic nitroalkanes, aliphatic nitroalcohols, and aliphatic nitroacids, have been widely used in the chemical industry because they are readily accessible and stable for syntheses of various organic compounds [2]. Most of the aliphatic nitrocompounds are not technically produced from biological sources, but 3-nitropropanol (3NPOH) and 3-nitropropionic acid (3NPA) can be extracted from *Astragalus*, *Coronilla*, and *Indigofera* genera of the *Leguminosae* family [3]. The most common sources of nitrocompounds in nature are glucose esters of nitropropionic acid and glycoside of nitro-propanol, 3-nitro-1-propyl-beta-D-glucopyranoside, collectively known as miserotoxin. This toxin was first isolated from *Astragalus oblongifolius* [4,5]. Moreover, 3NPA and 3NPOH are observed in fungi, such as *Penicillium* spp. and *Aspergillus* spp., as well as kernel of the fruit of the karaka tree (*Corynocarpus laevigatus*) [6]. Understanding the properties and functions of nitrocompounds may provide us with novel insights and strategies for future application of nitrocompounds in the animal industry. Therefore, we have reviewed the literature and highlight how alphatic nitrocompounds, including 3-nitropropanol, 2-nitro-1-propanol, nitroethane, and 2-nitroethanol, impact on animal production.

2. Inhibition of Pathogenic Bacteria, Ammonia and Methane Production

Previous studies have reported that nitrocompunds exhibit broad-spectrum antimicrobial activity both in vitro and in vivo [7–11]. The effects of nitrocompounds on pathogen inhibition have been reviewed and listed in Table 1. Jung et al. [12] and Dimitrijevic et al. [13] indicated that *Enterococcus faecalis* and *Listeria monocytogenes* were reduced in the medium containing 10 mM 2-nitro-1-propanol (2NPOH), whereas *Salmonella Typhimurium* and *Escherichia coli* were significantly inhibited by 2.5 mM 2NPOH. An unpublished test conducted in our lab also showed that 2NPOH (4 and 8 mM) and 2-nitroethanol (2NEOH) (8 mM) significantly inhibited growth of *Clostridium perfringens*. Moreover, 2NPOH, 2NEOH, nitroethane (NE), and 2-nitro-methyl-propionate have been reported to reduce *Campylobacter jejuni* and *Campylobacter coli* in culture of Bolton broth at pH 8.2, whereas 2NPOH are more capable inhibiting *Listeria monocytogenes* than 2NEOH and NE [13,14]. Additionally, Kim et al. [15] reported that 2NEOH, 2NPOH, and 3NPA have the potential to reduce uric acid-utilizing microorganisms isolated from poultry manure. The study also suggested that nitrocompounds had superior inhibitory effects compared to their acid and alcohol counterparts [15]. Furthermore, 2NPOH reduced *Listeria monocytogenes*, *Salmonella enterica* serovar Enteritidis, *Escherichia coli*, *Staphylococcus aureus*, and *Bacillus cereus* inoculated on Russian-type salad and corn-flour-based doughs [11,16,17].

Table 1. Summary of Antimicrobial ability of nitrocompounds in vitro and in vivo.

Nitrocompound	Dosage	Unit	Pathogens Inhibition	Reference
			In vitro	
2NEOH	10, 20	mM	*Campylobacter coli*	[14]
	10, 20	mM	*Campylobacter jejuni*	[14]
	8	mM	*Clostridium perfringens*	Unpublished data
	15	mM	*Listeria monocytogenes* strain 18	[13]
	50	mM	uric acid-utilizing microorganisms	[15]
2NMP	10, 20	mM	*Campylobacter jejuni*	[14]
2NPOH	5	%	*Bacillus cereus* [1]	[11]
	10, 20	mM	*Campylobacter coli*	[14]
	10, 20	mM	*Campylobacter jejuni*	[14]
	4, 8	mM	*Clostridium perfringens*	Unpublished data
	10	mM	*Enterococcus faecalis*	[12]
	2.5, 5, 10	mM	*Escherichia coli*	[12]
	0.5, 2, 5	%	*Escherichia coli* [1]	[11]
	10, 15	mM	*Listeria monocytogenes* strain 18	[13]
	50	mM	*Listeria monocytogenes* [2]	[17]
	0.5, 2, 5	%	*Salmonella enterica* serovar Enteritidis [1]	[11]
	2.5, 5, 10	mM	*Salmonella Typhimurium*	[12]
	0.5, 2, 5	%	*Staphylococcus aureus* [1]	[11]
	50	mM	uric acid-utilizing microorganisms	[15]
3NPA	50	mM	uric acid-utilizing microorganisms	[15]
NE	10, 20	mM	*Campylobacter coli*	[14]
	10, 20	mM	*Campylobacter jejuni*	[14]
	15	mM	*Listeria monocytogenes* strain 18	[13]
			In vivo/feces incubation	
2NEOH	20	mM	*Escherichia coli*	[18]
	20	mM	*Escherichia coli*	[18]
	44	mM	*Salmonella Typhimurium*	[19]
	13, 65, 130	mg/bird	*Salmonella Typhimurium* [2]	[8]
	44	mM	*Salmonella Typhimurium*	[19]
	20	mM	*Salmonella Typhimurium*	[18]
	100	mM	uric acid-utilizing microorganisms	[10]
3NPA	100	mM	uric acid-utilizing microorganisms	[10]
Ethyl-nitroacetate	44	mM	*Salmonella Typhimurium*	[19]
NE	12	mM	*Escherichia coli*	[19]
	44	mM	*Salmonella Typhimurium*	[19]
	20	mM	*Salmonella Typhimurium*	[18]
	12	mM	Total Coliforms	[19]

2NEOH, 2-Nitroethanol; 2NMP, 2-Nitro-methyl-porprionate; 2NPOH, 2-Nitro-1-propanol; 3NPA, 3-Nitropropionic acid; NE, Nitroethane.
[1] Inoculated on corn flour-based doughs; [2] Inoculated on Russian-type salad.

On the other hand, previous studies reported the effects of nitrocompounds on the microbial community in the animal intestine, manure or ruminal fluid. Jung et al. [8] demonstrated that a bird gavaged with 13, 65 and 130 mg 2NPOH inhibited *Salmonella Ty-*

phimurium and regulated volatile fatty acids in the cecal content, whereas broilers fed with 16.7 ppm 2NPOH significantly reduced ammonia nitrogen in feces [20]. It is suggested that nitrocompounds reduce ammonia production by inhibiting uric acid-utilizing microorganisms in animal manure [10,20]. Early work by Kim et al. [10] confirmed that 100 mM 2NPOH and 3NPA significantly suppressed uric acid-utilizing microorganisms isolated from poultry feces. Furthermore, Ruiz-Barrera et al. [19] observed that NE reduced *Escherichia coli* and total coliforms after incubation of layer hen manure for 24 h, whereas 2NEOH and 2NPOH reduced *Salmonella Typhimurium* in feces of 6-month-old poultry litter and manure collected from mature sows [18]. Additionally, nitrocompounds regulated immune responses in laying hens challenged with *Salmonoella* [7]. The previous study reported that *Salmonella* challenge increased gene expression of interferon-γ, interleukin-1B, and Toll-like receptor-4 in the ileum of laying hens, but 2NPOH downregulated these cytokines and numerically reduced *Salmonella* in the ceca [7].

Nitrocompounds not only inhibited pathogens and reduced ammonia production in poultry, but also decreased skatole levels in swine manure [21], as well as methane production in ruminants [22]. Zhang et al. [1] summarized the roles of nitrocompounds as methanogenic inhibitors in ruminant animals. Nitrocompounds were first evaluated in an in vitro study [23], indicating that methane production was inhibited by 2NPOH, NE, 2NEOH in the ruminal fluid collected from Holstein-Friesian cows. These results were in agreement with another in vitro study which suggested that NE, 2NPOH, and 2NEOH enhanced volatile fatty acids production and reduced methane formation in the broiler cecal content after 24 h inoculation [24]. Moreover, several in vivo studies have been conducted to confirm the methane-inhibiting ability of selected nitrocompounds. Anderson et al. [9] demonstrated that daily administration of 2NEOH and 2NPOH reduced methane production in mature ewes, whereas Gutierrez-Banuelos et al. [25] reported that NE inhibited methane-producing ability in the rumen and feces of steers. Furthermore, methane emissions and the ratio of acetate to propionate were linearly reduced in response to the increase in NE supplementation. To access the effect of NE on methanogenesis, the same research group conducted an in vitro test. The study showed that NE increased nitro-metabolizing bacteria, reduced methane production, but did not accumulate hydrogen levels in the ruminal fluid [26]. Apart from alphatic nitroalkanes, nitroalcohols, and nitroacids, another nitrocompound has been studied recently. It has been reported that 3-nitrooxypropanol (3NOP) also has methane-inhibitory effects in ruminants [22,27,28]. In addition, dairy cows could produce less methane production for 10 additional weeks after withdrawal of 3NOP [29]; thus, it suggested that 3NOP might be a potential feed additive acting as a methane inhibitor in ruminant animals [1].

3. Inhibition of *Eimeria* spp.

Interestingly, nitrocompounds act like monensin, in terms of both inhibiting methane production and suppressing parasite colonization in ruminants and chickens [23,30–35]. Teng et al. [32] demonstrated that 0.5 mg/mL of monensin and 0.5 mM 2NPOH and 2NEOH significantly inhibited development of sporozoites in the Madin-Darby bovine kidney cells. Moreover, dietary supplementation of 200 ppm 2NPOH reduced cecal lesion scores, as well as improved digestibility of energy in the birds challenged with *Eimeria* spp. [32]. However, 2NPOH did not improve intestinal permeability in a recent experiment [35]. In the ruminants, NE and monensin exhibit similar effects on inhibition of methane production [26,33]. Furthermore, monensin could further regulate butyrate formation, whereas NE did not show significant effects on production of ruminal volatile fatty acids [26]. A recent study was conducted to compare the effects of monensin and NE on digestibility and growth performance of lambs. The authors demonstrated that both monensin and NE did not improve digestibility of crude protein, organic matter, neutral detergent fiber, and acid detergent fiber [33]. However, lamb fed with NE had higher average daily gain and better feed conversion rate compared to the group administrated with monensin [33].

4. Toxicity of Nitrocompounds

Even though nitrocompounds induce several positive outcomes on inhibiting pathogens as well as reducing methane and ammonia production in the gastrointestinal tracts, the toxicity of these organic compounds has caused consumers' caution. Previous in vitro studies have reported that 0.4 to 3 mM is the range of cytotoxic threshold of various testing cells following exposure to nitrocompounds [32,36,37]. It should be noted that the toxicity thresholds of cells are much lower than the thresholds of pathogens (from 4 to 50 mM), indicating that animal cells are more vulnerable than pathogens to nitrocompounds.

Ingestion of *Astragalus* spp. may cause livestock poisoning in ruminant and monogastric animals [38]. Moreover, several reviews have concluded that miserotoxin of *Astragalus* spp. is less toxic to ruminants than nonruminants after oral administration [3,39]. Miserotoxin (3-nitro-1-propyl-beta-D-glucopyranoside) was isolated and characterized from *Astragalus* spp. by Stermitz et al. [5]. The concentrations of miserotoxin in *Astragalus* spp. vary from 2 to 6% [40]. Miserotoxin is relatively innocuous to animals compared to the pure nitrocompounds, such as 3NPA or 3NPOH. A previous study reported that the LD50 of miserotoxin to rats was greater than 2.5 g/kg, whereas the LD50 of 3NPOH was 77 mg/kg [41]. The symptoms of toxicity caused by miserotoxin or nitrocompounds are similar to methemoglobinemia, including depression of feed intake, a tendency to fall down, difficulty in breathing and head extension [3]. These observed reactions in animals are also classic symptoms of nitrite poisoning. However, nitrocompounds do not cause lethal levels of methemoglobinemia as nitrite does [3]. The toxicity of 3NPA and 3NPOH in humans and animals has been reviewed by [4,6]. The 3NPA does not exhibit chronic toxicity; furthermore, the acute toxicity of LD50 dose of 3NPA is between 60 to 120 mg/kg (oral challenge). Burdock et al. [6] also concluded that the acceptable daily intake of 3NPA should not be above 25 mg/kg/day for human.

Toxicity levels of nitroalkanes and nitroalcohols have been concluded in a previous review article. Smith et al. [39] indicated that the acute LD50 values for mice following intraperitoneal injection of nitromethane, NE, 1-nitropropane, 2-nitropropane, and 2NEOH were 110, 310, 250, 800, and 2100 mg/kg body weight, respectively. Moreover, rats under inhalation exposure to NE at 100 or 200 ppm for 2 years had no significant effects on body weights, hematology, nonneoplastic, and neoplastic pathology [42]. An early study also demonstrated that supplementation of less toxic nitrocompounds, such as NE, was capable to prevent *Astragalus* spp. poisoning in ruminants [43].

Less is known regarding toxicity of dietary supplementation of synthetic pure nitrocompounds on economic animals. Previous studies have not indicated any adverse effect of nitrocompounds on performance of laying hens and ovine [7,9]. However, Jung et al. [8] reported that 6-day-old broiler chickens gavaged with a single dose of 130 mg 2NPOH caused 30% mortalities, whereas 13 mg 2NPOH showed no apparent adverse effects. Moreover, dietary supplementation of 33 and 100 ppm 2NPOH and 2NEOH had no impacts on growth performance of young broiler chickens, but 200 ppm 2NPOH and 2NEOH addition resulted in decrease in body weight [20,32]. In summary, toxicity of nitrocompounds is diverse and is influenced by various factors, including animal species, ages, and types and doses of nitrocompounds. Little evidence of chronic toxicity caused by nitrocompounds was reported in previous studies; thus, further investigation is needed before the application of nitrocompounds in animal production.

5. Metabolism of Nitrocompounds

The metabolism of natural sources of nitrocompounds is illustrated in Figure 1. In ruminants, glycoside of nitro-propanol and glucose esters of nitropropionic acid are hydrolyzed by microbial β-glucosidase and esterase, respectively, in the rumen. The rate of hydrolyzation is estimated at 0.75 g mol/mL/h in ruminal fluid [44]. After liberating free 3NPA and 3NPOH, ruminal microbials rapidly metabolize 3NPOH to 3NPA, indicating that these nitrocompounds are equally toxic to animals [44,45]. Apart from metabolizing to 3NPA, 3NPOH is also oxidized to 3-aminopropanol, whereas 3NPA is further metabolized to 3-aminopropionic acid (β-alanine) in the rumen [46]. A previous study indicated that the metabolism of 3NPA was faster than 3NPOH, and the disappearance of 3NPOH was proceeded at a faster rate than that of 2NPOH [44]. It has been reported that the efficiency of ruminal metabolism was influenced by dietary treatments, such as supplementation of NE [43,47]. Dietary protein also contributed to the rate of microbial detoxification [43]. A recent study demonstrated that ruminal microbials could cleave 3NPOH and 3NPA to nitrite [48], and the nitrate will further be degraded to ammonia by rumen microorganisms [49]. However, nitrocompounds are not only metabolized to their respective amines, nitrite, and ammonia, but are also directly absorbed by reticulo-rumen in both sheep and cattle [50,51]. If the 3NPOH was not metabolized to 3-aminopropanol, 3NPA, or nitrate in the rumen, it might be further metabolized to 3NPA in the liver [52].

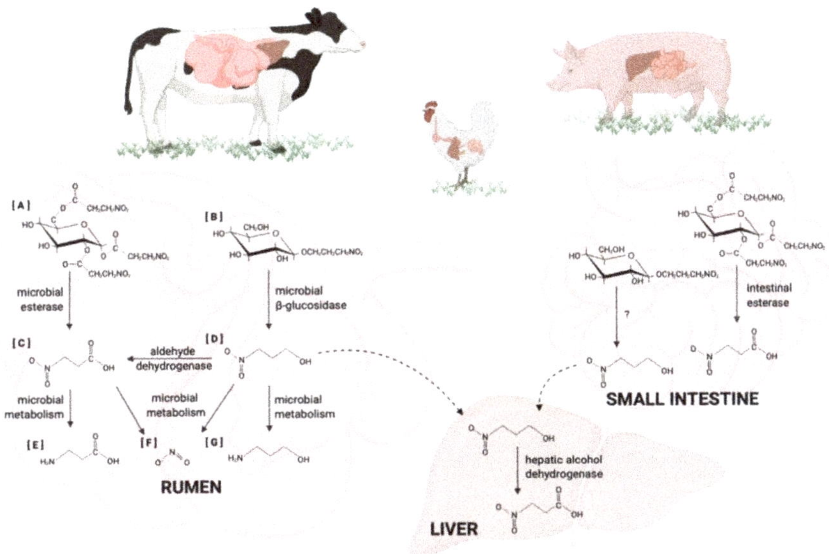

Figure 1. Metabolism of natural sources of nitrocompounds in ruminants and non-ruminants. (**A**) Glucose esters of 3-nitropropionic acid; (**B**) glycoside of 3-nitro-propanol (3-nitro-1-propyl-beta-D-glucopyranoside); (**C**) 3-nitropropionic acid; (**D**) 3-nitropropanol; (**E**) 3-aminopropionic acid (β-alanine); (**F**) nitrite; (**G**) 3-aminopropanol.

Unlike ruminants, monogastric animals, such as pigs and chickens, are not able to secrete β-glucosidase; thus, they absorb miserotoxin in the upper gastrointestinal tract [4] (Figure 1). Though little is known regarding how non-ruminants hydrolyze miserotoxin to liberate free 3NPOH after absorption, previous studies indicated that free 3NPOH is metabolized to 3NPA by aldehyde dehydrogenase and hepatic alcohol dehydrogenase [52,53]. Moreover, monogastric animals are able to metabolize glucose esters of nitropropionic acid by tissue esterase [4].

The metabolism of NE, 2NEOH, and 2NPOH in animals might not share the same pathway as 3NPOH and 3NPA. It has been suggested that NE is transformed to acetaldehyde and nitrite in animals [39]. The acetaldehyde might be oxidized to acetate by acetaldehyde dehydrogenase and further enter into tricarboxylic acid (TCA) cycles, but nitrite is critical to cause acute poisoning [39,54]. Zhang et al. [48] reported that 90% of NE could be degraded by microorganisms, whereas only 75% of 2NEOH and 60% of 2NPOH were metabolized in the ruminal fluid. Moreover, NE, 2NEOH, and 2NPOH might be degraded to ethylamine, amino-ethanol, and 2-amino-1-propnol, respectively, and these intermediates might be further metabolized to nitrite and ammonia by ruminal microbials. The authors also suggested that NE and 2NPOH produced more ammonia compared to 2NEOH [48].

Several ruminal bacteria are able to degrade nitrocompounds and metabolize nitrite anaerobically, including *Megasphaera* spp., *Coprococcus* spp., *Ramibacterium* spp., and *Selenomonas* spp. [49]. Moreover, a new group of anaerobic bacteria, *Denitrobacterium detoxificans*, was identified by Anderson et al. [55]. The author demonstrated that growth of *Denitrobacterium detoxificans* was supported by 3NPA, 3NPOH, 2NPOH, NE, and 2NEOH as electron acceptors, whereas hydrogen and formate served as electron donors. Furthermore, *Clostridium* spp. also has similar effects on the reduction of aliphatic nitrocompounds by hydrogenase and ferredoxin [56].

6. Mechanisms of Actions of Nitrocompounds

Three mechanisms of actions of nitrocompounds have been proposed in previous studies, explaining how nitrocompounds cause toxicity to animals and how nitrocompounds inhibit pathogens and methanogenesis [55,57–60]. The most common toxicity of nitrocompounds is associated with nitrite poisoning shown in Figure 2. After being reduced by ruminal microbial, nitrite acts as a strong reductant in the circulation which rapidly reduces ferrous (Fe^{2+}) iron in oxyhemoglobin (oxyHb) to ferric (Fe^{3+}) state also known as methemoglobin (metHb). The reaction between nitrate and oxyHb not only generates methemoglobin, but also produces hydrogen peroxide and nitrate. The hydrogen peroxide will initiate an autocatalytic propagation with metHb, forming a ferrylhemoglobin (ferrylHb)-radical. The ferrylHb-radical reduces back to metHb by generating two molecules of nitrogen dioxide from nitrite. The nitrogen dioxide can further oxidize oxyHb to ferrylHb-radical, leading to the unstoppable autocatalytic propagation. As the propagation will not be terminated until most of the nitrite in the circulation system is consumed [61], the serial reactions elevate metHb precipitously. Moreover, deoxyhemoglobin also reacts with nitrite, forming nitrosyl hemoglobin as the end product [62]. Both MetHb and Nitrosyl hemoglobin are incapable of carrying oxygen; thus, animals fed high levels of nitrocompounds or nitrite will fail to transport oxygen to tissue and result in death due to severe hypoxia [57].

It is speculated that the more nitrite generated from nitrocompounds, the stronger toxicity that might be observed. As it is discussed above, in the ruminants, nitrocompounds are degraded to various organic compounds by ruminal microorganisms, whereas monogastric animals only produce nitrite as the final product. If animals were fed the same amount of nitrocompounds, ruminants could generate less nitrite than non-ruminants do; thus, previous studies have concluded that miserotoxin is less toxic to ruminants than non-ruminants [39].

On the other hand, 3NPA could inhibit succinate dehydrogenase (SDH) and manipulates energy production of cells (Figure 3). SDH plays important roles in both TCA cycle and respiratory complex II, an enzyme involved in the electron transport chain [63]. SDH regulates oxidation of succinate to fumarate and the reduction of quinone to quinol in the membrane [63]. Hylin et al. [64] first reported the effects of 3NPA on SDH in the heart muscle of rat. It was proposed that 3NPA can act as a suicide inhibitor of SDH because chemical structure of 3NPA is similar to succinic acid, the substrate of SDH [58]. Coles et al. [59] further summarized how 3NPA inactivates the enzyme. In the initiating step, 3NPA is metabolized to 3-nitroacrylate, following with two electrons transferring to the flavin and generating reduced flavin adenine dinucleotide (FADH2) on the enzymes. The nucleophilic of a thiol group is later added to the double bound of 3-nitroacrylate, formatting a thioether on the SDH. Even though FADH2 can be oxidized by respiratory chain, the 3-nitroacrylate is not able to release from the enzyme anymore; thus, the effect of 3NPA on SDH is considered as an irreversible reaction. Moreover, a previous study reported that nitrocompounds could inhibit formate dehydrogenase, formate hydrogen lyase, and hydrogenase activity [60]. As SDH, formate dehydrogenase, and formate hydrogen lyase play important roles in energy metabolism in both eukaryotes and prokaryotes, it is concluded that nitrocompounds might impede energy production in pathogens and parasites by suppressing metabolism of formate and hydrogen as well as inhibiting the SDH involved in complex II and TCA cycle [65–67].

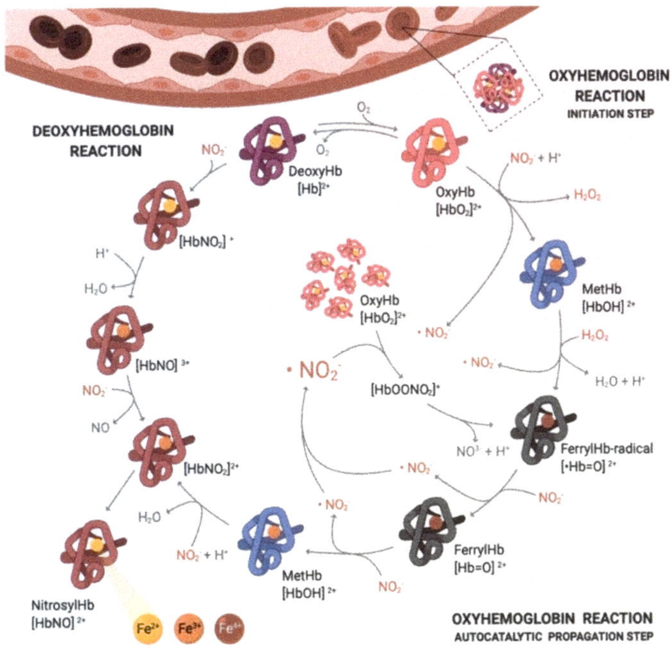

Figure 2. Mechanisms of actions of nitrite poisoning caused by nitrocompounds administration. Once nitrocompounds are metabolized to nitrite in the gastrointestinal tracts in animals, nitrite will interact with both oxyhemoglobin and deoxyhemoglobin, leading to failure of oxygen transportation. Nitrite could further initiate an autocatalytic propagation that keep oxidizing oxyhemoglobin to methemoglobin and end up with the formation of nitrosyl hemoglobin.

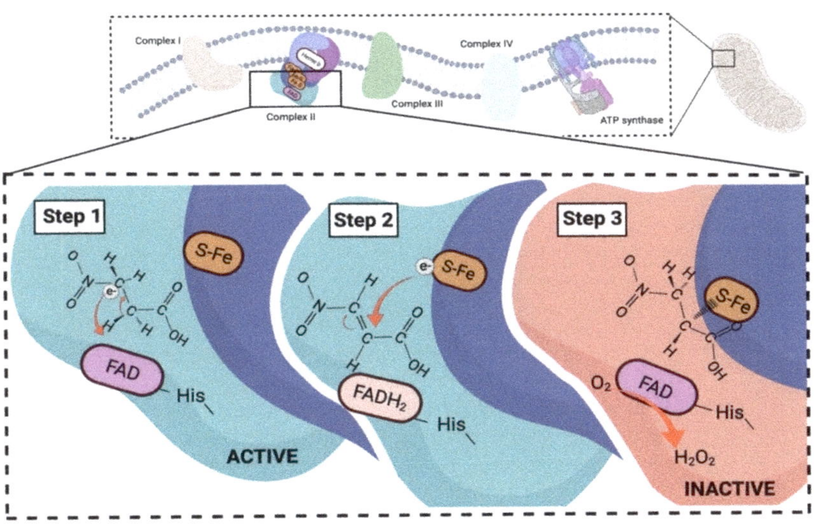

Figure 3. Inhibition of succinate dehydrogenase in electron transport chain by 3-nitropropionic acid (3NPA). 3NPA has a similar chemical structure as succinic acids which make it possible to attach on succinate dehydrogenase and irreversibly inactive the enzyme, causing failure of electron transportation in mitochondria.

The presence of *Denitrobacterium detoxificans* and the inhibition of formate dehydrogenase by nitrocompounds are the main mechanisms of actions that reduced methane production in the ruminants. *D. detoxificans* processes nitrocompounds metabolizing activity and causes the reduction of nitrite, nitroalkanes and nitroalcohols with oxidation of hydrogen and formate (Figure 4). In the reaction, hydrogen and formate are oxidized to water, carbon dioxide and hydrogen, whereas nitrite and nitrocompounds are reduced to ammonia and nitrite, respectively [68]. Zhang et al. [1] further concluded that nitrocompounds act as alternative electron acceptors, diverting the flow of reducing equivalent away from methanogenesis (Figure 4, [pathway 3]). On the contrary, other studies indicated that the inhibition of methane production was independent to the presence of *D. detoxificans* and the loss of reducing equivalents by the reduction of nitrocompounds [23,69]. Though competing reductants might not be the main mechanisms of actions of nitrocompounds, it should be noted that metabolizing nitrocompounds by *D. detoxificans*, indeed, spares the reducing equivalents from the production of methane [68].

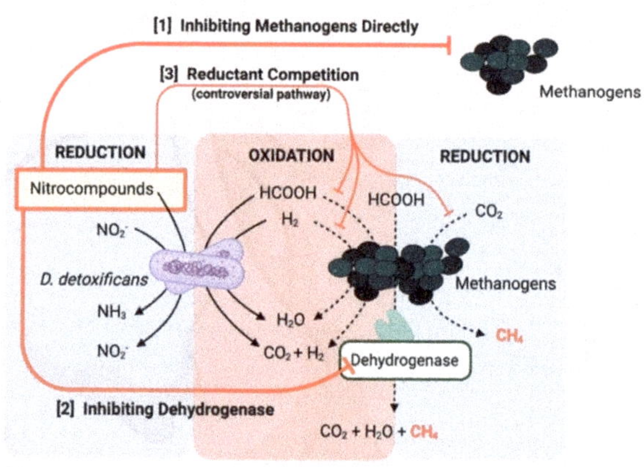

Figure 4. Possible mechanisms of actions of nitrocompounds withhold methanogenesis in ruminants. Nitrocompounds might directly inhibit methanogens [pathway 1] or suppress dehydrogenase which is an enzyme metabolizing formate to methane [pathway 2]. Metabolism of Nitrocompounds by D. detoxificans consumes reductant in the rumen, such as hydrogen and formate. Nitrocompounds might compete these reductants with carbon dioxides, indirectly reducing methane production. [pathway 3] Reductant competition is a possible mechanism of action of nitrocompounds, but it might not play the main role on inhibition of methanogenesis in ruminants.

It has been proposed that nitrocompounds reduce methane production by inhibiting methanogens directly or suppressing ability of formate dehydrogenase [60] (Figure 4, [pathway 1] and [pathway 2]). The ruminal methanogens are capable of metabolizing formate to carbon dioxide and hydrogen, which are rapidly oxidized to methane [70]. Additionally, some methanogens could degrade formate to carbon dioxides, water, and methane directly via formate dehydrogenase [71]. Approximately 18% of ruminal methane was formed from formate rather than carbon dioxide [72]; thus, inhibiting dehydrogenase for formate oxidation is a potential mechanism of action of nitrocompounds in withholding methanogenesis in the rumen.

7. Conclusions

Conjugates of 3NPOH and 3NPA in forages, and various synthetic nitrocompounds have been reviewed in the context of their effects on the inhibition of foodborne pathogens, parasites, methane, and ammonia production in economic animals. The toxicity, metabolism, and mechanisms of actions have been discussed in the current review to conclude the advantages and disadvantages of application of nitrocompounds in animal production. Previous studies have elucidated the properties of 3NPA and 3NPOH comprehensively because they are the only nitrocompounds observed from natural sources so far. It has been demonstrated that the toxicity of 3NPA and 3NPOH is associated with nitrite poisoning and inactivation of SDH. Nevertheless, little is known regarding the mode of actions and toxicity of 2NPOH, NE, and 2NEOH. Even though the effects of short-chain nitrocompounds on broiler chicken, laying hen, cattle, lamb and swine have been studied for decades, further research is needed to determine a range of safe dosages in order to use nitrocompounds as a novel strategy for the control of pathogens in animal production.

Author Contributions: Writing—original draft preparation and visualization, P.-Y.T.; supervision, review, and editing, W.K.K. All authors have read and agreed to the published version of the manuscript.

Funding: This research was funded by a cooperative agreement 58-6040-8-034 from United States Department of Agriculture-Agricultural Research Service.

Institutional Review Board Statement: Not applicable.

Data Availability Statement: No new data were created or analyzed in this study. Data sharing is not applicable to this article.

Acknowledgments: Special thanks to all members in Kim's lab. Figures are created with Biorender.

Conflicts of Interest: The authors declare no conflict of interest.

References

1. Zhang, Z.-W.; Cao, Z.-J.; Wang, Y.-L.; Wang, Y.-J.; Yang, H.-J.; Li, S.-L. Nitrocompounds as potential methanogenic inhibitors in ruminant animals: A review. *Anim. Feed Sci. Tech.* **2018**, *236*, 107–114. [CrossRef]
2. Shvekhgeimer, M.-G.A. Aliphatic nitro alcohols. Synthesis, chemical transformations and applications. *Russ. Chem. Rev.* **1998**, *67*, 35–68. [CrossRef]
3. Gustine, D.L. Aliphatic Nitro Compounds in Crownvetch: A Review. *Crop. Sci.* **1979**, *19*, 197–203. [CrossRef]
4. Anderson, R.C.; Majak, W.; Rassmussen, M.A.; Callaway, T.R.; Beier, R.C.; Nisbet, D.J.; Allison, M.J. Toxicity and metabolism of the conjugates of 3-nitropropanol and 3-nitropropionic acid in forages poisonous to livestock. *J. Agric. Food Chem.* **2005**, *53*, 2344–2350. [CrossRef]
5. Stermitz, F.R.; Norris, F.A.; Williams, M.C. Miserotoxin, new naturally occurring nitro compound. *J. Am. Chem. Soc.* **1969**, *91*, 4599–4600. [CrossRef]
6. Burdock, G.A.; Carabin, I.G.; Soni, M.G. Safety assessment of β-nitropropionic acid: A monograph in support of an acceptable daily intake in humans. *Food Chem.* **2001**, *75*, 1–27. [CrossRef]
7. Adhikari, P.; Cosby, D.E.; Cox, N.A.; Kim, W.K. Effect of dietary supplementation of nitrocompounds on Salmonella colonization and ileal immune gene expression in laying hens challenged with *Salmonella enteritidis*. *Poult. Sci.* **2017**, *96*, 4280–4286. [CrossRef] [PubMed]
8. Jung, Y.S.; Anderson, R.C.; Edrington, T.S.; Genovese, K.J.; Byrd, J.A.; Callaway, T.R.; Nisbet, D.J. Experimental use of 2-nitropropanol for reduction of *Salmonella typhimurium* in the Ceca of broiler chicks. *J. Food Prot.* **2004**, *67*, 1945–1947. [CrossRef] [PubMed]
9. Anderson, R.C.; Carstens, G.E.; Miller, R.K.; Callaway, T.R.; Schultz, C.L.; Edrington, T.S.; Harvey, R.B.; Nisbet, D.J. Effect of oral nitroethane and 2-nitropropanol administration on methane-producing activity and volatile fatty acid production in the ovine rumen. *Bioresour. Technol.* **2006**, *97*, 2421–2426. [CrossRef] [PubMed]
10. Kim, W.K.; Weeks, L.J.; Anderson, R.C.; Nisbet, D.J.; Dunkley, K.; Ricke, S.C. Effects of nitrocompounds on uric acid-utilizing microorganisms, nitrogen retention, and microbial community in laying hen manure. *J. Environ. Sci. Health B* **2009**, *44*, 403–406. [CrossRef]
11. Morente, E.O.; Abriouel, H.; Lopez, R.L.; Ben Omar, N.; Galvez, A. Antibacterial activity of carvacrol and 2-nitro-1-propanol against single and mixed populations of foodborne pathogenic bacteria in corn flour dough. *Food Microbiol.* **2010**, *27*, 274–279. [CrossRef] [PubMed]
12. Jung, Y.S.; Anderson, R.C.; Callaway, T.R.; Edrington, T.S.; Genovese, K.J.; Harvey, R.B.; Poole, T.L.; Nisbet, D.J. Inhibitory activity of 2-nitropropanol against select food-borne pathogens in vitro. *Lett. Appl. Microbiol.* **2004**, *39*, 471–476. [CrossRef]

13. Dimitrijevic, M.; Anderson, R.C.; Callaway, T.R.; Jung, Y.S.; Harvey, R.B.; Ricke, S.C.; Nisbet, D.J. Inhibitory Effect of Select Nitrocompounds on growth and survivability of *Listeria monocytogenes* in vitro. *J. Food Prot.* **2006**, *69*, 1061–1065. [CrossRef] [PubMed]
14. Horrocks, S.M.; Jung, Y.S.; Huwe, J.K.; Harvey, R.B.; Ricke, S.C.; Carstens, G.E.; Callaway, T.R.; Anderson, R.C.; Ramlachan, N.; Nisbet, D.J. Effects of short-chain nitrocompounds against *Campylobacter jejuni* and *Campylobacter coli* in vitro. *J. Food Sci.* **2007**, *72*, M50–M55. [CrossRef] [PubMed]
15. Kim, W.K.; Anderson, R.C.; Ratliff, A.L.; Nisbet, D.J.; Ricke, S.C. Growth inhibition by nitrocompounds of selected uric acid-utilizing microorganisms isolated from poultry manure. *J. Environ. Sci. Health B* **2006**, *41*, 97–107. [CrossRef]
16. Antonio, C.M.; Abriouel, H.; Lopez, R.L.; Omar, N.B.; Valdivia, E.; Galvez, A. Enhanced bactericidal activity of enterocin AS-48 in combination with essential oils, natural bioactive compounds and chemical preservatives against *Listeria monocytogenes* in ready-to-eat salad. *Food Chem. Toxicol.* **2009**, *47*, 2216–2223.
17. Cobo Molinos, A.; Lucas Lopez, R.; Abriouel, H.; Ben Omar, N.; Valdivia, E.; Galvez, A. Inhibition of *Salmonella enterica* Cells in deli-type salad by enterocin AS-48 in combination with other antimicrobials. *Probiotics Antimicrob. Proteins* **2009**, *1*, 85–90. [CrossRef]
18. Anderson, R.C.; Jung, Y.S.; Oliver, C.E.; Horrocks, S.M.; Genovese, K.J.; Harvey, R.B.; Callaway, T.R.; Edrington, T.S.; Nisbet, D.J. Effects of nitrate or nitro supplementation, with or without added chlorate, on *Salmonella enterica* serovar Typhimurium and *Escherichia coli* in swine feces. *J. Food Prot.* **2007**, *70*, 308–315. [CrossRef]
19. Ruiz-Barrera, O.; Anderson, R.C.; Hume, M.E.; Corrales-Millan, J.; Castillo-Castillo, Y.; Corral-Luna, A.; Guevara-Valdez, J.L.; Salinas-Chavira, J.; Rodriguez-Muela, C.; Arzola-Alvarez, C. Short chain nitrocompounds as a treatment of layer hen manure and litter; effects on in vitro survivability of Salmonella, generic *E. coli* and nitrogen metabolism. *J. Environ. Sci. Health B* **2017**, *52*, 23–29. [CrossRef]
20. Mowrer, J.E.; Sedlacek, P.; Kim, J.; Ritz, C.; Kim, W.K. Supplementation of nitrocompounds in broiler diets: Effects on bird performance, ammonia volatilization and nitrogen retention in broiler manure. *J. Environ. Sci. Health B* **2016**, *51*, 126–131. [CrossRef]
21. Beier, R.C.; Anderson, R.C.; Krueger, N.A.; Edrington, T.S.; Callaway, T.R.; Nisbet, D.J. Effect of nitroethane and nitroethanol on the production of indole and 3-methylindole (skatole) from bacteria in swine feces by gas chromatography. *J. Environ. Sci. Health B* **2009**, *44*, 613–620. [CrossRef] [PubMed]
22. Haisan, J.; Sun, Y.; Guan, L.L.; Beauchemin, K.A.; Iwaasa, A.; Duval, S.; Barreda, D.R.; Oba, M. The effects of feeding 3-nitrooxypropanol on methane emissions and productivity of Holstein cows in mid lactation. *J. Dairy Sci.* **2014**, *97*, 3110–3119. [CrossRef] [PubMed]
23. Anderson, R.C.; Callaway, T.R.; Van Kessel, J.A.; Jung, Y.S.; Edrington, T.S.; Nisbet, D.J. Effect of select nitrocompounds on ruminal fermentation; an initial look at their potential to reduce economic and environmental costs associated with ruminal methanogenesis. *Bioresour. Technol.* **2003**, *90*, 59–63. [CrossRef]
24. Saengkerdsub, S.; Kim, W.K.; Anderson, R.C.; Nisbet, D.J.; Ricke, S.C. Effects of nitro compounds and feedstuffs on in vitro methane production in chicken cecal contents and rumen fluid. *Anaerobe* **2006**, *12*, 85–92. [CrossRef]
25. Gutierrez-Banuelos, H.; Anderson, R.C.; Carstens, G.E.; Slay, L.J.; Ramlachan, N.; Horrocks, S.M.; Callaway, T.R.; Edrington, T.S.; Nisbet, D.J. Zoonotic bacterial populations, gut fermentation characteristics and methane production in feedlot steers during oral nitroethane treatment and after the feeding of an experimental chlorate product. *Anaerobe* **2007**, *13*, 21–31. [CrossRef]
26. Gutierrez-Banuelos, H.; Anderson, R.C.; Carstens, G.E.; Tedeschi, L.O.; Pinchak, W.E.; Cabrera-Diaz, E.; Krueger, N.A.; Callaway, T.R.; Nisbet, D.J. Effects of nitroethane and monensin on ruminal fluid fermentation characteristics and nitrocompound-metabolizing bacterial populations. *J. Agric. Food Chem.* **2008**, *56*, 4650–4658. [CrossRef] [PubMed]
27. Martinez-Fernandez, G.; Abecia, L.; Arco, A.; Cantalapiedra-Hijar, G.; Martin-Garcia, A.I.; Molina-Alcaide, E.; Kindermann, M.; Duval, S.; Yanez-Ruiz, D.R. Effects of ethyl-3-nitrooxy propionate and 3-nitrooxypropanol on ruminal fermentation, microbial abundance, and methane emissions in sheep. *J. Dairy Sci.* **2014**, *97*, 3790–3799. [CrossRef]
28. Reynolds, C.K.; Humphries, D.J.; Kirton, P.; Kindermann, M.; Duval, S.; Steinberg, W. Effects of 3-nitrooxypropanol on methane emission, digestion, and energy and nitrogen balance of lactating dairy cows. *J. Dairy Sci.* **2014**, *97*, 3777–3789. [CrossRef] [PubMed]
29. Hristova, A.N.; Oha, J.; Giallongoa, F.; Fredericka, T.W.; Harpera, M.T.; Weeksa, H.L.; Brancob, A.F.; Moatec, P.J.; Deightonc, M.H.; Williamsc, S.R.O.; et al. An inhibitor persistently decreased enteric methane emission from dairy cows with no negative effect on milk production. *Proc. Natl. Acad. Sci. USA* **2015**, *112*, 10663–10668. [CrossRef]
30. Chapman, H.D. Milestones in avian coccidiosis research: A review. *Poult. Sci.* **2014**, *93*, 501–511. [CrossRef] [PubMed]
31. Chapman, H.D.; Jeffers, T.K.; Williams, R.B. Forty years of monensin for the control of coccidiosis in poultry. *Poult. Sci.* **2010**, *89*, 1788–18001. [CrossRef] [PubMed]
32. Teng, P.Y.; Fuller, A.L.; Kim, W.K. Evaluation of nitro compounds as feed additives in diets of *Eimeria*-challenged broilers in vitro and in vivo. *Poult. Sci.* **2020**, *99*, 1320–1325. [CrossRef]
33. Zhang, Z.W.; Wang, Y.L.; Chen, Y.Y.; Wang, W.K.; Zhang, L.T.; Luo, H.L.; Yang, H.J. Nitroethanol in comparison with monensin exhibits greater feed efficiency through inhibiting rumen methanogenesis more efficiently and persistently in feedlotting lambs. *Animals* **2019**, *9*, 784. [CrossRef] [PubMed]
34. Bangoura, B.; Bardsley, K.D. Ruminant Coccidiosis. *Vet. Clin. North. Am. Food Anim. Pract.* **2020**, *36*, 187–203. [CrossRef]
35. Teng, P.Y.; Yadav, S.; Dos Santos, T.S.; Fuller, A.L.; Kim, W.K. 2-Nitro-1-propanol improved nutrient digestibility and oocyst shedding but not growth performance of *Eimeria*-challenged broilers. *Poult. Sci.* **2020**, *99*, 4314–4322. [CrossRef]

36. Paik, D.C.; Wen, Q.; Braunstein, R.E.; Trokel, S.L. Short chain aliphatic beta-nitro alcohols for corneoscleral cross-linking: Corneal endothelial toxicity studies. *J. Refract. Surg.* **2008**, *24*, S741–S747. [CrossRef]
37. Kim, M.; Takaoka, A.; Hoang, Q.V.; Trokel, S.L.; Paik, D.C. Pharmacologic alternatives to riboflavin photochemical corneal cross-linking: A comparison study of cell toxicity thresholds. *Investig. Ophthalmol. Vis. Sci.* **2014**, *55*, 3247–3257. [CrossRef] [PubMed]
38. Williams, M.C.; James, L.F. Toxicity of nitro-containing *Astragalus* to sheep and chicks. *J. Range Manag.* **1975**, *28*, 260–263. [CrossRef]
39. Smith, D.J.; Anderson, R.C. Toxicity and metabolism of nitroalkanes and substituted nitroalkanes. *J. Agric. Food Chem.* **2013**, *61*, 763–779. [CrossRef]
40. Majak, W.; Wikeem, B.M. Miserotoxin levels in fertilized *Astragalus miser* var. serotinus. *J. Range Manag.* **1986**, *39*, 130–131. [CrossRef]
41. Majak, W.; Pass, M.A.; Madryga, F.J. Toxicity of miserotoxin and its aglycone. *Toxicol. Lett.* **1983**, *19*, 171–178. [CrossRef]
42. Griffin, T.B.; Stein, A.A.; Coulston, F. Chronic inhalation exposure of rats to vapors of nitroethane. *Ecotoxicol. Environ. Saf.* **1988**, *16*, 11–24. [CrossRef]
43. Majak, W. Further enhancement of 3-nitropropanol detoxification by ruminal bactdria in cattle. *Can. J. Anim. Sci.* **1992**, *72*, 863–870. [CrossRef]
44. Majak, W.; Clark, L.J. Metabolism of aliphatic nitro compounds in bovine rumen fluid. *Can. J. Anim. Sci.* **1980**, *60*, 319–325. [CrossRef]
45. Muir, A.D.; Majak, W.; Pass, M.A.; Yost, G.S. Conversion of 3-nitropropanol (miserotoxin aglycone) to 3-nitropropionic acid in cattle and sheep. *Toxicol. Lett.* **1984**, *20*, 137–141. [CrossRef]
46. Anderson, R.C.; Rasmussen, M.A.; Allison, M.J. Metabolism of the plant toxins nitropropionic acid and nitropropanol by ruminal microorganisms. *Appl. Environ. Microbiol.* **1993**, *59*, 3056–3061. [CrossRef] [PubMed]
47. Majak, W.; Cheng, K.J.; Hall, J.W. Enhanced degradation of 3-nitropropanol by ruminal microorganisms. *J. Anim. Sci.* **1986**, *62*, 1072–1080. [CrossRef]
48. Zhang, Z.W.; Wang, Y.L.; Wang, W.K.; Chen, Y.Y.; Si, X.M.; Wang, Y.J.; Wang, W.; Cao, Z.J.; Li, S.L.; Yang, H.J. The Antimethanogenic nitrocompounds can be cleaved into nitrite by rumen microorganisms: A comparison of Nitroethane, 2-Nitroethanol, and 2-Nitro-1-propanol. *Metabolites* **2020**, *10*, 15. [CrossRef]
49. Majak, W.; Cheng, K.J. Identification of rumen bacteria that anaerobically degrade aliphatic nitrotoxins. *Can. J. Microbiol.* **1981**, *27*, 646–650. [CrossRef]
50. Majak, W.; Pass, M.A.; Muir, A.D.; Rode, L.M. Absorption of 3-nitropropanol (miserotoxin aglycone) from the compound stomach of cattle. *Toxicol. Lett.* **1984**, *23*, 9–15. [CrossRef]
51. Pass, M.A.; Majak, W.; Muir, A.D.; Yost, G.S. Absorption of 3-nitropropanol and 3-nitropropionic acid from the digestive system of sheep. *Toxicol. Lett.* **1984**, *23*, 1–7. [CrossRef]
52. Benn, M.; McDiarmid, R.; Majak, W. In-vitro biotransformation of 3-nitropropanol (miserotoxin aglycone) by horse liver alcohol dehydrogenase. *Toxicol. Lett.* **1989**, *47*, 165–172. [CrossRef]
53. Pass, M.A.; Muir, A.D.; Majak, W.; Yost, G.S. Effect of alcohol and aldehyde dehydrogenase inhibitors on the toxicity of 3-nitropropanol in rats. *Toxicol. Appl. Pharmacol.* **1985**, *78*, 310–315. [CrossRef]
54. Little, H.N. Oxidation of nitroethane by extracts from Neurospora. *J. Biol. Chem.* **1951**, *193*, 347–358. [CrossRef]
55. Anderson, R.C.; Rasmussen, M.A.; Jensen, N.S.; Allison, M.J. *Denitrobacterium detoxificans* gen. nov., sp. nov., a ruminal bacterium that respires on nitrocompounds. *Int. J. Syst. Evol. Microbiol.* **2000**, *50* (Pt. 2), 633–638. [CrossRef] [PubMed]
56. Angermaier, L.; Simon, H. On the reduction of aliphatic and aromatic nitro compounds by *Clostridia*, the role of ferredoxin and its stabilization. *Hoppe Seylers Z Physiol. Chem.* **1983**, *364*, 961–975. [CrossRef] [PubMed]
57. Shiono, H.; Yagi, Y.; Thongnoon, P.; Kurabayashi, N.; Chikayama, Y.; Miyazaki, S.; Nakamura, I. Acquired methemoglobinemia in anemic cattle infected with *Theileria sergenti*. *Vet. Parasitol.* **2001**, *102*, 45–51. [CrossRef]
58. Alston, T.A.; Mela, L.; Bright, H.J. 3-Nitropropionate, the toxic substance of *Indigofera*, is a suicide inactivator of succinate dehydrogenase. *Proc. Natl. Acad. Sci. USA* **1977**, *74*, 3767–3771. [CrossRef]
59. Coles, C.J.; Edmondson, D.E.; Singer, T.P. Inactivation of succinate dehydrogenase by 3-nitropropionate. *J. Biol. Chem.* **1979**, *254*, 5161–5167. [CrossRef]
60. Anderson, R.C.; Krueger, N.A.; Stanton, T.B.; Callaway, T.R.; Edrington, T.S.; Harvey, R.B.; Jung, Y.S.; Nisbet, D.J. Effects of select nitrocompounds on in vitro ruminal fermentation during conditions of limiting or excess added reductant. *Bioresour. Technol.* **2008**, *99*, 8655–8661. [CrossRef]
61. Keszler, A.; Piknova, B.; Schechter, A.N.; Hogg, N. The reaction between nitrite and oxyhemoglobin: A mechanistic study. *J. Biol. Chem.* **2008**, *283*, 9615–9622. [CrossRef]
62. Hathazi, D.; Mahut, S.D.; Scurtu, F.V.; Bischin, C.; Stanciu, C.; Attia, A.A.; Damian, G.; Silaghi-Dumitrescu, R. Involvement of ferryl in the reaction between nitrite and the oxy forms of globins. *J. Biol. Inorg. Chem.* **2014**, *19*, 1233–1239. [CrossRef] [PubMed]
63. Rasheed, M.; Tarjan, G. Succinate Dehydrogenase Complex: An Updated Review. *Arch. Pathol. Lab. Med.* **2018**, *142*, 1564–1570. [CrossRef]
64. Hylin, J.W.; Matsumoto, H. Inhibition of succinic dehydrogenase by 3-nitropropanoate. *Toxicol. Appl. Pharmacol.* **1964**, *6*, 168–171. [CrossRef]
65. Harada, S.; Inaoka, D.K.; Ohmori, J.; Kita, K. Diversity of parasite complex II. *Biochim. Biophys. Acta* **2013**, *1827*, 658–667. [CrossRef] [PubMed]
66. Ruiz-Herrera, J.; Garcia, L.G. Regulation of succinate dehydrogenase in *Escherichia coli*. *J. Gen. Microbiol.* **1972**, *72*, 29–35. [CrossRef]

67. Pecsi, I.; Hards, K.; Ekanayaka, N.; Berney, M.; Hartman, T.; Jacobs, W.R., Jr.; Cook, G.M. Essentiality of succinate dehydrogenase in *Mycobacterium smegmatis* and its role in the generation of the membrane potential under hypoxia. *mBio* **2014**, *5*. [CrossRef] [PubMed]
68. Anderson, R.C.; Huwe, J.K.; Smith, D.J.; Stanton, T.B.; Krueger, N.A.; Callaway, T.R.; Edrington, T.S.; Harvey, R.B.; Nisbet, D.J. Effect of nitroethane, dimethyl-2-nitroglutarate and 2-nitro-methyl-propionate on ruminal methane production and hydrogen balance in vitro. *Bioresour. Technol.* **2010**, *101*, 5345–5349. [CrossRef]
69. Anderson, R.C.; Rasmussen, M.A. Use of a novel nitrotoxin-metabolizing bacterium to reduce ruminal methane production. *Bioresour. Technol.* **1998**, *64*, 89–95. [CrossRef]
70. Tapio, I.; Snelling, T.J.; Strozzi, F.; Wallace, R.J. The ruminal microbiome associated with methane emissions from ruminant livestock. *J. Anim. Sci. Biotechnol.* **2017**, *8*, 7. [CrossRef]
71. Asanuma, N.; Iwamoto, M.; Hino, T. Formate metabolism by ruminal microorganisms in relation to methanogenesis. *Anim. Sci. Technol.* **1998**, *69*, 576–584.
72. Hungate, R.E.; Smith, W.; Bauchop, T.; Yu, I.; Rabinowitz, J.C. Formate as an intermediate in the bovine rumen fermentation. *J. Bacteriol.* **1970**, *102*, 389–397. [CrossRef] [PubMed]

animals

MDPI

Article

Date Palm Pollen Extract Avert Doxorubicin-Induced Cardiomyopathy Fibrosis and Associated Oxidative/Nitrosative Stress, Inflammatory Cascade, and Apoptosis-Targeting Bax/Bcl-2 and Caspase-3 Signaling Pathways

Samar S. Elblehi [1,*], Yasser S. El-Sayed [2], Mohamed Mohamed Soliman [3] and Mustafa Shukry [4,*]

1 Department of Pathology, Faculty of Veterinary Medicine, Alexandria University, Alexandria, Edfina 22758, Egypt
2 Department of Forensic Medicine and Toxicology, Faculty of Veterinary Medicine, Damanhour University, Damanhour 22511, Egypt; elsayed-ys@vetmed.dmu.edu.eg
3 Clinical Laboratory Sciences Department, Turabah University College, Taif University, P.O. Box 11099, Taif 21944, Saudi Arabia; mmsoliman@tu.edu.sa
4 Department of Physiology, Faculty of Veterinary Medicine, Kafrelsheikh University, Kafrelsheikh 33516, Egypt
* Correspondence: samar.saad@alexu.edu.eg (S.S.E.); mostafa.ataa@vet.kfs.edu.eg (M.S.)

Citation: Elblehi, S.S.; El-Sayed, Y.S.; Soliman, M.M.; Shukry, M. Date Palm Pollen Extract Avert Doxorubicin-Induced Cardiomyopathy Fibrosis and Associated Oxidative/ Nitrosative Stress, Inflammatory Cascade, and Apoptosis-Targeting Bax/Bcl-2 and Caspase-3 Signaling Pathways. *Animals* **2021**, *11*, 886. https://doi.org/10.3390/ani11030886

Academic Editors: Youssef A. Attia, Woo Kyun Kim, Nesrein Hashem and Maria de Olivera

Received: 3 February 2021
Accepted: 17 March 2021
Published: 20 March 2021

Simple Summary: The use of date palm pollen ethanolic extract (DPPE) is a conventional approach in improving the side-effects induced by Doxorubicin (DOX).DPPE mitigated DOX-induced body and heart weight changes and ameliorated DOX-induced elevated cardiac injury markers. In addition, serum cardiac troponin I concentrations (cTnI), troponin T (cTnT), and N-terminal NBP and cytosolic (Ca^{+2}) were amplified by alleviating the inflammatory and oxidative injury markers and decreasing histopathological lesions severity. DPPE decreased DOX-induced heart injuries by mitigating inflammation, fibrosis, and apoptosis through its antioxidant effect. To reduce DOX-induced oxidative stress injuries and other detrimental effects, a combined treatment of DPPE is advocated.

Abstract: Doxorubicin (DOX) has a potent antineoplastic efficacy and is considered a cornerstone of chemotherapy. However, it causes several dose-dependent cardiotoxic results, which has substantially restricted its clinical application. This study was intended to explore the potential ameliorative effect of date palm pollen ethanolic extract (DPPE) against DOX-induced cardiotoxicity and the mechanisms underlying it. Forty male Wistar albino rats were equally allocated into Control (CTR), DPPE (500 mg/kg bw for 4 weeks), DOX (2.5 mg/kg bw, intraperitoneally six times over 2 weeks), and DPPE + DOX-treated groups. Pre-coadministration of DPPE with DOX partially ameliorated DOX-induced cardiotoxicity as DPPE improved DOX-induced body and heart weight changes and mitigated the elevated cardiac injury markers activities of serum aminotransferases, lactate dehydrogenase, creatine kinase, and creatine kinase-cardiac type isoenzyme. Additionally, the concentration of serum cardiac troponin I (cTnI), troponin T (cTnT), N-terminal pro-brain natriuretic peptide (NT-pro BNP), and cytosolic calcium (Ca^{+2}) were amplified. DPPE also alleviated nitrosative status (nitric oxide) in DOX-treated animals, lipid peroxidation and antioxidant molecules as glutathione content, and glutathione peroxidase, catalase, and superoxide dismutase activities and inflammatory markers levels; NF-κB p65, TNF-α, IL-1β, and IL-6. As well, it ameliorated the severity of histopathological lesions, histomorphometric alteration and improved the immune-staining of the pro-fibrotic (TGF-β1), pro-apoptotic (caspase-3 and Bax), and anti-apoptotic (Bcl-2) proteins in cardiac tissues. Collectively, pre-coadministration of DPPE partially mitigated DOX-induced cardiac injuries via its antioxidant, anti-inflammatory, anti-fibrotic, and anti-apoptotic potential.

Keywords: cardiac injury markers; oxidative stress; histopathology; Bcl-2; Bax; TGF-β1; date palm (pollen extract); doxorubicin

1. Introduction

Doxorubicin (Adriamycin®), an anthracycline chemotherapeutic medication, has been effective against several types of malignancies since the 1960s [1,2]. It is the most valuable cytotoxic medication approved by oncologists in tandem with other anti-tumor medications or radiation and surgery [3]. It is highly potent and effective against solid tumors, i.e., breast, lung, bladder, gastrointestinal, thyroid, testicular, and ovarian carcinoma [4,5]. It is also used for treating hematological cancers, i.e., Hodgkin's and non-Hodgkin's lymphoma and pediatric leukemia [2,3]. Two proposed mechanisms for DOX antineoplastic effects have been reported [6]. The first one is through DNA chelation as DOX interacts with DNA, inhibits topoisomerase-II progression, and hinders DNA repair, which triggers DNA damage and cell death [7]. The second mechanism includes reactive oxygen species creation (ROS) and oxidative stress induction [8]. In vivo, DOX is metabolized into an unstable semiquinone, which is transformed back to DOX in a reaction that discharges ROS and reactive nitrogen species (RNS), causing lipid peroxidation, cell membrane, DNA, and proteins damages, and prompts apoptotic pathways of cell downfall to kill cancer cells [8,9]. Genes that can regulate this pathway include those involved with the oxidation outcome (xanthine oxidase, NADH dehydrogenases, and nitric oxide synthases) and those which disable free radicals, involving catalase (CAT), superoxide dismutase (SOD) and glutathione peroxidase (GPx) [9,10]. Nevertheless, such effects are not selective for cancer cells alone as the same mechanisms can also affect healthy cells [11,12].

It has been reported that DOX administration can induce structural and functional cardiac alterations, i.e., ventricular distension, diminished output, systolic and diastolic disturbance [13,14], congestive heart failure (CHF), left ventricular remodeling, and cardiomyopathy [15,16]. The detailed mechanisms behind DOX-induced cardiac injury have not been elucidated, but it is possibly involved with several paths. Previous studies reported that DOX-induced cardiotoxicity involved the production of oxidative ROS [17]. As DOX enters the body, it binds tightly to cardiolipin present in the inner mitochondrial sheath [18], accumulates in mitochondria, and affects the electron transport chain creation of ROS and RNS [19]. Later, they aggravate mitochondrial and cellular membrane damage and diminish the antioxidant defense system [20,21], subsequently leading to cell apoptosis [21].

Mitochondrial damage can also initiate intracellular Ca^{2+} imbalance [22], which further affects the apoptosis paths and causes myocardial cell death [23]. DOX also interferes with iron regulation [24], up-regulates NF-κB expression, which consequently causes the release of pro-inflammatory cytokines, i.e., tumor necrosis factor-alpha (TNF-α), interleukin-1 beta (IL-1β), and interleukin-6 (IL-6), and triggers vascular and cardiac inflammatory reaction [25] and exaggerates their downstream apoptotic pathways [26]. In addition, oxidative stress activates several pro-fibrogenic factors, which enhances the accumulation of extracellular matrix and development of cardiac fibrosis [27], remodeling [28], and eventual cardiac dysfunction [29]. DOX is still in use. To remain an efficient anticancer medication, it is essential to find appropriate new therapeutic agents to serve as adjuvants to mitigate DOX-induced cardiotoxicity. Several therapeutic strategies were developed to minimize DOX-induced oxidative injury, inflammation, DNA damage, and apoptosis. However, most of them interfere with DOX's therapeutic effects, limiting their clinical use for cardioprotection against DOX-induced cardiotoxicity [30] and up till now no specific effective and safe drugs have been found.

Phytomedicine is one of the strategies that focus on chemical substances naturally present in plants to improve health conditions and prevent, manage, and treat many diseases. Plant phenolics are natural antioxidant agents that embrace an electron that forms comparatively stable phenoxyl radicals and, consequently, interrupt the redox reactions within the cells [31]. They were also found to activate a cellular redox defense mechanism by stimulating endogenous antioxidant fractions [32] and keep the cells from xenobiotic oxidative stress, DNA impairment, and apoptosis [33]. Much attention has been given to using plant chemicals as a defensive strategy to resolve cardiotoxicity triggered by DOX [34–36].

Date palm pollen (DPP) is a powder formed from date palm (*Phoenix dactylifera* L.) male reproductive cells. It has been utilized by the initial Chinese and the primeval Egyptians as a regenerating factor and worldwide as a dietary supplement [37,38]. Yearly, approximately one thousand tons of DPP are created in Arabic areas [39]. DPP is rich in many health-promoting factors, i.e., flavonoids and volatile unsaturated fatty acids [40,41]; that have strong antioxidant properties in scavenging free radicals [41,42]. In addition, DPP has anti-inflammatory, anti-coccidial, aphrodisiac, anti-apoptotic actions, and is a hepatoprotective agent [43–46]. Egyptian DPP has been proven to have a vast range of nutritionally and biochemically bioactive constituents, i.e., essential and non-essential amino acids, nucleic acids, different carbohydrates, trace elements, minerals, and vitamins. It also contains important phenolic compounds, including gallic, caffeic, coumaric, cinnamic, ferulic acids, catechin, rutin, quercetin, and naringenin propyl gallate. It also contains saturated (arachidic, capric, lauric, myristic, palmitic, and stearic) and unsaturated (arachidonic, linoleic, linolenic, oleic and palmitoleic) fatty acids, ω3, ω6 [47] and a lot of enzymes and cofactors [38,48]. Furthermore, Egyptian DPP has estrogenic substances, i.e., estriol, estradiol (E2), and estrone, which were recognized to alleviate male subfertility problems through their gonadotrophic activity [49]. This study was intended to explore the potential ameliorative effect of date palm pollen ethanolic extract (DPPE) against DOX-induced cardiotoxicity and the mechanisms underlying it.

2. Materials and Methods

2.1. Chemicals, Kits, and Reagents

Doxorubicin HCl (Adricin®) injectable solution was procured from EIMC United Pharmaceuticals (Badr City, Cairo, Egypt). Commercially available kits for the measurement of ALT, AST, LDH, CK, CK-MB, GSH, GPx, CAT and SOD pursuits, and NO and MDA contents were obtained from Biodiagnostic Co. (Cairo, Egypt). Rat specific ELISA kits for cTnI and cTnT levels were gained from MyBiosource, Inc. (San Diego, CA, USA). ELISA kit for NT-proBNP was obtained from CUSABIO (Hubei, China). A commercially available colorimetric kit for Ca^{+2} was purchased from Elabscience Co. (Houston, TX, USA). Rats-specific ELISA kits for IL-6, IL-1β, and TNF-α were bought from BD Biosciences (San Jose, CA, USA). Masson's trichrome was purchased from (Sigma Aldrich, St. Louis, MI, USA). NF-κB p65 total ELISA Kit, hydroxyproline colorimetric assay kit, hematoxylin and eosin stain (H&E), rabbit polyclonal anti-TGF-β1 antibody (Product# ab25121), rabbit polyclonal anti-cleaved caspase-3 antibody (Product# ab4051) and rabbit monoclonal anti-Bax antibody E63 (Product# ab32503) were purchased from Abcam Co. (Cambridge Science Park, Cambridge, UK). Rabbit polyclonal anti-Bcl-2 antibody (Product# PA5-27094) was obtained from Thermo Fisher Scientific Co. (Waltham, MA, USA).

2.2. Date Palm Pollen Grains Collection and Ethanolic Extract Preparation

Date palm pollen grains were gathered from *Phoenix dactylifera* L. in March 2020 from El-Beheira, Egypt. They are authenticated at the Department of Botany, Faculty of Science, Alexandria University. After collection, the pollen grains were dissected from the bark and washed with water, dried with air, and ground at room temperature using a grinder to fine powder kept at 4 °C until use.

Two hundred grams of DPP powder was extracted twice with 1600 mL of 80% ethanol for 24 h at room temperature. The extract was filtered in a Buchner funnel and then centrifuged at 5000 radius centrifugation force (RCF) for 30 min. The obtained supernatant was evaporated at 40 °C in a rotary evaporator under vacuum till complete dryness; then, the final dry extract and stock solution was preserved in dark glass bottles in the refrigerator at 4 °C for further analysis. The DPPE was re-dispersed in distilled H_2O and orally intubated to treated rats using an intragastric tube at the time of experimentation.

2.3. Acute Oral Toxicity of DPPE "Median Lethal Dose, LD_{50}"

Acute toxicity trial was carried out following the guidelines of the Organization for Economic Co-operation and Development [50] to evaluate the acute oral hazard of DPPE. The 'Limit Test' in the up and down procedure (UDP) was conducted to reduce the overall animals' suffering. A maximum of 5 male Wister albino rats per group was administered sequentially with DPPE up to a test dose of 5000 mg/kg bw. Twenty-five adult male Wister albino rats (180 ± 10 g bw) were allocated randomly into 5 groups (5 rats each) and were acclimatized for 7 days. The rat groups were fasted overnight and the next morning, freshly prepared DPPE was orally administrated to groups 1–5 at doses of 1, 2, 3, 4, and 5 g/kg bw, respectively. The limit test involved dosing an animal with up to 5 g/kg bw. If the animal managed to survive, two extra animals were dosed. If both animals stayed alive, the LD_{50} was supposed to be higher than the limit dose, and the test was finished. The rats were observed every 2 h for 24 h, and again at 48 and 72 h to record any behavioral changes, signs of toxicity, and mortalities. The survived animals were observed for any delayed toxic signs or death for the next 14 days.

2.4. Animals Experimentation and Sampling

Forty adult male Wister albino rats weighing approximately 190 ± 10 g (10 weeks-old) were purchased from the Medical Research Institute, Alexandria University, Egypt. The rats were kept in stainless-steel boxes at controlled environment "temperature 25 ± 5 °C and humidity 55 ± 5%" with a 12 h light/dark cycle and free access to standard rat feed (El-Nasr Co., Cairo, Egypt) and water for 2 weeks before the experiment to follow-up normal growth and behavior. The animals were given humane treatment in compliance with the Institutional and National Procedures for the Care and Use of Experimental Animals (NIH). They were declared by the Local Committee of the Faculty of Veterinary Medicine, Alexandria University (Ethical Committee Approval Number: 2020/013/59) and ethical approval of Taif University (42-0081).

After acclimatization, the rats were randomly distributed into 4 equal groups (10 rats each). Group I (CTR) rats weighing approximately (192 ± 3.6) were orally intubated with 1 mL distilled water using a stomach tube daily for about 4 weeks. Group II (DPPE-treated) rats weighing approximately (190 ± 8.1) were orally intubated with DPPE at a dose of 0.5 g/kg bw daily for 4 weeks. Groups I and II were also intraperitoneally injected with 0.5 mL isotonic saline solution six times over the last two weeks of the experiment. Group III (DOX-treated) rats weighing approximately (195 ± 5.2) were orally intubated with 1 mL distilled water for 4 weeks and were DOX injected intraperitoneally at a dose of 2.5 mg/kg bw six times over the last two weeks of the experiment [51] with an accumulative dose of 15 mg/kg bw. Group IV (DPPE + DOX) weighing approximately (194 ± 4.5) rats obtained DPPE and DOX at the same dosage used in groups II and III. DPPE was administrated to rats an hour before DOX administration (Figure 1).

At the end of the experimentation, the animals were only allowed free access to water and fasted for 12 h. After that, they were weighed, and blood samples were obtained just before euthanasia from the retro-orbital plexus of the inner eye canthus under diethyl ether anesthesia. The collected blood was centrifuged for 10 min at 3000 rpm, and then the resulting sera samples were kept at −20 °C for further analysis. Subsequently, rats were euthanized by cervical dislocation. The heart was rapidly harvested, rinsed with saline, dried, weighted, and dissected. The cardiac specimens were immediately frozen and kept at −80 °C. In an ice-cold phosphate buffer saline, the frozen samples were thawed and homogenized. (0.1 M pH 7.4) utilizing a homogenizer with a Teflon pestle and then centrifuged at 5000× g for 15 min. at 4 °C. Aliquots of the supernatant were frozen at −80 °C for the chemical analysis. In neutral buffered formalin 10% solution, other heart specimens were immediately fixed for the histopathological and immunohistochemical assessment.

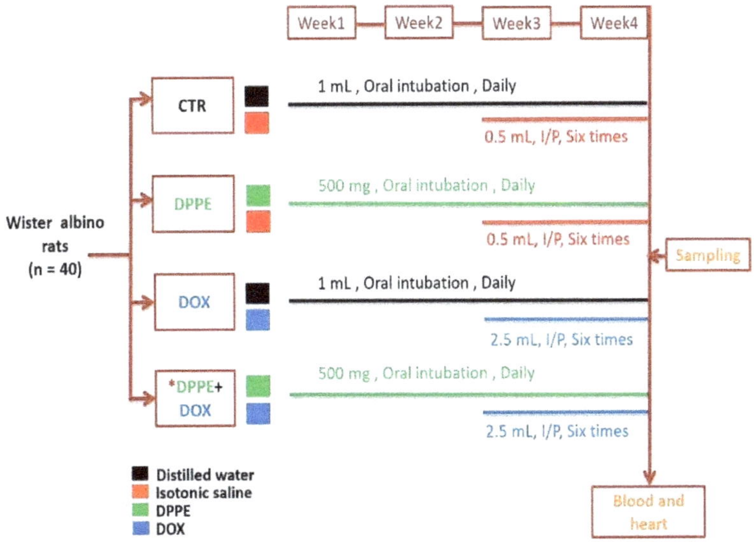

Figure 1. A schematic overview of the experimental protocol. CTR, Control group; DPPE, date palm pollen ethanolic extract-treated group; DOX, doxorubicin-treated group, DPPE + DOX, date palm pollen ethanolic extract- and doxorubicin treated group. * DPPE was given an hour before DOX administration.

2.5. Assessment of the Body, Heart, and Relative Heart Weights

At the end of the experimentation, in each rat, the body and heart weight were recorded. The relative heart weights (RHW) were estimated using the following formula:

$$\text{RHW} = \frac{\text{Heart weight (g)}}{\text{Bodyweight (g)}} \times 100$$

2.6. Assessment of Cardiac Injury Biomarkers and Cytosolic Calcium (Ca^{+2})

The serum ALT, AST, LDH, CK, and CK-MB levels were estimated as instructed by the manufacturers. The serum cTnI, cTnT, and NT-proBNP (a marker of heart failure) concentrations were also measured using the corresponding enzyme-linked immunosorbent assay (ELISA) kits using ELISA Plate Reader (Bio-Rad, Hercules, CA, USA). The supernatant obtained by centrifugation of cardiac tissue homogenate was used to evaluate the concentration of cytosolic Ca^{2+} using the Ca^{2+} colorimetric assay kit, as instructed by the manufacturer.

2.7. Estimation of Cardiac Nitro-Oxidative Stress and Lipid Peroxidation

The concentration of nitric oxide (NO) was assessed in the supernatants of the cardiac homogenates based on the enzymatic reduction of nitrate to nitrite. For nitrite detection, the colored azo dye product "Griess reaction" was spectrophotometrically monitored at 550 nm absorbance [52]. The levels of MDA [53] and GSH [54] and the activities of GPx [55], CAT [56], and SOD [57] were spectrophotometrically estimated in the cardiac tissue homogenates. The total protein content was also assessed [58].

2.8. Assessment of Inflammatory Markers

The cardiac total NF-κB p65, TNF-α, IL-1β, and IL-6 were measured in the supernatant obtained by centrifugation of cardiac tissue homogenate using the corresponding rat-specific ELISA kit following the manufacturer's protocols.

2.9. Estimation of Hydroxyproline Content

Briefly, about 100 mg of the right ventricle was homogenized in double-distilled water. In a tightened screw-capped polypropylene vial, the tissue homogenates were mixed with conc NaOH (10 N) and then boiled for an hour at 120 °C. The alkaline lysate was ice-cooled, neutralized to pH 7.0, and centrifuged to get off supernatants. The hydrolysates were hot air-dried, chloramine T-oxidized, and finally reacted with Ehrlich's reagent. The resultant colored product was measured at 560 nm absorbance, and the amount of hydroxyproline content was detected by comparing it with a standard curve [59].

2.10. Histopathological Assessment and Semi-Quantitative Scoring Approach

Cardiac samples were immediately fixed in phosphate-buffered formalin (10%, pH 7.4) after necropsy for 24 h, then were handled using the conventional paraffin embedding method. The 5 μm thick pieces were cut and placed on slides, deparaffinated in xylene, and rehydrated using decreasing concentrations of ethanol. One set of slides was hematoxylin and eosin (H&E)-stained for the routine histopathological setting. An additional set was Masson's trichrome-stained for detecting the amount and distribution of collagen fibers [60]. Stained sections were blindly examined using light microscopes and photographed using a digital camera at a magnification of 400× (Nikon Corporation Co., Ltd., Tokyo, Japan).

To convey the occurrence and severity of the histopathological lesions, a semi-quantitative scoring approach was used. In each animal group, seven H&E-stained slides (one slide/rat) were examined, and 10 random fields per slide were used for grading the various pathological lesions in a blinded fashion. The severity of pathological lesions was asessed according to the percentage of tissue affected in the entire section as None (−): normal histology with zero immersion of the inspected field, Mild (+): 5–25% of the tested field was involved, Moderate (++): 26–50% of the inspected field was involved, Severe (+++): >50% of the examined field was applied. The incidence represented the number of lesion rats per total examined.

2.11. Immunohistochemical Assessment

According to Hsu, et al. [61], the immunodetection was assessed using four overlapping paraffin-embedded cardiac tissue sections. Sections were sliced at 4 μm thicknesses utilizing a microtome and put-on slides that are positively charged. Then, the sections were deparaffinized, rehydrated in xylene, then in different graded ethanol solutions and underwent antigen repossession using sodium citrate buffer (10 mM, pH 6.0) in the microwave at 105 °C for 10 min. Then, the activity of endogenous peroxidase was inhibited with 3% H_2O_2 for 10 min.; the non-specific proteins were blocked with 5% goat serum for 30 min at room temperature. The cardiac tissue slices were washed thrice in Dako Tris-buffered saline (TBS) and then incubated with the specific rabbit primary antibodies: polyclonal anti-TGF-β1 (dilution 1/200), anti-cleaved caspase-3 (dilution1/100), monoclonal anti-Bax (dilution1/250), and anti-Bcl-2 (dilution 1/100) at 4 °C overnight. In the negative control sections, normal IgG was substituted for the primary antibodies at the same concentration and antibody species. Following PBS washing, the tissue sections were incubated for an hour with goat anti-rabbit biotin-labeled secondary antibody, rinsed in PBS for 2 min. Then the sections were incubated with streptavidin-horseradish peroxidase reagent (VECTASTAIN1 Elite ABC kit, Vector Laboratories, Inc., Burlingame, CA, USA) at 37 °C for 20 min then washed with rinsing buffer and incubated with 3,3-diaminobenzidine tetrahydrochloride (DAB Substrate Kit, Thermo Fischer Scientific, Rockford, IL, USA) as the chromogen to developed peroxidase reaction. The sections were finally objected to Mayer's Hematoxylin to augment the nuclear staining and mounted with di-poly cysteine xylene (DPX). All slides were assessed blindly and photographed using a digital camera.

2.12. Histomorphometric Assessment

The assessment was performed using H&E, Masson's trichrome, and immunostained cardiac sections (one section from each rat and seven per group). The digital images (ten different fields per section at ×400 magnification power) were blindly analyzed using image analysis software (ImageJ Version 1.47, National Institutes of Health, Bethesda, MD, USA, wayne@codon.nih.gov. The ten values were averaged in each animal, and the average was used as individual sampling data.

Using H&E-stained sections, the cross-sectional area of cardiomyocytes was assessed in the left ventricular wall. Fifteen cardiomyocytes with a visible nucleus and intact cellular membrane were selected per field for the measurement and analysis [62].

Using Masson's trichrome-stained sections [63], the collagen volume fraction (CVF %) and perivascular collagen area (PVCA %) percentage were estimated as per the following formulas:

$$\text{CVF } (\%) = \frac{\text{Collagen area}}{\text{Total area}} \times 100;$$

$$\text{PVCA } (\%) = \text{area occupied by the collagen/total area of the vessel section} \times 100\,.$$

Using images of immunostained slides, the area percentage (%) of TGF-β1, cleaved caspase-3, Bax, and Bcl-2 immunopositive cardiomyocytes were estimated as area percent (%) across ten different fields/sections [64].

2.13. Data Analysis

One-way Variance Analysis evaluated the numerical data estimation (ANOVA)test using SPSS data analysis software (Version 21; SPSS Inc., Chicago, IL, USA) and summarized it as means ± (SEM). Tukey's post-hoc test was used to ascertain the statistical difference between experimental groups. * $p < 0.05$ was set as statistically significant.

3. Results

3.1. Median Lethal Dose, Mortality, and Survival Rates

Rats of DPPE groups did not exhibit any behavioral changes, toxic side-effects, or even mortalities after 24 h and 14 days post-treatment. Thus, dosing was ceased at 5 g/kg bw. Consequently, the LD_{50} of DPPE was evaluated to be more than 5 g/kg bw. The CTR and DPPE group did not exhibit any mortality all over the experimental period. However, the DOX group showed a scruffy appearance and exhibited 30% mortality (three dead rats), and DPPE + DOX group exhibited 10% mortality (one dead rat) (the data not shown).

3.2. Body, Heart, and Relative Heart Weights and Myocyte Cross-Sectional Area

As demonstrated in Table 1, CTR and DPPE groups exhibited a statistically non-significant change in the body weight, heart weight, RHW, and myocyte cross-sectional area. Conversely, the DOX group displayed a significant reduction in body weight (≈0.86-fold) and a substantial rise in the heart weight, RHW, and the myocyte cross-sectional area (1.6, 1.7, and 1.52-fold, respectively) compared to CTR values. Nevertheless, DPPE + DOX group exhibited a non-substantial rise in body weight (≈1.08-fold) and a considerable reduction in heart weight (≈0.76-fold), RHW (≈0.71-fold), and the myocyte cross-sectional area (≈0.73-fold) compared to DOX group values. In comparison with the CTR group values, DPPE + DOX group expressed a non-significant reduction in body weight (≈0.93-fold), a significant rise in the heart weight (≈1.22-fold), a non-significant increase in the RHW (≈1.2-fold), and a substantial increasing in the cardiomyocyte cross-sectional area (≈1.22-fold).

Table 1. Effect of date palm pollen ethanolic extract (500 mg/kg bw/day) and/or doxorubicin (2.5 mg/kg bw/day) on body, heart, and relative heart weights myocyte cross-sectional area of control and treated rats.

Groups	Initial Body Weight (g)	Body Weight (g)	Heart Weight (g)	RHW (%)	Cardiomyocyte Cross-Sectional Area (μm^2)
CTR	194 ± 4.5	218.43 ± 4.86	0.85 ± 0.013	0.52 ± 0.07	138.03 ± 3.90
DPPE	195 ± 5.2	221.85 ± 6.51	0.87 ± 0.023	0.49 ± 0.06	135.40 ± 2.65
DOX	190 ± 8.1	189.14 ± 8.36 *	1.36 ± 0.047 *	0.88 ± 0.11 *	210.87 ± 5.26 *
DOX + DPPE	192 ± 3.6	204.43 ± 2.72	1.04 ± 0.061 †	0.63 ± 0.09	154.94 ± 2.06 †

Control (CTR); date palm pollen ethanolic extract (DPPE); doxorubicin-treated (DOX); doxorubicin and date palm pollen ethanolic extract-treated (DOX + DPPE) groups, and relative heart weight (RHW). (n = 7). Each value is the average of 7 observations. Values are mean ± standard error (SEM). Mean values within the same columns were statistically different from CTR and DPPE (* $p < 0.05$). CTR, DPPE and DOX († $p < 0.05$).

3.3. *Cardiac Injury Biomarkers and Cardiac Cytosolic Calcium (Ca^{2+})*

As demonstrated in Table 2, the CTR and DPPE groups disclosed a statistically non-significant change in the activities of ALT, AST, LDH, CK, CK-MP, and the levels of cTnI, cTnT, NT-pro BNP, and cardiac Ca^{2+}. Meanwhile, the DOX group exhibited a significant increase in ALT (≈1.52-fold), AST (≈1.42-fold), LDH (≈2.49-fold), CK (≈2.8-fold) and CK-MP (≈2.03-fold) activities, and cTnI (≈5-fold), cTnT (≈3.44-fold), NT-pro BNP (≈2.38-fold), and cardiac Ca^{2+}(≈1.74-fold) levels as compared to CTR values. In contrast, DPPE + DOX group demonstrated a significant reduction in ALT (≈0.79-fold), AST (≈0.85-fold) LDH (≈0.73-fold), CK (≈0.7-fold) and CK-MP (≈0.72-fold) activities, and cTnI (≈0.57-fold) fold, cTnT (≈0.54-fold), NT-pro BNP (≈0.63-fold) and cardiac Ca^{2+} (≈0.82-fold) levels, as compared to DOX values. In addition, this group showed a significant increase in the activities of ALT (≈1.2-fold), AST (≈1.22-fold), LDH (≈1.83-fold), CK (≈1.97-fold), and CK-MP (≈1.48-fold), and the levels of cTnI (≈2.89-fold), cTnT (≈1.84-fold) NT-pro BNP and cardiac Ca^{2+}(≈1.43-fold), as compared to the CTR values.

Table 2. Effect of date palm pollen ethanolic extract (500 mg/kg bw/day) and/or doxorubicin (2.5 mg/kg bw/day) on serum cardiac injury biomarkers and cardiac cytosolic calcium of control and treated rats.

Parameters	CTR	DPPE	DOX	DPPE + DOX
ALT (U/L)	69.26 ± 1.98	66.39 ± 4.25	105.09 ± 3.71 *	83.71 ± 3.04 †
AST (U/L)	143.99 ± 4.74	140.25 ± 5.32	203.97 ± 6.98 *	175.36 ± 6.27 †
LDH (U/L)	496.99 ± 18.28	485.76 ± 31.16	1237.62 ± 65.72 *	908.19 ± 9.61 †
CK (U/L)	218.63 ± 17.16	203.71 ± 5.81	612.92 ± 7.97 *	431.50 ± 17.14 †
CKMP (U/L)	458.99 ± 24.39	440.86 ± 32.52	934.56 ± 41.55 *	682.14 ± 37.44 †
CTnI (pg/mL)	0.37 ± 0.05	0.30 ± 0.05	1.85 ± 0.38 *	1.07 ± 0.13 †
CTnT (pg/mL)	0.91 ± 0.018	0.87 ± 0.03	3.15 ± 0.17 *	1.69 ± 0.15 †
NT- ProBNP (pg/mL)	63.45 ± 3.73	57.31 ± 2.089	151.17 ± 3.44 *	96.55 ± 4.02 †
Ca^{+2} (µg/g tissue)	32.80 ± 3.60	31.81 ± 2.95	77.35 ± 5.54 *	54.03 ± 5.17 †

Control (CTR); date palm pollen ethanolic extract (DPPE); doxorubicin-treated (DOX); doxorubicin and date palm pollen ethanolic extract-treated (DOX + DPPE) groups; alanine aminotransferase (ALT); aspartate aminotransferase (AST); lactate dehydrogenase (LDH); creatine kinase (CK); creatine kinase-cardiac type isoenzyme (CK-MB); cardiac troponin I (CTnI); cardiac troponin T (CTnT); N-terminal pro-brain natriuretic peptide (NT-proBNP) and cardiac cytosolic calcium (Ca^{+2}). (n = 7). Each value is the average of seven observations Values are mean ± standard error (SEM). Mean values within the same lines were statistically different from CTR and DPPE (* $p < 0.05$). CTR, DPPE and DOX († $p < 0.05$).

3.4. *Cardiac Nitro-Oxidative Stress and Lipid Peroxidation*

As demonstrated in Table 3, the CTR and DPPE groups revealed a non-significant change in the concentrations of NO, lipid peroxidation marker (MDA), and antioxidant parameters (GSH level and GPx, CAT, and SOD activities). Meanwhile, the DOX group disclosed a statistically substantial increase in the quantities of NO (≈3.61-fold), MDA (≈1.86-fold), and a significant reduction of the GSH level, and GPx, CAT, and SOD pursuits (≈0.46, 0.31, 0.54 and 0.31-fold, respectively) compared to the CTR values. Quite the opposite, the DPPE + DOX group demonstrated a considerable drop in the levels of NO

(≈0.58-fold) and MDA (≈0.69-fold) and a significant rise in the GSH level and GPX, CAT, and SOD activities (≈1.67, ≈2.18, ≈1.41 and ≈2.28-fold, respectively) equated to the DOX group values. Compared to the CTR group values, DPPE + DOX group showed a significant rise in the levels of NO (≈2.11–fold), MDA (≈1.29–fold), and a considerable reduction in the GSH level, and GPx, CAT, and SOD pursuits (≈ 0.77, 0.68, 0.77 and 0.72-fold, respectively).

Table 3. Effect of date palm pollen ethanolic extract (500 mg/kg bw/day) and/or doxorubicin (2.5 mg/kg bw/day) on cardiac nitro-oxidative stress lipid peroxidation in control and treated rats.

Parameters	CTR	DPPE	DOX	DPPE + DOX
NO (μmol/g tissue)	4.84 ± 0.93	3.32 ± 1.09	17.49 ± 2.45 *	10.25 ± 1.07 †
MDA (nmol/g tissue)	37.35 ± 2.41	34.48 ± 2.73	69.52 ± 3.52 *	48.3726 ± 2.29 †
GSH (mmol/g tissue)	25.27 ± 2.96	28.71 ± 1.65	11.73 ± 1.55 *	19.63 ± 0.85 †
GPx (U/g tissue)	17.23 ± 1.48	19.08 ± 2.21	5.42 ± 1.26 *	11.84 ± 1.48 †
CAT (U/g tissue)	34.61 ± 2.19	37.64 ± 2.38	18.92 ± 2.44 *	26.75 ± 2.70 †
SOD (U/g tissue)	13.71 ± 1.318	15.45 ± 1.21	4.31 ± 1.22 *	9.85 ± 1.51 †

Control (CTR); date palm pollen ethanolic extract (DPPE); doxorubicin-treated (DOX); doxorubicin date palm pollen ethanolic extract-treated (DOX + DPPE) groups; nitric oxide (NO); malondialdehyde (MDA); reduced glutathione (GSH); glutathione peroxidase (GPx); catalase (CAT) and superoxide dismutase (SOD). (n = 7). Each value is the average of 7 observations Values are mean ± standard error (SEM). Mean values within the same lines were statistically different from CTR and DPPE (* $p < 0.05$). CTR, DPPE and DOX († $p < 0.05$).

3.5. Inflammatory Markers

As shown in Table 4, the CTR and DPPE groups displayed a statistically non-significant change in the NF-κB p65, TNF-α, IL-1β, and IL-6 concentrations. Meanwhile, the DOX group exhibited a significant rise in the quantities of NF-κB p65 (≈2.63-fold), TNF-α (≈2.35-fold), IL-1β (≈1.74-fold), and IL-6 (≈1.85-fold) equated to CTR values. Conversely, DPPE + DOX group disclosed a significant reduction in the concentrations of NF-κB p65 (≈0.72-fold), TNF-α (≈0.63-fold), IL-1β (≈0.82-fold), and IL-6 (≈0.74-fold) compared to the DOX group values. Compared with CTR values, DPPE + DOX group exhibited a significant rise in the NF-κB p65 (≈1.9-fold), TNF-α (≈1.48-fold), IL-11β (≈1.43-fold), and IL-6 (≈1.36-fold) levels.

Table 4. Effect of date palm pollen ethanolic extract (500 mg/kg bw/day) and/or doxorubicin (2.5 mg/kg bw/day) on cardiac inflammatory biomarkers hydroxyproline content in control and treated rats.

Parameters	CTR	DPPE	DOX	DPPE + DOX
NF-κB p65 (ng/g tissue)	90.23 ± 3.5	83.65 ± 3.8	238.09 ± 22.5 *	172.11 ± 14.8 †
TNF-α (pg/g tissue)	34.29 ± 1.65	32.89 ± 1.89	80.75 ± 5.87 *	50.67 ± 4.81 †
IL-1β (pg/g tissue)	79.81 ± 1.6	81.54 ± 1.7	138.85 ± 3.23 *	114.24 ± 3.65 †
IL-6 (pg/g tissue)	47.20 ± 2.66	44.13 ± 1.71	87.63 ± 2.93 *	64.64 ± 2.21 †
Hydroxyproline (μg/g tissue)	22.67 ± 2.03	20.61 ± 1.92	46.33 ± 2.21 *	38.3 ± 2.87 †

Control (CTR); date palm pollen ethanolic extract (DPPE); doxorubicin-treated (DOX); doxorubicin date palm pollen ethanolic extract-treated (DOX + DPPE) groups; nuclear factor- kappa B (NF-κB p65); tumor necrosis factor-alpha (TNF-α), interleukin-1β (IL-1β) and interleukin-6 (IL-6). (n = 7). Each value is the average of seven observations Values are mean ± standard error (SEM). Mean values within the same lines were statistically different from CTR and DPPE (* $p < 0.05$). CTR, DPPE and DOX († $p < 0.05$).

3.6. Cardiac Hydroxyproline Content

As demonstrated in Table 4, the CTR and DPPE groups displayed a non-significant change in the hydroxyproline concentrations. Meanwhile, the DOX group exhibited a substantial rise of about 2-fold when equated with the CTR quantity. In contrast, the DPPE + DOX group displayed a significant reduction (≈0.83-fold) compared to DOX values and a considerable enhancement (≈1.6-fold) compared to the CTR value.

3.7. Histopathological Results and Lesions Scoring

Figure 2 demonstrated the histomorphological results of H&E-stained cardiac tissue sections. Table 5 also illustrated the prevalence and severity of the identified pathological lesions in various treatments.

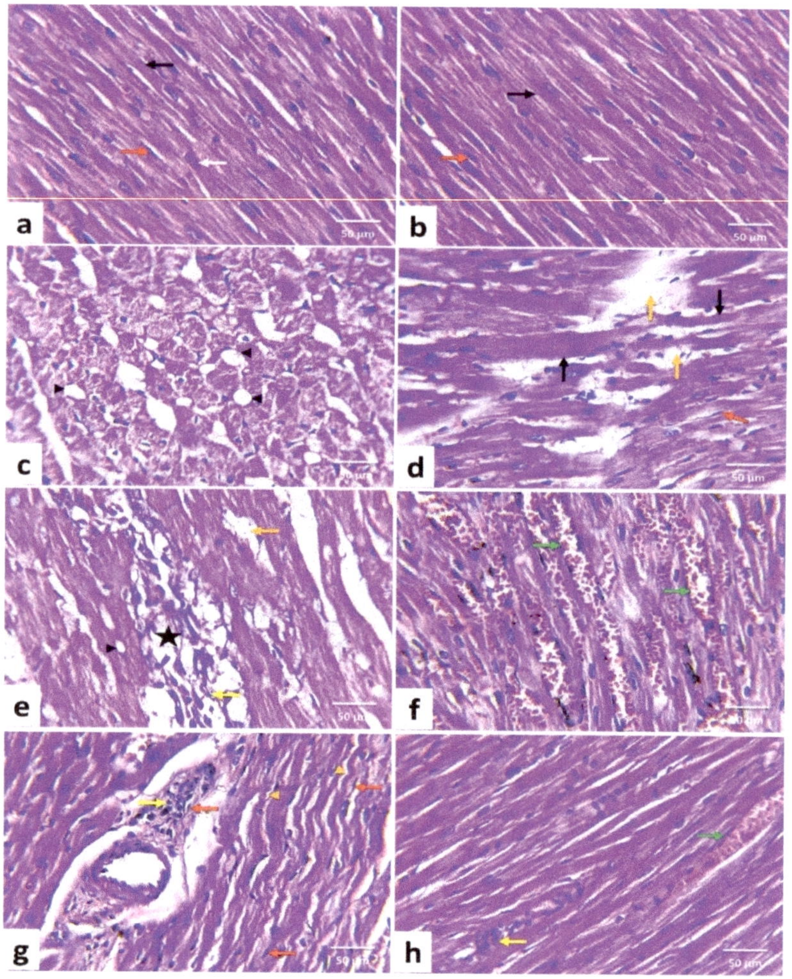

Figure 2. Histopathological changes of rats' cardiac tissues (H&E, ×400). A rat from the control (**a**) and a rat from date palm pollen ethanolic extract-treated (**b**) groups, respectively showing normal histoarchitecture of the cardiomyocytes with well-organized and branched cardiac myofibers (black arrow), centrally located oval nuclei (white arrow), and minimal interstitial connective tissue with few interstitial fibroblasts (red arrow) in between. Doxorubicin-treated rats (**c**–**g**) showing vacuolization of the sarcoplasm (black arrowhead), Zenker's necrosis (black arrow), wavy muscle fibers (orange arrowhead), loss of myofibrils (orange arrow), myocardial necrosis (star), mononuclear inflammatory cells infiltrations (yellow arrow), fibroblasts proliferation (red arrow), and hyperemic interstitial blood vessels (green arrow). DPPE + DOX-treated rat (**h**) showing marked improvement in muscle fibers striation, which almost looks like the control. However, minute areas of myocardial necrosis with inflammatory cells infiltration (yellow arrow) and hyperemic interstitial blood vessels (green arrow) are still evident. (n = 7). Each value is the average of seven observations.

Table 5. Incidence and severity of cardiac histopathological lesions in the examined cardiac tissues in the control group and after various treatments in male Wistar albino rats.

Groups (n = 7)	Lesion Severity	Loss of Muscular Striations	Myocardial Vacuolation	Myocardial Necrosis	Myofibrillar Loss	Interstitial Inflammatory Cells Infiltrations	Hyperemic Blood Vessels	Interstitial Edema	Interfibrillar Hemorrhage	Myocardial Fibrosis
CTR	None (−)	6 −85.71%	7 −100%	7 −100%	7 −100%	7 −100%	7 −100%	7 −100%	7 −100%	7 −100%
	Mild (+)	1 −14.28%	0 0%	0 0%	0 0%	0 0%	0 0%	0 0%	0 0%	0 0%
	Moderate (++)	0 0%	0 0%	0 0%	0 0%	0 0%	0 0%	0 0%	0 0%	0 0%
	Severe (+++)	0 0%	0 0%	0 0%	0 0%	0 0%	0 0%	0 0%	0 0%	0 0%
DPPE	None (−)	7 −100%	7 −100%	7 −100%	7 −100%	7 −100%	7 −100%	7 −100%	7 −100%	7 −100%
	Mild (+)	0 0%	0 0%	0 0%	0 0%	0 0%	0 0%	0 0%	0 0%	0 0%
	Moderate (++)	0 0%	0 0%	0 0%	0 0%	0 0%	0 0%	0 0%	0 0%	0 0%
	Severe (+++)	0 0%	0 0%	0 0%	0 0%	0 0%	0 0%	0 0%	0 0%	0 0%
DOX	None (−)	0 0%	0 0%	0 0%	0 0%	0 0%	0 0%	0 0%	0 0%	0 0%
	Mild (+)	0 0%	0 0%	0 0%	1 −14.28%	0 0%	0 0%	1 −14.28%	1 −14.28%	1 −14.28%
	Moderate (++)	2 −28.57%	6 −85.71%	5 −71.42%	6 −85.71%	7 −100%	5 −71.42%	6 −85.71	5 −71.42%	6 −85.71%
	Severe (+++)	5 (71.42%)	1 −14.28%	2 −28.57%	0 0%	0 0%	2 −28.57%	0 0%	1 −14.28%	0 0%
DPPE + DOX	None (−)	0 0%	0 0%	0 0%	1 −14.28%	0 0%	0 0%	1 −14.28%	2 −28.57%	0 0%
	Mild (+)	5 (71.42%)	5 −71.42%	4 −57.14%	4 −57.14%	2 −28.57%	5 (71.42%)	5 −71.42%	3 −42.85%	5 −71.42%
	Moderate (++)	2 −28.57%	2 −28.57%	3 −42.85%	2 −28.57%	5 −71.42%	2 −28.57%	1 −14.28%	2 −28.57%	2 −28.57%
	Severe (+++)	0 0%	0 0%	0 0%	0 0%	0 0%	0 0%	0 0%	0 0%	0 0%

Control (CTR); date palm pollen ethanolic extract (DPPE); doxorubicin-treated (DOX); doxorubicin and date palm pollen ethanolic extract-treated (DOX + DPPE) groups. The severity of pathological lesions in different groups of rats was determined according to the percentage of tissue affected as: None (−): normal histology with zero involvement of the examined field, Mild (+):5–25% of the examined field is involved, Moderate (++): 26–50% of the examined field is involved Severe (+++): ≥50% of the examined field is involved. Incidence is the number of rats with lesions per total examined. (n = 7).

Heart tissues from the CTR (Figure 2a) and DPPE (Figure 2b) groups revealed normal histoarchitecture with well-organized and branched cardiac myofibers. The cardiomyocytes were closely arranged with oval centrally located nuclei, eosinophilic cytoplasm, and cross striations. In addition, minimal interstitial connective tissue and few fibroblasts were noticed. Meanwhile, the DOX group exhibited moderate to severe histological alterations and high lesion scores (Table 5), wherein disoriented cardiac myofibers with wavy appearance were evident. Furthermore, myocardial degenerative changes such as sarcoplasmic vacuolization (Figure 2c), myofibrillar flocculation, and fragmentation were noticed. Many cardiomyocytes showed Zenker's degeneration. Meanwhile, others exhibited Zenker's necrosis. Myocytolysis (Figure 2d) and multifocal zones of myocardial necrosis combined with infiltrations of mononuclear inflammatory cells (Figure 2e) were obvious. Additionally, hyperemia of interfibrillar blood vessels (Figure 1f), perivascular edema with inflammatory cell infiltrations (Figure 2g), intramyocardial edema, fibrin deposition, and focal areas of hemorrhage were noticed. There were interfibrillar infiltrations of active fibroblasts with a hypertrophic nucleus and marked myocardial and perivascular fibrosis. On the contrary, the DPPE + DOX group displayed a marked enhancement in cardiac tissue structure and integrity. Nevertheless, they were not identical to the CTR limits. Compared with the DOX group, the previously noted lesions were less in severities and distribution in the DPPE + DOX group (Figure 2h, Table 5).

3.8. Masson's Trichrome Staining and Histomorphometric Findings

As illustrated in Figure 3, the cardiac tissue sections of the CTR (Figure 3a,e) and DPPE (Figure 3b,f) groups displayed normal spreading of greenish delicate collagen fibers in between the cardiomyocyte fibers and around the intramyocardial coronary vessels. They also revealed non-significant changes ($p > 0.05$) in the mean CVF% (Figure 3i) and PVCA% (Figure 3j). Conversely, the DOX group (Figure 3c,g) exhibited an apparent increase in the amount of collagen fiber deposition as well the mean CVF % ($\approx$8.5-Fold) and PVCA% ($\approx$3.09-fold) as compared to the CTR values (Figure 3i,j, respectively). However, the DPPE + DOX group revealed a marked reduction in collagen fiber deposition (Figure 3d,h). Meanwhile, the mean CVF % and PVCA% showed significant ($p > 0.05$) decrease with approximately 0.61-fold and 0.68-fold, respectively, when linked to DOX group values and significant ($p > 0.05$) increases with approximately 5.24 and 2.11-fold, correspondingly when equated to the CTR values (Figure 3i,j, respectively).

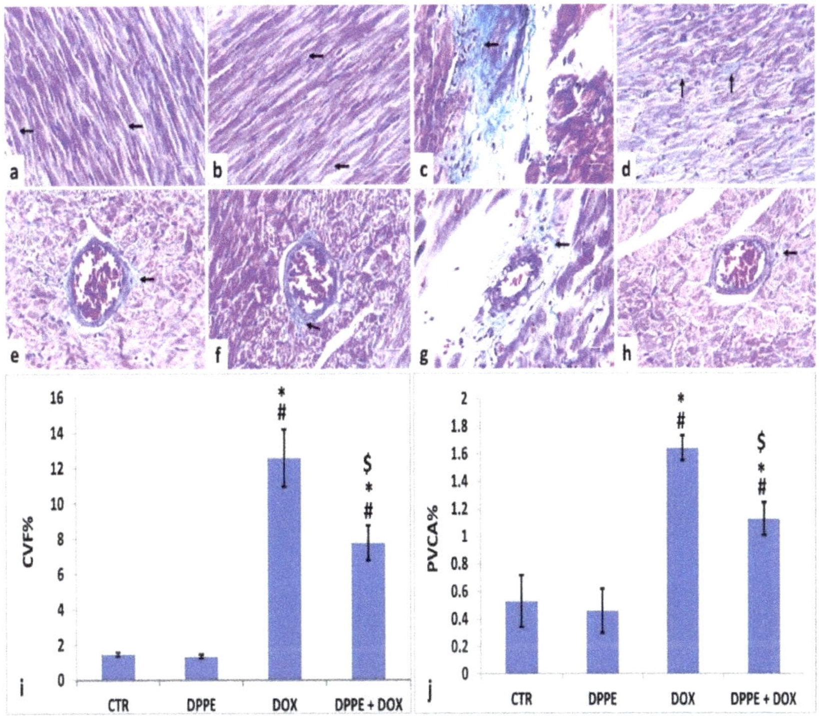

Figure 3. Histopathological changes of rats' cardiac tissues (Masson's trichrome, ×400, arrows: green stained collagen fibers). Rats from the control (**a**,**e**) and DPPE (**b**,**f**) groups, respectively showing scanty collagen fibers deposition. DOX treated rat (**c**,**g**) showing increased collagen deposition. DPPE + DOX-treated rat (**d**,**h**) showed a relative reduction in collagen deposition. Quantification of collagen volume fraction (CVF%), and perivascular collagen area (PVCA%) (**i**,**j**), respectively, ×400 across 10 different fields/section), $n = 7$ rat/group. Mean values were statistically differ from CTR ($^{\#}$ $p < 0.05$), DPPE (* $p < 0.05$), DOX ($^{\$}$ $p < 0.05$) groups.

3.9. Immunohistochemical Analysis

For the expression of TGF-β1 in the cardiac tissue of the control groups, the CTR (Figure 4a) and DPPE (Figure 4b) groups revealed a normal expression of TGF-β1 (brown color). Both groups showed no significant alterations in the mean immune-stained area % of TGF-β1 (Figure 4e). Conversely, DOX-treated rats exhibited a noticeable increase in TGF-β1 expression (Figure 4c) with a substantial ($p < 0.05$) rise of the mean immune-stained area% (≈2.6-fold), as associated with the CTR group value (Figure 4e). However, DPPE + DOX-treated rats exhibited a conspicuous decrease in TGF-β1 expression (Figure 4d) with a substantial ($p > 0.05$) decline in mean immune-stained area% (≈0.7-fold) as compared to the DOX group value and a significant ($p > 0.05$) increase (≈1.84-fold) concerning CTR group quantity (Figure 4e).

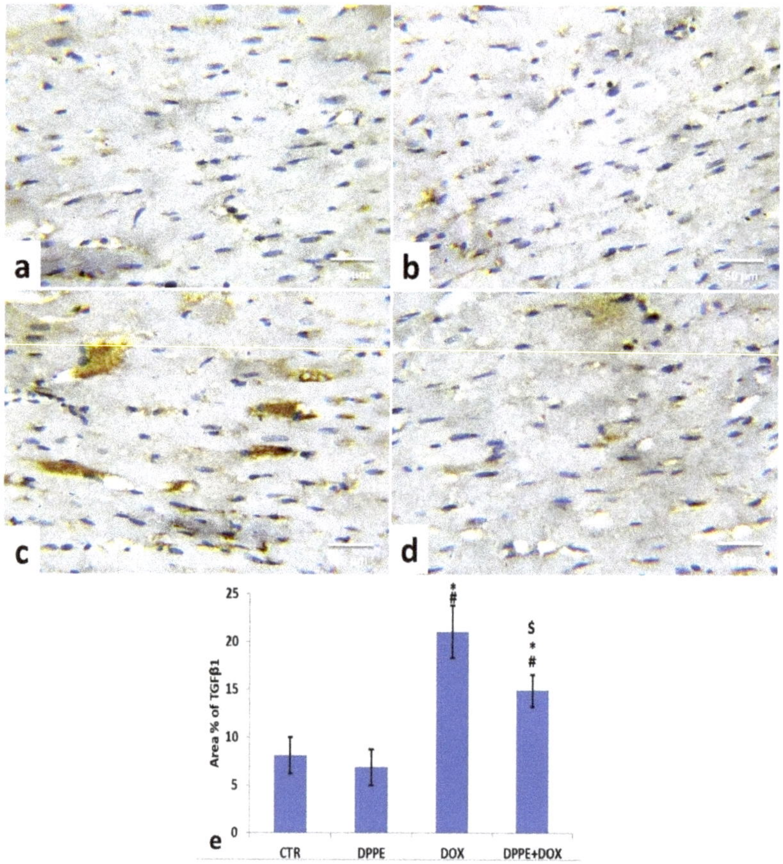

Figure 4. Immunohistochemical staining of transforming growth factor β1 (TGF-β1) in the experimental rats' cardiac cells (IHC, ×400). A control (**a**), DPPE-treated (**b**) DOX-treated (**c**) and DPP + DOX-treated (**d**) rats. (**e**) Quantification of TGF-β1 expression, the immunohistochemical staining of TGF-β1 was measured as area percent (%) across 10 different fields/section, $n = 7$ rat/group. Mean values were statistically differed from CTR ($^{\#}$ $p < 0.05$), DPPE (* $p < 0.05$), DOX ($^{\$}$ $p < 0.05$) group.

Referring to cleaved caspase-3, Bax, and Bcl-2 expressions, the control (Figures 5a, 6a and 7a) and DPPE (Figures 5b, 6b and 7b respectively) groups revealed weak representation of cleaved caspase-3 and Bax. In addition, both groups showed a robust expression of Bcl-2 with dispersed, intensely brown stained immune-reactive cardiomyocytes. There were no significant alterations in the mean area% of cleaved caspase-3, Bax, and Bcl-2 immunostained cells (Figures 5e, 6e and 7e, respectively). Meanwhile, DOX-treated rats' cardiac tissues displayed moderate to strong expression and immune-staining of cleaved caspase-3 (Figure 5c) and Bax (Figure 6c) and weak expression and immune-staining for Bcl-2 (Figure 7c). Relative to the CTR group values, the mean area% of immune-stained cells showed a substantial ($p > 0.05$) rise for cleaved caspase-3 (≈8.95-fold) and BAX (≈2.37-fold) and a significant decrease for Bcl-2 (≈0.3-fold) (Figures 5e, 6e and 7e, respectively).

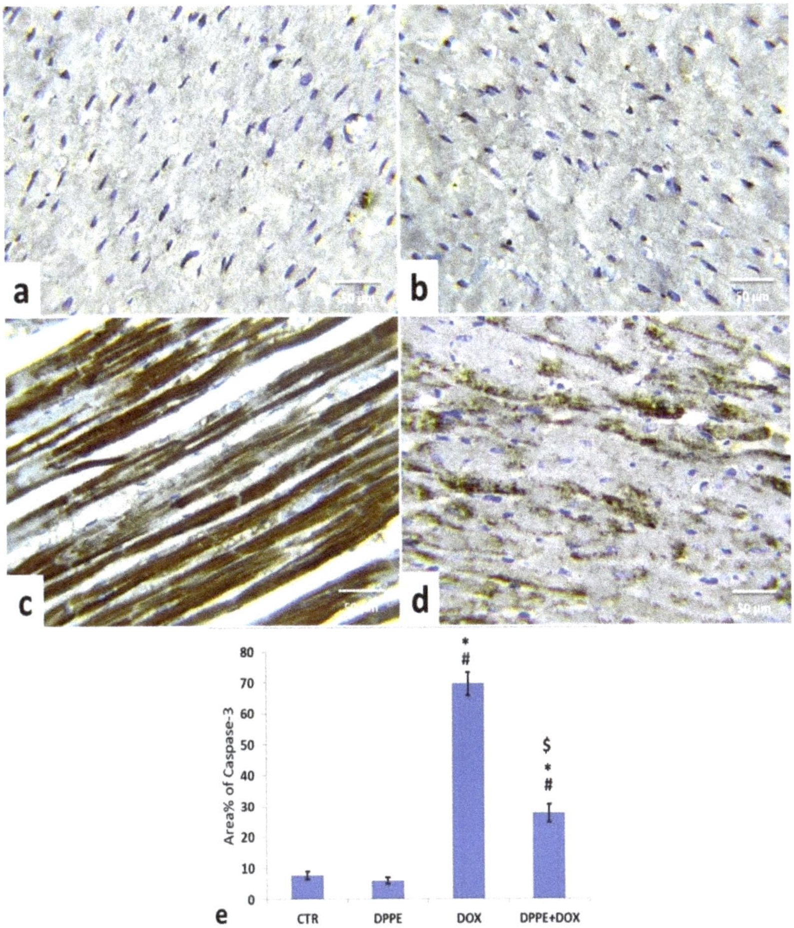

Figure 5. Immunohistochemical staining of cysteine aspartate specific protease-3 (cleaved caspase-3) in the cardiac cells of the experimental rats (IHC, ×400). A control (**a**), DPPE-treated (**b**) DOX-treated (**c**) and DPP + DOX-treated (**d**) rats. (**e**) Quantification of caspase-3 expression, the immunohistochemical staining of cleaved caspase-3 was measured as area percent (%) across 10 different fields/section, $n = 7$ rat/group. Mean values were statistically different from the CTR ($^{\#}$ $p < 0.05$), DPPE (* $p < 0.05$), and DOX ($^{\$}$ $p < 0.05$) groups.

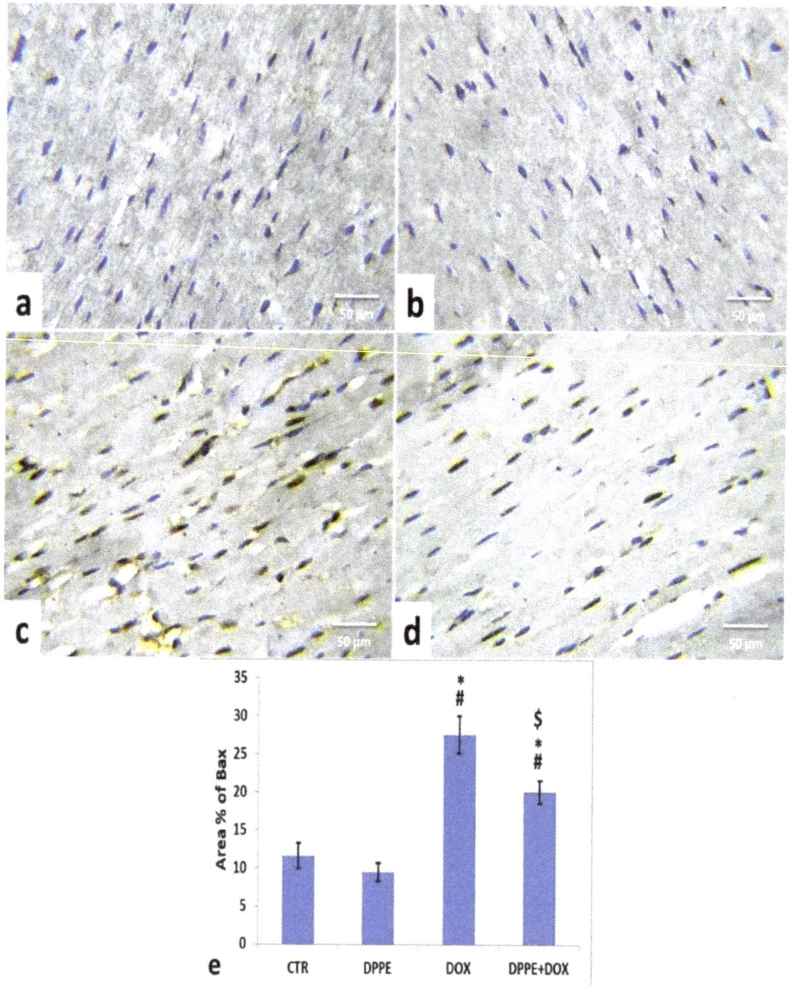

Figure 6. Immunohistochemical staining of Bcl2 associated X protein (Bax) in the experimental rats' cardiac cells (IHC, ×400). A control (**a**), DPPE-treated (**b**) DOX-treated (**c**) and DPP + DOX-treated (**d**) rats. (**e**) Quantification of Bax expression, the immunohistochemical staining of Bax was measured as area percent (%) across 10 different fields/section, $n = 7$ rat/group. Mean values were statistically different from the CTR ($^{\#}$ $p < 0.05$), DPPE (* $p < 0.05$), and DOX ($^{\$}$ $p < 0.05$) groups.

The cardiac tissues of DPPE + DOX-treated rats exhibited weak expression and immune-staining of cleaved caspase-3 (Figure 5d) and Bax (Figure 6d). It also showed moderate to strong expression and immune-staining of Bcl-2 (Figure 7d). Constantly, the mean immune-stained areas % in DPPE + DOX group showed a significant ($p < 0.05$) reduction for cleaved caspase-3 (≈0.39-fold) and Bax (≈0.72-fold) and a considerable increase for Bcl-2 (≈1.9-fold) as associated with the DOX treated group values (Figures 5e, 6e and 7e, respectively). Meanwhile, they showed a substantial ($p < 0.05$) surge for cleaved caspase-3 (≈3.55-fold) and Bax (≈1.72-fold) and a significant decrease for Bcl-2 (≈0.59-fold), as equated with the CTR group (Figures 5e, 6e and 7e, respectively).

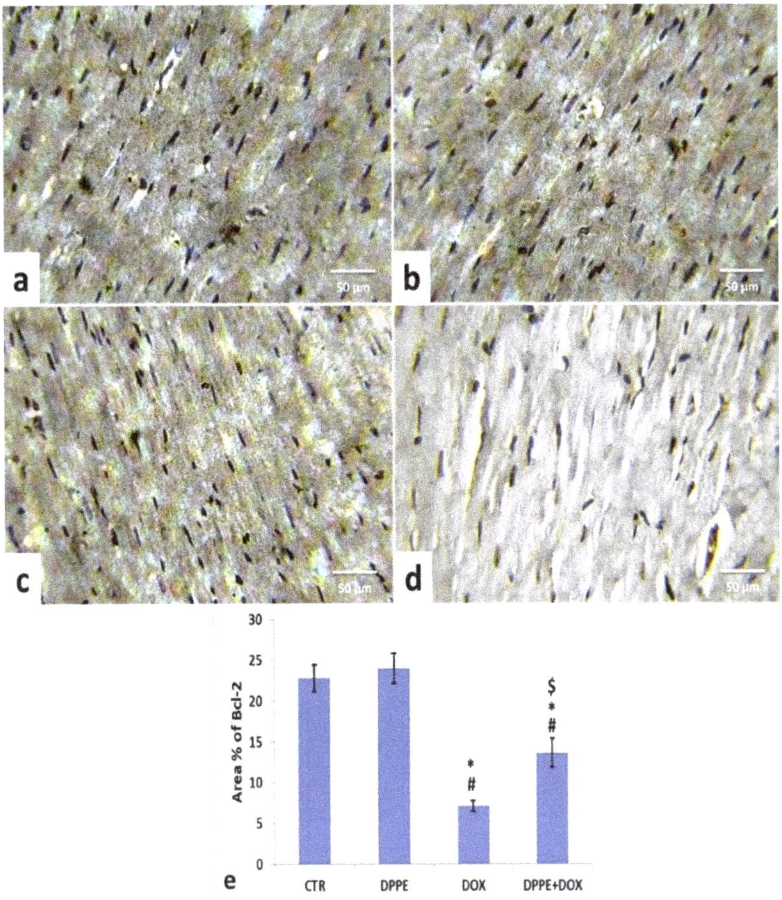

Figure 7. Immunohistochemical staining of B-cell lymphoma-2 (Bcl-2) in cardiac cells of the experimental rats (IHC, ×400). A control (**a**), DPPE-treated (**b**) DOX-treated (**c**) and DPP + DOX-treated (**d**) rats. (**e**) Quantification of Bcl-2 expression, the immunohistochemical staining of Bcl-2 was measured as area percent (%) across 10 different fields/section, $n = 7$ rat/group. Mean values were statistically different from the CTR ($^{\#}$ $p < 0.05$), DPPE (* $p < 0.05$), and DOX ($^{\$}$ $p < 0.05$) groups.

4. Discussion

Anthracyclines, including doxorubicin, play a crucial role in chemotherapy for the medication of numerous solid organ tumors and hematologic malignancy [12,16]. However, dose-based cardiotoxicity of anthracyclines is frequently reported to limit their therapeutic efficacy [5,65,66]. Since the DOX cardiotoxic effects are generally irreversible, searching for new protective approaches that could interrupt DOX-induced pathogenic events and confer protection against its cardiotoxicity should be developed [67]. Currently, the handout study is nearly the first to verify the meliorative potential of DPPE on DOX-induced cardiotoxicity. DOX cardiotoxicity is the ultimate obstacle to be solved to enhance its clinical usage [68]. In this work, rats were treated with 15 mg DOX/kg bw as a cumulative dose to mimic its chronic cardiotoxicity, as seen in clinical therapies [69]. Several pathways participate in DOX-induced cardiotoxicity.

However, the main mechanism involved is ROS generation, which causes peroxidation of lipids and depletion of antioxidant enzymes [35]. The cardiac tissue has many mitochondria because it needs much energy, making it more susceptible to DOX toxicity [67]. They have a high DOX affinity because their inner membrane encompasses cardiolipin, an anionic phospholipid with a high binding affinity to cationic DOX [67,70]. In mitochondria, cytochrome p450 reductase, xanthine oxidase, and NADPH dehydrogenase have to convert DOX into a semiquinone radical that interacts with molecular oxygen forming superoxide anion and an extra ROS [67]. Since free radicals act as the main contributor to DOX-induced cardiotoxicity, antioxidant compounds are known to be possible protective and therapeutic agents [71,72]. Dexrazoxane is the only synthetic medication used for cardiotoxicity prevention in clinical conditions [73,74]. DPP contains polyphenols and flavonoids with effective antioxidant and anti-inflammatory potentials that clarify its prospective use in many diseases. It also has antimicrobial, anti-coccidial, anti-apoptotic, and hepatoprotective potential [43,44,46]. Similarly, DPP is used as an anti-toxicant [42] and provides a cardio-preventive ability against isoproterenol-triggered myocardial infarction [42]. Its effect against DOX-induced cardiotoxicity has never been investigated.

The DOX group showed a scruffy appearance and exhibited 30% mortality. This result was parallel to those reported by Wu, et al. [75]. However, in DPPE + DOX-treated animals, there was relatively low mortality, reflecting that pre-cotreatment with DPPE might improve the survival of DOX-intoxicated rats. Additionally, the DOX persuaded a significant reduction in body weight [76,77], which might be due to decreased appetite, reduced protein synthesis, mucositis, and/or inadequate assimilation of nutrients [78–80]. The improved body weight in DPPE + DOX-treated animals mirrored the protective effects of DPPE. Cardiomyopathy caused by DOX is a shift from myocardial hypertrophy to heart failure [81]. In animals, myocardial hypertrophy is mainly assessed by measuring heart index weight [82]. In the current work, the DOX caused an increment in heart and relative heart weights and cardiomyocyte cross-sectional area, which indicated ventricular hypertrophy [75,81,83–85]. However, pre-cotreatment with DPPE revealed a reduction in the previous parameters, suggesting the ability of DPPE to maintain the normal integrity of cardiomyocytes. The cardiac enzymes ALT, AST, CK, CK-MB, LDH, CK, LDH, cTnI, and cTnT are a dynamic bioindicator of myocardial injury [35]. The DOX-induced cardiac damage was evident through the substantial rise of cardiac injury biomarkers: ALT, AST, LDH (not very specific biomarkers), CK, CK-MB, cTnI, and cTnT (more specific and sensitive biomarkers) activities, reflecting cardiomyocyte membrane disruption and extensive cardiomyocyte damage [59,76,86]. However, they were reduced following pre-cotreatment with DPPE, suggesting its competency to maintain the normal integrity of cardiac muscle and to inhibit DOX-induced myocardial damage [42]. An earlier study showed that antioxidant compounds could decrease cardiac function biomarkers in DOX-intoxicated rats [35].

The N-terminal pro-brain natriuretic peptide is a peptide produced to control blood pressure fluid equilibrium. It is liberated from the heart following ventricle volume expansion and/or pressure overload [87]. A large amount of NT-pro BNP is released into the blood during cardiac insufficiency, so it is considered a sensitive biomarker of congestive heart failure [88]. It is also a valuable predictor in patients with anthracycline chemotherapy as a vital biomarker of left ventricular dysfunction [89]. Herein, the DOX-induced, a dramatic increment of serum NT-pro BNP level, demonstrating that it can cause acute cardiac failure [35,90]. However, pre-cotreatment with DPPE lowered NT-pro BNP serum level, suggesting that DPPE may defend the heart from DOX-induced toxicity and cardiac damage.

The generation of large quantities of ROS and O_2^- in DOX metabolism results in DNA and mitochondrial injury, therefore enhancing lipid peroxidation in the cell membrane and increased MDA levels in cardiac cells [91]. In turn, free radicals released in response to DOX can interfere with the balance between oxidative and antioxidants agents, followed by depletion of the endogenous myocardial antioxidant compounds (GSH) and enzymes (GPx, SOD, and CAT) [77]. Herein, DOX-intoxicated rats exhibited increased oxidative and nitrosative stresses, indicated by an increment of MDA and NO, and a substantial decline in antioxidant enzyme activity [76,92]. The recorded decline in the antioxidant enzyme activity could be attributed to their utilization in the fight against oxidative stress [35]. Remarkably, pre-cotreatment with DPPE diminished MDA and NO levels, and improved antioxidant activity in cardiac tissue, suggesting the antioxidant ability of DPPE against DOX-induced oxidative/nitrosative stress. Furthermore, DPPE expressed antioxidant activity and defensive mechanisms by restoring oxidative stress/antioxidant balance in several toxic modules [40,49,93]. Many earlier studies of phytochemical or antioxidant elements have demonstrated their ability to reduce lipid peroxidation and improve the value of antioxidant markers in DOX cardiotoxicity [21,35,71,76,94,95]. The mechanism for protecting the DPPE may include de-activating potentially toxic metabolites and free radicals and potentiation of antioxidant paths [96]. Other investigators claimed that DPPE incorporates considerable amounts of flavonoids, phytosterols, and carotenoids [42,97], which are antioxidants with redox activities acting as reducing agents ROS/NOS quenchers [98].

It is well established that increased free radicals output with excitotoxicity and lipid peroxidation accelerates inflammatory conciliators' synthesis and thus activates the inflammatory response in the cardiac tissue [99]. NF-κB, a transcription factor, is involved in cell survival, inflammation, and immune responses. NF-κB p65 modulates the inflammatory responses, whereas its translocation to the nucleus enhances transcription of the pro-inflammatory cytokines, i.e., TNF-α, IL-1β, and IL-6 [100,101]. In turn, they provoke leukocyte infiltration into the myocardium and aggravate inflammatory injury [102]. Consistently, DOX stimulated NF-κB p65, TNF-α, and IL-1β production in cardiac tissues, which can also cause cardiomyocyte apoptosis by increasing Ca^{2+} store of heart muscle cells [26]. As well, DOX provoked a substantial rise in NF-κB level and correspondingly induced an increment in TNF-α and IL-1β quantities, reflecting enhanced inflammatory responses [71]. However, DPPE pre-cotreatment generated a substantial decline in NF-kB's cardiac contents and the related downstream pro-inflammatory cytokines, TNF-α and IL-1β, indicating its potential to suppress the initiated inflammatory cascade.

Disruption of Ca^{2+} homeostasis is another pathway implicated in DOX-generated cardiac toxicity. It is documented that DOX cardiotoxicity is accompanied by a high overload of Ca^{2+} in cardiomyocytes, resulting in inadequate contraction and interference with Ca^{2+} regulation, thus triggering ROS generation and leading to cell dysfunction [103–105]. In the present work, DPPE was found to mitigate oxidative stress-mediated Ca^{2+} overload in DOX-challenged cardiac tissues. This may be attributed to DPPE's antioxidant properties, which suppress ROS generation and consequently reduce such high Ca^{2+}. Moreover, ROS and O_2^- exaggerated cardiac dysfunction and mitochondrial damage induced in DOX therapy [17]. The formation of the Fe-anthracycline complex catalyzes the transformation of H_2O_2 to $OH^\bullet$ radicals, resulting in severe cytoskeleton injury and plasma membrane disruption followed by myofibril loss, sarcoplasmic reticulum dilation, and myocardial necrosis [24,106].

The findings of the cardiac tissues histopathological analysis supported the biochemical interpretations. Due to enlargement of the sarcoplasmic reticulum, DOX induced several degenerative changes in heart tissue, including myocardial hyalinization, and sarcoplasmic vacuolization [107], wavy myocardial fibers flocculation, and fragmentation. In addition, multifocal areas of myocardial necrosis, myofibrillar loss, inflammatory cell infiltrations, myocardial fibrosis, hemorrhage, vascular congestion, and interfibrillar edema were observed [71,74,100,108,109]. On the contrary, cardiac tissues of DPPE plus DOX-treated animals showed a marked improvement in cardiac tissue structure and integrity. According to the histomorphometric analysis, the previously declared lesions were less in incidence and severity. These remarks were comparable with other experiments that verified the potential of DPPE as anti-myocardial damage [42]. The literature reviewed that DPPE contains bioactive substances, including estradiol [110], stigmasterol [111], β-sitosterol [112], carotenoids lutein [113,114], δ-tocotrienol [115], and isorhamnetin [116], which have potential cardioprotective activities.

Fibrosis is a reparative reaction to DOX-induced cardiotoxicity [85]. The necrotic or apoptotic cardiomyocytes are replaced by overproduced collagen by fibroblast. However, it contributes to heart rigidity and instability [85,117]. DOX-induced cardiac fibrosis is based on the inflammatory and growth factors signaling paths regulated by TGF-β1. Increased oxidative stress and the subsequent antioxidant depletion and lipid peroxidation trigger tissue inflammation and necrosis and enhance tissue fibrogenesis progression [35,59,118]. TGF-β1 is another key factor in the regulation of collagen production in DOX-induced cardiomyopathy. The pro-fibrogenic cytokine, TGF-β1, is a proliferation-mediated fibrotic protein produced by cardiac myofibroblast and is responsible for cardiomyocyte hypertrophy, apoptosis, and fibrosis. It may stimulate cardiac fibroblastic hyperplasia, increased production of type I and III collagen fiber and fibronectin, and cause increased extracellular matrix and decreased extracellular matrix degradation through inhibition of collagen enzyme release [28,119]. In the current work, DOX elevated the cardiac hydroxyproline level, the major component of fibrillar collagen [59]. Additionally, DOX-induced remarkable hypertrophy of fibroblast increased collagen deposition, and fibrosis was further confirmed by Masson's trichrome staining of the cardiac tissues [120]. DOX also increased the TGF-β1 expression in cardiac tissues [85,121–123]. However, DPPE pre-cotreatment induced a marked reduction in collagen fiber deposition between cardiac muscle fibers and around the intramyocardial coronary vessels, hydroxyproline content, and TGF-β1 expression in cardiac tissues. Therefore, DPPE has a potential aptitude to maintain the normal integrity of cardiac muscle and inhibit DOX-induced myocardial damage, which attenuated fibrosis development via modulation of fibrogenic genes.

The major regulators of apoptosis are Bcl-2 family members, which involve pro-apoptotic protein (Bax, caspase-3) and anti-apoptotic (Bcl-2) proteins [74]. During apoptosis, the Bcl-2 expression declines, while Bax and caspase-3 expressions rise [124]. Bax stimulation ensures cell damage by forming a pore in the mitochondrial membrane, leading to poly (ADP-ribose) polymerase cleavage and mitochondrial cytochrome-c induction, which mediate apoptosis [125]. Meanwhile, the Bcl-2 inhibits apoptosis by inhibiting mitochondrial permeability transition [126] in cardiomyocytes protecting mitochondrial structure and function [124]. Caspases are essential parts of the apoptotic process. Opened mitochondrial pores lead to mitochondrial cytochrome-C release, and activated caspase-3 triggers proteolytic degradation of cellular components death [127]. Herein, the immunohistochemical staining of myocardial tissues showed that DOX caused an increase in the cleaved caspase-3 and Bax expressions and decreased Bcl-2 expression, which reflected apoptosis's role in DOX-induced cardiomyopathy [76,86,128]. Nevertheless, pre-cotreatment with DPPE reduced cleaved caspase-3 and Bax and enhanced Bcl-2 expressions, implying that DPPE's anti-apoptotic activity could conserve myocardial integrity and mitigate myocardial damage.

5. Conclusions

Collectively, for the first time from these observations, it is indicated that date palm pollen ethanolic extract displayed an effective cardioprotective potential against doxorubicin-induced cardiac myopathy. The antifibrotic and anti-apoptotic mechanisms of DPPE were attributed to suppressing cardiac oxidative/nitrosative damage, pro-inflammatory cytokines production, and fibrogenic and apoptotic gene expressions, thereby reducing myocardial myopathy and detrimental structural alterations. Accordingly, DPPE is highly recommended as an adjunct to avert the toxic side-effects caused by doxorubicin.

Author Contributions: Methodology, S.S.E.; software, S.S.E., Y.S.E.-S., M.M.S. and M.S.; validation, M.S., S.S.E. and M.M.S.; formal analysis, Y.S.E.-S., M.S. and S.S.E.; writing—original draft preparation, Y.S.E.-S., S.S.E., M.M.S. and M.S.; visualization, S.S.E., M.S., Y.S.E.-S. and M.M.S.; supervision. M.S. and M.M.S. All authors contributed to this work and agreed to participate in the paper. All authors have read and agreed to the published version of the manuscript.

Funding: This study was supported by Taif University Researchers Supporting Project (TURSP-2020/09), Taif University, Taif, Saudi Arabia.

Institutional Review Board Statement: This study was declared by the Local Committee of the Faculty of Veterinary Medicine, Alexandria University (Ethical Committee Approval Number: 2020/013/59) and ethical approval of Taif University (42-0081).

Informed Consent Statement: Not applicable.

Data Availability Statement: All data sets obtained and evaluated during the current study are available upon appropriate request from the corresponding author.

Acknowledgments: We appreciate and thank Taif University for financial support to Taif University Researchers Supporting Project (TURSP-2020/09), Taif University, Taif, Saudi Arabia.

Conflicts of Interest: The authors declared no competing conflict of interest.

Abbreviations

ALT	alanine transaminase
AST	Aspartate aminotransferase
Bax	Alanine aminotransferase
Bcl2	B-cell lymphoma 2
BW	body weight
CAT	Catalase
CHF	congestive heart failure
CK	creatine kinase
CK-MB	creatine kinase-cardiac type isoenzyme
CVF	collagen volume fraction
cTnI	Serum cardiac troponin I
cTnT	Troponin T
DOX	Doxorubicin
DPPE	date palm pollen ethanolic extract
GPx	glutathione peroxidase
GSH	reduced glutathione
IL-1β	interleukin-1 beta
IL-6	interleukin-6
LDH	lactate dehydrogenase
MDA	Malondialdehyde
NO	nitric oxide
NT-pro BNP	N-terminal pro-brain natriuretic peptide
PVCA	collagen area
RCF	radius centrifugation force
RNS	reactive nitrogen species
ROS	Redox Oxygen species
SOD	superoxide dismutase
TNF-α	tumor necrosis factor-alpha
UDP	up and down procedure

References

1. Hu, H.; Zhang, W.; Huang, D.; Yang, Q.; Li, J.; Gao, Y. Cardiotoxicity of anthracycline (ANT) treatment in children with malignant tumors. *Pediatr. Hematol. Oncol.* **2018**, *35*, 111–120. [CrossRef]
2. Yu, J.; Wang, C.; Kong, Q.; Wu, X.; Lu, J.J.; Chen, X. Recent progress in doxorubicin-induced cardiotoxicity and protective potential of natural products. *Phytomedicine* **2018**, *40*, 125–139. [CrossRef] [PubMed]
3. Nagiub, M.; Nixon, J.V.; Kontos, M.C. Ability of nonstrain diastolic parameters to predict doxorubicin-induced cardiomyopathy: A Systematic review with meta-analysis. *Cardiol. Rev.* **2018**, *26*, 29–34. [CrossRef] [PubMed]
4. Finn, N.A.; Findley, H.W.; Kemp, M.L. A switching mechanism in doxorubicin bioactivation can be exploited to control doxorubicin toxicity. *PLoS Comput. Biol.* **2011**, *7*, e1002151. [CrossRef] [PubMed]
5. Octavia, Y.; Tocchetti, C.G.; Gabrielson, K.L.; Janssens, S.; Crijns, H.J.; Moens, A.L. Doxorubicin-induced cardiomyopathy: From molecular mechanisms to therapeutic strategies. *J. Mol. Cell Cardiol.* **2012**, *52*, 1213–1225. [CrossRef] [PubMed]

6. Shafei, A.; El-Bakly, W.; Sobhy, A.; Wagdy, O.; Reda, A.; Aboelenin, O.; Marzouk, A.; El Habak, K.; Mostafa, R.; Ali, M.A.; et al. A review on the efficacy and toxicity of different doxorubicin nanoparticles for targeted therapy in metastatic breast cancer. *Biomed. Pharmacother.* **2017**, *95*, 1209–1218. [CrossRef]
7. Tewey, K.M.; Rowe, T.C.; Yang, L.; Halligan, B.D.; Liu, L.F. Adriamycin-induced DNA damage mediated by mammalian DNA topoisomerase II. *Science* **1984**, *226*, 466–468. [CrossRef]
8. Tokarska-Schlattner, M.; Zaugg, M.; Zuppinger, C.; Wallimann, T.; Schlattner, U. New insights into doxorubicin-induced cardiotoxicity: The critical role of cellular energetics. *J. Mol. Cell Cardiol.* **2006**, *41*, 389–405. [CrossRef]
9. Pawlowska, J.; Tarasiuk, J.; Wolf, C.R.; Paine, M.J.; Borowski, E. Differential ability of cytostatics from anthraquinone group to generate free radicals in three enzymatic systems: NADH dehydrogenase, NADPH cytochrome P450 reductase, and xanthine oxidase. *Oncol. Res.* **2003**, *13*, 245–252. [CrossRef]
10. Fogli, S.; Nieri, P.; Breschi, M.C. The role of nitric oxide in anthracycline toxicity and prospects for pharmacologic prevention of cardiac damage. *FASEB J.* **2004**, *18*, 664–675. [CrossRef]
11. Li, D.L.; Wang, Z.V.; Ding, G.; Tan, W.; Luo, X.; Criollo, A.; Xie, M.; Jiang, N.; May, H.; Kyrychenko, V.; et al. Doxorubicin blocks cardiomyocyte autophagic flux by inhibiting lysosome acidification. *Circulation* **2016**, *133*, 1668–1687. [CrossRef]
12. Menna, P.; Salvatorelli, E. Primary prevention strategies for anthracycline cardiotoxicity: A brief overview. *Chemotherapy* **2017**, *62*, 159–168. [CrossRef] [PubMed]
13. Chatterjee, K.; Zhang, J.; Honbo, N.; Karliner, J.S. Doxorubicin cardiomyopathy. *Cardiology* **2010**, *115*, 155–162. [CrossRef]
14. Pecoraro, M.; Del Pizzo, M.; Marzocco, S.; Sorrentino, R.; Ciccarelli, M.; Iaccarino, G.; Pinto, A.; Popolo, A. Inflammatory mediators in a short-time mouse model of doxorubicin-induced cardiotoxicity. *Toxicol. Appl. Pharmacol.* **2016**, *293*, 44–52. [CrossRef] [PubMed]
15. Ibrahim, D.M.; Radwan, R.R.; Fattah, S.M.A. Antioxidant and antiapoptotic effects of sea cucumber and valsartan against doxorubicin-induced cardiotoxicity in rats: The role of low dose gamma irradiation. *J. Photochem. Photobiol. B* **2017**, *170*, 70–78. [CrossRef]
16. McGowan, J.V.; Chung, R.; Maulik, A.; Piotrowska, I.; Walker, J.M.; Yellon, D.M. Anthracycline chemotherapy and cardiotoxicity. *Cardiovasc. Drugs Ther.* **2017**, *31*, 63–75. [CrossRef] [PubMed]
17. Pugazhendhi, A.; Edison, T.; Velmurugan, B.K.; Jacob, J.A.; Karuppusamy, I. Toxicity of doxorubicin (Dox) to different experimental organ systems. *Life Sci.* **2018**, *200*, 26–30. [CrossRef] [PubMed]
18. Goormaghtigh, E.; Huart, P.; Praet, M.; Brasseur, R.; Ruysschaert, J.M. Structure of the adriamycin-cardiolipin complex. Role in mitochondrial toxicity. *Biophys. Chem.* **1990**, *35*, 247–257. [CrossRef]
19. Priya, L.B.; Baskaran, R.; Huang, C.Y.; Padma, V.V. Neferine ameliorates cardiomyoblast apoptosis induced by doxorubicin: Possible role in modulating NADPH oxidase/ROS-mediated NFkappaB redox signaling cascade. *Sci. Rep.* **2017**, *7*, 12283. [CrossRef]
20. Ma, T.; Kandhare, A.D.; Mukherjee-Kandhare, A.A.; Bodhankar, S.L. Fisetin, a plant flavonoid ameliorates doxorubicin-induced cardiotoxicity in experimental rats: The decisive role of caspase-3, COX-II, cTn-I, iNOs and TNF-alpha. *Mol. Biol. Rep.* **2019**, *46*, 105–118. [CrossRef]
21. Zhang, X.; Hu, C.; Kong, C.Y.; Song, P.; Wu, H.M.; Xu, S.C.; Yuan, Y.P.; Deng, W.; Ma, Z.G.; Tang, Q.Z. FNDC5 alleviates oxidative stress and cardiomyocyte apoptosis in doxorubicin-induced cardiotoxicity via activating AKT. *Cell Death Differ.* **2020**, *27*, 540–555. [CrossRef] [PubMed]
22. Wallace, K.B. Adriamycin-induced interference with cardiac mitochondrial calcium homeostasis. *Cardiovasc. Toxicol.* **2007**, *7*, 101–107. [CrossRef] [PubMed]
23. Wenningmann, N.; Knapp, M.; Ande, A.; Vaidya, T.R.; Ait-Oudhia, S. Insights into doxorubicin-induced cardiotoxicity: Molecular mechanisms, preventive strategies, and early monitoring. *Mol. Pharmacol.* **2019**, *96*, 219–232. [CrossRef]
24. Minotti, G.; Recalcati, S.; Menna, P.; Salvatorelli, E.; Corna, G.; Cairo, G. Doxorubicin cardiotoxicity and the control of iron metabolism: Quinone-dependent and independent mechanisms. *Methods Enzymol.* **2004**, *378*, 340–361. [CrossRef]
25. Wang, X.; Wang, Q.; Li, W.; Zhang, Q.; Jiang, Y.; Guo, D.; Sun, X.; Lu, W.; Li, C.; Wang, Y. TFEB-NF-kappaB inflammatory signaling axis: A novel therapeutic pathway of dihydrotanshinone I in doxorubicin-induced cardiotoxicity. *J. Exp. Clin. Cancer Res.* **2020**, *39*, 93. [CrossRef]
26. Shaker, R.A.; Abboud, S.H.; Assad, H.C.; Hadi, N. Enoxaparin attenuates doxorubicin induced cardiotoxicity in rats via interfering with oxidative stress, inflammation and apoptosis. *BMC Pharmacol. Toxicol.* **2018**, *19*, 3. [CrossRef]
27. Zhang, N.; Wei, W.Y.; Li, L.L.; Hu, C.; Tang, Q.Z. Therapeutic potential of polyphenols in cardiac fibrosis. *Front. Pharmacol.* **2018**, *9*, 122. [CrossRef] [PubMed]
28. Sun, X.; Chen, G.; Xie, Y.; Jiang, D.; Han, J.; Chen, F.; Song, Y. Qiliqiangxin improves cardiac function and attenuates cardiac remodelling in doxorubicin-induced heart failure rats. *Pharm. Biol.* **2020**, *58*, 417–426. [CrossRef]
29. Borriello, M.; Iannuzzi, C.; Sirangelo, I. Pinocembrin protects from age-induced cytotoxicity and inhibits non-enzymatic glycation in human insulin. *Cells* **2019**, *8*, 385. [CrossRef]
30. Salzer, W.L.; Devidas, M.; Carroll, W.L.; Winick, N.; Pullen, J.; Hunger, S.P.; Camitta, B.A. Long-term results of the pediatric oncology group studies for childhood acute lymphoblastic leukemia 1984–2001: A report from the children's oncology group. *Leukemia* **2010**, *24*, 355–370. [CrossRef]

31. Pandey, K.B.; Rizvi, S.I. Plant polyphenols as dietary antioxidants in human health and disease. *Oxid. Med. Cell Longev.* **2009**, *2*, 270–278. [CrossRef] [PubMed]
32. Gupta, V.K.; Sharma, B. Role of phytochemicals in neurotrophins mediated regulation of Alzheimer's disease. *Int. J. Complement. Altern. Med.* **2017**, *7*, 00231. [CrossRef]
33. Singh, N.; Sharma, B. Biotoxins mediated DNA damage and role of phytochemicals in DNA protection. *Biochem. Mol. Biol. J.* **2018**, *4*, 5. [CrossRef]
34. Ayza, M.A.; Balasubramanian, R.; Berhe, A.H. Cardioprotective effect of *Croton macrostachyus* stem bark extract and solvent fractions on cyclophosphamide-induced cardiotoxicity in rats. *Evid. Based Complement. Altern. Med.* **2020**, *2020*, 8467406. [CrossRef] [PubMed]
35. Sandamali, J.A.N.; Hewawasam, R.P.; Jayatilaka, K.; Mudduwa, L.K.B. Cardioprotective potential of *Murraya koenigii* (L.) Spreng. leaf extract against doxorubicin-induced cardiotoxicity in rats. *Evid. Based Complement. Altern. Med.* **2020**, *2020*, 6023737. [CrossRef] [PubMed]
36. Wang, Y.; Chao, X.; Ahmad, F.U.D.; Shi, H.; Mehboob, H.; Hassan, W. *Phoenix dactylifera* protects against doxorubicin-induced cardiotoxicity and nephrotoxicity. *Cardiol. Res. Pract.* **2019**, *2019*, 7395239. [CrossRef] [PubMed]
37. Bahadir, A.; Kurucu, N.; Kadioglu, M.; Yenilme, E. The role of nitric oxide in doxorubicin-induced cardiotoxicity: Experimental study. *Turk. J. Haematol.* **2014**, *31*, 68–74. [CrossRef] [PubMed]
38. Hassan, H.M.M. Chemical composition and nutritional value of palm pollen grains. *Glob. J. Biotechnol. Biochem.* **2011**, *6*, 1–7.
39. Soliman, F.A.; Soliman, A. The gonad stimulating potency of date palm pollen grains. *Experientia* **1958**, *14*, 92–93. [CrossRef] [PubMed]
40. El-Kashlan, A.M.; Nooh, M.M.; Hassan, W.A.; Rizk, S.M. Therapeutic potential of date palm pollen for testicular dysfunction induced by thyroid disorders in male rats. *PLoS ONE* **2015**, *10*, e0139493. [CrossRef]
41. Kroyer, G.; Hegedus, N. Evaluation of bioactive properties of pollen extracts as functional dietary food supplement. *Innov. Food Sci. Emerg. Technol.* **2001**, *2*, 171–174. [CrossRef]
42. Daoud, A.; Ben Mefteh, F.; Mnafgui, K.; Turki, M.; Jmal, S.; Ben Amar, R.; Ayadi, F.; ElFeki, A.; Abid, L.; Rateb, M.E.; et al. Cardiopreventive effect of ethanolic extract of Date Palm Pollen against isoproterenol induced myocardial infarction in rats through the inhibition of the angiotensin-converting enzyme. *Exp. Toxicol. Pathol.* **2017**, *69*, 656–665. [CrossRef]
43. Bentrad, N.; Gaceb-Terrak, R.; Benmalek, Y.; Rahmania, F. Studies on Chemical Composition and Antimicrobial Activities of Bioactive Molecules from Date Palm (Phoenixdactylifera L.) Pollens and Seeds. *Afr. J. Tradit. Complement. Altern. Med.* **2017**, *14*, 242–256. [CrossRef]
44. Elberry, A.A.; Mufti, S.T.; Al-Maghrabi, J.A.; Abdel-Sattar, E.A.; Ashour, O.M.; Ghareib, S.A.; Mosli, H.A. Anti-inflammatory and antiproliferative activities of date palm pollen (*Phoenix dactylifera*) on experimentally-induced atypical prostatic hyperplasia in rats. *J. Inflamm* **2011**, *8*, 40. [CrossRef] [PubMed]
45. Abedi, A.; Parviz, M.; Karimian, S.M.; Rodsari, H.R.S. Aphrodisiac activity of aqueous extract of *Phoenix dactylifera* pollen in male rats. *Adv. Sex. Med.* **2013**, *3*, 28–34. [CrossRef]
46. Metwaly, M.S.; Dkhil, M.A.; Al-Quraishy, S. Anti-coccidial and anti-apoptotic activities of palm pollen grains on *Eimeria papillata*-induced infection in mice. *Biologia* **2014**, *69*, 254–259. [CrossRef]
47. El-Kholy, W.M.; Soliman, T.N.; Darwish, A.M.G. Evaluation of date palm pollen (*Phoenix dactylifera* L.) encapsulation, impact on the nutritional and functional properties of fortified yoghurt. *PLoS ONE* **2019**, *14*, e0222789. [CrossRef] [PubMed]
48. Abbas, F.; Ateya, A. Estradiol, esteriol, estrone and novel flavonoids from date palm pollen. *Aust. J. Appl. Sci. Res.* **2011**, *7*, 606–614.
49. El-Neweshy, M.S.; El-Maddawy, Z.K.; El-Sayed, Y.S. Therapeutic effects of date palm (*Phoenix dactylifera* L.) pollen extract on cadmium-induced testicular toxicity. *Andrologia* **2013**, *45*, 369–378. [CrossRef]
50. OECD. Acute oral toxicity-up-and-down procedure. In *OECD Guidelines for the Testing of Chemicals*; OECD Publishing: Paris, France, 2001; pp. 1–26. [CrossRef]
51. Arafa, M.H.; Mohammad, N.S.; Atteia, H.H.; Abd-Elaziz, H.R. Protective effect of resveratrol against doxorubicin-induced cardiac toxicity and fibrosis in male experimental rats. *J. Physiol. Biochem.* **2014**, *70*, 701–711. [CrossRef]
52. Tarpey, M.M.; Wink, D.A.; Grisham, M.B. Methods for detection of reactive metabolites of oxygen and nitrogen: In vitro and in vivo considerations. *Am. J. Physiol. Regul. Integr. Comp. Physiol.* **2004**, *286*, R431–R444. [CrossRef] [PubMed]
53. Ohkawa, H.; Ohishi, N.; Yagi, K. Assay for lipid peroxides in animal tissues by thiobarbituric acid reaction. *Anal. Biochem.* **1979**, *95*, 351–358. [CrossRef]
54. Beutler, E.; Duron, O.; Kelly, B.M. Improved method for the determination of blood glutathione. *J. Lab. Clin. Med.* **1963**, *61*, 882–888.
55. Paglia, D.E.; Valentine, W.N. Studies on the quantitative and qualitative characterization of erythrocyte glutathione peroxidase. *J. Lab. Clin. Med.* **1967**, *70*, 158–169. [PubMed]
56. Aebi, H. Catalase in vitro. *Methods Enzymol.* **1984**, *105*, 121–126. [CrossRef]
57. Nishikimi, M.; Appaji, N.; Yagi, K. The occurrence of superoxide anion in the reaction of reduced phenazine methosulfate and molecular oxygen. *Biochem. Biophys. Res. Commun.* **1972**, *46*, 849–854. [CrossRef]
58. Bradford, M.M. A rapid and sensitive method for the quantitation of microgram quantities of protein utilizing the principle of protein-dye binding. *Anal. Biochem.* **1976**, *72*, 248–254. [CrossRef]

59. El-Agamy, D.S.; El-Harbi, K.M.; Khoshhal, S.; Ahmed, N.; Elkablawy, M.A.; Shaaban, A.A.; Abo-Haded, H.M. Pristimerin protects against doxorubicin-induced cardiotoxicity and fibrosis through modulation of Nrf2 and MAPK/NF-kB signaling pathways. *Cancer Manag. Res.* **2019**, *11*, 47–61. [CrossRef]
60. Bancroft, J.D.; Layton, C. The hematoxylin and eosin, connective and mesenchymal tissues with their stains. In *Bancroft s Theory and Practice of Histological Techniques*, 7th ed.; Kim, S., Suvarna, C.L., Eds.; Churchill Livingstone: Philadelphia, PA, USA, 2013; pp. 173–186.
61. Hsu, S.M.; Raine, L.; Fanger, H. The use of antiavidin antibody and avidin-biotin-peroxidase complex in immunoperoxidase technics. *Am. J. Clin. Pathol.* **1981**, *75*, 816–821. [CrossRef]
62. Fan, Z.D.; Zhang, L.; Shi, Z.; Gan, X.B.; Gao, X.Y.; Zhu, G.Q. Artificial microRNA interference targeting AT(1a) receptors in paraventricular nucleus attenuates hypertension in rats. *Gene Ther.* **2012**, *19*, 810–817. [CrossRef]
63. Ye, F.; He, J.; Wu, X.; Xie, J.; Chen, H.; Tang, X.; Lai, Z.; Huang, R.; Huang, J. The regulatory mechanisms of Yulangsan MHBFC reversing cardiac remodeling in rats based on eNOS-NO signaling pathway. *Biomed. Pharmacother.* **2019**, *117*, 109141. [CrossRef]
64. Elblehi, S.S.; El Euony, O.I.; El-Sayed, Y.S. Apoptosis and astrogliosis perturbations and expression of regulatory inflammatory factors and neurotransmitters in acrylamide-induced neurotoxicity under omega3 fatty acids protection in rats. *Neurotoxicology* **2020**, *76*, 44–57. [CrossRef]
65. Mitry, M.A.; Edwards, J.G. Doxorubicin induced heart failure: Phenotype and molecular mechanisms. *Int. J. Cardiol. Heart Vasc.* **2016**, *10*, 17–24. [CrossRef]
66. Nebigil, C.G.; Desaubry, L. Updates in anthracycline-mediated cardiotoxicity. *Front. Pharmacol.* **2018**, *9*, 1262. [CrossRef]
67. Angsutararux, P.; Luanpitpong, S.; Issaragrisil, S. Chemotherapy-induced cardiotoxicity: Overview of the roles of oxidative stress. *Oxid Med. Cell Longev.* **2015**, *2015*, 795602. [CrossRef]
68. Govender, J.; Loos, B.; Marais, E.; Engelbrecht, A.M. Melatonin improves cardiac and mitochondrial function during doxorubicin-induced cardiotoxicity: A possible role for peroxisome proliferator-activated receptor gamma coactivator 1-alpha and sirtuin activity? *Toxicol. Appl. Pharmacol.* **2018**, *358*, 86–101. [CrossRef] [PubMed]
69. Xiao, J.; Sun, G.B.; Sun, B.; Wu, Y.; He, L.; Wang, X.; Chen, R.C.; Cao, L.; Ren, X.Y.; Sun, X.B. Kaempferol protects against doxorubicin-induced cardiotoxicity in vivo and in vitro. *Toxicology* **2012**, *292*, 53–62. [CrossRef] [PubMed]
70. Geisberg, C.A.; Sawyer, D.B. Mechanisms of anthracycline cardiotoxicity and strategies to decrease cardiac damage. *Curr. Hypertens Rep.* **2010**, *12*, 404–410. [CrossRef]
71. Jardan, Y.A.B.; Ansari, M.A.; Raish, M.; Alkharfy, K.M.; Ahad, A.; Al-Jenoobi, F.I.; Haq, N.; Khan, M.R.; Ahmad, A. Sinapic acid ameliorates oxidative stress, inflammation, and apoptosis in acute doxorubicin-induced cardiotoxicity via the NF-kappaB-mediated pathway. *Biomed. Res. Int.* **2020**, *2020*, 3921796. [CrossRef]
72. Dos Santos, J.M.; Alfredo, T.M.; Antunes, K.A.; da Cunha, J.; Costa, E.M.A.; Lima, E.S.; Silva, D.B.; Carollo, C.A.; Schmitz, W.O.; Boleti, A.P.A.; et al. *Guazuma ulmifolia* Lam. decreases oxidative stress in blood cells and prevents doxorubicin-induced cardiotoxicity. *Oxid. Med. Cell Longev.* **2018**, *2018*, 2935051. [CrossRef] [PubMed]
73. Cvetkovic, R.S.; Scott, L.J. Dexrazoxane: A review of its use for cardioprotection during anthracycline chemotherapy. *Drugs* **2005**, *65*, 1005–1024. [CrossRef]
74. Zhang, Q.L.; Yang, J.J.; Zhang, H.S. Carvedilol (CAR) combined with carnosic acid (CAA) attenuates doxorubicin-induced cardiotoxicity by suppressing excessive oxidative stress, inflammation, apoptosis and autophagy. *Biomed. Pharmacother.* **2019**, *109*, 71–83. [CrossRef]
75. Wu, R.; Yao, P.A.; Wang, H.L.; Gao, Y.; Yu, H.L.; Wang, L.; Cui, X.H.; Xu, X.; Gao, J.P. Effect of fermented *Cordyceps sinensis* on doxorubicin-induced cardiotoxicity in rats. *Mol. Med. Rep.* **2018**, *18*, 3229–3241. [CrossRef] [PubMed]
76. Al-Taee, H.; Azimullah, S.; Meeran, M.F.N.; Almheiri, M.K.A.; Al Jasmi, R.A.; Tariq, S.; Ab Khan, M.; Adeghate, E.; Ojha, S. beta-caryophyllene, a dietary phytocannabinoid attenuates oxidative stress, inflammation, apoptosis and prevents structural alterations of the myocardium against doxorubicin-induced acute cardiotoxicity in rats: An in vitro and in vivo study. *Eur. J. Pharmacol.* **2019**, *858*, 172467. [CrossRef]
77. Wu, R.; Mei, X.; Wang, J.; Sun, W.; Xue, T.; Lin, C.; Xu, D. Zn(ii)-curcumin supplementation alleviates gut dysbiosis and zinc dyshomeostasis during doxorubicin-induced cardiotoxicity in rats. *Food Funct.* **2019**, *10*, 5587–5604. [CrossRef] [PubMed]
78. Herman, E.H.; Zhang, J.; Chadwick, D.P.; Ferrans, V.J. Comparison of the protective effects of amifostine and dexrazoxane against the toxicity of doxorubicin in spontaneously hypertensive rats. *Cancer Chemother. Pharmacol.* **2000**, *45*, 329–334. [CrossRef]
79. Jenkins, G.R.; Lee, T.; Moland, C.L.; Vijay, V.; Herman, E.H.; Lewis, S.M.; Davis, K.J.; Muskhelishvili, L.; Kerr, S.; Fuscoe, J.C.; et al. Sex-related differential susceptibility to doxorubicin-induced cardiotoxicity in B6C3F1 mice. *Toxicol. Appl. Pharmacol.* **2016**, *310*, 159–174. [CrossRef]
80. Sahu, B.D.; Kumar, J.M.; Kuncha, M.; Borkar, R.M.; Srinivas, R.; Sistla, R. Baicalein alleviates doxorubicin-induced cardiotoxicity via suppression of myocardial oxidative stress and apoptosis in mice. *Life Sci.* **2016**, *144*, 8–18. [CrossRef] [PubMed]
81. Ahmed, L.A.; El-Maraghy, S.A. Nicorandil ameliorates mitochondrial dysfunction in doxorubicin-induced heart failure in rats: Possible mechanism of cardioprotection. *Biochem. Pharmacol.* **2013**, *86*, 1301–1310. [CrossRef]
82. Kuzman, J.A.; O'Connell, T.D.; Gerdes, A.M. Rapamycin prevents thyroid hormone-induced cardiac hypertrophy. *Endocrinology* **2007**, *148*, 3477–3484. [CrossRef]
83. Zhang, Y.; Ma, C.; Liu, C.; Wei, F. Luteolin attenuates doxorubicin-induced cardiotoxicity by modulating the PHLPP1/AKT/Bcl-2 signalling pathway. *Peer J.* **2020**, *8*, e8845. [CrossRef] [PubMed]

84. Karagiannis, T.C.; Lin, A.J.; Ververis, K.; Chang, L.; Tang, M.M.; Okabe, J.; El-Osta, A. Trichostatin A accentuates doxorubicin-induced hypertrophy in cardiac myocytes. *Aging* **2010**, *2*, 659–668. [CrossRef] [PubMed]
85. Lin, H.; Zhang, J.; Ni, T.; Lin, N.; Meng, L.; Gao, F.; Luo, H.; Liu, X.; Chi, J.; Guo, H. Yellow wine polyphenolic compounds prevents doxorubicin-induced cardiotoxicity through activation of the Nrf2 signalling pathway. *J. Cell Mol. Med.* **2019**, *23*, 6034–6047. [CrossRef] [PubMed]
86. Zare, M.F.R.; Rakhshan, K.; Aboutaleb, N.; Nikbakht, F.; Naderi, N.; Bakhshesh, M.; Azizi, Y. Apigenin attenuates doxorubicin induced cardiotoxicity via reducing oxidative stress and apoptosis in male rats. *Life Sci.* **2019**, *232*, 116623. [CrossRef]
87. Folsom, A.R.; Nambi, V.; Bell, E.J.; Oluleye, O.W.; Gottesman, R.F.; Lutsey, P.L.; Huxley, R.R.; Ballantyne, C.M. Troponin T, N-terminal pro-B-type natriuretic peptide, and incidence of stroke: The atherosclerosis risk in communities study. *Stroke* **2013**, *44*, 961–967. [CrossRef] [PubMed]
88. Desai, A.S.; Lewis, E.F.; Li, R.; Solomon, S.D.; Assmann, S.F.; Boineau, R.; Clausell, N.; Diaz, R.; Fleg, J.L.; Gordeev, I.; et al. Rationale and design of the treatment of preserved cardiac function heart failure with an aldosterone antagonist trial: A randomized, controlled study of spironolactone in patients with symptomatic heart failure and preserved ejection fraction. *Am. Heart J.* **2011**, *162*, 966–972.e910. [CrossRef] [PubMed]
89. Argun, M.; Uzum, K.; Sonmez, M.F.; Ozyurt, A.; Derya, K.; Cilenk, K.T.; Unalmis, S.; Pamukcu, O.; Baykan, A.; Narin, F.; et al. Cardioprotective effect of metformin against doxorubicin cardiotoxicity in rats. *Anatol. J. Cardiol.* **2016**, *16*, 234–241. [CrossRef] [PubMed]
90. Li, G.; Li, W.R.; Jin, Y.G.; Jie, Q.Q.; Wang, C.Y.; Wu, L. Tetrandrine attenuated doxorubicin-induced acute cardiac injury in mice. *Biomed. Res. Int.* **2020**, *2020*, 2616024. [CrossRef]
91. Quagliariello, V.; Vecchione, R.; Coppola, C.; Di Cicco, C.; De Capua, A.; Piscopo, G.; Paciello, R.; Narciso, V.; Formisano, C.; Taglialatela-Scafati, O.; et al. Cardioprotective effects of nanoemulsions loaded with anti-inflammatory nutraceuticals against doxorubicin-induced cardiotoxicity. *Nutrients* **2018**, *10*, 1304. [CrossRef]
92. Haybar, H.; Goudarzi, M.; Mehrzadi, S.; Aminzadeh, A.; Khodayar, M.J.; Kalantar, M.; Fatemi, I. Effect of gemfibrozil on cardiotoxicity induced by doxorubicin in male experimental rats. *Biomed. Pharmacother.* **2019**, *109*, 530–535. [CrossRef]
93. Mohamed, N.A.; Ahmed, O.M.; Hozayen, W.G.; Ahmed, M.A. Ameliorative effects of bee pollen and date palm pollen on the glycemic state and male sexual dysfunctions in streptozotocin-Induced diabetic wistar rats. *Biomed. Pharmacother.* **2018**, *97*, 9–18. [CrossRef] [PubMed]
94. Alam, M.F.; Khan, G.; Safhi, M.M.; Alshahrani, S.; Siddiqui, R.; Moni, S.S.; Anwer, T. Thymoquinone ameliorates doxorubicin-induced cardiotoxicity in Swiss Albino mice by modulating oxidative damage and cellular inflammation. *Cardiol. Res. Pract.* **2018**, *2018*, 1483041. [CrossRef]
95. Baniahmad, B.; Safaeian, L.; Vaseghi, G.; Rabbani, M.; Mohammadi, B. Cardioprotective effect of vanillic acid against doxorubicin-induced cardiotoxicity in rat. *Res. Pharm. Sci.* **2020**, *15*, 87–96. [CrossRef]
96. Ohta, S.; Fujimaki, T.; Uy, M.M.; Yanai, M.; Yukiyoshi, A.; Hirata, T. Antioxidant hydroxycinnamic acid derivatives isolated from Brazilian bee pollen. *Nat. Prod. Res.* **2007**, *21*, 726–732. [CrossRef]
97. Hassan, W.; El-kashlan, A.M.; Mohamed, N. Egyptian date palm pollen ameliorates testicular dysfunction induced by cadmium chloride in adult male rats. *Am. Sci. J.* **2012**, *8*, 659–669.
98. Caldwell, C.R. Alkylperoxyl radical scavenging activity of red leaf lettuce (*Lactuca sativa* L.) phenolics. *J. Agric. Food Chem.* **2003**, *51*, 4589–4595. [CrossRef]
99. Bai, Z.; Wang, Z. Genistein protects against doxorubicin-induced cardiotoxicity through Nrf-2/HO-1 signaling in mice model. *Environ. Toxicol.* **2019**, *34*, 645–651. [CrossRef]
100. Alanazi, A.M.; Fadda, L.; Alhusaini, A.; Ahmad, R.; Hasan, I.H.; Mahmoud, A.M. Liposomal resveratrol and/or carvedilol attenuate doxorubicin-induced cardiotoxicity by modulating inflammation, oxidative stress and S100a1 in rats. *Antioxidants* **2020**, *9*, 159. [CrossRef]
101. Hafez, H.M.; Hassanein, H. Montelukast ameliorates doxorubicin-induced cardiotoxicity via modulation of p-glycoprotein and inhibition of ROS-mediated TNF-alpha/NF-kappaB pathways. *Drug Chem. Toxicol.* **2020**, *5*, 1–12. [CrossRef] [PubMed]
102. Hu, C.; Zhang, X.; Zhang, N.; Wei, W.Y.; Li, L.L.; Ma, Z.G.; Tang, Q.Z. Osteocrin attenuates inflammation, oxidative stress, apoptosis, and cardiac dysfunction in doxorubicin-induced cardiotoxicity. *Clin. Trans. Med.* **2020**, *2*, 124. [CrossRef]
103. Asensio-Lopez, M.C.; Soler, F.; Pascual-Figal, D.; Fernandez-Belda, F.; Lax, A. Doxorubicin-induced oxidative stress: The protective effect of nicorandil on HL-1 cardiomyocytes. *PLoS ONE* **2017**, *12*, e0172803. [CrossRef]
104. Xiong, C.; Wu, Y.Z.; Zhang, Y.; Wu, Z.X.; Chen, X.Y.; Jiang, P.; Guo, H.C.; Xie, K.R.; Wang, K.X.; Su, S.W. Protective effect of berberine on acute cardiomyopathy associated with doxorubicin treatment. *Oncol. Lett.* **2018**, *15*, 5721–5729. [CrossRef] [PubMed]
105. Ansari, M.A.; Raish, M.; Ahmad, A.; Alkharfy, K.M.; Ahmad, S.F.; Attia, S.M.; Alsaad, A.M.S.; Bakheet, S.A. Sinapic acid ameliorate cadmium-induced nephrotoxicity: In vivo possible involvement of oxidative stress, apoptosis, and inflammation via NF-kappaB downregulation. *Environ. Toxicol. Pharmacol.* **2017**, *51*, 100–107. [CrossRef] [PubMed]
106. Saad, S.Y.; Najjar, T.A.; Al-Rikabi, A.C. The preventive role of deferoxamine against acute doxorubicin-induced cardiac, renal and hepatic toxicity in rats. *Pharmacol. Res.* **2001**, *43*, 211–218. [CrossRef]
107. Carvalho, F.S.; Burgeiro, A.; Garcia, R.; Moreno, A.J.; Carvalho, R.A.; Oliveira, P.J. Doxorubicin-induced cardiotoxicity: From bioenergetic failure and cell death to cardiomyopathy. *Med. Res. Rev.* **2014**, *34*, 106–135. [CrossRef] [PubMed]

108. Mattila, M.; Soderstrom, M.; Ailanen, L.; Savontaus, E.; Savontaus, M. The Effects of neuropeptide Y overexpression on the mouse model of doxorubicin-induced cardiotoxicity. *Cardiovasc. Toxicol.* **2020**, *20*, 328–338. [CrossRef] [PubMed]
109. Menon, S.; Lawrence, L.; Sivaram, V.P.; Padikkala, J. Oroxylum indicum root bark extract prevents doxorubicin-induced cardiac damage by restoring redox balance. *J. Ayurveda Integr. Med.* **2019**, *10*, 159–165. [CrossRef] [PubMed]
110. Lagranha, C.J.; Deschamps, A.; Aponte, A.; Steenbergen, C.; Murphy, E. Sex differences in the phosphorylation of mitochondrial proteins result in reduced production of reactive oxygen species and cardioprotection in females. *Circ. Res.* **2010**, *106*, 1681–1691. [CrossRef]
111. Li, C.; Liu, Y.; Xie, Z.; Lu, Q.; Luo, S. Stigmasterol protects against Ang II-induced proliferation of the A7r5 aortic smooth muscle cell-line. *Food Funct.* **2015**, *6*, 2266–2272. [CrossRef] [PubMed]
112. Lei, L.; Wang, X.; Huang, W.; Liu, Y.; Zheng, F.; Ma, K.Y.; Li, Y.M.; Wang, L.; Man, S.W.; Zhang, C.; et al. Cholesterol side chain analogs but not its ether analogs possess cholesterol-lowering activity. *Food Funct.* **2015**, *6*, 630–634. [CrossRef]
113. Zou, Z.; Xu, X.; Huang, Y.; Xiao, X.; Ma, L.; Sun, T.; Dong, P.; Wang, X.; Lin, X. High serum level of lutein may be protective against early atherosclerosis: The Beijing atherosclerosis study. *Atherosclerosis* **2011**, *219*, 789–793. [CrossRef]
114. Zou, Z.Y.; Xu, X.R.; Lin, X.M.; Zhang, H.B.; Xiao, X.; Ouyang, L.; Huang, Y.M.; Wang, X.; Liu, Y.Q. Effects of lutein and lycopene on carotid intima-media thickness in Chinese subjects with subclinical atherosclerosis: A randomised, double-blind, placebo-controlled trial. *Br. J. Nutr.* **2014**, *111*, 474–480. [CrossRef] [PubMed]
115. Wong, W.Y.; Ward, L.C.; Fong, C.W.; Yap, W.N.; Brown, L. Anti-inflammatory gamma- and delta-tocotrienols improve cardiovascular, liver and metabolic function in diet-induced obese rats. *Eur. J. Nutr.* **2017**, *56*, 133–150. [CrossRef]
116. Ibarra, M.; Perez-Vizcaino, F.; Cogolludo, A.; Duarte, J.; Zaragoza-Arnaez, F.; Lopez-Lopez, J.G.; Tamargo, J. Cardiovascular effects of isorhamnetin and quercetin in isolated rat and porcine vascular smooth muscle and isolated rat atria. *Planta Med.* **2002**, *68*, 307–310. [CrossRef]
117. Timm, K.N.; Tyler, D.J. The role of AMPK activation for cardioprotection in doxorubicin-induced cardiotoxicity. *Cardiovasc. Drugs Ther.* **2020**, *34*, 255–269. [CrossRef] [PubMed]
118. Hamza, A.A.; Ahmed, M.M.; Elwey, H.M.; Amin, A. *Melissa officinalis* protects against doxorubicin-induced cardiotoxicity in rats and potentiates its anticancer activity on MCF-7 cells. *PLoS ONE* **2016**, *11*, e0167049. [CrossRef]
119. Wang, B.; Hao, J.; Jones, S.C.; Yee, M.S.; Roth, J.C.; Dixon, I.M. Decreased Smad 7 expression contributes to cardiac fibrosis in the infarcted rat heart. *Am. J. Physiol. Heart Circ. Physiol.* **2002**, *282*, H1685–H1696. [CrossRef] [PubMed]
120. Cao, Y.-Y.; Li, K.; Li, Y.; Tian, X.-T.; Ba, H.-X.; Wang, A.; Li, X.-H. *Dendrobium candidum* aqueous extract attenuates isoproterenol-induced cardiac hypertrophy through the ERK signalling pathway. *Pharm. Biol.* **2020**, *58*, 176–183. [CrossRef]
121. El-Said, N.T.; Mohamed, E.A.; Taha, R.A. Irbesartan suppresses cardiac toxicity induced by doxorubicin via regulating the p38-MAPK/NF-kappaB and TGF-beta1 pathways. *Naunyn Schmiedebergs Arch. Pharm.* **2019**, *392*, 647–658. [CrossRef]
122. Sun, Z.; Lu, W.; Lin, N.; Lin, H.; Zhang, J.; Ni, T.; Meng, L.; Zhang, C.; Guo, H. Dihydromyricetin alleviates doxorubicin-induced cardiotoxicity by inhibiting NLRP3 inflammasome through activation of SIRT1. *Biochem. Pharmacol.* **2020**, *175*, 113888. [CrossRef]
123. Xu, S.; Wang, Y.; Yu, M.; Wang, D.; Liang, Y.; Chen, Y.; Liao, C.; Xie, Z.; Zhao, B.; Han, J.; et al. LongShengZhi capsule inhibits doxorubicin-induced heart failure by anti-oxidative stress. *Biomed. Pharmacother.* **2020**, *123*, 109803. [CrossRef]
124. Deavall, D.G.; Martin, E.A.; Horner, J.M.; Roberts, R. Drug-induced oxidative stress and toxicity. *J. Toxicol.* **2012**, *2012*, 645460. [CrossRef] [PubMed]
125. Radhiga, T.; Rajamanickam, C.; Sundaresan, A.; Ezhumalai, M.; Pugalendi, K.V. Effect of ursolic acid treatment on apoptosis and DNA damage in isoproterenol-induced myocardial infarction. *Biochimie* **2012**, *94*, 1135–1142. [CrossRef]
126. Zamzami, N.; Marchetti, P.; Castedo, M.; Zanin, C.; Vayssiere, J.L.; Petit, P.X.; Kroemer, G. Reduction in mitochondrial potential constitutes an early irreversible step of programmed lymphocyte death in vivo. *J. Exp. Med.* **1995**, *181*, 1661–1672. [CrossRef] [PubMed]
127. Li, Y.; Wang, B.; Zhou, C.; Bi, Y. Matrine induces apoptosis in angiotensin II-stimulated hyperplasia of cardiac fibroblasts: Effects on Bcl-2/Bax expression and caspase-3 activation. *Basic Clin. Pharm. Toxicol.* **2007**, *101*, 1–8. [CrossRef] [PubMed]
128. Zaki, S.M.; Algaleel, W.A.; Imam, R.A.; Abdelmoaty, M.M. Mesenchymal stem cells pretreated with platelet-rich plasma modulate doxorubicin-induced cardiotoxicity. *Hum. Exp. Toxicol.* **2019**, *38*, 857–874. [CrossRef] [PubMed]

animals

MDPI

Article

Effects of Exercise Combined with Undenatured Type II Collagen on Endurance Capacity, Antioxidant Status, Muscle Lipogenic Genes and E3 Ubiquitin Ligases in Rats

Cemal Orhan [1], Emre Sahin [1], Besir Er [2], Mehmet Tuzcu [2], Andrey P. Lopes [3], Nurhan Sahin [1], Vijaya Juturu [4] and Kazim Sahin [1,*]

1 Animal Nutrition Department, Veterinary Faculty, University of Firat, 23119 Elazig, Turkey; corhan@firat.edu.tr (C.O.); esahin@bingol.edu.tr (E.S.); nsahin@firat.edu.tr (N.S.)
2 Division of Biology, Science Faculty, Firat University, 23119 Elazig, Turkey; beshir.er@hotmail.com (B.E.); mtuzcu@firat.edu.tr (M.T.)
3 Department of Development & Innovation, Lonza, Rio de Janeiro 22793, Brazil; andrey.lopes@lonza.com
4 Department of Research & Development, Lonza, Morristown, NJ 07960, USA; vijaya.juturu@lonza.com
* Correspondence: nsahinkm@yahoo.com; Tel.:+90-532-7473506 or +90-424-2370000 (ext. 3938)

Simple Summary: Undenatured type II collagen (UCII), a collagen product that modulates the immune system by oral tolerance, has become a novel alternative agent to support skeletal system health. The current study explored the impact of UCII on endurance capacity, oxidative stress, inflammation, and antioxidant defense markers in exercised rats. UCII supplementation decreased serum lactate, malondialdehyde, inflammatory marker levels (TNF-α) and improved antioxidant status and lipid metabolism in training rats.

Citation: Orhan, C.; Sahin, E.; Er, B.; Tuzcu, M.; Lopes, A.P.; Sahin, N.; Juturu, V.; Sahin, K. Effects of Exercise Combined with Undenatured Type II Collagen on Endurance Capacity, Antioxidant Status, Muscle Lipogenic Genes and E3 Ubiquitin Ligases in Rats. *Animals* **2021**, *11*, 851. https://doi.org/10.3390/ani11030851

Academic Editors: Woo Kyun Kim and Raffaella Rossi

Received: 1 March 2021
Accepted: 15 March 2021
Published: 17 March 2021

Abstract: The current study aimed to investigate the effect of exercise combined with undenatured type II collagen (UCII) administration on endurance capacity, lipid metabolism, inflammation, and antioxidant status in rats. Twenty-one male Wistar albino rats were divided into three groups as follows: (1) Sedentary control, (2) Exercise (E), (3) Exercise + UCII (4 mg/kg BW/day; E + UCII). The findings showed that the exhaustive running time in the UCII group was significantly prolonged compared to that of the non-supplemented group ($p < 0.001$). When compared to the control group, total serum cholesterol (TC, $p < 0.05$) and triglyceride (TG, $p < 0.05$) levels decreased, while creatinine kinase (CK) levels increased in the E group ($p < 0.001$). Serum creatinine kinase levels were reduced in the E + UCII group compared to the E group ($p < 0.01$). Serum lactate, myoglobin ($p < 0.01$), and osteocalcin levels ($p < 0.01$) increased significantly in exercised rats compared to sedentary control rats, while serum lactate ($p < 0.01$) and myoglobin ($p < 0.0001$) levels decreased in the E + UCII group compared to control. Additionally, UCII supplementation caused significant increases in antioxidant enzyme activities [SOD ($p < 0.01$) and GSH-Px ($p < 0.05$)] and decreases in malondialdehyde (MDA) and tumor necrosis factor (TNF-α) levels ($p < 0.001$). Muscle lipogenic protein (SREBP-1c, ACLY, LXR, and FAS) levels were lower in the E + UCII group than in other groups. In addition, UCII supplementation decreased muscle MAFbx, MuRF-1, myostatin and increased MyoD levels in exercised rats. Moreover, the E + UCII group had lower muscle inflammatory markers [TNF-α ($p < 0.0001$) and IL-1β ($p < 0.01$)] than the control group. These results suggest exercise combined with UCII (4 mg/kg BW/day) modulates lipid, muscle, and antioxidant status in rats.

Keywords: exercise; endurance; undenatured type II collagen; anti-inflammatory; antioxidants; immune response

1. Introduction

Exercise is essential for improving functional capacity, cognitive function and preventing chronic diseases [1]. Many studies have demonstrated the importance of regular exercise in stimulating numerous metabolic health benefits such as metabolic syndrome, hypertension, muscle metabolism, and the antioxidant system [2–6]. Regular exercise strengthens muscle and bone quality around the joint, helps control weight, combat fatigue, and improves life [7]. Hurley et al. [8] reported reduced pain with an inevitable decline of 6%, equivalent to a 1.25-point reduction (improvement) in a review article of 9 clinical studies. In the same report, five exercise studies on health-related quality of life using the 36-item Short Form reported substantial social function benefits with an absolute percent of 7.9% [8]. Regular exercise can stimulate glycosaminoglycan content in cartilage and prevent cartilage damage, including loss of extracellular matrix, inflammation development, and osteophyte formation [9,10].

Regular exercise has a vital role in regulating metabolic function, such as carbohydrate and lipid metabolism [11]. Metabolism disorders caused by insufficient exercise or unbalanced nutrition can induce obesity, diabetes [12], lipogenic genes [sterol regulatory element-binding protein 1 (SREBP-1), liver X receptors (LXR), ATP-citrate lyase (ACL), and fatty acid synthase (FAS)] defects [13], weakness of antioxidant status and inflammation [14]. On the other hand, few studies have found that endurance training increases energy expenses and can trigger lipolytic hormones to facilitate post-exercise energy expenditure [15].

Muscle protein synthesis and degradation are regulated by several signaling pathways, such as the mammalian target of rapamycin (mTOR) and the ubiquitin-proteasome pathways [16]. E3 ubiquitin ligases, including muscle ring finger 1 (MuRF1) and muscle atrophy F-box protein (MAFbx), targets muscle proteins as substrate. During cachexia, MuRF1 and MAFbx are expressed explicitly in atrophying skeletal muscle and mediate muscle protein degradation [17]. Lokireddy et al. [18] demonstrated MAFbx/atrogin-1 activates the breakdown of the myogenic transcription factors MyoD, while Cohen et al. [19] suggested MuRF1 is mainly involved in the degradation of myofibrillar proteins such as myosin heavy chain protein and myosin-binding protein C. Function loses in joints may reduce muscle strength, subsequently muscle mass. Depending on inflammatory inductors, tibialis anterior and quadriceps muscles probably undergo atrophy during knee osteoarthritis (OA), associated with increased MuRF1 expression [20]. Because joint health is closely related to muscle activity and mass [21,22], muscle integrity and joint health can affect each other and provides better endurance capacity in exercise.

The undenatured form of type II collagen (UCII) is safe, non-toxic, and has high antigenic specificity. It is derived from chicken sternum cartilage and is a powdered, glycosylated, and shelf-resistant component [23]. UCII modulates the immune system by oral tolerance, has become a novel alternative agent to support skeletal muscle system health over the past two decades [23,24]. Small amounts of UCII taken orally interact with gut-associated lymphoid tissue, where the naive T cells (Th0) transform into T regulatory cells (Treg) targeting the type II collagen [25]. When these specific Treg cells encounter type II collagen, recognized as an antigen by the immune system, they prevent autoimmune reactions by reducing killer T cell attacks on joint cartilage and stimulates anti-inflammatory cytokine production [25]. Subjects receiving UCII supplementation presented a vital improvement in knee extension compared to placebo [26]. One study reported that UCII increased daily activities, improved quality of life, and no side effects [27]. In a multicenter, randomized study comparing the efficacy and tolerability of UCII and glucosamine hydrochloride plus chondroitin sulfate for six months, the WOMAC score was improved with UCII compared with GC [28]. Previous studies have shown that small UCII doses modulate joint health in arthritis [27]. Tong et al. [29] demonstrated that ingestion of microgram amounts of UCII reduced inflammatory cytokine levels and possibly served to reduce both the incidence and severity of arthritis.

Although there are several inconsistent studies regarding the effects of exercise on inflammation and antioxidant enzyme activities, regular exercise can enhance functional capacity by balancing oxidation processes through increasing resistance against oxidative stress and accelerating recovery from its harmful effects [1,4,5,30]. Several studies have shown that exercise can regulate the mitochondrial antioxidant enzymes and the activity of DNA repair enzymes [31] by reducing the accumulation of lipid peroxidation markers [32] such as MDA in skeletal muscle cells [33]. On the other hand, Yan et al. [34] reported that UCII, a nutritional supplement, improves the antioxidant capacity of the body by increasing SOD activity and decreasing MDA content. They also stated that UCII could alleviate inflammation by regulating inflammatory cytokine levels. Many dietary supplements are widely used to improve performance and reduce muscle fatigue and/or possible damage during physical exercises [5,6,33]. Although there are reports about the positive effects of treadmill exercise on inflammation and antioxidant enzymes, no studies investigate the effects of exercise combined with UCII on endurance capacity, lipid metabolism, oxidative stress, and inflammation markers, lipogenic proteins, and E3 ubiquitin ligases. Hence, we investigated the effects of exercise combined with UCII supplementation on performance, inflammation including IL-1β, TNF-α, oxidative stress, and lipogenic proteins (SREBP-1c, ACLY, LXRs, FAS), and E3 ubiquitin ligases (MAFbx, MuRF-1) in regularly exercised rats.

2. Materials and Methods

2.1. Animals

Twenty-one male Wistar Albino rats (8 weeks old, 180 ± 20 g) were obtained from the Firat University (Elazig, Turkey) and housed in cages at 22 ± 2 °C and controlled lighting (12 h light and 12 h dark). The Firat University of Animal Ethical Committee, Elazig, Turkey (2019/139–206) approved all the study processes.

2.2. Experimental Design

Rats were divided into three groups as follows: (i) Sedentary control (n = 7), (ii) Exercise (E, n = 7), (iii) Exercise+ UCII [(E + UCII, n = 7); 4 mg/kg BW/day)]. UC-II® formulation (lot number#1808021) as the powder was provided by Lonza Consumer Health Inc., Morristown, NJ, USA. UCII and physiological saline as a placebo were given daily by gavage before exercise during the trial period (8 weeks). According to the FDA, the animal dose was calculated by converting the human equivalent dose (HED) [35].

2.3. Exercise Procedure

The exercise was done on the treadmill (Commat Limited, Ankara, Turkey), which comprises a motivation grid at its rear end that provides an electric shock if the animal places in our previous studies [33]. Rats in exercise ran on the treadmill 25 m/min, 45 min/day, and five days per week for eight weeks [36]. This model offers adaptations to the cardiovascular system, comprising the heart representative's physiological remodeling with improved O_2 intake, enhancement of cardiac contractile function, and calcium utilization [36,37]. Exhaustion time and average distance run were noted at the end of each training session.

2.4. Sample Collection

After decapitation with cervical dislocation under anesthesia, blood and gastrocnemius muscle were taken. Serum samples were taken to biochemical gel tubes after centrifugation. The muscle samples were quickly removed and stored at −80 °C for further analysis. Tissue was homogenized within 10 min in 10 volumes of cold Tris 10 mM (pH 7.4). The homogenates were then centrifuged to give the low-speed supernatant fraction used for analysis.

2.5. Biochemical Analysis

Serum samples were analyzed for serum glucose, lipid profile, liver enzymes (AST, ALT), blood urea-N, and creatinine levels by the biochemical analyzer (Samsung Electronics Co., Suwon, Korea). Serum lactate (Cayman Chemical Co., Ann Arbor, MI, USA), myoglobin, and osteocalcin concentrations (MyBioSource, San Diego, CA, USA) were measured by ELISA (Elx-800, Bio-Tek Instruments Inc., Winooski, VT, USA) according to the manufacturer's instructions. The intra- and interassay coefficients of variation for lactate, myoglobin, and osteocalcin kits were <15%. Serum cartilage oligomeric matrix protein (COMP), interleukin 1β, (IL-1β), IL-6, and tumor necrosis factor (TNF-α) levels were also analyzed with ELISA kits (MyBioSource, San Diego, CA, USA) according to the manufacturer's instructions.

The malondialdehyde (MDA) level in samples was detected by HPLD (Shimadzu, Tokyo, Japan) using a UV-vis SPD-10 AVP detector and C18 ODS-3, 5 μm, 4.6 mm × 250 mm column. Antioxidant enzymes (SOD, CAT, GSHPx) were assessed by commercially available kits (Cayman Chemical, Ann Arbor, MI, USA) according to the manufacturer's process.

2.6. Western Blot Method

Muscle SREBP-1c, ACLY, LXRs, FAS, MAFbx, MuRF-1, MyoD, Myostatin, IL-1β, TNF-α, and NCAM levels were determined using the Western blot technique [27]. The muscle homogenates were prepared in ice-cold lysis buffer. SDS-PAGE sample buffer containing 2% β-mercaptoethanol was added to the supernatant. Twenty micrograms of protein were electrophoresed and then transferred into nitrocellulose membranes (Schleicher and Schuell Inc., Keene, NH, USA). Nitrocellulose blots blocked with 1% bovine serum albumin in PBS for one hour prior to administration of the primary antibodies (SREBP-1c, ACLY, LXRs, FAS, MAFbx, MuRF-1, MyoD, Myostatin, TNF-α, IL-1β, and NCAM) (Abcam, Cambridge, UK) that were diluted (1:1000) in the same buffer containing 0.05% Tween-20. Protein loading was checked using an antibody against α-actin (A5316; Sigma Aldrich, St. Louis, MO, USA). Bands were analyzed densitometrically using an image analysis system (Image J; National Institute of Health, Bethesda, MA, USA).

2.7. Statistical Analysis

Data were noted as mean ± SE. The sample size is based on a power of 85% to achieve a p-value of 0.05. Seven animals were tested to get the significance of the treatments per group. Analyses were done with the SPSS software program (IBM SPSS, Version 22.0; Chicago, IL, USA). Significance was detected with a one-way ANOVA followed by a post-hoc Tukey test and determined as significant for probability values less than $p < 0.05$.

3. Results

3.1. Performance and Serum Analyses

Significant body weight changes were observed between the E group and the sedentary control group ($p < 0.0001$ for control vs. E; $p < 0.01$ for control vs. E + UCII; Figure 1A). Exhaustion time increased in the exercise and E + UCII groups compared to the control group ($p < 0.0001$; Figure 1B). Additionally, training combined with UCII increased the exhaustion time by 18.7% in exercised rats ($p < 0.01$). Moreover, while there was no difference between E + UCII and control groups, significant decreases were observed in serum cholesterol (4.0%) and triglyceride (6.0%) in group E compared to the control group ($p < 0.05$). Serum creatinine kinase concentration was higher in the E ($p < 0.0001$) and E + UCII groups compared to the control group ($p < 0.01$). There was no statistical change in liver and kidney function tests for all groups ($p > 0.05$; Table 1).

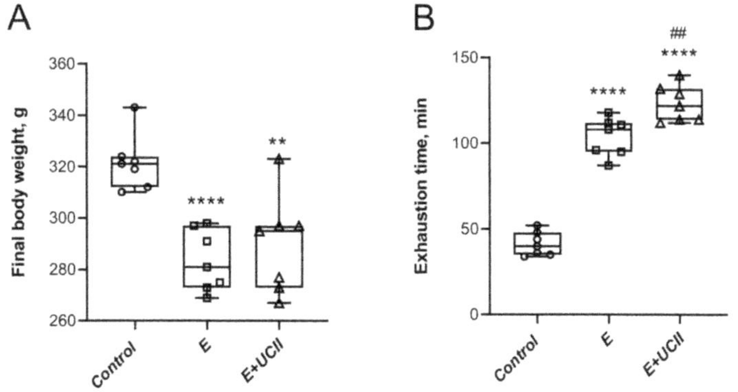

Figure 1. The effect of UCII on body weight (**A**) and exhaustion time (**B**) in exercised rats. Box and Whisker plots show median, min, and max values. ANOVA and Tukey's post-hoc test were used to compare the results among different treatment groups. Statistical Scheme ** $p < 0.01$; **** $p < 0.0001$ compared to the control group and ## $p < 0.01$ compared to the exercise group. UCII, Undenatured type II collagen. Control (Circle): no exercise and no UCII, E (Square): exercised rats, E + UCII (triangle), exercised rats receiving 4 mg/kg BW/day UCII.

Table 1. The effects of UCII on serum parameters in exercised rats ($n = 7$).

Items	Control	Exercise (E)	E + UCII
Glucose, mg/dL	108.29 ± 3.19	102.14 ± 1.77	102.43 ± 1.65
TC, mg/dL	98.51 ± 1.66	92.06 ± 1.57 *	95.70 ± 1.48
Triglyceride, mg/dL	103.51 ± 1.17	96.58 ± 1.74 *	102.40 ± 1.82 #
TP, g/dL	6.56 ± 0.19	6.59 ± 0.11	6.52 ± 0.17
Albumin, g/dL	3.43 ± 0.10	3.51 ± 0.09	3.53 ± 0.12
Globulin, g/dL	3.09 ± 0.12	3.29 ± 0.10	3.19 ± 0.10
ALT, U/L	97.29 ± 4.65	98.43 ± 5.42	95.43 ± 3.00
AST, U/L	118.43 ± 6.21	116.86 ± 6.28	114.43 ± 7.75
TBil, mg/dL	0.24 ± 0.01	0.24 ± 0.01	0.24 ± 0.01
CK, IU/L	125.80 ± 1.75	193.60 ± 2.86 ****	179.98 ± 2.81 ****, ##
Creatine, mg/dL	0.48 ± 0.03	0.47 ± 0.02	0.48 ± 0.04
BUN, mg/dL	20.47 ± 0.77	20.86 ± 0.23	19.44 ± 0.60

Data are expressed as mean± SE. ANOVA and Tukey's post-hoc test were used to compare the results among different treatment groups. Statistical significance between groups is shown by: * $p < 0.05$; **** $p < 0.0001$ compared as control group and, # $p < 0.05$, ## $p < 0.01$ compared as exercise Group. UCII, Undenatured type II collagen; TC, total cholesterol; TG, triglyceride; TP, total protein; ALT, alanine transaminase; AST, aspartate transaminase; TBil, total bilirubin; CK, creatinine kinase; BUN, blood urea nitrogen. Control: no exercise and no UCII, E: exercised rats, E + UCII, exercised rats receiving 4 mg/kg BW/day UCII.

3.2. *Inflammatory and Cartilage Markers*

Concentrations of lactate in serum increased by 2.1 and 1.8 fold in the E group and E + UCII group compared to the control group ($p < 0.0001$ for both; Figure 2A). It was lower by 13.8 in the E + UCII group than the E group ($p < 0.01$). Similarly, serum myoglobin concentrations in the E group and E + UCII group were 1.9 and 1.7 folds higher than in the control group. However, myoglobin concentrations in the E + UCII group were 13.6% lower than in the E group ($p < 0.001$; Figure 2B). Serum osteocalcin levels increased by 14.2% and 15.4% in the E ($p < 0.0001$) and E + UCII ($p < 0.001$) groups compared to the control group (Figure 2C), while osteocalcin levels did not change between E and E + UCII groups ($p > 0.05$). Serum COMP, IL-1β, and IL-6 concentrations were similar among groups ($p > 0.05$; Figure 2D–F). While there was no difference between the control group and E group in terms of serum TNF-α concentration ($p > 0.05$), it was 12.2% lower in the E + UCII group than in the control group ($p < 0.05$; Figure 2G).

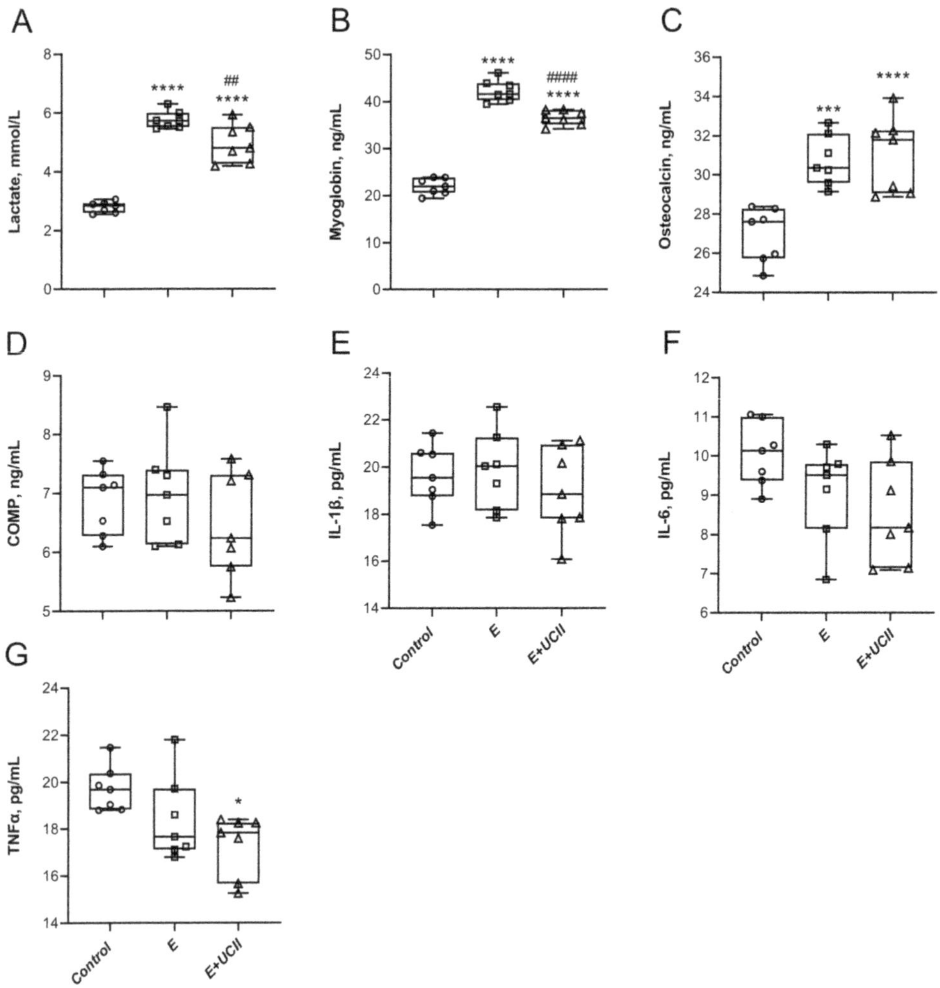

Figure 2. The effect of UCII on the serum lactate (**A**), myoglobin (**B**), osteocalcin (**C**), COMP (**D**), IL-1β (**E**), IL-6 (**F**), and TNFα (**G**) levels in exercised rats. Box and Whisker plots show median, min, and max values. ANOVA and Tukey's post-hoc test were used to compare the results among different treatment groups. Statistical significance between groups is shown by: * $p < 0.05$, *** $p < 0.001$, **** $p < 0.0001$ compared as control group and, ## $p < 0.01$, #### $p < 0.01$ com-pared as exercise Group. UCII, Undenatured type II collagen; COMP, cartilage oligomeric matrix protein; IL-1β, interleukin-1β, IL-6, interleukin-6; TNFα, tumor necrosis factor-alpha. Control (Circle): no exercise and no UCII, E (Square): exercised rats, E + UCII (triangle), exercised rats receiving 4 mg/kg BW/day UCII.

3.3. Oxidative Stress and Antioxidant Properties

A significant decrease in serum and muscle MDA levels of 10.3% and 12.7% was observed in group E compared to control (Figure 3A,B; $p < 0.01$ for both). Besides, serum and muscle MDA levels were 14.3% and 15.0% lower in the E + UCII group than the control group ($p < 0.001$ for both). Serum SOD, CAT, and GSHPx (Figure 3C–E) activities improved in the E and E + UCII groups compared to the control group. Serum SOD and GSHPx levels increased by 12.1% and 11.9% in the E + UCII group compared to the E group ($p < 0.01$ and $p < 0.05$).

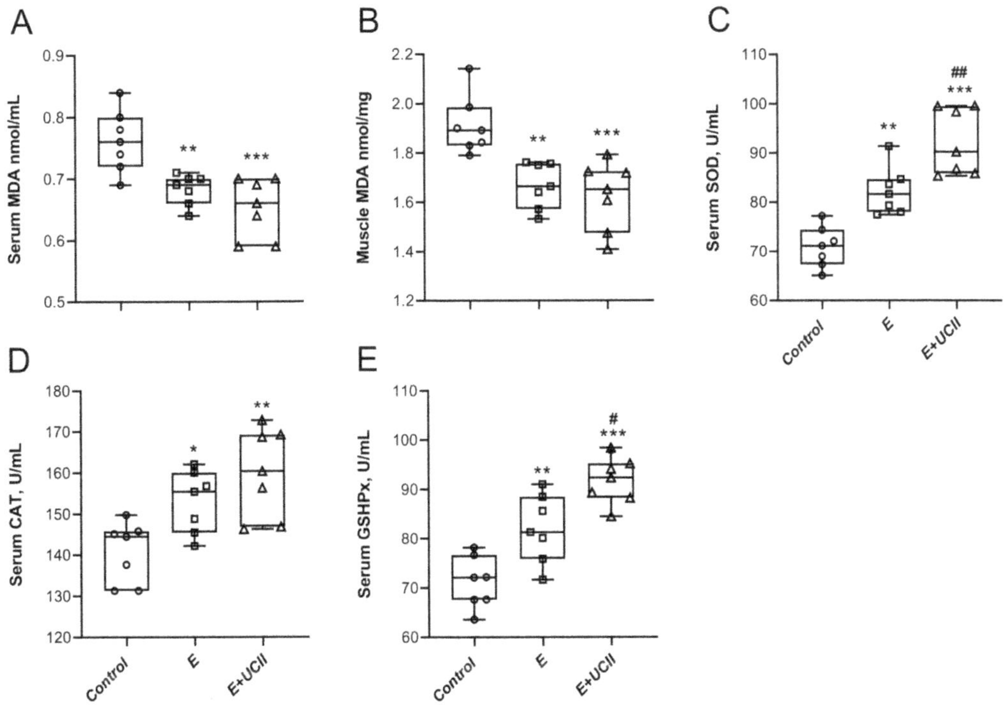

Figure 3. The effects of UCII on the serum MDA (**A**), muscle MDA (**B**), serum SOD (**C**), CAT (**D**), and GSHPx (**E**) levels in exercised rats. Box and Whisker plots show median, min, and max values. ANOVA and Tukey's post-hoc test were used to compare the results among different treatment groups. Statistical significance between groups is shown by: * $p < 0.05$, ** $p < 0.01$, *** $p < 0.001$ compared as control group and # $p < 0.05$, ## $p < 0.01$ compared as exercise Group. UCII, Undenatured type II collagen; MDA, malondialdehyde; SOD, superoxide dismutase, CAT, catalase; GSHPx, glutathione peroxidase, Control (Circle): no exercise and no UCII, E (Square): exercised rats, E + UCII (triangle), exercised rats receiving 4 mg/kg BW/day UCII.

3.4. Muscle Proteins and Inflammatory Cytokines

While exercise alone did not alter the muscle SREBP-1c level, with UCII supplementation, the muscle SREBP-1c was reduced by 35.1% compared to the control ($p < 0.0001$; Figure 4A). Muscle SREBP-1c in the E + UCII group decreased by 27.8% compared to the E group ($p < 0.0001$). In the E + UCII and E groups, muscle ACLY levels reduced by 29.8% and 27.0% ($p < 0.0001$) and did not change at ACLY levels between the E and E + UCII groups ($p > 0.05$; Figure 4B).

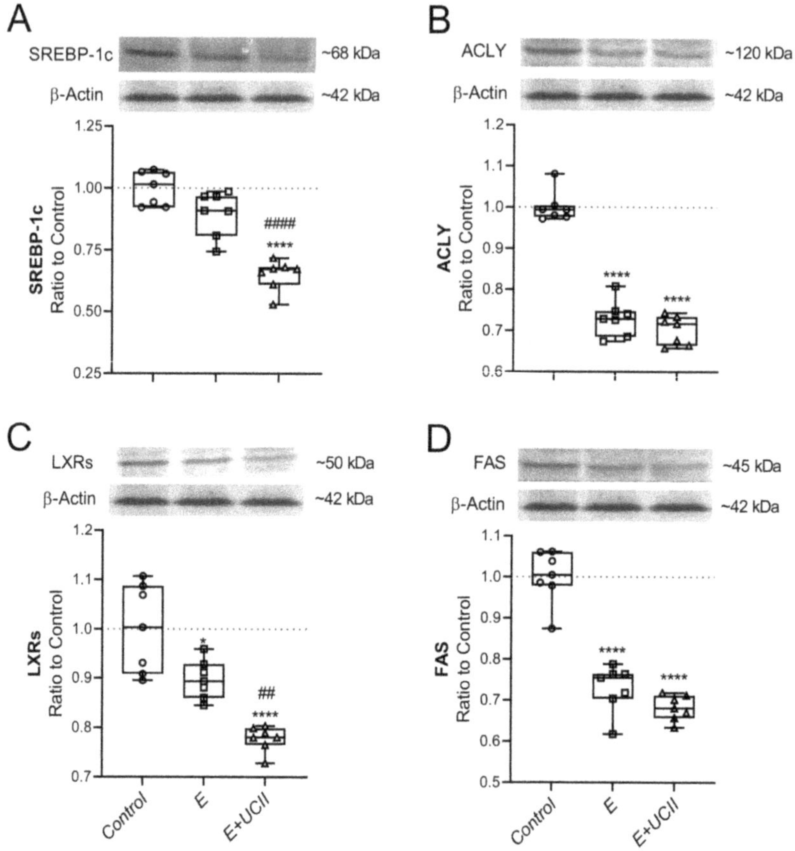

Figure 4. The effect of UCII on the muscle SREBP-1c (**A**), ACLY (**B**), LXRs (**C**), and FAS (**D**) levels in exercised rats. The densitometric analysis of the relative intensity according to the control group of the western blot bands (Figure S1) was performed with β-actin normalization to ensure equal protein loading. Blots were repeated at least three times ($n = 3$), and a representative blot is shown. Box and Whisker plots show median, min, and max values. Data are expressed as a ratio of the control set at 1.0. ANOVA and Tukey's post-hoc test were used to compare the results among different treatment groups. Statistical significance between groups is shown by: * $p < 0.05$; **** $p < 0.0001$ compared as control group and, ## $p < 0.01$; #### $p < 0.0001$ compared as exercise group. E: Exercise; UCII, Undenatured type II collagen; SREBP-1c, Sterol regulatory element-binding protein 1c; ACLY, ATP citrate lyase; LXRs, Liver X receptors; FAS, Fatty acid synthase, Control (Circle): no exercise and no UCII, E (Square): exercised rats, E + UCII (triangle), exercised rats receiving 4 mg/kg BW/day UCII.

LXRs level was lower in E and E + UCII groups than the control group ($p < 0.05$ and $p < 0.0001$; Figure 4C). Moreover, the LXRs level in the E + UCII group was lower than the E group ($p < 0.01$). The exercise alone and exercise combined with UCII supplementation decreased muscle FAS levels ($p < 0.0001$ for both; Figure 4D). However, the muscle FAS levels of E and E + UCII groups were similar ($p > 0.05$; Figure 4D).

Muscle MAFbx, MuRF-1, and Myostatin levels decreased in the E group compared to the control group ($p < 0.001$, $p < 0.0001$ $p < 0.05$, respectively; Figure 5A–C). Muscle MAFbx, MuRF-1, and Myostatin levels were lower in the E + UCII group compared to the control group ($p < 0.0001$, $p < 0.0001$ $p < 0.001$, respectively). However, muscle MAFbx and MuRF1 levels in the E + UCII group were lower than the E group ($p < 0.001$ and $p < 0.0001$). Muscle

MyoD levels increased 2.0 and 2.1 fold in E and E + UCII groups compared to the control group ($p < 0.0001$ for both; Figure 5D).

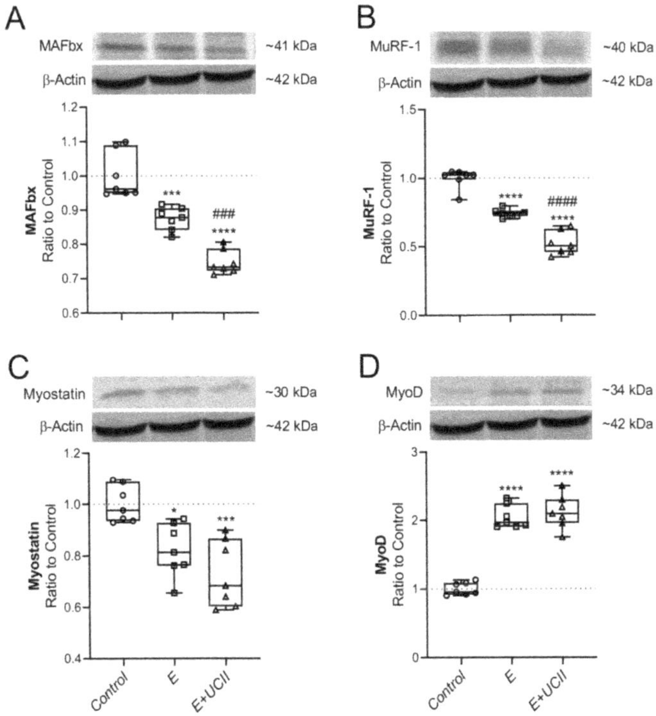

Figure 5. The effect of UCII on the muscle MAFbx (**A**), MuRF-1 (**B**), Myostatin (**C**), and MyoD (**D**) levels in exercised rats. The densitometric analysis of the relative intensity according to the control group of the western blot bands (Figure S2) was performed with β-actin normalization to ensure equal protein loading. Blots were repeated at least three times (n = 3), and a representative blot is shown. Box and Whisker plots show median, min, and max values. Data are expressed as a ratio of the control set at 1.0. ANOVA and Tukey's post-hoc test were used to compare the results among different treatment groups. Statistical significance between groups is shown by: * $p < 0.05$; *** $p < 0.001$; **** $p < 0.0001$ compared as control group and, ### $p < 0.001$; #### $p < 0.0001$ compared as exercise group. E: Exercise; UCII, Undenatured type II collagen; MAFbx, Muscle atrophy F-box; MuRF-1, Muscle RING-finger protein-1; MyoD, Myogenic differentiation factor, Control (Circle): no exercise and no UCII, E (Square): exercised rats, E + UCII (triangle), exercised rats receiving 4 mg/kg BW/day UCII.

Supplementing UCII to exercised rats resulted in a 28.6% reduction in TNF-α levels compared to control rats ($p < 0.0001$; Figure 6A), and similarly, the exercise group reduced TNF-α levels by 18.6 compared to the control groups ($p < 0.0001$). The decrease in muscle TNF levels was more in the E + UCII group than the E group ($p < 0.05$). While muscle IL-1β level did not change in the E group ($p > 0.05$), it decreased by 25.4% in the E + UCII group compared to the control group. Muscle NCAM level neither changed in the E + UCII group nor the E group compared to the control group ($p > 0.05$ for both; Figure 6C).

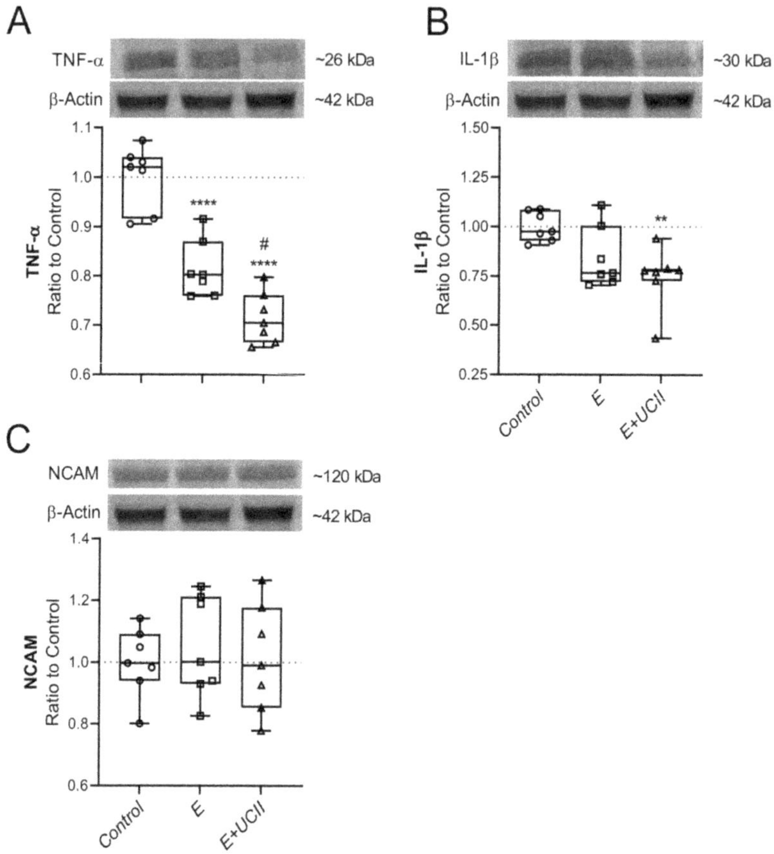

Figure 6. The effect of UCII on the muscle TNF-α (**A**), IL-1β (**B**), and NCAM (**C**) levels in exercised rats. The densitometric analysis of the relative intensity according to the control group of the western blot bands (Figure S3) was performed with β-actin normalization to ensure equal protein loading. Blots were repeated at least three times (n = 3), and a representative blot is shown. Box and Whisker plots show median, min, and max values. Data are expressed as a ratio of the control set at 1.0. ANOVA and Tukey's post-hoc test were used to compare the results among different treatment groups. Statistical significance between groups is shown by: ** $p < 0.01$; **** $p < 0.0001$ compared to the control group and, # $p < 0.05$ compared to the exercise group. E: Exercise; UCII, Undenatured type II collagen; TNF-α, Tumor necrosis factor-α; IL-1β, Interleukin-1β; NCAM, Neural cell adhesion molecules, Control (Circle): no exercise and no UCII, E (Square): exercised rats, E + UCII (triangle), exercised rats receiving 4 mg/kg BW/day UCII.

4. Discussion

This study investigated the efficacy and possible mechanisms of UCII supplementation associated with endurance capacity, lipid metabolism, cartilage markers, and modulation of cartilage markers and antioxidants, and reduced oxidative stress and reduced oxidative stress, regulated muscle lipogenic and E3 Ubiquitin ligases proteins, as well as inflammatory markers including IL-1β and TNF-α in exercised rats.

The mechanism of action of UCII would provide new insights on the development of natural anti-inflammatory properties and immune tolerance and new therapeutic approaches for joint health. The clinically authorized laboratory confirms that the active epitopes in UCII are resistant to digestion and preserve the undenatured 3D structure required to interact with Peyer's patches and induce oral tolerance. Upon consumption, UCII is thought to be taken up by the Peyer's patches to trigger immune cells. It transforms naive T cells into T regulatory (Treg) cells that specifically target type-2 collagen. Then, Treg cells migrate through the circulation. When they identify type-2 collagen in articular cartilage, these collagen-specific regulatory T cells slow down the inflammatory cytokines' production by secreting anti-inflammatory mediators such as TGF-β, IL-4, and IL-10. This action supports anti-inflammatory and cartilage-protecting signaling pathways that prevent the immune system from damaging and erosion joint cartilage while promoting cartilage repair and regeneration. Next, UCII provides relief of joint damage symptoms, which is recognized as oral tolerance and modulating inflammatory pathways [24,25]. Importantly, muscular weakness is one of the most predisposing factors in the progression of OA symptoms. Inflammatory factors lead to presynaptic reflex inhibition and changes in the neuromuscular junctions; thus, muscle atrophy and muscle weakness may be observed after inflammatory status [20]. The current study indicates various modulating pathways to inhibit muscle damage and inflammation, as shown in Figures 1–6.

IL-1β, considered a pronociceptive cytokine, maybe primarily antinociceptive in disease states characterized by thermal hyperalgesia [38]. It independently causes inflammatory responses and catabolic consequences and is combined with other mediators to the articular cartilage and other articular elements. IL-6, considered a cytokine, strongly stimulates immunity and improves inflammatory reaction [38]. According to Haseeb and Haqqi [39], the immune system's participation in OA's development and progression is critical in disease pathogenesis. Mueller and Tuan [40] reported that cytokines might disturb the catabolism and anabolism progressions, most vital in tissues subjected to high mechanical load, including human joints. Thus, there is a progressive articular cartilage degeneration, and this degeneration includes both inflammation and degradation and production processes, which together lead to a gradual loss of joint function and pain. For the first time, the present study results showed that UCII significantly reduced lactate, myoglobin levels and increased osteocalcin levels in exercised rats, suggesting that it could improve endurance capacity. However, there are no studies to compare the lactate, myoglobin, and osteocalcin levels obtained in the present study in exercised rats. Moreover, exercise combined with UCII supplementation did not affect serum COMP, IL-1β, IL-6 in rats. However, UCII supplementation significantly reduced TNFα levels in exercised rats, suggesting that it alleviates inflammation. A recent study conducted by Varney et al. [41] in dogs indicates that UCII supplementation reduces inflammation after exercise, thus supporting our findings. In addition, UCII supplementation for osteoarthritis and healthy subjects demonstrated that 40 mg once a day of UCII supplementation reported efficacy in supporting joint comfort, mobility, flexibility [26,28], and companion animal support joint health and function [23]. Bagi et al. [42] reported the weight-bearing preservation of injured leg capacity and the cancellous bone integrity, suggesting potential for preventing worsening of articular cartilage damage with UCII supplementation the OA rat model. Besides the anti-inflammatory function of UCII, it is prolonged exhaustion compared to the unsupplemented group depending on its joint protecting properties.

Exercise triggers reactive oxygen species formation that acts as significant mediators and cellular adaptations, modulation of antioxidant protection, and repair of oxidative damage [43]. The present study data are consistent with the earlier studies showing that regular training increases antioxidant capacity, possibly adaptation to increased oxidative stress [4,5,44]. In addition, we found that exercise combined with UCII supplementation increased the activities of antioxidant enzymes (SOD and GSHPx) in rats. It indicates UCII may help to improve the antioxidant capacity in exercised rats. Moreover, decreased muscle TNF-α and IL-1β levels support that UCII has anti-inflammation activity on muscle in exercised rats. Though there is no literature to compare the effects of UCII on antioxidant levels in rats that were exercised, similar findings were obtained in OA rats. For example, Yan et al. [34] reported that UCII increased serum SOD activity in OA rats.

SREBP-1c, directly stimulated by LXR, induces fatty acid and triglyceride production by upregulating some lipogenic genes, including FAS and ACLY [45]. In skeletal muscle, lipid utilization increases with exercise, and muscle SREBP-1c protein levels tend to elevate after training [46]. However, SREBP actions may be altered with the type of training and energy intake. For example, Jeong et al. [47] determined that 8-week low-fat diet and exercise program reduces muscle SREBP-1c and TNF-α levels in high-fat diet (HFD) fed C57BL/6J mice. Likewise, muscle LXR levels were reduced in HFD fed rats after regular exercise (4 weeks) [48]. Additionally, Smith et al. [49] observed a reduction of muscle SREBP-1c mRNA levels in men at the end of the six-month endurance exercise. In contrast, de Souza Cordeiro et al. [50] reported that gastrocnemius muscle SREBP-1c expression increased after 10-week aerobic exercise in rats. Long-term exercise and caloric restriction could also increase the SREBP-1c levels independently in gastrocnemius and soleus muscles in rats [51]. SREBP-1c, ACLY, FAS, and LXRs protein levels were reduced after exercise in the current study. UCII supplementation further lowered SREBP-1c and LXRs levels compared to the exercised group. Unfortunately, the effects of exercise on muscle lipid metabolism still unclear, and we could not reach any paper to discuss the impact of UCII on muscle lipid metabolism.

Muscle wasting is inevitably related to aging, and, lately, it has been revealed in patients with OA [52]. The ubiquitin ligase MAFbx shows a critical role in muscle loss through regulating MyoD degradation. MAFbx expression, enhanced by myostatin and inflammatory disorders like OA, in muscle inhibits MyoD, promoting muscle regeneration [53]. In mice, mRNA expression and protein levels of MAFbx decreased in muscle after exercise, whereas MyoD protein levels increased [54]. Moreover, MuRF1 plays a vital role in muscle remodeling and triggers muscle protein degradation via ubiquitination [55]. Following repeated resistance exercise, MuRF-1 and MAFbx levels could be inhibited in human muscle, promoting muscle regeneration [56]. Recently, Zeng et al. [57] proved that E3 ubiquitin ligases such as MuRF1 and Atrogin-1 decreased after different types of exercise in aged rats. Similarly, we found that exercise reduced MAFbx, MuRF-1, and myostatin levels in muscle, whereas MyoD levels increased. UCII treatment decreased MAFbx and MuRF-1 compared to the exercise group. Although there is inadequate data on the effect of UCII on muscle functions, these results showed that UCII might regulate muscle metabolism by regulating MAFbx, MuRF-1, and myostatin.

Our study has a limitation, primarily related to the lack of data in the UCII-receiving sedentary control group alone. We could not show whether the use of UCII in healthy non-exercised animals improves the measured parameters. In addition, how UCII regulates the expression of anti-inflammatory cytokines inhibits the synthesis of inflammatory cytokines also needs in healthy animals in further study. However, the primary goal of this study is to investigate the effects of UCII combined with exercise in rats. We suggest that training combined with UCII supplementation may improve lipid, muscle, and antioxidant status in the exercised rat model; however, more in-depth studies are needed to confirm it. On the other hand, animal studies are models based on indications and effects; they help human studies move on. Based on the results, human studies should be planned. These results are not to exaggerate to humans. Further human double-blind studies should be designed in exercise or sporting activities.

5. Conclusions

These results suggest UCII supplementation with exercise modulates lipid, muscle, and antioxidant status in the exercised rat model. The action of anti-inflammatory cytokines inhibits the synthesis of inflammatory cytokines in muscle. These results also contribute significantly to expanding the academic community's knowledge of the increase in UCII benefits combined with physical exercise by modulating inflammatory markers and antioxidant status. Furthermore, clinical studies are needed to demonstrate the effects of exercise combined with UCII in large animals and humans.

Supplementary Materials: The following are available online at https://www.mdpi.com/2076-2615/11/3/851/s1, Figure S1. Full immunoblots related to Figure 4 (SREB-1C (A), ACLY (B), LXRs (C), FAS (D); β-Actin (E)); Figure S2. Full immunoblots related to Figure 5 (MAFbx (A), MuRF-1 (B), Myostatin (C), Myo D (D), and β-Actin (E)); Figure S3. Full immunoblots related to Figure 6 (TNF-α (A), IL1-β (B), NCAM (C) and β-Actin (D)).

Author Contributions: Conceptualization, K.S.; methodology, K.S., N.S., C.O.; validation, K.S., N.S., C.O., and M.T.; formal analysis, N.S., C.O., E.S., B.E., and M.T.; investigation, E.S., B.E. and M.T.; data curation, C.O., N.S., E.S., B.E.; writing—original draft preparation, A.P.L.; K.S., and V.J.; writing—review and editing, K.S., V.J., and A.P.L.; supervision, K.S.; project administration, K.S.; funding acquisition, K.S. All authors have read and agreed to the published version of the manuscript.

Funding: This project was supported by Lonza Consumer Health (Morristown, NJ, USA) and the Turkish Academy of Sciences (KS).

Institutional Review Board Statement: The study was conducted according to the guidelines of the Declaration of Helsinki, and approved by the experiment was conducted under the protocol approved by the Firat University Animal Ethical Committee, Elazig, Turkey (2019/139-206).

Data Availability Statement: The data presented in this study are available within the article.

Acknowledgments: The authors thank Lonza Consumer Health (Morristown, NJ, USA) for financial support. This work was also supported in part by the Turkish Academy of Sciences (KS). This article was presented at the Clinical Congress of Rheumatology on 8 October 2020.

Conflicts of Interest: V.J. and A.P.L. are an employee of Lonza Consumer Health (Morristown, NJ, USA, and Rio de Janeiro, Brazil). Other authors declare no conflict of interest.

References

1. Steele, J.; Fisher, J.; Skivington, M.; Dunn, C.; Arnold, J.; Tew, G.; Batterham, A.M.; Nunan, D.; O'Driscoll, J.M.; Mann, S.; et al. A higher effort-based paradigm in physical activity and exercise for public health: Making the case for a greater emphasis on resistance training. *BMC Public Health* **2017**, *17*, 300. [CrossRef] [PubMed]
2. Elokda, A.S.; Nielsen, D.H. Effects of exercise training on the glutathione antioxidant system. *Eur. J. Cardiovasc. Prev. Rehabil.* **2007**, *14*, 630–637. [CrossRef]
3. Kemmler, W.; Von Stengel, S.; Engelke, K.; Kalender, W.A. Exercise decreases the risk of metabolic syndrome in elderly females. *Med. Sci. Sports Exerc.* **2009**, *41*, 297–305. [CrossRef] [PubMed]
4. Rosety-Rodriguez, M.; Rosety, I.; Fornieles-Gonzalez, G.; Diaz-Ordonez, A.J.; Camacho, A.; Rosety, M.A.; Pardo, A.; Rosety, M.; Alvero, R.; Ordonez, F.J. A 6-week training program increased muscle antioxidant system in elderly diabetic fatty rats. *Med. Sci. Monit.* **2012**, *18*, BR346–BR350. [CrossRef]
5. Silva, E.P., Jr.; Borges, L.S.; Mendes-da-Silva, C.; Hirabara, S.M.; Lambertucci, R.H. l-Arginine supplementation improves rats' antioxidant system and exercise performance. *Free Radic. Res.* **2017**, *51*, 281–293. [CrossRef]
6. Sahin, K.; Orhan, C.; Tuzcu, M.; Sahin, N.; Erten, F.; Juturu, V. Capsaicinoids improve consequences of physical activity. *Toxicol. Rep.* **2018**, *5*, 598–607. [CrossRef]
7. Musumeci, G. Effects of exercise on physical limitations and fatigue in rheumatic diseases. *World J. Orthop.* **2015**, *6*, 762–769. [CrossRef] [PubMed]
8. Hurley, M.; Dickson, K.; Hallett, R.; Grant, R.; Hauari, H.; Walsh, N.; Stansfield, C.; Oliver, S. Exercise interventions and patient beliefs for people with hip, knee or hip and knee osteoarthritis: A mixed methods review. *Cochrane Database Syst. Rev.* **2018**, *4*, CD010842. [CrossRef] [PubMed]
9. Iijima, H.; Aoyama, T.; Ito, A.; Tajino, J.; Yamaguchi, S.; Nagai, M.; Kiyan, W.; Zhang, X.; Kuroki, H. Exercise intervention increases expression of bone morphogenetic proteins and prevents the progression of cartilage-subchondral bone lesions in a post-traumatic rat knee model. *Osteoarthr. Cartil.* **2016**, *24*, 1092–1102. [CrossRef]
10. Smith, J.K. Exercise as an adjuvant to cartilage regeneration therapy. *Int. J. Mol. Sci.* **2020**, *21*, 9471. [CrossRef] [PubMed]
11. Thyfault, J.P.; Bergouignan, A. Exercise and metabolic health: Beyond skeletal muscle. *Diabetologia* **2020**, *63*, 1464–1474. [CrossRef] [PubMed]
12. Hemmingsen, B.; Gimenez-Perez, G.; Mauricio, D.; Roque, I.F.M.; Metzendorf, M.I.; Richter, B. Diet, physical activity or both for prevention or delay of type 2 diabetes mellitus and its associated complications in people at increased risk of developing type 2 diabetes mellitus. *Cochrane Database Syst. Rev.* **2017**, *12*, CD003054. [CrossRef] [PubMed]
13. Solinas, G.; Boren, J.; Dulloo, A.G. De novo lipogenesis in metabolic homeostasis: More friend than foe? *Mol. Metab.* **2015**, *4*, 367–377. [CrossRef] [PubMed]
14. Booth, F.W.; Roberts, C.K.; Laye, M.J. Lack of exercise is a major cause of chronic diseases. *Compr. Physiol.* **2012**, *2*, 1143–1211. [PubMed]

15. Irving, B.A.; Davis, C.K.; Brock, D.W.; Weltman, J.Y.; Swift, D.; Barrett, E.J.; Gaesser, G.A.; Weltman, A. Effect of exercise training intensity on abdominal visceral fat and body composition. *Med. Sci. Sports Exerc.* **2008**, *40*, 1863–1872. [CrossRef]
16. Yoon, M.S. mTOR as a Key Regulator in Maintaining Skeletal Muscle Mass. *Front. Physiol.* **2017**, *8*, 788. [CrossRef]
17. Glass, D.J. Signaling pathways perturbing muscle mass. *Curr. Opin. Clin. Nutr. Metab. Care* **2010**, *13*, 225–229. [CrossRef] [PubMed]
18. Lokireddy, S.; Wijesoma, I.W.; Sze, S.K.; McFarlane, C.; Kambadur, R.; Sharma, M. Identification of atrogin-1-targeted proteins during the myostatin-induced skeletal muscle wasting. *Am. J. Physiol. Cell Physiol.* **2012**, *303*, C512–C529. [CrossRef]
19. Cohen, S.; Brault, J.J.; Gygi, S.P.; Glass, D.J.; Valenzuela, D.M.; Gartner, C.; Latres, E.; Goldberg, A.L. During muscle atrophy, thick, but not thin, filament components are degraded by MuRF1-dependent ubiquitylation. *J. Cell Biol.* **2009**, *185*, 1083–1095. [CrossRef]
20. Cunha, J.E.; Barbosa, G.M.; Castro, P.; Luiz, B.L.F.; Silva, A.C.A.; Russo, T.L.; Vasilceac, F.A.; Cunha, T.M.; Cunha, F.Q.; Salvini, T.F. Knee osteoarthritis induces atrophy and neuromuscular junction remodeling in the quadriceps and tibialis anterior muscles of rats. *Sci. Rep.* **2019**, *9*, 6366. [CrossRef] [PubMed]
21. Heiden, T.L.; Lloyd, D.G.; Ackland, T.R. Knee joint kinematics, kinetics and muscle co-contraction in knee osteoarthritis patient gait. *Clin. Biomech.* **2009**, *24*, 833–841. [CrossRef] [PubMed]
22. Van der Esch, M.; Steultjens, M.; Knol, D.L.; Dinant, H.; Dekker, J. Joint laxity and the relationship between muscle strength and functional ability in patients with osteoarthritis of the knee. *Arthritis Rheum.* **2006**, *55*, 953–959. [CrossRef] [PubMed]
23. Gencoglu, H.; Orhan, C.; Sahin, E.; Sahin, K. Undenatured Type II Collagen (UC-II) in Joint Health and Disease: A Review on the Current Knowledge of Companion Animals. *Animals* **2020**, *10*, 697. [CrossRef] [PubMed]
24. Bagchi, D.; Misner, B.; Bagchi, M.; Kothari, S.C.; Downs, B.W.; Fafard, R.D.; Preuss, H.G. Effects of orally administered undenatured type II collagen against arthritic inflammatory diseases: A mechanistic exploration. *Int. J. Clin. Pharmacol. Res.* **2002**, *22*, 101–110.
25. Park, K.S.; Park, M.J.; Cho, M.L.; Kwok, S.K.; Ju, J.H.; Ko, H.J.; Park, S.H.; Kim, H.Y. Type II collagen oral tolerance; mechanism and role in collagen-induced arthritis and rheumatoid arthritis. *Mod. Rheumatol.* **2009**, *19*, 581–589. [CrossRef]
26. Lugo, J.P.; Saiyed, Z.M.; Lau, F.C.; Molina, J.P.; Pakdaman, M.N.; Shamie, A.N.; Udani, J.K. Undenatured type II collagen (UC-II(R)) for joint support: A randomized, double-blind, placebo-controlled study in healthy volunteers. *J. Int. Soc. Sports Nutr.* **2013**, *10*, 48. [CrossRef] [PubMed]
27. Crowley, D.C.; Lau, F.C.; Sharma, P.; Evans, M.; Guthrie, N.; Bagchi, M.; Bagchi, D.; Dey, D.K.; Raychaudhuri, S.P. Safety and efficacy of undenatured type II collagen in the treatment of osteoarthritis of the knee: A clinical trial. *Int. J. Med. Sci.* **2009**, *6*, 312–321. [CrossRef]
28. Lugo, J.P.; Saiyed, Z.M.; Lane, N.E. Efficacy and tolerability of an undenatured type II collagen supplement in modulating knee osteoarthritis symptoms: A multicenter randomized, double-blind, placebo-controlled study. *Nutr. J.* **2016**, *15*, 14. [CrossRef] [PubMed]
29. Tong, T.; Zhao, W.; Wu, Y.Q.; Chang, Y.; Wang, Q.T.; Zhang, L.L.; Wei, W. Chicken type II collagen induced immune balance of main subtype of helper T cells in mesenteric lymph node lymphocytes in rats with collagen-induced arthritis. *Inflamm. Res.* **2010**, *59*, 369–377. [CrossRef]
30. Radak, Z.; Chung, H.Y.; Koltai, E.; Taylor, A.W.; Goto, S. Exercise, oxidative stress and hormesis. *Ageing Res. Rev.* **2008**, *7*, 34–42. [CrossRef] [PubMed]
31. Bo, H.; Kang, W.; Jiang, N.; Wang, X.; Zhang, Y.; Ji, L.L. Exercise-induced neuroprotection of hippocampus in APP/PS1 transgenic mice via upregulation of mitochondrial 8-oxoguanine DNA glycosylase. *Oxid. Med. Cell. Longev.* **2014**, *2014*, 834502. [CrossRef]
32. Sallam, N.; Laher, I. Exercise Modulates Oxidative Stress and Inflammation in Aging and Cardiovascular Diseases. *Oxid. Med. Cell. Longev.* **2016**, *2016*, 7239639. [CrossRef]
33. Sahin, K.; Pala, R.; Tuzcu, M.; Ozdemir, O.; Orhan, C.; Sahin, N.; Juturu, V. Curcumin prevents muscle damage by regulating NF-kappaB and Nrf2 pathways and improves performance: An in vivo model. *J. Inflamm. Res.* **2016**, *9*, 147–154.
34. Yan, Z.; Zhao, H.; Liu, A.; Liu, S.; Zou, G.; Wang, H. The Effects of Undenatured Type II Collagen on Inflammatory Mediators and Oxidative Stress in an Osteoarthritis Rat Model. *IOP Conf. Ser. Earth Environ. Sci.* **2020**, *598*, 012067. [CrossRef]
35. Shin, J.W.; Seol, I.C.; Son, C.G. Interpretation of animal dose and human equivalent dose for drug development. *Korean J. Intern. Med.* **2010**, *31*, 1–7.
36. Feng, R.; Wang, L.; Li, Z.; Yang, R.; Liang, Y.; Sun, Y.; Yu, Q.; Ghartey-Kwansah, G.; Sun, Y.; Wu, Y.; et al. A systematic comparison of exercise training protocols on animal models of cardiovascular capacity. *Life Sci.* **2019**, *217*, 128–140. [CrossRef]
37. Rivas-Estany, E.; Sixto-Fernandez, S.; Barrera-Sarduy, J.; Hernandez-Garcia, S.; Gonzalez-Guerra, R.; Stusser-Beltranena, R. Effects of long-term exercise training on left ventricular function and remodeling in patients with anterior wall myocardial infarction. *Arch. Cardiol. Mex.* **2013**, *83*, 167–173.
38. Wojdasiewicz, P.; Poniatowski, L.A.; Szukiewicz, D. The role of inflammatory and anti-inflammatory cytokines in the pathogenesis of osteoarthritis. *Mediat. Inflamm.* **2014**, *2014*, 561459. [CrossRef] [PubMed]
39. Haseeb, A.; Haqqi, T.M. Immunopathogenesis of osteoarthritis. *Clin. Immunol.* **2013**, *146*, 185–196. [CrossRef]
40. Mueller, M.B.; Tuan, R.S. Anabolic/Catabolic balance in pathogenesis of osteoarthritis: Identifying molecular targets. *PM&R* **2011**, *3*, S3–S11.

41. Varney, J.L.; Fowler, J.W.; Coon, C.N. PSVI-33 Undenatured type II collagen mitigates inflammation and cartilage degeneration in healthy untrained Labrador retrievers after exercise. *J. Anim. Sci.* **2020**, *98*, 313–314. [CrossRef]
42. Bagi, C.M.; Berryman, E.R.; Teo, S.; Lane, N.E. Oral administration of undenatured native chicken type II collagen (UC-II) diminished deterioration of articular cartilage in a rat model of osteoarthritis (OA). *Osteoarthr. Cartil.* **2017**, *25*, 2080–2090. [CrossRef] [PubMed]
43. Giles, L.V.; Koehle, M.S. The health effects of exercising in air pollution. *Sports Med.* **2014**, *44*, 223–249. [CrossRef]
44. Powers, S.K.; Ji, L.L.; Leeuwenburgh, C. Exercise training-induced alterations in skeletal muscle antioxidant capacity: A brief review. *Med. Sci. Sports Exerc.* **1999**, *31*, 987–997. [CrossRef] [PubMed]
45. Porstmann, T.; Santos, C.R.; Griffiths, B.; Cully, M.; Wu, M.; Leevers, S.; Griffiths, J.R.; Chung, Y.L.; Schulze, A. SREBP activity is regulated by mTORC1 and contributes to Akt-dependent cell growth. *Cell Metab.* **2008**, *8*, 224–236. [CrossRef] [PubMed]
46. Ikeda, S.; Miyazaki, H.; Nakatani, T.; Kai, Y.; Kamei, Y.; Miura, S.; Tsuboyama-Kasaoka, N.; Ezaki, O. Up-regulation of SREBP-1c and lipogenic genes in skeletal muscles after exercise training. *Biochem. Biophys. Res. Commun.* **2002**, *296*, 395–400. [CrossRef]
47. Jeong, J.H.; Park, H.G.; Lee, Y.R.; Lee, W.L. Moderate exercise training is more effective than resveratrol supplementation for ameliorating lipid metabolic complication in skeletal muscle of high fat diet-induced obese mice. *J. Exerc. Nutr. Biochem.* **2015**, *19*, 131–137. [CrossRef]
48. Yu, Q.; Xia, Z.; Liong, E.C.; Tipoe, G.L. Chronic aerobic exercise improves insulin sensitivity and modulates Nrf2 and NFkappaB/IkappaBalpha pathways in the skeletal muscle of rats fed with a high fat diet. *Mol. Med. Rep.* **2019**, *20*, 4963–4972.
49. Smith, I.J.; Huffman, K.M.; Durheim, M.T.; Duscha, B.D.; Kraus, W.E. Sex-specific alterations in mRNA level of key lipid metabolism enzymes in skeletal muscle of overweight and obese subjects following endurance exercise. *Physiol. Genom.* **2009**, *36*, 149–157. [CrossRef]
50. De Souza Cordeiro, L.M.; Mario, E.G.; Moreira, C.C.L.; Rodrigues, A.H.; Wanner, S.P.; Soares, D.D.; Botion, L.M.; Ferreira, A.V.M. Aerobic training induces differential expression of genes involved in lipid metabolism in skeletal muscle and white adipose tissues. *J. Cell Biochem.* **2019**, *120*, 18883–18893. [CrossRef]
51. Nadeau, K.J.; Ehlers, L.B.; Aguirre, L.E.; Moore, R.L.; Jew, K.N.; Ortmeyer, H.K.; Hansen, B.C.; Reusch, J.E.; Draznin, B. Exercise training and calorie restriction increase SREBP-1 expression and intramuscular triglyceride in skeletal muscle. *Am. J. Physiol. Endocrinol. Metab.* **2006**, *291*, E90–E98. [CrossRef] [PubMed]
52. Larsson, L.; Degens, H.; Li, M.; Salviati, L.; Lee, Y.I.; Thompson, W.; Kirkland, J.L.; Sandri, M. Sarcopenia: Aging-Related Loss of Muscle Mass and Function. *Physiol. Rev.* **2019**, *99*, 427–511. [CrossRef] [PubMed]
53. Silva, J.M.S.; Alabarse, P.V.G.; Teixeira, V.O.N.; Freitas, E.C.; de Oliveira, F.H.; Chakr, R.; Xavier, R.M. Muscle wasting in osteoarthritis model induced by anterior cruciate ligament transection. *PLoS ONE* **2018**, *13*, e0196682. [CrossRef] [PubMed]
54. Okada, A.; Ono, Y.; Nagatomi, R.; Kishimoto, K.N.; Itoi, E. Decreased muscle atrophy F-box (MAFbx) expression in regenerating muscle after muscle-damaging exercise. *Muscle Nerve* **2008**, *38*, 1246–1253. [CrossRef]
55. Koyama, S.; Hata, S.; Witt, C.C.; Ono, Y.; Lerche, S.; Ojima, K.; Chiba, T.; Doi, N.; Kitamura, F.; Tanaka, K.; et al. Muscle RING-finger protein-1 (MuRF1) as a connector of muscle energy metabolism and protein synthesis. *J. Mol. Biol.* **2008**, *376*, 1224–1236. [CrossRef] [PubMed]
56. Mascher, H.; Tannerstedt, J.; Brink-Elfegoun, T.; Ekblom, B.; Gustafsson, T.; Blomstrand, E. Repeated resistance exercise training induces different changes in mRNA expression of MAFbx and MuRF-1 in human skeletal muscle. *Am. J. Physiol. Endocrinol. Metab.* **2008**, *294*, E43–E51. [CrossRef] [PubMed]
57. Zeng, Z.; Liang, J.; Wu, L.; Zhang, H.; Lv, J.; Chen, N. Exercise-Induced Autophagy Suppresses Sarcopenia Through Akt/mTOR and Akt/FoxO3a Signal Pathways and AMPK-Mediated Mitochondrial Quality Control. *Front. Physiol.* **2020**, *11*, 583478. [CrossRef] [PubMed]

animals

Article

Effect of a Phytogenic Feed Additive on Growth Performance, Nutrient Digestion, and Immune Response in Broiler-Fed Diets with Two Different Levels of Crude Protein

Jinquan Wang [1], Shengchen Su [1], Chasity Pender [2], Raj Murugesan [2,3], Basharat Syed [3] and Woo Kyun Kim [1,*]

1 Department of Poultry Science, University of Georgia, Athens, GA 30602, USA; jq.wang@uga.edu (J.W.); su1@uga.edu (S.S.)
2 BIOMIN America Inc., Overland Park, KS 66210, USA; chasity.pender@biomin.net (C.P.); raj.murugesan@biomin.net (R.M.)
3 BIOMIN Holding GmbH, 3130 Getzersdorf, Austria; basharat.syed@biomin.net
* Correspondence: wkkim@uga.edu

Simple Summary: The rising concerns on antibiotics resistance from using antibiotics in animal production has resulted in an increase in researches on antibiotic alternatives. A phytogenic feed additive from a blend of extracts of oregano, cinnamon, citrus peel, and fructooligosaccharides was evaluated in the present study. The objective of the present study is not only to evaluate the effect of phytogenic feed additive on broiler performance, but also to explore the potential mode of actions through immune response, digestive enzyme activities, nutrient transporter gene expressions and nutrient digestibility. Supplementation of phytogenic feed additives improved broiler FCR through stimulating ileum immunity.

Citation: Wang, J.; Su, S.; Pender, C.; Murugesan, R.; Syed, B.; Kim, W.K. Effect of a Phytogenic Feed Additive on Growth Performance, Nutrient Digestion, and Immune Response in Broiler-Fed Diets with Two Different Levels of Crude Protein. *Animals* **2021**, *11*, 775. https://doi.org/10.3390/ani11030775

Academic Editor: Raffaella Rossi

Received: 12 February 2021
Accepted: 8 March 2021
Published: 11 March 2021

Abstract: The aim of this experiment was to evaluate the effect of a phytogenic feed additive (PFA) on growth performance and nutrient digestibility of broilers fed corn and soybean meal-based diets containing two different levels of crude protein. A 2 × 2 completely randomized factorial arrangement (eight replicates/treatment, 30 birds/replicate) was conducted with a positive control (PC) and negative control (NC) containing crude protein at standard or reduced by 1.5% (equivalent to a reduction of 15 g/kg), respectively, and supplementation of PFA at 0 or 125 ppm of diet. There were no significant interactions found between PFA and CP levels in the current study. Main effect analysis showed that during 0–42 d of age NC diets decreased body weight gain ($p < 0.05$), but increased feed intake ($p < 0.05$) and feed conversion ratio (FCR, $p < 0.01$), whereas supplementation of PFA resulted in a lower FCR ($p < 0.01$). The ileal nutrient digestibility was reduced ($p < 0.05$) in the broilers fed a reduced protein diet at 21 d compared to the standard protein level group, but there were no effects for PFA levels. Similarly, supplementing PFAs showed no effects on digestive enzyme (Alkaline phosphatase, amylase, and lipase) activity in jejunal digesta and jejunal brush border enzyme (maltase, sucrase, and aminopeptidase) activity. Supplementation of PFA downregulated ($p < 0.05$) the mRNA expressions of cytochrome P450 1A and interleukin 6 in the ileum but had no effects on nutrient transporter genes in the jejunum. In conclusion, supplementation of PFA reduced broiler FCR during the whole grow-out period and positively regulated the immune responses in the ileum.

Keywords: broiler; dietary protein level; phytogenic feed additive; growth performance; immune response

1. Introduction

The ban of using antibiotics as growth promoters have spurred research into using plant-derived compounds named phytogenic feed additives (PFAs). The use of PFAs, categorized as sensory and flavoring compounds by European Union legislation (EC 1831/2003), from herbs or spices as antibiotic alternatives are generally recognized as safe [1]. The botanical constituents used in broiler diets as a single compound or multiple cocktails exhibit growth-promoting, immune-regulatory, antimicrobial, stimulating nutrient digestibility and antioxidant properties [2–6].

Dietary protein is a major contributing factor in driving feed cost. Reducing the crude protein level in the broiler diet has recently attracted much attention since it could reduce the feed cost and nitrogen excretion [7,8]. The previous study demonstrated the inclusion of PFA stimulated the nutrient digestibility and small intestine villus height of broilers fed a corn soybean diet [4], however, it remains unknown for the effects of PFA in a low protein diet, and there might be an interaction between PFA and crude protein level.

The PFA in the present study is a blend of extracts of oregano, cinnamon, and citrus peel, and fructooligosaccharides, which has been reported to have a positive result in the performance or metabolism of turkey poults, piglets, aquatic animals, and dairy calves [9–12]. In broilers, PFA in a diet with reduced metabolizable energy and crude protein reduced plasma cholesterol and improved plasma and meat total antioxidant capacity, gut microbiota, Toll-like signaling molecules, and tight junction genes of broilers [13,14]. Furthermore, it has been reported that PFA also improved the immune system and gut health of broilers infected with *Clostridium perfringens* [15]. In a recent study, supplementation of PFA showed a trend towards improving the livability and performance of broilers challenged with *Eimeria tenella* [16]. Whether PFA can affect nutrient digestion and the immune response of broilers fed a reduced protein diet are unclear.

It is hypothesized that PFA can beneficially modulate anabolism and immunity in broilers fed a low protein diet. The objective of the present study is to evaluate the effects of PFA and dietary protein levels on growth performance, nutrient digestibility and transportation, and the immunity of broilers.

2. Materials and Methods

The experimental protocol was reviewed and approved by the University of Georgia Institutional Animal Care and Use Committee (A2019 08-022).

2.1. Experimental Treatments

The experiment is a 2 × 2 completely randomized factorial arrangement with two levels of dietary crude protein, standard protein (positive control, PC) following Cobb500 Performance and Nutritional Guide [17] or 1.5% reduction (negative control, NC) and inclusion of PFA (Digestarom® BIOMIN Holding GmbH, Getzersdorf, Austria) at 125 ppm or not. The experimental diets were iso-caloric and formulated based on corn and soybean meal. Broilers were fed throughout a 42-day production period, and no antibiotic growth promoters were used in the diet. PC and NC diets (Table 1) were mixed using a horizontal mixer (Davis Double Ribbon Mixer, Bonner Springs, KS, USA) for 12 min. Then PFA was mixed with 5 kg of either PC or NC diet to create a premix before adding to the mixer for the final treatment diets. All the diets were fed as a mash form. Birds were fed starter, grower, and finisher diet from 0–14, 15–28, and 29–42 d of age, respectively.

Table 1. Feed formulation and composition of the experimental diets.

Item	0–14 d		15–28 d		29–42 d	
	PC [1]	NC [1,2]	PC [1]	NC [1,2]	PC [1]	NC [1,2]
			Ingredient (%)			
Corn	58.17	58.20	62.75	62.25	64.52	69.00
Soybean meal	36.42	33.27	31.58	28.21	29.36	25.49
DCP [1]	1.57	1.57	1.44	1.47	1.24	1.26
Soybean oil	1.70	2.56	2.19	2.54	3.00	2.35
Limestone	1.18	1.18	1.13	1.14	1.05	1.07
Common salt	0.30	0.30	0.30	0.30	0.30	0.30
DL-methionine	0.25	0.24	0.22	0.19	0.18	0.16
Premix [1,3]	0.25	0.25	0.25	0.25	0.25	0.25
L-lysine-HCL	0.08	0.09	0.08	0.09	0.02	0.05
Sand	0	2.21	0	1.48	0	0
			Calculated nutrient [1] (%)			
ME [1] (kcal/kg)	3008	3008	3086	3086	3160	3160
Crude protein	22.00	20.50	20.00	18.50	19.00	17.50
Dig-Lysine	1.18	1.10	1.05	0.97	0.95	0.88
Dig-Methionine	0.58	0.55	0.52	0.48	0.47	0.43
Dig-TSAA	0.88	0.83	0.80	0.74	0.74	0.68
Dig-Threonine	0.78	0.73	0.71	0.65	0.67	0.62
Ca	0.90	0.90	0.84	0.84	0.76	0.76
Non-phytate P	0.45	0.45	0.42	0.42	0.38	0.38
			Analyzed nutrient			
Crude protein, %	22.24	20.81	20.36	18.25	19.15	17.88
GE [1], kcal/kg	3815	3803	3986	3957	4016	4004

[1] DCP, dicalcium phosphate; ME, metabolizable energy; TSAA, total sulfur amino acid; PC, positive control containing standard crude protein; NC, negative control containing reduced crude protein by 1.5% (equivalent to a reduction of 15 g/kg), GE, gross energy.[2] The amino acids ratio to lysine remained the same as positive control. [3] Provided per kg of diet: vitamin A (retinyl acetate), 8000 IU; cholecalciferol, 1000 IU; vitamin E (DL-tocopheryl acetate), 20 IU; vitamin K, 0.5 mg; thiamin, 2.0 mg; riboflavin, 8.0 mg; d-pantothenic acid, 10 mg; niacin, 35 mg; pyridoxine, 3.5 mg; biotin, 0.18 mg; folic acid, 0.55 mg; vitamin B12, 0.010 mg; manganese, 120 mg; iodine, 0.70 mg; iron, 100 mg; copper, 8 mg; zinc, 100 mg; and selenium, 0.30 mg.

2.2. Birds Husbandry and Sample Collection

A total of 960 Cobb500 male broiler chicks were obtained on the day of the hatch from a hatchery at Cleveland, GA and randomly allocated into an environmentally controlled house with 32-floor pens (length, 1.52 m; width, 1.22 m; height, 0.61 m) with 30 birds each located at the Poultry Research Center of University of Georgia. All birds were individually weighed and grouped prior to the allocation to ensure an equal initial bodyweight for all pens. The birds were managed as described previously by Wang et al. (2020) [18]. Feed and body weight were measured weekly to determine body weight gain (BWG), feed intake (FI), and feed conversion ratio (FCR). On 21 and 42 d, 10 birds from each pen (replicate) were randomly selected and euthanized by cervical dislocation to collect ileal digesta (obtained from the Meckel's diverticulum to 1 cm before the ileal–cecal junction) for nutrient digestibility. Jejunum and ileum without contents (rinsed out using phosphate buffer saline) from one bird per pen in treatments PC and PC + PFA at 125 ppm were collected and stored in −80 °C for mRNA expression analysis. The rinsed jejunal digesta were collected in centrifuge tubes and the brush border enzymes were gently scrubbed from rinsed jejunum using a microscope slide. Jejunal digesta and brush border mucous were collected at 21 and 42 d for digestive enzyme activity. Soluble proteins from jejunal digesta and brush border mucous were extracted with 0.01 M PBS at pH 7.2. The samples were centrifuged at 4 °C, and an aliquot of the supernatant was used for future analyses. Soluble protein was determined by the Braford method using Bio-Rad protein assay kits (BioRad, Hercules, CA, USA).

2.3. Chemical Analysis

Chromic (III) oxide (Sigma-Aldrich., St. Louis, MO, USA) was used as a marker for nutrient digestibility in the diet and ileal digesta were analyzed following the method by Williams et al. (1962) [19]. Nitrogen content in feed and digesta was determined using the LECO system as indicated by AOAC International [20] performed at the Agricultural Experimental Station Chemical Laboratories, University of Missouri. Gross energy values in feed and digesta were determined using a calorimeter (IKA C1 Compact Bomb Calorimeter, IKA-Werke., Staufen, Germany).

Calculation:

Nutrient digestibility was calculated using the following equation:

$$\text{Nutrient digestibility} = [1 - (\text{Ci}/\text{Co}) \times (\text{No}/\text{Ni})] \times 100$$

where: Ci is the concentration of chromium in the diet; Co is the concentration of chromium in the ileal digesta or feces; Ni is the concentration of the nutrient in the diet; No is the concentration of the nutrient in the ileal digesta or feces; all values were expressed as a percentage of dry matter.

2.4. Isolation of mRNA and RT-qPCR

Total mRNA was extracted from jejunal and ileal tissue using Trizol reagent (Invitrogen, Thermo Fisher Scientific, Pittsburgh, PA, USA) according to the manufacturer's instruction. RNA quantity and purity were determined using a Nanodrop 1000 spectrophotometer (Thermo Fisher Scientific). The cDNA was synthesized from total RNA (1000 ng) using the high-capacity cDNA reverse transcription kit (Thermo Fisher Scientific) and was diluted to 5 ng/µL for RT-qPCR (Real-time polymerase chain reaction) analysis.

Jejunal samples were used to detect nutrient transporter genes including excitatory amino acid transporters (Eaat3), peptide transporter 1 (Pept1), glucose transporter 5 (Glut5) and sodium-glucose transporter 1 (Sglt1) and ileal samples for interleukin (IL) 6, IL-8, heme oxygenase-1 (HO-1), cytochrome P450 isoform 1A1 (CYP1A1), and UDP-glucuronosyltransferases isoform 1A1 (UGT1A1) were analyzed in ileum samples. Glyceraldehyde 3-phosphate dehydrogenase was chosen as a reference gene. The information of primers is shown in Table 2. qPCR was performed on an Applied Biosystems StepOnePlus™ (Thermo Fisher Scientific, Waltham, MA, USA) with iTaq™ Universal SYBR Green Supermix (BioRad, Hercules, CA, USA) using the following conditions for all genes: 95 °C for 10 min followed by 40 cycles of 95 °C for 15 s and 60 °C for 60 s. Samples were run in duplicate and relative gene expression data were analyzed using the $2-\Delta\Delta$Ct method [21]. The mean ΔCt of PC was used to calculate the $\Delta\Delta$Ct value.

2.5. Enzyme Activity Assay

Alkaline phosphatase activity in jejunal digesta is determined using an alkaline phosphatase assay kit (ab83369, Abcam, Cambridge, MA, USA). Alkaline phosphatase activity is reported as U/mg of protein. One unit of alkaline phosphatase is defined as the amount of enzyme causing the hydrolysis of one micromole of para-Nitrophenylphosphate per minute at pH 9.6 and 25 °C. Amylase activity in jejunal digesta is measured using EnzChek Ultra Amylase Assay Kit (E33651, Molecular Probes, Eugene, OR, USA). Amylase activity is reported as U/mg of protein. One unit of amylase is the amount of enzyme will generate 1 micromole of glucose from corn starch. Amylase from Bacillus sp. was used as the standard. Lipase activity in jejunal digesta is determined using Lipase Activity Assay Kit (MAK047, Sigma-Aldrich, St. Louis, MO, USA) and is reported as mU/mg of protein. One unit of lipase is the amount of enzyme that will generate 1.0 µmole of TNB per minute at 42 °C.

Table 2. Information of primers for quantitative real-time PCR.

Gene Name	Primer (5′→3′)		Length	Reference Sequence
	Forward	Reverse		
Nutrient transporters				
Eaat3	tgctgctttggattcagtgt	agcaatgactgtagtgcagaagtaatatatg	79	XM_424930.5
Pept1	cccctgaggaggatcactggtt	caaaagagcagcagcaacga	66	NM_204365.1
Glut5	ttgctggctttgggttgtg	ggaggttgagggccaaagtc	60	XM_417596.5
Sglt1	gccgtggccagggctta	caataacctgatctgtgcaccagta	71	NM_001293240.1
Immunity				
CYP1A1	gcttcaaccccaacagctac	gtgttcatgttcaccacgct	118	NM_205147.1
IL-6	ataaatcccgatgaagtgg	ctcacggtcttctccataaa	146	NM_204628.1
IL-8	cgttcagcgattgaactccg	ctgccttgtccagaattgcc	211	NM_205018.1
HO-1	cacgagttcaagctggtcac	ctgcagctccatcggaaaat	120	NM_205344.1
UGT1A1	ccaacctgccaaagaacgtg	ccctcgtaaacaccgtgtga	115	XM_015289249.1
GAPDH	tcagcagcaggcttcactac	gctaaggctgtggggaaagt	161	NM_204305.1

CYP1A1, cytochrome P450 family 1 subfamily A member 1; Eaat3, excitatory amino acid transporter 3; GAPDH, glyceraldehyde 3-phosphate dehydrogenase; Glut5, fructose transporter 5; HO-1, heme oxygenase 1; IL, interleukin; Pept1, peptide transporter 1; SEM, standard error of the mean; Sglt1, sodium glucose linked transporter 1; UGT1A1, UDP-glucuronosyltransferases isoform 1A1.

Jejunal brush border enzyme activities (maltase, sucrase and leucine aminopeptidase) were analyzed. For maltase and sucrase, jejunal mucous were homogenized in 100 mM mannitol 2 mM HEPES/KOH (pH 6.5), then centrifuged at 2200× g for 10 min. Aliquots of the supernatant were stored at −20 °C. Maltase and sucrase activities were analyzed according to Dahlqvist methods [22] with modification for microplate (Corning Costar 3631, Corning, NY, USA) assay at 42 °C for 20 min incubation. Sucrase or maltase unit were described as mg of glucose generated from sucrose or maltose per minute per mg of protein. The background glucose content from maltose and sucrose was measured and deducted from the reading. For leucine aminopeptidase, 15 μL of undiluted homogenized tissue was incubated with 135 μL 1 μmole L-leucine-p-nitroanilide 42 °C per 0.01 mole of PBS for 30 min. We used 4-nitroaniline (Fisher AC18069-1000, Hampton, NH, USA) as a standard. Change of absorbance was detected on a Molecular Devices microplate reader at O. D. 405 mM [23] (Sun, 2007). One unit of aminopeptidase N is defined as the hydrolysis of 1 μmol of the substrate in one minute at 42 °C, pH 7.0.

2.6. Statistical Analysis

Growth performance and nutrient digestibility data were analyzed using a two-way ANOVA model and gene expression and enzyme activity data were analyzed using a one-way ANOVA model as a completely randomized design using the GLM procedure of SAS 9.4 [24]. Significant level was set at $p < 0.05$ and tendency at $0.05 \leq p < 0.10$. Each pen was regarded as an experimental unit. The least square means were reported in the results. Treatment means were further separated using Tukey's multiple range test when the interaction is presented.

3. Results and Discussions

Because there were no significant interactions ($p > 0.1$) found in the current study, our results and discussion were focused on the main effect of crude protein and PFA on broiler performance, nutrient digestibility, and gene expression of nutrient transporter and immunity.

3.1. Growth Performance

Birds in all the pens maintained general health throughout the trial. The effect of crude protein level and phytogenic feed additive on the growth performance of broilers is shown in Table 3. Main effects showed that diets with low crude protein level had trended to decrease (p = 0.076) BWG and increased (p = 0.089) FI, but FCR was significantly increased during 0–21 (p = 0.032) and 0–42 d of age (p < 0.01). Supplementation of PFA reduced FCR during 0–21 d (p = 0.042) and the whole grow out period (p = 0.034). There were no interactions between dietary protein and PFA on the growth performance.

Table 3. Effects of crude protein level and phytogenic feed additive on the growth performance of broilers.

Item	BWG (g/bird)			FI (g/bird)			FCR		
	0–14 d	0–28 d	0–42 d	0–14 d	0–28 d	0–42 d	0–14 d	0–28 d	0–42 d
Main effect of dietary protein level									
PC [1,2]	324	1235	2488	457	1805	4110	1.409	1.420 [B]	1.665 [B]
NC [1,2]	322	1207	2424	460	1882	4194	1.430	1.510 [A]	1.731 [A]
SEM [2]	2.91	14.63	22.03	4.65	28.5	35.7	0.018	0.021	0.010
Main effect of PFA level									
0	323	1223	2434	458	1838	4185	1.425	1.479 [A]	1.724 [A]
125 ppm	324	1218	2478	459	1856	4157	1.420	1.456 [B]	1.683 [B]
SEM [2]	2.97	14.1	24.5	4.53	31.28	32.3	0.020	0.016	0.013
Treatments									
PC [1,2] + 0	321	1235	2479	452	1777	4085	1.411	1.390	1.668
NC [1,2] + 0	328	1212	2498	461	1885	4204	1.427	1.506	1.760
PC [1,2] + PFA [2]	325	1235	2390	463	1833	4127	1.432	1.444	1.662
NC [1,2] + PFA [2]	320	1202	2458	457	1879	4184	1.370	1.515	1.702
SEM [2]	2.38	10.5	18.0	2.99	22.5	23.9	0.018	0.021	0.010
p-value									
Protein level	0.639	0.211	0.076	0.637	0.108	0.089	0.475	0.032	<0.010
PFA [2]	0.808	0.820	0.221	0.911	0.702	0.984	0.984	0.042	0.034
Interaction	0.170	0.821	0.484	0.294	0.500	0.525	0.885	0.584	0.104

Means within a column with different superscripts differ: [A,B] p < 0.05. [1] standard crude protein levels were 22.0, 20.0 and 19.0% at 0–14, 15–28 and 29–42 d of age, respectively; reduced protein group were reduced crude protein by 1.5%. [2] PC, positive control; NC, negative control; PFA, phytogenic feed additive, Digestarom® P.E.P.at 125 ppm; SEM, standard error of the mean with n = 8.

In the present study, reducing the dietary protein at 1.5% showed a negative effect on growth by decreasing BWG and increasing FCR and FI, particularly in the finisher phase. Dietary protein is potentially used for muscle growth and immunity in animals [25]. Broilers in the present study attempted to increase FI to meet their protein and amino acid requirements for growth and maintenance needs. Whereas a reduced protein diet could be insufficient in supporting both roles even with the increased FI. It is plausible that PFA in the diet may reduce the immunity cost on energy and protein [4]. The nutrients are redirected to animal growth thus broilers fed diet contain PFA had a better FCR. Further studies are necessary on the mechanism of PFA on broiler immunity and performance. Studies reported that reduction of crude protein by 1.7% decreased BWG while the FCR maintained the same [26], but a 2% reduction in a miscellaneous meal diet reduced both FI and BWG in broilers [5]. Dietary protein level was reported as a significant positive contributor for broiler BWG and feed efficiency by Pesti. (2009) from summarizing 26 types of research [27], however, the protein quality from the ingredients, reduction level and rearing environments are played as cofactors on the dietary protein level and broiler performance.

The PFA is a blend of citrus peel, cinnamon, oregano and fructooligosaccharides. During the finisher and whole grow-out period in the present study, the increased FCR was induced by reduced dietary protein, whereas PFA supplementation improved the broiler efficiency This is in agreement with the previous study that supplementation of

PFA improved performance and offset the negative effect of a low protein diet [28]. Sadek et al. (2014) and Murugesan et al. (2015) reported PFA improved the feed efficiency of broilers [4,29]. Paraskeuas et al. (2016) found that the reduction of dietary protein by 1.32% and ME by 0.8 MJ/kg increased FCR, but PFA addition did not affect the growth performance of broilers [13]. The response of phytogenic compounds on broiler performance may be contributed by variable sources of plants and processing methods of extracting those active compounds. The recent studies are more likely to show the beneficial effects on growth performance, which may be due to the improvement in extraction and refinement of those active compounds [12–14].

3.2. Nutrient Digestibility

The effect of crude protein level and phytogenic feed additive on the growth performance of broilers is shown in Table 4. Low dietary protein diets decreased ($p = 0.047$) crude protein digestibility at 21 d (Table 4), but PFA addition did not cause significant differences in the digestibility of dry matter, crude protein, and ileal digestible energy among the treatments at 21 and 42 d of age in the present study.

Table 4. Effects of crude protein level and phytogenic feed additive on the nutrient digestibility of broilers.

Item	DM (%)		CP (%)		IDE (kcal/kg)	
	21 d	42 d	21 d	42 d	21 d	42 d
			The main effect of dietary protein level			
PC [1,2]	71.8	71.1	81.2 [A]	79.7	2837	2878
NC [1,2]	70.6	72.1	78.2 [B]	78.0	2791	2838
SEM [2]	1.13	0.75	0.96	0.89	51.0	34.1
			The main effect of PFA level			
0	71.0	71.6	79.6	78.9	2807	2853
125 ppm	71.4	71.6	79.8	78.8	2821	2863
SEM [2]	1.12	0.76	1.03	0.95	50.6	34.3
			Treatments			
PC [1,2] + 0	71.2	71.4	80.9	80.0	2834	2849
NC [1,2] + 0	70.9	71.9	78.4	77.9	2781	2857
PC [1,2] + PFA [2]	72.4	70.9	81.4	79.4	2840	2826
NC [1,2] + PFA [2]	70.4	72.3	78.1	78.2	2801	2899
			p-value			
PFA	0.827	0.997	0.910	0.925	0.8596	0.846
CP level	0.476	0.389	0.047	0.226	0.5432	0.426
Interaction	0.622	0.666	0.789	0.761	0.9296	0.516

Means within a column with different superscripts differ: [A,B] $p < 0.05$. [1] standard crude protein levels were 22.0, 20.0 and 19.0% at 0–14, 15–28 and 29–42 d of age, respectively. [2] PC, positive control; NC, negative control; DM, dry matter; IDE, ileal digestible energy; PFA, phytogenic feed additive, Digestarom® P.E.P. at 125 ppm; SEM, standard error of the mean with $n = 8$.

The activities of digestive enzymes, including amylase, lipase, aminopeptidase, sucrose, maltase, and alkaline phosphatase were not different at 21 and 42 d in the present study (Table 5).

Table 5. Effect of phytogenic feed additive (PFA) on the jejunal digesta and brush border enzyme activities of broilers.

Item	Jejunal Digesta			Brush Border Enzyme		
	Alkaline Phosphatase U/mg	Amylase U/mg	Lipase mU/mg	Maltase mg/mg	Sucrase mg/mg	Aminopeptidase µmoL/mg
			21 d of age			
Control [2]	1.549	4.912	23.172	3.236	2.458	0.202
PFA [1]	1.109	4.348	22.963	4.161	2.456	0.195
SEM [1]	0.291	0.158	1.564	0.25	0.15	0.009
p-value	0.289	0.475	0.950	0.115	0.996	0.693
			42 d of age			
Control [2]	1.381	4.764	22.560	2.983	2.011	0.098
PFA [1]	1.048	4.489	17.500	2.769	2.077	0.101
SEM [1]	0.231	0.161	2.246	0.33	0.38	0.011
p-value	0.602	0.8812	0.275	0.659	0.845	0.485

[1] PFA, phytogenic feed additive, Digestarom® P.E.P. added at 125 ppm of diet; SEM, standard error of the mean with $n = 8$. [2] Containing standard crude protein levels were 22.0, 20.0 and 19.0% at 0–14, 15–28 and 29–42 d of age, respectively.

In the present study, the decreased protein digestibility in the low protein diet negatively affected BWG and FCR, while supplementation of PFA at 125 mg/kg had no effect on digestibility. Zumbaugh et al. (2020) [12] reported that inclusion of a similar blend of PFA at 1 g/kg in the diet with a 1.5% reduction of dietary protein improved BWG, but did not affect FCR, protein intake, and protein digestibility of turkey poults. Meanwhile, there are studies that reported feeding PFAs to broilers improved crude protein digestibility and AMEn [4,30,31]. The different results of feeding PFAs on poultry species may associate with mechanisms of PFAs. The PFAs generally affect broilers through the mechanism of anti-oxidation, stimulating digestive enzymes, or regulating the immune response. Additionally, there was no effect of PFA on digestive enzymes for broilers fed a standard protein level diet in the present study, and similar results were found when turkey poults were fed PFA except a reduction of aminopeptidase activity [12]. The different results in nutrient digestibility and digestive enzymes could be related to the PFA dose and poultry species, which deserves further study.

3.3. Gene Expression of Transporter and Immunity

The present study further detected the mRNA profiles of nutrient transporters and immune parameters in treatments PC and PFA at 125 ppm (Table 6). In contrast with PC, the mRNA expressions of Eaat3, Pept1, Glut5, and Sglt1 in the jejunum were not influenced by supplementing PFA at 21 and 42 d of age. Ileal immune parameters CYP1A1 ($p = 0.030$) and IL-6 ($p = 0.037$) were downregulated by PFA at 21 d, but not for IL-8, HO-1, UGA1A1 at 21 and 42 d.

Table 6. Effect of PFA on the gene expression of nutrient transporters and immune parameters of broilers.

Item	Nutrient Transporter				Immunity				
	Eaat3 [1]	Pept1 [1]	Glut5 [1]	Sglt1 [1]	CYP1A [1]	IL-6 [1]	IL-8 [1]	HO-1 [1]	UGAT1A1 [1]
21 d of age									
Control [2]	1.09	1.04	1.07	1.04	1.42 [A]	1.16 [A]	1.11	1.01	1.03
PFA [1]	0.73	0.71	0.92	0.86	0.38 [B]	0.57 [B]	1.19	0.99	0.86
SEM [1]	0.12	0.12	0.17	0.14	0.25	0.15	0.26	0.09	0.1
p-value	0.193	0.151	0.721	0.583	0.032	0.037	0.507	0.881	0.279
42 d of age									
Control [2]	1.18	1.2	1.14	1.24	0.91	1.34	1.22	1.03	1.02
PFA [1]	0.82	0.75	0.83	0.78	1.03	1.04	1.35	1.07	1.1
SEM [1]	0.22	0.23	0.18	0.26	0.33	0.38	0.32	0.12	0.12
p-value	0.429	0.119	0.135	0.402	0.945	0.544	0.699	0.821	0.679

[A,B] means within a column with different superscripts trend to be different ($p < 0.05$). [1] Eaat3, excitatory amino acid transporter 3; Peptl, peptide transporter 1; Glut5, fructose transporter 5; Sglt1, sodium glucose linked transporter 1; CYP1A1, cytochrome P450 family 1 subfamily A member 1; HO-1, heme oxygenase 1; IL, interleukin; UGT1A1, UDP-glucuronosyltransferases isoform 1A1; PFA, phytogenic feed additive, Digestarom® P.E.P. added at 125 ppm of diet; SEM, standard error of the mean with n = 8. [2] Containing standard crude protein levels were 22.0, 20.0 and 19.0% at 0–14, 15–28 and 29–42 d of age, respectively.

The unaffected transporters in broilers fed the PFA diet in the present study, coupled with unchanged digestive enzyme activity and most nonsignificant digestibility parameters, indicate that PFA as a phytogenic nutraceutical is not enough to trigger significant differences in nutrient digestion, transportation, and assimilation for broilers. Likely, in turkeys, these transporters were not influenced by the PFA diet [12].

Importantly, in the present study, PFA deregulated ileal CYP1A1 and IL-6, indicating that PFA may reduce inflammation and redirect nutrients towards growth in broilers. Expression of CYP1A1 is a sensitive indicator for certain immune cell loss and susceptibility to enteric infection [32,33]. IL-6 acts as a pro-inflammatory cytokine especially in the smooth muscle cell [34]. PFA protected the intestinal barrier by upregulating the tight junction protein gene in broilers [14]. The beneficial immune response and intestinal barrier may be the main contributors for the improvement in feed efficiency by PFA. It is known that immune response cost energy and protein [35], thus, the down-regulation of immunity genes in the present study may partially explain the improved FCR with PFA addition. Due to the less demand on immunity, more nutrients are possibly redirected to growth. Recent studies found that phytochemicals modulated intestinal endogenous bactericidal peptides and muscle physicochemical property [36,37], and whether PFA has an effect on these aspects needs further study.

4. Conclusions

In conclusion, dietary supplementation of PFA improved feed conversion ratio and ileum immune gene expression in the present study. However, the nutrient digestibility, nutrient transporter gene expression, and digestive enzyme activities were not influenced for broiler-fed diet supplemented with PFAs. The beneficial effects of PFA on broiler performance could be linked with the antioxidant and anti-inflammatory compounds from PFA-related plants.

Author Contributions: Conceptualization, C.P., R.M., B.S., and W.K.K.; methodology and data collection, and formal analysis, J.W. and S.S.; writing—original draft preparation, J.W.; writing—review and editing, C.P., R.M., B.S., and W.K.K.; All authors have read and agreed to the published version of the manuscript.

Funding: This research was funded by Biomin America Inc with grant number 03082017.

Institutional Review Board Statement: The study was conducted according to the guidelines and approved by The Institutional Animal Care and Use Committee of University of Georgia (A2019 08-022).

Informed Consent Statement: Not applicable.

Data Availability Statement: Data available on request due to privacy restrictions.

Acknowledgments: This article was funded by the Biomin America Inc. Therefore, the authors appreciate the technical and financial support of the Biomin America Inc.

Conflicts of Interest: The authors declare no conflict of interest. The funders had no role in the design of the study; in the collection, analyses, or interpretation of data; in the writing of the manuscript, or in the decision to publish the results.

References

1. The European Parliament and the Council of the European Union. Regulation No 1831/2003 of the European Parliament and Council of 22 September 2003 on additives for use in animal nutrition. *Off. J. Eur. Union* **2003**, *268*, 29–43.
2. Jamroz, D.; Wiliczkiewicz, A.; Wertelecki, T.; Orda, J.; Skorupińska, J. Use of active substances of plant origin in chicken diets based on maize and locally grown cereals. *Br. Poult. Sci.* **2005**, *46*, 485–493. [CrossRef]
3. Hong, J.-C.; Steiner, T.; Aufy, A.; Lien, T.-F. Effects of supplemental essential oil on growth performance, lipid metabolites and immunity, intestinal characteristics, microbiota and carcass traits in broilers. *Livest. Sci.* **2012**, *144*, 253–262. [CrossRef]
4. Murugesan, G.R.; Syed, S.B.; Haldar, S.; Pender, C. Pender. Phytogenic feed additives as an alternative to antibiotic growth promoters in broiler chickens. *Front. Vet. Sci.* **2015**, *2*, 21.
5. Liu, N.; Wang, J.; Gu, K.; Deng, Q.; Wang, J. Effects of dietary protein levels and multienzyme supplementation on growth performance and markers of gut health of broilers fed a miscellaneous meal based diet. *Anim. Feed. Sci. Technol.* **2017**, *234*, 110–117. [CrossRef]
6. Liu, N.; Wang, J.; Liu, Z.; Chen, Y.; Wang, J.P. Tetramethylpyrazine attenuates necrotic enteritis by reducing gut oxidative stress, inflammation, opportunistic bacteria and endotoxins of broilers. *Euro. Poult. Sci.* **2018**, *82*, 233.
7. Bregendahl, K.; Sell, J.L.; Zimmerman, D.R. Effect of low-protein diets on growth performance and body composition of broiler chicks. *Poult. Sci.* **2002**, *81*, 1156–1167. [CrossRef]
8. Chalova, V.; Kim, J.; Patterson, P.; Ricke, S.; Kim, W. Reduction of nitrogen excretion and emissions from poultry: A review for conventional poultry. *Worlds Poult. Sci. J.* **2016**, *72*, 509–520. [CrossRef]
9. Kroismayr, A.; Schedle, K.; Sehm, J.; Pfaffl, M.W.; Plitzner, C.; Foissy, H.; Ettle, T.; Mayer, H.; Schreiner, M.; Windisch, W. Effects of antimicrobial feed additives on gut microbiology and blood parameters of weaned piglets. *Bodenkultur* **2008**, *59*, 111–120.
10. Akbarian-Tefaghi, M.; Ghasemi, E.; Khorvash, M. Performance, rumen fermentation and blood metabolites of dairy calves fed starter mixtures supplemented with herbal plants, essential oils or monensin. *J. Anim. Physiol. Anim. Nutr.* **2018**, *102*, 630–638. [CrossRef]
11. Peterson, B.; Peatman, E.; Ourth, D.; Waldbieser, G. Phytogenic feed-additive effects on channel catfish rhamnose-binding lectin levels, and susceptibility to Edwardsiella ictaluri. *Dis. Aquat. Org.* **2018**, *129*, 99–106. [CrossRef]
12. Zumbaugh, C.; Murugesan, G.; Wong, E.; Syed, B.; Persia, M. Evaluation of a phytogenic feed additive on performance, nutrient digestion, and absorption in turkey poults. *Anim. Feed. Sci. Technol.* **2020**, *267*, 114575. [CrossRef]
13. Paraskeuas, V.; Fegeros, K.; Palamidi, I.; Theodoropoulos, G.; Mountzouris, K.C. Phytogenic Administration and Reduction of Dietary Energy and Protein Levels Affects Growth Performance, Nutrient Digestibility and Antioxidant Status of Broilers. *J. Poult. Sci.* **2016**, *53*, 264–273. [CrossRef] [PubMed]
14. Paraskeuas, V.V.; Mountzouris, K.C. Modulation of broiler gut microbiota and gene expression of Toll-like receptors and tight junction proteins by diet type and inclusion of phytogenics. *Poult. Sci.* **2019**, *98*, 2220–2230. [CrossRef] [PubMed]
15. Mitsch, P.; Zitterl-Eglseer, K.; Köhler, B.; Gabler, C.; Losa, R.; Zimpernik, I. The effect of two different blends of essential oil components on the proliferation of Clostridium perfringens in the intestines of broiler chickens. *Poult. Sci.* **2004**, *83*, 669–675. [CrossRef]
16. Zheng, L.; Bielke, J.; Ramirez, S.; Pender, C.; Tacconi, A.; Murugessan, G.R.; Bielke, L. Using yucca and/or phytogenics on alleviating negative effects of necrotic enteritis in broilers. *Poult. Sci. Ann. Meet.* **2019**, *93*, 186.
17. Cobb Vantress Inc. *Cobb500 Broiler Performance & Nutrition Supplement Guide*; Cobb Vantress Inc.: Siloam Springs, AR, USA, 2018.
18. Wang, J.; Choi, H.; Kim, W. Effects of dietary energy level and 1,3-diacylglycerol on growth performance and carcass yield in broilers. *J. Appl. Poult. Res.* **2020**, *29*, 665–672. [CrossRef]
19. Williams, C.H.; David, D.J.; Iismaa, O. The determination of chromic oxide in faeces samples by atomic absorption spectrophotometry. *J. Agric. Sci.* **1962**, *59*, 381–385. [CrossRef]
20. AOAC. *Official Method of Analysis of the Association of Official Analytical Chemists*; AOAC International: Arlington, TX, USA, 1996.
21. Livak, K.J.; Schmittgen, T.D. Analysis of relative gene expression data using real-time quantitative PCR and the 2-ΔΔCt method. *Methods* **2001**, *25*, 402–408. [CrossRef]
22. Dahlqvist, A. Method for assay of intestinal disaccharidases. *Anal. Biochem.* **1964**, *7*, 18–25. [CrossRef]
23. Sun, X. *Effect of Corn Quality and Enzyme Supplementation on Broiler Performance, Gastrointestinal Enzyme Activity, Nutrient Retention, Intestinal Mucin, and Jejunal Gene Expression*; Virginia Tech: Montgomery, VA, USA, 2007.
24. SAS Institute. *SAS®User's Guide for Personal Computer*; SAS Institute Inc.: Cary, NC, USA, 2008.
25. Wu, G.; Wu, Z.; Dai, Z.; Yang, Y.; Wang, W.; Liu, C.; Wang, B.; Wang, J.; Yin, Y. Dietary requirements of "nutritionally non-essential amino acids" by animals and humans. *Amino Acids* **2013**, *44*, 1107–1113. [CrossRef]

26. Ullrich, C.; Langeheine, M.; Brehm, R.; Taube, V.; Siebert, D.; Visscher, C. Influence of Reduced Protein Content in Complete Diets with a Consistent Arginine–Lysine Ratio on Performance and Nitrogen Excretion in Broilers. *Sustainability* **2018**, *10*, 3827. [CrossRef]
27. Pesti, G.M. Impact of dietary amino acid and crude protein levels in broiler feeds on biological performance. *J. Appl. Poult. Res.* **2009**, *18*, 477–486. [CrossRef]
28. Pender, C.; Ramirez, S.; Murugesan, G.R.; Sobotik, B.E.; Archer, G.S. Evaluating the effects of a novel phytogenic feed additive on performance of broilers fed a low or standard protein diet. *Poult. Sci. Ann. Meet.* **2020**, *82*, 161.
29. Sadek, K.; Ahmed, H.; Ayoub, M.; Elsabagh, M. Evaluation of Digestarom and thyme as phytogenic feed additives for broiler chickens. *Eur. Poult. Sci.* **2014**, *78*. [CrossRef]
30. Mountzouris, K.; Paraskevas, V.; Tsirtsikos, P.; Palamidi, I.; Steiner, T.; Schatzmayr, G.; Fegeros, K. Assessment of a phytogenic feed additive effect on broiler growth performance, nutrient digestibility and caecal microflora composition. *Anim. Feed. Sci. Technol.* **2011**, *168*, 223–231. [CrossRef]
31. Pirgozliev, V.; Mansbridge, S.C.; Rose, S.P.; Lillehoj, H.S.; Bravo, D. Immune modulation, growth performance, and nutrient retention in broiler chickens fed a blend of phytogenic feed additives. *Poult. Sci.* **2019**, *98*, 3443–3449. [CrossRef] [PubMed]
32. Schiering, C.; Wincent, E.; Metidji, A.; Iseppon, A.; Li, Y.; Potocnik, A.J.; Omenetti, S.; Henderson, C.J.; Wolf, C.J.H.C.R.; Nebert, D.W.; et al. Feedback control of AHR signalling regulates intestinal immunity. *Nat. Cell Biol.* **2017**, *542*, 242–245. [CrossRef]
33. Mescher, M.; Haarmann-Stemmann, T. Modulation of CYP1A1 metabolism: From adverse health effects to chemoprevention and therapeutic options. *Pharmacol. Ther.* **2018**, *187*, 71–87. [CrossRef] [PubMed]
34. Deng, Q.; Shi, H.; Luo, Y.; Zhao, H.; Liu, N. Effect of dietary Lactobacilli mixture on Listeria monocytogenes infection and virulence property in broilers. *Poult. Sci.* **2020**, *99*, 3655–3662. [CrossRef]
35. Klasing, K. Nutrition and the immune system. *Br. Poult. Sci.* **2007**, *48*, 525–537. [CrossRef] [PubMed]
36. Ding, K.; Jiang, Q.; Wang, J.; Liu, N.; Zhang, F. Effect of tetramethylpyrazine on growth performance, Campylobacter jejuni carriage and endogenous antimicrobial peptides in rabbits. *Czech J. Anim. Sci.* **2019**, *64*, 465–471. [CrossRef]
37. Wang, J.; Liu, N.; Zhang, F. Tetramethylpyrazine Protects Oxidative Stability and Gelation Property of Rabbit Myofibrillar Proteins. *Food Sci. Anim. Resour.* **2019**, *39*, 623–631. [CrossRef] [PubMed]

animals

MDPI

Article

Zinc and/or Selenium Enriched Spirulina as Antioxidants in Growing Rabbit Diets to Alleviate the Deleterious Impacts of Heat Stress during Summer Season

Fawzia Hassan [1], Samia Mobarez [2], Manal Mohamed [2], Youssef Attia [3], Aml Mekawy [4] and Khalid Mahrose [5,*]

1 Department of By-products Utilization Research, Animal Production Research Institute, Agricultural Research Center 12618, El-Dokki, Giza 12611, Egypt; fawzia_amer@yahoo.com
2 Department of Poultry Nutrition Research, Animal Production Research Institute, Agricultural Research Center 12618, El-Dokki, Giza 12611, Egypt; samiamostafa116@gmail.com (S.M.); manalsaudy@yahoo.com (M.M.)
3 Department of Agriculture, Faculty of Environmental Sciences, King Abdulaziz University, Jeddah 21589, Saudi Arabia; yaattia@kau.edu.sa
4 Poultry Production Department, Agriculture College, Damietta University, Damietta 34516, Egypt; amlmekawyaa@gmail.com
5 Animal and Poultry Production Department, Faculty of Technology and Development, Zagazig University, Zagazig 44511, Egypt
* Correspondence: ostrichkhalid@zu.edu.eg; Tel.: +20-100-127-8452

Citation: Hassan, F.; Mobarez, S.; Mohamed, M.; Attia, Y.; Mekawy, A.; Mahrose, K. Zinc and/or Selenium Enriched Spirulina as Antioxidants in Growing Rabbit Diets to Alleviate the Deleterious Impacts of Heat Stress during Summer Season. *Animals* **2021**, *11*, 756. https://doi.org/10.3390/ani11030756

Academic Editors: Raffaella Rossi and Ilias Giannenas

Received: 1 February 2021
Accepted: 27 February 2021
Published: 10 March 2021

Simple Summary: Heat stress in summer season impairs growth and causes heat-induced physiological stress in rabbits. Zinc acts as an antioxidant stress agent by inhibition of oxidation of macromolecules such as DNA as well as inhibition of inflammatory response, eventually resulting in the down-regulation of reactive oxygen species production. Selenium is a powerful biological anti-oxidant mineral. Spirulina is comparatively confined extreme protein (55–65%) and comprised all important amino acids, has wellbeing assistances, immuno-stimulatory influences and antiviral activity and ensured the capability to diminish heat stress impacts. In the current work, effects of dietary supplemental zinc and/ or selenium enriched spirulina (100 mg Zn-Sp/kg diet, 0.5 mg Se-Sp/kg diet or 100 mg Zn-Sp+ 0.5 mg Se-Sp, respectively) as antioxidants on growth performance, nutrient digestibility, plasma biochemicals and antioxidant status of New Zealand White growing rabbits under summer conditions were evaluated. The findings showed that the supplemented diets enhanced growth performance of rabbits at marketing, hot carcass weight, dressing percentage, high density lipoprotein cholesterol and total antioxidant capacity and reduced thio-barbituric acid reactive substances. Finally, dietary supplementation of 100 mg Zn-Sp, 0.5 mg Se-Sp or their combination could improve growth performance, nutrients digestibility and antioxidant status of heat stressed growing rabbits.

Abstract: Effects of dietary supplemental zinc and/ or selenium enriched spirulina (Zn-Sp, Se-Sp and Zn-Se-Sp, respectively) as antioxidants on growth performance, nutrient digestibility, plasma biochemicals and antioxidant status of growing rabbits under summer conditions were evaluated. A total of 160 New Zealand White male rabbits at six-weeks-old were randomly assigned to four groups. The first group received untreated diet (control). The other groups received diet supplemented with 100 mg Zn-Sp/kg diet, 0.5 mg Se-Sp/kg diet or 100 mg Zn-Sp+ 0.5 mg Se-Sp, respectively. The findings showed that the supplemented diets enhanced growth performance of rabbits at marketing. Rabbits fed Zn-Sp exhibited high dry and organic matter digestibilities while those fed Zn-Sp and Zn-Se-Sp diet supplemented achieved high crude protein digestibility. Rabbits fed diet supplemented with Zn-Se-Sp gave the highest hot carcass weight when competed with their counterparts. Zn-Sp and Zn-Se-Sp supplemented diets tended to promote dressing percentage. Low concentrations of plasma total cholesterol, LDL-cholesterol and VLDL-cholesterol were recorded by Se-Sp and Zn-Se-Sp groups. Rabbits fed Se-Sp, Zn-Se-Sp had the greatest HDL, plasma TAC and catalase and the lowest TBARs. Conclusively, dietary supplementation of 100 mg Zn-Sp, 0.5 mg Se-Sp or their combination could improve growth performance, nutrients digestibility and antioxidant status of heat stressed growing rabbits.

Keywords: antioxidants; growth; heat stress; rabbit; selenium; spirulina; zinc

1. Introduction

The environmental and nutritional factors are affecting the intensive rabbit production [1,2]. Rabbits play an increasingly important role in meat production throughout the world [3]. Growing rabbits are very susceptible to high temperature and the heat anxiety in summer season is correlated with reduces in growth performance and increases in mortality [2,4]. Oxidative stress refers to the imbalance between free radicals production and the ability of the antioxidant defense system of the body to detoxify or impair oxidative damage to DNA, proteins, and lipids [5].

Zinc (Zn) is a component of more than 300 enzymes and more than 2000 transcriptional factors and is involved in the biosynthesis of nucleic acids and in cell division processes [6,7]. Practical commercial rabbit diets include a broad range of zinc levels (40–140 mg/kg). Growing rabbits respond positively to 100 mg zinc/kg diets in terms of improving body weight gain (BWG) and feed conversion ratio (FCR) [8]. Besides, zinc acts as an antioxidant stress agent by inhibition of oxidation of macromolecules such as DNA and proteins as well as inhibition of inflammatory response, eventually resulting in the down-regulation of reactive oxygen species production [5].

Selenium (Se) is a powerful biological antioxidant mineral. It can control several vital biological processes [9,10]. Moreover, Se is an integral component of at least 25 selenoproteins and serving as an essential co-factor in the antioxidant enzyme system. The intake of Se in productive animals affects nutrient utilization, productive performance, antioxidative mechanism, reproductive function, hormone metabolism and responses of the immune system [11–13].

Spirulina platensis have been exhausted for several years as nourishment for people and animals owing to the outstanding nutritious profile and great carotenoid substance. Spirulina is comparatively confined extreme protein (55–65%) and comprised all important amino acids [14,15], has wellbeing assistances [16], immuno-stimulatory influences and antiviral activity [15,17] and ensured the capability to diminish heat stress impacts [18].

The purpose of this study was to investigate the effects of dietary supplementation of Zn- and/ or Se-enriched Spirulina or their combination on growth performance, plasma biochemicals and antioxidant status of growing New Zealand White rabbits (NZW) during summer season (Julie and August).

2. Materials and Methods

The current study was performed at Borg-El Arab, Alexandria Governorate, Animal Production Research Institute, Agricultural Research Center, Ministry of Agriculture, Egypt.

Spirulina platensis (*Arthrospira platensis*) was obtained from Agricultural Microbiology Department, National Research Centre (NRC), Giza, Egypt. The principles of the cultivation system and modification of culture medium by the addition of inorganic zinc and selenium sources were described [19]. The biomass concentration was 1 g of dry mass/L. Resultant Zn-enriched Spirulina contained 100.17 mg Zn for each 1 gm dry algae. Selenium enriched algae is produced by growing strain of Spirulina platensis, algae containing 1 mg Se/g algae.

A total number of hundred and sixty weaned New Zealand White (NZW) male rabbits, 6-weeks old (average initial live body weight; 744.79 ± 17.56 g), were randomly assigned to four experimental groups (40 rabbits/each; 5 replicates, 4 rabbits/each). Rabbits were housed an open-sided house in individual cages (60 × 40 × 24 cm). Feed and water were offered ad libitum throughout the experimental period (6–14 weeks of age). Environmental temperature and RH were noted daily, and then averages of temperature, RH and temperature-humidity index (THI) for two months (Julie and August) were estimated [20] expending the next principle: THI = db °C-{(0.31−0.31RH)(db °C−14)}, where db °C is the dry bulb temperature in Celsius and RH is the relative humidity %; the assessed estimates of THI were categorized [20] as follows: < 22.2 is lack of heat anxiety, 22.2− < 23.2 is reasonable heat anxiety, 23.3− < 25.5 is acute heat anxiety, and 25.5 or more is extremely acute heat anxiety. Four pelleted diets were formulated and nutrients requirements were adjusted [21] as shown in Table 1. The first experimental group received untreated diet (control). The second, third and fourth experimental groups received diet supplemented with 100 mg Zn-Sp/kg diet, 0.5 mg Se-Sp/kg diet, 100 mg Zn-Sp+ 0.5 mg Se-Sp, respectively. Body weight (BW) and feed intake (FI) were recorded weekly and then BWG and FCR were computed.

Table 1. Feed ingredients and chemical composition of rabbit basal diet (%DM basis).

Feed Ingredient	(%)	Nutrient Composition (%DM Basis)	
Soybeanmeal (44%CP)	19.30	Dry matter (DM)	88.90
Barley	17.10	Organic matter (OM)	90.70
Wheat bran	24.88	Crude protein (CP)	17.57
Yellow corn	7.00	Crude fiber (CF)	13.01
Clover hay	25.00	Ether extract (EE)	2.01
Molasses	3.00	Nitrogen free extract (NFE)	58.11
Limestone	1.08	Ash	9.30
Di-calcium phosphate	1.71	NDF	29.85
DL-Methionine	0.28	ADF	17.02
Sodium chloride	0.35	ADL	3.44
Vit.-Min. premix [a]	0.30	Methionine [b]	0.68
Total	100	Lysine [c]	0.99
		Calcium	1.27
		Available Phosphours	0.55
		Digestible energy (Kcal/Kg DM) [d]	2599.49
		Zn (mg/kg DM) [e]	107.81
		Se (mg/kg dM) [f]	0.82

[a] Vit. And Min. premix per kg contains: Vit A 6000 IU; Vit D3450 IU; Vit E 40 mg; Vit K3 1 mg; Vit B1 1 mg; Vit B2 3 mg; Niacin 180 mg; Vit B6 39 mg; Vit B12 2.5 mg; Pantothenic acid 10 mg; biotin 10 mg; folic acid 2.5 mg; choline chloride 1200 mg; Manganese 15 mg; Zinc 60 mg; Iron 38 mg; Copper 5 mg; Selenium 0.1 mg; Iodine 0.2 mg; Selenium 0.05 mg; ([b,c,e,f]): Calculated on the basis of the ingredients composition. ([d]) Digestible energy (DE) was calculated according to Lebas [21] using the following equation: $DE = 15.627 + 0.000982\ CP^2 + 0.0040\ EE^2 - 0.0114\ MM^2 - 0.169\ ADF \pm 1.250$ MJ/kg DM. DM = Dry matter; CP = %crude protein in DM; EE = % ether extract (lipids) in DM; MM =% minerals (ash) in DM; ADF = % acid detergent fibre in DM; CF = % crude fibre in DM.

At the end of the experimental period, digestibility trial was carried out on ten rabbits per group. Rabbits were housed individually in metabolic cages (1825 mm height × 1370 mm length × 840 depth including feeders × 1210 width when the door is opened) which allowed for the collection of feces and urine separately for five consecutive days collection according to European reference method for rabbit digestion trials [22]. The experimental diets were offered daily and fresh water was provided all times. During the collection period, feces were collected every 24 h for 5 consecutive days, daily FI and feces excreted were accurately determined. Feces of each animal were dried, ground and stored until analysis. Digestible energy (DE, Kcal/Kg diet) was calculated as follow: TDN (Total Digestible Nutrients) × 44.3 [23].

Chemical analyses of both experimental diets and feces were [24] for determining moisture, crude protein (CP), crude fiber (CF), ether extract (EE), nitrogen free extract (NFE), ash. Calcium and Zinc were determined by atomic absorption spectrophotometer and phosphorous was determined colorimetrically using spectrophotometer (3300 perken Elmer, California, United States).

At the end of the experimental period, six male rabbits from each group were randomly taken, fasted for 12 h, individually weighed and immediately slaughtered. Slaughter procedure and carcass analysis were carried out [25]. After complete bleeding, pelt, viscera and tail were removed and then the carcass and giblets (liver, heart, and kidney) were weighed. Dressing percentage included relative weights of the carcass, giblets and head were estimated. Blood samples (5 mL from each rabbit) were collected at slaughtering time (during bleeding) in heparinized glass tubes. Blood plasma was separated by centrifugation at 3000 rpm for 15 min. The collected plasma was stored at −20 °C until assay. Plasma total protein, albumin, total cholesterol, LDL, HDL-cholesterol, aspartate aminotransferase (AST) and alanine aminotransferase (ALT) were colorimetrically determined using commercial kits (acquired from Bio-diagnostic, Giza, Egypt), according to the manufacturers' instructions. Plasma total protein, albumin, cholesterol, LDL-cholesterol, HDL-cholesterol and trans-aminases were determined [26–31]. Plasma globulin values were obtained by subtracting albumin values from the corresponding total protein values. The albumin/globulin ratio was calculated. Zinc and selenium concentrations in plasma were determined by using atomic absorption analysis. Plasma total antioxidant capacity (T-AOC) was measured [32]. Thiobarbituric acid reactive substances (TBARS), superoxide dismutase enzyme (SOD), catalase (CAT) and glutathione peroxidase enzyme (GSH-Px) were verified by colorimetric techniques using a commercially obtainable kit (Bio-diagnostic., Cairo, Egypt).

The obtained data were statistically analyzed by one-way analysis of variance using the general linear model procedure of SAS® Software Statistical Analysis [33]. Differences among treatment means were tested by Duncan's Multiple Range-Test [34].

3. Results

Temperature and relative humidity values in this work ranged between 33.74 to 34.82 °C and 83.83 to 84.39%, correspondingly. Estimates of THI extended amid 32.57 to 33.79 through the investigational interval, representing exposure of NZW growing rabbits to extremely acute heat anxiety.

Growth performance evaluation of rabbits during the different experimental periods is presented in Table 2. Groups of rabbits consumed diets supplemented with Zn- and Se-enriched SP or their combination had significantly ($p < 0.01$) greater BW at 14 weeks of age ($p = 0.005$) and BWG during 6–14 weeks of age ($p = 0.006$). Rabbits fed diet supplemented with Se-Sp achieved the greatest ($p < 0.05$) BWG when compared with the control, and without significant variations with the other groups of supplementation, during 6–10 weeks of age. However, rabbits fed diet supplemented with Zn-Se-Sp exhibited better ($p < 0.05$) BWG than those of the control group during 10–14 weeks of age. Average of FI was decreased ($p < 0.05$) in rabbits fed diet supplemented with Se-Sp and Zn-Se-Sp compared to the control group during 10–14 weeks of age. There was insignificant difference in average FI among the tested groups through 6–10 and 6–14 weeks of age. Regarding FCR, the rabbit groups fed the supplemented diets presented inferior ($p < 0.01$ and 0.001) FCR than the control group through 6–14 weeks and 10–14 weeks of age.

Table 2. Effect of dietary Zn-Sp, Se-Sp and their combination on the growth performance of growing rabbits.

Items	Control	Supplementation (mg/kg DM)			SEM	*p*-Value
		Zn-Sp	Se-Sp	Zn-Se-Sp		
Average body weight, (g)						
at 6 week	745.73	745.33	745.60	746.00	31.27	1.000
at 10 week	1460.66	1515.66	1551.33	1499.33	30.79	0.22
at 14 week	2034.33 ^{b}	2169.66 ^{a}	2199.33 ^{a}	2200.33 ^{a}	30.13	0.005
Average daily weight gain (g/rabbit/day)						
6–10 week	25.53 ^{b}	27.51 ^{ab}	28.77 ^{a}	26.90 ^{ab}	0.81	0.05
10–14 week	20.48 ^{b}	23.35 ^{ab}	23.14 ^{ab}	25.03 ^{a}	1.06	0.03
6–14 week	23.01 ^{b}	25.43 ^{a}	25.96 ^{a}	25.97 ^{a}	0.66	0.006
Average daily feed intake (g/rabbit/day)						
6–8 week	99.04	97.65	96.16	98.78	4.59	0.96
10–14 week	99.98 ^{a}	96.33 ^{ab}	86.98 ^{c}	91.21 ^{bc}	2.41	0.002
6–14 week	98.78	96.99	91.57	94.99	2.89	0.34
Feed conversion (g feed/g gain)						
6–8 week	3.87	3.55	3.34	3.67	0.19	0.24
10–14 week	4.88 ^{a}	4.12 ^{b}	3.75 ^{b}	3.64 ^{b}	0.22	0.0005
6–14 week	4.29 ^{a}	3.81 ^{b}	3.52 ^{b}	3.65 ^{b}	0.14	0.002

Mean values with the same letter within the same row did not differ significantly ($p > 0.05$). SEM: Standard Error of Means.

Results found in Table 3 showed that rabbits fed Zn-Sp recorded significantly ($p < 0.05$) higher digestibility of DM and OM compared to the control group. Rabbits fed diet supplemented with Zn-Sp and Zn-Se-Sp had higher ($p < 0.05$) CP digestibility than the control group. Besides, EE and NFE digestibilities were greater ($p < 0.05$) in rabbits fed diets supplemented with Zn-Sp and Zn-Se-Sp compared to the control group. There were non-significant variations in DM, OM, CP, EE and NFE between the supplemented diets. On the other hand, the different supplementations had no significant effect on CF digestibility. Data of the nutritive values including DCP, TDN and DE illustrated insignificant alterations were noticed between all groups under investigation (Table 3).

Table 3. Effect of dietary supplemental Zn-Sp, Se-algae and their combination on digestibility and nutritive value of experimental rabbit diets.

Supplementation (mg/kg DM)						
Item	Control	Zn-Sp	Se-Sp	Zn-Se-Sp	±SEM	*p*-Value
Digestibility (%)						
DM	62.38 ^{b}	73.23 ^{a}	67.77 ^{ab}	70.86 ^{ab}	2.65	0.09
OM	64.26 ^{b}	74.75 ^{a}	69.71 ^{ab}	72.71 ^{ab}	2.55	0.08
CP	66.23 ^{b}	75.25 ^{a}	70.94 ^{ab}	73.83 ^{a}	2.51	0.13
CF	49.25	62.05	54.07	55.27	4.72	0.35
EE	71.42 ^{b}	77.92 ^{a}	73.28 ^{ab}	78.02 ^{a}	1.53	0.03
NFE	67.41 ^{b}	78.11 ^{a}	73.66 ^{ab}	76.96 ^{a}	2.60	0.07
Nutritive Values						
DCP (g)	11.78	13.02	12.28	12.77	0.46	0.23
TDN (%)	64.61	72.93	67.56	71.06	2.88	0.31
DE(kcal/kg)	2862.22	3230.79	2992.91	3147.95	127.75	0.25

Means in the same row with different superscripts are significantly different ($p < 0.05$). SEM: Standard Error of Means.

The impacts of Zn, Se-enriched SP and or combination on carcass traits of rabbits are displayed in Table 4. All carcass traits, except for pre-slaughter weight and spleen percentage, were significantly changed ($p < 0.05$, 0.01 and 0.001) due to the tested supplementations. Growing rabbits fed diet supplemented with Zn-Se-Sp had the highest

($p < 0.05$) hot carcass weight when competed with their counterparts. The addition of Zn-Sp and Zn-Se-Sp in diets tended to promote ($p < 0.01$) dressing % when compared with the control. Rabbits fed diet containing Se-Sp and Zn-Se-Sp had lower ($p < 0.05$) liver % than the control group, while rabbits fed diets supplemented with Se-Sp and the control group had higher ($p < 0.05$) heart % as compared to those fed diet containing Zn-Se-Sp. However, rabbits fed diets supplemented with Zn-Sp had higher ($p < 0.05$) kidneys % compared to rabbits fed Zn-Se-Sp diet. Rabbits of the control group had greater ($p < 0.05$) giblets % and non-edible parts than those fed diet containing Zn-Se-SP. The rabbits fed diet supplemented with Zn-Se-Sp had worthier ($p < 0.05$) total edible parts than the control group. Meat of rabbits fed diets supplemented with Se-SP and Zn-Se-Sp had lower ($p < 0.05$) values of EE than those of the control. Regarding Zn and Se content of meat, rabbits consumed diets supplemented with Zn-Sp and Zn-Se-Sp had greater ($p < 0.001$) Zn content of meat than those of the control and those fed Se-Sp diet. Meat of growing rabbits fed Zn-Se-Sp and Se-Sp presented higher ($p < 0.001$) content of Se than the control and those fed Zn-Sp diets (Table 4).

Table 4. Effect of different experimental diets on carcass traits and meat chemical composition of growing rabbits.

	Supplementation (mg/kg DM)					
Items	**Control**	**Zn-Sp**	**Se-Sp**	**Zn-Se-Sp**	**±SEM**	***p*-Value**
		Carcass traits				
Pre-slaughter weight (g)	2000.0	2078.33	2046.67	2181.00	27.59	0.21
Hot carcass weight (g)	1070.56 ^{c}	1251.93 ^{b}	1229.67 ^{b}	1366.43 ^{a}	32.67	0.001
Dressing %	53.52 ^{b}	60.27 ^{a}	60.13 ^{ab}	65.68 ^{a}	1.73	0.01
Liver weight %	3.19 ^{a}	2.73 ^{ab}	2.46 ^{b}	2.29 ^{b}	0.19	0.05
Heart weight %	0.38 ^{a}	0.36 ^{ab}	0.37 ^{a}	0.31 ^{b}	0.01	0.06
Kidney weight %	0.64 ^{ab}	0.67 ^{a}	0.59 ^{ab}	0.56 ^{b}	0.02	0.01
Spleen%	0.11	0.12	0.12	0.18	0.02	0.25
Edible Giblets [1] %	4.21 ^{a}	3.76 ^{ab}	3.45 ^{ab}	3.17 ^{b}	0.22	0.05
Total edible parts [2] %	57.73 ^{b}	64.03 ^{ab}	63.59 ^{ab}	68.85 ^{a}	1.86	0.01
Non- edible parts %	42.27 ^{a}	35.97 ^{ab}	39.38 ^{ab}	34.30 ^{b}	1.77	0.01
		Meat chemical composition (%)				
DM	27.87 ^{a}	27.32 ^{ab}	26.85 ^{b}	26.91 ^{ab}	0.26	0.07
CP	22.12	23.14	22.80	22.84	0.31	0.19
EE	3.67 ^{a}	2.70 ^{b}	2.49 ^{bc}	2.65 ^{c}	0.24	0.02
Ash	2.09	1.48	1.56	1.42	0.22	0.19
		Zn and Se content in meat				
Zn (mg/100 g)	0.95 ^{b}	1.57 ^{a}	0.96 ^{b}	1.58 ^{a}	0.009	0.0001
Se (μg/g)	0.094 ^{b}	0.097 ^{b}	0.46 ^{a}	0.47 ^{a}	0.003	0.0001

Mean values with the same letter within the same row did not differ significantly ($p > 0.05$); [1] Edible Giblets, % = (Liver+ kidneys + heart)/ Pre-slaughter weight (g) × 100; [2] Total edible parts, % = (Carcass weight + edible giblets weight)/Pre-slaughter weight (g) × 100.

As shown in Table 5, plasma total protein levels were higher ($p < 0.05$) in rabbits fed diet supplemented with Zn-Se-Sp compared to the control group. Meanwhile, the same group and those fed Se-Sp recorded higher ($p < 0.05$) globulin levels compared to the control group. While, no significant effect on albumin, A/G ratio, AST and ALT group were observed among all the tested groups and the control group. Additionally, lower plasma total cholesterol, LDL-cholesterol and VLDL-cholesterol concentrations were observed for rabbits given Se-Sp and Zn-Se- Sp diets in comparison with rabbits given the control diet. In the opposite direction, Se-Sp, Zn-Se-Sp were greater ($p < 0.05$) in HDL than the control group. Plasma lipids and triglycerides levels were significantly decreased ($p < 0.05$) with dietary supplementation of Zn-Sp, Se-Sp and Zn-Se-Sp.

Table 5. Effect of dietary supplemental Zn-Sp, Se-Sp and their combination on plasma biochemicals and antioxidative status of the experimental growing rabbits.

Items	Supplementation (mg/kg DM)				±SEM	p-Value
	control	Zn-Sp	Se-Sp	Zn-Se-Sp		
Plasma biochemicals						
Total protein (g/dL)	6.20 b	6.68 ab	6.90 ab	7.33 a	0.22	0.02
Albumin (g/dL)	3.72	3.93	3.25	3.65	0.25	0.31
Globulin (g/dL)	2.47 b	2.75 ab	3.65 a	3.68 a	0.31	0.03
Albumin / Globulin ratio	1.51	1.43	0.89	0.99	0.20	0.10
AST	33.37	34.81	37.20	35.07	2.05	0.63
ALT	44.00	44.50	43.37	46.12	1.51	0.62
Total lipids (mg/dL)	292.24 a	244.60 b	225.80 b	232.30 b	7.04	0.0001
Triglycerides (mg/dL)	73.44 a	67.85 ab	60.44 b	59.15 b	2.81	0.01
Total cholesterol (mg/dL)	107.74 a	94.17 ab	86.40 b	80.36 b	6.39	0.05
HDL (mg/dL)	24.42 c	27.03 bc	28.44 b	32.53 a	1.13	0.002
LDL (mg/dL)	68.63 a	53.57 ab	45.87 b	36.00 b	6.33	0.02
VLDL (mg/dL)	14.68 a	13.57 ab	12.09 b	11.83 b	0.56	0.011
Antioxidative status						
T-AOC (mmol/L)	0.59 c	0.95 bc	1.07 b	1.97 a	0.14	0.0001
TBARs	6.47 a	6.08 a	5.17 b	2.95 c	0.15	0.0001
GSH-Px (U/L)	0.92 b	1.07 b	1.20 b	2.00 a	0.16	0.0020
Catlase (U/L)	495.07 b	571.78 ab	635.73 a	623.48 a	30.91	0.0285
SOD (U/L)	28.98 c	33.29 bc	40.43 ab	45.57 a	2.68	0.0075

Mean values with the same letter within the same row did not differ significantly ($p > 0.05$). Total antioxidant capacity (mmol/L), TBARS, thiobarbituric acid reactive substances, Glutathione peroxidase (U/L) and Superoxide dismutase (U/L).

Data in Table 5 postulated a significant ($p < 0.05$) increase in plasma T-AOC and catalase levels in rabbits fed Se-Sp and Zn-Se-Sp compared to the control group. On the other hand, a significant decrease ($p < 0.05$) in TBARs was noticed in rabbits fed diet containing Se-Sp and Zn-Se-Sp compared to the control group and Zn-Sp diet. Rabbits fed diet supplemented with Zn-Se-Sp tended to increase ($p < 0.05$) GSH-Px level compared to the other tested groups. As well, the SOD concentration was found to be significantly higher in the rabbits fed diet included Zn-Se-Sp than in rabbits given Zn-Sp containing diets and those fed the control diet. Moreover, SOD levels were higher ($p < 0.05$) in Se-Sp group than those of the control one.

4. Discussion

The present study demonstrated that Zn- or Se-enriched Sp or their combination improved BW at marketing, BWG and FCR of growing rabbits, while, these supplementations decreased average FI. Similar to the present findings, Hassan et al. [35] reported that dietary Zn-Sp supplementation at levels of 50, 75 and 100 mg/kg diet caused an increase in marketing BW and improved BWG and FCR. In this regard, Hassan et al. [36] stated that dietary supplementation of Se-algae at 0.05, 0.1, 0.2, 0.4 and 0.5 mg had a positive impact on growth performance of growing rabbits. On contrary, Hosny et al. [37] found non-significant impact on BW and FI of rabbits fed diet including 0.3 mg organic Se/kg diet.

The present study also revealed that the combination effect of Zn and Se-enriched Spirulina showed better growth performance compared to using each alone, this improvement may be due to greater bio-efficacy of Zn-Sp. It could provide more Zn for absorption and resulted in improved growth performance (Hassan et al. [35]. Besides, zinc acts as an antioxidant stress agent by inhibition of oxidation of macromolecules such as DNA and proteins (Prasad and Bao [5]. As well, a potential for better growth may be the profile of organic compounds of Spirulina in the Se-algae (Larsen et al. [38]. Such enhancement of supplementation may be due to the synergetic effects of organic Zn and Se and they have a potential nutritive value as feed additives for growing rabbits under summer heat stress conditions. Furthermore, confirmation is associated with the use of Spirulina which can improve the growth performance because it has some natural constituents such as phycocyanin, beta-carotene, tocopherols, linolenic acid, minerals, vitamins and phenolic compounds that had been shown to have strong antioxidant properties with promote growth and maintain health (Michalak and Mahrose [15] and Farag et al. [16]. It has very high content of macro and micronutrients, essential amino acids, proteins, lipids, vitamins, minerals and anti-oxidants Soni et al. [39]. Furthermore, it is a strong antioxidant due to the presence of high content of antioxidant phenols or flavonoids Gabr et al. [40].

The dietary Zn-Sp supplementation resulted in an escalation in DM, OM. Zn-Se-Sp improved CP, EE and NFE digestibility compared to the control group. In this direction, Hassan et al. [36] found that rabbits fed diets supplemented with Zn-Sp at 100 mg/kg diet led to a rise in all nutrient digestibilities. Our results suggested that supplemental Zn-Sp or Zn-Se-SP to the growing rabbits improved nutrients digestibility. This improvement may be related to the role of zinc in metabolism, whereas zinc participates actively in protein synthesis, carbohydrate and lipid metabolism Chrastinová et al. [41] as well, organic Zn has been considered as an alternative to inorganic Zn in the diets of rabbits and broilers due to its better absorption and efficiency Hassan et al. [35]. Furthermore, organic Se is metabolized much more efficiently the inorganic Se forms and could be efficiently utilized for synthesis of selenoproteins under stress conditions Qazi et al. [42]. The positive effect of Zn-Se-Sp on the digestibility of nutrients may be revealed that Se improved the antioxidative status of rabbits Hassan et al. [43]. Moreover, Zn has an oxidative activity Prasad and Bao [5] which reduces the oxidative capacity and improve animal health.

The effect of Zn-Se-Sp was positive in hot carcass weight, dressing% and total edible parts compared to the control group. The current results are in line with those of Hassan et al. [36] who stated that rabbits on diets supplied with Se-algae at level of 0.5 mg/kg diet increased hot carcass weight, dressing% and total edible parts%. Moreover, Hassan et al. [35] indicated that Zn-Sp supplementation at 50, 75 and 100 mg/kg diet increased hot carcass weight, dressing and total edible parts% compared with the rabbits fed the control diet. On the contrary, Selim et al. [8] postulated that rabbits fed diets containing 50, 100, 200 or 400 mg zinc oxide /kg diet did not change carcass traits. Moreover, no significant effect of Se-algae addition on carcass yield of rabbits was noticed (Marounek et al. [44].

Regarding meat composition, our results agreed with the findings obtained by Hassan et al. [43] who mentioned that dietary addition of Se-algae in rabbit diets at level of 0.2 mg/kg diet decreased EE content of meat. Similarly, Hassan et al. [36] reported that 0.5 mg Se algae/kg diet decreased EE content of rabbit meat. Marounek et al. [44] stated that supplemental Se in the rabbit diet had no impact on CP content of meat. However, Se content in the hind leg meat was found to be increased to dietary Se-alga at 0.2 mg/kg diet [43]. Comparable findings were reported by Marounek et al. [44] and Amer et al. [45] who found that Se-yeast supplementation in rabbit diet deposited in the meat and improved meat quality.

The present study showed a positive influence of the supplemented diets on plasma biochemistry indices. These outcomes are consistent with those of Hassan et al. [36] who mentioned that rabbit consumed diet including 50, 70 and 100 mg/kg had high total protein and HDL. In addition, the dietary Zn- and Se-enriched Sp or their combination had no significant effect of A/G ratio, albumin levels and activities of AST and ALT. Similarly, Hassan et al. [35] concluded that Zn-Sp supplementation did not impact AST, ALT, albumin and A/G ratio. Hassan et al. [43] revealed an increase in plasma total protein concentration as a result of Se-algae addition at level of 0.2 mg/kg diet. Likewise, El-Kholy et al. [46] stated that the addition of different forms of Se led to increases in total protein and globulin levels and did not change ALT and AST activities. The reduction in plasma total cholesterol, LDL-cholesterol and VLDL-cholesterol levels in rabbits fed the supplemented diets may be due to that Zn inhibits the lipolysis in adipose tissues, reduces free fatty acid release into the circulation and its availability to the liver and excessive lipoprotein synthesis (Dieck et al. [47]. Besides Zinc contribution to insulin secretion and action, Zinc directly affects lipid metabolism then increased free fatty acid flux to the liver which stimulates the assembly and secretion of vLDL resulting in hypertriglyceridemia (Ranasinghe et al. [48]. On the other hand, our findings matched with El-Kholy et al. [46] who observed that rabbits received either organic or inorganic Se forms had lower total cholesterol and LDL levels than those of the control group. Moreover, Hassan et al. [35] showed that supplemental Zn-Sp at 75 mg/kg diet decreased in serum total cholesterol and LDL concentration.

It is noteworthy that the supplemental Zn-Se-Sp had a potential antioxidant effect on the rabbits under high temperature and was associated with the lower of TBARs, the higher TAC, GSH, SOD and catalase activities. So it may protect the tissues against oxidative damage which included protein and fat oxidation of growing rabbits under hot conditions. The present results are in agreement with the findings of Alissa et al. [49] who showed that plasma TBARs concentration was reduced by zinc supplementation (0.5%, *w/w*) in rabbit diets and suggested that zinc was associated with a reduction in plasma lipid peroxides. Similar findings have been also reported by Zhang et al. [50] who indicated that rabbits fed a diet containing 0.24 mg/kg Se had the greatest serum GSH-Px and CAT activities. Moreover, there was an increase in the serum T-AOC concentration due to Zn-Sp supplementation at levels of 50, 75 and 100 mg/kg diet [35]. The addition of organic Se at 0.3 mg/diet increased glutathione peroxidase activity in rabbits [37]. The findings reported by Prasad and Bao [5] strongly suggested that zinc reduces oxidative stress and ROS-mediated inflammatory responses, and that zinc acts as a potent agent by inhibition of ROS production and inflammation. Whereas, Zn has a potential role as an antioxidative stress agent, and a pro-antioxidant effect or protective effect against oxidative stress in biological system [5]. In addition, Se positivity affects the antioxidative status of rabbits; this effect has been attributed to selenium which is an essential constituent of GSH-Px [9,10,12]. Glutathione peroxidase helps in protecting cellular membranes from oxidative damage which resulted in enhancing the growth performance of rabbits under hot conditions [51]. Recent literature have shown that Spirulina has an antioxidant, immunomodulatory, anti-inflammatory, antiviral, and antimicrobial activity in various experimental animals [15,40,52]. In this respect, Park et al. [52] suggested that dietary Spirulina supplementation at levels of 0.25, 0.5, 0.75, or 1.0% in broiler diets caused an increase in the serum SOD, and GSH-Px activities. As observed in this study, supplementing rabbit diets with Zn- and Se-enriched SP enhanced the antioxidative status as they are an efficient scavenger of free radicals [53,54].

5. Conclusions

It is clear from the present study, that supplementation of 100 mg Zn-Sp, 0.5 gm Se-Sp and or their combination could improve growth performance, nutrients digestibility and antioxidant status of heat stressed growing rabbits.

Author Contributions: Conceptualization, F.H. and K.M.; methodology, F.H.; software, S.M.; validation, F.H., M.M. and A.M.; formal analysis, S.M.; investigation, M.M.; resources, F.H.; data curation, F.H.; writing—original draft preparation, F.H. and A.M.; writing—review and editing, K.M. and Y.A.; visualization, A.M.; supervision, K.M.; project administration, F.H. All authors have read and agreed to the published version of the manuscript.

Funding: This research received no external funding.

Institutional Review Board Statement: The study was conducted according to the guidelines of the Declaration of Helsinki, and approved by the Ethics Committee of Animal Production Research Institute, Giza, Egypt.

Data Availability Statement: The data that support the findings of this study are available on request from the corresponding author. The data are not publicly available due to privacy or ethical restrictions.

Conflicts of Interest: The authors declare no conflict of interest.

References

1. Hassan, F.; Mahrose, K.M.; Basyony, M.M. Effects of grape seed extract as a natural antioxidant on growth performance, carcass characteristics and antioxidant status of rabbits during heat stress. *Arch. Anim. Nutr.* **2016**, *70*, 141–154. [CrossRef] [PubMed]
2. Farghly, M.F.A.; Mahrose, K.M.; Mahmoud, G.B.; Ali, R.M.; Daghash, W.; Metwally, K.A.; Abougaba, M.S. Lighting programs as an appliance to improve growing New Zealand white rabbit's performance. *Int. J. Biometeorol.* **2020**, *64*, 1295–1303. [CrossRef] [PubMed]
3. Biagini, D.; Gascoa, L.; Rosatoc, R.; Peirettib, P.G.; Gaib, F.; Lazzaronia, C.; Montonerid, C.; Ginepro, M. Compost-sourced substances (SBO) as feedstuff additives in rabbit production. *Anim. Feed Sci. Technol.* **2016**, *214*, 66–76. [CrossRef]
4. Marai, I.F.M.; Habeeb, A.A.M.; Gad, A.E. Rabbits' productive, reproductive and physiological performance traits as affected by heat stress: A review. *Livest. Prod. Sci.* **2002**, *78*, 71–90. [CrossRef]
5. Prasad, A.S.; Bao, B. Review, Molecular Mechanisms of Zinc as a Pro-Antioxidant Mediator: Clinical Therapeutic Implications. *Antioxidants* **2019**, *8*, 164. [CrossRef]
6. Chasapis, C.T.; Ntoupa, P.A.; Spiliopoulou, C.A.; Stefanidou, M.E. Recent aspects of the effects of zinc on human health. *Arch. Toxicol.* **2020**, *94*, 1443–1460. [CrossRef]
7. Mateos, G.G.; Garla-Rebollar, P.; de Blas, C. Minerals, vitamins and additives. In *Nutrition of the Rabbits*, 3rd ed.; De Blas, C., Wiseman, J., Eds.; CABI Publishing: Wallingford, UK, 2020; pp. 126–133.
8. Selim, N.; Abdel-Khalek, A.M.; Gad, S.M. Effect of supplementation zinc, magnesium or iron on performance and some physiological traits of growing rabbits. *Asia J. Poult. Sci.* **2012**, *6*, 23–30. [CrossRef]
9. Abd El-Hack, M.E.; Mahrose, K.H.M.; Arif, M.; Chaudhry, M.T.; Saadeldin, I.M.; Saeed, M.; Soomro, R.N.; Abbasi, I.H.R.; Rehman, Z.U. Alleviating the environmental heat burden on laying hens by feeding on diets enriched with certain antioxidants (vitamin E and selenium) individually or combined. *Environ. Sci. Pollut. Res.* **2017**, *24*, 10708–10717. [CrossRef]
10. Abd El-Hack, M.E.; Mahrose, K.H.; Askar, A.A.; Alagawany, M.; Arif, M.; Saeed, M.; Abbasi, F.; Soomro, R.N.; Siyal, F.A.; Chaudhry, M.T. Single and combined impacts of vitamin A and selenium in diet on productive performance, egg quality, and some blood parameters of laying hens during hot season. *Biol. Trace Elem. Res.* **2017**, *177*, 169–179. [CrossRef]
11. Cardoso, B.R.; Roberts, B.R.; Bush, A.I.; Hare, D.J. Selenium, selenoproteins and neurodegenerative diseases. *Metallomics* **2015**, *7*, 213–1228. [CrossRef]
12. Zoidis, E.; Seremelis, I.; Kontopoulos, N.; Danezis, G.P. Selenium-dependent antioxidant enzymes: Actions and properties of selenoproteins. A review. *Antioxidants* **2018**, *7*, 66. [CrossRef]
13. Li, M.; Zhang, Y.; Li, S. Effects of selenium deficiency on testis development and autophagy in chicks. *Ital. J. Anim. Sci.* **2020**, *19*, 753–761. [CrossRef]
14. Hajati, H.; Zaghari, M. *Spirulina Platensis in Poultry Nutrition;* Cambridge Scholars Publishing Lady Stephenson Library: Newcastle upon Tyne, UK, 2019.
15. Michalak, I.; Mahrose, K.M. Seaweeds, intact and processed, as a valuable component of poultry feed. *J. Mar. Sci. Eng.* **2020**, *8*, 620. [CrossRef]
16. Farag, M.R.; Alagawany, M.; Abd El-Hack, M.E.; Dhama, K. Nutritional and healthical aspects of *Spirulina* (*arthrospira*) for poultry, animals and human. *Int. J. Pharmacol.* **2015**, *12*, 36–51. [CrossRef]
17. Khan, M.; Shobha, J.C.; Mohan, I.K.; Naidu, M.U.R.; Sundaram, C.; Singh, P.K.; Kutala, V.K. Protective effect of *Spirulina* against doxorubicin-induced cardiotoxicity. *Phytother. Res.* **2005**, *19*, 1030–1037. [CrossRef] [PubMed]
18. Hajati, H.; Zaghari, M.; Oliveira, H.C. *Arthrospira* (*Spirulina*) *Platensis* can be considered as a probiotic alternative to reduce heat stress in laying Japanese quails. *Braz. J. Poult. Sci.* **2020**, *22*, 1–8. [CrossRef]
19. Gaese, H. Chemical composition and potential application of *Spirulina platensis* biomass. *Int. J. Agric. Environ.* **2012**, *32*, 32–40.
20. Marai, I.F.M.; Ayyat, M.S.; Abd El-Monem, U.M. Growth performance and reproductive traits at first parity of New Zealand white female rabbits as affected by heat stress and its alleviation under Egyptian conditions. *Trop. Anim. Health Prod.* **2001**, *33*, 1–12. [CrossRef]
21. Lebas, F. Feeding strategy for small and medium scale rabbit. In Proceedings of the 3rd Conference of Asian Rabbit Production Association, Bali, Indonesia, 27–29 August 2013; pp. 1–15.
22. Perez, J.M.; Lebas, F.; Gidenne, T.; Maertens, L.; Xiccato, G.; Parigi-Bini, R.; Dalla Zotte, A.; Cossu, M.E.; Carazzolo, A.; Villamide, M.J.; et al. European reference method for in vivo determination of diet digestibility in rabbits. *World Rabbits Sci.* **1995**, *3*, 41–43.
23. Schneider, B.H.; Flatt, W.P. *The Evaluation of Feed through Digestibility Experiments;* University of Georgia Experiments, University of Georgia Press: Athens, GA, USA, 1975; p. 423.
24. Association of Official Analytical Chemists (AOAC). *Official Methods of Analysis*, 17th ed.; AOAC: Washington, DC, USA, 2000.
25. Blasco, A.; Ouhayoun, J. Harmonization of criteria and terminology in rabbit meat research. *World Rabbit Sci.* **1996**, *4*, 93–99. [CrossRef]
26. Orsonneau, J.L.; Douet, P.; Massoubre, C.; Lustenberger, P.; Bernard, S. An improved pyrogallol red-molybdate method for determining total urinary protein. *Clin. Chem.* **1989**, *35*, 2233–2236. [CrossRef]
27. Doumas, B.T.; Watson, W.A.; Biggs, H.G. Albumin standards and the measurement of serum albumin with bromcresol green. *Clin. Chem. Acta* **1971**, *31*, 87–96. [CrossRef]

28. Richmond, W. Preparation and properties of a cholesterol oxidase from *Nocardia* sp. and its application to the enzymatic assay of total cholesterol in serum. *Clin. Chem.* **1973**, *19*, 1350–1356. [CrossRef]
29. Reitman, S.; Frankel, S. A calorimetric method for the determination of serum glutamic oxaloacetic and glutamic pyruvic transaminase. *Am. J. Clin. Path.* **1957**, *28*, 56–63. [CrossRef]
30. Lopez-Virella, M.F.; Stone, P.; Ellis, S.; Colwell, J.A. Cholesterol determination in high-density lipoproteins separated by three different methods. *Clin. Chem.* **1977**, *23*, 882–884. [CrossRef]
31. Wieland, H.; Seidel, D. A simple specific method for precipitation of low density lipoproteins. *J. Lipid Res.* **1983**, *24*, 904–909. [CrossRef]
32. Koracevic, D.; Koracevic, G.; Djordjevic, V.; Andrejevic, S.; Cosic, V. Method for the measurement of antioxidant activity in human fluids. *J. Clin. Pathol.* **2001**, *54*, 356–361. [CrossRef]
33. SAS. User's guide statistic (Release 9.1). In *Statistical Analysis System*; Institute Inc.: Cary, NC, USA, 2002.
34. Duncan, D.B. Multiple range and multiple F-tests. *Biometrics* **1955**, *11*, 1–42. [CrossRef]
35. Hassan, F.A.; Amin, F.H.; Ali, W.A.; El-Aasar, T.A.; Azoz, A.A. Effect of dietary addition of zinc enriched *Spirulina* on growth and physiological performance in growing rabbits. *Egypt. J. Rabbit Sci.* **2016**, *26*, 155–173.
36. Hassan, F.A.; Hoballah, E.M.; Basyony, M.M.; El-Medany, S.A. Effect of dietary selenium enriched micro-algae supplementation on growth performance and anti-oxidative status of rabbits under high ambient temperature in summer season. *Egypt. J. Nutr. Feeds* **2015**, *18*, 229–244. [CrossRef]
37. Hosny, N.S.; Hashem, N.M.; Morsy, A.S.; Abo-elezz, Z.R. Effects of organic selenium on the physiological response, blood metabolites, redox status, semen quality, and fertility of rabbit bucks kept under natural heat stress conditions. *Front. Vet. Sci.* **2020**, *7*, 290. [CrossRef]
38. Larsen, E.H.; Hansen, M.; Fan, T.; Vahl, M. Speciation of selenoamino acids, selenonium ions and inorganic selenium by ion exchange HPLC with mass spectometric detection and its application to yeast and algae. *J. Anal. Atom. Spectrom.* **2001**, *16*, 1403–1408. [CrossRef]
39. Soni, R.A.; Sudhakar, K.; Rana, R.S. Review *Spirulina* from growth to nutritional product: A review. *Trends Food Sci. Technol.* **2017**, *69*, 157–171. [CrossRef]
40. Gabr, G.A.; El-Sayed, S.M.; Hikal, M.S. Antioxidant activities of phycocyanin: A bioactive compound from *Spirulina platensis*. *J. Pharm. Res. Int.* **2020**, *32*, 73–85. [CrossRef]
41. Chrastinová, L'.; Čobanová, K.; Chrenková, M.; Poláčiková, M.; Formelová, Z.; Lauková, A.; Ondruška, L'.; Pogány Simonová, M.; Strompfová, V.; Bučko, O.; et al. High dietary levels of zinc for young rabbits. *Slovak J. Anim. Sci.* **2015**, *48*, 57–63.
42. Qazi, I.H.; Angel, C.; Yang, H.; Zoidis, E.; Pan, B.; Wu, Z.; Ming, Z.; Zeng, C.; Meng, Q.; Han, H.; et al. Role of selenium and selenoproteins in male reproductive function: A review of past and present evidences. *Antioxidants* **2019**, *8*, 268. [CrossRef] [PubMed]
43. Hassan, F.A.; Abdel-Azeem, N.M.; Abdel-Rahman, S.M.; Amin, H.F.; Abdel-Mawla, L.F. Effect of dietary organic selenium supplementation on growth performance, carcass characteristics and antioxidative status of growing rabbits. *World Vet. J.* **2019**, *9*, 16–25. [CrossRef]
44. Marounek, M.; Dokoupilová, A.; Volek, Z.; Hoza, I. Quality of meat and selenium content in tissues of rabbits fed diets supplemented with sodium selenite, selenized yeast and selenized algae. *World Rabbit Sci.* **2009**, *17*, 207–212. [CrossRef]
45. Amer, S.A.; Omar, A.E.; Abd El-Hack, M.E. Effects of selenium- and chromium-enriched diets on growth performance, lipid profile, and mineral concentration in different tissues of growing rabbits. *Biol. Trace Elem. Res.* **2018**, *187*, 92–99. [CrossRef]
46. El-Kholy, K.H.; Tag El-Deen, H.T.; Abd-El-Lateif, A.I.; Mekaouy, A.I. Effects of dietary selenium sources on metabolic, enzymatic and immunoglobulin serum profiles in growing rabbits. *Pak. J. Nutr.* **2019**, *18*, 430–436. [CrossRef]
47. Dieck, H.; Döring, F.; Fuchs, D.; Roth, H.-P.; Daniel, H. Transcriptome and proteome analysis identifies the pathways that increase hepatic lipid accumulation in Zinc-deficient rats. *J. Nutr.* **2005**, *135*, 199–205. [CrossRef]
48. Ranasinghe, P.; Wathurapatha, W.S.; Ishara, M.H.; Jayawardana, R.; Galappatthy, P.; Katulanda, P.; Constantine, G.R. Effects of Zinc supplementation on serum lipids: A systematic review and meta-analysis. *Nutr. Metab.* **2015**, *12*, 26–38. [CrossRef] [PubMed]
49. Alissa, E.M.; Bahijri, S.M.; Lamb, D.J.; Gordon, A.; Ferns, G.A.A. The effects of coadministration of dietary copper and zinc supplements on atherosclerosis, antioxidant enzymes and indices of lipid peroxidation in the cholesterol-fed rabbit. *Int. J. Exp. Pathol.* **2004**, *85*, 265–275. [CrossRef] [PubMed]
50. Zhang, Y.; Zhu, S.; Wang, X.; Wang, C.; Li, F. The effect of dietary selenium levels on growth performance, antioxidant capacity and glutathione peroxidase 1 (GSHPx1) mRNA expression in growing meat rabbits. *Anim. Feed Sci. Technol.* **2011**, *169*, 259–264. [CrossRef]
51. Dalle Zotte, A.; Szendrő, Z. The role of rabbit meat as functional food, a review. *Meat Sci.* **2011**, *88*, 319–331. [CrossRef] [PubMed]
52. Park, J.H.; Lee, S.I.; Kim, I.H. Effect of dietary *Spirulina* (*Arthrospira*) *platensis* on the growth performance, antioxidant enzyme activity, nutrient digestibility, cecal microflora, excreta noxious gas emission, and breast meat quality of broiler chickens. *Poult. Sci.* **2018**, *97*, 2451–2459. [CrossRef] [PubMed]
53. Attia, Y.A.; Abd El-Hamid, A.E.; Zeweil, H.S.; Qota, E.M.; Bovera, F.; Monastra, M.; Sahledom, M.D. Effect of dietary amounts of organic and inorganic Zinc on productive and physiological traits of white peckin ducks. *Animal* **2013**, *7*, 695–700. [CrossRef]
54. Attia, Y.A.; Abdalah, A.A.; Zeweil, H.S.; Bovera, F.; Tag El-Din, A.A.; Araft, M.A. Effect of inorganic or organic selenium supplementation on productive performance, egg quality and some physiological traits of dual purpose breeding hens. *Czech J. Anim. Sci.* **2010**, *55*, 505–519. [CrossRef]

animals

Review

Effects of Dietary Fiber on Nutrients Utilization and Gut Health of Poultry: A Review of Challenges and Opportunities

Amit Kumar Singh and Woo Kyun Kim *

Department of Poultry Science, University of Georgia, Athens, GA 30602, USA; aksingh@uga.edu
* Correspondence: wkkim@uga.edu

Simple Summary: The inclusion of agricultural co-products has been increased to utilize the nutrients in these products available at low cost, but inherently, it adds a high dietary fiber content in the poultry diets. The use of exogenous feed enzymes along with advancements in feed milling, feed formulation, and processing of these non-conventional ingredients to improve their digestibility and utilization have played an emphatic role in boosting their use globally. Despite such developments, the presence of a high level of dietary fibers (DF) acting in an anti-nutritive manner still poses challenges in poultry feeding. Various isolated forms of fiber or feed enzymes to break DF into fermentable substrates are being used extensively to provide potential prebiotics to support beneficial gut microbiota or probiotics to improve the gut health of poultry raised without antibiotic growth promoters (AGP). This review reports and discusses the existing challenges in feeding high-DF feed ingredients to poultry and the opportunities that are available to improve the nutritive value of such non-conventional feed ingredients by adopting various technologies.

Abstract: Many fibrous ingredients incorporated in poultry feed to reduce production costs have low digestibility and cause poor growth in poultry. However, all plant-based fibers are not equal, and thus exert variable physiological effects on the birds, including but not limited to, digestibility, growth performance, and microbial fermentation. Several types of fibers, especially oligosaccharides, when supplemented in poultry diets in isolated form, exhibit prebiotic effects by enhancing beneficial gut microbiota, modulating gut immunity, boosting intestinal mucosal health, and increasing the production of short-chain fatty acids (SCFA) in the gut. Recently, poultry producers are also facing the challenge of limiting the use of antibiotic growth promoters (AGP) in poultry feed. In addition to other alternatives in use, exogenous non-starch polysaccharides digesting enzymes (NSPase) and prebiotics are being used to provide substrates to support the gut microbiome. We also conducted a meta-analysis of different studies conducted in similar experimental conditions to evaluate the variability and conclusiveness in effects of NSPase on growth performance of broilers fed fibrous ingredients. This review presents a holistic approach in discussing the existing challenges of incorporating high-fiber ingredients in poultry feed, as well as strategies to fully utilize the potential of such ingredients in improving feed efficiency and gut health of poultry.

Keywords: antinutrient; enzyme; fermentation; fiber; gut health; microbiota; meta-analysis; poultry; prebiotic

Citation: Singh, A.K.; Kim, W.K. Effects of Dietary Fiber on Nutrients Utilization and Gut Health of Poultry: A Review of Challenges and Opportunities. *Animals* **2021**, *11*, 181. https://doi.org/10.3390/ani11010181

Received: 4 December 2020
Accepted: 11 January 2021
Published: 14 January 2021

Publisher's Note: MDPI stays neutral with regard to jurisdictional claims in published maps and institutional affiliations.

1. Introduction

The poultry diet is balanced for metabolizable energy and protein through the incorporation of several ingredients and additives. The cost of many cereal grains and legumes used in making poultry feed is increasing due to the growing markets utilizing them as food and fuel [1]. To counter this, alternative feedstuffs such as agricultural co-products, including wheat middlings, soy hulls, oil cakes, distillers dried grains and solubles (DDGS), and sugar beet pulp, etc., are regularly incorporated in poultry feed. However, these co-products inherently contain a high proportion of dietary fiber (DF) comprising of non-starch polysaccharides (NSP), lignin, and other indigestible plant-based carbohydrates [2]. Poultry lacks endogenous enzymes required for the breakdown of these NSP. The NSP fractions include cellulose and non-cellulosic polysaccharides (NCP), and the NCP portion further consists of pectic polysaccharides and hemicellulose [2,3]. DF is further divided into soluble and insoluble fibers based on their aqueous solubility. Furthermore, a non-digestible fraction of starch that remains resistant to enzymatic digestion is termed as 'resistant starch' and has been reported to possess physiological functions similar to other DF [4].

Recently, there has been an increase in the trend of incorporating DF and oligosaccharides in poultry diet to supply substrates for beneficial gut microbes [5]. Previous research on the characteristics of fiber demonstrates both opportunities and challenges in enhancing healthy and efficient poultry production. Fiber can act as an antinutrient, as it often encapsulates nutrients in cell walls of plant-based feed ingredients, negatively influences viscosity of digesta, and impacts mineral absorption through the chelating properties of some fiber moieties [6,7]. It has been noted that DF is utilized by microbes in the lower gut to produce short-chain fatty acids (SCFA) as fermentation metabolites [8]. These SCFA are utilized by the intestinal enterocytes for growth and are transported to the liver to produce ATP. However, because of their low digestibility, NSP reduces the apparent metabolizable energy (AME) value of feed, and consequently increases the viscosity of digesta, which adversely affects the digestibility of other nutrients [5]. Thus, preprocessing and enzyme supplementation to increase the digestibility of fiber will also improve utilization of other nutrients in feed and will increase fermentable resources for the gut microbes [5,9,10]. The fermentable substrate can range from complex fragments to simple oligomers that could serve as prebiotics if they could selectively enhance the population of beneficial bacteria leading to immunomodulation and improved gut health [11–13].

2. Composition and Properties of DF

The generic term 'fiber' in nutrition is broadly used for a diverse group of complex carbohydrate fractions of NSP, oligosaccharides and resistant starch, and polyphenolic compound lignin. The proximate analysis system developed by Weende Experiment Station in Germany classified carbohydrates in feed into a more digestible component called nitrogen-free extract (NFE) and a less digestible fibrous component called crude fiber (CF) [14]. The CF is still used in poultry feed formulation and it is determined as the organic residue of acid and alkali digestion, and it fails to account for total fractions of NSP. The other analysis process using neutral and acid detergents by Van Soest [15] categorized fiber into neutral detergent fiber (NDF) comprising of cellulose, hemicellulose and lignin, and acid detergent fiber (ADF), largely consisting of cellulose and lignin. The Van Soest detergent fiber system is also affected by unreliability and falls short of accounting for all NSP in the poultry feed ingredients. The term DF is more associated with the physiological effect and method of determination of fiber component in the feed. The DF in feed is determined either using enzymatic-gravimetric methods adopted by the Association of Official Analytical Chemists for total, soluble, and insoluble DF [16], or by the Upsala and Englyst method that quantifies each monosaccharide converted to aldol acetates and measured using chromatography and spectrophotometry [17,18]. The DF primarily consists of carbohydrates polymers such as cellulose, hemicelluloses, pectins, mucilage, gums, β-glucans, oligosaccharides and resistant starch, and associated substances like lignin [19]. Cellulose is the major component of the plant cell wall and consists of a linear chain of up to 10,000 glucose monomer units per molecule linked by β (1→4) glycosidic bonds. Hemicellulose is a heterogeneous group of chemicals that also include both linear and branched chain of monomers other than glucose. Pectins are gel-forming polysaccharides that are mostly found in the outer skin of rind of fruits and vegetables and consists of polymers of galacturonic acids interspersed with rhamnose and branched chain of pentoses and hexoses. β-glucans are polysaccharides of variable sizes that consist of glucose polymer linked via β-(1→3) and β-(1→6) or via β-(1→4) and β-(1→3) glycosidic bonds. Resistant starch is a homopolysaccharide of glucose that is resistant to digestion by endogenous enzymes and is categorized into various types based on its physical inaccessibility, granular form, retrogradation, and chemical modification.

The physiochemical properties of DF include solubility, water-holding capacity, viscosity and gelation, binding ability, bulking ability, and fermentability [19,20]. Based on the dissolving characteristics of DF, they are either soluble (e.g., pectins and gums) or insoluble (e.g., lignin and cellulose). Notably, DF can hold water in void spaces or hydrophilic sites and the amount of water retained is defined as its water-holding capacity [2]. Viscosity is the property of liquid to resist flow due to internal friction. Viscosity is a proportional relationship between the flow of the fluid and the force directed on it and it relates to DF, where some polysaccharides physically entangle and mix with fluids, thicken, and form gel [21]. Besides, DF can also entrap and bind some bile acids, form bulk due to water holding, and increase fermentation metabolites by being broken down and utilized by the gut microbes. In poultry feeding, soluble DF is desired for enhanced action of gut microbes but there is also an increase of the unstirred water layer on the intestinal mucosa in case of viscous fiber that decreases the efficiency of nutrient absorption.

3. Antinutritive Effect of DF in Poultry

DF from different cereals such as wheat, rye, and barley, etc., either in insoluble or soluble forms, can exert an antinutritive effect in poultry by depressing AME, starch digestibility, nitrogen retention, and other nutrient utilization, leading to poor growth performance [7]. Despite the positive attributes of DF, the inclusion of a high level of fiber is limited either because certain NSP can bind bile acids, fats, or cholesterol, and cause lipid malabsorption, leading to low AME value of feed and poor growth [22–24].

The viscosity of digesta is one of the major factors impacting digestibility. It is thought that higher viscosity interferes with efficient nutrient diffusion, subsequently reducing their breakdown and transport by endogenous enzymes at the mucosal surface [6,25]. The antinutritive effect of pentosans such as arabinoxylans and arabinogalactan also depends on the degree of polymerization, which in turn increases their viscosity [7]. When intact arabinoxylans (30 g/kg) and depolymerized arabinoxylans (30 g/kg) were added to the broiler diet in Choct and Annison's [7] study, the ileal viscosity compared with water increased from 1.2 in the control to 2.2 in the depolymerized arabinoxylan, and it increased to 3.0 in the intact arabinoxylan-added diets. In the same study, it was found that when 35 g/kg of arabinoxylan was added to the diet, the digesta viscosity increased by more than two times. It has been reported that the addition of soluble NSP such as arabinoxylan in broiler diet can increase the endogenous loss of amino acids and depress the ileal digestibility of protein [3]. In a study on broiler chicken conducted by Kluth and Rodehutscord [26], it was found that inevitable endogenous loss of CP in low-fiber (CF = 30 g/kg) diet was 11.7 g/kg dry matter intake (DMI), while in the high-fiber (CF = 80 g/kg) group it was 16.3 g/kg. This loss for lysine and methionine was 0.4 and 0.17 g/kg respectively, in low-fiber diet and 0.59 and 0.19 g/kg respectively, in high-fiber diet. Angkanaporn et al. [27] found that adding 15 g/kg of a wheat pentosan (arabinoxylan) decreased the average apparent amino acid digestibility by 17% and increased the average endogenous amino acid loss by 23.5 g/kg DMI. This provides a relatively favorable environment for the establishment of fermentative microbiota in the upper gut, which may not typically reside there in high numbers [6,7]. Moreover, fermentation occurring at the site of the upper gut is not beneficial for the host, as it yields relatively low amounts of energy compared to typical enzymatic digestion and nutrient absorption by the host [28]. Jørgensen et al. [29] reported that NSP fermentation could only contribute up to 3–4% energy of ME intake. However, soluble NSP is more easily digested than insoluble NSP and some of these soluble fibers, such as inulin and wheat dextrin, would not reduce the digestibility of other nutrients, as they do not increase the viscosity of digesta [22]. When a wheat and barley-based diet was supplemented with 7.5 g inulin by replacing an equal amount of wheat in a 28-day broiler study by Rodriguez et al. [30], the viscosity of the jejunal digesta reduced from 1.83 in control to 1.30 millipascal seconds (mPa s). If the viscosity of digesta is managed by using feed additives, then soluble fibers can be better utilized by poultry due to reduced interference of the movement of the digesta and improved diffusion of digestive enzymes to the substrates [7].

The insoluble fiber present in poultry feed causes less viscosity than soluble fiber and has low fermentability due to its limited accessibility by the action of microbial or host enzymes. The insoluble fiber binds water by surface tension or hydrogen bonds in the pores of its matrix, and the quantity of water it can bind depends upon its swelling characteristics or water-holding capacity [2]. As such, a poultry diet containing a higher amount of insoluble fiber can increase the transit rate of digesta and passage of nutrients in the lower gastrointestinal tract due to this higher water-holding capacity [19]. However, in poultry, it is reported that coarse particles can delay the transit of digesta in the gizzard and thus increase the exposure of substrates to the digestive enzymes [31]. Some authors suggest that it is not the water-holding capacity of insoluble DF, but rather it is the mechanical stimulation, that leads to excess mucus secretion and increased peristalsis in response to coarse particles that increase motility and decrease digesta transit or retention time [32,33]. In a pig study by Wilfart et al. [34], it was found that the addition of around 0.8% insoluble fiber in the diet reduced the mean retention time of digesta in the total tract by 9 h when the solid-phase marker was used. Moreover, encapsulation of other nutrients by fiber in the cell wall of plant-based feed also reduces the utilization of nutrients and limits their digestibility in several feedstuffs [5,35]. Downstream consequences of high NSP feed ingredients with high water retention properties, such as wheat, barley, oats, cassava, and rye, include wet droppings and increased moisture content of litter. In turn, these conditions lead to poor foot-pad quality and increased ammonia volatilization [36,37]. Together, these data suggest that high NSP feed ingredients may adversely affect poultry health both directly and indirectly.

It is not well-defined how DF would decrease the bioavailability of minerals and vitamins, but the adsorption property of DF is expected to reduce the utilization of these nutrients by the host. The presence of higher levels of phytate associated with fiber increases the excretion of endogenous minerals in broilers [38]. Cowieson et al. [38] reported that in a precision feeding assay on 6-week-old female broilers, feeding of 1 g phytic acid increased endogenous excretion of calcium by 69%, iron by 31%, sodium by 300%, and sulfur by 47%. It is established that phytate present in most plant-based fibers can strongly bind phosphorus and divalent cations such as zinc, copper, calcium, and magnesium, thus reducing their absorption and disturbing their homeostasis in the body [39,40].

4. The Beneficial Effect of DF in Poultry Nutrition and Gut Health

4.1. Effects of Fibrous Diet on Nutrient Utilization and Ammonia Emission

The dietary protein and amino acids that escape host digestion are subject to fermentation by gut microbes. Approximately half of these nitrogen sources are metabolized to uric acid and ammonia in the gut, thus depriving nourishment and increasing toxicity in the host [41]. The uric acid is subsequently volatilized to ammonia in the litter by the microbes, which causes respiratory discomfort to the birds and poses a major public health concern [42]. Although carbohydrates are the preferred substrate for energy metabolism by the gut microbes, depletion of carbohydrate substrates causes specific groups of putrefactive and proteolytic microbes to turn to residual protein breakdown and shift fermentation from saccharolytic to proteolytic [43,44]. Besides emitting odorous sulfur compounds and ammonia, fermentation of protein also yields other harmful metabolites such as amines, phenols, and indoles [45]. During the microbial fermentation of DF, nitrogen sources such as ammonia are also utilized for bacterial protein synthesis, which may reduce its emission [46]. It follows that some reports suggest that the inclusion of a higher amount of fermentable DF, combined with the reduction in crude protein in the diet of chicken, has been reported to reduce ammonia emission [47]. The authors stated that the addition of 10% corn-DDGS in the corn-SBM control diet, decreased 7-day cumulative manure ammonia emission from 3.9 g/kg of manure DM to 1.9 g/kg of manure DM, a reduction by 51%. It can be summarized that fermentable fiber provides energy for microbial protein synthesis and prevents the fermentation of undigested protein into ammonia.

4.2. Poultry Gut Microbiome and Its Modulation by DF

The gut microbiome is being regarded as an essential and integral part of the gastrointestinal tract (GIT) ecosystem, which functions as an additional organ and contributes to various aspects ranging from nutrient utilization to improved health status and immune modulation in the host [48–51]. The GIT of poultry is the shelter for a diverse community of microorganisms which comprises over 900 species of bacteria, along with some protozoa, fungi, yeast, and viruses, collectively referred to as microbiome or microbiota that assist the host in breakdown and utilization of consumed feeds [52,53]. The microbiome is present throughout the GIT of poultry from crop to colon, with their population increasing gradually along the distal intestine, and the vast majority reside in the caecum and colon, ranging from 10^{11} to 10^{12} colony forming unit (CFU)/g of luminal content [52]. Various bacterial species reside in different microhabitats of the GIT, ranging from the lumen to mucus and mucosal linings, and are found in significant quantity and diversity [54,55]. A normal process of mucus secretion, epithelial turnover, and peristaltic movements occurring in the GIT is expected to distribute the subsets of the luminal microbiome to the mucus and mucosal surfaces [50]. Rinttilä and Apajalahti [56] reviewed that the GIT environment in chicks is more aerobic initially and is first colonized by facultative aerobic bacteria, such as *Enterobacteriaceae*, *Lactobacillus*, and *Streptococcus*. Later, the GIT gradually transitions to anaerobic, subsequently inducing outgrowth of obligate anaerobes in the growing chicks. The lower gut microbiome depends on the residual digesta and intestinal secretions for deriving nutrients and energy for their growth [57]. The normal microbiome of the lower gut does not compete with the host for nutrients as they utilize the residual feed, salvaging a considerable proportion of energy for the host through fermentation, and precluding colonization of pathogenic and putrefactive bacteria [50]. The composition of this microbiome initially depends on the inoculum passed from the breeder hen as well as the surrounding environmental condition for the chicks during hatch, and later gets modified with age, diet type, and the intestinal environment of the birds [51,52,55,58]. Thus, with the growth of birds, diet serves as one of the strongest determinants of microbial diversity and colonization in the gut.

The bacterial population is the component of the microbiome of major interest for poultry nutritionists because of its role in fermentation and being a target of various AGPs activity. In the ceca of chicken, many families of bacteria, including *Lachnospiraceae*, *Ruminococcaceae*, and *Veillonellaceae*, belonging to the order *Clostridiales*, are non-pathogenic, produce SCFA (lactate, propionate, and butyrate), and are characteristically different than pathogenic *Clostridium perfringens* [56]. In addition to the normal gut microbiome in poultry, several probiotic strains are also added in the poultry feed to enhance the population of known beneficial microbes to prevent dysbiosis or to limit the use of AGP typically implemented to reduce the load of pathogenic microbes. In general, the beneficial bacteria are associated with the promotion of gut maturation and integrity, modulation of the immune response of the host, and persistent antagonism against the pathogen colonization in the gut [49]. However, for practical use as probiotics, beneficial bacteria should fulfill specific criteria: possess antimicrobial activity, adhere to the mucosal lining, have phenotypic and genotypic stability, resist lysosomal destruction and AGP in feed, tolerate acid and bile, and utilize carbohydrates [50].

The inclusion of fermentable DF in the poultry feed supports the growth and establishment of beneficial microbes and probiotic bacteria by providing them substrates for extracting energy and fueling their metabolism. As discussed previously, all DF included in feed are not the same, and those providing benefits to the host by selectively stimulating the growth of beneficial and commensal bacteria in the gut are defined as prebiotics [59]. However, some prebiotics can directly stimulate the immune system and bind the pathogen to facilitate their removal [60,61]. The pathogen could bind with the feed oligosaccharides, mimicking host cell receptors instead of adhering to the host cell surface oligosaccharides, and ultimately get flushed out of the GIT. For example, galactooligosaccharides (GOS) have been found to prevent adhesion of enteropathogenic *Escherichia coli* (EPEC) in human intestinal cells in cell culture and mannan oligosaccharides (MOS) has been reported to decrease the population of *Salmonella* in broiler chicks [61,62]. In general, the most common prebiotics are small fragments of carbohydrates which are oligosaccharides of fructose, xylose, mannose, or galactose, etc., although inulin, raffinose, and resistant starch are used as well [8,11]. The potential of oligosaccharides to modify the intestinal microbiota in poultry is dose-dependent [63], and it has been reported in some studies that oligosaccharides or different fibers with a lower degree of polymerization are more thoroughly broken down through fermentation [12]. Inclusion of DF in poultry diet can effectively support cellulolytic and beneficial bacteria, including *Lactobacillus* and *Bifidobacterium*, and enhance the production of SCFA, and a combined effect would prevent digestive disturbances and wet litter [23,50,64]. Beneficial bacteria present as a part of the gut microbiome also produce metabolites such as bacteriocins that provide protective activity against pathogenic bacteria in poultry [50,54]. Compared with the probiotics added in poultry feed, ingredients containing fiber with potential prebiotic effects provide the advantage of stimulating such commensal and beneficial microbes that are normally present in the host GIT [65]. The degradation of fiber in high-fiber diets could also enhance the population of fiber fermenting microbes, including bacteria of genus *Lactobacillus*, family *Ruminococcoceae*, and family *Lachnospiraceae* [66]. Thus, the fermentable DF can modulate the gut microbiome and promote the growth of beneficial bacteria that would be required to improve broiler performance in the absence of AGP in the diet. However, more research is required to understand the interaction of different components of DF with microbes in a dynamic and competitive gut environment.

4.3. Microbial Fermentation of DF

The principal metabolites of DF fermentation are SCFA, which mainly include acetate, propionate, butyrate, lactate and succinate, H_2O, and gases (CO_2, H_2, and CH_4), along with the accumulation of bacterial cell biomass [25]. The fermentable carbohydrates of hexose and pentose sugar monomers are converted to pyruvate through glycolysis and the pentose phosphate pathway, which can then be converted to lactate, propionate (via succinate), acetate, and butyrate (via acetyl-Co-A) [57]. Acetate is the major SCFA produced in poultry GIT, followed by either propionate or butyrate based on the diet type and site of the GIT [66–68]. The production of SCFA depends on the availability of the fermentable substrates, and a high-fiber diet does not always increase SCFA, as Walugembe et al. [69] reported that reducing NDF from 15% to 10% increased cecal butyrate production by 37% on average in both broiler and layer birds. In addition, other SCFAs including valerate, isobutyrate, and isovalerate are produced in trace amounts in the poultry GIT [66,67]. Besides SCFA, lactic acid is also produced in significant quantities, with the highest concentration in the ileum, followed by the crop, gizzard, and ceca [68]. It has been reported that 95–99% of the total SCFA produced in the gut is rapidly absorbed in the gut lumen before it can reach the rectum of non-ruminant animals [4,25]. The ceca in poultry serve as the major site for fermentation and production of SCFA and methane [70]. The SCFA produced can provide up to 5% to 15% of daily metabolizable energy for the maintenance energy requirement of birds [56].

It has also been noted that probiotic bacteria such as *Lactobacillus* increase the production of butyrate in chicken, which has been attributed to the fact that there may be cross-feeding of lactate to the butyrate-producing bacteria [71]. The proportion of production of different SCFA compared to the total production varies based on the type of ingredients and the microbiota dynamics in the gut [44,72]. The dietary feeding of *Lactobacillus plantarum B1* in the finisher phase and during the total period of 6 weeks in broilers has been noticed by Peng et al. [73] to increase the production of propionate by more than 27.5% and total SCFA by more than 30.5% compared with the control diet. In broilers fed the diet supplemented with 16,000 birchwood xylan unit (BXU)/kg of xylanase enzyme for 42 days, Lee et al. [74] observed that the enzyme increased the cecal acetate, propionate, and butyrate by more than 20%, 30%, and 40% respectively, while it decreased the production of branched-chain SCFA by a small amount. Rehman et al. [75] reported that supplementing broilers' diet with 1% inulin did not influence total cecal SCFA at day 42 but increased the proportion of butyrate to 15.6% from 11.7% in the control. More favorably, the supply of DF is essential in maintaining saccharolytic fermentation and it can influence intestinal physiology indirectly through its metabolic products.

4.4. Role of SCFA on Gut Health of Poultry

In addition to supplying energy, SCFA also contributes to the normal functioning of the lower GIT by acting on the intestinal musculature and vasculature and through their impact on the metabolism of enterocytes and colonocytes [76,77]. Among all SCFA, butyrate has received the most attention due to its nutritional properties for intestinal epithelial cells and its inhibitory effect on pathogenic bacteria in the gut. The presence of SCFA in the GIT can affect both gut motility and ionic absorption [76]. During a surgical catheterization experiment in dogs, it was observed that acetic acid was a more potent stimulant for ileal motility based on propulsive motor events, while propionic acid was less effective and butyric acid tended to decrease ileal contraction [78]. In an in vitro study on rat colon by Binder and Mehta [79], it was concluded that stimulation of Na and Cl absorption was greater in response to mucosal butyrate than that of propionate and acetate. The enteral nutrition of SCFA, including acetate, but preferentially butyrate and propionate, could lead to cecal crypt proliferation and act as luminal trophic factors on the cecal epithelium. Butyrate is the preferred source of energy for the enterocyte, where it is readily absorbed via passive diffusion and recognized to regulate the differentiation and proliferation of these cells [80]. In addition to supporting the growth of villi, butyrate can also suppress the invasion of epithelial cells by pathogens [81]. Fernández-Rubio et al. [81] reported that feeding of 0.92 g/kg sodium butyrate as a supplement in a standard broiler diet to orally challenged chicks reduced *Salmonella* Enteritidis at day 42 in the ceca (>10^8 CFU/g in 8 birds in control vs. >10^8 CFU/g in 1 bird in sodium butyrate group) and crops (10^7–10^8 CFU/g in 12 birds in control vs. 10^7–10^8 CFU/g in 2 birds in sodium butyrate group). Butyric acid fed to chickens in the form of impregnated microbeads in the feed in a study by Van Immerseel et al. [82] significantly reduced the colonization of *Salmonella* Enteritidis in the caecum but not in the spleen and liver.

Birds vaccinated against coccidiosis and receiving butyric acid can obtain additional benefits of maintenance of intestinal villi structures and better performance when challenged with coccidiosis [83]. The authors in the study of Leeson et al. [83] observed that villi to crypt ratio increased from 5.3 to 5.9 and final week weight gain increased by 25% when 0.2% butyric acid was supplemented in the diet of the birds challenged with coccidia. It has been understood that SCFA reduces intestinal pH and limits the growth of acid-sensitive pathogenic bacteria like *Enterobacteriaceae* by exhausting their H^+ATPase pump. This occurs as SCFA overcome the proton motive force across the bacterial cell membrane, enter the cell in undissociated form, and cause damage by dissociating in the neutral cytoplasm of the bacterial cells [84]. Moreover, with increased production of SCFA, pH is reduced, which causes ionization of ammonia and reduces its absorption in the hind gut [43–45]. Therefore, it can be inferred that production of SCFA by lower gut microbes does not only salvage the energy from the undigested nutrients, but it also improves the intestinal health of poultry.

4.5. Effect of DF on the Gut Histomorphometry, Integrity, and Immune Response

The epithelial cells of the mucosa that are responsible for the absorption of nutrients exist in a dynamic state, dying and shedding regularly, and quickly replenished by the new cells generated from crypts. The area and length of the villi, the depth of the crypts, and the ratio of the villi to the crypts provide a measure of absorption efficiency and gut health status [64,85]. The effect of DF on mucosal morphology in poultry is not well-established, but it has been known to affect cell turnover based on its physicochemical characteristics and inclusion level in the diet of birds in different growth phases [6,64]. The effect of fiber on intestinal histomorphological status is variable: a reduction in villi height has been reported in chickens fed viscous ingredients such as citrus pectin and gum xanthan, while insoluble fiber has been reported to favor villi development [6,25]. In contrast, Andoh et al. [86] observed that the ingestion of pectin increased both villus height and crypt depth in rats. Inclusion of lignin and MOS in the diet of broiler chicken can potentially increase villus height and goblet cell number and thereby may enhance feed efficiency [64]. When MOS was included in the diet at 0.2% until day 21 and 0.1% from day 21 to day 42 in the study by Baurhoo et al. [64], it increased the number of goblet cells per villus from 61 in control to 118 in the jejunum of broilers at day 42. Supplementation of 0.5% fructooligosaccharides (FOS) in broiler diets for a 21 day study by Shang et al. [87] revealed that FOS significantly increased villi height by 24% and mucosa thickness by 26% in ileum compared with control. An increase of 134% in duodenal villi height was also discovered by Ashraf et al. [88] in heat-stressed birds when fed 0.5% MOS prebiotics. The addition of MOS has been reported to increase villi height and goblet cell number, as well as enhance gut integrity in chickens [64]. Thus, the feeding of specific components of DF may stimulate GIT mucosa to increase villi length and surface area for better nutrient absorption, leading to a subsequent higher growth rate in chickens [6]. DF, or its degraded fragments that increase SCFA, especially butyrate, are expected to increase villus height/crypt depth ratio and improve the absorptive capacity of the intestine [4]. Importantly, the enhanced digestive and absorptive performance in response to the increased surface area of villi is associated with increased production of brush border enzymes and higher availability of nutrient transporters [89], thus providing further capacity for nutrient uptake.

Besides their contribution in nutrient utilization in poultry, intestinal epithelium and its mucus secretion also play an important role in defense against pathogens. The epithelial cells are connected by various junctional complex that consists of tight junctions, adherens junctions, gap junction, and desmosomes. The tight junction proteins block the paracellular pathway and regulate intestinal permeability, while claudin, occludin, and junctional adhesion molecule (JAM) family are other crucial transmembrane proteins that associate with peripheral scaffolding proteins, such as ZO family that anchors strands to the actin component of epithelial cells [90,91]. The abnormal changes triggered by pathogens can impair the functions of these tight junction proteins during inflammation and cause increased intestinal leakage [92]. It is also interesting to note that the addition of xylanase enzyme to the wheat-based diet increases expression of mRNA of the tight junction gene occludin, in the ileum of chickens with mucosal barrier impaired by *Clostridium perfringens* [93]. It is also plausible to state that the production of butyrate by fermentation of fiber can enhance intestinal epithelial barrier function via upregulation of tight junction protein claudin-1, and induce ZO-1 and occludin redistribution [77,94]. The increase in abundance of beneficial bacteria such as *Lactobacillus* by prebiotics can also play an important role in regulating intestinal tight junction protein and enhance epithelial barrier function [95]. The proper regulation of tight junctions is important as it can also be affected by immune cells such as tumor necrosis factor (TNF) and interferon gamma (IFNγ) and dysregulation in mucosal immune homeostasis can lead to barrier dysfunction and onset to other diseases [96].

The digesta matrix containing the nutrients from the feed are in intimate contact with the immune system in the gut (gut-associated lymphoid tissue, GALT), which is necessary for proper functioning and development of the immune components and antigen-presenting cells [13,97,98]. Compared to other specific nutrients, the impact of DF on immunity is less explored. It has been suggested in several studies that the bacteria and their components can stimulate and activate immune cells of GALT [48,49,60], suggesting that aspects of DF likely influence immunity by proxy. More directly, SCFA such as butyrate that is produced during fermentation can increase the activity of phagocytic cells and spare glutamine to be used by lymphocytes as a source of energy [13]. Rezaei et al. [10] observed that feeding of 0.5% and 1.0% oligosaccharide extract from palm kernel expeller and coconut flour increased immunoglobulin A (IgA) by 85% and 141% respectively, in 3-week broilers, which could provide some protection against pathogens such as *Salmonella*. FOS supplementation in hen diets significantly enhanced IgA secretion and Toll-like receptor-4 in the intestine and reduced *Salmonella* colonization in the ceca of laying hens [9]. Furthermore, the inclusion of 0.5% FOS in the diet of chicken could also increase IgM and IgG titers in the plasma [60]. Dietary supplementation of the yeast cell wall that is rich in β-glucan has also been found to increase mucosal IgA secretion, increase humoral as well as cell-mediated immunity, and potentially acts as an adjuvant to enhance the immune response against coccidiosis [99]. It is also noteworthy to mention that inclusion of DDGS that contains a high level of NSP and yeast β-glucan has been described to increase IgA, IgG, and gene expression of IL-4 and IL-6 in broilers [100]. In another study on broilers, it was observed that feeding of equal proportions of sugar beet pulp along with rice hulls at a 3% inclusion increased antibody titer against Newcastle disease virus by 100% [101]. The supplementation of oligosaccharides has also been reported to reduce the count of heterophils in chicken, which is suggestive of its stress-relieving action [10,54]. Stress in poultry is also known to adversely affect epithelial integrity, gut permeability, and immune response, and has been reviewed and studied elsewhere [102,103]. Thus, the inclusion of fermentable DF in the diet of poultry could enhance mucosal health, improve immune regulation, and modify the luminal environment for better absorption of nutrients.

5. Effects of Enzymatic Degradation and Processing on the Utilization of DF

The lack of information about the chemical composition of DF incorporated in several studies makes it difficult to compare the effects of physiochemical properties of these DFs on its nutritional value in poultry diet. Additionally, the poultry diet contains variable ingredients that are expected to result in differences in the enzymatic digestion and microbial fermentation of DF in different feed matrix. Exogenous feed enzymes can reduce the bacterial colonization in the ileum by reducing the nutrients available for fermentation [68,104]. Feed enzymes can provide benefits to the birds by releasing more nutrients for utilization by the host while providing degraded products such as oligomers of polysaccharide substrates for utilization by the cecal microbes for the production of SCFA [105]. Multi-carbohydrase enzyme supplementation in a wheat-based diet improves nutrient utilization, reduces digesta viscosity, and mitigates the negative impact of *Clostridium perfringens* challenge in broilers [104]. de Vries et al. [106] reviewed that the processing of fibrous feed ingredients by hammer and roller milling can increase the solubility of NSP-fraction and enhance the coefficient of digestibility in poultry. The authors also stated based on several studies that the application of feed enzyme to the ingredients subjected to hydrothermal processing can increase the digestibility of fiber fraction up to 1.5–6 times compared with that of unprocessed diets. There are limited processing techniques in use in poultry feed production to improve the utilization of DF but pelleting and micronizing have been reported to increase the action of pentosanase on fibrous diet [107]. Exogenous NSPase, phytase, and xylanase can increase the bioavailability of several nutrients affected by high-fiber content in feed and concurrently provide degraded fiber fragments and oligosaccharides for utilization by the gut microbiome [5,105]. These NSPase enzymes can decrease digesta viscosity and alleviate the deleterious effect of viscous fiber on the intestinal mucosa in poultry [35]. The use of such exogenous enzymes is thus an efficient method for removing the nutrient encapsulating effect of plant cell walls and generating biologically active oligomers for sustaining the gut microbiome and limiting the use of AGPs for maintaining gut health of poultry. With existing challenges of the antinutritional effect of DF, further research is warranted to explore the opportunities of enhancing the utilization of agricultural co-products through but not limited to chemical, enzymatic, irradiation, and milling techniques.

6. Scope of Improvement in Poultry Productivity by Exogenous Fiber-Degrading Enzymes

6.1. Meta-Analysis of the Effect of NSPase on Growth Performance of Broilers Fed Fibrous Diet

We summarized the effect of NSPase enzymes on average daily gain (ADG) and feed conversion ratio (FCR) of commercial broilers fed wheat, rye, and barley-based fibrous diet. To identify relevant studies, we searched the web of science core collection and google scholar for literature in accordance with the Preferred Reporting Items for Systematic Reviews and Meta-Analyses (PRISMA) 2009 checklist and population, intervention, comparison, outcome, and study type (PICOS) based search strategy. The search was targeted within the title, abstract, and indexed keywords using the following terms: xylanase or enzyme *, chicken * or broiler *, growth or performance, fiber or fibre, or NSP or polysaccharides. After combining all search results within two decades and removing duplicates, the collection was narrowed to 125 research articles. From this collection, we set exclusion criteria as protease enzyme use and in vitro trials, and inclusion criteria as randomized animal study with wheat, barley, and rye as fiber sources. We filtered 35 eligible studies involving broilers and analyzed 17 studies with comparable trial duration ranging from 20 to 25 days, while the longer and shorter period studies were excluded. Data from control and treatment groups were extracted and the effect size was calculated as the difference in means (Table 1). All the parameters included in the meta-analysis were on the same scale (unit of measure) and had the same outcome (continuous), so the standardized mean difference estimates (Cohen's d or Hedges' g) were not used. Review Manager v5.4, RevMan [108], was used for statistical analysis and generation of forest plots using random-effects model to accommodate for higher heterogeneity among studies.

In all the studies analyzed, xylanase was used as NSPase enzyme either alone or in combination with other carbohydrases. Few studies evaluated the effect of xylanase in conjunction with phytase included in the basal diet. Overall, the supplementation of NSPase enzyme seems to improve ADG in broilers from 20 to 25-days of age by 2.5 g/day and decrease FCR by 6 points compared to fibrous control diet. Compared to more recent studies, research outcomes from previous decades show higher effect sizes for ADG in response to enzyme treatments (Figure 1). This trend was not obvious in case of summary of effects on FCR (Figure 2). It is understandable that ample progress has been made in birds management and feed manufacturing technology which could have helped birds to grow better in recent times [5,106]. There might be a limit on improvement in performance in response to NSPase enzymes when birds are already able to perform to their potential. However, the effectiveness of such exogenous feed enzymes and their combination should also be evaluated in terms of gut health improvement, which would be more detectable under challenged and unhygienic rearing conditions [5,105]. Xylanase is just one among the other NSPase enzymes such as glucanase, amylase, phytase, cellulase, and mannanase, etc. Data from more studies need to be compared for conclusive interpretation on the effect of enzyme combination and types of ingredients used in poultry feed. Moreover, there are other various parameters such as carcass yield, meat quality, litter quality, and digestibility, etc., that would also need to be considered apart from the enzyme activity in feed for evaluating the efficiency of NSPase enzymes. Nonetheless, the meta-analysis of the effect of xylanase on ADG and FCR provides a quantitative assessment of its efficacy across different studies where broilers were fed a fibrous diet. The improvement in growth performance of broilers in response to exogenous enzymes is based on underlying mechanism of improvement in the digestibility of nutrients [109,110]. Thus, further statistical analysis of the results of multiple studies are required to ascertain the effect of NSPase enzymes on the ileal digestibility of the components of NSP that corresponds to the improvement in the growth of broilers.

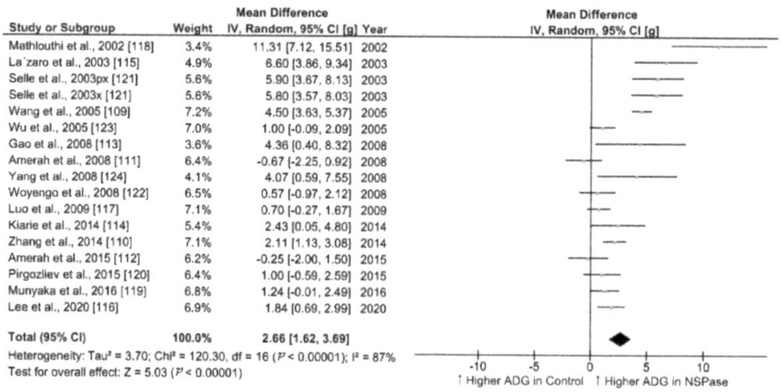

Figure 1. Forest plot of comparison: Effect of non-starch polysaccharides-degrading enzymes on average daily gain (g/day) of broilers fed either wheat or barley or rye-based diets during a trial duration of 20–25 days. Abbreviations: ADG: average daily gain; px: phytase + xylanase; x: xylanase; IV: inverse-variance; CI: confidence interval; Tau^2: tau-squared; Chi^2: chi-squared; df: degrees of freedom; I^2: I-squared statistic; Z: Z-test statistic. The vertical line corresponding to the value of zero (0) in the plot is the line of no effect. The hyphen (-) represents a negative mean difference (effect size) and corresponds to a higher average daily gain in control compared with NSPase group. The squared green boxes represent the point estimates and the width of the horizontal lines extending from the squared green boxes represent the 95% confidence interval of the individual study. The mid-point of the green boxes represents the mean effect estimate and the area of the boxes represent the corresponding weight given to each study. The diamond at the bottom of the studies represents the summary estimate and confidence interval of all studies combined. The points on the vertical axis of the diamond represents the overall effect and the width of the diamond represents the 95% confidence interval. The upward arrow in the label of the plot visually represents a higher average daily gain.

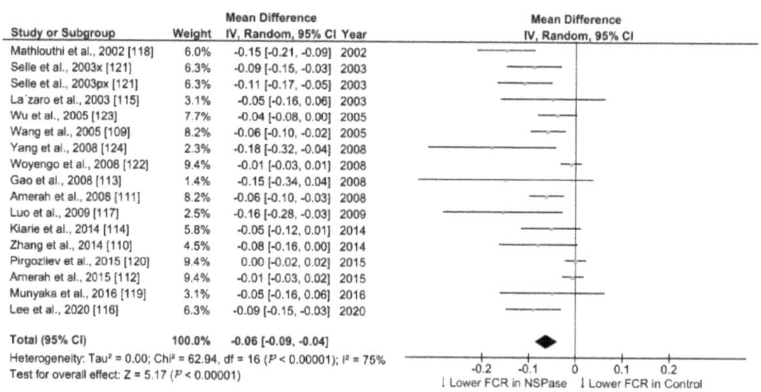

Figure 2. Forest plot of comparison: Effect of non-starch polysaccharides degrading enzymes on feed conversion ratio in broilers fed either wheat or barley or rye-based diets during a trial duration of 20–25 days. Abbreviations: FCR: feed conversion ratio; px: phytase + xylanase; x: xylanase; IV: inverse-variance; CI: confidence interval; Tau^2: tau-squared; Chi^2: chi-squared; df: degrees of freedom; I^2: I-squared statistic; Z: Z-test statistic. The vertical line corresponding to the value of zero (0) in the plot is the line of no effect. The hyphen (-) represents a negative mean difference (effect size) and corresponds to a lower feed conversion ratio in NSPase group compared with control. The squared green boxes represent the point estimates and the width of the horizontal lines extending from the squared green boxes represent the 95% confidence interval of the individual study. The mid-point of the green boxes represents the mean effect estimate and the area of the boxes represent the corresponding weight given to each study. The diamond at the bottom of the studies represents the summary estimate and confidence interval of all studies combined. The points on the vertical axis of the diamond represents the overall effect and the width of the diamond represents the 95% confidence interval. The downward arrow in the label of the plot visually represents a lower feed conversion ratio.

Table 1. Fiber source, enzyme combination, and summary statistics of research outcomes of selected research papers evaluating the effects of non-starch polysaccharides-degrading enzymes on average daily gain and feed conversion ratio in broilers.

						ADG in NSPase		ADG in Control		FCR in NSPase		FCR in Control	
S.N.	Study	Days	Fiber Source	Enzyme	Rep (N)	Mean, g	±SEM	Mean, g	±SEM	Mean	±SEM	Mean	±SEM
1	Amerah et al. 2008 [111]	21	w 66.5%	xyl	6	46.5	0.57	47.2	0.57	1.35	0.013	1.41	0.013
2	Amerah et al. 2015 [112]	21	w > 60%, s > 5%, rp 0–6%	xyl + gluc	8	49.6	0.63	49.9	0.63	1.34	0.008	1.35	0.008
3	Gao et al. 2008 [113]	21	w 60%	xyl + gluc + cel + pec	4	37.3	1.57	32.9	1.27	1.58	0.070	1.73	0.065
4	Kiarie et al. 2014 [114]	21	w 60%, wb 9%	xyl + basal phy	6	37.2	0.86	34.8	0.86	1.37	0.023	1.42	0.023
5	La'zaro et al. 2003 [115]	25	r 50%	xyl + gluc	7	38.3	0.99	31.7	0.99	1.66	0.040	1.71	0.040
6	Lee et al. 2020 [116]	20	w 10–20%	xyl	10	36.8	0.42	34.9	0.42	1.47	0.021	1.56	0.021
7	Luo et al. 2009 [117]	21	w 40%	xyl	5	27.9	0.35	27.2	0.35	1.56	0.046	1.71	0.046
8	Mathlouthi et al. 2002 [118]	21	w 40%, b 20%	xyl + gluc	12	38.7	1.21	27.4	1.77	1.50	0.009	1.65	0.030
9	Munyaka et al. 2016 [119]	21	w 42%, b 5%, r 5%, wm 2%	xyl + gluc	7	40.6	0.45	39.3	0.45	1.12	0.040	1.17	0.040
10	Pirgozliev et al. 2015 [120]	21	w 63%	xyl	6	39.4	0.57	38.4	0.57	1.35	0.008	1.35	0.008
11	Selle et al. 2003 [121]	24	w 70%	xyl + phy	8	44.3	0.81	38.4	0.81	1.46	0.021	1.57	0.021
12	Selle et al. 2003 [121]	24	w 70%	xyl	8	44.2	0.81	38.4	0.81	1.48	0.021	1.57	0.021
13	Wang et al. 2005 [109]	21	w 70%	xyl + gluc	6	41.4	0.32	36.9	0.32	1.53	0.013	1.59	0.013
14	Woyengo et al. 2008 [122]	21	w 58%	xyl main effect	8	44.3	0.56	43.7	0.56	1.26	0.008	1.27	0.008
15	Wu et al. 2005 [123]	21	w 66%	xyl	6	38.3	0.39	37.3	0.39	1.34	0.015	1.38	0.015
16	Yang et al. 2008 [124]	21	w 62.4%	xyl	8	56.9	1.26	52.9	1.26	1.65	0.049	1.83	0.049
17	Zhang et al. 2014 [110]	21	w 60%	xyl	6	37.7	0.35	35.6	0.35	1.52	0.030	1.60	0.030

Abbreviations: S.N. serial number; Rep: the number of replicates, NSP: non-starch polysaccharide; ADG: average daily gain; FCR: feed conversion ratio; N: replicate; SEM: pooled standard error of mean; w: wheat; b: barley; r: rye; wb: wheat bran; wm: wheat middlings; rp: rapeseed; s: sunflower; xyl: xylanase; gluc: glucanase; phy: phytase; cel: cellulase; pec: pectinase.

6.2. *Significance of Exogenous Fiber-Degrading Enzymes during Disease Challenge in Poultry*

The efficacy of exogenous enzymes becomes more important during disease challenge conditions in poultry flock when the digestive and immune system of the birds are in a compromised state. The disease condition deteriorates the performance and reduces the efficiency of feed utilization that can further increase the cost of production. Amerah et al. [125] found that in a wheat-based basal diet, xylanase supplementation (2000 U/kg of feed) increased weight gain by 16% and reduced FCR by 6% at day 42 in *Salmonella enterica* serovar Heidelberg (5×10^5 CFU/mL) challenged broilers. In the same study, xylanase supplementation also reduced the *Salmonella*-positive cecal samples from 32.5% in the challenged control to 12.5%. Sun et al. [126] mentioned that the enzyme complex containing xylanase, glucanase, and mannanase as major components supplemented at 500 mg/kg diet decreased *Clostridium perfringens* from 3.66 to 3.48 log CFU/g of ileal digesta, increased body weight by 4%, improved FCR by 3%, increased villus height by 8%, and villus height to crypt depth ratio by 11% in 3-week broilers. Likewise, in a study by Jia et al. [127] on broiler chickens challenged with *Clostridium perfringens*, the supplementation of carbohydrase enzyme complex at 1 kg/ton of feed (supplying 60 U cellulase, 1400 U pectinase, 1200 U xylanase, 800 U glucanase, 500 U mannanase, 30 U galactanase, and other minor enzyme activities per kilogram of diet) reduced the feed conversion ratio by 5–6% in wheat- and flaxseed-containing diets. The mixture of feed enzymes can also be used in combination with direct-fed microbials to improve feed utilization and compensate for the damage and performance loss if occurred due to a coccidial challenge [128]. Jackson et al. [129] supplemented 100 million units of β-mannanase per ton of feed of broilers subjected to necrotic enteritis using a *Eimeria* sp. and *Clostridium perfringens* model. The authors reported that the intestinal lesion score was decreased by 16% on day 14, weight gain was increased by 14%, and FCR was improved by 11% on day 21. In a 39-day broiler study, Choct et al. [130] reported that the inclusion of 2.5 g/kg xylanase enzyme in a wheat-based diet reduced the number of ileal and cecal population of *Clostridium perfringens* to an insignificant level. Bortoluzzi et al. [131] confirmed that the addition of β-mannanase at 400 mg/kg in the diet increased *Lactobacillus* and *Ruminococcaceae* and reduced *Bacteroides* in the ceca of 21-day broilers regardless of the *Eimeria* challenge. Thus, an ideal enzyme or enzyme blend can reduce digesta viscosity, increase available energy, improve nutrient utilization, provide a health benefit, and reduce environmental pollution [132]. Therefore, such exogenous enzymes can prove effective in circumstances where AGPs are not desired. Further, it would be interesting to focus on research to elucidate the host–immune–diet–microbiome interactions to realize the benefits of additives like prebiotic DF and feed enzymes during restricted use of AGPs. Moreover, the supplementation of potent NSPases along with dietary fiber could improve productive performance and gut health of poultry, and thus increase profitability in both healthy and disease-affected flocks.

7. Conclusions

Several agricultural co-products are mixed in poultry diet to reduce the cost of feed production. The increase in fiber content of feed from these alternative ingredients limits their inclusion owing to their low digestibility and antinutrient properties. However, there has been further progress in understanding the additional roles of the fiber component of diet in modulating gut microbiome, stimulating immunity, and promoting gut integrity. This advancement in knowledge has made the inclusion of fiber in poultry feeding a matter of further interest and due consideration. Furthermore, the departure of poultry producers from absolute dependence on AGP has also diverted focus towards exploiting the existing properties of feed components and non-AGP additives to achieve similar levels of performance. Unfortunately, no such alternatives have provided a comparable and consistent improvement over AGP. However, the addition of targeted DF components integrated with proper processing techniques and the application of exogenous enzymes can be utilized to maximize the benefits of DF additives while reducing their antinutrient properties, resulting in more efficient and profitable poultry production.

Author Contributions: Conceptualization and methodology, A.K.S. and W.K.K.; data curation and original draft preparation, A.K.S.; supervision, W.K.K., review and editing, A.K.S. and W.K.K. All authors have read and agreed to the published version of the manuscript.

Funding: Not applicable.

Institutional Review Board Statement: Not applicable.

Data Availability Statement: Not applicable.

Acknowledgments: We would like to thank Brett Marshall for helping with the manuscript preparation and technical assistance.

Conflicts of Interest: The authors declare no conflict of interest.

References

1. Muscat, A.; de Olde, E.M.; de Boer, I.J.M.; Ripoll-Bosch, R. The battle for biomass: A systematic review of food-feed-fuel competition. *Glob. Food Secur.* **2019**. [CrossRef]
2. Thebaudin, J.Y.; Lefebvre, A.C.; Harrington, M.; Bourgeois, C.M. Dietary fibers: Nutritional and technological interest. *Trends Food Sci. Technol.* **1997**, *8*, 41–48. [CrossRef]
3. Choct, M.; Dersjant-Li, Y.; McLeish, J.; Peisker, M. Soy oligosaccharides and soluble non-starch polysaccharides: A review of digestion, nutritive and anti-nutritive effects in pigs and poultry. *Asian Austral J. Anim.* **2010**, *23*, 1386–1398. [CrossRef]
4. Topping, D.L.; Clifton, P.M. Short-chain fatty acids and human colonic function: Roles of resistant starch and nonstarch polysaccharides. *Physiol. Rev.* **2001**, *81*, 1031–1064. [CrossRef]
5. Aftab, U.; Bedford, M.R. The use of NSP enzymes in poultry nutrition: Myths and realities. *Worlds Poult. Sci. J.* **2018**, *74*, 277–286. [CrossRef]
6. Bederska-Łojewska, D.; Świątkiewicz, S.; Arczewska-Włosek, A.; Schwarz, T. Rye non-starch polysaccharides: Their impact on poultry intestinal physiology, nutrients digestibility and performance indices—A review. *Ann. Anim. Sci.* **2017**, *17*, 351–369. [CrossRef]
7. Choct, M.; Annison, G. Anti-nutritive effect of wheat pentosans in broiler chickens: Roles of viscosity and gut microflora. *Br. Poult. Sci.* **1992**, *33*, 821–834. [CrossRef]
8. Teng, P.Y.; Kim, W.K. Review: Roles of prebiotics in intestinal ecosystem of broilers. *Front. Vet. Sci* **2018**, *5*, 245. [CrossRef]
9. Adhikari, P.; Cosby, D.E.; Cox, N.A.; Franca, M.S.; Williams, S.M.; Gogal, R.M., Jr.; Ritz, C.W.; Kim, W.K. Effect of dietary fructooligosaccharide supplementation on internal organs Salmonella colonization, immune response, ileal morphology, and ileal immunohistochemistry in laying hens challenged with Salmonella enteritidis. *Poult. Sci.* **2018**, *97*, 2525–2533. [CrossRef]
10. Rezaei, S.; Faseleh Jahromi, M.; Liang, J.B.; Zulkifli, I.; Farjam, A.S.; Laudadio, V.; Tufarelli, V. Effect of oligosaccharides extract from palm kernel expeller on growth performance, gut microbiota and immune response in broiler chickens. *Poult. Sci.* **2015**, *94*, 2414–2420. [CrossRef]
11. Patterson, J.A.; Burkholder, K.M. Application of prebiotics and probiotics in poultry production. *Poult. Sci.* **2003**, *82*, 627–631. [CrossRef] [PubMed]
12. Saminathan, M.; Sieo, C.C.; Kalavathy, R.; Abdullah, N.; Ho, Y.W. Effect of prebiotic oligosaccharides on growth of *Lactobacillus* strains used as a probiotic for chickens. *Afr. J. Microbiol. Res.* **2011**, *5*, 57–64.
13. Schley, P.D.; Field, C.J. The immune-enhancing effects of dietary fibers and prebiotics. *Br. J. Nutr.* **2002**, *87* (Suppl. 2), S221–S230. [CrossRef] [PubMed]
14. AOAC. *Official Methods of Analysis*, 18th ed.; Association of Official Analytical Chemists: Arlington, VA, USA, 2006.
15. Van Soest, P.J. Use of detergents in the analysis of fibrous feeds. I. Preparation of fiber residues of low nitrogen content. *J. Assoc. Off. Agric. Chem.* **1963**, *46*, 825–829.
16. McCleary, B.V.; DeVries, J.W.; Rader, J.I.; Cohen, G.; Prosky, L.; Mugford, D.C.; Champ, M.; Okuma, K. Determination of insoluble, soluble, and total dietary fiber (CODEX definition) by enzymatic-gravimetric method and liquid chromatography: Collaborative study. *J. AOAC Int.* **2012**, *95*, 824–844. [CrossRef]
17. Englyst, H.N.; Quigley, M.E.; Hudson, G.J. Determination of dietary fibre as non-starch polysaccharides with gas–liquid chromatographic, high-performance liquid chromatographic or spectrophotometric measurement of constituent sugars. *Analyst* **1994**, *119*, 1497–1509. [CrossRef]
18. Theander, O.; Åman, P.; Westerlund, E.; Andersson, R.; Pettersson, D. Total dietary fiber determined as neutral sugar residues, uronic acid residues, and Klason lignin (the Uppsala method): Collaborative study. *J. AOAC Int.* **1995**, *78*, 1030–1044. [CrossRef]
19. Mudgil, D.; Barak, S. Composition, properties and health benefits of indigestible carbohydrate polymers as dietary fiber: A review. *Int. J. Biol. Macromol.* **2013**, *61*, 1–6. [CrossRef]
20. Holscher, H.D. Dietary fiber and prebiotics and the gastrointestinal microbiota. *Gut Microbes* **2017**, *8*, 172–184. [CrossRef]
21. Dikeman, C.L.; Fahey, G.C., Jr. Viscosity as related to dietary fiber: A review. *Crit. Rev. Food Sci. Nutr.* **2006**, *46*, 649–663. [CrossRef]
22. Chutkan, R.; Fahey, G.; Wright, W.L.; McRorie, J. Viscous versus nonviscous soluble fiber supplements: Mechanisms and evidence for fiber-specific health benefits. *J. Am. Acad. Nurse Pract.* **2012**, *24*, 476–487. [CrossRef] [PubMed]
23. Mateos, G.G.; Jiménez-Moreno, E.; Serrano, M.P.; Lázaro, R.P. Poultry response to high levels of dietary fiber sources varying in physical and chemical characteristics. *J. Appl. Poult. Res.* **2012**, *21*, 156–174. [CrossRef]
24. Topping, D. Cereal complex carbohydrates and their contribution to human health. *J. Cereal Sci.* **2007**, *46*, 220–229. [CrossRef]
25. Montagne, L.; Pluske, J.R.; Hampson, D.J. A review of interactions between dietary fiber and the intestinal mucosa, and their consequences on digestive health in young non-ruminant animals. *Anim. Feed Sci. Technol.* **2003**, *108*, 95–117. [CrossRef]
26. Kluth, H.; Rodehutscord, M. Effect of inclusion of cellulose in the diet on the inevitable endogenous amino acid losses in the ileum of broiler chicken. *Poult. Sci.* **2009**, *88*, 1199–1205. [CrossRef] [PubMed]
27. Angkanaporn, K.; Choct, M.; Bryden, W.L.; Annison, E.F.; Annison, G. Effects of wheat pentosans on endogenous amino acid losses in chickens. *J. Sci. Food Agric.* **1994**, *66*, 399–404. [CrossRef]
28. Jamroz, D.; Jakobsen, K.; Bach Knudsen, K.E.; Wiliczkiewicz, A.; Orda, J. Digestibility and energy value of non-starch polysaccharides in young chickens, ducks and geese, fed diets containing high amounts of barley. *Comp. Biochem. Physiol. A Mol. Integr. Physiol.* **2002**, *131*, 657–668. [CrossRef]
29. Jørgensen, H.; Zhao, X.-Q.; Bach Knudsen, K.E.; Eggum, B.O. The influence of dietary fibre source and level on the development of the gastrointestinal tract, digestibility and energy metabolism in broiler chickens. *Br. J. Nutr.* **1996**, *75*, 379–395. [CrossRef]

30. Rodriguez, M.L.; Rebole, A.; Velasco, S.; Ortiz, L.T.; Trevino, J.; Alzueta, C. Wheat- and barley-based diets with or without additives influence broiler chicken performance, nutrient digestibility and intestinal microflora. *J. Sci. Food Agric.* **2012**, *92*, 184–190. [CrossRef]
31. Kheravii, S.K.; Swick, R.A.; Choct, M.; Wu, S.-B. Coarse particle inclusion and lignocellulose-rich fiber addition in feed benefit performance and health of broiler chickens. *Poult. Sci.* **2017**, *96*, 3272–3281. [CrossRef]
32. Capuano, E. The behavior of dietary fiber in the gastrointestinal tract determines its physiological effect. *Crit. Rev. Food Sci. Nutr.* **2017**, *57*, 3543–3564. [CrossRef] [PubMed]
33. Le Goff, G.; van Milgen, J.; Noblet, J. Influence of dietary fibre on digestive utilization and rate of passage in growing pigs, finishing pigs and adult sows. *Anim. Sci.* **2002**, *74*, 503–515. [CrossRef]
34. Wilfart, A.; Montagne, L.; Simmins, H.; Noblet, J.; van Milgen, J. Digesta transit in different segments of the gastrointestinal tract of pigs as affected by insoluble fibre supplied by wheat bran. *Br. J. Nutr.* **2007**, *98*, 54–62. [CrossRef] [PubMed]
35. Choct, M.; Kocher, A. *Non-Starch Carbohydrates: Digestion and Its Secondary Effects in Monogastrics*; Nutrition Society of Australia: Crows Nest, NSW, Australia, 2000; pp. 31–38.
36. Liu, Z.; Wang, L.; Beasley, D.; Oviedo, E. Effect of moisture content on ammonia emissions from broiler litter: A laboratory study. *J. Atmos. Chem.* **2007**, *58*, 41–53. [CrossRef]
37. Shepherd, E.M.; Fairchild, B.D. Footpad dermatitis in poultry. *Poult. Sci.* **2010**, *89*, 2043–2051. [CrossRef]
38. Cowieson, A.J.; Acamovic, T.; Bedford, M.R. The effects of phytase and phytic acid on the loss of endogenous amino acids and minerals from broiler chickens. *Br. Poult. Sci.* **2004**, *45*, 101–108. [CrossRef]
39. Kumar, V.; Sinha, A.K.; Makkar, H.P.S.; Becker, K. Dietary roles of phytate and phytase in human nutrition: A review. *Food Chem.* **2010**, *120*, 945–959. [CrossRef]
40. Oatway, L.; Vasanthan, T.; Helm, J.H. Phytic acid. *Food Rev. Int.* **2001**, *17*, 419–431. [CrossRef]
41. Latshaw, J.D.; Zhao, L. Dietary protein effects on hen performance and nitrogen excretion. *Poult. Sci.* **2011**, *90*, 99–106. [CrossRef]
42. Kim, W.K.; Patterson, P.H. Effect of minerals on activity of microbial uricase to reduce ammonia volatilization in poultry manure. *Poult. Sci.* **2003**, *82*, 223–231. [CrossRef]
43. Piva, A.; Panciroli, A.; Meola, E.; Formigoni, A. Lactitol enhances short-chain fatty acid and gas production by swine cecal microflora to a greater extent when fermenting low rather than high fiber diets. *J. Nutr.* **1996**, *126*, 280–289. [CrossRef] [PubMed]
44. Jha, R.; Leterme, P. Feed ingredients differing in fermentable fiber and indigestible protein content affect fermentation metabolites and faecal nitrogen excretion in growing pigs. *Animal* **2012**, *6*, 603–611. [CrossRef] [PubMed]
45. Jha, R.; Berrocoso, J.F.D. Dietary fiber and protein fermentation in the intestine of swine and their interactive effects on gut health and on the environment: A review. *Anim. Feed Sci. Technol.* **2016**, *212*, 18–26. [CrossRef]
46. Metges, C.C. Contribution of microbial amino acids to amino acid homeostasis of the host. *J. Nutr.* **2000**, *130*, 1857S–1864S. [CrossRef]
47. Roberts, S.A.; Xin, H.; Kerr, B.J.; Russell, J.R.; Bregendahl, K. Effects of dietary fiber and reduced crude protein on ammonia emission from laying-hen manure. *Poult. Sci.* **2007**, *86*, 1625–1632. [CrossRef]
48. Ivanov, I.I.; Honda, K. Intestinal commensal microbes as immune modulators. *Cell Host Microbe* **2012**, *12*, 496–508. [CrossRef]
49. Littman, D.R.; Pamer, E.G. Role of the commensal microbiota in normal and pathogenic host immune responses. *Cell Host Microbe* **2011**, *10*, 311–323. [CrossRef]
50. Sánchez, B.; Delgado, S.; Blanco-Míguez, A.; Lourenço, A.; Gueimonde, M.; Margolles, A. Probiotics, gut microbiota, and their influence on host health and disease. *Mol. Nutr. Food Res.* **2017**, *61*, 1600240. [CrossRef]
51. Stanley, D.; Hughes, R.J.; Moore, R.J. Microbiota of the chicken gastrointestinal tract: Influence on health, productivity and disease. *Appl. Microbiol. Biotechnol.* **2014**, *98*, 4301–4310. [CrossRef]
52. Apajalahti, J.; Kettunen, A.; Graham, H. Characteristics of the gastrointestinal microbial communities, with special reference to the chicken. *Worlds Poult. Sci. J.* **2004**, *60*, 223–232. [CrossRef]
53. Wei, S.; Morrison, M.; Yu, Z. Bacterial census of poultry intestinal microbiome. *Poult. Sci.* **2013**, *92*, 671–683. [CrossRef]
54. Shang, Y.; Kumar, S.; Oakley, B.; Kim, W.K. Chicken gut microbiota: Importance and detection technology. *Front. Vet. Sci* **2018**, *5*, 254. [CrossRef] [PubMed]
55. Kubasova, T.; Kollarcikova, M.; Crhanova, M.; Karasova, D.; Cejkova, D.; Sebkova, A.; Matiasovicova, J.; Faldynova, M.; Pokorna, A.; Cizek, A.; et al. Contact with adult hen affects development of caecal microbiota in newly hatched chicks. *PLoS ONE* **2019**, *14*, e0212446. [CrossRef] [PubMed]
56. Rinttilä, T.; Apajalahti, J. Intestinal microbiota and metabolites—Implications for broiler chicken health and performance. *J. Appl. Poult. Res.* **2013**, *22*, 647–658. [CrossRef]
57. Macfarlane, G.T.; Gibson, G.R. Carbohydrate fermentation, energy transduction and gas metabolism in the human large intestine. In *Gastrointestinal Microbiology: Gastrointestinal Ecosystems and Fermentations*; Mackie, R.I., White, B.A., Eds.; Springer: Boston, MA, USA, 1997; Volume 1, pp. 269–318.
58. Kers, J.G.; Velkers, F.C.; Fischer, E.A.J.; Hermes, G.D.A.; Stegeman, J.A.; Smidt, H. Host and environmental factors affecting the intestinal microbiota in chickens. *Front. Microbiol.* **2018**, *9*, 235. [CrossRef] [PubMed]
59. Gibson, G.R.; Roberfroid, M.B. Dietary modulation of the human colonic microbiota: Introducing the concept of prebiotics. *J. Nutr.* **1995**, *125*, 1401–1412. [CrossRef] [PubMed]
60. Janardhana, V.; Broadway, M.M.; Bruce, M.P.; Lowenthal, J.W.; Geier, M.S.; Hughes, R.J.; Bean, A.G. Prebiotics modulate immune responses in the gut-associated lymphoid tissue of chickens. *J. Nutr.* **2009**, *139*, 1404–1409. [CrossRef] [PubMed]

61. Shoaf, K.; Mulvey, G.L.; Armstrong, G.D.; Hutkins, R.W. Prebiotic galactooligosaccharides reduce adherence of enteropathogenic *Escherichia coli* to tissue culture cells. *Infect. Immun.* **2006**, *74*, 6920–6928. [CrossRef] [PubMed]
62. Spring, P.; Wenk, C.; Dawson, K.A.; Newman, K.E. The effects of dietary mannaoligosaccharides on cecal parameters and the concentrations of enteric bacteria in the ceca of salmonella-challenged broiler chicks. *Poult. Sci.* **2000**, *79*, 205–211. [CrossRef]
63. Jung, S.J.; Houde, R.; Baurhoo, B.; Zhao, X.; Lee, B.H. Effects of galacto-oligosaccharides and a *Bifidobacteria lactis*-based probiotic strain on the growth performance and fecal microflora of broiler chickens. *Poult. Sci.* **2008**, *87*, 1694–1699. [CrossRef]
64. Baurhoo, B.; Phillip, L.; Ruiz-Feria, C.A. Effects of purified lignin and mannan oligosaccharides on intestinal integrity and microbial populations in the ceca and litter of broiler chickens. *Poult. Sci.* **2007**, *86*, 1070–1078. [CrossRef]
65. Snel, J.; Harmsen, H.J.M.; Van der Wielen, P.W.J.J.; Williams, B.A. Dietary strategies to influence the gastrointestinal microflora of young animals, and its potential to improve intestinal health. In *Nutrition and Health on the Gastrointestinal Tract*; Blok, M.C., Vahl, H.A., de Lange, L., van de Braak, A.E., Hemke, G., Hessing, M., Eds.; Wageningen Academic Publishers: Wageningen, NL, USA, 2002; pp. 37–69.
66. Singh, A.K. In Ovo and Post-Hatch Nutritional Programming to Improve Broiler Performance and Gut Health. Ph.D. Thesis, University of Hawai'i at Manoa, Ann Arbor, MI, USA, 2019. Available online: https://scholarspace.manoa.hawaii.edu/bitstream/10125/63507/1/Singh_hawii_0085A_10392.pdf (accessed on 1 December 2020).
67. Dunkley, K.D.; Dunkley, C.S.; Njongmeta, N.L.; Callaway, T.R.; Hume, M.E.; Kubena, L.F.; Nisbet, D.J.; Ricke, S.C. Comparison of in vitro fermentation and molecular microbial profiles of high-fiber feed substrates incubated with chicken cecal inocula. *Poult. Sci.* **2007**, *86*, 801–810. [CrossRef] [PubMed]
68. Józefiak, D.; Kaczmarek, S.; Rutkowski, A.; Józefiak, A.; Jensen, B.B.; Engberg, R.M. Fermentation in broiler chicken gastrointestinal tract as affected by high dietary inclusion of barley and β-glucanase supplementation. *J. Anim. Feed Sci.* **2005**, *14*, 695. [CrossRef]
69. Walugembe, M.; Hsieh, J.C.F.; Koszewski, N.J.; Lamont, S.J.; Persia, M.E.; Rothschild, M.F. Effects of dietary fiber on cecal short-chain fatty acid and cecal microbiota of broiler and laying-hen chicks. *Poult. Sci.* **2015**, *94*, 2351–2359. [CrossRef] [PubMed]
70. Ricke, S.C.; Woodward, C.L.; Kwon, Y.M.; Kubena, L.F.; Nisbet, D.J. Limiting avian gastrointestinal tract *Salmonella* colonization by cecal anaerobic bacteria, and a potential role for methanogens. In *Preharvest and Postharvest Food Safety: Contemporary Issues and Future Directions*; Beier, R.C., Pillai, S.D., Phillip, T.D., Eds.; Blackwell Publishing: Ames, IA, USA, 2004; p. 141.
71. Meimandipour, A.; Shuhaimi, M.; Soleimani, A.F.; Azhar, K.; Hair-Bejo, M.; Kabeir, B.M.; Javanmard, A.; Muhammad Anas, O.; Yazid, A.M. Selected microbial groups and short-chain fatty acids profile in a simulated chicken cecum supplemented with two strains of *Lactobacillus*. *Poult. Sci.* **2010**, *89*, 470–476. [CrossRef]
72. Reichardt, N.; Vollmer, M.; Holtrop, G.; Farquharson, F.M.; Wefers, D.; Bunzel, M.; Duncan, S.H.; Drew, J.E.; Williams, L.M.; Milligan, G.; et al. Specific substrate-driven changes in human faecal microbiota composition contrast with functional redundancy in short-chain fatty acid production. *ISME J.* **2018**, *12*, 610–622. [CrossRef]
73. Peng, Q.; Zeng, X.F.; Zhu, J.L.; Wang, S.; Liu, X.T.; Hou, C.L.; Thacker, P.A.; Qiao, S.Y. Effects of dietary Lactobacillus plantarum B1 on growth performance, intestinal microbiota, and short chain fatty acid profiles in broiler chickens. *Poult. Sci.* **2016**, *95*, 893–900. [CrossRef]
74. Lee, S.A.; Apajalahti, J.; Vienola, K.; González-Ortiz, G.; Fontes, C.M.G.A.; Bedford, M.R. Age and dietary xylanase supplementation affects ileal sugar residues and short chain fatty acid concentration in the ileum and caecum of broiler chickens. *Anim. Feed Sci. Technol.* **2017**, *234*, 29–42. [CrossRef]
75. Rehman, H.; Hellweg, P.; Taras, D.; Zentek, J. Effects of dietary inulin on the intestinal short chain fatty acids and microbial ecology in broiler chickens as revealed by denaturing gradient gel electrophoresis. *Poult. Sci.* **2008**, *87*, 783–789. [CrossRef]
76. Scheppach, W. Effects of short chain fatty acids on gut morphology and function. *Gut* **1994**, *35*, S35–S38. [CrossRef]
77. Yan, H.; Ajuwon, K.M. Butyrate modifies intestinal barrier function in IPEC-J2 cells through a selective upregulation of tight junction proteins and activation of the Akt signaling pathway. *PLoS ONE* **2017**, *12*, e0179586. [CrossRef] [PubMed]
78. Kamath, P.S.; Hoepfner, M.T.; Phillips, S.F. Short-chain fatty acids stimulate motility of the canine ileum. *Am. J. Physiol. Gastrointest. Liver Physiol.* **1987**, *253*, G427–G433. [CrossRef] [PubMed]
79. Binder, H.J.; Mehta, P. Short-chain fatty acids stimulate active sodium and chloride absorption in vitro in the rat distal colon. *Gastroenterology* **1989**, *96*, 989–996. [CrossRef]
80. Fukunaga, T.; Sasaki, M.; Araki, Y.; Okamoto, T.; Yasuoka, T.; Tsujikawa, T.; Fujiyama, Y.; Bamba, T. Effects of the soluble fiber pectin on intestinal cell proliferation, fecal short chain fatty acid production and microbial population. *Digestion* **2003**, *67*, 42–49. [CrossRef]
81. Fernández-Rubio, C.; Ordonez, C.; Abad-González, J.; Garcia-Gallego, A.; Honrubia, M.P.; Mallo, J.J.; Balana-Fouce, R. Butyric acid-based feed additives help protect broiler chickens from *Salmonella* Enteritidis infection. *Poult. Sci.* **2009**, *88*, 943–948. [CrossRef]
82. Van Immerseel, F.; Fievez, V.; de Buck, J.; Pasmans, F.; Martel, A.; Haesebrouck, F.; Ducatelle, R. Microencapsulated short-chain fatty acids in feed modify colonization and invasion early after infection with *Salmonella* enteritidis in young chickens. *Poult. Sci.* **2004**, *83*, 69–74. [CrossRef]
83. Leeson, S.; Namkung, H.; Antongiovanni, M.; Lee, E.H. Effect of butyric acid on the performance and carcass yield of broiler chickens. *Poult. Sci.* **2005**, *84*, 1418–1422. [CrossRef]
84. van Der Wielen, P.W.; Biesterveld, S.; Notermans, S.; Hofstra, H.; Urlings, B.A.; van Knapen, F. Role of volatile fatty acids in development of the cecal microflora in broiler chickens during growth. *Appl. Environ. Microbiol.* **2000**, *66*, 2536–2540. [CrossRef]
85. Shyer, A.E.; Tallinen, T.; Nerurkar, N.L.; Wei, Z.; Gil, E.S.; Kaplan, D.L.; Tabin, C.J.; Mahadevan, L. Villification: How the gut gets its villi. *Science* **2013**, *342*, 212–218. [CrossRef]

86. Andoh, A.; Bamba, T.; Sasaki, M. Physiological and anti-inflammatory roles of dietary fiber and butyrate in intestinal functions. *JPEN J. Parenter. Enter. Nutr.* **1999**, *23*, S70–S73. [CrossRef]
87. Shang, Y.; Regassa, A.; Kim, J.H.; Kim, W.K. The effect of dietary fructooligosaccharide supplementation on growth performance, intestinal morphology, and immune responses in broiler chickens challenged with *Salmonella* Enteritidis lipopolysaccharides. *Poult. Sci.* **2015**, *94*, 2887–2897. [CrossRef]
88. Ashraf, S.; Zaneb, H.; Yousaf, M.S.; Ijaz, A.; Sohail, M.U.; Muti, S.; Usman, M.M.; Ijaz, S.; Rehman, H. Effect of dietary supplementation of prebiotics and probiotics on intestinal microarchitecture in broilers reared under cyclic heat stress. *J. Anim. Physiol. Anim. Nutr. (Berl.)* **2013**, *97* (Suppl. 1), 68–73. [CrossRef]
89. Amat, C.; Planas, J.M.; Moreto, M. Kinetics of hexose uptake by the small and large intestine of the chicken. *Am. J. Physiol.* **1996**, *271*, R1085–R1089. [CrossRef] [PubMed]
90. Ulluwishewa, D.; Anderson, R.C.; McNabb, W.C.; Moughan, P.J.; Wells, J.M.; Roy, N.C. Regulation of tight junction permeability by intestinal bacteria and dietary components. *J. Nutr.* **2011**, *141*, 769–776. [CrossRef]
91. Zihni, C.; Mills, C.; Matter, K.; Balda, M.S. Tight junctions: From simple barriers to multifunctional molecular gates. *Nat. Rev. Mol. Cell Biol.* **2016**, *17*, 564. [CrossRef]
92. Teng, P.-Y.; Yadav, S.; de Souza Castro, F.L.; Tompkins, Y.H.; Fuller, A.L.; Kim, W.K. Graded Eimeria challenge linearly regulated growth performance, dynamic change of gastrointestinal permeability, apparent ileal digestibility, intestinal morphology, and tight junctions of broiler chickens. *Poult. Sci.* **2020**, *99*, 4203–4216. [CrossRef] [PubMed]
93. Liu, D.; Guo, S.; Guo, Y. Xylanase supplementation to a wheat-based diet alleviated the intestinal mucosal barrier impairment of broiler chickens challenged by *Clostridium perfringens*. *Avian Pathol.* **2012**, *41*, 291–298. [CrossRef]
94. Wang, H.B.; Wang, P.Y.; Wang, X.; Wan, Y.L.; Liu, Y.C. Butyrate enhances intestinal epithelial barrier function via up-regulation of tight junction protein claudin-1 transcription. *Dig. Dis. Sci.* **2012**, *57*, 3126–3135. [CrossRef] [PubMed]
95. Karczewski, J.; Troost, F.J.; Konings, I.; Dekker, J.; Kleerebezem, M.; Brummer, R.J.; Wells, J.M. Regulation of human epithelial tight junction proteins by *Lactobacillus plantarum* in vivo and protective effects on the epithelial barrier. *Am. J. Physiol. Gastrointest. Liver Physiol.* **2010**, *298*, G851–G859. [CrossRef] [PubMed]
96. Turner, J.R. Intestinal mucosal barrier function in health and disease. *Nat. Rev. Immunol.* **2009**, *9*, 799–809. [CrossRef]
97. Wu, D.; Lewis, E.D.; Pae, M.; Meydani, S.N. Nutritional modulation of immune function: Analysis of evidence, mechanisms, and clinical relevance. *Front. Immunol.* **2019**, *9*, 3160. [CrossRef] [PubMed]
98. Cunningham-Rundles, S.; Lin, H.; Ho-Lin, D.; Dnistrian, A.; Cassileth, B.R.; Perlman, J.M. Role of nutrients in the development of neonatal immune response. *Nutr. Rev.* **2009**, *67*, S152–S163. [CrossRef] [PubMed]
99. Gómez-Verduzco, G.; Cortes-Cuevas, A.; López-Coello, C.; Ávila-González, E.; Nava, G.M. Dietary supplementation of mannan-oligosaccharide enhances neonatal immune responses in chickens during natural exposure to *Eimeria* spp. *Acta Vet. Scand.* **2009**, *51*, 11. [CrossRef]
100. Min, Y.N.; Li, L.L.; Liu, S.K.; Zhang, J.; Gao, Y.P.; Liu, F.Z. Effects of dietary distillers dried grains with solubles (DDGS) on growth performance, oxidative stress, and immune function in broiler chickens. *J. Appl. Poult. Res.* **2015**, *24*, 23–29. [CrossRef]
101. Sadeghi, A.; Toghyani, M.; Gheisari, A. Effect of various fiber types and choice feeding of fiber on performance, gut development, humoral immunity, and fiber preference in broiler chicks. *Poult. Sci.* **2015**, *94*, 2734–2743. [CrossRef]
102. Lara, L.J.; Rostagno, M.H. Impact of heat stress on poultry production. *Animals* **2013**, *3*, 356–369. [CrossRef]
103. Vicuña, E.A.; Kuttappan, V.A.; Galarza-Seeber, R.; Latorre, J.D.; Faulkner, O.B.; Hargis, B.M.; Tellez, G.; Bielke, L.R. Effect of dexamethasone in feed on intestinal permeability, differential white blood cell counts, and immune organs in broiler chicks. *Poult. Sci.* **2015**, *94*, 2075–2080. [CrossRef]
104. Jia, W.; Slominski, B.A.; Bruce, H.L.; Blank, G.; Crow, G.; Jones, O. Effects of diet type and enzyme addition on growth performance and gut health of broiler chickens during subclinical *Clostridium perfringens* challenge. *Poult. Sci.* **2009**, *88*, 132–140. [CrossRef]
105. Bedford, M.R.; Apajalahti, J. Microbial interactions in response to Exogenous enzyme utilization. In *Enzymes in Farm Animal Nutrition*; Bedford, M.R., Partrige, G.G., Eds.; CAB International: Wallingford, UK, 2001; pp. 299–314.
106. De Vries, S.; Pustjens, A.M.; Schols, H.A.; Hendriks, W.H.; Gerrits, W.J.J. Improving digestive utilization of fiber-rich feedstuffs in pigs and poultry by processing and enzyme technologies: A review. *Anim. Feed Sci. Technol.* **2012**, *178*, 123–138. [CrossRef]
107. Teitge, D.A.; Campbell, G.L.; Classen, H.L.; Thacker, P.A. Heat pretreatment as a means of improving the response to dietary pentosanase in chicks fed rye. *Can. J. Anim. Sci.* **1991**, *71*, 507–513. [CrossRef]
108. Review Manager (RevMan). *Computer Program, Version 5.4*; The Cochrane Collaboration: London, UK; The Nordic Cochrane Center: Copenhagen, Denmark, 2020.
109. Wang, Z.R.; Qiao, S.Y.; Lu, W.Q.; Li, D.F. Effects of enzyme supplementation on performance, nutrient digestibility, gastrointestinal morphology, and volatile fatty acid profiles in the hindgut of broilers fed wheat-based diets. *Poult. Sci.* **2005**, *84*, 875–881. [CrossRef]
110. Zhang, L.; Xu, J.; Lei, L.; Jiang, Y.; Gao, F.; Zhou, G.H. Effects of xylanase supplementation on growth performance, nutrient digestibility and non-starch polysaccharide degradation in different sections of the gastrointestinal tract of broilers fed wheat-based diets. *Asian Austral. J. Anim* **2014**, *27*, 855. [CrossRef] [PubMed]
111. Amerah, A.M.; Ravindran, V.; Lentle, R.G. Influence of wheat hardness and xylanase supplementation on the performance, energy utilisation, digestive tract development and digesta parameters of broiler starters. *Anim. Prod. Sci.* **2008**, *49*, 71–78. [CrossRef]

112. Amerah, A.M.; Van de Belt, K.; van Der Klis, J.D. Effect of different levels of rapeseed meal and sunflower meal and enzyme combination on the performance, digesta viscosity and carcass traits of broiler chickens fed wheat-based diets. *Animal* **2015**, *9*, 1131–1137. [CrossRef] [PubMed]
113. Gao, F.; Jiang, Y.; Zhou, G.H.; Han, Z.K. The effects of xylanase supplementation on performance, characteristics of the gastrointestinal tract, blood parameters and gut microflora in broilers fed on wheat-based diets. *Anim. Feed Sci. Technol.* **2008**, *142*, 173–184. [CrossRef]
114. Kiarie, E.; Romero, L.F.; Ravindran, V. Growth performance, nutrient utilization, and digesta characteristics in broiler chickens fed corn or wheat diets without or with supplemental xylanase. *Poult. Sci.* **2014**, *93*, 1186–1196. [CrossRef]
115. Lazaro, R.; Garcia, M.; Medel, P.; Mateos, G.G. Influence of enzymes on performance and digestive parameters of broilers fed rye-based diets. *Poult. Sci.* **2003**, *82*, 132–140. [CrossRef]
116. Lee, S.H.; Hosseindoust, A.; Laxman Ingale, S.; Rathi, P.C.; Yoon, S.Y.; Choi, J.W.; Kim, J.S. Thermostable xylanase derived from *Trichoderma citrinoviride* increases growth performance and non-starch polysaccharide degradation in broiler chickens. *Br. Poult. Sci.* **2020**, *61*, 57–62. [CrossRef]
117. Luo, D.; Yang, F.; Yang, X.; Yao, J.; Shi, B.; Zhou, Z. Effects of xylanase on performance, blood parameters, intestinal morphology, microflora and digestive enzyme activities of broilers fed wheat-based diets. *Asian Austral. J. Anim.* **2009**, *22*, 1288–1295. [CrossRef]
118. Mathlouthi, N.; Mallet, S.; Saulnier, L.; Quemener, B.; Larbier, M. Effects of xylanase and β-glucanase addition on performance, nutrient digestibility, and physico-chemical conditions in the small intestine contents and caecal microflora of broiler chickens fed a wheat and barley-based diet. *Anim. Res.* **2002**, *51*, 395–406. [CrossRef]
119. Munyaka, P.M.; Nandha, N.K.; Kiarie, E.; Nyachoti, C.M.; Khafipour, E. Impact of combined beta-glucanase and xylanase enzymes on growth performance, nutrients utilization and gut microbiota in broiler chickens fed corn or wheat-based diets. *Poult. Sci.* **2016**, *95*, 528–540. [CrossRef] [PubMed]
120. Pirgozliev, V.; Rose, S.P.; Pellny, T.; Amerah, A.M.; Wickramasinghe, M.; Ulker, M.; Rakszegi, M.; Bedo, Z.; Shewry, P.R.; Lovegrove, A. Energy utilization and growth performance of chickens fed novel wheat inbred lines selected for different pentosan levels with and without xylanase supplementation. *Poult. Sci.* **2015**, *94*, 232–239. [CrossRef] [PubMed]
121. Selle, P.; Ravindran, V.; Ravindran, G.; Pittolo, P.; Bryden, W. Influence of phytase and xylanase supplementation on growth performance and nutrient utilisation of broilers offered wheat-based diets. *Asian-australas. J. Anim. Sci.* **2003**, *16*, 394–402. [CrossRef]
122. Woyengo, T.A.; Guenter, W.; Sands, J.S.; Nyachoti, C.M.; Mirza, M.A. Nutrient utilisation and performance responses of broilers fed a wheat-based diet supplemented with phytase and xylanase alone or in combination. *Anim. Feed Sci. Technol.* **2008**, *146*, 113–123. [CrossRef]
123. Wu, Y.; Lai, C.; Qiao, S.; Gong, L.; Lu, W.; Li, D. Properties of Aspergillar xylanase and the effects of xylanase supplementation in wheat-based diets on growth performance and the blood biochemical values in broilers. *Asian Austral. J. Anim.* **2005**, *18*, 66–74. [CrossRef]
124. Yang, Y.; Iji, P.A.; Kocher, A.; Mikkelsen, L.L.; Choct, M. Effects of xylanase on growth and gut development of broiler chickens given a wheat-based diet. *Asian Austral. J. Anim.* **2008**, *21*, 1659–1664. [CrossRef]
125. Amerah, A.M.; Mathis, G.; Hofacre, C.L. Effect of xylanase and a blend of essential oils on performance and *Salmonella* colonization of broiler chickens challenged with *Salmonella* Heidelberg. *Poult. Sci.* **2012**, *91*, 943–947. [CrossRef]
126. Sun, Q.; Liu, D.; Guo, S.; Chen, Y.; Guo, Y. Effects of dietary essential oil and enzyme supplementation on growth performance and gut health of broilers challenged by *Clostridium perfringens*. *Anim. Feed Sci. Technol.* **2015**, *207*, 234–244. [CrossRef]
127. Jia, W.; Slominski, B.A.; Bruce, H.L.; Nyachoti, C.M.; Jones, R.O. Enzyme addition facilitates the post-disease compensatory growth of broiler chickens challenged with *Clostridium perfringens*. *Can. J. Anim. Sci.* **2009**, *89*, 369–381. [CrossRef]
128. Dersjant-Li, Y.; Gibbs, K.; Awati, A.; Klasing, K.C. The effects of enzymes and direct fed microbial combination on performance and immune response of broilers under a coccidia challenge. *J. Appl. Anim. Nutr.* **2016**, *4*. [CrossRef]
129. Jackson, M.E.; Anderson, D.M.; Hsiao, H.Y.; Mathis, G.F.; Fodge, D.W. Beneficial effect of β-mannanase feed enzyme on performance of chicks challenged with *Eimeria* sp. and *Clostridium perfringens*. *Avian Dis.* **2003**, *47*, 759–763. [CrossRef] [PubMed]
130. Choct, M.; Sinlae, M.; Al-Jassim, R.A.M.; Pettersson, D. Effects of xylanase supplementation on between-bird variation in energy metabolism and the number of *Clostridium perfringens* in broilers fed a wheat-based diet. *Aust. J. Agric. Res.* **2006**, *57*, 1017–1021. [CrossRef]
131. Bortoluzzi, C.; Scapini, L.B.; Ribeiro, M.V.; Pivetta, M.R.; Buzim, R.; Fernandes, J.I.M. Effects of β-mannanase supplementation on the intestinal microbiota composition of broiler chickens challenged with a coccidiosis vaccine. *Livest. Sci.* **2019**, *228*, 187–194. [CrossRef]
132. Khattak, F.M.; Pasha, T.N.; Hayat, Z.; Mahmud, A. Enzymes in poultry nutrition. *J. Anim. Plant Sci.* **2006**, *16*, 1–7.

MDPI
St. Alban-Anlage 66
4052 Basel
Switzerland
Tel. +41 61 683 77 34
Fax +41 61 302 89 18
www.mdpi.com

Animals Editorial Office
E-mail: animals@mdpi.com
www.mdpi.com/journal/animals

www.ingramcontent.com/pod-product-compliance
Lightning Source LLC
Chambersburg PA
CBHW061549220926
51850CB00005B/290
9783036589695